Essential Cytometry Methods

Reliable Lab Solutions

Essential Cytometry Methods

Reliable Lab Solutions

Edited by

Zbigniew Darzynkiewicz

Brander Cancer Research Institute
New York Medical College
Valhalla, NY 10595

J. Paul Robinson

Purdue University Cytometry Laboratories
Bindley Bioscience Center
West Lafayette, IN 47907-2057

Mario Roederer

Immunology Laboratory
Vaccine Research Center
National Institute of Allergy and Infectious Diseases
National Institute of Health
Bethesda, Maryland 20892

AMSTERDAM • BOSTON • HEIDELBERG • LONDON
NEW YORK • OXFORD • PARIS • SAN DIEGO
SAN FRANCISCO • SINGAPORE • SYDNEY • TOKYO
Academic Press is an imprint of Elsevier

Academic Press is an imprint of Elsevier
Linacre House, Jordan Hill, Oxford OX2 8DP, UK
30 Corporate Drive, Suite 400, Burlington, MA 01803, USA
525 B Street, Suite 1900, San Diego, CA 92101-4495, USA
32 Jamestown Road, London NW1 7BY, UK

First edition 2009

ISBN: 978-0-12-375045-7

For information on all Academic Press publications
visit our website at www.elsevierdirect.com

Printed and bound by CPI Group (UK) Ltd, Croydon, CR0 4YY

Transferred to Digital Print 2012

CONTENTS

v

PART II   Cellular DNA Content Analysis

PART III   Cell Proliferation and the Death Assays

PART IV   Cell Surface Immunophenotyping

Part V   Cytogenetics/Chromatin Structure

27. Telomere Length Measurements Using Fluorescence *In Situ*
Hybridization and Flow Cytometry

Gabriela M. Baerlocher and Peter M. Lansdorp

28. Sperm Chromatin Structure Assay: DNA Denaturability

Donald Evenson and Lorna Jost

PART VI   Cell Physiology Assays

29. Cell Membrane Potential Analysis

Howard M. Shapiro

PART VII   Detection of Microorganisms and Pathogens

CONTRIBUTORS

Numbers in parentheses indicate the pages on which the authors' contributions begin.

David Ambrozak (183), Immunology Laboratory, Vaccine Research Center, National Institute of Allergy and Infectious Diseases, National Institute of Health, Bethesda, Maryland 20892

Nicholas Ashbolt (803), Australian Water Technologies, Science and Environment, Sydney, New South Wales 2114, Australia

Kenneth A. Ault (781), Maine Medical Center Research Institute, South Portland, Maine 04106

Gabriela M. Baerlocher (603), Department of Hematology and Department of Clinical Research, University Hospital Bern, 3010 Bern, Switzerland, and Terry Fox Laboratory, British Columbia Cancer Agency, Vancouver, British Columbia, Canada V5Z 1L3

Nicole Baumgarth (577), Center for Comparative Medicine, University of California, Davis, California 95616

Elżbieta Bedner (463), Department of Pathology, Pomeranian School of Medicine, Szczecin, Poland

Michael J. Boyer (673), Department of Medical Oncology, Royal Prince Alfred Hospital, Sydney, New South Wales 2050, Australia

Marcel P. Bruchez (129), Carnegie Mellon University, Department of Chemistry and Molecular Biosensor and Imaging Center, 4400 Fifth Ave., Pittsburgh, PA 15213

Wayne O. Carter (713), Hill's Pet Food, Topeka, KS

Ib Jarle Christensen (223), Finsen Laboratory, Rigshospitalet, DK-2100 Copenhagen, Denmark

Zbigniew Darzynkiewicz (205, 307, 437, 463), Braner Cancer Research Institute, New York Medical College, Valhalla, New York 10595

Frank Dolbeare (331), Biology and Biotechnology Program, Lawrence Livermore National Laboratory, Livermore, California 94550

Elmar Endl (397), Division of Molecular Immunology, Research Center Borstel, D-23845 Borstel, Germany

Donald Evenson (635), Olson Biochemistry Laboratories, Department of Chemistry, South Dakota State University, Brookings, South Dakota 57007

David W. Galbraith (109, 251), Department of Plant Sciences, Institute for Biomedical Science and Biotechnology, University of Arizona, Tucson, Arizona 85721

Johannes Gerdes (397), Division of Molecular Immunology, Research Center Borstel, D-23845 Borstel, Germany

Jianping Gong (437), Cancer Research Institute, New York Medical College, Valhalla, New York 10595

Wojciech Gorczyca (541), Genzyme Genetics (New York Laboratory), New York, New York 10019

Andreas Grützkau (143), Deutsches Rheuma-Forschungszentrum, 10117 Berlin, Germany

Ronald Hamelik (235), Department of Pathology, Miller School of Medicine, University of Miami, Miami, Florida 33101

Jhagvaral Hasbold (373), The Centenary Institute of Cancer Medicine and Cell Biology, Sydney, Australia

David W. Hedley (241, 673), Departments of Medicine and Pathology, Ontario Cancer Institute/Princess Margaret Hospital, Toronto, Ontario, Canada M4X 1K9

Philip D. Hodgkin (373), Medical Foundation, University of Sydney, and The Centenary Institute of Cancer Medicine and Cell Biology, Sydney, Australia, and Discipline of Pathology, Faculty of Health Science, The University of Tasmania, Hobart, Australia

Christiane Hollmann (397), Division of Molecular Immunology, Research Center Borstel, D-23845 Borstel, Germany

Kevin L. Holmes (29, 183), Chief, Flow Cytometry Section Research Technologies Branch, NIAID, NTH, DHHS Bethesda, Maryland 20892, and Immunology Laboratory, Vaccine Research Center, National Institute of Health, Bethesda, Maryland 20892

Michael S. Janes (129), Invitrogen Corporation—Molecular Probes®, Labeling and Detection Technologies, Eugene, Oregon 97402

Chris J. Janse (865), Laboratory of Parasitology, University of Leiden, 2300 RC Leiden, The Netherlands

Peter Østrup Jensen (757), Department of Clinical Microbiology, Rigshospitalet, DK-2100 Copenhagen, Denmark

Lorna Jost (635), Olson Biochemistry Laboratories, Department of Chemistry, South Dakota State University, Brookings, South Dakota 57007

Carl H. June (687), Department of Immunobiology, Naval Medical Research Institute, Bethesda, Maryland 20889

Kathryn L. Kellar (161), Division of Scientific Resources, National Center for Preparedness, Detection and Control of Infectious Diseases, Centers for Disease Control and Prevention, Atlanta, Georgia 30333

Mariam Klouche (725), Laborzentrum Bremen, Friedrich-Karl-Strasse 22, D-28205 Bremen, Germany

Richard A. Koup (183), Immunology Laboratory, Vaccine Research Center, National Institute of Allergy and Infectious Diseases, National Institute of Health, Bethesda, Maryland 20892

Awtar Krishan (235), Department of Pathology, Miller School of Medicine, University of Miami, Miami, Florida 33101

Peter M. Lansdorp (603), Department of Medicine, University of British Columbia, Vancouver, British Columbia, Canada V5Z 4E3, and Terry Fox Laboratory, British Columbia Cancer Agency, Vancouver, British Columbia, Canada V5Z 1L3

Larry M. Lantz (29), Flow Cytometry Section, National Institute of Allergy and Infectious Diseases, National Institutes of Health, Bethesda, Maryland 20892

Jacob Larsen (757), Department of Clinical Pathology, Næstved Hospital, 4700 Næstved, Denmark

Jørgen K. Larsen (757), Borgergade 30III, DK-1300 Copenhagen K, Denmark

Xun Li (437), Cancer Research Institute, New York Medical College, Valhalla, New York 10595

A. Bruce Lyons (373), Discipline of Pathology, Faculty of Health Science, The University of Tasmania, Hobart, Australia

János Matkó (55), Department of Immunology, Eötvös Loránd University, Budapest H-1117, Hungary

Birgit Mechtold (143), Institut für Genetik, Universität zu Köln, Germany

Stefan Miltenyi (143), Miltenyi Biotec GmbH, 51429 Bergisch Gladbach, Germany

Jane Mitchell (781), Maine Medical Center Research Institute, South Portland, Maine 04106

Joe Narai (803), Commonwealth Centre for Laser Applications, Macquarie University, Sydney, New South Wales 2109, Australia

Padma Kumar Narayanan (713), Amgen Inc., Bothell, Washington, and Department of Basic Medical Sciences, School of Veterinary Medicine, Purdue University, West Lafayette, Indiana 47907

Kary L. Oakleaf (129), Invitrogen Corporation—Molecular Probes®, Labeling and Detection Technologies, Eugene, Oregon 97402

Peggy L. Olive (417), Department of Medical Biophysics, British Columbia Cancer Research Centre, Vancouver, British Columbia, Canada V5Z 1L3

Kerry G. Oliver (161), Radix BioSolutions Ltd., Georgetown, Texas 78626

David R. Parks (205), Department of Genetics, Stanford University, Stanford, California 94305

Stephen P. Perfetto (183), Immunology Laboratory, Vaccine Research Center, National Institute of Allergy and Infectious Diseases, National Institute of Health, Bethesda, Maryland 20892

Eckhard Pflüger (143), Miltenyi Biotec GmbH, 51429 Bergisch Gladbach, Germany

Robert H. Pierce (769), Department of Pathology, Wright-Patterson Medical Center, Wright-Patterson Air Force Base, Dayton, Ohio 45433

Martin Poot (769), Department of Pathology, University of Washington, Seattle, Washington 98195

Peter S. Rabinovitch (275, 687), Department of Pathology, University of Washington, Seattle, Washington 98195

Andreas Radbruch (143), Deutsches Rheuma-Forschungszentrum, 10117 Berlin, Germany

Bartek Rajwa (3), Purdue University, West Lafayette, Indiana 47907

J. Paul Robinson (3, 713, 837), Weldon School of Biomedical Engineering, Purdue University, West Lafayette, Indiana 47907, and Department of Basic Medical Sciences, School of Veterinary Medicine, Purdue University Cytometry Laboratories, West Lafayette, Indiana 47907

Mario Roederer (183, 205), Immunology Laboratory, Vaccine Research Center, National Institute of Allergy and Infectious Diseases, National Institute of Health, Bethesda, Maryland 20892

Gregor Rothe (725), Laborzentrum Bremen, Friedrich-Karl-Strasse 22, D-28205 Bremen, Germany

Ingrid Schmid (183), Department of Hematology/Oncology, David Geffen School of Medicine, University of California at Los Angeles, Los Angeles, California 90095

Jules R. Selden (331), Department of Safety Assessment, Merck Research Laboratories, West Point, Pennsylvania 19486

Howard M. Shapiro (657), 283 Highland Avenue, West Newton, Massachusetts 02465-2513

Carleton C. Stewart (487), Laboratory of Flow Cytometry, Roswell Park Cancer Institute, Buffalo, New York 14263

Sigrid J. Stewart (487), Laboratory of Flow Cytometry, Roswell Park Cancer Institute, Buffalo, New York 14263

Jennifer Sturgis (3), Purdue University, West Lafayette, Indiana 47907

János Szöllősi (55), Department of Biophysics and Cell Biology, Cell Biophysics Research Group of the Hungarian Academy of Sciences, University of Debrecen, Debrecen H-4012, Hungary

Nicholas H. A. Terry (353), Departments of Experimental Radiation Oncology, M. D. Anderson Cancer Center, The University of Texas, Houston, Texas 77030

Andreas Thiel (143), Berlin-Brandenburg Center for Regenerative Therapies, Charité, 13353 Berlin, Germany

Frank Traganos (463), Brander Cancer Research Institute, New York Medical College, Valhalla, New York 10595

Sorina Tugulea (541), Genzyme Genetics (New York Laboratory), New York, New York 10019

Duncan Veal (803), School of Biological Sciences, Macquarie University, Sydney, New South Wales 2109, Australia

György Vereb (55), Department of Biophysics and Cell Biology, Cell Biophysics Research Group of the Hungarian Academy of Sciences, University of Debrecen, Debrecen H-4012, Hungary

Graham Vesey (803), School of Biological Sciences, Macquarie University, Sydney, New South Wales 2109, Australia

Philip H. Van Vianen (865), Laboratory of Parasitology, University of Leiden, 2300 RC Leiden, The Netherlands

Lars L. Vindeløv (223), Department of Haematology, Rigshospitalet, DK-2100 Copenhagen, Denmark

R. Allen White (353), Department of Biomathematics, M. D. Anderson Cancer Center, The University of Texas, Houston, Texas 77030

Keith Williams (803), School of Biological Sciences, Macquarie University, Sydney, New South Wales 2109, Australia

Brent Wood (521), Department of Laboratory Medicine, University of Washington, Seattle, Washington 98195

Xingyong Wu (129), Quantum Dot Corporation, Department of Chemistry and Molecular Biosensor and Imaging Center, 4400 Fifth Ave., Pittsburgh, PA 15213

PREFACE

Two hundred and sixteen individual chapters dedicated to different cytometric methodologies were published since 1990 in the six cytometry-committed volumes of the series of *Methods in Cell Biology* (MCB; Volumes 33, 41, 42, 63, 64, and 75). The editors of these volumes attempted to assemble chapters describing the most widely used methods of flow- and quantitative image-cytometry. The chapters outlined principles of these methods, their applications, advantages, as well as possible pitfalls in their use, and presented them to the forum of cell biologists. Within the series of MCB, these volumes received wide readership, high citation rates, and were valuable in promoting cytometric techniques among cell biologists across different fields.

The exceptionally high interest in the MCB chapters on cytometry prompted the Publisher to propose a special edition of the "*Essential Cytometry Methods*" within the framework of the new series of volumes defined "*Reliable Lab Solutions.*" This volume presents the chapters describing the most frequently used methods among those presented in the previous volumes. The chapters were selected based on high frequency of citations and relevance of the methodology. Since most these methods are still widely used, such an edition is contemporary and will be of use to many, particularly to young investigators who are starting to use the methods of cytometry in their research.

Authors of this volume were asked to update their chapters by providing a short foreword to the original text and make corrections in the text, if needed. The update highlights in brief progress in the methodology, novel reagents, new applications, and sister methodologies developed since the original publication. Additional references essential for the presentation of the update are included. Because some authors, particularly of the chapters early published, in MCB volumes 33, 41, and 42, could not be reached, their chapters remain without the update. Some chapters were updated by the Editors who are familiar with the methodology described in them.

Applications of cytometric methods have had a tremendous impact on research in various fields of cell and molecular biology, immunology, microbiology, and medicine. We hope that this volume will be of help to many researchers who need these methods in their investigation, stimulate application of the methodology in new areas, and promote further progress in science.

Zbigniew Darzynkiewicz
Mario Roederer
J. Paul Robinson

PART I

Fluorochromes/General Techniques

CHAPTER 1

Principles of Confocal Microscopy

J. Paul Robinson, Jennifer Sturgis, and Bartek Rajwa

Purdue University
West Lafayette
Indiana 47907

I. Brief History of Microscope Development

Microscopy techniques have taken over 200 years to mature into technologies capable of the measurements now possible using confocal microscopy. Prior to 1800, production microscopes using simple lens systems were of higher resolution

ESSENTIAL CYTOMETRY METHODS
3
DOI: 10.1016/B978-0-12-375045-7.00001-5

than compound microscopes despite the chromatic and spherical aberrations present in the double convex lens design. Microscopy did not really prosper until W. H. Wollaston made a significant improvement to the simple lens in 1812. Soon after, Brewster improved upon this design in 1820 and in 1827 Giovanni Battista Amici introduced the first matched achromatic microscope. Key features of Amici's design were the recognition of the importance of cover glass thickness and the development of the concept of "water immersion." Then Carl Zeiss and Ernst Abbe introduced oil-immersion systems by developing oils that matched the refractive index of glass. By 1886, Dr Otto Schott formulated glass lenses that allowed for color correction, and produced the first "apochromatic" objectives. Just after the turn of the century, Köhler illumination revolutionized bright-field microscopy. This discovery has been considered one of the most significant developments in microscopy prior to the electronic age.

Later developments such as the use of phase-contrast illumination, Nomarski illumination, and epi-illumination have each had significant impact on cell biology. In recent years, the advent of confocal microscopy has changed the way cell biologists prepare and examine material because we now have more options. Using this technology, the biologist can pinpoint the location of labeled molecules (e.g., a growth factor) in relatively thick specimens. This allows us to identify the organelle or location for the synthesis of the molecule. It is also possible to reconstruct the 3D structure of many cells, organs, and even small organisms quite accurately. Such information has given us a great deal of insight into the structure and function of many biological systems. This chapter will discuss the basic principles of confocal microscopy. Detailed texts on the subject include (Cox and Sheppard, 1983; Hibbs, 2004; Matsumoto, 2002; Paddock, 1999; Pawley, 2006).

II. Development of Confocal Microscopy

Marvin Minsky, then at Harvard University, filed the first patent for the concept of a confocal microscope in 1957 (Minsky, 1988). Ten years later, the first analog confocal microscope was built by Mojmir Petran and Maurice D. Egger (Egger and Petran, 1967). However, the origins of confocal optics go back to a microscopic spectrophotometer made by Hiroto Naora in the immediate postwar years. His first publication on this subject in English appeared in *Science* 58 years ago (Naora, 1951). A laser was first used as the light source for a confocal microscope by Davidovits and Egger (1969). In 1983, Cox and Sheppard recognized the value of a computer to collect and store confocal images (Cox and Sheppard, 1983), and the first commercial confocal microscopes based on the design of Brad Amos (Bio-Rad MRC500) appeared in 1987. Many scientists contributed to the enhancement and practical application of the technology (Amos, 1988; Amos *et al.*, 1987; Brakenhoff *et al.*, 1979, 1985; Carlsson *et al.*, 1985; White *et al.*, 1987).

The term *confocal* in the context of biological microscopy probably was for the first time used by Brakenhoff and others in 1979 (Brakenhoff *et al.*, 1979).

It describes an optical platform in which the illumination is confined to a diffraction-limited spot in the specimen and the detection is similarly confined by placing an aperture (a pinhole) in front of the detector in a position optically conjugate to the focused spot (Amos and White, 2003).

A confocal microscope achieves crisp images of structures even within thick tissue specimens by a process known as optical sectioning. The image source is primarily the photon emission from fluorescent molecules within or attached to structures within the object being sectioned. An alternative to fluorescence emission, reflectance, is discussed later. A point source of laser light illuminates the back focal plane of the microscope objective and is subsequently focused to a diffraction-limited spot within the specimen. Within this spot fluorescent molecules are excited and emit light in all directions. However, the emitted light is refocused in the objective image plane and any out-of-focus light is essentially removed from the image by passing the light through a pinhole aperture, so only a thin optical section of the specimen is formed. The effective removal of out-of-focus light by the aperture creates an essentially background-free image—as opposed to the traditional fluorescent microscope which includes all of this *out-of-focus* light. The comparison between traditional fluorescence microscopy and confocal microscopy is demonstrated in Fig. 1. As the diameter of the pinhole is reduced, the amount of light collected from the specimen is reduced, as is the "thickness" of the optical section. This effectively decreases the resolution of the images obtained. The resolution of a point light source is defined by the circular Airy diffraction pattern with a central bright region and outer dark ring formed on the image plane. The radius of this central bright region is defined as $r_{Airy} = 0.61\lambda/NA$, where λ is the wavelength of the excitation source and NA is the numerical aperture of the objective lens. To increase the signal and decrease the background light, it is necessary to decrease the pinhole to a size slightly less than r_{Airy}; a correct adjustment can decrease the background light by a factor of 10^3 over conventional fluorescence microscopy. Thus, achieving the correct pinhole diameter is crucial for achieving maximum resolution in a thick specimen. This becomes a tradeoff, however, between *optimizing* axial resolution (optimum = $0.7r_{Airy}$) and lateral resolution (optimum = $0.3r_{Airy}$).

While the image collection optics removes the background light and creates a nice clean section, it is important to realize that the entire image is still being bathed in excitation light. By the time a thick section has been imaged a number of times, the *impact* of substantial photobleaching must be considered.

III. Image Formation in Confocal Microscopy

There are several methods for achieving a confocal image. The most common method scans the point source of light (a laser beam) over the sample using a pair of galvanometer mirrors. One galvanometer scans in the X direction and the other in the Y direction. The emitted fluorescence traverses the reverse pathway, is

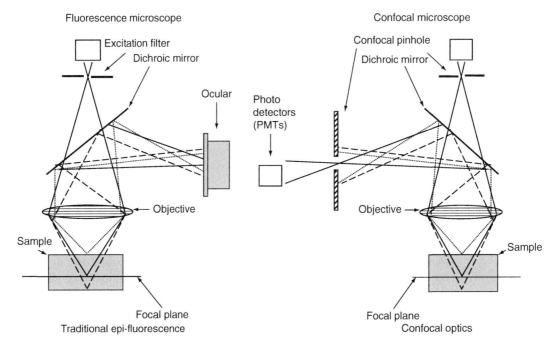

Fluorescence microscope

Confocal microscope

Fig. 1 The light pathway in conventional fluorescence microscopy versus confocal microscopy. It should be noted that the principle of confocality is that the emission signal recorded will be exclusive for each focal plane within the specimen that is imaged. This feature is essentially the role of the confocal iris as described in the text.

separated from the excitation source by a beam-splitting dichroic mirror, and is reflected to a photomultiplier tube, amplifying the signal. After passing through an analog-to-digital converter, the signal is displayed as a sequential raster scan of the image. Depending on the desired measurements within the imaging requirement it is possible to collect very small scan ranges from 50×50 points (or even smaller) up to rather large scanning areas with as many as 4096×4096 points. Most current systems utilize 16-bit ADCs, allowing an effective image of 1024×1024 pixels or more with at least 256 gray levels. Some confocal microscopes can collect high-speed images at video rates, and use 8- or 12-bit ADCs, such as the Zeiss Live, which can collect up to 1536×1536 pixels, also for several channels, with continuously variable scanning speed up to 120 frames/s with 512×512 pixels; there are faster modes with smaller frames (e.g., 505 frames/s with 512×100 pixels, 1010 frames/s with 512×50 pixels) and an ultrafast line scan mode with $>60,000$ lines/s. Some instruments achieve fast scanning by slit scanning. Just because an instrument can collect more points (and thus higher resolution) does not necessarily mean it is useful. For example, if the time required to collect a very large image is excessive, there might be severe photobleaching making the collection of no value. A single scan of a 4096×4096 image might take several seconds. To collect a relatively small number of sections (50) with signal averaging of 3 scans per image would take many minutes, an impractical time constraint with many biological specimens. Regardless, the perfect image could take several runs to acquire, and so the "high-resolution" mode is less practical for 3D imaging than the commercial literature might suggest.

Frequently, practical operation of the confocal microscope will be image collection at a size using the fastest possible point scanning available on the instrument to achieve a quality signal without photobleaching. Once the imaging area is selected, the top and bottom (in the Z-axis) of the image sections are identified; if desired, the image collection parameters can be changed at this point to obtain higher resolution. Electronic magnification is one of the most useful components of the confocal collection system and is universally available on all microscopes. The principle of electronic magnification is that the imaging area is reduced, but the number of pixels in the collection area remains constant. This effectively magnifies the image. However, it is generally not possible to magnify the image beyond the point where the Nyquist criterion (2.3f) is exceeded, since beyond this is considered empty magnification—although there are cases where "super-resolution" is possible (Plášek and Reischig, 1998). An important point to consider is that the power delivered to the specimen increases with the square of the magnification. Therefore, a zoom factor of 2 places four times the laser power onto the object. This could cause serious bleaching or physically heat the specimen beyond a reasonable level. An example of electronic magnification is shown in Fig. 2. The primary advantage is that one can view a larger field of the sample and zoom in to areas of particular interest using the zoom feature.

Investigators have demonstrated two-photon excitation in which a fluorophore simultaneously absorbs two photons each having half the energy—and twice the wavelength–normally required to raise the molecule to its excited state. A significant advantage of this system is that only the fluorophore molecules in the focal plane are excited, as this is the only area with sufficient light intensity. The higher wavelengths used mean that considerably less background noise is collected and the efficiency of imaging thick specimens is significantly increased. Those probes requiring UV-excitation can be excited by means of two-photon excitation (Sako *et al.*, 1997), which may have the advantage of causing less tissue damage (particularly when imaging live cells, as compared to using a UV-excitation source); this is still subject to verification, although the evidence appears to support this notion. Multiphoton microscopy has decided advantages in imaging to a greater depth. While resolution in most thin ($<70\ \mu m$) tissues is actually worse than in conventional confocal microscopy, it is actually improved in thick tissues, where conventional confocal microscopy is unable to image well at all.

A. Benefits of Confocal Microscopy

A now-familiar tool in the research laboratory, confocal microscopy has a number of significant advantages over conventional fluorescence microscopy:

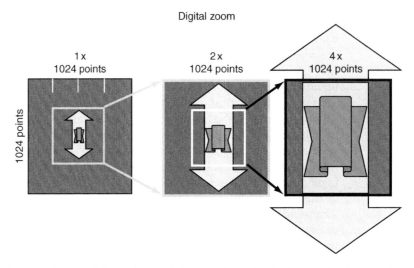

Fig. 2 The principles of electronic zoom is based on the notion that by reducing the area of the scan, but not reducing the number of points within the scan, the image will be electronically magnified. This is shown in the cartoon above. The first box represents a 1024 × 1024 matrix. If a box half this size is now imaged using 1024 × 1024 points, the resultant image will be magnified by a factor of 2. If this is repeated the resultant image will be magnified by a factor of 4. These of course depend on the fact that the image is oversampled. The impact is that the area of the object to be imaged is reduced as demonstrated by the arrow-like objects in the cartoon.

1. Improved Resolution in the "*x-y*" Plane

Because the effective resolution of optical instruments is limited by diffraction of light, Fraunhofer diffraction on a circular aperture can be used to model the imaging process (Born and Wolf, 1999). The intensity distribution $I(v)$ produced by diffraction is proportional to

$$I(v) \propto \left[\frac{J_1(v)}{v}\right]^2 \tag{1}$$

where $J_1(x)$ is a Bessel function of the first kind and v is a coordinate which is related to the transverse distance in the focal plane d by

$$v = \frac{2\pi}{\lambda} d \, \sin\theta \tag{2}$$

Airy diffraction pattern or *Airy disk* is the term used to identify the intensity distribution, and the Airy disc describes the two-dimensional intensity point spread function (PSF) of the system; θ is the half angle of the cone of light converging to an illuminated spot or diverging from one; λ is the wavelength.

The resolution r defined by Rayleigh is the required separation of two objects such that their diffraction pattern shows a measurable drop in intensities between these two points. When the distance to the first dark fringe in the diffraction pattern is treated as a measurement of the resolution in a Rayleigh sense, this detectable drop is approximately 26%. A simpler measure is the FWHM criterion which relies on the measurement of the Airy disk at half its maximum height. For the conventional microscope this is about the same as the radius of the first minimum (Amos, 2000):

$$r = \frac{0.61\lambda}{n\,\sin\theta} = \frac{0.61\lambda}{\mathrm{NA}}, \quad r_{\mathrm{FWHM}} = \frac{0.5\lambda}{n\,\sin\theta} = \frac{0.5\lambda}{\mathrm{NA}} \tag{3}$$

where n is the index of refraction of the medium; NA, the product of $\sin\theta$ and the former value, is called the *numerical aperture*.

The use of a small confocal pinhole in a confocal microscope causes the photomultiplier to function as a *point detector*. Therefore, when identical optics are used for illumination and observation, the function describing intensity distribution becomes a product of two identical Airy disks, and is thus proportional to

$$I(v) \propto \left[\frac{J_1(v)}{v}\right]^2 \left[\frac{J_1(v)}{v}\right]^2 = \left[\frac{J_1(v)}{v}\right]^4 \tag{4}$$

The resolution limit r, defined again as the width of the Airy disk at half its maximum height, derived from the function above differs from the resolution limit for a classical microscope and is expressed by the following equation:

$$r_{\mathrm{FWHM}}^{\mathrm{conf}} = \frac{0.44\lambda}{\mathrm{NA}} \tag{5}$$

The confocal system has an amplitude point spread function (PSF) which is narrower than that of the corresponding nonconfocal system (such as a fluorescence microscope) by factor of 0.72, as measured between the half-power points (the FWHM criterion). Indeed, comparing Eqs. (3) and (5) it is evident that confocal resolution in the "x-y" plane (measured using the FWHM criterion) is increased over that achievable by conventional fluorescence microscopy. However, the distance from peak to the first zero (being the conventional Rayleigh criterion) is unchanged (Sheppard and Choudhury, 1977). The resolution improvement clearly varies for different imaging modes (fluorescence, backscattered light, and transmitted light).

2. Improvement of Signal-to-Noise Ratio Through Rejection of Out-of-Focus Light from Planes Other Than the Plane of Focus

Confocal microscopes use only light eminating from the volume of the object conjugate to the detector and the source. Once the background light has been reduced, the full resolution available from the optics may be utilized. The confocal diffraction pattern has much less energy outside the central peak than does the wide-field pattern. Hence, a bright object near a dim one is less likely to contribute background light to reduce image contrast. Thus a dim object resolved in a confocal laser scanning microscope from the Rayleigh perspective can be really *seen* by the observer as being resolved. This translates to the fact that the rejection of out-of-focus background results in an improved signal-to-noise ratio. It has been shown that the reduction of background in a point-scanning confocal system could provide a signal-to-noise ratio that is an order of magnitude better than that of a conventional microscope. The spinning-disk confocal instruments have a signal-to-noise ratio that is greater than that of the conventional microscope by a factor of 2–3 (Sandison and Webb, 1994). Confocal optics also effective discriminate diffuse scattering from planes away that are remote from the focal plane (Sheppard and Wilson, 1978).

3. 3D Imaging Capability: Optical Sectioning through Thick Specimens

For a conventional microscope the intensity variation as a point object is displaced along the axis is given by

$$I(u) \propto \left(\frac{\sin(u/4)}{u/4} \right)^2 \tag{6}$$

where u is a normalized optical coordinate related to the axial distance z by

$$u = \frac{2\pi}{\lambda} z \sin^2\theta \tag{7}$$

However, in a confocal microscope this intensity is squared (Sheppard and Wilson, 1978):

$$I(u) \propto \left(\frac{\sin(u/4)}{u/4} \right)^4 \tag{8}$$

The consequence of this effect is that the axial extent of the confocal microscope PSF is about 30% smaller compared to the conventional microscope PSF. Therefore, it is possible to express the axial resolutions for conventional and confocal microscopes as (Jonkman and Stelzer, 2002)

$$r_z = \frac{2\lambda n}{\text{NA}^2}, \ r_z^{\text{conf}} \approx \frac{1.4\lambda n}{\text{NA}^2} \tag{9}$$

If the depth of field is now defined in terms of the reduction in maximum intensity for a point image, then it is clear that the depth of field for confocal microscope is only slightly reduced relative to that of a conventional microscope (Sheppard and Matthews, 1987; Sheppard, 1988). Therefore, the improvement in axial resolution (Eq. (9)) does not explain the optical sectioning capabilities of confocal microscopy. However, if we consider the variation in the integrated intensity for the image of a point source (which shows the total power in the image) we will see why using a confocal microscope can discriminate against parts of the object that do not fall within the focal plane.

The integrated intensity of the light emanating from any one point in the specimen is almost unchanged in a conventional microscope as the object moves away from the focus. However, the confocal integrated intensity PSF is a maximum in the focal plane. Therefore, in contrast to conventional fluorescent microscopes, confocal systems collect signal only from fluorochromes located in the immediate location of the focus (Jonkman and Stelzer, 2002; Sheppard and Wilson, 1978). The axial imaging properties of confocal system are degraded by the presence of spherical aberration, which occurs when focusing with a high-NA, oil-immersion objective into a biological aqueous environment.

4. Depth Perception in Z-Sectioned Images

Reconstruction techniques are used to reconstruct an image of the fluorescence emission of the specimen through the entire depth of the specimen. While confocal instruments faithfully image structures with dimensions as small as subcellular organelles up to whole tissue preparations, there are some technical constraints that limit the maximal thickness of the objects which can be imaged. To obtain 3D images which closely represent the geometry of the sample, the light path through the sample must be as short as possible, since imaging artifacts like astigmatism, spherical aberration, and intensity attenuation increase with path length (Hell *et al.*, 1993). By using advanced digital reconstruction techniques which are readily available, it is possible to extract image information which is not easily accessible by simply presenting individual sections.

5. Electronic Magnification Adjustment

By reducing the scanned area of the excitation source, but retaining the effective resolution, it is possible to magnify the image electronically. This has a number of advantages over conventional microscopy.

B. Excitation Sources

The most common light sources for confocal microscopes are lasers. The acronym LASER stands for light amplification by stimulated emission of radiation. Most lasers on conventional confocal microscopes are continuous wave lasers (CW) and are either gas, dye, or solid-state lasers. Argon-ion (Ar) lasers are the most popular gas lasers, followed by either krypton-ion (Kr) or a mixture of argon and krypton (Kr-Ar) or helium and neon (He-Ne). The small argon-ion lasers used in confocal microscopy produce 10–500 mW TEM_{00} mode at 488 nm. They are compact and air cooled, stabilize in less than 15 min after being turned on, and exhibit low amplitude noise. Krypton-argon-ion lasers contain a mixture of argon and krypton gases. They provide both the strong blue and green emissions of argon lasers, and additionally the red and yellow lines of the krypton-ion transitions at 647.1 and 568.2 nm. Helium-neon lasers, which are available now with 543.5-nm lines, are also popular because of their low cost, reliability, and compactness.

Helium-cadmium (He-Cd) can also used in confocal microscopy. The He-Cd can provide UV lines at 325 or 441 nm, although use of the 325 nm line is very difficult in most microscopes because of loss of signal transmission at wavelengths below 350 nm. The most common source of UV excitation for the confocal microscope is the argon-ion laser, which can emit 350–363-nm UV light. The traditional UV-excitation source for UV/vis confocal systems has been the Coherent "Enterprise," which of course requires a source of chilled water. A listing of the frequently used probes for confocal microscopy, together with the laser lines required, is shown in Table I. The power necessary to excite fluorescent molecules at a specific wavelength can be calculated. For instance, consider 1 mW of power at 488 nm focused via a 1.25-NA objective to a Gaussian spot whose radius at $1/e^2$ intensity is 0.25 μm. The peak intensity at the center will be 10^{-3} W [$\pi(0.25 \times 10^{-4}$ cm)^2] = 5.1×10^5 W/cm^2 or 1.25×10^{24} photons/(cm s) (Shapiro, 1995). If FITC were the fluorochrome used in such a system, 63% of its molecules would be in an excited state and 37% in the ground state at any one point in time (Shapiro, 2003). This would be sufficient to obtain efficient excitation of this probe. For optimal confocal microscopy, the power delivered to the fluorescent probe must be sufficient to saturate the fluorescent molecules in the specimen.

Diode lasers have been commercially available since 1962, but have only recently been capable of producing sufficient output power and beam quality to be used in imaging systems. Unfortunately, the cheap mass-produced diodes known from CD

Table I
Probes for Proteins

Probe	Excitation	Emission
FITC	488	525
PE	488	525
APC	630	650
PerCP™	488	680
Cascade Blue	360	450
Coumarin-phalloidin	350	450
Texas Red™	610	630
Tetramethylrhodamine	550	575
CY3 (indotrimethinecyanines)	540	575
CY5 (indopentamethinecyanines)	640	670

and DVD players emit in red and have little utility for biological confocal microscopy. Among red diodes only the 635-nm AlGaInP laser is useful in standard confocal systems.

Alternatives for providing shorter wavelengths (440, 405, 375 nm) are available in blue diode lasers. Currently, blue and violet gallium nitride (GaN) laser diodes are manufactured by only a few companies using two main technologies: by growing GaN crystals on dissimilar materials like sapphire (Nichia Chemicals, Japan) and silicon carbide (Cree, Durham, NC), or by utilizing a unique approach of extremely high pressure to grow GaN crystals on GaN substrates (Unipress Top-GaN, Warsaw, Poland). Blue diode lasers found their way to confocal microscopy almost immediately after being introduced (Girkin and Ferguson, 2000). Coherent Vioflame/Radius (Coherent Inc., Santa Clara, CA) or iFLEX-2000 (Point Source, Southampton, UK) are examples of commercial lasers utilizing the new GaN diodes.

Most confocal microscopes are designed around conventional microscopes, with the modification of the light source, which can be one of several lasers. For most cell biology studies, arc lamps are not adequate sources of illumination for confocal microscopy. When using multiple laser beams, it is vital to expand the laser beams using a beam-expander telescope so that the back focal aperture of the objective is always completely filled. The beam widths from several different lasers must also be matched if simultaneous excitation is required. The most important feature in selecting the laser line is the absorption maximum of the fluorescent probe.

C. Nipkow Disk Scanners

Instead of scanning the sample with a laser beam, similar effects can be achieved using multiple pinholes arranged in a raster pattern. The most commonly used pattern—the Nipkow disk—consists of a series of rectangular perforated holes arranged in an Archimedes spiral. There are significant advantages in using a Nipkow disk in a scanning confocal microscope because it is possible to view the sample in real time through the eyepiece. However, there are significant problems with spinning-disk microscope, such as low illumination efficiency. Despite this, such systems are very popular and useful in the area of live-cell imaging.

Until recently, confocal microscopy technology did not offer reliable and inexpensive systems for real-time observations. This problem was solved by Tanaami *et al.*, who improved the original design by Petran. The new spinning-disk instruments utilize two disks instead of one. The upper disk consists several thousand very small microlenses. When light illuminates this disk, the microlens focuses the light onto the lower disk, where several thousands pinholes are arranged with the identical pattern. The light passing through each pinhole is aimed by the objective lens at a spot on the specimen. Light from the specimen passes back through the objective lens and pinholes of the first disk, and is reflected by a beam splitter to a CCD camera. The upper disk containing the microlenses and the lower disk containing the pinholes are physically connected and rotated together by an electrical motor, thus raster-scanning the specimen (Tanaami *et al.*, 2002). This design marketed by Yokogawa has been implemented by a number of commercial manufacturers.

D. Structured Illumination, Programmable Array Microscopes (PAMs): Alternative Confocal Technologies

Programmable array microscopes (PAMs) are a family of microscope systems in which a spatial light modulator is placed in an image plane of the microscope and used to generate patterns of illumination and/or detection (Hanley *et al.*, 1999). An example of spatial light modulation would be the Digital Micromirror Device (DMD) (Texas Instruments). The DMD is a semiconductor-based "light switch" array of thousands of individually addressable, tiltable mirror pixels. DMD technology is widely used as a spatial light modulator for projectors. In a microscope, micromirrors of DMDs can be used to create a pattern of reflection pinholes for highly parallel light collection. Microscopes using DMDs to create confocality have been reported but the idea still awaits commercialization (Hanley *et al.*, 1998; Liang *et al.*, 1997).

Instruments for excellent resolution imaging can also be designed using the principles of structured illumination. This concept proposes to change the illumination system of the microscope to project a single spatial-frequency grid pattern onto the object. A microscope utilizing such an illumination model would image efficiently only that portion of the object for which the grid pattern is in focus. The resultant image would have the unwanted grid pattern superimposed; however, this can be removed in real time, permitting acquisitions of optically sectioned images from a conventional wide-field microscope (Neil *et al.*, 1997).

E. Spectral Imaging Instruments

Multispectral imaging has been available in the remote sensing field for over 40 years. However, only recently was it introduced to standard confocal scanning instruments, becoming one of the most important advances in the field of biological imaging. This is an important innovation for a number of very good reasons. First, spectral overlap or crosstalk can be difficult to eliminate even in the simplest systems where two or three fluorophores are used simultaneously. Second, spectral fingerprints of intrinsic or introduced fluorophores can reveal information about the physiological processes inside live cells. Third, spectral imaging can enhance other sophisticated techniques like FRET or dye ratio imaging (Berg, 2004; Dickinson *et al.*, 2001). A number of commercial systems are now readily available.

IV. Useful Fluorescent Probes for Confocal Microscopy

The essential requirement for a fluorescent molecule is an appropriate excitation source. Since most lasers can be successfully used in confocal systems, the number of fluorescent probes available for use in confocal microscopy is very broad. A series of tables is provided detailing the properties of fluorescent probes for proteins (Table I), for intracellular organelles (Table II), for nucleic acids (Table III), for ions (Table IV), and for measuring intracellular changes in oxidation state (Table V). The excitation properties of each probe depend upon its chemical composition. Ideal fluorescent probes will have high quantum yield, large Stokes shift, and nonreactivity with the molecules to which they are bound. It is vital to match the absorption maximum of each probe to the appropriate laser excitation line. For fluorochrome combinations, it is desirable to have fluorochromes with similar absorption peaks but significantly different emission peaks, enabling use of a single excitation source. It is common in confocal microscopy to use two, three, or even four distinct fluorescent molecules simultaneously, although it is usually necessary to image each probe independently and combine the images postcollection.

Table II
Organelle Specific Stains

Probe	Specificity	Excitation	Emission
BODIPY	Golgi	505	511
NBD	Golgi	488	525
DPH	Lipid	350	420
TMA-DPH	Lipid	350	420
Rhodamine 123	Mitochondria	488	525
DiO	Lipid	488	500
diI-C_n(5)	Lipid	550	565
diO-C_n(3)	Lipid	488	500

Table III
Nucleic Acid Dyes

Hoechst 33342		350	460
DAPI		350	470
PI		530	620
Acridine Orange		500	520
TOTO-1		514	530
TOTO-3		640	660
Thiazole Orange (vis)		450	480

Table IV
Probes for Ionic Fluxes

Molecule/Probe	Excitation (nm)	Emission (nm)
Calcium/Indo-1	351	405, >460
Magnesium/Mag-Indo-1	351	405, >460
Calcium/Fluo-3	488	525
Calcium/Fura-2	363	>500
Calcium/Calcium Green	488	515
Pl A/Acyl Pyrene	351	405, >460

Table V
Probes for Oxidative States

Probe	Oxidant	Excitation	Emission
DCFH-DA	(H_2O_2)	488	525
HE	(O_2^-)	488	590
DHR 123	(H_2O_2)	488	525
DAF-2	(NO)	488	538

DCFH-DA, dichlorofluorescin diacetate; HE, hydroethidine; DHR-123, dihydrorhodamine 123; DAF-2, 4,5-diaminofluorescein diacetate.

A. Fluorochrome Photobleaching

Photobleaching is defined as the irreversible destruction of an excited fluorophore by light. Uneven bleaching throughout the thickness of a specimen will bias the detection of fluorescence, causing a significant problem in confocal microscopy. Methods for countering photobleaching include shorter scan times, high magnification, high-NA objectives, and wide emission filters as well as reduced excitation intensity. A number of "antifade" reagents are available; unfortunately, many are not compatible with viable cells. In the absence of an antifade reagent, FITC in particular is very susceptible to photobleaching.

B. Antifade Reagents

Many quenchers act by reducing oxygen concentration to prevent formation of excited species of oxygen. Antioxidants such as propyl gallate, hydroquinone, and *p*-phenylenediamine can be used for fixed specimens but are not useful for live cell studies. Quenching fluorescence in live cells is possible using either systems with reduced O_2 concentration or singlet-oxygen quenchers such as carotenoids (50 mM crocetin or etretinate in cell cultures), ascorbate, imidazole, histidine, cysteamine, reduced glutathione, uric acid, or trolox (vitamin E analog). Photobleaching can be calculated for a particular fluorochrome to determine the maximum scan time possible for that molecule. For example, the most commonly used fluorescent probe, FITC, bleaches with a quantum efficiency Q_b of 3×10^{-5}. A standard laser intensity would pump 4.4×10^{23} photons/(cm s) and FITC would be bleached with a rate constant of 4.2×10^3 s^{-1}. After 240 μs of irradiation, only 37% of the molecules would remain. In a single plane, 16 scans would cause 6–50% bleaching (Tsien and Waggoner, 1990). An excellent source for information on photobleaching is the excellent chapter on the subject by Diaspro *et al.* (2006).

V. Applications of Confocal Microscopy

A. Cell Biology

The applications in cell biology are expanding on a daily basis, in no small part owing to a new generation of simple-to-use confocal microscopes that have been designed to remove the technical difficulties previously associated with operating these instruments. Currently, one of the more frequent applications is cell tracking using green fluorescent protein (GFP), a naturally occurring protein from the jellyfish *Aequorea victoria* that fluoresces when excited by UV or blue light (Jordan *et al.*, 1999; Moerner *et al.*, 1999; Sullivan and Shelby, 1998). A gene for a fluorescent protein such as GFP can be transfected into cells so that subsequent replication of the organism carries with it the fluorescent reporter molecule, providing a valuable tool for tracking the presence of that protein in developing tissue or differentiated cells. This is particularly useful for identifying regulatory genes in developmental biology, and for identifying the biological impact of alterations to normal growth and development processes.

In almost any application, multiple fluorescent wavelengths can be detected simultaneously. If a UV/vis confocal microscope is available, Hoechst 33342 (420 nm), FITC (525 nm), and Texas Red (630 nm) can be simultaneously collected to create a three-color image, providing excellent information regarding the location of the labeled molecules and the structures they identify, and the relationships between them. Figure 3 shows an example of multiple materials being imaged by confocal microscopy.

B. Microscopy of Living Cells

Evaluation of live cells using confocal microscopy presents some difficult challenges. One is the need to maintain a stable position while imaging a live cell. For example, a viable respiring cell may be constantly changing shape, preventing a finely resolved 3D-image reconstruction. Fluorescent probes must be found which are not toxic to the cell. Figure 4 presents an example of cells attached to an extracellular matrix. In this image, the cells can be accurately identified and enumerated, and their relative locations within the matrix determined as well.

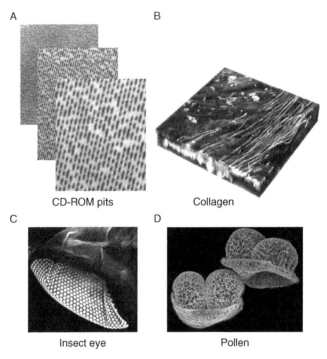

Fig. 3 Various materials imaged with a confocal microscope. This composite image shows (A) the pits from a CD-ROM imaged by confocal microscopy. (B) A 3D reconstruction of extracellular matrix using approximately 100 sections imaged on a Biorad 1024 confocal microscope. Collagen bundles are clearly visible—and while the resolution on the x–z plane is not high, the 3D reconstruction provides ample information regarding the nature of the materials. (C) A reconstruction of the eye of a fly. (D) Pine tree pollen showing brightest pixel display.

This figure demonstrates the effectiveness of confocal microscopy as a qualitative and quantitative tool for creating a 3D-image reconstruction of live cells. Figure 5 is another example of 3D imaging of live cells, in this case endothelial cells growing on glass in a tissue culture dish. Thirty image sections were taken 0.2 μm apart; the image plane presented shows an x–z plane with the cells attached to the cover glass. Figure 6 is a cartoon showing how attached cells might be imaged using a line scanning confocal microscope, for example.

C. Calcium Imaging

Confocal microscopy can be used for evaluation of physiological processes within cells. Examples are changes in cellular pH, changes in free Ca^{2+} ions, and changes in membrane potential and oxidative processes within cells. One of the most successful methods for evaluating these phenomena is emission ratioing in real time. Usually the molecules under study are excited at one wavelength but emit

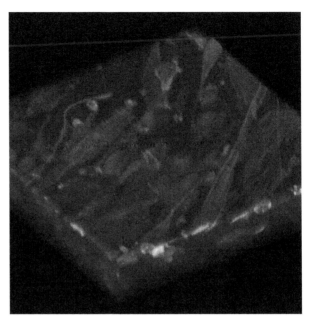

Fig. 4 Extracellular matrix materials imaged with a confocal microscope. In this image, the cells can be accurately enumerated and their relative locations within the matrix determined as well. The central nuclei are stained with Hoechst (UV excited) and the rest of the collagen is shown as the autofluorescent signal. This image shows a 3D reconstruction of approximately 100 sections imaged on a Biorad 1024 confocal microscope. Collagen bundles are clearly visible—and while the resolution on the x–z plane is not high, the 3D reconstruction provides ample information regarding the nature of the material. (See Plate no. 1 in the Color Plate Section.)

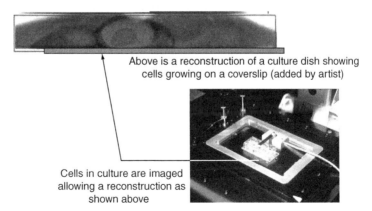

Fig. 5 3D imaging of live endothelial cells growing on glass in a tissue culture dish. Thirty image sections were taken 0.2 μm apart; the image plane presented shows an x–z plane with the cells attached to the cover glass. The bar underneath the cells was added to show where the cover slip would normally be since this image is a reconstruction of the x–z plane from data collected in the x–y planes. Below is a photo of the culture dish sitting on the inverted stage of the confocal microscope. (See Plate no. 2 in the Color Plate Section.)

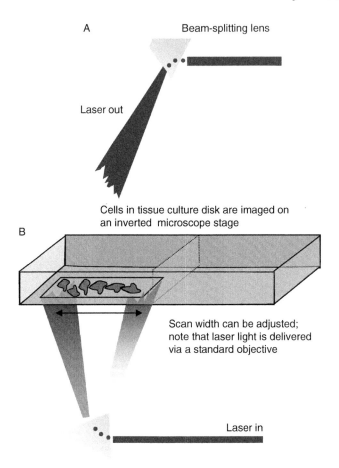

Fig. 6 A cartoon showing an example of how attached cells can be analyzed using confocal microscopy. In this case, the cells are scanned via a line scanning microscope, which is faster than point scanners; however, the principle of evaluating cultured cells is identical using both confocal technologies.

at two wavelengths depending on the change in properties of the molecule. Changes in cellular pH can be identified using SNARF-1 (Edwards *et al.*, 1998) or BCECF (Stephano and Gould, 1997; Yip and Kurtz, 1995), which is excited at 488 nm and emits at 525 and 590 nm. The ratio of 590/525 signals reflects the intracellular pH. Calcium changes can be detected using INDO-1, which can be excited at 350 nm (Niggli *et al.*, 1994; Sako *et al.*, 1997). INDO-1 can bind Ca^{2+}, and the fluorescence of the bound molecule is preferentially at the lower emission wavelength; the ratio of emission signals at 400 nm/525 nm reflects the concentration of Ca^{2+} in the cell.

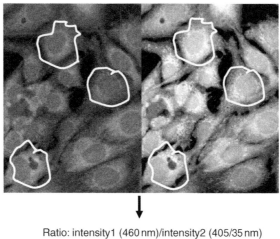

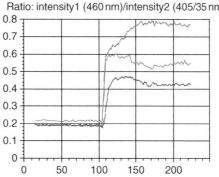

Ratio: intensity1 (460 nm)/intensity2 (405/35 nm)

Fig. 7   Fast kinetic imaging is a common application in confocal microscopy. While the intent is not to evaluate 3D or tissue depth, the confocal microscope lends itself very well to kinetic imaging. In this case, it is calcium imaging using Indo-1, which requires a UV excitation source. Here, the changes in fluorescence of cells loaded with calcium-sensitive dye were measuring using a confocal microscope and calcium-ratioing software. The same regions in each wavelength were measured, and the relative changes were recorded and plotted.

Rapid changes in Ca^{2+} can be detected by kinetic imaging   taking a series of images at both emission wavelengths in quick succession. An example is shown in Fig. 7.

D. Cell Adhesion Studies

One early example of the power of confocal microscopy was the study of chondrocytes essentially *in vivo* (Errington *et al.*, 1997). In addition, studies of osteoblastic cell adhesion have been performed using confocal microscopy. Investigators in those studies were interested in the cell attachment and release mechanisms of human osteoblasts to orthopedic devices used for bone or joint replacement (Shah *et al.*, 1999).

E. Colocalization Studies

One of the routine uses for confocal microscopy is the colocalization of or distribution of molecules produced within living organisms. For example, studies of the distribution of HMG-I protein, a high-mobility group protein which interacts *in vitro* with the minor groove of AT-rich B-DNA, have demonstrated that it is found exclusively in the nucleus (Amirand *et al.*, 1998). Other examples of colocalization have been shown in studies of the TR6 protein produced by equine herpes virus. Confocal microscopy was able to determine that the IR6 protein of wild-type RacL11 virus colocalizes with nuclear lamins very late in infection, whereas the mutant IR6 protein encoded by the RacM24 strain did not colocalize with the lamin proteins (Osterrieder *et al.*, 1998).

Similarly, colocalization studies using confocal microscopy have recently determined that gene 1 products associated with murine hepatitis virus (MHV) are directly associated with the viral RNA synthesis. Confocal microscopy revealed that all the viral proteins detected by these antisera colocalized with newly synthesized viral RNA in the cytoplasm, particularly in the perinuclear region of infected cells. Several cysteine and serine protease inhibitors—E64d, leupeptin, and zinc chloride—inhibited viral RNA synthesis without affecting the localization of viral proteins, suggesting that the processing of the MHV gene 1 polyprotein is tightly associated with viral RNA synthesis. Dual labeling with antibodies specific for cytoplasmic membrane structures showed that RNA and MHV gene 1 products colocalized with the Golgi apparatus in HeLa cells. However, in murine 17CL-1 cells, the viral proteins and viral RNA did not colocalize with the Golgi apparatus but, instead, partially colocalized with the endoplasmic reticulum (Shi *et al.*, 1999). It is fair to say that despite the many alternative technologies available, only confocal microscopy, via its unique ability to create accurate 3D structural representations of cells and their organelles, was able to demonstrate the location of the viral RNA.

F. Fluorescence Resonance Energy Transfer (FRET)

FRET is a distance-dependent photophysical interaction between two molecules in which the excitation energy of one molecule (the donor) can be transferred nonradiatively to the other. For FRET to occur, molecules must be in close proximity (a maximum distance of 100 Å), and the absorption spectrum of the recipient molecule and the emission spectrum of the donor must overlap (Emptage, 2001; Wouters *et al.*, 2001). FRET has been used in both spectrophotometry and microscopy for many years; however, only recently with the introduction of fluorescent proteins which form convenient donor-acceptor FRET pairs has the method become universal.

G. Fluorescence Recovery After Photobleaching (FRAP)

Fluorescence recovery after photobleaching (FRAP) is a measure of the dynamics of the chemical changes in a fluorescent molecule within an object such as a cell. A small area of the cell is bleached by exposure to an intense laser beam, and the recovery of fluorescent species in the bleached area is measured. The recovery time (t) can be calculated from the equation $t = W^2/4D$, where W is the diameter of the bleached spot and D is the diffusion coefficient of the fluorescent molecule under study. An alternative technique can measure interactions between cells; one of two attached cells is bleached and the recovery of the pair as a whole is monitored. To satisfactorily perform FRAP experiments, it is necessary to be able to park the laser beam over a particular cell, or part of a cell, and effectively bleach the fluorescence of this component of the cells. A number of studies designed to explore multistep signal-transducing events can be performed using FRAP apparatus—for example, tracking G-protein segments to localized regions of the cell (Kwon *et al.*, 1994). FRAP has also been used to determine macromolecular diffusion of biological polymers (Gribbon and Hardingham, 1998). The fluorescence loss in photobleaching (FLIP) technique requires monitoring of one site and bleaching of another site in the sample. If there is an exchange of fluorescent molecules between two regions, an increase of fluorescence in the bleached site will occur, whereas the other region should gradually exhibit loss of fluorescence (McNally and Smith, 2002).

VI. Conclusions

Confocal microscopy has reached the point that instruments are now effective and inexpensive compared to the early 1990s when commercial technologies were introduced. More complex systems such as UV microscopes are still relatively rare; however, these may eventually be superseded by the multiphoton microscopes now becoming almost turnkey in operation. For routine one-, two-, or three-color fluorescence microscopy where 3D is important, conventional confocal microscopes will considerably surpass multiphoton resolution and will be around for many more years. A new breed of low-cost instruments is now available, making this technology a simple tool requiring no significant technical or engineering knowledge. New fluorescent dyes are also available, increasing the applications for confocal microscopy. In addition, one of the most important innovations has been the expansion of analytical software for analysis and graphical presentation of confocal imaging data. For example, software packages are now very powerful

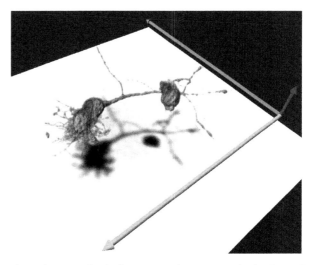

Fig. 8 A neuron imaged on a confocal microscope and reconstructed using Image Pro Plus software. By using advanced software packages, highlights such as shadows, color effects, and rotations are easily obtainable from 3D image stacks. (See Plate no. 3 in the Color Plate Section.)

and fast compared to just a few years ago. The ability to represent materials graphically is a vital and growing part of confocal microscopy. An example of one such graphic is presented in Fig. 8. Finally, new applications are being developed that utilize reflectance (Brightman *et al.*, 2000) or autofluorescence, which while previously considered a problem can be used a tool for extracting 3D information from within thick tissue. Despite the fact that more sophisticated techniques like two-photon microscopy, four-pi microscopy, and others have become available, confocal microscopy, owing to its simplicity and affordable price, as well as to progress in laser technology and in computing, remains the most widely used tool for three-dimensional optical imaging of cells and tissues. The future of confocal microscopy is most likely going to be heavily influenced by the availability of moderately low cost "personal" confocal instruments, and expansion of highly automated HTS/HCS instruments utilizing confocal optics.

References

Amirand, C., Viari, A., Ballini, J. P., Rezaei, H., Beaujean, N., Jullien, D., Kas, E., and Debey, P. (1998). Three distinct sub-nuclear populations of HMG-I protein of different properties revealed by co-localization image analysis. *J. Cell Sci.* **111**(23), 3551–3561.

Amos, W. B. (1988). Results obtained with a sensitive confocal scanning system designed for epifluorescence. *Cell Motil. Cytoskeleton* **10**(1-2), 54–61.

Amos, W. B. (2000). Instruments for fluorescence imaging. *In* "Protein Localization by Fluorescent Microscopy: A Practical Approach" (V. J. Allan, ed.), pp. 67–108. Oxford University Press, Oxford.

Amos, W. B., and White, J. G. (2003). How the confocal laser scanning microscope entered biological research. *Biol. Cell* **95**(6), 335–342.

Amos, W. B., White, J. G., and Fordham, M. (1987). Use of confocal imaging in the study of biological structures. *Appl. Opt.* **26**(16), 3239–3243.

Berg, R. H. (2004). Evaluation of spectral imaging for plant cell analysis. *J. Microsc.* **214**(Pt 2), 174–181.

Born, M., and Wolf, E. (1999). "Principles of Optics: Electromagnetic Theory of Propagation, Interference and Diffraction of Light." 7th expanded edn., Cambridge University Press, Cambridge.

Brakenhoff, G. J., van der Voort, H. T. M., van Spronsen, E. A., Linnemans, W. A., and Nanninga, N. (1985). Three-dimensional chromatin distribution in neuroblastoma nuclei shown by confocal scanning laser microscopy. *Nature* **317**, 748–749.

Brakenhoff, G. J., Blom, P., and Barends, P. (1979). Confocal scanning light-microscopy with high aperture immersion lenses. *J. Microsc. Oxf.* **117**(Nov.), 219–232.

Brightman, A. O., Rajwa, B. P., Sturgis, J. E., McCallister, M. E., Robinson, J. P., and Voytik-Harbin, S. L. (2000). Time-lapse confocal reflection microscopy of collagen fibrillogenesis and extracellular matrix assembly *in vitro*. *Biopolymers* **54**(3), 222–234.

Carlsson, K., Danielsson, P. E., Lenz, R., Liljeborg, A., Majlof, L., and Aslund, N. (1985). Three-dimensional microscopy using a confocal laser scanning microscope. *Opt. Lett.* **10**(2), 53–55.

Cox, I. J., and Sheppard, C. J. R. (1983). Scanning optical microscope incorporating a digital framestore and microcomputer. *Appl. Opt.* **22**(10), 1474–1478.

Davidovits, P., and Egger, M. D. (1969). Scanning laser microscope. *Nature* **223**(5208), 831.

Diaspro, A., Chirico, G., Usai, C., Ramoino, P., and Dobrucki, J. (2006). "Photobleaching. Handbook of Biological Confocal Microscopy," 3rd edn., Springer, New York.

Dickinson, M. E., Bearman, G., Tille, S., Lansford, R., and Fraser, S. E. (2001). Multi-spectral imaging and linear unmixing add a whole new dimension to laser scanning fluorescence microscopy. *Biotechniques* **31**(6), 1272, 1274–1276, 1278.

Edwards, L. J., Williams, D. A., and Gardner, D. K. (1998). Intracellular pH of the mouse preimplantation embryo: Amino acids act as buffers of intracellular pH. *Hum. Reprod.* **13**(12), 3441–3448.

Egger, M. D., and Petran, M. (1967). New reflected-light microscope for viewing unstained brain and ganglion cells. *Science* **157**(3786), 305–307.

Emptage, N. J. (2001). Fluorescent imaging in living systems. *Curr. Opin. Pharmacol.* **1**(5), 521–525.

Errington, R. J., Fricker, M. D., Wood, J. L., Hall, A. C., and White, N. S. (1997). Four-dimensional imaging of living chondrocytes in cartilage using confocal microscopy: A pragmatic approach. *Am. J. Physiol. Cell Physiol.* **272**(3), 1040–1051.

Girkin, J. M., and Ferguson, A. I. (2000). Confocal microscopy using an InGaN violet laser diode at 406 nm. *Opt. Express* **7**(10), 336–341.

Gribbon, P., and Hardingham, T. E. (1998). Macromolecular diffusion of biological polymers measured by confocal fluorescence recovery after photobleaching. *Biophys. J.* **75**(2), 1032–1039.

Hanley, Q. S., Verveer, P. J., and Jovin, T. M. (1998). Optical sectioning fluorescence spectroscopy in a programmable array microscope. *Appl. Spectrosc.* **52**(6), 783–789.

Hanley, Q. S., Verveer, P. J., Gemkow, M. J., Arndt-Jovin, D., and Jovin, T. M. (1999). An optical sectioning programmable array microscope implemented with a digital micromirror device. *J. Microsc.* **196**(Pt 3), 317–331.

Hell, S., Reiner, G., Cremer, C., and Stelzer, E. H. K. (1993). Aberrations in confocal fluorescence microscopy induced by mismatches in refractive-index. *J. Microsc. Oxf.* **169**, 391–405.

Hibbs, A. R. (2004). "Confocal Microscopy for Biologists." Kluwer Academic/Plenum Publishers, New York.

Jonkman, J. E. N., and Stelzer, E. H. K. (2002). Resolution and contrast in confocal and two-photon microscopy. *In* "Confocal and Two-Photon Microscopy Foundations, Applications, and Advances" (A. Diaspro, ed.), pp. 101 126. Wiley-Liss, New York.

Jordan, K., Solan, J. L., Dominguez, M., Sia, M., Hand, A., Lampe, P. D., and Laird, D. W. (1999). Trafficking, assembly, and function of a connexin43-green fluorescent protein chimera in live mammalian cells. *Mol. Biol. Cell* **10**, 2033–2050.

Kwon, G., Axelrod, D., and Neubig, R. R. (1994). Lateral mobility of tetramethylrhodamine (TMR) labelled G protein alpha and beta gamma subunits in NG 108-15 cells. *Cell Signal.* **6**(6), 663.

Liang, M. H., Stehr, R. L., and Krause, A. W. (1997). Confocal pattern period in multiple-aperture confocal imaging systems with coherent illumination. *Opt. Lett.* **22**(11), 751–753.

Matsumoto, B. (2002). "Cell Biological Applications of Confocal Microscopy," 2nd edn., Academic Press, San Diego.

McNally, J., and Smith, C. L. (2002). Photobleaching by confocal microscopy. *In* "Confocal and Two-Photon Microscopy Foundations, Applications, and Advances" (A. Diaspro, ed.), pp. 525–538. Wiley-Liss, New York.

Minsky, M. (1988). Memoir on inventing the confocal scanning microscope. *Scanning* **10**, 128–138.

Moerner, W. E., Peterman, E. J. G., Brasselet, S., Kummer, S., and Dickson, R. M. (1999). Optical methods for exploring dynamics of single copies of green fluorescent protein. *Cytometry* **36**(3), 232–238.

Naora, H. (1951). Microspectrophotometry and cytochemical analysis of nucleic acids. *Science* **114**(2959), 279–280.

Neil, M. A. A., Juskaitis, R., and Wilson, T. (1997). Method of obtaining optical sectioning by using structured light in a conventional microscope. *Opt. Lett.* **22**(24), 1905–1907.

Niggli, E., Piston, D. W., Kirby, M. S., Cheng, H., Sandison, D. R., Webb, W. W., and Lederer, W. J. (1994). A confocal laser scanning microscope designed for indicators with ultraviolet excitation wavelengths. *Am. J. Physiol. Cell Physiol.* **266**(1), 303–310.

Osterrieder, N., Neubauer, A., Brandmuller, C., Kaaden, O. R., and O'Callaghan, D. J. (1998). The equine herpesvirus 1 IR6 protein that colocalizes with nuclear lamins is involved in nucleocapsid egress and migrates from cell to cell independently of virus infection. *J. Virol.* **72**(12), 9806–9817.

Paddock, S. W. (1999). "Confocal Microscopy: Methods and Protocols." Humana Press, Totowa, NJ.

Pawley, J. B. (2006). "Handbook of Biological Confocal Microscopy," 3rd edn., Plenum Press, New York.

Plášek. J., and Reischig, J. (1998). Transmitted-light microscopy for biology: A physicist's point of view. Parts I and II. *Proc. R. Microsc. Soc.* **33**, 121–127.

Sako, Y., Sekihata, A., Yanagisawa, Y., Yamamoto, M., Shimada, Y., Ozaki, K., and Kusumi, A. (1997). Comparison of two-photon excitation laser scanning microscopy with UV-confocal laser scanning microscopy in three-dimensional calcium imaging using the fluorescence indicator Indo-1. *J. Microsc.* **185**(1), 9–20.

Sandison, D. R., and Webb, W. W. (1994). Background rejection and signal-to-noise optimization in confocal and alternative fluorescence microscopes. *Appl. Opt.* **33**(4), 603–615.

Shah, A. K., Sinha, R. K., Hickok, N. J., and Tuan, R. S. (1999). High-resolution morphometric analysis of human osteoblastic cell adhesion on clinically relevant orthopedic alloys. *Bone* **24**(5), 499–506.

Shapiro, H. M. (1995). "Practical Flow Cytometry," 3rd edn., Wiley-Liss, New York.

Shapiro, H. M. (2003). "Practical Flow Cytometry." 4th edn., Wiley-Liss, New York.

Sheppard, C. J. R. (1988). Super-resolution in confocal imaging. *Optik* **80**(2), 53–54.

Sheppard, C. J. R., and Choudhury, A. (1977). Image-formation in scanning microscope. *Opt. Acta* **24**(10), 1051–1073.

Sheppard, C. J. R., and Matthews, H. J. (1987). Imaging in high-aperture optical-systems. *J. Opt. Soc. Am. A Opt. Image Sci. Vis.* **4**(8), 1354–1360.

Sheppard, C. J. R., and Wilson, T. (1978). Depth of field in scanning microscope. *Opt. Lett.* **3**(3), 115–117.

Shi, S. T., Schiller, J. J., Kanjanahaluethai, A., Baker, S. C., Oh, J. W., and Lai, M. M. C. (1999). Colocalization and membrane association of murine hepatitis virus gene 1 products and *de novo*-synthesized viral RNA in infected cells. *J. Virol.* **73**(7), 5957–5969.

Stephano, J. L., and Gould, M. C. (1997). The intracellular calcium increase at fertilization in Urechis caupo oocytes: Activation without waves. *Dev. Biol.* **191**(1), 53–68.

Sullivan, K. F., and Shelby, R. D. (1998). Using time-lapse confocal microscopy for analysis of centromere dynamics in human cells. *Green Fluorescent Proteins* **58**, 183.

Tanaami, T., Otsuki, S., Tomosada, N., Kosugi, Y., Shimizu, M., and Ishida, H. (2002). High-speed 1-frame/ms scanning confocal microscope with a microlens and Nipkow disks. *Appl. Opt.* **41**(22), 4704–4708.

Tsien, R. Y., and Waggoner, A. (1990). Fluorophores for confocal microscopy: Photophysics and photochemistry. *In* "Handbook of Biological Confocal Microscopy" (J. B. Pawley, ed.), pp. 153–161. Plenum, New York.

White, J. G., Amos, W. B., and Fordham, M. (1987). An evaluation of confocal versus conventional imaging of biological structures by fluorescence light microscopy. *J. Cell Biol.* **105**(1), 41–48.

Wouters, F. S., Verveer, P. J., and Bastiaens, P. I. H. (2001). Imaging biochemistry inside cells. *Trends Cell Biol.* **11**(5), 203–211.

Yip, K. P., and Kurtz, I. (1995). NH3 permeability of principal cells and intercalated cells measured by confocal fluorescence imaging. *Am. J. Physiol. Renal Physiol.* **269**(4), 545–550.

CHAPTER 2

Protein Labeling with Fluorescent Probes

Kevin L. Holmes and Larry M. Lantz

Flow Cytometry Section
National Institute of Allergy and Infectious Diseases
National Institutes of Health
Bethesda, Maryland 20892

I. Update

This chapter describes the chemistry involved in fluorochrome antibody conjugation in order to provide a better understanding of the processes involved and also to provide tips to avoid common pitfalls in these procedures. Although the chapter focuses on conjugation of fluorescent probes to antibodies, these procedures can easily be adapted to protein and peptide conjugation as well. This chapter will guide you through the chemistry involved in protein conjugation in an easy to read format. Although many of the reagents necessary to perform these

DOI: 10.1016/B978-0-12-375045-7.00002-7

procedures are now available in kit form, a thorough knowledge of the chemistry of the reaction will help the investigator avoid many pitfalls and to adjust the experiment to their specific needs.

A perusal of the referenced procedures at the end of this chapter demonstrates that the vast majority of these protocols have been in use for a number of years. The major change in this area of science is the virtual explosion of fluorochromes available for conjugation (FITC, biotin, Cy5, Cy3, APC, PE, Alexa 488, Alexa 547, Qdots to name a few). Additionally, many companies now offer conjugation kits with detailed instructions to guide you thorough the conjugation process.

With a few exceptions (i.e., PE, APC, Qdots, Zenon labeling), the majority of fluorochrome conjugation kits available today utilize fluorescent dyes containing an organic functional group, most commonly, NHS esters, isothiocyanates, or sulfonyl-cholorides. Labeling proteins with organic dyes are simple one step protocols. Because of their ease of operation, labeling kits utilizing these labeling procedures have seen a large increase in the past several years. But the reaction rates are highly pH-dependent and the organic reactants, especially NHS esters and sulfonyl chlorides, hydrolyze rapidly in the aqueous, highly alkaline environment needed for these conjugations. A new organic functional group has recently been marketed by Invitrogen of the activated ester type termed SDP. It is a derivative of NHS esters yet is reported to be hydrolytically stable in aqueous environments. The major pitfalls to avoid in organic dye conjugations are the use of expired reagents, hydrolysis of the diluted activated fluorochrome, and the addition of reactants at incorrect molar ratios.

To avoid pitfalls:

- Fresh reagents are key. Reagents older than 2 weeks can cause a drastic reduction in labeling efficiency. These reactions are routinely performed at pH 8.4 at 25 °C and any drift in pH or ionic strength may have detrimental effects on the outcome of the reaction.
- Addition of reactants in correct molar ratios.
- Some of the newer NHS ester reagents and now soluble in aqueous solvents and must be used immediately to prevent hydrolysis of the reactive probe.

APC, PE, and Qdot labeling utilize heterobifuntional linkers which contain two different reactive groups to covalently bind the fluorochrome to the antibody or protein. Many of the kits available today prepare the antibody for labeling by reducing existing disulfide linked cysteines in the hinge region of the antibody creating sulfhydryls that react with a maleimide group of the heterobifunctional linker.

To avoid pitfalls:

- All of these reactive groups are volatile and will degrade in aqueous solution in very short time frames as noted earlier.
- Addition of reactants in correct molar ratios.
- These procedures will take a minimum of 4–5 h and the investigator must allocate enough time to perform the experiment in a single day.
- As stated earlier, fresh reagents are key to a productive outcome.

One drawback of this protocol is that the protein must have endogenous disulfide linked cysteines. Although intact antibodies do have endogenous cysteines, many proteins do not or the reduction of these cysteines may affect the secondary or tertiary structure of the protein. In these instances, sulfhydryls may be needed be added to the protein as described later in this chapter.

The majority of this chapter details covalent conjugation of antibodies or proteins with fluorochrome probes. Another, noncovalent conjugation procedure has been introduced in recent years known as Zenon labeling. The Zenon labeling method uses a labeled Fab fragment directed against the Fc portion of the antibody to be labeled. Advantages of this procedure are its short preparation time and the ability to label small (microgram) amounts of antibodies. Disadvantages of this procedure are that it is not a covalent conjugation of the fluorochrome to the antibody and only intact antibodies may be labeled using this method. Also, since this is an antigen-antibody reaction, this method cannot be used for other proteins, peptides, or antibody fragments.

Pitfalls:

- Zenon labeled antibody complexes must be used within 30–60 mins.
- When used in multiparameter labeling protocols, fixation with aldehyde-based fixatives following staining is recommended to prevent transfer of the Zenon label between antibodies.
- Addition of reactants in correct molar ratios.
- Labeled material cannot be stored.

Two other pitfalls that are common to all the labeling techniques are

- Failure to titrate the fluorochrome antibody conjugate.
- The choice of an incorrect fluorochrome for your experiment.

Regardless of the efficiency of the labeling reaction or the brightness of the fluorochrome used, if too much labeled antibody is added, nonspecific binding is increased and if too little is added, a reduced signal will result. A fairly large majority of the problems that investigators have in flow cytometry and immunochemistry can be traced to a failure to properly titrate the labeled antibody or protein. Also, before choosing a fluorochrome, the investigator should research the available literature or contact the operator of the flow cytometer, confocal microscope, or fluorescent microscope to determine the correct fluorochrome or fluorochrome combinations that are appropriate for their applications. For example, combinations of PE-Cy5 and APC are extremely difficult to compensate using flow cytometers and FITC will photobleach very quickly when used in confocal microscopy.

This chapter delves into the chemistry involved in the fluorochrome antibody conjugation to give you a better understanding of the processes involved and will help you avoid common pitfalls in these procedures There have been many excellent reviews of bioconjugation chemistry in general (Brinkley, 1992;

Hermanson, 1996a; Wong, 1991a) and detailed methods of fluorochrome labeling (Hardy, 1986; Haugland, 1995; Haugland and You, 1995; Holmes *et al.*, 1997; Johnson and Holborow, 1986). The reader is directed to the latter sources for specific protocols in fluorochrome conjugation.

II. Introduction

Conjugation of fluorescent molecules to proteins is a subset of the much larger field of bioconjugation chemistry. It has nevertheless developed into an important field itself, since the initial descriptions of the use of fluorescently labeled antibodies in tissue sections (Coons and Kaplan, 1950; Coons *et al.*, 1942). With the advent of monoclonal antibodies and flow cytometry, the value of labeling proteins and, in particular, antibodies with fluorescent molecules has become quite evident in the biomedical and basic research communities. The following is a concise overview of the very large field of fluorochrome bioconjugation, including a discussion of the basis of the procedures, optimization of the conjugation, and fluorochrome specific topics. The conjugation of fluorochromes to antibody molecules is emphasized, but the same principles can be applied to other large proteins. There have been many excellent reviews of bioconjugation chemistry in general (Brinkley, 1992; Hermanson, 1996a; Wong, 1991a) and detailed methods of fluorochrome labeling (Hardy, 1986; Haugland, 1995; Haugland and You, 1995; Holmes *et al.*, 1997; Johnson and Holborow, 1986). The reader is directed to the latter sources for specific protocols in fluorochrome conjugation.

III. Labeling of Proteins with Organic Fluorescent Dyes

A. Conjugation Chemistry

The coupling of fluorescent probes to proteins, in particular antibodies, can be best accomplished by an understanding of the chemistry involved in the conjugation. Proteins are polymers of amino acids that contain various side chains. These side chains are utilized as reactive groups to attach dyes and fluorochromes. The reactivity of the protein/antibody will, therefore, be determined by the amino acid composition and the sequence location of the individual amino acids in the three-dimensional structure of the molecule. Thus, the nonpolar hydrophobic amino acids (glycine, alanine, valine, leucine, isoleucine, methionine, proline, phenylalanine, and tryptophan) are usually found on the interior of the protein and unavailable for modification. Amino acids with ionizable side chains (arginine, aspartic acid, glutamic acid, cysteine, histidine, lysine, and tyrosine) and with polar groups (glutamine, serine, and threonine) are usually located on the protein surface and are available for modification (Wong, 1991b). Protein modification reactions are nucleophilic substitution reactions (Fig. 1). In this type of reaction, a nucleophile

(Nu:) with a lone pair of electrons attacks an electron deficient (electrophilic) center, resulting in the displacement of a leaving group (X:). The nucleophile in protein modification is the amino acid side chain. The electrophilic center is most commonly a carbon atom in which a more electronegative atom, such as oxygen, has been attached. The relative reactivity is determined by the nucleophilicity of the amino acid side chain. For protein modification with fluorochromes, there are two general classes of agents that are most commonly used—acylating and alkylating agents (Fig. 2).

In alkylation, an alkyl group is transferred to the nucleophile (amino acid side chain for proteins), and in acylation, an acyl group is bonded. Figure 3 shows a listing of commonly used reactive groups, the functional group on the protein or antibody that is targeted, the linkage formed, and the dyes that are available that use this linkage for conjugation. The acylating agents are amine reactive and form amide, thiourea, or sulfonamides. The alkylating agents most commonly used are the chlorinated s-triazine and N-maleimide groups. Chlorinated triazines react with amines forming amino triazines with loss of a chloride ion. Maleimides react primarily with sulfhydryl groups at neutral pH to form a thioether bond with proteins. Sulfhydryls already present on the protein, such as cytsteines, can be utilized or sulfhydryls can be added to the protein with thiolating reagents such as N-succinimidyl 3-(2-pyridyldithio)propionate (SPDP) or N-succinimidyl

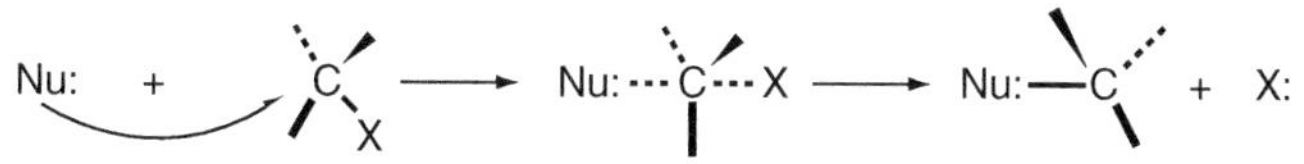

Fig. 1 Schematic representation of a nucleophilic substitution reaction. In this type of reaction, a nucleophile (Nu:) with a lone pair of electrons attacks an electron-deficient (electrophilic) center, resulting in a covalent coupling of the nucleophile and the electrophile along with the displacement of a leaving group (X:).

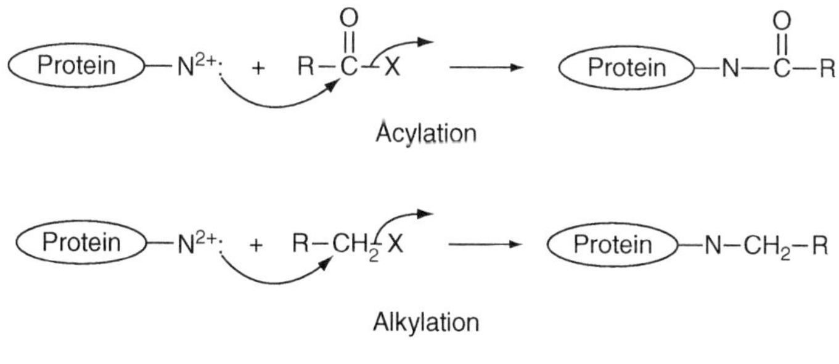

Fig. 2 Schematic representation of alkylation and acylation reactions. In acylation, an active carbonyl group undergoes addition to the amino acid side chain, and in alkylation, an alkyl group is transferred to the nucleophile. Both reactions result in the displacement of a leaving group (X:) as shown in Fig. 1.

Protein functional group	Reactive group on dye	Linkage formed	Dyes available with these reactive groups
 Protein—NH₂ Primary amine	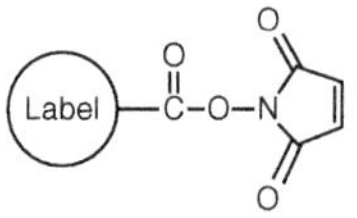 NHS-ester	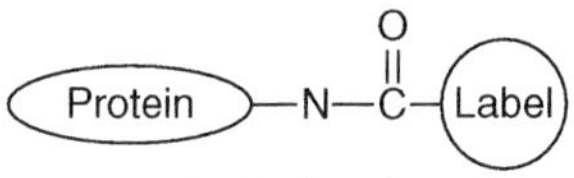 Amide bond	Fluorescein, AMCA, carboxy-fluorescein, biotin, cyanine and alexa dyes, rhodamine,
 Protein—NH₂ Primary amine	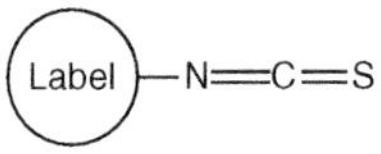 Isothio-cyanate	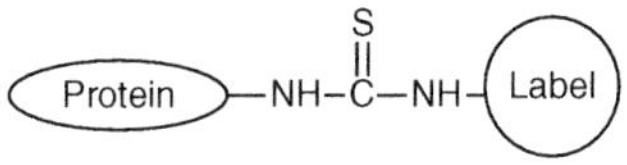 Thiourea	Fluorescein, rhodamine, oregon green
 Protein—NH₂ Primary amine	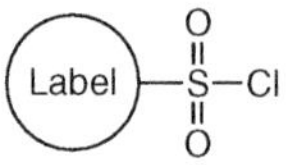 Sulfonyl halide	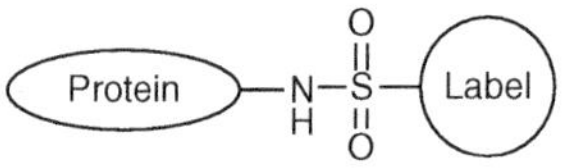 Sulfonamide bond	Texas red, lissamine rhodamine B, sulfonyl chloride
 Protein—NH₂ Primary amine	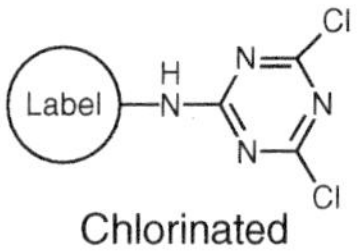 Chlorinated S-triazines	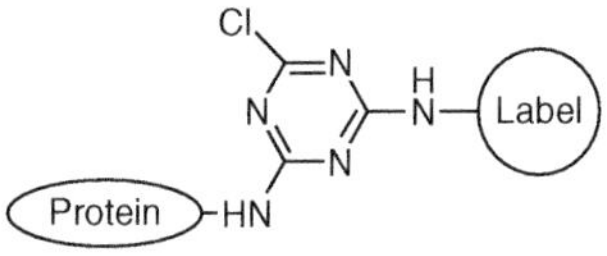 Amino triazines	5-DTAF
 Protein—SH Sulfhydryl	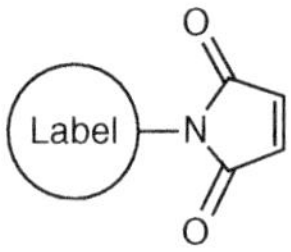 Maleimide	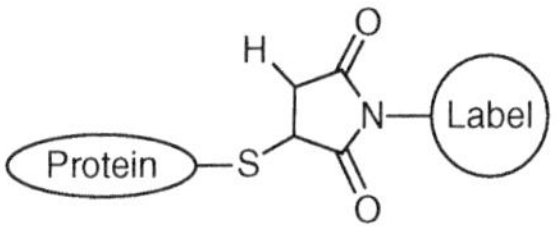 Thioether bond	Biotin, fluorescein, alexa dyes, texas red, rhodamine, oregon green

Fig. 3 Reactive groups and reactions of commonly used fluorescent dyes. Listing of commonly used reactive groups for protein labeling, the functional group on the protein or antibody that is targeted, the linkage formed, and the dyes that are available that use this linkage for conjugation.

S-acetylthioacetate (SATA). These may provide an alternative to amine reactive agents, when it has been found that this coupling results in a loss of biological activity (Imam, 1979).

B. Practical Considerations: Optimizing Conditions for Labeling

When performing coupling of fluorochromes to antibodies, the procedure used to couple is generally less dependent on the fluorophore and more dependent on the reactive moiety attached to it. Therefore, all dyes having N-hydroxysuccinimide (NHS) ester reactive groups, for example, can utilize the same conjugation protocol. The differences in the reaction conditions between the commonly used reactive groups depend on the characteristics of the functional group(s) that is targeted as well as of the reactive group on the fluorochrome.

1. pH and Buffers

The rate of the labeling reaction is governed by several factors, but is dependent on the nucleophilicity of the amino acid side chains, which in turn is dependent on their pK_a (Table I). Ionizable groups in proteins such as carboxylic and amine groups exist either in protonated or unprotonated forms. The degree of protonation is dependent on their pK_a and the pH. Carboxylic groups at pH values above their pK_a will be unprotonated and carry a negative charge. At pH values below their pK_a, carboxylic groups will be protonated and carry no charge. Amine groups, however, are protonated and positively charged at pH values below their pK_a, and they are unprotonated and neutrally charged at pH values above their pK_a. Because protonation decreases nucleophilicity, pH will affect the rate and specificity of the reaction. This can be used to direct reactivity of agents toward particular groups. For example, at neutral pH N-ethylmaleimide reacts more readily to the sulfhydryl group of cysteine (having a pK_a of 8.5–8.8) than with the ε-amino group of lysine (having a pK_a of 9.11–10.7). It is therefore important to choose a buffer that has a buffering capacity within the pH range that is optimal for the reaction. Isothiocyanate labeling which occurs optimally at a pH of 9–9.5 requires a borate or carbonate buffer, whereas N-ethylmaleimide coupling, which occurs at near neutral pH requires a phosphate buffer. In addition, amine-containing buffers such as Tris should not be used because of competition between the amine groups in the buffer and the amino acid side chains of the protein. It should be noted, however, that attempts to selectively target particular amino

Table I
pK_a of Amino Acid Functional Groups[a]

Functional group	pKa in free amino acids	pKa in proteins
ε-Amino	10.5–10.8	9.11–10.7
α-Amino	8.8–10.8	6.72–8.14
Sulfhydryl	8.3–8.4	8.5–8.8
α-Carboxyl	1.8–2.6	3.1–3.7

[a]Data compiled from Wong (1991b) and Botelho and Gurd (1989).

acid groups *exclusively*, using pH alone, is not possible. This is because the microenvironmental effects of the protein, in which the group exists, modify the pK_a of amino acid groups. This means that there may be significant overlap in the pK_a ranges of the different reactive amino acid side chains, and, therefore, overlap in the nucleophilicity of these groups.

2. Protein Concentration and Hydrolysis

The speed and degree of substitution of the coupling reaction is dependent on the concentration of the protein. This is primarily due to hydrolysis of the acylating (or alkylating) agent, because water competes as a nucleophile with the amino acid side chains. This is especially noticeable with NHS esters, which are more reactive with amines at alkaline pH, but also show an increased rate of hydrolysis with increasing pH. Studies performed on NHS ester compounds indicate the half-life of hydrolysis for a homobifunctional NHS-ester is 4–5 h at pH 7 and 0 °C in aqueous environments free of primary amines. This half-life decreases to 1 h at pH 8 and 25 °C, and 10 min at pH 8.6 and 4 °C (Cuatrecasas and Parikh, 1972; Lomant and Fairbanks, 1976; Staros *et al.*, 1986). Therefore, a pH range of 7.5–8.5 has been recommended (Haugland and You, 1995). Protein concentrations of at least 1 mg/ml are desirable; higher concentrations give higher rates of reaction and higher fluorophore/protein (*F/P*) ratios. The *F/P* ratio is the quantitative measure of the level of fluorophore modification of the protein, is determined spectrophotometrically, and is expressed as moles of fluorophore/moles of protein. (For a more detailed discussion of this topic, see Brinkley, 1992; Haugland, 1995.) Brinkley (1992) recommends concentrations of 7.5–15 mg/ml. Ideally, the protein concentration should be kept constant to ensure reproducible results. In this regard, protein concentration is more important than total amount of protein when labeling small amounts of antibody.

3. Degree of Substitution and Activity/Quenching Effects

The final *F/P* ratio achieved in fluorochrome labeling is dependent on the factors previously listed, that is, pH and protein concentration, but also time and temperature. In general, most reactions can be performed at room temperature, with some reactions showing more or less sensitivity to temperature changes (Wong, 1991c). It has been suggested that a convenient procedure is to add the fluorochrome to a stirred protein solution in an ice-bath and allow the bath to warm to room temperature over a period of about 2 h (Brinkley, 1992). Most protocols are designed to provide an optimal *F/P* within usually 1–2 h. Longer incubation time will provide increased *F/P* ratios, but the final *F/P* ratio desired may vary, dependent on the dye and the antibody or protein that is being labeled. It is best to systematically vary conditions, usually either time of incubation or amount of dye added to the protein solution, to achieve optimal labeling. Optimal labeling will be below the maximum substitution possible for three reasons. First, unless the antibody-combining site is somehow protected (Imam, 1979) it may be labeled, and the degree of substitution is limited by the desire to retain the biological activity of the antibody. Second, the intensity of fluorescence of the conjugate will reach a maximum and then decrease with increasing F/P ratios due to fluorescence quenching effects (Der-Balian *et al.*, 1988; Haugland, 1996). Third, increases in F/P may result in higher nonspecific binding of the antibody. For example, fluorescein isothiocyanate (FITC) shows increased negative charge with increases in F/P, resulting in binding to positively charged cellular ions, similar to the binding of eosin dyes (The and Feltkamp, 1970a). Therefore, a balance must be achieved between obtaining the highest degree of substitution that is consistent with the preservation of activity and reduction of nonspecific binding.

C. Specific Organic Dyes

1. Fluorescein Derivatives

Since its introduction as a fluorescent label for immune serum, FITC has been used extensively for immunofluorescent techniques (Riggs *et al.*, 1958). FITC continues to be the most widely used inorganic fluorochrome for labeling antibodies and owing to its long history of use has been studied extensively, particularly for labeling of antibody proteins. The conditions for optimal labeling of FITC are determined by the requirements of the isothiocyanate reactive group. In general, optimal labeling of proteins with FITC is achieved when conditions include high pH, temperature, and protein concentration. FITC is soluble in aqueous solutions but may be more completely dissolved in DMSO, resulting in more predictable results (Goding, 1976). Below pH 9, FITC reacts primarily with the α-amino group of the N-terminal amino acid, thus limiting the degree of substitution; however, at pH values above their pH 9, the degree of substitution increases due to reaction with ε-amino groups of lysine (Maeda, 1969). Maximal fluorescein to protein ratios are obtainable at 37 °C, but 25 °C incubation provides optimal results (The and Feltkamp, 1970b). High protein concentrations allow high F/P ratios in a shorter amount of time, compared with low concentrations (The and Feltkamp, 1970b). Optimally, concentrations of 25 mg/ml are desirable, but may be impractical; proteins ranging from 1 to 10 mg/ml can be labeled with an increase in incubation time.

A modification of fluorescein, dichlorotriazinylaminofluorescein (5-DTAF), has been used to label carbohydrates and protein, including antibodies (Der-Balian *et al.*, 1988). The chlorinated S-triazines are highly reactive with nucleophiles and bind to α- and ε-amino groups of proteins. The conjugates appear stable, but may show changes with time, attributable to hydrolysis of the remaining, relatively inert, chloro group (Zuk *et al.*, 1979).

Another modification of fluorescein, carboxyfluorescein succinimidyl ester, known as CFSE or FAM, utilizes the reactive succinimidyl ester to form very stable protein conjugates (Zuk *et al.*, 1979). This derivative has also been modified with a seven-atom aminohexanoyl spacer group between the FAM fluorphore and the succinimidyl ester in an attempt to reduce quenching effects seen with increased *F/P* ratios (Haugland, 1996).

2. Biotin/Avidin

Although biotin is not a fluorochrome, its extensive use as a label in combination with fluorochrome-labeled avidin necessitates a discussion of its properties. Biotin is a small molecule (MW 244) which serves as an intermediate carrier of carbon dioxide in carboxylating enzymes. Its usefulness in immunochemistry and flow cytometry, however, is the very high affinity ($K_a = 10^{15}$ M^{-1}) binding observed with the egg white protein avidin. Avidin is a tetramer, consisting of four subunits with a combined molecular mass of 67,000–68,000. Each avidin molecule contains four binding sites for biotin. Avidin is positively charged at neutral pH, having an isoelectric point of 10.5. Because of its positive charge and the presence of the oligosaccharides mannose and *N*-acetylglucosamine, avidin has been shown to bind nonspecifically to negatively charged molecules and to carbohydrate-binding proteins on cell surfaces. Avidin has also been shown to bind nonspecifically to the cytoplasmic granules of mast cells (Bussolati and Gugliotta, 1983). These and other observations have led to the more widespread use of streptavidin, particularly for fluorescence microscopy and flow cytometry.

Streptavidin, isolated from *Streptomyces avidinii*, lacks carbohydrate, reducing the potential for possible protein interaction, and it has an isoelectric point of 5–6, which lowers the overall charge of the molecule. Although it has been suggested that these characteristics of streptavidin will eliminate many of the problems associated with nonspecific binding of avidin, this may not be true in all instances. Indeed, streptavidin has been shown to contain the tripeptide sequence Arg-Tyr-Asp (RYD) that mimics the binding sequence of fibronectin Arg-Gly-Asp (RGD), a universal recognition domain of the extracellular matrix that promotes cell adhesion (Alon, 1990). Additionally, the higher isoelectric point of avidin, compared with streptavidin, may be irrelevant since it may be altered by conjugation with fluorochromes. In this regard, FITC conjugation to protein has been shown to increase their negative electrical charge (The and Feltkamp, 1970a). Another disadvantage to streptavidin is that the biotin-binding cleft is different than avidin, and may require a longer spacer arm for biotinylation to achieve optimum binding. These disadvantages have prompted the introduction and use of a chemically deglycosylated form of avidin, known as variously as NeutraLite or NeutrAvidin. This form has a near neutral isoelectric point but maintains the biotin-binding affinity of native avidin (Hiller *et al.*, 1987).

The use of biotin as reagents has been reviewed (Wilchek and Bayer, 1990), as well as their use in bioassays (Wilchek and Bayer, 1988) and their methods of coupling to antibodies (Haugland and You, 1995).

Several active biotin derivatives have been produced to biotinylate proteins and glycoproteins. They are summarized in Table II. The most popular biotin derivative is the NHS ester for labeling the *ε*-amino groups of lysine.

Table II
Biotin Derivatives

Protein functional group	Biotin derivative
Amine	*N*-Hydroxysuccinimide ester (NHS ester)
Tyrosyl, histidyl	*p*-Diazobenzoyl-biocytin
Sulfhydryl	3-(*N*-Maleimidopropionyl)biocytin, iodoacetyl-LC-biotin
Carboxyl	Biocytin hydrazide/carbodiimide

There are water-soluble sulfo-NHS forms also available; these do not offer advantages over NHS biotin for labeling antibodies but have been used to restrict labeling to the cell surface in immunoprecipitation protocols (Lantz and Holmes, 1995). Biocytin hydrazide has been used to label carbohydrates or glycoproteins, such as the carbohydrate groups of the Fc region of antibodies. Biocytin is an adduct of biotin and lysine (*N*-ε-biotinyl-L-lysine), but in this application biocytin hydrazide labeling was found to be inferior to labeling via the ε-amino groups (Diamandis and Christopoulos, 1991; Gretch *et al.*, 1987).

The biotin molecule is small and dependent on the location of the functional group on the protein to which biotin conjugates; the avidin-binding site may be inaccessible to avidin. For this reason, derivatives of biotin have been made that are known as long chain or biotin-X (or XX) which include one or two seven-atom aminohexanoic spacers attached to the carboxyl group of biotin. These separate the protein-binding site from the avidin-binding site, greatly enhancing the efficiency of formation of the avidin-binding complex by reducing steric hindrance. As noted earlier, it has been suggested that biotin-binding cleft of streptavidin is different than avidin and may require a longer spacer group for optimal binding (Haugland, 1996; Haugland and You, 1995). Haugland and You (1995) found that the use of two aminohexanoic acid spacers in biotin resulted in higher titers in enzyme-linked immunosorbent assays.

The previous discussion emphasizes the need to be knowledgeable of the reagents used when performing flow cytometric analysis or immunohistochemistry with the biotin-avidin system. The choice of use of avidin, NeutrAvidin, streptavidin, or biotin with aminohexanoic spacer groups may depend on the specific application or cell type being analyzed. When commercially prepared reagents are used, it may be difficult to determine which biotin is used for conjugations. The major antibody supply companies vary in whether they provide their antibodies coupled with biotin, its long-chain version, or both, whereas most supply streptavidin-fluorochrome conjugates.

3. Texas Red and Rhodamine

Some of the first dyes to be used as second labels in combination with FITC (or in other multicolor applications) were the sulforhodamine dyes Lissamine rhodamine B and Texas Red (sulforhodamine 101; Titus *et al.*, 1982). Both of these dyes are sulfonyl chlorides, which will form amide bonds with amino groups of the protein. They are not group specific, however, and due to a high rate of hydrolysis at the high pH (pH 9–9.5) required for conjugation to aliphatic amines, they require low-temperature (4 °C) conditions for conjugation (Lefevre *et al.*, 1996). In particular, sulfonyl chlorides will form conjugates with tyrosine, histidine, and cysteine (Wong, 1991b), which are unstable and require subsequent removal with hydroxylamine to ensure stability with storage (Brinkley, 1992). These deficiencies result in a lack of reproducibility in conjugations and an inability to successfully label some antibody classes or species (Titus *et al.*, 1982). New derivatives of Texas Red and Lissamine rhodamine B incorporate a 6-aminohexanoic spacer between the fluorophore and a NHS ester group (Lefevre *et al.*, 1996). This modification allows labeling under more mild conditions, with less precipitation of proteins and an increased fluorescence yield (Lefevre *et al.*, 1996).

4. Cyanine Dyes

Cyanine dyes are synthetic dyes that depending on their structure cover the spectrum form UV to IR. They were first synthesized over 100 years ago and were originally used, and still are, in photography to increase the sensitivity range of photographic emulsions. They are also used in CD-R and DVD-R media but due to their chemical instability, cyanine disks are unsuitable for archival use.

The cyanine dyes have high extinction coefficients ($>100,000$ mol^{-1} cm^{-1}), moderate quantum yields, and are photostable, and thus they are extremely useful as fluorescent probes. A major advantage of these dyes is that the spectral properties can be easily manipulated by altering the heterocyclic nucleus and the number of double bonds in the polymethine chain. These dyes can be synthesized with excitation maxima ranging from below 500 nm to beyond 750 nm and have been of great interest to the biomedical field for a number of years because of their unique physical properties.

Ernst first proposed the nomenclature for the naming of the dyes in 1989 (Ernst *et al.*, 1989). This nomenclature is nonstandard, as it gives no hint of their chemical structures. In the original paper, the number designated the count of the methanes and the side chains were ignored. Thus, various structures were designated as Cy2, Cy3, Cy5, and Cy7 by this basic nomenclature in the literature.

Cy2, Cy3, Cy5, and their derivatives are the major cyanine dyes in use today in biomedical applications. They are water-soluble flourescent dyes of the cyanine dye family.

Cy2 conjugates have maximum adsorption/excitation around 492 nm and fluoresce (510 nm) in the green region of the visible spectrum like FITC (520 nm) but are more photostable, less sensitive to pH changes, and more fluorescent in organic mounting media than FITC.

Cy3 conjugates are excited maximally at 550 nm, with peak emission at 570 nm. Cy3 has been used with fluorescein for double labeling. Cy3 can also be paired with Cy5 for multiple labeling when using a confocal microscope.

Cy5 conjugates are excited maximally at 650 nm and fluoresce maximally at 670 nm. Cy5 can be used with a variety of other fluorophores for multiple labeling due to a wide separation of its emission from that of shorter wavelength-emitting fluorophores. A significant advantage of using Cy5 over other fluorophores is the low autofluorescence of biological specimens in this region of the spectrum. However, because of its emission maximum at 670 nm, Cy5 cannot be seen well by eye, and it cannot be excited optimally with a mercury lamp. Therefore, it is not recommended for use with conventional epifluorescence microscopes. It is most commonly visualized with a confocal microscope equipped with an appropriate laser for excitation and a far-red detector.

Cy7 conjugates can be excited maximally at 755 nm, with peak emission at 7850 nm.

The Cy3, Cy5, and Cy7 dyes were further modified in their structure (Mujumdar *et al.*, 1996; Southwick *et al.*, 1990) to yield up to 10 different cyanine dyes each with separate spectral properties. At the present, the Cy2, Cy3, Cy 3.5, Cy 5, Cy 5.5, and Cy 7 are the most widely used cyanine dyes. Cy 7 has been mostly used for the PE\Cy 7 and APC-Cy7 tandem conjugates that will be discussed later.

Cy 2, Cy3, Cy 3.5, Cy 5, and Cy 5.5 are available in a choice of three labeling chemistries to enable labeling via amine groups (using NHS ester dyes), thiol groups (using maleimide dyes), or aldehyde groups (using hydrazide dyes). They all exhibit pH tolerance (pH 3–10), good aqueous solubility which eliminates need for organic solvents in assay buffers, as well as excellent photostability.

5. Alexa Fluor Dyes

A relatively recent addition to the family of organic dyes are a group of fluorochromes named the Alexa Fluor dyes (Panchuk-Voloshina *et al.*, 1999). These organic dyes cover the entire spectrum from the UV to red to long wavelength dyes. Their names denote the approximate excitation maxima of the dyes and are Alexa 350, Alexa 430, Alexa 488, Alexa 532, Alexa 546, Alexa 568, Alexa 594, Alexa 555, Alexa 565, Alexa 633, Alexa 647, Alexa 660, Alexa 680, Alexa 700, and Alexa 750. These are all sulfonated derivatives of well known and often used fluorochromes, aminocoumarin (Alexa 350, Alexa 430), rhodamine (Alexa 488, Alexa 532, Alexa 546, Alexa 568, Alexa 594, Alexa 633), or the Cy3 or Cy5 type carbocyanines (Alexa 555, Alexa 565, Alexa 633, Alexa 647, Alexa 660, Alexa 680, Alexa 700, Alexa 750). Sulfonation increases the polarity of molecules and decreases the tendency for them to form aggregates (Mujumdar, 1993). Sulfonation also introduces negative charges into the fluorochromes making them more hydrophilic and ring sulfonation increases the brightness of many dyes including the carbocyanines Cy3 and Cy5 (Wessendorf and Brelje, 1992) and some pyrenes, such as Cascade Blue (Whitaker *et al.*, 1991).

All of these Alexa dyes are very stable in pH ranges from 4 to 9 and have fluorescent intensities generally brighter than comparable dyes. These dyes are also very photostable and resist photobleaching making them extremely useful in immunocytochemistry applications (Berlier *et al.*, 2003; Panchuk-Voloshina *et al.*, 1999). All of these dyes are available as NHS ester derivatives allowing conjugation to proteins at relatively mild conditions without organic solvents.

IV. Labeling of Antibodies with Zenon Probes

The majority of this section discusses covalent conjugation of antibodies or proteins with fluorochrome probes. Another, non-covalent conjugation procedure that has been gaining popularity in recent years is known as Zenon labeling.

A. Background

The Zenon labeling method uses a labeled FAb fragment directed against the Fc portion of the antibody to be labeled. These kits are available in a variety of fluorophore, biotin, or enzyme conjugates for use in mouse and human IgG antibodies and to a lesser extent in rabbit and goat IgG antibodies.

There are several advantages to this labeling protocol. The labeling is rapid and quantitative with labeling complexes ready for cellular staining within 5–15 min. Zenon labeled samples are compatible with single and multicolor applications and several Zenon labeled complexes can be used together in a single multicolor protocol. It is not necessary to remove exogenous proteins from the antibody preparation before Zenon labeling, and a single Zenon labeling only requires 1 μg of antibody. Zenon labeled antibody complexes have shown good reactivity in both flow cytometry and immunohistochemistry protocols (Airey *et al.*, 2004; Ono and Freed, 2004; Petrovic *et al.*, 2004).

There are some shortcomings to this labeling protocol. Primarily, this is a noncovalent labeling system. Since it is noncovalent modification, labeled antibodies should be used within 30 min to 1 h and cannot be stored for subsequent use. Also when used in multiparameter labeling protocols, fixation with aldehyde-based fixatives following staining is recommended to prevent transfer of the Zenon label between antibodies (Haugland, 2002). This may not be compatible with some applications. Also, the Zenon labeling system is only compatible with antibody applications and cannot be used for labeling other proteins, peptides, or antibody fragments. At this point, Zenon kits are available in a limited number of fluorophore, biotin, or enzyme conjugates for use in most mouse and human IgG antibodies and to a lesser extent in rabbit IgG and goat IgG antibodies. There are some combinations that are yet to be available and may not be offered for very specific fluorophore or antibody combinations. Also, these Zenon labeled antibody complexes has been reported to have lower fluorescent signals as in traditional methods, but may be acceptable since the fluorescent background is very low.

B. Principle of Zenon Labeling

The Zenon antibody labeling system allows noncovalent coupling of fluorochromes or enymes to unconjugated mouse, rabbit, and goat IgG primary antibodies. This method uses directly labeled FAb particles from antibodies directed against the Fc regions of the target antibody. The fluorochrome linked FAb is mixed with the target antibody and allowed to incubate forming fluorochrome-FAb-antibody complexes. These complexes can be used immediately

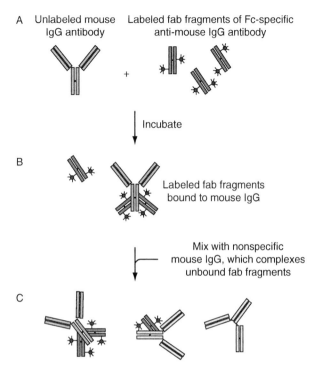

How Zenon Technology works:

Fig. 4 Schematic representation of Zenon Technology. Labeled FAb fragments are mixed with their taget IgG molcules and incubated. Residual Zenon labeling reagent is adsorped to avoid cross-reactivity.

for single-color labeling. In multicolor applications, it is essential to adsorb any residual fluorochrome linked FAb to avoid cross-reactivity. This is accomplished by mixing a solution of soluble nonspecific IgG with the complex. After a short incubation, the resulting complex is relatively stable and can be used for both single and multicolor labeling as if it were a conventional direct conjugate (Fig. 4) (Haugland, 2002).

In summary, the Zenon labeling system offers a quick, convenient, cost effective method of labeling antibodies. It is especially helpful when antibody is in limited supply, or in a mixture of other proteins, or when the investigator needs the labeled antibody immediately. It is not without its limitations but it provides the investigator another tool for flow cytometric and immunohistochemical analysis.

C. Choice of Reactive Group

The availability of a large variety of fluorochrome derivatives provides the researcher with the opportunity to tailor their conjugations to their specific needs. The choice of the derivative is dependent on several factors: (1) The number of functional groups available on the protein. The most commonly used method of protein modification involves coupling through the aliphatic amines, the N-terminal α-amines, and the ε-amino group of lysine. This is due primarily to their high reactivity and to their abundance in proteins. Although some proteins lack N-terminal α-amines (i.e., cytochrome *c* and ovalbumin; Brinkley, 1992), the majority possess both α- and ε-amines. This is true for immunoglobulin G (IgG) molecules, which are composed of four polypeptide chains and approximately 90 lysine residues (Nakagawa *et al.*, 1972). (2) As previously stated, the pH of the reaction is controlled using the appropriate buffer system. However, for reactive groups requiring more alkaline pH, some proteins may be more sensitive to these conditions. For example, IgM molecules are unstable at alkaline pH, and, therefore, necessitate the use of derivatives that can be used at neutral pH such as NHS esters (Haugland and You, 1995). (3) The most variable portion of the antibody molecule is the antigen-combining site, which implies that there will be variability in susceptibility of the antigen-combining site to be labeled during conjugation. Therefore, it may be necessary to utilize a different functional group for conjugation if it is determined that activity cannot be preserved with a particular reactive group. This is obviously of more importance when labeling monoclonal rather than polyclonal antibody preparations. For example, as stated earlier, if conjugation directed toward primary amines (i.e., NHS ester) results in loss of activity, reactive groups targeting sulfhydryls (i.e., *N*-ethylmaleimide derivatives at neutral pH) may achieve labeling with preservation of activity. In this regard, one strategy employed is to cleave the antibodies by reduction at their disulfide groups in the hinge region, using 2-mercaptoethylamine. This generates two heavy plus light chain fragments containing free sulfhydryls that are removed from the antigen-combing site and available for conjugation (Hermanson, 1996a). This may be particularly advantageous since the frequency of free thiols in proteins can be relatively low.

A new organic functional group has recently been marketed (Invitrogen) of the activated ester type termed SDP(sulfodicholorphenol). It is a derivative of NHS esters yet is reported to be hydrolytically stable in aqueous environments and will allow the investigator to better control the degree of labeling and use less organic flourochrome to achieve the optimal labeling. This organic functional group is only available in the Alexa 488 activated dye at the date of this publcation. The authors have not tested this reagent, but it would be subject to the same constraints of all of the organic fluorescent probes (e.g. use of fresh reagents and addition of reagents in correct molar ratio).

V. Labeling of Proteins with Phycobiliproteins

A. Background

Phycobiliproteins are water-soluble fluorescent pigment proteins of the photosynthetic machinery of cyanobacteria and eukaryotic algae. They function as accessory or antenna pigments for the collection of light energy. These pigments absorb light energy of the visible spectrum in wavelengths that are poorly absorbed by chlorophyll. The phycobiliproteins in most algae are arranged in subcellular structures, known as phycobilisomes. These structures optimize the capture and transfer of light energy. All of the phycobiliproteins absorb light energy directly, and, through a series of energy transfer intermediates, transfer this light energy to the photosynthetic reaction core (phycoerythrin to phycocyanin to allophycocyanin to chlorophyll *a*) (Glazer, 1985). When the individual phycobiliproteins are purified and isolated, these proteins become highly fluorescent because there are no nearby molecules to act as energy acceptors. Phycobilisomes provide 30–50% of the total light-harvesting capacity of cyanobacteria and red alga cells (Glazer, 1982).

Phycobiliproteins are classified on the basis of their absorbance maxima into three major groups—the phycoerythrins, the phycocyanins, and allophycocyanin. Absorption maxima for the phycoerythrins lie between 490 and 570 nm whereas absorption maxima for the phycocyanins and allophycocyanin lie between 610 and 665 nm.

Phycobiliproteins are composed of a number of subunits, each consisting of a protein backbone to which linear tetrapyrrole chromophores are covalently bound. All phycobiliproteins contain either phycocyanobilin or phycoerythrobilin chromophores, and may also contain one of three minor bilins: phycourobilin, cryptobilin, or the 697-nm bilin.

Three phycobiliproteins are commonly used as fluorescent labels, R-phycoerythrin (R-PE), B-phycoerythrin (B-PE), and allophycocyanin (APC), due to their high quantum yields and absorbance/fluorescent properties.

The two most commonly used phycoerythrins, R-PE and B-PE, are composed of three types of subunits, α, β, and γ. The subunit structure of both R-PE and B-PE is $(\alpha\beta)_6\gamma$, producing a protein with a molecular weight of 240,000 Da. The R- and B-prefixes refer to conventional nomenclature indicating the type of organism from which the pigment was originally isolated. These prefixes have now evolved to denote the shape of the absorbance curve of the purified phycoerythrin (PE) as more recent evidence has determined that differences in PE structure are not species specific. R-PE and B-PE are the most intensely fluorescent of the phycobiliproteins, with quantum efficiencies of 90% or more (Glazer, 1985).

Allophycocyanin is composed of two dissimilar polypeptide subunits, α and β, each containing one covalently bound phycocyanobilin chromophore (Glazer, 1982). In nature, allophycocyanin exists as a trimer $(\alpha\beta)_3$ producing a protein of approximately 104,000 Da. These allophycocyanin trimers readily dissociate into monomers $(\alpha\beta)$ on dilution to very low protein concentration (<30 ng/ml), under acidic conditions, or on exposure to chaotropic salts. Disassociated subunits typically exhibit changes in both absorbance and fluorescence spectra. For example, APC trimers typically display absorbance maxima at 650 nm, whereas the monomer has absorbance maxima of 620 nm. Techniques have been developed to cross-link APC that stabilize the trimeric structure and preserve its absorbance and fluorescence properties (Yeh *et al.*, 1987). Dissociation of the subunit structure of the phycoerythrins has not been observed under typical laboratory conditions.

B. Principle of Coupling and the Use of Heterobifunctional Reagents

The isolation and use of phycobiliproteins from algae and cyanobacteria has revolutionized flow cytometry by providing dyes with high quantum yields and extinction coefficients, as well as large Stokes shifts, permitting their use in multi-color applications. Labeling of macromolecules with phycobiliprotein derivatives can provide absorbance coefficients 30-fold higher than with small synthetic fluorophores (Zola *et al.*, 1990). In addition, the external environment does not easily affect these fluorophores. They are not readily quenched by conjugation to another molecule, and their fluorescence is independent of pH or ionic strength. Also, their excellent stability and solubility in aqueous solutions allows conjugation reactions under mild conditions conducive with protein integrity.

Conjugation of these fluorophores to antibodies requires the use of cross-linking reagents known as heterobifunctional cross-linkers. This is a multistep process in which reactive groups are introduced or activated on each protein before the conjugation process is begun. In addition, coupling of two proteins with hetero-bifunctional cross-linking agents will produce a mixture of polypeptide species, the number depending on the cross-linker chosen. Therefore, the conjugated product may require purification prior to use. This makes the bioconjugation of these dyes more complex than other organic dyes.

Heterobifunctional conjugation reagents contain two different reactive groups that can be used to target different functional species on proteins or other macro-molecules. These reagents typically require a two- or three-step process depending on the cross-linker utilized. This allows greater control of the F/P ratio and may also be used to site-direct a conjugation reaction toward particular functional group of the target molecules. For example, as stated earlier, targeting sulfhydryls may be desirable when conjugation with ε-amino groups compromises activity of the antibody molecule. The low abundance of free sulfhydryls on antibody mole-cules is less problematic when performing phycobiliprotein:antibody conjugations, in contrast to organic dye:antibody conjugations. This is because the high quan-tum yields of the phycobiliproteins provide optimal results when an approximately 1:1 molar conjugation ratio is achieved. An added advantage of targeting sulfhy-dryl and carbohydrate molecules within the antibody is the spatial separation of the phycobiliprotein from the functional binding sites within the conjugate. This lessens the potential of loss of activity due to steric hindrance effects of the attached phycobiliprotein.

Several heterobifunctional agents are available for protein:protein conjugation. These have been reviewed extensively, and only the most commonly used reagents will be discussed here (Hermanson, 1996b; Wong, 1991d). Typically, the hetero-bifunctional reagent contains an amine-reactive moiety and a sulfhydryl-reactive group separated by a variable length spacer arm. Often the amine-reactive group is an NHS ester, whereas the sulfhydryl-reactive group can be one of several different reactive groups. Frequently, one protein is first modified with the most reactive or most labile end of the cross-linker. For example, with a NHS ester-maleimide heterobifunctional linker (i.e., succinimidyl-4-(*N*-maleimidomethyl)cyclohexane-1-carboxylate, SMCC), one protein is initially bound to the cross-linker via its ε-amino groups, using the NHS ester end. Excess heterobifunctional cross-linker is removed from this reaction by gel filtration, and the maleimide reactive group is then utilized to couple the second protein to the initial reaction complex through activated sulfhydryls. The NHS ester group is utilized first because it is much more labile in aqueous solution.

1. Types of Heterobifunctional Cross–Linkers

Two groups of heterobifunctional cross-linkers are commonly used for coupling phycobiliproteins to immunoglobulins. Both classes utilize ε-amino residues of lysine for protein coupling but differ primarily in the type of sulfhydryl-reactive group presented (Figs. 5 and 6).

The first group, which includes SPDP, 4-succinimidyloxycarbonyl-α-methyl-α-(2-pyridylditio)toluene (SMPT), and SATA, introduces a reactive or activatable sulfhydryl into the protein (Fig. 5). These inserted sulfhydryl residues can then be used to form disulfide bonds with introduced or endogenous sulfhydryls of the second protein. The resulting disulfide bond linking the phycobiliprotein and the immunoglobulin when using SPDP and SATA is, therefore, labile under reducing conditions. However, with the presence of the aromatic ring in the SPDP derivative, SMPT sterically hinders the disulfide sufficiently to increase its *in vivo* half-life (Thorpe *et al.*, 1987). Several lines of evidence suggest that short chain cross-

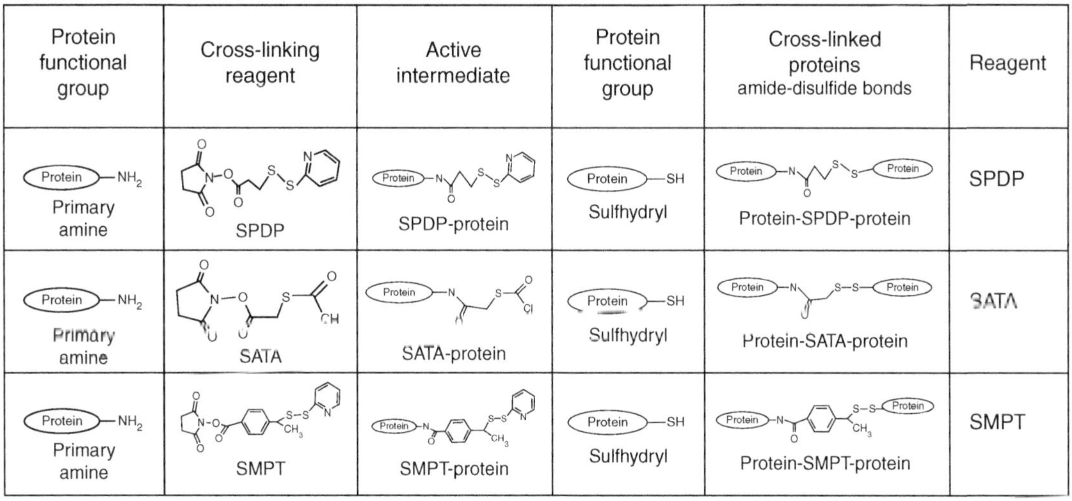

Protein functional group	Cross-linking reagent	Active intermediate	Protein functional group	Cross-linked proteins amide-disulfide bonds	Reagent
Protein—NH₂ Primary amine	SPDP	SPDP-protein	Protein—SH Sulfhydryl	Protein-SPDP-protein	SPDP
Protein—NH₂ Primary amine	SATA	SATA-protein	Protein—SH Sulfhydryl	Protein-SATA-protein	SATA
Protein—NH₂ Primary amine	SMPT	SMPT-protein	Protein—SH Sulfhydryl	Protein-SMPT-protein	SMPT

Fig. 5 Reactions of selected short chain heterobifunctional cross-linking reagents. This group of heterobifunctional cross-linking reagents introduces a reactive or activated sulfhydryl into the protein. These inserted sulfhydryl residues can then be used to form disulfide bonds with introduced or endogenous sulfhydryls of the second protein.

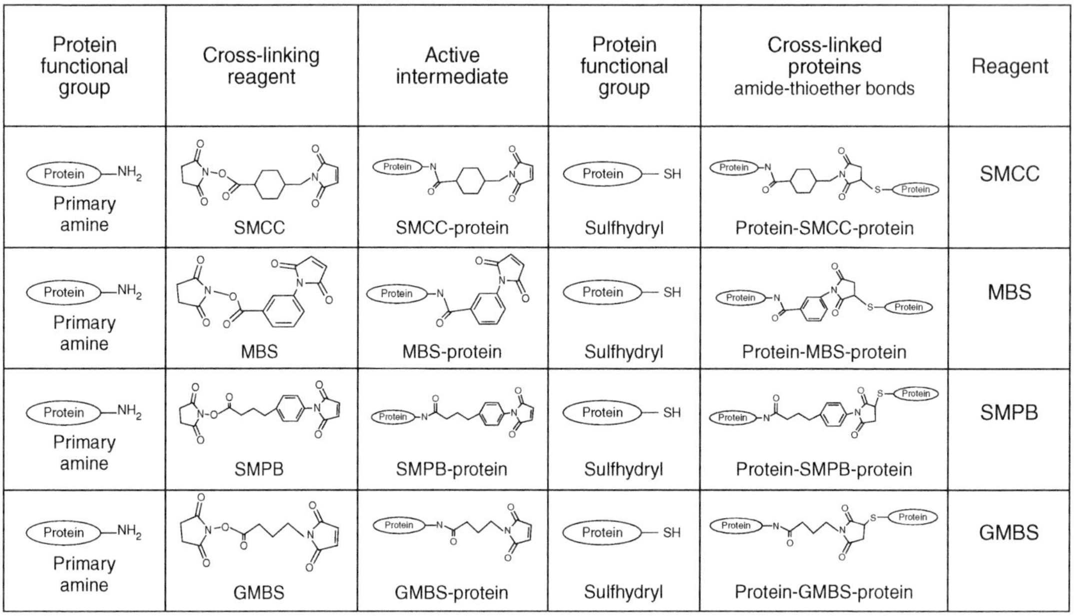

Protein functional group	Cross-linking reagent	Active intermediate	Protein functional group	Cross-linked proteins amide-thioether bonds	Reagent
Protein—NH₂ Primary amine	SMCC	SMCC-protein	Protein—SH Sulfhydryl	Protein-SMCC-protein	SMCC
Protein—NH₂ Primary amine	MBS	MBS-protein	Protein—SH Sulfhydryl	Protein-MBS-protein	MBS
Protein—NH₂ Primary amine	SMPB	SMPB-protein	Protein—SH Sulfhydryl	Protein-SMPB-protein	SMPB
Protein—NH₂ Primary amine	GMBS	GMBS-protein	Protein—SH Sulfhydryl	Protein-GMBS-protein	GMBS

Fig. 6 Reactions of selected amino- and sulfhydryl-directed heterobifunctional cross-linking reagents. This group of heterobifunctional reagents contains an NHS ester on one end and a maleimide group on the other. The NHS-ester can react with primary amines in macromolecules producing an amide bond, and the maleimide reacts with sulfhydryl groups to form thioether bonds as described earlier (Fig. 3).

linkers are the least immunogenic (Boeckler *et al.*, 1996). However, increasing the length of the spacer arm between conjugated proteins and antibodies results in improved antibody binding, presumably due to decreased steric hindrance (Bieniarz *et al.*, 1996). Additionally, because many sulfhydryl residues are located below the surface of the protein in more hydrophobic regions, longer spacer arms may induce a more efficient conjugation. Furthermore, the generation of disulfide bonds is intrinsically much less specific than the generation of thioether bonds. The production of phycobiliprotein:phycobiliprotein and immunoglobulin: immunoglobulin dimers is more probable, thus reducing the yield of functional product (Brinkley, 1992). Despite these shortcomings, SPDP is probably the most popular of all the amino- and sulfhydryl-directed heterobifunctional reagents. It has been used in the preparation of bispecific antibodies (Bode *et al.*, 1989), the production of immunotoxins (Bjorn *et al.*, 1986), as well as the generation of B-phycoerythrin-allophycocyanin conjugates (Glazer and Stryer, 1983).

The second group of amino- and sulfhydryl-directed heterobifunctional reagents includes SMCC, *m*-maleimidobenzoyl-*N*-hydroxysuccinimide ester (MBS), *N*-(γ-malemidobutyryloxy)succinimide ester (GMBS), and succinimidyl-4-(*p*-maleimidophenyl)butyrate (SMPB) (Fig. 6). All of these reagents contain a NHS ester on one end and a maleimide group on the other. The NHS ester can react with primary amines in macromolecules producing an amide bond, and the maleimide reacts with sulfhydryl groups to form thioether bonds as described earlier (Fig. 3). The primary difference among the members of this group of reagents is the spacer arm separating the two reactive groups. SMCC is probably the most stable compound of this group (Hermanson, 1996b). This stability is probably due to the location of the maleimide group away from the aromatic ring structure. Proteins modified with SMCC form relatively stable long-lived maleimide-activated intermediates and may be freeze-dried with minimal loss of activity (Ishikawa et al., 1983). MBS is less stable than SMCC due to the location of the aromatic ring directly adjacent to the maleimide group. Nevertheless, MBS has enjoyed much popularity, probably because it was one of the first NHS ester-maleimide heterobifunctional reagents. SMPB is an analog of MBS containing an extended spacer arm. The maleimide group is nonetheless immediately adjacent to an aromatic ring. Both MBS and SMPB are much more labile than SMCC, and the maleimide-activated protein intermediate should be desalted and mixed with the sulfhydryl-containing molecule quickly to prevent hydrolysis of the maleimide reactive component. The maleimide group of GMBS is adjacent to an aliphatic spacer providing this heterobifunctional reagent better stability than MBS or SMPB. It is not as stable as SMCC, and maleimide-activated intermediates of GMBS should be immediately mixed with the corresponding sulfhydryl-containing molecule after purification to achieve optimal results. This class of amino- and sulfhydryl-directed heterobifunctional reagents has been utilized in cross-linking of alkaline phosphatase and human IgG F(ab')$_2$ fragments, for generating phycobiliprotein-antibody conjugates, and for generating immunotoxin conjugates (Hardy et al., 1983; Holmes et al., 1995; Mahan et al., 1987; Myers et al., 1989).

2. Choice of Heterobifunctional Cross-Linker

The addition of the NHS ester-maleimide heterobifunctional reagents to a protein at neutral pH causes the nucleophilic attack of the NHS ester reactive group by ε-amino groups of lysine and results in the formation of an amide bond between the protein and the heterobifunctional linker. As discussed earlier (Section III.B.1), at neutral pH, the ε-amino groups of lysine are relatively unreactive toward the maleimide group of the heterobifunctional cross-linker. On addition of this protein-cross-linker complex to a second protein, which contains a sulfhydryl molecule, the maleimide group undergoes alkylation forming a thioether bond, and the two proteins are coupled. The sulfhydryl group on the second protein can either be added exogenously to the protein, or endogenous cysteines can be activated. One example of the exogenous source of sulfhydryl molecules is the first group of heterobifunctional reagents discussed earlier (i.e., SPDP, SMPT, or SATA). The activation of endogenous sulfhydryls often involves the addition of reducing agents (i.e., 2-mercaptoethanol or dithiothreitol) that reduce disulfide-bonded cysteines.

These alkylation events are random events that occur wherever lysine residues are present on the surface of the antibody. The majority of heterobifunctional cross-linker reagents, whether they add sulfhydryl (Fig. 5) or maleimide (Fig. 6) groups are directed toward lysine. As stated earlier, because these residues can be located within the antigen-combining sites of the antibody, conjugation may yield antibodies with lower activity. Because of this, site-specific methods of labeling antibodies away from the antigen-combining areas have been pursued. One method of site-directed labeling of antibodies is through the reduction of interchain disulfide bonds and the subsequent acylation with maleimide-containing hetero-bifunctional reagents. One important feature of the labeling of endogenous cysteine molecules is that they are generally localized away from the antigen combing sites (del Rosario *et al.*, 1990). These considerations may be of particular interest when attaching a large macromolecule such as a phycobiliprotein, in contrast to an organic dye such as FITC or biotin.

Pursuant to the previous discussion, a relatively new class of heterobifunctional cross-linking reagents is available that targets carbohydrate groups of proteins. This new class of cross-linker contains a carbonyl active group on one end and a sulfhydryl reactive group on the other end. The carbonyl reactive group is a hydrazide group that can form hydrazone bonds with aldehyde residues. Aldehyde residues are produced by the oxidation of the carbohydrate molecules with sodium periodate. The sulfhydryl reactive groups are of two types. The first type, characterized by 4-(4-*N*-maleimido-phenyl)butyric acid hydrazide (MPBH) and 4-(*N*-maleimidomethyl)cyclohexane-1-carboxyl-hydrazide hydrochloride (M_2C_2H), contains a maleimide group. The principal difference between these molecules is that the maleimide group of M_2C_2H is adjacent to an aliphatic hexane ring, analogous to SMCC, and the maleimide is expected to be more stable in aqueous solutions. The maleimide group of MPBH is adjacent to an aromatic phenyl group. 3-(2-Pyridyldithio)propionyl hydrazide (PDPH) contains a pyridyl disulfide group, similar to SPDP, which on reduction with dithiothreitol (DTT) forms a sulfhydryl reactive group. Because carbohydrate molecules are generally found on the Fc portion of immunoglobulins, coupling of the carbonyl reactive group to an antibody may help to preserve antigenic activity.

The coupling of phycobiliproteins to antibodies with heterobifunctional cross-linking reagents is generally not totally efficient. Unconjugated antibody and phycobiliprotein, as well as overlabeled species, must be removed to ensure optimal fluorescence signal. Unconjugated antibody will reduce the effective titer of the resulting conjugate while unconjugated phycobiliprotein and overconjugated proteins (antibodies or phycobiliproteins) may contribute to excessive background. Gel filtration is the most effective technique to separate the individual peaks. Good results are obtained using Bio-Gel A-1.5 (Bio-Rad, Hercules, CA) gel filtration medium or ion-exchange chromatography using hydroxyapatite.

3. Tandem Conjugate Dyes

The demand for fluorochromes that can be used in simultaneous multicolor applications has resulted in the development of dyes that utilize fluorescence resonance energy transfer to achieve a high Stokes shift (Glazer and Stryer, 1983).

These are known as tandem conjugate dyes, and the most widely used are phycoerythrin-cyanine 5 (PE-Cy5) (Lansdorp *et al.*, 1991; Shih *et al.*, 1993; Waggoner *et al.*, 1993) and phycoerythrin-Texas Red (PE-TR). Procedures for constructing tandem conjugate dyes and their coupling to antibodies basically follow the guidelines outlined in this chapter. The organic dye is first conjugated to the phycobiliprotein as described in Section II. The phycobiliprotein-dye complex is then conjugated to the antibody using a heterobifunctional cross-linking reagent as outlined earlier.

VI. Conclusion

The ability to label proteins or antibodies with fluorochrome dyes empowers the researcher with the tools to investigate a myriad of biological questions. Foremost is the ability to couple antibodies with organic dyes and phycobiliproteins for use in flow cytometry and imaging. A reflection of the usefulness of these reagents is the growth of commercially available fluorescent-labeled antibodies in more recent years. However, as new monoclonal antibodies are made or new dyes discovered, it is important to understand the relatively simple chemistries involved in fluorescent conjugation. Successful conjugations can be achieved through the use of published protocols and by following the guidelines presented here. The most important fact to remember, however, is that optimal conjugation may require slight modification of conditions for each protein labeled. As detailed earlier, this is due to the variation in amino acid sequence and three-dimensional structure of individual proteins.

References

Airey, J. A., Almeida-Porada, G., Colletti, E. J., Porada, C. D., Chamberlain, J., Movsesian, M., Sutko, J. L., and Zanjani, E. D. (2004). Human mesenchymal stem cells form Purkinje fibers in fetal sheep heart. *Circulation* **109**, 1401–1407.

Alon, R. (1990). Streptavidin contains an RYD sequence which mimics the RGD receptor domain of fibronectin. *Biochem. Biophys. Res. Commun.* **170**, 1236–1241.

Berlier, J. E., Rothe, A., Buller, G., Bradford, J., Gray, D. R., Filanoski, B. J., Telford, W. G., Yue, S., Liu, J., Cheung, C. Y., Chang, W., Hirsch, J. D., *et al.* (2003). Quantitative comparison of long-wavelength Alexa Fluor dyes to Cy dyes: Fluorescence of the dyes and their bioconjugates. *J. Histochem. Cytochem.* **51**, 1699–1712.

Bieniarz, C., Husain, M., Barnes, G., King, C. A., and Welch, C. J. (1996). Extended length heterobifunctional coupling agents for protein conjugations. *Bioconjug. Chem.* **7**, 88–95.

Bjorn, M. J., Groetsema, G., and Scalapino, L. (1986). Antibody-*Pseudomonas* exotoxin A conjugates cytotoxic to human breast cancer cells *in vitro*. *Cancer Res.* **46**, 3262–3267.

Bode, C., Runge, M. S., Branscomb, E. E., Newell, J. B., Matsueda, G. R., and Haber, E. (1989). Antibody-directed fibrinolysis. An antibody specific for both fibrin and tissue plasminogen activator. *J. Biol. Chem.* **264**, 944–948.

Boeckler, C., Frisch, B., Muller, S., and Schuber, F. (1996). Immunogenicity of new heterobifunctional cross-linking reagents used in the conjugation of synthetic peptides to liposomes. *J. Immunol. Methods* **191**, 1–10.

Botelho, L. H., and Gurd, F. R. N. (1989). Amino acids and proteins. *In* "Practical Handbook of Biochemistry and Molecular Biology" (G. D. Fasman, ed.), pp. 359–366. CRC Press, Boca Raton, FL.

Brinkley, M. (1992). A brief survey of methods for preparing protein conjugates with dyes, haptens, and cross-linking reagents. *Bioconjug. Chem.* **3,** 2–13.

Bussolati, G., and Gugliotta, P. (1983). Nonspecific staining of mast cells by avidin-biotin-peroxidase complexes (ABC). *J. Histochem. Cytochem.* **31,** 1419–1421.

Coons, A. H., and Kaplan, M. H. (1950). Localization of antigen in tissue cells: II. Improvements in a method for the detection of antigen by means of fluorescent antibody. *J. Exp. Med.* **91,** 1–13.

Coons, A. H., Creech, H. J., Jones, R. N., and Berliner, E. (1942). Demonstration of pneumococcal antigen in tissues by the use of fluorescent antibody. *J. Immunol.* **45,** 159–170.

Cuatrecasas, P., and Parikh, I. (1972). Adsorbents for affinity chromatography. Use of N-hydroxysuccinimide esters of agarose. *Biochemistry* **11**(12), 2291–2299.

del Rosario, R. B., Wahl, R. L., Brocchini, S. J., Lawton, R. G., and Smith, R. H. (1990). Sulfhydryl site-specific cross-linking and labeling of monoclonal antibodies by a fluorescent equilibrium transfer alkylation cross-link reagent. *Bioconjug. Chem.* **1,** 51–59.

Der-Balian, G. P., Kameda, N., and Rowley, G. L. (1988). Fluorescein labeling of Fab' while preserving single thiol. *Anal. Biochem.* **173,** 59–63.

Diamandis, E. P., and Christopoulos, T. K. (1991). The biotin-(strept)avidin system: Principles and applications in biotechnology. *Clin. Chem.* **37,** 625–636.

Ernst, L. A., Gupta, R. K., Mujumdar, R. B., and Waggoner, A. S. (1989). Cyanine dye labeling reagents for sulfhydryl groups. *Cytometry* **10**(1), 3–10.

Glazer, A. N. (1982). Phycobilisomes: Structure and dynamics. *Annu. Rev. Microbiol.* **36,** 173–198.

Glazer, A. N. (1985). Light harvesting by phycobilisomes. *Annu. Rev. Biophys. Biophys. Chem.* **14,** 47–77.

Glazer, A. N., and Stryer, L. (1983). Fluorescent tandem phycobiliprotein conjugates. Emission wavelength shifting by energy transfer. *Biophys. J.* **43,** 383–386.

Goding, J. W. (1976). Conjugation of antibodies with fluorochromes: Modifications to the standard methods. *J. Immunol. Methods* **13,** 215–226.

Gretch, D. R., Suter, M., and Stinski, M. F. (1987). The use of biotinylated monoclonal antibodies and streptavidin affinity chromatography to isolate herpesvirus hydrophobic proteins or glycoproteins. *Anal. Biochem.* **163,** 270–277.

Hardy, R. R. (1986). Purification and coupling of fluorescent proteins for use in flow cytometry. *In* "Handbook of Experimental Immunology" (D. M. Weir, L. A. Herzenberg, and C. Blackwell, eds.), pp. 31.1–31.12. Blackwell, Boston.

Hardy, R. R., Hayakawa, K., Parks, D. R., and Herzenberg, L. A. (1983). Demonstration of B-cell maturation in X-linked immunodeficient mice by simultaneous three-colour immunofluorescence. *Nature* **306,** 270–272.

Haugland, R. P. (1995). Coupling of monoclonal antibodies with fluorophores. *Methods Mol. Biol.* **45,** 205–221.

Haugland, R. P. (1996). Fluorophores and their amine-reactive derivatives. *In* "Handbook of Fluorescent Probes and Research Chemicals," (M. T. Z. Spence, eds.), p. 19. Molecular Probes, Inc., Eugene, OR.

Haugland, R. P. (2002). Section 7.2—Zenon technology—Versatile reagents for immunolabeling. *In* "Handbook of Fluorescent Probes and Research Chemicals," (M. T. Z. Spence, eds.), p. 202. Molecular Probes, Inc., Eugene, OR.

Haugland, R. P., and You, W. W. (1995). Coupling of monoclonal antibodies with biotin. *Methods Biol.* **45,** 223–233.

Hermanson, G. T. (1996a). Antibody modification and conjugation. *In* "Bioconjugate Techniques," p. 463. Academic Press, San Diego.

Hermanson, G. T. (1996b). Heterobifunctional cross-linkers. *In* "Bioconjugate Techniques," pp. 228–286. Academic Press, San Diego.

Hiller, Y., Gershoni, J. M., Bayer, E. A., and Wilchek, M. (1987). Biotin binding to avidin. Oligosaccharide side chain not required for ligand association. *Biochem. J.* **248,** 167–171.

Holmes, K. L., Fowlkes, B. J., Schmid, I., and Giorgi, J. V. (1995). Preparation of cells and reagents for flow cytometry. *In* "Current Protocols in Immunology" (J. E. Coligan, A. M. Kruisbeck, D. H. Margulies, E. M. Shevach, and W. Strober, eds.), pp. 5.3.1–5.3.23. Wiley, New York.

Holmes, K. L., Lantz, L. M., and Russ, W. (1997). Conjugation of fluorochromes to monoclonal antibodies. *In* "Current Protocols in Cytometry" (J. P. Robinson, Z. Darzynkiewicz, P. N. Dean, A. Orfao, P. S. Rabinovitch, C. C. Stewart, H. J. Tanke, and L. L. Wheeless, eds.), pp. 4.2.1–4.2.12. Wiley, New York.

Imam, S. A. (1979). Labelling of specific antibodies with fluorescein isothiocyanate with protection of the antigen-binding site [proceedings]. *Biochem. Soc. Trans.* **7**, 1013–1014.

Ishikawa, E., Imagawa, M., Hashida, S., Yoshitake, S., Hamaguchi, Y., and Ueno, T. (1983). Enzyme-labeling of antibodies and their fragments for enzyme immunoassay and immunohistochemical staining. *J. Immunoassay* **4**, 209–327.

Johnson, G. D., and Holborow, E. J. (1986). Preparation and use of fluorochrome conjugates. *In* "Handbook of Experimental Immunology" (D. M. Weir, L. A. Herzenberg, and C. Blackwell, eds.), pp. 28.1–28.21. Blackwell, Boston.

Lansdorp, P. M., Smith, C., Safford, M., Terstappen, L. W., and Thomas, T. E. (1991). Single laser three color immunofluorescence staining procedures based on energy transfer between phycoerythrin and cyanine 5. *Cytometry* **12**, 723–730.

Lantz, L. M., and Holmes, K. L. (1995). An improved nonradioactive cell surface labelling technique for immunoprecipitation. *BioTechniques* **18**, 56–60.

Lefevre, C., Kang, H. C., Haugland, R. P., Malekzadeh, N., and Arttamangkul, S. (1996). Texas Red-X and rhodamine Red-X, new derivatives of sulforhodamine 101 and lissamine rhodamine B with improved labeling and fluorescence properties. *Bioconjug. Chem.* **7**, 482–489.

Lomant, A. J., and Fairbanks, G. (1976). Chemical probes of extended biological structures: Synthesis and properties of the cleavable protein cross-linking reagent [^{35}S]dithiobis(succinimidyl propionate). *J. Mol. Biol.* **104**(1), 243–261.

Maeda, H. (1969). Reaction of fluorescein-isothiocyanate with proteins and amino acids. I. Covalent and non-covalent binding of fluorescein-isothiocyanate and fluorescein to proteins. *J. Biochem. (Tokyo)* **65**, 777–783.

Mahan, D. E., Morrison, L., Watson, L., and Haugneland, L. S. (1987). Phase change enzyme immunoassay. *Anal. Biochem.* **162**, 163–170.

Mujumdar, R. B., Ernst, L. A., Mujumdar, S. R., Lewis, C. J., and Waggoner, A. S. (1993). Cyanine dye labeling reagents: sulfoindocyanine succinimidyl esters. *Bioconj. Chem.* **4**, 105–111.

Mujumdar, S. R., Mujumdar, R. B., Grant, C. M., and Waggoner, A. S. (1996). Cyanine-labeling reagents: Sulfobenzindocyanine succinimidyl esters. *Bioconjug. Chem.* **7**, 356–362.

Myers, D. E., Uckun, F. M., Swaim, S. E., and Vallera, D. A. (1989). The effects of aromatic and aliphatic maleimide crosslinkers on anti-CD5 ricin immunotoxins. *J. Immunol. Methods* **121**, 129–142.

Nakagawa, Y., Capetillo, S., and Jirgensons, B. (1972). Effect of chemical modification of lysine residues on the conformation of human immunoglobulin G. *J. Biol. Chem.* **247**, 5703–5708.

Ono, A., and Freed, E. O. (2004). Cell-type-dependent targeting of human immunodeficiency virus type 1 assembly to the plasma membrane and the multivesicular body. *J. Virol.* **78**, 1552–1563.

Panchuk-Voloshina, N., Haugland, R. P., Bishop-Stewart, J., Bhalgat, M. K., Millard, P. J., Mao, F., Leung, W. Y., and Haugland, R. P. (1999). Alexa dyes, a series of new fluorescent dyes that yield exceptionally bright, photostable conjugates. *J. Histochem. Cytochem.* **47**, 1179–1188.

Petrovic, S., Barone, S., Xu, J., Conforti, L., Ma, L., Kujala, M., Kere, J., and Soleimani, M. (2004). SLC26A7: A basolateral Cl-/HCO3-exchanger specific to intercalated cells of the outer medullary collecting duct. *Am. J. Physiol. Renal. Physiol.* **286**, F161–F169.

Riggs, J. L., Seiwald, R. J., Burckhalter, J. H., Downs, C. M., and Metcalf, T. G. (1958). Isothiocyanate comopounds as fluorescent labeling agents for immune serum. *Am. J. Pathol.* **34**, 1081–1091.

Shih, C. C., Bolton, G., Schy, D., Lay, G., Campbell, D., and Huang, C. M. (1993). A novel dye that facilitates three-color analysis of PBMC by flow cytometry. *Ann. NY Acad. Sci.* **677**, 389–395.

Southwick, P. L., Ernst, L. A., Tauriello, E. W., Parker, S. R., Mujumdar, R. B., Mujumdar, S. R., Clever, H. A., and Waggoner, A. S. (1990). Cyanine dye labeling reagents-carboxymethylindocyanine succinimidyl esters. *Cytometry* **11**, 418–430.

Staros, J. V., Wright, R. W., and Swingle, D. M. (1986). Enhancement by *N*-hydroxysulfosuccinimide of water-soluble carbodiimidemediated coupling reactions. *Anal. Biochem.* **156**, 220–222.

The, T. H., and Feltkamp, T. E. (1970a). Conjugation of fluorescein isothiocyanate to antibodies. II. A reproducible method. *Immunology* **18**, 875–881.

The, T. H., and Feltkamp, T. E. (1970b). Conjugation of fluorescein isothiocyanate to antibodies. I. Experiments on the conditions of conjugation. *Immunology* **18**, 865–873.

Thorpe, P. E., Wallace, P. M., Knowles, P. P., Relf, M. G., Brown, A. N., Watson, G. J., Knyba, R. E., Wawrzynczak, E. J., and Blakey, D. C. (1987). New coupling agents for the synthesis of immunotoxins containing a hindered disulfide bond with improved stability *in vivo*. *Cancer Res.* **47**, 5924–5931.

Titus, J. A., Haugland, R., Sharrow, S. O., and Segal, D. M. (1982). Texas Red, a hydrophilic, red-emitting fluorophore for use with fluorescein in dual parameter flow microfluorometric and fluorescence microscopic studies. *J. Immunol. Methods* **50**, 193–204.

Waggoner, A. S., Ernst, L. A., Chen, C. H., and Rechtenwald, D. J. (1993). PE-CY5. A new fluorescent antibody label for three-color flow cytometry with a single laser. *Ann. NY Acad. Sci.* **677**, 185–193.

Wessendorf, M. W., and Brelje, T. C. (1992). Which fluorophore is brightest? A comparison of the staining obtained using fluorescein, tetramethylrhodamine, lissamine rhodamine, Texas Red and cyanine 3.18. *Histochemistry* **98**, 81–85.

Whitaker, J. E., Haugland, R. P., Moore, P. L., Hewitt, P. C., Resse, M., and Haugland, R. P. (1991). Cascade Blue derivatives: Water soluble, reactive,blue emission dyes evaluated as fluorescent labels and tracers. *Anal. Biochem.* **198**, 119–130.

Wilchek, M., and Bayer, E. A. (1988). The avidin-biotin complex in bioanalytical applications. *Anal. Biochem.* **171**, 1–32.

Wilchek, M., and Bayer, E. A. (1990). Biotin-containing reagents. *Methods Enzymol.* **184**, 123–138.

Wong, S. S. (1991a). Chemistry of Protein Conjugation and Cross-Linking. CRC Press, Boca Raton, FL.

Wong, S. S. (1991b). Reactive groups of proteins and their modifying agents. *In* "Chemistry of Protein Conjugation and Cross-Linking," pp. 7–48. CRC Press, Boca Raton, FL.

Wong, S. S. (1991c). Procedures, analysis, and complications. *In* "Chemistry of Protein Conjugation and Cross-Linking," pp. 209–220. CRC Press, Boca Raton, FL.

Wong, S. S. (1991d). Heterobifunctional cross-linkers. *In* "Chemistry of Protein Conjugation and Cross-Linking," pp. 147–194. CRC Press, Boca Raton, FL.

Yeh, S. W., Ong, L. J., Clark, J. H., and Glazer, A. N. (1987). Fluorescence properties of allophycocyanin and a crosslinked allophycocyanin trimer. *Cytometry* **8**, 91–95.

Zola, H., Neoh, S. H., Mantzioris, B. X., Webster, J., and Loughnan, M. S. (1990). Detection by immunofluorescence of surface molecules present in low copy numbers. High sensitivity staining and calibration of flow cytometer. *J. Immunol. Methods* **135**, 247–255.

Zuk, R. F., Rowley, G. L., and Ullman, E. F. (1979). Fluorescence protection immunoassay: A new homogeneous assay technique. *Clin. Chem.* **25**, 1554–1560.

CHAPTER 3

Cytometry of Fluorescence Resonance Energy Transfer

György Vereb,★ János Matkó,† and János Szöllősi★

★Department of Biophysics and Cell Biology
Cell Biophysics Research Group of the Hungarian Academy of Sciences
University of Debrecen
Debrecen H-4012, Hungary

†Department of Immunology
Eötvös Loránd University
Budapest H-1117, Hungary

I. Introduction

With the onset of modern proteomics, hundreds of pairs of cellular proteins that are capable of interacting with each other *in vitro* have been identified. However, the extent to which this possibility of interaction reflects their behavior in living cells is not clear. Nonetheless, it is abundantly clear that these protein-protein interactions are crucial in both maintaining the stable "resting" state of cells and

driving activation processes to eventually give cellular responses to external stimuli. Hence, it becomes more and more important to be able to detect such interactions *in situ* inside or on the surface of cells. Fluorescence techniques are widely used
to quantify molecular parameters of various biochemical and biological processes
in vivo because of their inherent sensitivity, specificity, and temporal resolution. The
combination of fluorescence spectroscopy with flow and image cytometry provided a
solid basis of rapid and continuous improvements of these technologies. A major
asset in studying molecular-level interactions was the application of fluorescence
resonance energy transfer (FRET) to cellular systems. In fact, the application of
FRET has been steadily expanding and doubled, based on PubMed searches, in the
past 5 years. FRET is a special phenomenon in fluorescence spectroscopy during
which energy is transferred from an excited donor molecule to an acceptor molecule
under favorable spectral, proximity, and orientational conditions. Applying donor-
and acceptor-labeled antibodies, lipids, and various types of fluorescent proteins
(such as green fluorescent protein (GFP) analogs), the FRET technique can be
used to determine intermolecular and intramolecular distances of cell surface components in biological membranes and molecular associations within live cells. For
reviews, see Szöllősi *et al.* (1998, 2002), Selvin (2000), Jares-Erijman and Jovin (2003,
2006), Scholes (2003), Szöllősi and Alexander (2003), Vogel *et al.* (2006), Szöllősi
et al. (2006), and van Royen *et al.* (2009). With the help of FRET, one can measure
molecular dimensions and determine them in functioning live cells, providing information that would be impossible to obtain with other classic approaches such as with
electron microscopic methods.

Excellent reviews are available on the applicability of FRET to biological
systems, as well as descriptions and comparison of various approaches (Bastiaens
and Squire, 1999; Berney and Danuser, 2003; Clegg, 1995, 2002; Gordon *et al.*,
1998; Matkó and Edidin, 1997; Periasamy *et al.*, 2008; Schmid and Birbach, 2007;
Szöllősi *et al.*, 1998; Vereb *et al.*, 2003).

On the following sections, we describe three flow cytometric FRET (FCET) and
three image cytometric FRET approaches in technical detail (Bastiaens *et al.*, 1996;
Nagy *et al.*, 1998a,b, 2002, 2006; Sebestyen *et al.*, 2002; Szöllősi *et al.*, 1984; Tron
et al., 1984; Vereb *et al.*, 1997, 2004). We have selected these approaches because
either we have played seminal roles in their elaboration or we have used these
techniques extensively in biological systems. Also, the methods detailed have proved
to be rather straightforward and robust in answering biological questions about
molecular associations *in situ* in/on the cell. After a brief introduction to the theory
of FRET, we explain details and relevant protocols of these FRET techniques. We
also provide references to free software solutions that aid the evaluation of data
obtained in these experiment (Roszik *et al.*, 2008, 2009; Szentesi *et al.*, 2004, 2005)
This is followed by several demonstrative FRET applications used by us or others.
Specific advantages and limitations of FRET approaches are also discussed, together with perspectives and advances that may provide new approaches for the detection of FRET in the future. The reader will also be referred to most relevant papers
and reviews—both classic and very recent—on FRET methodologies.

II. Theory of FRET

In FRET, an excited fluorescent dye, called a *donor*, donates energy to an acceptor dye, a phenomenon first described *correctly* by Förster (1946), as a dipole-dipole resonance energy transfer mechanism. For the process to occur, a set of conditions have to be fulfilled:

- The emission spectrum of the donor has to overlap with the excitation spectrum of the acceptor. The larger the overlap, the higher the rate of FRET is.
- The emission dipole vector of the donor and the absorption dipole vector of the acceptor need to be close to parallel. The rate of FRET decreases as the angle between the two vectors increases. In biological situations where molecules are free to move (rotate), we generally assume that dynamic averaging takes place; that is, during its excited lifetime, the donor assumes many possible steric positions, among them those that can yield an effective transfer of energy.
  - The distance between the donor and acceptor should be between 1 and 10 nm.

This latter phenomenon is the basis of the popularity of FRET in biology: The distance over which FRET occurs is small enough to characterize the proximity of possibly interacting molecules; under special circumstances, it even provides quantitative data on exact distances and information on the spatial orientation of molecules or their domains. Hence, the very apropos term from Stryer and Haugland (1967), who equated FRET with a "spectroscopic ruler."

Figure 1 illustrates some major aspects and qualities of the FRET process. The usual term for characterizing the efficiency of FRET is E, which is the ratio of excited state molecules relaxing by FRET to the total number of excited molecules. The energy transfer rate is dependent on the negative sixth power of the distance R between the donor and the acceptor, resulting in a sharply dropping curve when plotting E against R, centered around R_0, the distance where $E = 0.5$, that is, where there is a 50% chance that the energy of the excited donor will be transferred to the acceptor. As separation between the donor and acceptor increases, E decreases, and at $R = 2R_0$, E is already getting negligible. Conversely, as the distance reaches values below 1 nm, strong ground-state interactions or transfer by exchange interactions become dominant at the expense of FRET (Dexter, 1953).

When looking at the Jablonski scheme of FRET (Fig. 1), it is easy to see that its occurrence has profound consequences on the fluorescence properties of both the donor and the acceptor. An additional de-excitation process is introduced in the donor, with rate constant k_t, in addition to fluorescent (k_f) and nonfluorescent relaxations (k_{nf}). This decreases the fluorescence lifetime and the quantum efficiency of the donor ($Q_{D(+A)} < Q_D$), rendering it less fluorescent. The decrease in donor fluorescence (often termed *donor quenching*) can be one of the most easily measured spectroscopic characteristic that indicates the occurrence of FRET. Additionally, because the acceptor is excited as a result of FRET, acceptors that are fluorescent will emit photons (proportional to their quantum efficiency) also when FRET occurs. This is called *sensitized emission* and can also be a sensitive measure of FRET.

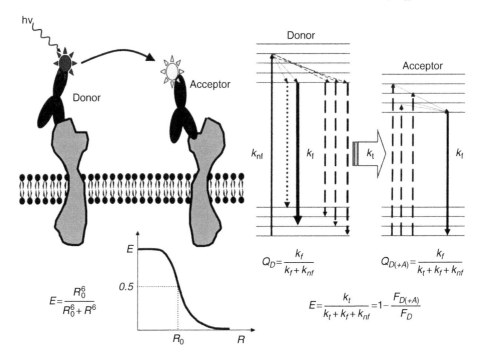

Fig. 1 The fluorescence resonance energy transfer (FRET) primer. FRET occurs when two suitable dye molecules that serve as donor and acceptor in the process come into sufficiently close molecular proximity of each other, the donor emission spectrum overlaps with the acceptor absorption spectrum and the spatial orientation of the respective emission, and absorption dipoles is favorable. The efficiency of FRET, E, is the function of the inverse sixth power of donor-acceptor separation, and thus, FRET occurs only over short R distances in the range of 1–10 nm, distributed around R_0, which marks 50% efficiency for a given donor-acceptor pair. The requirement for molecular proximity makes the phenomenon ideal for detecting association or aggregation of various macromolecules even in functioning cells. The Jablonski diagram for FRET shows that this process is an additional de-excitation pathway (rate constant $= k_t$) for the excited donor that otherwise can relax by radiative (k_f) and nonfluorescent (k_{nf}) processes. Henceforth, the Q_D quantum efficiency of the donor decreases in the presence of FRET ($Q_{D(+A)}$). It ensues that measuring donor fluorescence, a parameter proportional to Q, in the presence and absence of acceptor is one way to assess FRET efficiency.

III. Measuring FRET

The various approaches that have been used to quantitate FRET can be categorized based on the spectrofluorometric parameter detected and whether the donor or the acceptor is investigated:

- Measurements based on intensity changes upon FRET.

- Measuring donor quenching based on donor fluorescence with and without the acceptor.
  - Based on samples labeled with and without acceptor (Turcatti *et al.*, 1996).
  - Based on photobleaching the acceptor (Bastiaens *et al.*, 1996; Roszik *et al.*, 2008; Vereb *et al.*, 1997).
  - Based on reversibly switching on and off the acceptor (Giordano *et al.*, 2002).
- Measuring sensitized acceptor emission (Suzuki, 2000).
- Measuring both donor quenching and sensitized acceptor emission.
  - Measuring integrated emission in spectral bands through filters (Nagy *et al.*, 1998b; Tron *et al.*, 1984).
  - Measuring and fitting emission spectra (Clegg *et al.*, 1992; Megias *et al.*, 2009).
- Measurements of decreased donor lifetime.
  - Direct measurement of donor lifetime with and without the acceptor (Bastiaens *et al.*, 1996; Clegg *et al.*, 2003).
  - Estimation of donor lifetime based on photobleaching rate of the donor (Gadella and Jovin, 1995; Jovin and Arndt-Jovin, 1989; Vereb *et al.*, 1997).
  - Estimation of donor lifetime based on fluorescence anisotropy (Matkó *et al.*, 1993; Piston and Rizzo, 2008).
- Measurements of increased acceptor photobleaching rate (Mekler, 1994).
- Measurements of decreased fluorescence anisotropy of identical fluorophores (Clayton *et al.*, 2002; Lidke *et al.*, 2003; Szabó *et al.*, 2008).

These techniques vary greatly in their applicability, sensitivity, accuracy, and robustness as was demonstrated in a review by Berney and Danuser (2003). The authors systematically compared eight methods for calculating FRET efficiency and seven methods for calculating FRET indices. Using Monte Carlo simulation, representative sets of FRET efficiencies and indices were validated in different experimental settings. It is highly recommended to consult the results of this paper when choosing the right FRET approach for a special experimental system (Berney and Danuser, 2003).

FRET can be measured both in microscopic imaging and in flow cytometry. FCET carries the advantage of examining large cell populations in a short time and provides a FRET efficiency value averaged over the population or on a cell-by-cell basis, but averaged over each cell. The microscopic approaches have the ability to provide subcellular detail and the possibility to correlate FRET values with other biological information gained from fluorescent labeling, on a pixel-by-pixel basis. However, the sample size is restricted, because data acquisition and processing are rather time consuming. Also, biological variation may influence the composition of the cell population selected for imaging, which may or may not be advantageous, depending on our understanding of selection criteria.

In a review about FRET imaging, Jares-Erijman and Jovin (2003) classified 22 approaches to quantifying FRET in a systematic way. Six of these techniques have not been tested in a biological system yet. The characterization is based on the fact that in exploiting FRET in image cytometry, two fundamental challenges should be faced. First, the formalism must be appropriate for quantifying FRET under conditions of arbitrary, generally unknown, intramolecular and/or intermolecular stoichiometries, distributions and microenvironments of donor and acceptor. Second, to provide temporal resolution, continuous methods of observation of FRET are needed in most studies of live cells. The techniques fall in two major groups: group I includes 18 methods and is based on donor quenching and/or acceptor sensitization; group II, having only four methods, is based on measuring emission anisotropy of either the donor or the acceptor. Of the 22, 8 techniques can be applied in flow cytometry as well (Jares-Erijman and Jovin, 2003).

A. Flow Cytometric FRET

1. Donor-Quenching FRET

If one quickly and easily needs to estimate the efficiency of FRET between two epitopes, the easiest way is to label one of them with a donor and the other with an acceptor fluorophore, bound to monoclonal antibodies (mAbs) or their Fab fragments. Suitable donor-acceptor pairs are mostly selected based on the availability of conjugated antibodies and excitation lines in the flow cytometer. The classic dye pair used for FRET is fluorescein and tetramethylrhodamine (TR). Other spectrally similar dyes such as Alexa fluor 488 or Cy2 can be used in place of fluorescein, and TR can be replaced by Cy3 or Alexa fluor 546. Suitable pairs of visible fluorescent proteins (such as cyan fluorescent protein, CFP and yellow fluorescent protein, YFP) fused to the target proteins are also applicable, but their usage is mostly advised after the cellular system has already been explored using specific antibodies.

When applied for the measurement of donor quenching, only the fluorescence of fluorescein (the donor) is measured, and thus, the 488 Ar laser line most commonly available in flow cytometers is adequate. We need to measure the *background-corrected* fluorescence from a donor-only-labeled sample (F_D) and from a sample double labeled with donor and acceptor ($F_{D(+A)}$). FRET efficiency is calculated as

$$E = 1 - \frac{F_{D(+A)}}{F_D} \tag{1}$$

Since F_D and $F_{D(+A)}$ are measured on distinct samples, only the histogram means from the two populations can be considered here. This is one of the main disadvantages of the method, so it is only suggested as a quick and rough estimate of whether FRET and thus molecular proximity occurs. However, it is a quite useful approach when signals are low compared to background/autofluorescence.

There is always one additional control to make, and that is to check for competition of the antibody carrying the acceptor with donor labeling. This should be done with the unlabeled antibody against the "acceptor" epitope, and any decrease of donor fluorescence caused by adding the unlabeled antibody should be attributed to competition rather than donor quenching caused by FRET. Needless to say, competition between labeling antibodies is likely also a sign of molecular proximity, albeit not as readily quantitated as FRET.

2. Classic FCET Based on Both Donor Quenching and Acceptor Sensitization

Although the measurement of FRET-induced donor quenching does not need either dual-laser instruments or complicated evaluation, quenching cannot be used for cell-by-cell data analysis of FRET efficiency. In the classic approach introduced for the cell-by-cell measurement of FRET (Damjanovich et al., 1983; Szöllősi et al., 1984; Tron et al., 1984), either a dual-laser flow cytometer with Argon ion lasers tuned to 488 and 514 nm or a single laser device modified with a prism to split these two lines during multiline operation can be used. The method, later adapted to digital microscopy as well (Nagy et al., 1998b), is based on the following model.

In a sample labeled with both the donor and the acceptor, the intensity measured when exciting the donor (say at 488 nm) and detecting donor emission (e.g., with a band-pass filter centered around 540 nm) can come from the donor (I_D), but also from the acceptor if that is excited at 488 nm and its emission spectrum overlaps with the detection bandpass of the donor. In the ideal situation, which is the case for the fluorescein-TR dye pair, the acceptor TR is not detected in the donor excitation/emission regimen. (As a point of caution, Cy3 as an acceptor is less red shifted than TR, so using it as an acceptor in place of TR requires either different filters or a correction for crosstalk in the model of calculation.)

If FRET is present, the unquenched donor intensity I_D is diminished to the $1 - E$ fraction of its original value:

$$I_1(\text{ex}:D, \text{em}:D) = I_D(1 - E) \tag{2}$$

The emission measured in the acceptor emission channel (>590 nm) has two possible sources of excitation. We can generate fluorescence not only by exciting at 514 nm (see I_3 in Eq. (9)) but also by exciting at 488. In this latter case, we have two sources of emission. The donor will emit light above 590 nm, and this is proportional to the quantity of donor dyes, I_D, its quenching by FRET ($1 - E$) and a factor S_1 that accounts for the ratio of signals generated by the same amount of donor molecules in the acceptor and the donor emission channels:

$$S_1 = \frac{I_2(\text{donor})}{I_1(\text{donor})} \tag{3}$$

$$I_2(\text{ex}:D,\text{em}:A) = I_D(1-E)S_1 + I_A S_2 + I_D E\alpha \tag{4}$$

The second two components of this I_2 emission originate from the acceptor. If the acceptor were to emit in the acceptor channel after excited by its proper (514 nm) excitation wavelength (ex: A, em: A), the intensity measured in the absence of FRET would be I_A. Exciting the acceptor at the donor wavelength (ex: D) is also possible, but it will be less efficient, the factor of proportionality being S_2:

$$S_2 = \frac{I_2(\text{acceptor})}{I_3(\text{acceptor})} \tag{5}$$

This factor can be obtained from measuring the I_2 and I_3 intensities for samples that are labeled with acceptor only.

The emission of acceptor has yet another source in addition to excitation at ex: $D = 488$ nm. It gains energy from the donor via FRET, resulting in an emission component $I_D E\alpha$. The signal arising from sensitized emission of the acceptor is proportional to the number of excitation quanta transferred from the donor to the acceptor and is therefore proportional to $I_D E$. The α factor defines the detection sensitivity of fluorescence from an excited acceptor molecule with respect to the sensitivity to detect fluorescence from an excited donor molecule and can be determined from the ratio of the following signals: the I_2 signal arising from N pieces of excited acceptor molecules and the I_1 signal arising from the same number of excited donor molecules. It depends on the fluorescence quantum yields of the used dyes (Q_D for the donor and Q_A for the acceptor), and the overall detection efficiencies in the donor (η_D) and the acceptor (η_A) detection channels for photons with wavelength distribution of the donor and the acceptor emission, respectively:

$$\alpha = \frac{Q_A \eta_A}{Q_D \eta_D} \tag{6}$$

Experimentally, α can be determined by comparing the I_1 signal of a donor-only-labeled sample to the I_2 signal of a sample labeled with acceptor only:

$$\alpha = \frac{I_{2A}}{I_{1D}} \cdot \frac{B_D}{B_A} \cdot \frac{L_D}{L_A} \cdot \frac{\varepsilon_D}{\varepsilon_A} \tag{7}$$

where B denotes the mean number of receptors per cell labeled by the corresponding antibody, L stands for the mean number of dye molecules attached to an antibody molecule, and ε is the molar extinction coefficient of the dyes at 488 nm; subscripts A and D refer to acceptor and donor, respectively. The ratios of B, L, and ε correct for the different number of photons absorbed by the acceptor- and donor-only-labeled samples. The α factor has to be determined in each experiment, because it depends on the instrument setup and the dye pair. In the equation for α, there is a contribution from the sample labeled only with acceptor excited at the donor wavelength. Usually, this fluorescence intensity is rather small, thus giving the main error source in the calculations. To decrease this error, α should be determined using a protein that is abundant on our cells and recalculated for the actual antibodies used in the experiment.

The fluorescence quantum yields of the dyes may depend on the type of antibody they are attached to and even on the labeling ratio L, thereby affecting the value of α. The α factor determined for a given donor-acceptor antibody pair can be used for other antibody pairs labeled with the same dyes, provided its value is corrected for the differences in the quantum yields:

$$\alpha_2 = \alpha_1 \frac{Q_{A,2}}{Q_{A,1}} \frac{Q_{D,1}}{Q_{D,2}} \tag{8}$$

where subscript 1 refers to the antibody pair for which α has been determined previously, and subscript 2 refers to the new antibody pair.

With this said, we finally progress to analyzing the signal in the acceptor channel detected after exciting specifically the acceptor:

$$I_3(\text{ex}:A, \text{em}:A) = I_D(1-E)S_3 + I_A + I_D E\alpha S_x \tag{9}$$

The first component is the contribution of quenched donor emission, $I_D(1-E)$, to this channel, which is weighted by the factor S_3, obtained on donor-only-labeled samples similarly to S_1.

$$S_3 = \frac{I_3(\text{donor})}{I_1(\text{donor})} \tag{10}$$

The second component is the acceptor emission without sensitization, I_A itself, and the third component is the sensitized acceptor emission. This is the same as in Eq. (4) for I_2, except an S_x factor has to be introduced to correct for exciting the donor at the acceptor's optimum wavelength rather than at its own (as for I_2):

$$S_x = \frac{I_3(\text{donor})}{I_2(\text{donor})} \tag{11}$$

Incidentally, S_x can be expressed from S_3 and S_1 that were already introduced, so we can reformulate Eq. (9) as

$$I_3(\text{ex}:A, \text{em}:A) = I_D(1-E)S_3 + I_A + I_D E\alpha \frac{S_3}{S_1} \tag{12}$$

As already hinted at, the factors S_1, S_2, S_3, and α can be determined from measuring the intensities I_1, I_2, and I_3 for donor- and acceptor-only-labeled samples. Now, it only remains to measure the intensities I_1, I_2, and I_3 on samples labeled with both donor and acceptor (then, of course, subtract the appropriate fluorescence values measured on unlabeled control cells), and calculate, from the list-mode data file, an "*A*" value for each cell:

$$A = \frac{1}{\alpha}\left[\frac{(I_2 - S_2 I_3)}{\left(1 - \frac{S_3 S_2}{S_1}\right)I_1} - S_1 \right] \tag{13}$$

From Eqs. (2), (4), and (12), it follows that the FRET efficiency *E* is related intimately to *A*:

$$E = \frac{A}{1 + A} \tag{14}$$

And hence FRET efficiency on a cell-by-cell basis can be calculated.

3. FCET with Cell-by-Cell Autofluorescence Correction

Both the accuracy and the reproducibility of FRET measurements are compromised if the ligands for the fluorescently labeled probes are expressed at low levels. In such cases, the contribution of autofluorescence may be significant relative to the specific signal. To improve the applicability of FRET in low signal-to-noise systems, one strategy is to apply a red-shifted donor-acceptor pair such as Cy3 and Cy5, or Alexa fluor 546 and Alexa fluor 633. This approach is particularly useful because cellular autofluorescence is higher in the blue and green spectral regions than in the red edge of the visible spectrum (Loken *et al.*, 1987). Furthermore, a cell-by-cell correction for autofluorescence can be achieved using yet another fluorescence channel in addition to those specific for the donor and the acceptor mentioned in Section III.A.2 (Sebestyen *et al.*, 2002).

The mathematical model can be developed based on that described for FCET earlier, with the following modifications. Because Cy3 and Cy5 are used as the donor-acceptor pair, the 488 nm line of an argon ion laser and the 635-nm line of a red diode laser can efficiently excite the donor and acceptor dyes, respectively. The emission maximum of Cy3 (565 nm) falls in the range of the *FL2* detection channel (e.g., through an 585 nm band-pass filter), whereas the maximum of Cy5 emission (667 nm) is in the range of the *FL3* (670-nm long-pass filter) and *FL4* (661-nm band-pass filter) channels. The *FL1* channel, which does not overlap with the emission spectrum of Cy5 and has very little overlap with the spectrum of Cy3, can be used for detecting autofluorescence. Thus, *FL2*, *FL3*, and *FL4* will replace I_1, I_2, and I_3 of Section III.A.2, and for each of these intensities, an additional contributing component, that from autofluorescence, is also considered. *FL1* is introduced as the channel specific for autofluorescence, and the contribution of autofluorescence to each of the other channels is estimated from weight factors B_2,

B_3, and B_4, similar to the S_1, S_2, and S_3 factors already used. Clearly, the contribution of fluorescent labels to the autofluorescence channel also needs to be considered and consequentially further S factors—S_5 and S_6—need to be introduced. Furthermore, the acceptor Cy5 may contribute to the donor-specific $FL2$ signal depending on the optical setup, so its contribution has to be taken into account using an S_4 factor. The equation system that evolves is as follows:

$$FL1(488, 530) = AF + I_D(1 - E) \cdot S_5 + I_A \cdot S_6 + I_D \cdot E \cdot \alpha \cdot \frac{S_6}{S_2} \tag{15}$$

$$FL2(488, 585) = AF \cdot B_2 + I_D(1 - E) + I_A \cdot S_4 + I_D \cdot E \cdot \alpha \cdot \frac{S_4}{S_2} \tag{16}$$

$$FL3(488, > 670) = AF \cdot B_3 + I_D(1 - E) + S_1 + I_A \cdot S_2 + I_D \cdot E \cdot \alpha \tag{17}$$

$$FL4(635, 661) = AF \cdot B_4 + I_D(1 - E) + S_3 + I_A + I_D \cdot E \cdot \alpha \cdot \frac{S_3}{S_1} \tag{18}$$

The numbers in brackets refer to the excitation and detection wavelengths. The autofluorescence (native to $FL1$), the unquenched donor fluorescence (native to $FL2$), and the directly excited acceptor fluorescence (native to $FL4$) are denoted by AF, I_D, and I_A. S_1, S_3, and S_5 are determined using samples labeled only with Cy3:

$$S_1 = \frac{FL3}{FL2} \tag{19}$$

$$S_3 = \frac{FL4}{FL2} \tag{20}$$

$$S_5 = \frac{FL1}{FL2} \tag{21}$$

S_2, S_4, and S_6 are determined on cells labeled only with Cy5.

$$S_2 = \frac{FL3}{FL4} \tag{22}$$

$$S_4 = \frac{FL2}{FL4} \tag{23}$$

$$S_6 = \frac{FL1}{FL4} \tag{24}$$

B_2, B_3, and B_4 are determined on unlabeled cells:

$$B_2 = \frac{FL2}{FL1} \tag{25}$$

$$B_3 = \frac{FL3}{FL1} \tag{26}$$

$$B_4 = \frac{FL4}{FL1} \tag{27}$$

When using the filters and dichroic mirrors available in the BD FACSCalibur, S_3, S_4, and S_6 are zero, which simplifies Eqs. (15)–(18). Solving the simplified equations yields the following expression for the parameter $A = E/(1 - E)$:

$$\frac{FL1 \cdot (B_2 S_1 + B_4 S_2 - B_3) + FL2 \cdot (B_3 S_5 - S_1 - B_4 S_2 S_5) + FL3 \cdot (1 - B_2 S_5) + FL4 \cdot S_2 \cdot (B_2 S_5 - 1)}{\alpha (FL2 - FL1 \cdot B_2)}$$

$$\tag{28}$$

Using this formula, we can easily calculate the value of E for each cell either in a spreadsheet of imported list-mode data or in a specific program where gating and determination of the S, B, and α factors is considerably easier. An example showing the improvement resulting from this approach in the case of relatively low receptor expression is depicted in Fig. 2.

4. Sample Experiment for FCET Measurement

Two molecular epitopes on the surface of cells are chosen for testing their nanometer-scale proximity. They are labeled with Cy3 and Cy5 conjugated dyes. Samples are made where both, either, or none of the epitopes are labeled. The samples are run in a flow cytometer and the appropriate emission intensities are collected in list-mode files. Then the samples can be analyzed using a custom-made program (Szentesi *et al.*, 2004; available to the public from http://www.freewebs.com/cytoflex) to assess whether the labeled epitopes are in each other's molecular proximity. The analysis yields a cell-by-cell distribution of FRET efficiency values. Such analysis can also be performed in a spreadsheet table if one prefers.

In general, the following samples are necessary:

1. Unlabeled cells.
2. Cells with one epitope labeled using a Cy3 conjugated antibody (or alternatives like Alexa fluor 546 or 555).

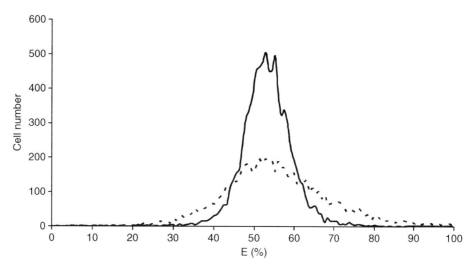

Fig. 2 Improving fluorescence resonance energy transfer (FRET) analysis using cell-by-cell auto-fluorescence correction. Distribution histograms of FRET efficiency are produced from flow cytometric measurements on breast cancer cells expressing relatively low numbers (~40,000/cell) of ErbB2 molecules. Two epitopes on ErbB2 were tagged with Alexa fluor 546 and Alexa fluor 647 conjugated monoclonal Fab fragments. Consequently, FRET distribution histograms peak at about 50%, indicating efficient intramolecular energy transfer. Cell-by-cell autofluorescence correction (*solid line*) vastly improves the dispersion of data as compared to analysis with subtraction of an average autofluorescence value (*dashed line*), but the mean remains unchanged. This approach is especially useful when signals from the specific label are low relative to autofluorescence.

3. Cells with the same epitope labeled using a Cy5 conjugated antibody (or alternatives like Alexa fluor 633 or 647).

4. Cells with the other epitope labeled using a Cy3 (or alternative) conjugated antibody.

5. Cells with the other epitope labeled using a Cy5 (or alternative) conjugated antibody.

6. Cells double labeled with the pair of antibodies Cy3 on epitope 1 and Cy5 on epitope 2.

7. Cells double labeled with the pair of antibodies Cy5 on epitope 1 and Cy3 on epitope 2.

This set allows for measuring FRET in both directions, which is useful if expression levels of the two epitopes are not comparable. If a positive control with known molecular interactions is known, it is advisable to have a set of samples of the same cell type similar to those listed earlier. For example, if we want to know about the molecular interaction of major histocompatibility complex (MHC) class I and class II on a lymphocytic cell, we need the following:

1. Unlabeled cells.
2. Cells with MHC class I labeled using a Cy3 conjugated antibody.
3. Cells with MHC class I labeled using a Cy5 conjugated antibody.
4. Cells with MHC class II labeled using a Cy3 conjugated antibody.
5. Cells with MHC class II labeled using a Cy5 conjugated antibody.
6. Cells double labeled with anti-MHC-I-Cy3 and anti-MHC-II-Cy5 antibodies.
7. Cells double labeled with anti-MHC-I-Cy5 and anti-MHC-II-Cy3 antibodies.

   As for the positive control, we know that MHC-I binds the β_2-microglobulin ($\beta 2m$), so we make further samples:

8. Cells with $\beta 2m$ labeled using a Cy3 conjugated antibody (this replaces MHC-II as the donor, when MHC-I is the acceptor, and the reason to make $\beta 2m$ the donor is that surely all $\beta 2m$ is bound to an MHC-I, but not all MHC-I may bind $\beta 2m$).
9. Cells double labeled with anti-$\beta 2m$-Cy3 and anti-MHC-I-Cy5.
   Additional information needed for the calculations is as follows:

- The dye/protein molar ratio of all antibodies used.
- The molar absorption coefficients of all dyes used.
- The quantum efficiencies of the dyes used, if any of the labels give intensity less than three to five times the background.

a. Materials
Fluorochromes

- Cyanine dyes Cy3 and Cy5 or Alexa fluor dyes Alexa 546/555 and Alexa 633/647 as donor and acceptor, respectively. (The Alexa fluor 546–647 pair appears optimal for most cytometers—see Horváth *et al.*, 2005).

Cells

- Lymphocytic cells in suspension.
- Adherent cells should be detached (trypsinized, or treated with collagenase) before washing and labeling.

Solutions

- PBS (for washing)
- PBS + 0.1% BSA (for labeling)
- PBS + 1% formaldehyde (for fixation)
- Antibodies W6/32 against MHC-I, L368 against the $\beta 2m$, L243 against MHC-II

b. Sample Preparation
Harvesting Cells

- 75-ml flasks of adherent cells are trypsinized and medium with FCS is added to stop the trypsin.
- Cells are left to recover 20 min in the flask. It has been determined that after gentle trypsinization, most cell surface proteins are either unchanged or totally recovered within 20–30 min.
- One needs to determine labeling intensity as a function of time after trypsinization for the particular proteins examined.

Labeling

- Wash cells with ice-cold PBS and centrifuge suspension.
- Repeat washing.
- Add 1 million cells per sample tube and store on ice.
- Label cells with (usually) 5–50 μg/ml final concentration of antibodies (should be above saturating concentration that was determined previously) in 50 μg total volume of PBS-0.1% BSA mixture for 15–30 min on ice.
- Wash cells twice with ice-cold PBS and centrifuge suspension.
- Fix cells with PBS + 1% formaldehyde in 500–1000 μl.
- Store samples in refrigerator or cold room until measurement.

Before measurement, resuspend the cells with gentle shaking, and if upon examination in the microscope clumps are detected, run the suspension through a fine sieve.

Always examine labeled cells dropped on a microscopic slide in the fluorescence microscope to verify proper cellular position (e.g., membrane) of the label.

c. Instrument Settings

- Use sample 1 (background) as a negative control.
- Set FSC and SSC in linear mode to see your population on the scatter plot (*SSC/FSC* dot plot).
- Set *FL1, FL2, FL3,* and *FL4* in logarithmic mode.
- Set *FL1, FL2, FL3,* and *FL4* voltages so that mean fluorescence intensities are about 10.
- Run sample 2 (donor) and adjust *FL1, FL2,* and *FL3* voltages so that mean fluorescence intensities are in the 10^2–10^4 range.
- Run sample 4 (acceptor) and adjust *FL3* and *FL4* voltages so that mean fluorescence intensities are in the 10^2–10^4 range.
- Save instrument settings.
- *Draw the following plots*:
- *FSC/SSC*

- *FL1*, *FL2*, *FL3*, and *FL4* for intensity check
- *FL2/FL3*, *FL2/FL4*, and *FL3/FL4* for correlation check
- Run sample 1 and draw a gate on the *FSC/SSC* dot plot around viable cells
- Format plots so that only R1-gated events will appear
- Define statistics window to show the mean fluorescence intensities of all histograms from the R1-gated events
- Set the machine to acquire 20,000 events
- Run samples from 1 to 9.

d. Analysis

During data analysis, it is advised to follow a general scheme. First, one needs to determine the mean background intensities and the autofluorescence correction factors. Then calculate the α factor and spectral overspill parameters (*S* factors) from the acceptor- and donor-labeled samples. With these parameters in hand, now the energy transfer efficiency can be determined on a cell-by-cell basis.

B. Image Cytometric FRET

When measuring FRET in a flow cytometer, one should always keep in mind that FRET efficiency is calculated from fluorescence intensities that are averaged over each cell. Thus, the FCET technique inherently excludes the possibility of gaining information about the subcellular distribution of molecular proximities. Thus, the idea of measuring FRET in microscopic imaging has been introduced quite early in the joint history of cytometry and FRET (Jovin and Arndt-Jovin, 1989). All possible approaches to imaging FRET—both those already implemented and those only in the proposal stage—are reviewed in Jares-Erijman and Jovin (2003, 2006). We concentrate on three approaches that can be implemented using rather conventional imaging equipment and that have been applied extensively to answer biological questions. The common feature in these techniques is that they exploit the change of fluorescence intensity caused by the occurrence of FRET. A fourth approach that has successfully been used in cell biology is fluorescence lifetime imaging (FLIM) (Bastiaens and Jovin, 1996; Bastiaens and Squire, 1999; Gadella *et al.*, 1994; Grecco *et al.*, 2009; Haj *et al.*, 2002; Shcherbo *et al.*, 2009; Van Munster and Gadella, 2004). For an excellent overview, see Clegg *et al.* (2003).

As for intensity-based approaches, the first two measure donor fluorescence, whereas the third is similar in principle to the ratiometric technique described for FCET in section III.A.3. When we observe fluorescently stained microscopic samples, one labeled with donor and the other with both donor and acceptor, the samples differ in two basic features (Fig. 3). First, those labeled with both donor and acceptor will show (on average) less intense donor fluorescence because of quenching by the acceptor. Second, excited state reactions resulting in irreversible photobleaching of the donor—a phenomenon usually most annoying to

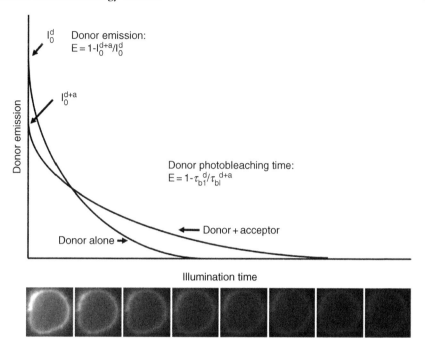

Fig. 3 Fluorescence resonance energy transfer (FRET) measured from donor parameters. Comparing donor fluorescence of donor with donor- and acceptor-labeled samples in the microscope yields two basic differences, each exploitable for the determination of FRET efficiency. On a single observation, lower fluorescence intensity in the donor- and acceptor-labeled sample because of "donor quenching" will be apparent. During continuous or repeated observations, photobleaching of the dye can occur (see also image sequence on lower inset—MHC-I labeled on human lymphocyte), the rate of which is inversely proportional to the excited state lifetime of the donor. Both these phenomena can be exploited to assess the change of donor lifetime upon the occurrence of the acceptor and thus the efficiency of FRET.

microscopists—will occur at different rates. This latter is related to the decreased excited state lifetime of the donor in the presence of the acceptor, which confers some degree of "protection" from photobleaching to the donor, thereby lengthening the time constant of photobleaching. Both of these features can be exploited for the measurement of FRET efficiency, and because excited state reactions are central to both, these techniques are often called *photobleaching FRET* (pbFRET).

1. Donor Photobleaching FRET

As depicted in Fig. 3, a fluorescently labeled cellular sample is subjected to photobleaching under continuous observation in the fluorescence microscope. The rate of photobleaching depends on the concentration of free radicals, the illumination flux, and the excited state lifetime. Keeping the former two constant, one is able to deduce the relation between the photobleaching time constant τ_{bl}, which is inversely proportional to the excited state lifetime τ, and the E efficiency of FRET (Bastiaens and Jovin, 1998; Jovin and Arndt-Jovin, 1989):

$$E = 1 - \frac{\tau^{D+A}}{\tau^{D}} = 1 - \frac{\tau_{bl}^{D}}{\tau_{bl}^{D+A}} \tag{29}$$

Here, the D and D + A upper indices refer to donor-only- and donor-plus-acceptor-labeled samples, respectively. It ensues that for this measurement approach, we need to prepare two samples, one with the epitope in question labeled with donor and the other with both donor and acceptor. Both samples are observed and imaged in the microscope using donor-specific optical filters, taking a sequence of images until the donor is totally bleached and hence its intensity reaches background. Intensity usually decays according to an exponential function to which a pixel-by-pixel fit can be made. Because the control (donor only) bleaching time constants originate from a separate sample, one needs to average these time constants and use the average as a reference τ_{bl}^{D} value for the calculation of a pixel-by-pixel FRET efficiency according to Eq. (29). These E values can then be evaluated with respect to subcellular localization, or they can be pooled to form distribution histograms. A program for the evaluation of donor photobleaching FRET image series has been created in LabView and is avaliable as a runtime executable (Szentesi *et al.*, 2005).

The method carries inherent advantages and disadvantages. Although subcellular distribution of E is derived from the measurement, this distribution should be judged carefully, because pixel-by-pixel variations of bleaching times can also result from variations of the local molecular environment, oxygenation, and previous bleaching of neighboring cells or even pixels. For this reason, the method offers more reliable results in full-field microscopy, whereas confocal laser scanning microscopes (LSMs) tend to fare worse in implementing this approach. Nonetheless, donor pbFRET is relatively simple to implement and is rather sensitive; FRET efficiencies of 2–4% can be reasonably well measured if labeling is good and images are free of noise. Adherent cells are the best targets for investigations with this approach, similar to other image cytometric FRET measurements. However, suspension cells and cell lines can also be measured after making the cells adhere to a substrate, either by sedimentation onto poly-L-lysine or collagen-coated coverslips or by using a cytocentrifuge.

Some disadvantages and pitfalls should also be considered when choosing and implementing this method. Primarily, the measurements are not self-controlled in the conventional sense, so care should be taken to execute bleaching sequences alternately between the donor-only- and the donor-plus-acceptor-labeled samples—even more so because fluctuations of temperature, illumination light intensity, and oxygenation greatly influence the photobleaching rate. A dye that is easily photobleached should be chosen as the donor, for example, fluorescein. This will hopefully minimize movement artifacts that hinder the proper exponential fitting. Should such artifacts persist, the microscope needs to be checked for mechanical stability, and cells for proper adherence. Sequential images can be corrected for registration, but in the case of a long sequence (and about 30 images are necessary for a good fitting), this may be cumbersome even if using a simple fast Fourier transform (FFT)-based algorithm.

Special consideration ought to be given to the actual kinetics of photobleaching. This mostly depends on the number and nature of various excited state reactions the donor can undergo. Fluorescein, for example, shows a rather complex behavior in this respect (Song *et al.*, 1995), so although it offers the advantage of fast bleaching, a multiexponential fitting may be necessary to obtain the proper bleaching time constants. In practice, quite often a double exponential works very well. Here, to be able to easily compare pixels or cells, an amplitude weighted average can be calculated from the two bleaching time constants (Nagy *et al.*, 1998a; Szabó *et al.*, 1995). Another consideration is the initial bleaching that occurs during the adjustment of the microscope, localization of the spot to measured, and focusing. If the bleaching is not monoexponential, an overestimation of the time constant (and underestimation of E) may result. Furthermore, it is most necessary to choose a highly photostable acceptor, because photobleaching of the acceptor after excitation via FRET will primarily destroy the nearest acceptors and eliminate FRET between the tightest donor-acceptor pairs, leading to an underestimation of E. A sign that hints at such a possibility is the relative stability or unexpected increase of donor fluorescence in the initial phases of the bleaching curve.

Comparison between FCET measurements and the donor photobleaching image cytometric approach revealed that consistently higher transfer values are obtained with the pbFRET method. This overestimation was independent of the pixel size: pixel sizes as large as a cell gave similar results to those obtained with smaller ones, reinforcing the view that energy transfer values are independent of fluorescence intensity in the samples. Some of this discrepancy can be attributed to the different weighting of energy transfer values in the pbFRET and the ratiometric energy transfer methods. Using Monte Carlo simulation, Nagy *et al.* (1998b) demonstrated that this overestimation in pbFRET is proportional to the heterogeneity in the pixel-by-pixel FRET efficiency values. Therefore, discrepancies between FRET efficiency values obtained with pbFRET and ratiometric approaches should be interpreted with caution (Nagy *et al.*, 1998b).

2. Acceptor Photobleaching FRET

Some of the pitfalls of donor pbFRET can be addressed by using the acceptor pbFRET technique (Bastiaens and Jovin, 1998; Bastiaens *et al.*, 1996, 1997; Vereb *et al.*, 1997). Most importantly, the acceptor bleaching is a reasonably simple technique that offers the unique advantage of self-controlled measurements on a pixel-by-pixel basis. The idea is to label a sample with both donor and acceptor, measure the fluorescence of the donor, and after having used an acceptor-specific excitation wavelength targeted at a given region of interest (ROI), measure it again to assess the increase in donor fluorescence. This is easily done in a (confocal) LSM. The energy transfer efficiency E can be calculated as

$$E_{(i,j)} = 1 - \frac{F^{\mathrm{I}}_{\mathrm{D}(i,j)}}{\gamma \cdot F^{\mathrm{II}}_{\mathrm{D}(i,j)}}, \tag{30}$$

where $F^{\mathrm{I}}_{\mathrm{D}(i,j)}$ and $F^{\mathrm{II}}_{\mathrm{D}(i,j)}$ are the background-subtracted donor fluorescence values of the (i, j)th pixel in the entire image before (I) and after (II) photobleaching the acceptor. In the denominator, γ is a correction factor that takes into consideration the photobleaching of the donor during the whole protocol. It can be calculated as

$$\gamma = \frac{\langle F^{\mathrm{ref,I}}_{\mathrm{D}(i,j)} \rangle}{\langle F^{\mathrm{ref,II}}_{\mathrm{D}(i,j)} \rangle}, \tag{31}$$

where $\langle F^{\mathrm{ref,I}}_{\mathrm{D}(i,j)} \rangle$ and $\langle F^{\mathrm{ref,II}}_{\mathrm{D}(i,j)} \rangle$ are the mean, background-subtracted donor intensities in pixels (i, j) above threshold of a reference sample labeled with donor only, before (I) and after (II) running an identical acceptor photobleaching protocol. This constant must be determined for each experimental setting.

In addition to determining this constant, the following controls and measures of caution need to be implemented. (1) During the image acquisition protocol, we must confirm that the acceptor was totally bleached, or apply a correction for incomplete bleaching. (2) Using an acceptor-only-labeled specimen, the whole protocol should be run, and no change in the donor channel should be seen. This usually indicates that there is no acceptor crosstalk into the donor channel before bleaching, and that there is no fluorescent photoproduct formed from the bleached acceptor that crosstalks into the donor channel afterward. However, as these two processes may be independently present, it is best recommended to correct for both. (3) It is recommended to leave the acceptor intact in part of the image and generate a FRET efficiency histogram from that area alone to see whether after applying the appropriate correction (γ), the histogram mean is zero. (4) If the photobleached acceptor can still absorb in the spectral range of donor emission, "dark transfer" can occur, which does not manifest in sensitized acceptor fluorescence, but nonetheless quenches the donor. (5) It follows from the points above that the ideal dye pair for acceptor pbFRET is made of an extremely photostable donor with high absorption and quantum efficiency, and an acceptor that is easy to bleach, does not accept FRET after bleaching, and does not spectrally interfere with the donor before or after bleaching. With the proper choice of filters, the indocarbocyanine dyes Cy3 and Cy5 reasonably well match these criteria. (6) Donor images have to be checked (and, if necessary, corrected) for proper spatial registration before implementing pixel-by-pixel calculations. Failure to do so will usually cause edge effects that grossly distort FRET distribution histograms. (7) Depending on the noise in our raw images, it may help to do a Gaussian filtering possibly using a filter size (usually 3×3) that is still less than the optical resolving power of our microscope. (8) From the prebleach donor and acceptor images an acceptor/donor ratio image should also be formed and its correlation with the FRET image checked. It is expected that higher acceptor/donor ratios can yield higher E values without decreasing donor-acceptor separation.

We have recently overviewed these issues and have created a freely available Image J Java plugin for the facile evaluation of acceptor photobleaching data (Roszik *et al.*, 2008). AccPbFRET allows pixel-wise calculation and detailed, ROI-based analysis of FRET efficiencies including semiautomatic analysis of large image sets. It uses a complex correction algorithm that solves or obviates most of the issues raised above: In addition to correcting for unwanted photobleaching of the donor, crosstalk of the acceptor and/or its photoproduct to the donor channel, as well as partial photobleaching of the acceptor are taken into account. Registraion of images is automatic, and optional Gaussian filtering is also available. The acceptor photobleaching technique can be easily implemented using an LSM, and the image processing that yields a 2D map of FRET efficiency is not too complicated. However, in addition to the several controls that one has to perform, the somewhat low sensitivity of the method is also a disadvantage. In our experience, E values higher than 5% can be determined safely with a well-set LSM system and proteins that are expressed at more than about 20,000 per cell.

The great advantage of self-controlled measurements can be further exploited by combining this method with the donor bleaching method. Briefly, the ROI is divided into two halves: In one half the acceptor is bleached and a map of FRET is calculated. Then in the whole ROI the donor bleaching kinetics is measured, and a map of FRET is calculated in the region with the acceptor still present, based on the average bleaching time constant measured in the area already void of acceptor. The histogram and subcellular distribution of E in the two half-ROIs can then be compared. Those interested should consult Bastiaens and Jovin (1998) for details.

3. Intensity–Based Ratiometric FRET

Both donor and acceptor photobleaching approaches carry the inherent drawback of irreversibly destroying fluorophores in the sample. It follows that measurements cannot be repeated at different times in the same sample, which may well be a problem if the time course of interactions is to be followed in a self-controlled manner. Strategies have already been suggested to reversibly switch on or off the donor or the acceptor (Jares-Erijman and Jovin, 2003). Lifetime imaging of the donor in the constant presence of the acceptor, using a sample without acceptor as a reference has been successful to follow in time intracellular protein interactions (Haj *et al.*, 2002; Legg *et al.*, 2002). Among the relatively less complicated intensity-based measurement modalities, the microscopic intensity-based ratiometric FRET (MI-FRET) offers similar advantages. This approach has been elaborated to exploit acceptor sensitization for estimating FRET efficiency on a pixel-by-pixel basis without losing temporal resolution (Nagy *et al.*, 1998b). The method originally used fluorescein as donor and rhodamine as acceptor, but it can be applied to other donor and acceptor pairs (e.g., Cy3 and Cy5) with slight modification.

To achieve quantitative FRET measurements, the autofluorescence of cells needs to be assessed on a pixel-by-pixel basis. Because the autofluorescence of cells has a wide excitation and emission spectrum and is fairly stable, it is possible to define ratios (B_1, B_2, and B_3) of fluorescence intensities recorded with the fluorescein (I_1), energy transfer (I_2), and rhodamine (I_3) filter sets to the fluorescence intensity (I_{Bg}) called *background intensity* in this terminology, recorded with ultraviolet (UV) excitation at (360-nm) and 450-nm emission

$$B_x = \frac{I_x}{I_{Bg}}, \quad x = 1, 2, 3. \tag{32}$$

These autofluorescence correction ratios are determined with unlabeled cells. Calculation of B_x values is not carried out on a pixel-by-pixel basis, but the pixel values are summed in the images, especially because the fluorescence intensities in single pixels are sometimes very low. With the B factors in hand, the following compensation is carried out on a pixel-by-pixel basis:

$$I_x(\text{without cellular autofluorescence}) = I_x(\text{with cellular autofluorescence}) - B_x I_{Bg}, \tag{33}$$

where I_x stands for I_1, I_2, and I_3 (fluorescein, energy transfer, and rhodamine intensities, respectively), I_{Bg} is the UV-excited background, and B_x is the ratio described earlier. This is valid because our fluorescent dyes have negligible absorption in the relevant UV range. This correction eliminates the contribution of autofluorescence from the I_1, I_2, and I_3 intensities.

In microscopy, intensity-based ratiometric determination of energy transfer usually relies on the measurement of sensitized acceptor emission, by exciting the donor (fluorescein) and measuring the fluorescence emission from the acceptor (rhodamine, I_2 in Eq. (35)). This, however, is complicated by the fact that I_2 has a background component ($B_2 I_{Bg}$) and a contribution from the direct emission of the donor ($I_F(1 - E)S_1$) and the acceptor ($I_R S_2$) even with the best designed filter sets. To overcome this problem, four independent images of the same field should be measured. The relationship between the background image [$I_{Bg}(360, 450)$], fluorescein image (I_1), rhodamine image (I_3), energy transfer image (I_2), and the FRET efficiency (E) are presented in the following equations:

$$I_1(490, 535) = B_1 I_{Bg} + I_F(1 - E) + I_R S_4 \tag{34}$$

$$I_2(490 > 590) = B_2 I_{Bg} + I_F(1 - E)S_1 + I_R S_2 + I_F E\alpha \tag{35}$$

$$I_3(546 > 590) = B_3 I_{Bg} + I_F(1 - E)S_3 + I_R \tag{36}$$

In Eqs. (34)–(36), I_F is the unquenched fluorescein emission, and I_R is the direct rhodamine emission. The numbers in parentheses refer to the wavelengths of excitation and emission. The S_1–S_4 factors characterize the spectral overlap between different channels. S_1 and S_3 are determined using samples labeled with fluorescein only according to the following equations (Szöllősi *et al.*, 1984; Tron *et al.*, 1984):

$$S_1 = \frac{I_2}{I_1} \tag{37}$$

$$S_3 = \frac{I_3}{I_1} \tag{38}$$

Fortunately, using a standard fluorescein and rhodamine filter and dichroic mirror setup, S_3 is usually zero, making Eqs. (34)–(36) simpler. In spite of this, we retained this designation to make our equations compatible with those of the flow cytometric approach (see Section III.A.2). S_2 and S_4 are determined on samples labeled with rhodamine only according to the following equations:

$$S_2 = \frac{I_2}{I_3} \tag{39}$$

$$S_4 = \frac{I_1}{I_3} \tag{40}$$

The fluorescence intensities in Eqs. (37)–(40) represent summed fluorescence intensities from the same part of the images. For example, to calculate S_1, some bright cells are selected, and the intensity of these cells is summed in the energy transfer (I_2) and the fluorescein (I_1) image. The proportionality α factor is the ratio of the fluorescence intensity of a given number of excited rhodamine molecules measured in the I_2 channel to the fluorescence intensity of the same number of excited fluorescein molecules detected in the I_1 channel. This is constant for a given experimental setup and a particular pair of donor and acceptor conjugated ligands and must be measured for every defined case (Nagy *et al.*, 1998b).

Although photobleaching is useful in pbFRET, it presents a problem in the case of intensity-based ratiometric measurements. When the energy transfer values are calculated using the intensity-based approach, correction for photobleaching should be taken into account. Because the bleaching rate of rhodamine is very low with usual full-field illumination intensities, this correction needs to be applied only for the fluorescein and the energy transfer images. Fluorescein and energy transfer images are sequentially recorded. When calculating the energy transfer in an image, relevant fluorescein intensity is computed as the average of the previous and the next fluorescein image. This linear approximation of bleaching is reasonable only for short exposure times. From these images, the FRET efficiency can be calculated on a pixel-by-pixel basis using two different approaches.

First, FRET can be calculated in a similar way to the flow cytometric approach, which has already been discussed in Section III.A.2. In this case, $I_R S_2$ is subtracted from the energy transfer image, and the remaining $I_F E\alpha$ is divided by $\alpha(I_1 - B_1 I_{Bg} - I_R S_4) = I_F(1 - E)\alpha$. This gives $E/(1 - E)$, which is termed β in Eq. (41)

$$\frac{I_2 - B_2 I_{Bg} - (I_1 - B_1 I_{Bg})S_1 + (I_3 - B_3 I_{Bg})(S_1 S_4 - S_2)}{\alpha(I_1 - B_1 I_{Bg} - (I_3 - B_3 I_{Bg})S_4)} = \beta \qquad (41)$$

In the denominator, we have the I_1 image corrected for background and rhodamine contribution multiplied by α. In the numerator, we have the I_2 image corrected for background and fluorescein contribution $(I_2 - B_2 I_{Bg} - [I_1 - B_1 I_{Bg}]S_1)$ and the background-corrected I_3 image multiplied by $(S_1 S_4 - S_2)$.

Second, E can also be calculated according to the classic equation, which is modified slightly so that it can be applied for microscopes not using single-wavelength excitation, but band-pass filters:

$$\frac{F_{AD}}{F_A} = 1 + \frac{A_D c_D}{A_A c_A} E, \qquad (42)$$

where F_A is the direct acceptor fluorescence (in our case $I_R S_2$) measured at the same excitation wavelength as the sensitized emission, A_D and A_A are the integrated absorptions of the donor and the acceptor in the fluorescein excitation range, respectively, and c_D and c_A are the concentrations of the donor and the acceptor, respectively. Solving this equation gives

$$\left(\frac{I_2 - B_2 I_{Bg} - (I_1 - B_1 I_{Bg} - (I_3 - B_3 I_{Bg})S_4)S_1}{(I_3 - B_3 I_{Bg})S_2} - 1 \right) \frac{A_A c_A}{A_D c_D} = E. \qquad (43)$$

To be able to calculate E according to Eq. (43), we need to calculate the absorption ratio (A_A/A_D) and the pixel-by-pixel acceptor/donor ratio (c_A/c_D). Determination of these parameters is not trivial, and a detailed experimental approach is given by Nagy *et al.* (1998b).

The intensity-based ratiometric FRET approach uses accurate background correction and subtraction of spectral overlaps from the energy transfer channel. This measurement results in images similar to those obtained by the donor photobleaching method. Important advantages of the ratiometric method are that it requires much less image-storing capacity and it is less sensitive to environmental factors than photobleaching. Photophysics of bleaching is far from being understood and sometimes pbFRET results give anomalously high values, especially when cell surface proteins are very dense. One drawback of the ratiometric microscopic technique is that for accurate determination of FRET, it requires higher fluorescence intensity than pbFRET does, because sensitized emission is only a small fraction of the I_2 intensity in the FRET image. The method is also very sensitive to precise determination of the S factors and the pixel-by-pixel donor/acceptor ratio, which also demands high levels of fluorescence intensity. A further drawback of the technique is that the absorption ratio is instrument dependent, and when lamps are used for excitation instead of lasers, this absorption ratio may be altered as the spectral characteristics of the lamp change with time (Nagy *et al.*, 1998b).

In order to facilitate the easy, pixel-by-pixel evaluation of ratiometric FRET images, we have created a freely available Image J plugin (Roszik *et al.*, 2009). RiFRET allows the user to calculate FRET from image stacks, that is from 3D datasets, making possible the calculation and visualization of three-dimensional formations in cells or tissues, or creating two-dimensional FRET maps as a function of time. Semiautomatic processing for larger datasets is also included in the program. All bleed-through corrections are included in the calculations, together with the calculation of these correction constants. Furthermore, three methods to determine the α factor for calibrating FRET efficiency calculations are included. These are optimized either for fluorophore-labeled antibodies or for fluorescent protein fusion products of fixed or varying donor-acceptor ratios. The reader is referred to Roszik *et al.* (2009) for further details and is encouraged to use this program that step-by-step seemlessly guides through the analysis process without losing the user in mathematical details.

4. Sample Experiment for Image Cytometric FRET Measurement

a. Materials

1. Prepare 1% formaldehyde solution in PBS. This solution should be made fresh.

2. Prepare fluorescently conjugated mAbs according to the method described in Szöllősi *et al.* (1989), Nagy *et al.* (1998b), and Sebestyen *et al.* (2002). The concentration of the stock solution should be in the range of 0.5–1.0 mg/ml. The dye/protein labeling ratio for donor- or acceptor-labeled antibody should preferentially be between 2 and 4. Store stock solutions in the presence of 0.1% NaN_3 at 4 °C.

3. Before fluorescent labeling of cells, remove antibody aggregates by centrifuging the diluted fluorescently conjugated antibody solution (100 μg/ml) at 100,000 $\times$ g for 30 min.

b. Sample Preparation

1. Deposit cells on microscope slide by growing them on the slide or by attaching suspended cells to polylysine-coated slide as described earlier (Damjanovich *et al.*, 1995).

2. Label cells with fluorescently tagged mAbs at saturating concentration (1-2 μM depending on the type of antibody) for 30 min on ice in the dark. The incubation volume should be 80 μl to cover the cells at a 15 $\times$ 15 mm^2 area. At least four samples should be prepared: unlabeled cells, cells labeled with donor only, cells labeled with acceptor only, and cells labeled with both donor and acceptor (FRET sample). For the double-labeled sample, donor and acceptor conjugated antibodies should be mixed in the needed concentration ratio before addition to the cells.

3. The excess mAb should be removed by washing cells in PBS twice.

4. Cells should be fixed in 1% formaldehyde solution in PBS for at least 1 h.

5. Mount cells under a coverslip by adding a drop of a solution containing 90% glycerol in PBS and seal the preparation by melted paraffin. Alternatively, a commercial antifade mounting medium (e.g., Vectashield, Vector laboratories, Inc., Burlingame, California) can be used. When intensity-based and pbFRET are compared on the same set of samples, avoid using antifade solution.

c. Fluorescence Measurements and Analysis

1. Choose the appropriate excitation, emission, and dichroic filters according to your donor and acceptor dye.

2A. For donor pbFRET, take a donor-only- and a donor-plus-acceptor-labeled sample. Take a sequence of images of the donor in both samples until donor is bleached to background. Finally take an image of the acceptor in the double-labeled sample.

2B. For acceptor pbFRET, use a double-labeled sample and a donor-only-labeled sample. (For the sake of simplicity, we assume that for the Cy3-Cy5 pair, all aforementioned controls are done and negative.) Run the following protocol on both samples: take donor image, take acceptor image, bleach acceptor in consecutive scans, take acceptor image to confirm it is totally bleached, and take donor image again.

2C. For ratiometric FRET, collect images (I_1, I_2, and I_3 and I_{Bg}) for all samples (donor, acceptor, and double labeled). For autofluorescence correction, one extra set of images is to be taken from nonlabeled cells. For each sample, take the autofluorescence (I_{Bg}) image first, then the acceptor (I_3), the donor (I_1), and the FRET (I_2) images. If there is photobleaching, take the FRET and the donor images, in this order, again.

3. Correct all the images for camera dark current by subtracting images taken with closed shutters.

4. Because of the filter changes (especially for dichroic filters), check images for image registration. Correct images for any pixel shift.

5. Correct images for the background caused by the fluorescence of the optical elements of the microscope, using cell-free areas of the images.

5A. For donor pbFRET, do a pixel-by-pixel monoexponential or biexponential fit for both donor-only- and donor-plus-acceptor sample sequences. Calculate average bleaching time constant from the donor-only image and determine E for every pixel in the donor-plus-acceptor sample using Eq. (29).

5B. For acceptor pbFRET, determine the γ correction factor according to Eq. (31) from images of the donor-only sample. Then calculate E for every pixel using γ and applying Eq. (30) to images of the double-labeled sample.

5C. For ratiometric FRET, analyze unlabeled cells first, determining the spectral overlap factors of autofluorescence. Then analyze donor-only and acceptor-only samples. Using all four images, correct the images for autofluorescence first. Then determine the spectral overlap factors for the donor spectrum from the donor-only sample and for the acceptor spectrum from the acceptor-only sample. Consecutively, analyze all four images of the double-labeled samples, and correct for autofluorescence, then for donor emission and acceptor emission. After all the corrections, calculate the FRET efficiency on a pixel-by-pixel basis. If there is photobleaching during image recording, use the average of donor and FRET images that were taken in alternated sequence.

C. Limitations of FRET

Although FRET can provide very useful information about molecular proximity and associations, it has its own limitations. The most serious drawback of FRET is that it has restricted capacity in determining absolute distances because FRET efficiency depends not only on the actual distance between the donor and acceptor, but also on their relative orientation (κ^2). It is still quite good at determining relative distances, namely, whether the two labels are getting closer or farther on a certain stimulus/effect. Even when measuring relative distances, care must be taken to ensure that the orientation factor (κ^2) does not change between the two systems to be compared. If the fluorescent dye is attached to an antibody or Fab fragment via a carbon linker having 6–12 carbon atoms, the linker often allows relatively free rotation of the dye, which minimizes uncertainty of κ^2.

Another problem is that FRET has a very sharp distance dependence, making it difficult to measure relatively long distances because the signal gets very weak. At the same time, energy transfer tends to occur on an all or none basis; if the donor and acceptor are within $1.63 \times R_0$ distance, energy transfer is detectable, and if they are farther apart, energy is transferred with very little efficiency. Because of this sharp decrease, the absence of FRET is not direct proof of the absence of molecular proximity between the epitopes investigated. Also, absence of FRET can be caused by sterical hindrance even for neighboring molecules or protein domains. On the other hand, presence of FRET to any appreciable extent above the experimental error of measurement is strong evidence of molecular interactions.

Indirect immunofluorescent labeling strategies may be applied to FRET measurements if suitable fluorophore-conjugated mAbs are not available or as an approach to enhancing the specific fluorescence signal. In such cases, special attention should be paid to the fact that the size of the antibody complexes used affects the measured FRET efficiency values. Application of a larger antibody complex causes a decrease in FRET efficiency as a result of the geometry of the antibody complexes, because when antibody or F(ab′) complexes become larger, the actual distance between the donor and acceptor fluorophores increases (Sebestyen *et al.*, 2002). This explains the decrease of FRET efficiency when fluorescent secondary F(ab′)$_2$ fragments are used on both the donor and the acceptor side as opposed to using direct-labeled primary antibodies. It also explains a further decrease in FRET efficiency when intact fluorescent secondary antibodies are used. Such findings underline the notion that FRET values cannot be compared directly to each other if they are obtained using different labeling strategies (Sebestyen *et al.*, 2002).

To increase the signal, we can use phycoerythrin (PE)- or allophycocyanine (APC)-labeled antibodies, because PE and APC have exceptional brightness. However, the size of these molecules is comparable to or even greater than whole antibodies. Because of steric limitations, the measurable FRET efficiency values can be low at the border of detection limit (Batard *et al.*, 2002). Nonetheless, it should be noted that even these low FRET efficiency values might have a biological meaning, because the accuracy of measurements is greatly improved by the high level of specific signals. Appropriate positive and negative controls can help only make the decision of whether given molecules are associated or not on the basis of the measured FRET efficiency values (Batard *et al.*, 2002).

Another possibility is to optimize the donor-acceptor pair formed from conventional fluorescent dyes. We have measured normalized fluorescence intensity values and normalized energy transfer efficiencies, spectral overlap integrals, and calculated crucial dye- and instrument-dependent parameters for all matching pairs of the seven most used indocarbocyanine and Alexa fluorophores on three commercial flow cytometers. The most crucial parameter in determining the applicability of the donor-acceptor pairs was the normalized fluorescence intensity and the least important one was the spectral overlap. On the basis of available laser lines, the optimal dye pair for all three cytometers is the Alexa546-Alexa647 pair, which produces high energy transfer efficiency values and has the best spectral characteristics with regard to laser excitation, detection of emission, and spectral overlap (Horváth *et al.*, 2005).

When studying cells labeled with donor and acceptor conjugated mAbs, we must perform averaging at different levels. The first averaging follows from the random conjugation of the fluorescent label. An additional averaging is brought about by the actual distribution of separation distances between the epitopes labeled with mAbs. This multiple averaging, an inevitable consequence of the nonuniform stoichiometry, explains why the goals of FRET measurements are so uniquely different in the case of purified molecular systems on the one hand, and in the case of *in situ*-labeled membrane or cytoplasmic molecules on the other hand. In the former case, FRET efficiency values can be converted into absolute distances, whereas in the latter, relative distances and their changes are investigated.

Calculation of distance relationships from energy transfer efficiencies is easy in the case of a single-donor/single-acceptor system if the localization and relative orientation of the fluorophores are known. If the FRET measurements are performed on the cell surface or inside the cell, many molecules might not be labeled at all; many could be single without a FRET pair, and others may be in smaller groups of hetero-oligomers, creating higher rates of FRET than expected from stand-alone pairs. A large number of epitopes binding the acceptor increases E simply by increasing the rate of transfer rather than actually meaning that the two epitopes investigated are closer to each other. If, in these cases, reversing the labeling still results in large E values, the proximity can be considered verified. However, if reversing the ratio of donor/acceptor to more than 1 makes FRET disappear, chances are that we have previously seen random colocalizations because of the high number of acceptors. In addition, if cell membrane components are investigated, a 2D restriction applies to the labeled molecules. Analytical solutions for randomly distributed donor and acceptor molecules and numerical solutions for nonrandom distribution have been elaborated by different groups (Dewey and Hammes, 1980; Snyder and Freire, 1982; Wolber and Hudson, 1979). To differentiate between random and nonrandom distributions, energy transfer efficiencies have to be determined at different acceptor concentrations. However, the donor/acceptor ratios should be in the range of 0.1–10.0 (Berney and Danuser, 2003) if we want to obtain reliable FRET measurements. Outside this range, noise and data irreproducibility propagate unfavorably, rendering accurate FRET efficiency calculations impossible. Berney and Danuser (2003) also suggest that to obtain stable FRET measurements, energy transfer has to be observed in the FRET channel, that is, by excitation of the donor and a measurement of the acceptor emission. Methods estimating FRET from the donor signal only in the presence and absence of acceptor are less robust. Donor-acceptor dye pairs should be chosen with the maximal spectral overlap, although this will increase the crosstalk between the detection channels. However, spectral overlaps can be corrected on the basis of samples labeled with donor or acceptor only (Berney and Danuser, 2003).

IV. Applications

A. Mapping Receptor Oligomerization

FRET technology has two major advantages: It can provide spatial information about molecular proximities/interactions in a distance range not covered by conventional optical microscopies (i.e., the range below the ∼250-nm diffraction limit of resolution); additionally FRET data have a unique specificity for the two labels (donor and acceptor dyes or fluorescent proteins) sitting on specified sites of two labeled macromolecules. This technique, thus, provides a special tool for *in situ*, semiquantitative mapping of oligomerization, or clustering of cell membrane molecules (proteins or lipids) in live cells, without any significant distortion of cellular integrity often experienced, for example, in electron microscopy because of the special sample preparation procedures. Since the introduction and adaptation of the FRET methodology to cellular studies in the early 1980s, the method gradually became very popular in cell biology, immunology, and neurobiology.

In the following section, we focus on several selected biological questions concerning homotropic and heterotropic associations of MHC molecules as antigen-presenting entities, colocalization of antigen receptors (e.g., T-cell receptor, TCR) and coreceptors (e.g., CD4 and CD8) on lymphoid cells or functional homotropic and heterotropic associations of several other cell surface receptors in distinct cell types first assessed and characterized by FRET. We pay special attention to specific technical details of various experimental strategies and the lessons and problems that have arisen during these investigations.

1. Clustering of Cell Surface MHC Molecules

The very first targets of cellular FRET studies were, among others, the class I and class II MHC molecules, because they are expressed on all nucleated human cells (MHC-I) or their expression is linked to and regulated by pathological processes, such as viral/bacterial infection (MHC-II). In addition, in FRET studies these molecules also proved favorable from the technical point of view, as MHC-I expression on distinct cell types is usually relatively high ($>10^5$ copies/cell) and the MHC-II level is also significantly upregulated by a number of pathogenic stimuli. Moreover, a large panel of mAbs reactive to different epitopes on their extracellular domains are also available. Thus, MHC glycoproteins served as ideal molecules for adaptation and optimization of FRET measurements to the live cell's surface. Early FRET studies (mostly FCET analysis) on the lateral membrane organization of MHC-I and MHC-II molecules (Bene *et al.*, 1994; Chakrabarti *et al.*, 1992; Matkó *et al.*, 1994, 1995; Mátyus *et al.*, 1995; Szöllősi *et al.*, 1989, 1996) in different cells (B and T lymphocytes, lymphomas, hybridomas, fibroblasts, etc. of mouse and human origin), in accordance with findings by other methods—lateral diffusion (Edidin *et al.*, 1991), single particle tracking (Smith *et al.*, 1999), and immunobiochemical analysis (Triantafilou *et al.*, 2000)—unequivocally demonstrated their inherent ability to form dimers and/or higher order oligomers at the cell surface. In addition, FRET experiments revealed several heterotropic interactions of MHC-I molecules with various other cell surface receptors of diverse function, such as the insulin receptor (Liegler *et al.*, 1991), the interleukin-2 (IL-2) receptor (Matkó *et al.*, 2002; Szöllősi *et al.*, 1987; Vereb *et al.*, 2000), the epidermal growth factor (EGF) receptor (Schreiber *et al.*, 1984), or the transferrin receptor (Mátyus *et al.*, 1995). These heterotropic MHC-I interactions revealed originally by FRET were confirmed later by other methods or functional assays (e.g., Ramalingam *et al.*, 1997).

Donor pbFRET and FCET were both essentially useful in exploring further functional aspects of MHC-I clustering in antigen-presenting cells. In works by Chakrabarti *et al.* (1992) and Bodnár *et al.* (2003), regulation of cell surface MHC-I clusters by exogenous β2m (light chain) was revealed on B lymphoblast antigen-presenting cells, in relation to its effect on the functional activity of cytotoxic effector T lymphocytes. In the experiments, *homo-FRET* between intact MHC-I molecules, as well as *hetero-FRET* between β2m-free MHC-I heavy chains (FHC), and intact MHC-I were both analyzed and found sensitive to the level/binding of exogenous β2m (Bodnár *et al.*, 2003). The high hetero-FRET efficiency, in accordance with the high degree of colocalization of FHC and intact heterodimer (α chain + β2m light chain) shown by high resolution scanning near-field optical microscopy (SNOM), suggested that the β2m-FHC are likely nucleating/initiating the large MHC-I clusters, because their recombination by exogenous β2m resulted in a remarkable β2m-concentration-dependent decrease in the efficiency of both homo-FRET and hetero-FRET, simultaneously with a decreased antigen presentation function.

Several important technical lessons could also be drawn from these basic FRET studies on MHC organization: (1) The FRET efficiency (E) measured for a given donor-acceptor dye pair coupled to antibodies (either whole immunoglobulin G [IgG] or Fab fragment) against the light (β2m) and heavy (α) chains of MHC-I, respectively, may serve as a *cell type-* and *cell cycle-independent internal standard for FRET measurements with comparative goals* (e.g., studies on different cells derived from experimental animals and measurements at different days or if the same cells/cell type is analyzed in different laboratories). This standard can be applied to most of the nucleated cells, because they all express MHC-I at their surface. (2) The efficiency of FRET between the heavy and light chains of MHC-I may be, however, asymmetrical in some cells: E was slightly lower when the heavy chain of intact MHC-I molecules was tagged by donor label and the β2m light chain by acceptor than in the case of reversed tagging (e.g., on B and T lymphoma cells with high FHC expression). The reason for this is likely the aforementioned mixed clustering of FHC and intact heterodimeric forms of MHC-I molecules, resulting also in an unequal number of accessible heavy and light chain epitopes on intact MHC-I. Therefore, we propose to use FRET from the β2m light chain (donor) to the α heavy chain (acceptor) for *internal standard*. (3) These data also demonstrate that using the same donor-acceptor combination on the same cells in different days, FRET efficiency was highly reproducible. Using the same conditions, Bodnár *et al.* (2003) experienced similarly good agreement between the results of FCET and microscopic pbFRET measurements. (4) The efficiency of FRET for a given fixed pair of antibody-tagged epitopes of MHC on a particular cell was, however, very sensitive to changes in the donor or acceptor dyes or in the tagging carrier (whole antibody or Fab fragment). Changing either the donor or the acceptor dye may result in an altered spectral overlap (R_0) or an altered dynamic orientational averaging (κ^2), both critically influencing the measurable efficiency of FRET, besides the actual distance of the donor and acceptor dipolar dyes (R). Some recent analogs of the classic fluorescein and rhodamine dyes (e.g., succinylated derivatives SFX or TAMRA-X, respectively, Molecular Probes, Eugene, OR) were developed with spacer arms, which favors their full dynamic orientational averaging when bound to antibodies and simultaneously reduces their aspecific binding (for review of these questions, see Matkó and Edidin, 1997).

A novel and intriguing aspect of FRET studies with MHC molecules was reported recently by Gáspár *et al.* (2001), who tried to fit *in situ* FRET data (on the MHC-I homotropic association and the relative spatial arrangement of membrane-bound MHC-I epitopes to the surface of the plasma membrane) to the X-ray crystallographic model of the given MHC-I molecule (human leukocyte antigen A2, HLA-A2), as well as other counteracting molecules, such as CD8 and TCR. This approach was based on the analysis of FRET efficiency (by FCET) between distinct structurally identified epitopes of MHC-I molecules relative to each other and between a given MHC-I epitope relative to the surface of the cell membrane labeled by fluorochrome-conjugated lipids (e.g., Bodipy-PC), using a modified Stern-Volmer analysis (Yguerabide, 1994). The structural models for the applied antibody and its Fab fragment were available and helped to define their docking site on MHC heavy chain. Fitting X-ray diffraction-derived structural models of macromolecules to each other while having informative *in situ* FRET data on their molecular interactions and positions relative to the plane of the membrane in hand proved to be a very powerful approach to generate models, approaching the physiological conditions for supramolecular clusters of MHC at cellular contacts, such as immunological synapses (Bromley *et al.*, 2001). Among the several thousands of computer-simulated possibilities for different molecular geometries/orientations of MHC-I molecules in homo-oligomers, only one to two matched the *in situ* FRET data measured on live cells (Gáspár *et al.*, 2001). Thus, besides giving a realistic model for supramolecular activation clusters at the contact site of antigen-presenting cells and cytotoxic T lymphocytes (CTLs), this approach provides a strategic template for studies on modeling the 3D structure of supramolecular receptor complexes appearing at the surface (or at a contact surface) of live cells, with a goal of designing drugs with effects on these molecular interactions.

It should be noted here that reliable models are available for analysis of FRET data obtained for *transverse (vertical) distance distributions* (Yguerabide, 1994), as well as for analysis of *lateral homotransfer (oligomerization)* in the case of membrane-bound proteins (Runnels and Scarlata, 1995).

As another special application of cellular FRET, *intercellular FRET* experiments should also be mentioned here. In this work, Bacso *et al.* (1996) labeled different molecular species (e.g., MHC-I and intercellular adhesion molecule-1, [ICAM-1] on antigen-presenting cells, CD8, and leukocyte function-associated antigen-1 [LFA-1] on T cells) with proper donor and acceptor dyes on antigen-presenting cells and T cells, respectively, and then measured intercellular FRET between these labeled molecules after initiating conjugation (synapse formation) of these cells. These FRET measurements had to be implemented inevitably with microscopic FRET, because formation of cell-cell contacts also had to be detected simultaneously. These intercellular pbFRET experiments (kinetic measurement of donor photobleaching) contributed important quantitative data to our view of immunological synapses. Namely, in the cytotoxic T-cell synapses the molecule pairs in point contact (e.g., CD8/MHC-I) displayed detectable nanometer scale FRET efficiency ($E = 17$–19%), while the longer adhesion molecules (e.g., ICAM-1/

LFA-1 pair) forming focal contacts and spanning larger distance range (≥ 25 nm) did not ($E < 1.5\%$). These FRET results were later confirmed by several pieces of evidence from confocal and electron microscopic analysis of immunological synapses (Monks et al., 1998; Wulfing et al., 1998). It is noteworthy that the FRET efficiency histogram for the CD8/MHC-I molecule pair obtained from this pbFRET analysis was broad and heterogenous ($E = 0–30\%$), suggesting the existence of molecular associations with varying tightness at the contact zone of antigen-presenting cells-T-cell conjugates.

Finally, we note that a "shorter spectroscopic ruler" alternative of FRET, the long-range electron transfer (LRET) technique (detecting electron transfer between fluorescent donors and nonfluorescent electron acceptor spin label radicals), was also successfully applied in characterizing cell surface MHC associations, in good quantitative agreement with FRET data on the same cells (for details and review, see Matkó et al., 1992, 1995; Matkó and Edidin, 1997).

2. Antigen Receptor–Coreceptor Interactions in T–Cell Immunology

Early FCET (Mittler et al., 1989) and line-scanning microscopic pbFRET (Szabó et al., 1995) studies convincingly demonstrated that the antigen receptor of T cells (i.e., TCR) physically associates with the CD4 coreceptor molecules on helper T cells upon ligation of TCR. That this association is dynamic was nicely confirmed by measuring FRET between fluoresceinated anti-CD4 and rhodaminated anti-CD3 at physiological ($37\,^{\circ}$C) and low ($4\,^{\circ}$C) temperatures. The efficiency of FRET at $37\,^{\circ}$C was closely 10%, whereas at $4\,^{\circ}$C it decreased to less than 1% (Mittler et al., 1989), suggesting that this association requires a dynamic redistribution of interacting molecules in the membrane. Using cells with mutant CD4 (lacking the cytoplasmic domain), the absence of FRET nicely demonstrated that this domain of CD4 is essential for its interaction with TCR. Further, FRET analysis of this question revealed that cytoplasmic CD4 motifs that bind p56[lck] are critical in binding CD4 to TCR (Collins et al., 1992).

These studies turned our attention to a critical point of FRET analysis. Care should be taken in the choice of the antibody used to tag our receptor. Here, for example, the whole anti-CD3 IgG molecule (145-2c11) induced molecular redistribution and microaggregation (involving TCR, CD4, and p56[lck]) in the plasma membrane of T cells. Therefore, to avoid such artifacts, for labeling of TCR, its rhodaminated F(ab')$_2$ fragment should be used. Hence, we propose to check all antibodies against multichain immune-recognition receptors (MIRRs) for such an activating/stimulating effect before their use as a simple fluorescent tag for FRET.

For further analyzing the role of CD4 coreceptors in antigen-specific T-cell activation, Zal et al. (2002) introduced a digital imaging deconvolution microscopic FRET approach. They applied a special labeling technique to attach donor and acceptor dyes to the TCR-CD3 complex and CD4 molecules. The CD3ζ chain was fused with a GFP mutant, ECFP (donor, Clontech) through a six amino acid linker, and the CD4 molecules were fused at their intracellular terminus, through a five amino acid linker, with another GFP mutant, EYFP (acceptor, Clontech). The cyan and yellow variants of GFP provide a good spectral overlap for FRET measurements, although their emissions also show some overlap (crosstalk). The double-labeled T cells were analyzed for FRET in the absence or presence of antigen-presenting cells in thermostated chambers, using an inverted fluorescence microscope equipped with a sensitive cooled CCD camera and computer-controlled motorized stage using an interactive control software driving also the 3D deconvolution (*z*-axis optical slicing) program. The images of the double-labeled cells were recorded in three optical channels of the microscope and then compensated for crosstalk between the channels, and finally the donor-normalized FRET information was derived based on earlier procedures described for FCET or FRET-microscopy (Gordon *et al.*, 1998; Tron *et al.*, 1984). Generating a 30-min time lapse of the corrected FRET signal, it was shown that TCR and CD4 rapidly accumulated into the contact area of the antigen-presenting cell and T cell (synapse) and were localized in close proximity (exhibiting high FRET), strictly depending on the nature of the antigen peptide. Only agonist peptide ligands promoted the TCR-CD4 interaction, whereas the antagonist ligands inhibited it (for more details, see Zal *et al.*, 2002). Several technical lessons emerge from this study: (1) FRET analysis should preferably be made on nondeconvoluted images, although they are of lower *z*-resolution, because of the possible nonlinearity of the deconvolution procedure; (2) cell areas where the donor (CFP) fluorescence level is less than 10 times noise level, as well as areas where the acceptor to donor fluorescence ratio is beyond the 1:1–3:1 range are recommended for exclusion from the analysis; and (3) using such imaging FRET analysis, the donor-normalized FRET signal is not a real FRET efficiency, but a number proportional to FRET efficiency. With the availability of user-friendly free software (Roszik *et al.*, 2009), the calculation of actual, corrected FRET images is highly recommended.

3. Oligomerization of Diverse Membrane Proteins

The different variations of FRET measurements (pbFRET, FCET, MI-FRET, FLIM, BRET (bioluminescence resonance energy transfer), described later, etc.) have all been extensively applied in studies of oligomerization, heterotropic interactions, and membrane compartmentation of various receptors, adhesion, accessory, or signal molecules in the plasma membrane of a number of human or murine cell types. Just to mention several examples, the cell surface distribution of FcεRI (IgE receptor) and its relationship to MAFA (a potential regulatory molecule) was nicely analyzed in detail on mast cells, using the donor pbFRET approach: The IgE receptor was found partially associated with MAFA, depending on FcεRI oligomerization/cross-linking by antigen (allergen), forming this way a platform for negative regulation of mast cell activation by MAFA (Jurgens *et al.*, 1996). *FRET* and *BRET* were both successfully applied for a number of G-protein-coupled hormone receptors (GPCR) to reveal their functionally critical dimerization (Eidne *et al.*, 2002; Milligan *et al.*, 2003). As an interesting observation, the human thyrotropin receptor (thyroid-stimulating hormone receptor, TSHR) was found to be naturally inhibited by its higher oligomeric surface distribution, whereas it became active when these forms dissociated to dimers or monomers, upon binding their ligand (thyroid-stimulating hormone, TSH). These conclusions were drawn from FRET measurements between the GFP-TSHR fusion construct (donor) and a TSHR-Myc tag construct + anti-Myc-Cy3 (acceptor), from MI-FRET analysis, where the FRET efficiency signal fell to less than 1% in ligated cells from about a 20% value in unstimulated cells (Latif *et al.*, 2002).

As new research areas taking an interest in the application of FRET, we must mention *neurobiology* and *intracellular organelle research*. This might be due to both the increasingly wide scale of microscopic FRET techniques and the technical development of microscopes. For example, subunit assembly and its regulation by nicotine of the $\alpha_4\beta_2$-nicotinic acetylcholine receptor in neurons was successfully characterized by MI-FRET measurements, using CFP and YFP constructs of the individual subunits after transfection to midbrain neurons (Nashmi *et al.*, 2003). A novel intracellular version of FRET—detected by FLIM between labeled CD44 (hyaluronan receptor) and ezrin (cytoskeletal linker protein)—revealed a new PKC-regulated mechanism by which CD44-ezrin coupling and the directional cell motility is controlled (Legg *et al.*, 2002). FLIM detection of FRET between GFP-fusion constructs of EGF receptor or platelet-derived growth factor (PDGF) receptor (transmembrane receptor tyrosine kinases, RTKs) and Cy3 antibody-labeled protein tyrosine phosphatase 1B (PTP1B) clearly demonstrated that the activated RTKs meet the PTP1B after their internalization, mainly on the surface of the endoplasmic reticulum, suggesting that the scenes of their activation and inactivation are spatially and temporally isolated within the cells (Haj *et al.*, 2002).

B. Defining Plasma Membrane Microdomains (Lipid Rafts)

The plasma membranes of most mammalian cells exhibit small microdomains (called *lipid rafts*) that are enriched in glycosphingolipids (GSLs), sphingomyelin (SM), and cholesterol (CHOL), while the polyunsaturated phospholipids are very rare in these domains. Lipid rafts are also rich in glycosyl-phosphatidylinositol (GPI)-anchored proteins. These domains were first defined chemically as detergent-resistant, light buoyant density membrane fractions, and then were characterized as diffusion barriers or copatches of fluorescently labeled lipids and proteins (Brown and London, 1998; Edidin, 2001; Jacobson and Dietrich, 1999; Matkó and Szöllősi, 2002; Simons and Ikonen, 1997; Subczynski and Kusumi, 2003; Vereb *et al.*, 2003).

FRET was also introduced into the studies directed to define existence, composition, and size (a highly controversial question) of lipid rafts. The starting strategy was to investigate homo-FRET between labeled GPI-anchored proteins—supposed to be enriched in raft microdomains—as a function of their cell surface density. First, applying an indirect approach, Varma and Mayor (1998) detected fluorescence *anisotropy* of GPI-anchored folate receptors (FR-GPI) in CHO cells with microscopy, using a fluorescent folate analog (N^α-pteroyl-N^ε-(4'-fluorescein-thiocarbamoyl)-L-lysine, PLF) for labeling. For comparison, chimeric transmembrane forms of FR (FR-TRM) were also investigated. Their working hypothesis was that if the FR-GPI was randomly distributed at the surface of CHO cells, the measured fluorescence anisotropy should depend on the actual surface density of FR (controlled by transfection-directed expression), due to concentration depolarization (homo-FRET) effects above a critical molecular proximity (within the 10-nm Förster-distance). In contrast, if the FR-GPI molecules are compartmented/concentrated in lipid microdomains, no density-dependent change is expected in the anisotropy, because only the number of microdomains increases with increased expression levels, but the mutual proximity of FR molecules does not. Interestingly, they detected density-dependent anisotropy in the case of the FR-TRM form, whereas the physiological FR-GPI form showed a constant anisotropy over a wide range of expression levels, consistent with the microdomain-localization of FR-GPI. This was further confirmed by the observed density dependence of FR-GPI anisotropy in cells, where the membrane was depleted of CHOL (stabilizer of rafts) before anisotropy measurements. The size of an individual domain was estimated from these measurements as being smaller than 70 nm (containing <50 GPI-anchored molecules) (Varma and Mayor, 1998).

Conflicting with these data, Kenworthy and Edidin (1998) measured homo-FRET between GPI-anchored 5'-nucleotidase proteins, using Cy3 (donor)- and Cy5 (acceptor)-conjugated antibodies, in MDCK cells, by the acceptor-photobleaching method. A relatively strong correlation was found between the homo-FRET efficiency and the cell surface density of 5'-nucleotidase. These data suggest that either the GPI-anchored 5'-nucleotidase is not constitutively compartmented in raft microdomains or their microdomains are not stable, that is, have a short lifetime. Further addressing this question, Kenworthy *et al.* (2000) analyzed homo-FRET between fluorescent cholera toxin B-labeled GSL components, GM_1 gangliosides, and their hetero-FRET with GPI-anchored proteins on three cell lines together with the dependence of FRET on cell surface densities of acceptor-labeled lipid or protein species. Surprisingly, in most cell types even the GM_1 ganglioside (considered a major lipid constituent of raft microdomains) showed density-dependent homo-FRET, indicative of its weak, hardly detectable clustering. This might mean that either raft microdomains occupy only a very small fraction of the cell surface area on the investigated cell types or the domains of the investigated GPI proteins are only transient (short lifetime) assemblies. This might be, however, highly cell specific, because the amount of cell surface raft constituents (e.g., GSL) may vary with cell type or differentiation, at least in lymphoid cells (e.g., Tuosto *et al.*, 2001). Nevertheless, most microdomains that we can see in the fluorescence microscope are likely composed of aggregates of small elementary raft microdomains, either upon biological signaling of the cell or just upon modulation of raft size by the procedure and agents used to label raft marker components.

Another often used FRET strategy to investigate raft (microdomain) composi-
tions on live cells is to measure hetero-FRET by microscopy (or by FCET) between
a selected fluorescently labeled membrane protein (receptor) and raft markers
proteins (e.g., GPI-anchored proteins) or GM_1 gangliosides. Such an approach
was used to show that a large fraction of CD25 (α chain of the IL-2 receptor) is
constitutively raft associated in lymphoma/leukemia T cells, using CD48 as a raft
marker. In these FRET experiments, *Fab* fragments were used to label the two
proteins, to minimize possible modulation of raft size by cross-linking with biva-
lent antibodies. Although on these T cells CD25 was expressed at a higher level
(twofold) than CD48, the FRET efficiency was similar in both directions (donor on
CD25 or on CD48). This, together with the pronounced decrease of FRET
efficiency upon depleting membrane CHOL, suggests that the two proteins are
colocalized within common microdomains (Matkó and Szöllősi, 2002; Matkó
et al., 2002; Vereb *et al.*, 2000).

C. Receptor Tyrosine Kinases

The type I family of transmembrane RTKs comprises four members: EGFRs
(EBFR or ErbB1), ErbB2 (HER2 or Neu), ErbB3, and ErbB4 (Citri *et al.*, 2003;
Nagy *et al.*, 1999b; Vereb *et al.*, 2002; Yarden and Sliwkowski, 2001). Within a
given tissue, these receptors are rarely expressed alone but are found in various
combinations. Members of the family form homoassociations and heteroassociations at the cell surface. Their ligands belong to three groups: EGF-like ligands
bind only to ErbB1, heregulin-like (or neuregulin-like) ligands bind only to ErbB3
and ErbB4, and EGF- and neuregulin-like ligands bind ErbB1 and ErbB4. ErbB3
shares growth factor-binding specificity with ErbB4 but lacks intrinsic kinase
activity. ErbB2 is an orphan receptor; no soluble physiological ligand specific to
ErbB2 has been detected. Despite this, ErbB2 participates actively in ErbB receptor combinations, and receptor complexes including the ErbB2-ErbB3 dimer that
appears to be more potent than any other combination (Citri *et al.*, 2003; Nagy
et al., 1999b; Sliwkowski *et al.*, 1999; Vereb *et al.*, 2002; Yarden and Sliwkowski,
2001). Thus, both ligand and receptor expression can vary by tissue and will, in
part, determine the specificity and the potency of cellular signals.

Molecular scale physical associations among ErbB family members have been
studied by classic biochemical (Sliwkowski *et al.*, 1994; Tzahar *et al.*, 1996), molecular
biological, and biophysical methods (Gadella and Jovin, 1995; Nagy *et al.*, 1998a,b,
2002). When isolated from cells, members of the ErbB family self-associate (homo-
associate) and associate with other family members (heteroassociate) (Tzahar *et al.*,
1996). However, experiments on isolated proteins are inherently unable to detect
interactions in cellular environments *in vivo* and *in situ* and cannot detect heterogeneity within or among cells. FRET measurements detected dimerization of ErbB1
receptors in fixed (Gadella and Jovin, 1995; Gadella *et al.*, 1994) and living cells
(Gadella *et al.*, 1994). FRET was also applied to monitor the association pattern of

ErbB2 in breast tumor cells (Nagy *et al.*, 1998a,b, 2002). First, we applied FCET measurements (Szöllősi *et al.*, 1984) to reveal cell-to-cell heterogeneity within a cell population. Classic FCET (Section III.A.2) and autofluorescence-corrected FCET (Section III.A.3) gave practically the same results, because the expression level of ErbB2 molecules is so high on these cell lines (SKBR-3, BT474, MDA453) that autofluorescence is practically negligible compared to the signal.

We estimated the association state of ErbB2 and assessed how it was affected by EGF treatment in breast tumor cell lines by measuring FRET between fluorescent mAbs or Fab fragments. There was considerable homoassociation of ErbB2 and heteroassociation of ErbB2 with EGFR in quiescent breast tumor cells. ErbB2 homoassociation was enhanced by EGF treatment in SKBR-3 cells and in the BT474 subline BT474M1 with high tumorigenic potential, whereas the original BT474 line was resistant to this effect. These differences correlated well with EGFR expression. Because we calculated these measurements in flow cytometry, we obtained one single FRET efficiency value for each cell analyzed. To reveal heterogeneity in the homoassociation pattern of ErbB2 within a single cell, we had to use one of the microscopic FRET approaches. We have used donor pbFRET microscopy to visualize FRET efficiency within single cells with spatial resolution limited only by diffraction in the optical microscope (Bastiaens *et al.*, 1996; Nagy *et al.*, 1998b; Vereb *et al.*, 1997). This allows detailed analysis of the spatial heterogeneity of molecular interactions. At first, we applied the donor pbFRET in wide-field microscopy and revealed extensive pixel-by-pixel heterogeneity in ErbB2 homoassociation (Nagy *et al.*, 1998a). In our measurements, ErbB2 homo-association was also heterogeneous in unstimulated breast tumor cells; and membrane domains with ErbB2 homoassociation had mean diameters of less than 1 μm (Nagy *et al.*, 1998a, 1999a). It was not clear whether the domain size was imposed by the optical resolution limit of wide-field microscopy in the X-Y plane or whether it originated from the actual size of ErbB2 aggregates. To refine the size estimate of domains containing ErbB2 molecules, we turned to SNOM and later combined FRET measurements with confocal microscopy.

To improve spatial resolution, first we studied the cell surface distribution of ErbB2 RTKs using SNOM (Kirsch *et al.*, 1996; Monson *et al.*, 1995; Vereb *et al.*, 1997). This technique is not limited by diffraction optics and can readily image objects in the 0.1- to 1.0-μm range, including submicrometer lipid and protein clusters in the plasma membrane (Edidin, 1997). ErbB2 was concentrated in irregular membrane patches, with a mean diameter of approximately 500 nm, containing up to 1000 ErbB2 molecules in nonactivated SKBR-3 and MDA453 human breast tumor cells. The mean cluster diameter increased to 600–900 nm when SKBR-3 cells were treated with EGF, heregulin, or a partially agonistic anti-ErbB2 antibody. The increase in cluster size was inhibited by an EGFR-specific tyrosine kinase inhibitor. Because the domain size was well within the resolution limit of the confocal microscopy (250–300 nm in the X-Y plane), we were able to confirm the SNOM results with a confocal LSM (CLSM) (Nagy *et al.*, 1999a).

We then implemented acceptor pbFRET in a CLSM equipped with three lasers, to study the correlation between the association of ErbB2 with ErbB2 and ErbB3 and the local density of these RTKs. From fluorescence intensities generated in two optical channels, we were able to determine the FRET efficiency values on a pixel-by-pixel basis, while the third laser beam provided the signal for measuring the expression level of ErbB3. The homoassociation of ErbB2 correlated positively with the local concentration of ErbB2 but negatively with ErbB3 local density. This negative correlation suggests that ErbB2-ErbB3 heterodimers compete with ErbB2 homodimers, and therefore, a high number of ErbB3 molecules can disassemble ErbB2 homodimers (Nagy *et al.*, 1999a).

We also investigated the colocalization of lipid rafts (Section IV.B) and ErbB2 clusters using similar approaches. The size of lipid rafts was identified with either fluoresceinated or Cy5-labeled subunit B of cholera toxin (CTX-B), which binds to the GSL GM_1 ganglioside. Once again, we used the signal from one laser beam to monitor the lipid rafts and signals from the other two laser beams to reveal the homoassociation pattern of ErbB2. In this case, we applied both the donor and the acceptor pbFRET, and they gave very similar results. Our observations suggest that like ErbB1, ErbB2 is localized mostly in lipid rafts. However, there is a negative correlation between ErbB2 homoassociation and the local density of the lipid raft marker CTX-B. This environment could alter the association properties of ErbB2, similar to our findings regarding other membrane receptors (Vereb *et al.*, 2000). Because stimulating ErbB2 increases the size of ErbB2 clusters (Nagy *et al.*, 1999a) and lipid rafts (Nagy *et al.*, 2002), the amount of ErbB2 concentrated in rafts is very likely related to the function of the protein. Localization of ErbB2 in lipid rafts is dynamic, because it can be dislodged from rafts by cholera toxin-induced raft cross-linking (Nagy *et al.*, 2002). The association properties and biological activity of ErbB2 expelled from rafts differ from those inside rafts. For example, 4D5 (parent murine version of trastuzumab)-mediated internalization of ErbB2 is blocked in cholera toxin-pretreated cells, whereas its antiproliferative effect is not. These results emphasize that alterations in the local environment of ErbB2 strongly influence its association properties, which are reflected in its biological activity and in its behavior as a target for therapy (Nagy *et al.*, 2002).

Finally, we have also used intramolecular and epitope-to membrane bilayer FRET measurements to facilitate the understanding of receptor structure and function at the molecular level. A molecular model was built for the nearly full-length ErbB2 dimer based on the X-ray or nuclear-magnetic resonance structures of extracellular, transmembrane, and intracellular domains. The extracellular domain was positioned above the cell membrane based on the distance determined from FRET. Favorable dimerization interactions were predicted for the extracellular, transmembrane, and protein kinase domains of ErbB2, which may act in a coordinated fashion in ErbB2 homodimerization, and also in heterodimers of ErbB2 with other members of the ErbB family (Bagossi *et al.*, 2005).

━━━━━━ **V. Perspectives**

Advances in developing new fluorescent probes, instrumentation, and methodologies have greatly improved the applicability of FRET to the systematic exploration of the localization, translocation, and association of signaling proteins in living or intact cells.

The newly developed fluorescent probes provide high sensitivity and great versatility while minimally perturbing the cell under investigation. The appearance of quantum dots was followed by their rapid spreading in biomedical applications owed to their great photostability and extreme brightness (for review, see Xing and Rao, 2008). Applications restricted to or optimal with a single laser excitation benefitted a lot from their wide excitation bandwidth and narrow emission spectrum. They also proved to be applicable as FRET donors offering a good separation of donor excitation from emission, thereby allowing for maximizing the overlap integral, or multiplexing FRET systems (Algar and Krull, 2008; Grecco *et al.*, 2004; McGrath and Barroso, 2008).

Genetically encoded reporter constructs that are derived from GFPs and red fluorescent proteins are leading a revolution in the real-time visualization and tracking of various cellular events (Aoki *et al.*, 2004; Scarlata and Dowal, 2004; Zhang *et al.*, 2002) and have brought the Nobel prize to the pioneers of their field. Some advances include the continued development of "passive" markers for the measurement of biomolecule expression and localization in live cells and "active" indicators for monitoring more complex cellular processes such as small molecule messenger dynamics, enzyme activation, and protein-protein interactions (Lippincott-Schwartz and Patterson, 2003; Lippincott-Schwartz *et al.*, 2003). A review published by Subramaniam *et al.* (2003) provides an excellent summary of the photophysical properties of GFPs and red fluorescent proteins and their application in quantitative microscopy. The plethora of novel spectral variants is reviewed in Shaner *et al.* (2007). New fluorescent protein variants have been developed that form optimal FRET pairs (Rizzo *et al.*, 2006; Shcherbo *et al.*, 2009; Shimozono and Miyawaki, 2008). Calibration constructs (Domingo *et al.*, 2007; Koushik *et al.*, 2006) of these protein pairs and calibration protocols (Nagy *et al.*, 2005) have also been created to aid the qualitative measurement of molecular interactions. Mutant proteins, which are capable of photoconversion from the nonfluorescent to a stable fluorescent form, have also been described (Chudakov *et al.*, 2003a,b; Patterson and Lippincott-Schwartz, 2002). These proteins can be used for precise *in vivo* photolabeling to track the movements of cells, organelles, and proteins and as acceptor in FRET experiments (Demarco *et al.*, 2006).

An ingenious yet somewhat involved route to assessing molecular interactions is bimolecular fluorescence complementation (BIFC) wherein the two potential interaction partners are each fused to half of a fluorescent protein, with the hope that upon interaction of the partners also these two halves would reconstitute into a functional fluorophor (Horváth *et al.*, 2009; Kerppola, 2006). An extension of this approach combines BIFC and FRET to measure the establishment of ternary complexes in living cells, by first complementing from two halves each a donor and an acceptor fluorescent protein, and then quantitating FRET occuring between them as all four investigated proteins assemble (Shyu *et al.*, 2008).

In addition to the development of versatile probes, another aspect that greatly enhanced the use of FRET in biological systems is the evolution of algorithms and software for the facile evaluation of FRET data. Chen *et al.* (2006) has introduced the concept of the ratio of sensitized acceptor emission to donor fluorescence quenching (G factor) and the ratio of donor/acceptor fluorescence intensity for equimolar concentrations in the absence of FRET (k factor) and developed a method for their determination which greatly simplifies quantitative FRET measurement in living cells as it does not require cell fixation, acceptor photobleaching, protein purification, or specialized equipment for determining fluorescence spectra or lifetime. Free programs have been developed for step-by-step guided evaluation of photobleaching and ratiometric microscopic FRET measurments (Roszik *et al.*, 2008, 2009; Szentesi *et al.*, 2005). A new approach termed 3D-FRET stoichiometry reconstruction (3DFSR) has been suggested and implemented for reconstructing 3D distributions of bound and free fluorescent molecules (Hoppe *et al.*, 2008).

One of the most exotic versions of FRET techniques is the photochromic FRET (pcFRET), when diheteroarylethenes are used as acceptors (Giordano *et al.*, 2002; Song *et al.*, 2002). In pcFRET, the fluorescent emission of the donor is modulated by cyclical transformations of a photochromic acceptor. Light induces a reversible change in the structure and, concomitantly, in the absorption properties of the acceptor. The corresponding variation in the overlap integral (and, thus, critical transfer distance R_0) between the two states provides the means for reversibly switching the process of FRET on and off, allowing direct and repeated evaluation of the relative changes in the donor fluorescence quantum yield. Diheteroaryl-lethenes demonstrate excellent stability in aqueous media, an absence of thermal back reactions, and negligible fatigue. pcFRET is applicable for monitoring the time course of changes in FRET efficiencies, that is, the temporal dynamics of protein translocation and association processes (Giordano *et al.*, 2002). Building SNAP tags into target proteins fused to GFP allows their *in situ* labeling with the photoswitchable probe, nitrobenzospiropyran (NitroBIPS), which substantially improves the sensitivity of detection to <1% FRET efficiency. Through orthogonal optical control of the colorful merocyanine and colorless spiro states of the NitroBIPS acceptor, donor fluorescence can be measured both in the absence and presence of FRET allowing the assesment of changes in molecular conformations (Mao *et al.*, 2008).

The various kinds of image cytometric FRET (i.e., donor pbFRET, acceptor pbFRET, and intensity-based ratiometric FRET) can be applied either in widefield microscopy or in laser-scanning confocal microscopy (Bastiaens and Jovin, 1998; Nagy *et al.*, 1998a,b, 2002; Vereb *et al.*, 1997). The major limitation of the widefield microscopy technique is that fluorescence light below and above the focal plane contributes to the useful signal, thereby degrading the quality of the image.

Confocal microscopy can overcome this limitation because of its capability to reject signals from outside the focal plane (Elangovan *et al.*, 2003; Nagy *et al.*, 2002). This capability provides a significant improvement in vertical resolution and allows for the use of serial optical sectioning of the living cells. The disadvantage of confocal microscopy is potentially serious photobleaching and photodamage of light-sensitive fluorophores, which can be detrimental during optical sectioning. Two-photon FRET microscopy can eliminate this problem; because two-photon excitation occurs only in the focal volume, the detected fluorescence signal is exclusively in-focus light. In addition, because two-photon excitation uses longer wavelength light, it is less damaging to living cells, thereby limiting problems associated with fluorophore photobleaching and photodamages, as well as intrinsic autofluorescence of cellular components (Elangovan *et al.*, 2003). The advantages of two-photon excitation can be further improved when it is combined with FLIM. Advantages of FLIM approaches are that lifetime measurements are less sensitive to changes in probe concentration and photobleaching. In addition, FLIM can discriminate fluorescence coming from different dyes that have similar absorption and emission characteristics but show a difference in fluorescence lifetime. The fluorescent lifetimes and relative concentrations of free and interacting molecules can be reliably estimated using the newly available global analysis approach of time correlated single photon counting FRET-FLIM data, even if the SNR is low (Grecco *et al.*, 2009). Also the combination of spectral ratiometric imaging of ECFP/Venus and high-speed FLIM-FRET of TagRFP/mPlum can increase the spectral bandwidth available and provide robust imaging of multiple FRET sensors within the same cell (Grant *et al.*, 2008). The concept of a minimal fraction of donor molecules involved in FRET (mf(D)), which can be obtained without fitting procedures and is derived directly from FLIM data has also improved our ability of interpreting FLIM-FRET experiments. It constitutes an interesting quantitative parameter for live cell studies because it is related to the minimal relative concentration of interacting proteins. For multilifetime donors mf (D) can possibly be the only quantitative determinant of FRET processes (Padilla-Parra *et al.*, 2008). In two-photon FRET-FLIM, the reduction in donor lifetime in the presence of acceptor reveals the dimerization of protein molecules and determines more precisely the distance between the donor and the acceptor. This methodology allows for studying the dynamic behavior of protein-protein interactions in living cells and tissues (Chen and Periasamy, 2004).

Another development of FRET modalities is when FRET measurements are combined with spectral imaging. Spectral imaging and linear unmixing extends the possibilities to discriminate distinct fluorophores with highly overlapping emission spectra and, thus, the possibilities of multicolor imaging. Spectral imaging also offers advantages for fast multicolor time-lapse microscopy and FRET measurements in living samples, as well as estimation of the fraction of free, non-FRETting fluorophors (Megias *et al.*, 2009; Wlodarczyk *et al.*, 2008; Zimmermann *et al.*, 2003).

The sensitivity of FRET can be extended to the single-molecule level by detecting FRET between a single-donor fluorophore and a single-acceptor fluorophore.

Single-pair FRET (spFRET) provides a unique means of observing conformational fluctuations and interactions of molecules at a single-molecule resolution even in the native, *in vivo* context of a living cell (Ha, 2001; Heyduk, 2002). Single molecule FRET has also been extended to measuring the interaction of three molecules through three fluorophores (Clamme and Deniz, 2005). The method stems in an earlier approach that allowed measuring in parallel three mutually dependent energy transfer processes between the fluorescent labels, such as cyan, yellow, and monomeric red fluorescent proteins in living cells (Galperin *et al.*, 2004).

A less involved procedure, two-sided FRET (tsFRET) explores the relationship of any two pairs of molecular entities out of three randomly chosen molecules. Using this method, labeling ErbB2 molecules with a mixture of Cy3 and Cy5 conjugated antibodies and $\beta1$ integrin with FITC conjugated antibody, first ErbB2 homoassociation was determined in an acceptor photobleaching protocol, followed by donor photobleaching of FITC to assess ErbB2-integrin heteroassociation. Interestingly, these measurements revealed a pixel-by-pixel anticorrelation of ErbB2 homoassociation and $\beta1$-integrin-ErbB2 heteroassociation, suggesting a competition for dimerization partners between ErbB2 and integrins (Fazekas *et al.*, 2008).

Another new FRET modality is the homotransfer or energy migration FRET (emFRET). This approach exploits fluorescence polarization measurements in flow cytometry, in wide-field LSMs, and in cLSMs, or in the form of anisotropy fluorescence lifetime imaging microscopy (rFLIM). These methods permit the assessment of rotational motion, association, and proximity of cellular proteins *in vivo*. They are particularly applicable to probes generated by fusions of visible fluorescence proteins and are capable of monitoring homoassociations of various signaling proteins (Clayton *et al.*, 2002; Lidke *et al.*, 2003; Tramier *et al.*, 2003). A version of this method can be adapted to flow cytometry and has recently been used to quantitatively characterize the large-scale association of ErbB1 and ErbB2 on the surface of breast tumor cells (Szabó *et al.*, 2008).

The sensitivity of the FRET can be greatly enhanced when bioluminescence is used as a source of excited donor. In BRET, one protein is fused to Renilla luciferase and the other protein to a mutant of GFP (Xu *et al.*, 1999). The luciferase can be activated by addition of its substrate, and if the proteins in question interact, resonance energy transfer occurs between the excited luciferase and the mutant GFP. BRET can be detected by monitoring the fluorescence signal emitted by the mutant GFP. By choosing the proper luciferase/GFP mutant combinations, BRET can be used to measure protein-protein interactions *in vitro* and *in vivo*. BRET is perfectly suited for cell-based proteomics applications, including receptor research and mapping signal transduction pathways (Devost and Zingg, 2003; Issad *et al.*, 2003; Xu *et al.*, 1999).

SNOM provides many interesting imaging possibilities, which can be further increased by combining it with FRET. The simplest combination is when instead of the classic microscope, SNOM is used to detect FRET between dye molecules bound to cell surfaces (Kirsch *et al.*, 1999). In the other approach, the acceptor dye of a FRET pair is attached to the tip of a near-field fiberoptic probe. Light exiting the SNOM probe, which is not absorbed by the acceptor dye, excites the donor dye introduced into a sample. As the tip approaches the sample containing the donor dye, energy transfer from the excited donor to the tip-bound acceptor produces a red-shifted fluorescence. By monitoring this red-shifted acceptor emission, one can observe a dramatic reduction in the sample volume probed with an uncoated SNOM tip (Vickery and Dunn, 1999). Local fluorescence probes based on CdSe semiconductor nanocrystals have already been tested as FRET-SNOM sources with the prospect of single-molecule resolution (Shubeita *et al.*, 2003). It should be noted that in the last two examples the FRET phenomenon is used only for enhancing the resolution of the SNOM technique. Another approach to breaking the resolution limit is to take FRET to TIRF (total internal reflection fluorescence) microscopy. The imaging system developed for this purpose (Paar *et al.*, 2008) enables screening of large numbers of cells under TIRF illumination combined with FRET imaging, thereby providing the means to record FRET between membrane-associated proteins labeled with a donor-acceptor pair. The system is capable of performing live-FRET scanning on stoichiometric FRET constructs reaching throughput of up to 1000 cells/s at the optical resolution limit. A comparison with confocal microscopy shows that TIRFM offers a 4.2-fold advantage over confocal microscopy in detecting contributions from membrane-localized proteins.

VI. Conclusions

A new and critically important strategic approach of today's cell biology is what we usually call "nanobiotechnology," that is, the efforts to resolve spatial details of living cells at nanometer resolution and to follow the kinetics of biochemical reactions or molecular interactions at a nanosecond (or even shorter) time scale. As we can see from this chapter, the dipole resonance energy transfer mechanism described by Förster (1946) became by now successfully adopted to cellular systems and with the past few years' enormous technical developments in laser and microscope technologies, as well as in ultrafast electronics, it became one of the most powerful approaches in detecting interactions between different molecular constituents of cells. Classic modalities of FRET measurements (FCET, pbFRET, MI-FRET/RiFRET) allowed us to assess many intriguing and important questions about the molecular architecture of the plasma membrane of cells and about its dynamic heterogeneity (microdomain structure). Advances in the development of fluorescent probes with special design for FRET (the natural fluorescent proteins, GFP family, different tandem fluorophores, quantum dots, etc.) and the novel innovative microscopic FRET modalities (confocal, two-photon microscopy, time-resolved FLIM-FRET, etc.) allow us now to explore interactions of signal molecules in the cytoplasm or at intracellular organelles. These, together with some more complicated approaches successfully evaluated in biological applications (e.g., FRET combined with spectral imaging, to allow the detection of multiple interactions at the same time or FRET combined with SNOM or TIRF to offer spatial resolution exceeding the diffraction limit), will certainly revolutionize nanobiology research.

Acknowledgments

The authors have been supported by the following research grants: OTKA K62648, K75752, and K68763; EU FP6 LSHBCT-2004-503467, LSHC-CT-2005-018914, and MRTN-CT-2005-019481.

References

Algar, W. R., and Krull, U. J. (2008). Quantum dots as donors in fluorescence resonance energy transfer for the bioanalysis of nucleic acids, proteins, and other biological molecules. *Anal. Bioanal. Chem.* **391**, 1609–1618.

Aoki, K., Nakamura, T., and Matsuda, M. (2004). Spatio-temporal regulation of Rac1 and Cdc42 activity during nerve growth factor-induced neurite outgrowth in PC12 cells. *J. Biol. Chem.* **279**, 713–719.

Bacso, Z., Bene, L., Bodnár, A., Matkó, J., and Damjanovich, S. (1996). A photobleaching energy transfer analysis of CD8/MHC-I and LFA-1/ICAM-1 interactions in CTL-target cell conjugates. *Immunol. Lett.* **54**, 151–156.

Bagossi, P., Horváth, G., Vereb, G., Szöllősi, J., and Tőzsér, J. (2005). Molecular modeling of nearly full-length ErbB2 receptor. *Biophys. J.* **88**, 1354–1363.

Bastiaens, P. I. H., and Jovin, T. M. (1996). Microspectroscopic imaging tracks the intracellular processing of a signal transduction protein: Fluorescent-labeled protein kinase C beta I. *Proc. Natl. Acad. Sci. USA* **93**, 8407–8412.

Bastiaens, P. I. H., and Jovin, T. M. (1998). Fluorescence resonance energy transfer microscopy. *In* "Cell Biology: A Laboratory Handbook" (J. E. Celis, ed.), 2nd edn., **Vol. 3**. Academic Press, New York.

Bastiaens, P. I. H., and Squire, A. (1999). Fluorescence lifetime imaging microscopy: Spatial resolution of biochemical processes in the cell. *Trends Cell Biol.* **9**, 48–52.

Bastiaens, P. I. H., Majoul, I. V., Verveer, P. J., Söling, H. D., and Jovin, T. M. (1996). Imaging the intracellular trafficking and state of the AB_5 quaternary structure of cholera toxin. *EMBO J.* **15**, 4246–4253.

Bastiaens, P. I. H., Majoul, I. V., Verveer, P. J., Söling, H. D., and Jovin, T. M. (1997). Imaging the intracellular trafficking and state of the AB_5 quaternary structure of cholera toxin. *EMBO J.* **15**, 4246–4253.

Batard, P., Szöllosi, J., Luescher, I., Cerottini, J. C., MacDonald, R., and Romero, P. (2002). Use of phycoerythrin and allophycocyanin for fluorescence resonance energy transfer analyzed by flow cytometry: Advantages and limitations. *Cytometry* **48**, 97–105.

Bene, L., Balázs, M., Matkó, J., Most, J., Dierich, M. P., Szöllosi, J., and Damjanovich, S. (1994). Lateral organization of the ICAM-1 molecule at the surface of human lymphoblasts: A possible model for its co-distribution with the IL-2 receptor, class I and class II HLA molecules. *Eur. J. Immunol.* **24**, 2115–2123.

Berney, C., and Danuser, G. (2003). FRET or no FRET: A quantitative comparison. *Biophys. J.* **84**, 3992–4010.

Bodnár, A., Bacso, Z., Jenei, A., Jovin, T. M., Edidin, M., Damjanovich, S., and Matkó, J. (2003). Class I HLA oligomerization at the surface of B cells is controlled by exogenous beta(2)-microglobulin: Implications in activation of cytotoxic T lymphocytes. *Int. Immunol.* **15**, 331–339.

Bromley, S. K., Burack, W. R., Johnson, K. G., Somersalo, K., Sims, T. N., Sumen, C., Davis, M. M., Shaw, A. S., Allen, P. M., and Dustin, M. L. (2001). The immunological synapse. *Annu. Rev. Immunol.* **19**, 375–396.

Brown, D. A., and London, E. (1998). Structure and origin of ordered lipid domains in biological membranes. *J. Membr. Biol.* **164,** 103–114.

Chakrabarti, A., Matkó, J., Rahman, N. A., Barisas, B. G., and Edidin, M. (1992). Self-association of class I major histocompatibility complex molecules in liposome and cell surface membranes. *Biochemistry* **31,** 7182–7189.

Chen, Y., and Periasamy, A. (2004). Characterization of two-photon excitation fluorescence lifetime imaging microscopy for protein localization. *Microsc. Res. Technol.* **63,** 72–80.

Chen, H., Puhl, H. L., 3rd, Koushik, S. V., Vogel, S. S., and Ikeda, S. R. (2006). Measurement of FRET efficiency and ratio of donor to acceptor concentration in living cells. *Biophys. J.* **91,** L39–41.

Chudakov, D. M., Belousov, V. V., Zaraisky, A. G., Novoselov, V. V., Staroverov, D. B., Zorov, D. B., Lukyanov, S., and Lukyanov, K. A. (2003a). Kindling fluorescent proteins for precise *in vivo* photolabeling. *Nat. Biotechnol.* **21,** 191–194.

Chudakov, D. M., Feofanov, A. V., Mudrik, N. N., Lukyanov, S., and Lukyanov, K. A. (2003b). Chromophore environment provides clue to "kindling fluorescent protein" riddle. *J. Biol. Chem.* **278,** 7215–7219.

Citri, A., Skaria, K. B., and Yarden, Y. (2003). The deaf and the dumb: The biology of ErbB-2 and ErbB-3. *Exp. Cell Res.* **284,** 54–65.

Clamme, J. P., and Deniz, A. A. (2005). Three-color single-molecule fluorescence resonance energy transfer. *Chemphyschem* **6,** 74–77.

Clayton, A. H., Hanley, Q. S., Arndt-Jovin, D. J., Subramaniam, V., and Jovin, T. M. (2002). Dynamic fluorescence anisotropy imaging microscopy in the frequency domain (rFLIM). *Biophys. J.* **83,** 1631–1649.

Clegg, R. M. (1995). Fluorescence resonance energy transfer. *Curr. Opin. Biotechnol.* **6,** 103–110.

Clegg, R. M. (2002). FRET tells us about proximities, distances, orientations and dynamic properties. *J. Biotechnol.* **82,** 177–179.

Clegg, R. M., Murchie, A. I., Zechel, A., Carlberg, C., Diekmann, S., and Lilley, D. M. (1992). Fluorescence resonance energy transfer analysis of the structure of the four-way DNA junction. *Biochemistry* **31,** 4846–4856.

Clegg, R. M., Holub, O., and Gohlke, C. (2003). Fluorescence lifetime-resolved imaging: Measuring lifetimes in an image. *Methods Enzymol.* **360,** 509–542.

Collins, T. L., Uniyal, S., Shin, J., Strominger, J. L., Mittler, R. S., and Burakoff, S. J. (1992). p56lck association with CD4 is required for the interaction between CD4 and the TCR/CD3 complex and for optimal antigen stimulation. *J. Immunol.* **148,** 2159–2162.

Damjanovich, S., Tron, L., Szöllösi, J., Zidovetzki, R., Vaz, W. L., Regateiro, F., Arndt-Jovin, D. J., and Jovin, T. M. (1983). Distribution and mobility of murine histocompatibility H-2Kk antigen in the cytoplasmic membrane. *Proc. Natl. Acad. Sci. USA* **80,** 5985–5989.

Damjanovich, S., Vereb, G., Schaper, A., Jenei, A., Matkó, J., Starink, J. P., Fox, G. Q., Arndt-Jovin, D. J., and Jovin, T. M. (1995). Structural hierarchy in the clustering of HLA class I molecules in the plasma membrane of human lymphoblastoid cells. *Proc. Natl. Acad. Sci. USA* **92,** 1122–1126.

Demarco, I. A., Periasamy, A., Booker, C. F., and Day, R. N. (2006). Monitoring dynamic protein interactions with photoquenching FRET. *Nat. Methods* **3,** 519–524.

Devost, D., and Zingg, H. H. (2003). Identification of dimeric and oligomeric complexes of the human oxytocin receptor by co-immunoprecipitation and bioluminescence resonance energy transfer. *J. Mol. Endocrinol.* **31,** 461–471.

Dewey, T. G., and Hammes, G. G. (1980). Calculation on fluorescence resonance energy transfer on surfaces. *Biophys. J.* **32,** 1023–1035.

Dexter, D. L. (1953). A theory of sensitized luminescence in solids. *J. Chem. Phys.* **21,** 836–850.

Domingo, B., Sabariegos, R., Picazo, F., and Llopis, J. (2007). Imaging FRET standards by steady-state fluorescence and lifetime methods. *Microsc. Res. Tech.* **70,** 1010–1021.

Edidin, M. (1997). Lipid microdomains in cell surface membranes. *Curr. Opin. Struct. Biol.* **7,** 528–532.

Edidin, M. (2001). Shrinking patches and slippery rafts: Scales of domains in the plasma membrane. *Trends Cell Biol.* **11,** 492–496.

Edidin, M., Kuo, S. C., and Sheetz, M. P. (1991). Lateral movements of membrane glycoproteins restricted by dynamic cytoplasmic barriers. *Science* **254**, 1379–1382.

Eidne, K. A., Kroeger, K. M., and Hanyaloglu, A. C. (2002). Applications of novel resonance energy transfer techniques to study dynamic hormone receptor interactions in living cells. *Trends Endocrinol. Metab.* **13**, 415–421.

Elangovan, M., Wallrabe, H., Chen, Y., Day, R. N., Barroso, M., and Periasamy, A. (2003). Characterization of one- and two-photon excitation fluorescence resonance energy transfer microscopy. *Methods* **29**, 58–73.

Fazekas, Z., Petrás, M., Fábián, A., Pályi-Krekk, Z., Nagy, P., Damjanovich, S., Vereb, G., and Szöllősi, J. (2008). Two-sided fluorescence resonance energy transfer for assessing molecular interactions of up to three distinct species in confocal microscopy. *Cytometry A* **73**, 209–219.

Förster, T. (1946). Energiewanderung und Fluoreszenz. *Naturwissenschaften* **6**, 166–175.

Gadella, T. W., Jr., and Jovin, T. M. (1995). Oligomerization of epidermal growth factor receptors on A431 cells studied by time-resolved fluorescence imaging microscopy. A stereochemical model for tyrosine kinase receptor activation. *J. Cell Biol.* **129**, 1543–1558.

Gadella, T. W., Jr., Clegg, R. M., and Jovin, T. M. (1994). Fluorescence lifetime imaging microscopy: Pixel-by-pixel analysis of phase-modulation data. *Bioimaging* **2**, 139–159.

Galperin, E., Verkhusha, V. V., and Sorkin, A. (2004). Three-chromophore FRET microscopy to analyze multiprotein interactions in living cells. *Nat. Methods* **1**, 209–217.

Gáspár, R., Jr., Bagossi, P., Bene, L., Matkó, J., Szöllősi, J., Tozser, J., Fesus, L., Waldmann, T. A., and Damjanovich, S. (2001). Clustering of class I HLA oligomers with CD8 and TCR: Three-dimensional models based on fluorescence resonance energy transfer and crystallographic data. *J. Immunol.* **166**, 5078–5086.

Giordano, L., Jovin, T. M., Irie, M., and Jares-Erijman, E. A. (2002). Diheteroarylethenes as thermally stable photoswitchable acceptors in photochromic fluorescence resonance energy transfer (pcFRET). *J. Am. Chem. Soc.* **124**, 7481–7489.

Gordon, G. W., Berry, G., Liang, X. H., Levine, B., and Herman, B. (1998). Quantitative fluorescence resonance energy transfer measurements using fluorescence microscopy. *Biophys. J.* **74**, 2702–2713.

Grant, D. M., Zhang, W., McGhee, E. J., Bunney, T. D., Talbot, C. B., Kumar, S., Munro, I., Dunsby, C., Neil, M. A., Katan, M., and French, P. M. (2008). Multiplexed FRET to image multiple signaling events in live cells. *Biophys. J.* **95**, L69–71.

Grecco, H. E., Lidke, K. A., Heintzmann, R., Lidke, D. S., Spagnuolo, C., Martinez, O. E., Jares-Erijman, E. A., and Jovin, T. M. (2004). Ensemble and single particle photophysical properties (two-photon excitation, anisotropy, FRET, lifetime, spectral conversion) of commercial quantum dots in solution and in live cells. *Microsc. Res. Tech.* **65**, 169–179.

Grecco, H. E., Roda-Navarro, P., and Verveer, P. J. (2009). Global analysis of time correlated single photon counting FRET-FLIM data. *Opt. Express.* **17**, 6493–6508.

Ha, T. (2001). Single-molecule fluorescence resonance energy transfer. *Methods* **25**, 78–86.

Haj, F. G., Verveer, P. J., Squire, A., Neel, B. G., and Bastiaens, P. I. (2002). Imaging sites of receptor dephosphorylation by PTP1B on the surface of the endoplasmic reticulum. *Science* **295**, 1708–1711.

Heyduk, T. (2002). Measuring protein conformational changes by FRET/LRET. *Curr. Opin. Biotechnol.* **13**, 292–296.

Hoppe, A. D., Shorte, S. L., Swanson, J. A., and Heintzmann, R. (2008). Three-dimensional FRET reconstruction microscopy for analysis of dynamic molecular interactions in live cells. *Biophys. J.* **95**, 400–418.

Horváth, G., Petrás, M., Szentesi, G., Fábián, A., Park, J. W., Vereb, G., and Szöllősi, J. (2005). Selecting the right fluorophores and flow cytometer for fluorescence resonance energy transfer measurements. *Cytometry A* **65**, 148–157.

Horváth, G., Young, S., and Latz, E. (2009). Toll-like receptor interactions imaged by FRET microscopy and GFP fragment reconstitution. *Methods Mol. Biol.* **517**, 33–54.

Issad, T., Boute, N., Boubekeur, S., Lacasa, D., and Pernet, K. (2003). Looking for an insulin pill? Use the BRET methodology! *Diabetes Metab.* **29**, 111–117.

Jacobson, K., and Dietrich, C. (1999). Looking at lipid rafts? *Trends Cell Biol.* **9**, 87–91.

Jares-Erijman, E. A., and Jovin, T. M. (2003). FRET imaging. *Nat. Biotechnol.* **21**, 1387–1395.

Jares-Erijman, E. A., and Jovin, T. M. (2006). Imaging molecular interactions in living cells by FRET microscopy. *Curr. Opin. Chem. Biol.* **10**, 409–416.

Jovin, T. M., and Arndt-Jovin, D. J. (1989). FRET microscopy: Digital imaging of fluorescence resonance energy transfer. Application in cell biology. *In* "Cell Structure and Function by Micro-spectrofluorimetry" (E. Kohen and J. G. Hirschberg, eds.). Academic Press, San Diego, CA.

Jurgens, L., Arndt-Jovin, D., Pecht, I., and Jovin, T. M. (1996). Proximity relationships between the type I receptor for Fc epsilon (Fc epsilon RI) and the mast cell function-associated antigen (MAFA) studied by donor photobleaching fluorescence resonance energy transfer microscopy. *Eur. J. Immunol.* **26**, 84–91.

Kenworthy, A. K., and Edidin, M. (1998). Distribution of a glycosylphosphatidylinositol-anchored protein at the apical surface of MDCK cells examined at a resolution of <100 Å using imaging fluorescence resonance energy transfer. *J. Cell Biol.* **142**, 69–84.

Kenworthy, A. K., Petranova, N., and Edidin, M. (2000). High-resolution FRET microscopy of cholera toxin B-subunit and GPI-anchored proteins in cell plasma membranes. *Mol. Biol. Cell* **11**, 1645–1655.

Kerppola, T. K. (2006). Design and implementation of bimolecular fluorescence complementation (BiFC) assays for the visualization of protein interactions in living cells. *Nat. protoc.* **1**, 1278–1286.

Kirsch, A., Meyer, C., and Jovin, T. M. (1996). Integration of optical techniques in scanning probe microscopes: The scanning near-field optical microscope (SNOM). *In* "Proceedings of NATO Advanced Research Workshop: Analytical Use of Fluorescent Probes in Oncology" Miami, FL, Oct. 14–18, 1995 (E. Kohen and J. G. Kirschberg, eds.). Plenum Press, New York.

Kirsch, A. K., Subramaniam, V., Jenei, A., and Jovin, T. M. (1999). Fluorescence resonance energy transfer detected by scanning near-field optical microscopy. *J. Microsc.* **194**, 448–454.

Koushik, S. V., Chen, H., Thaler, C., Puhl, H. L., 3rd, and Vogel, S. S. (2006). Cerulean, Venus, and Venus Y67C FRET reference standards. *Biophys. J.* **91**, L99–L101.

Latif, R., Graves, P., and Davies, T. F. (2002). Ligand-dependent inhibition of oligomerization at the human thyrotropin receptor. *J. Biol. Chem.* **277**, 45059–45067.

Legg, J. W., Lewis, C. A., Parsons, M., Ng, T., and Isacke, C. M. (2002). A novel PKC-regulated mechanism controls CD44 ezrin association and directional cell motility. *Nat. Cell Biol.* **4**, 399–407.

Lidke, D. S., Nagy, P., Barisas, B. G., Heintzmann, R., Post, J. N., Lidke, K. A., Clayton, A. H., Arndt-Jovin, D. J., and Jovin, T. M. (2003). Imaging molecular interactions in cells by dynamic and static fluorescence anisotropy (rFLIM and emFRET). *Biochem. Soc. Trans.* **31**, 1020–1027.

Liegler, T., Szöllősi, J., Hyun, W., and Goodenow, R. S. (1991). Proximity measurements between H-2 antigens and the insulin receptor by fluorescence energy transfer: Evidence that a close association does not influence insulin binding. *Proc. Natl. Acad. Sci. USA* **88**, 6755–6759.

Lippincott-Schwartz, J., and Patterson, G. H. (2003). Development and use of fluorescent protein markers in living cells. *Science* **300**, 87–91.

Lippincott-Schwartz, J., Altan-Bonnet, N., and Patterson, G. H. (2003). Photobleaching and photo-activation: Following protein dynamics in living cells. *Nat. Cell Biol.* Suppl. S7–S14.

Loken, M. R., Keij, J. F., and Kelley, K. A. (1987). Comparison of helium-neon and dye lasers for the excitation of allophycocyanin. *Cytometry* **8**, 96–100.

Mao, S., Benninger, R. K., Yan, Y., Petchprayoon, C., Jackson, D., Easley, C. J., Piston, D. W., and Marriott, G. (2008). Optical lock-in detection of FRET using synthetic and genetically encoded optical switches. *Biophys. J.* **94**, 4515–4524.

Matkó, J., and Edidin, M. (1997). Energy transfer methods for detecting molecular clusters on cell surfaces. *Methods Enzymol.* **278**, 444–462.

Matkó, J., and Szöllősi, J. (2002). Landing of immune receptors and signal proteins on lipid rafts: A safe way to be spatio-temporally coordinated? *Immunol. Lett.* **82**, 3–15.

Matkó, J., Ohki, K., and Edidin, M. (1992). Luminescence quenching by nitroxide spin labels in aqueous solution: Studies on the mechanism of quenching. *Biochemistry* **31**, 703–711.

Matkó, J., Jenei, A., Mátyus, L., Ameloot, M., and Damjanovich, S. (1993). Mapping of cell surface protein-patterns by combined fluorescence anisotropy and energy transfer measurements. *J. Photochem. Photobiol. B* **19**, 69–73.

Matkó, J., Bushkin, Y., Wei, T., and Edidin, M. (1994). Clustering of class I HLA molecules on the surfaces of activated and transformed human cells. *J. Immunol.* **152**, 3353–3360.

Matkó, J., Jenei, A., Wei, T., and Edidin, M. (1995). Luminescence quenching by long range electron transfer: A probe of protein clustering and conformation at the cell surface. *Cytometry* **19**, 191–200.

Matkó, J., Bodnár, A., Vereb, G., Bene, L., Vamosi, G., Szentesi, G., Szöllősi, J., Gáspár, R., Horejsi, V., Waldmann, T. A., and Damjanovich, S. (2002). GPI-microdomains (membrane rafts) and signaling of the multi-chain interleukin-2 receptor in human lymphoma/leukemia T cell lines. *Eur. J. Biochem.* **269**, 1199–1208.

Mátyus, L., Bene, L., Heiligen, H., Rausch, J., and Damjanovich, S. (1995). Distinct association of transferrin receptor with HLA class I molecules on HUT-102B and JY cells. *Immunol. Lett.* **44**, 203–208.

McGrath, N., and Barroso, M. (2008). Quantum dots as fluorescence resonance energy transfer donors in cells. *J. Biomed. Opt.* **13**, 031210.

Megias, D., Marrero, R., Martinez Del Peso, B., Garcia, M. A., Bravo-Cordero, J. J., Garcia-Grande, A., Santos, A., and Montoya, M. C. (2009). Novel lambda FRET spectral confocal microscopy imaging method. *Microsc. Res. Tech.* **72**, 1–11.

Mekler, V. M. (1994). A photochemical technique to enhance sensitivity of detection of fluorescence resonance energy transfer. *Photochem. Photobiol.* **59**, 615–620.

Milligan, G., Ramsay, D., Pascal, G., and Carrillo, J. J. (2003). GPCR dimerisation. *Life. Sci.* **74**, 181–188.

Mittler, R. S., Goldman, S. J., Spitalny, G. L., and Burakoff, S. J. (1989). T-cell receptor-CD4 physical association in a murine T-cell hybridoma: Induction by antigen receptor ligation. *Proc. Natl. Acad. Sci. USA* **86**, 8531–8535.

Monks, C. R., Freiberg, B. A., Kupfer, H., Sciaky, N., and Kupfer, A. (1998). Three-dimensional segregation of supramolecular activation clusters in T cells. *Nature* **395**, 82–86.

Monson, E., Merritt, G., Smith, S., Langmore, J. P., and Kopelman, R. (1995). Implementation of an NSOM system for fluorescence microscopy. *Ultramicroscopy* **57**, 257–262.

Nagy, P., Bene, L., Balázs, M., Hyun, W. C., Lockett, S. J., Chiang, N. Y., Waldman, F., Feuerstein, B. G., Damjanovich, S., and Szöllősi, J. (1998a). EGF-induced redistribution of erbB2 on breast tumor cells: Flow and image cytometric energy transfer measurements. *Cytometry* **32**, 120–131.

Nagy, P., Vamosi, G., Bodnár, A., Lockett, S. J., and Szöllősi, J. (1998b). Intensity-based energy transfer measurements in digital imaging microscopy. *Eur. Biophys. J.* **27**, 377–389.

Nagy, P., Jenei, A., Kirsch, A. K., Szöllősi, J., Damjanovich, S., and Jovin, T. M. (1999a). Activation-dependent clustering of the erbB2 receptor tyrosine kinase detected by scanning near-field optical microscopy. *J. Cell Sci.* **112**, 1733–1741.

Nagy, P., Jenei, A., Damjanovich, S., Jovin, T. M., and Szolosi, J. (1999b). Complexity of signal transduction mediated by ErbB2: Clues to the potential of receptor-targeted cancer therapy. *Pathol. Oncol. Res.* **5**, 255–271.

Nagy, P., Vereb, G., Sebestyen, Z., Horvath, G., Lockett, S. J., Damjanovich, S., Park, J. W., Jovin, T. M., and Szöllősi, J. (2002). Lipid rafts and the local density of ErbB proteins influence the biological role of homo- and heteroassociations of ErbB2. *J. Cell Sci.* **115**, 4251–4262.

Nagy, P., Bene, L., Hyun, W. C., Vereb, G., Braun, M., Antz, C., Paysan, J., Damjanovich, S., Park, J. W., and Szöllősi, J. (2005). Novel calibration method for flow cytometric fluorescence resonance energy transfer measurements between visible fluorescent proteins. *Cytometry A* **67**, 86–96.

Nagy, P., Vereb, G., Damjanovich, S., Mátyus, L., and Szöllősi, J. (2006). Measuring FRET in flow cytometry and microscopy. *Curr. Protoc. Cytom.*, 8. Chapter 12, Unit 12

Nashmi, R., Dickinson, M. E., McKinney, S., Jareb, M., Labarca, C., Fraser, S. E., and Lester, H. A. (2003). Assembly of alpha4beta2 nicotinic acetylcholine receptors assessed with functional fluorescently labeled subunits: Effects of localization, trafficking, and nicotine-induced upregulation in clonal mammalian cells and in cultured midbrain neurons. *J. Neurosci.* **23**, 11554–11567.

Paar, C., Paster, W., Stockinger, H., Schutz, G. J., Sonnleitner, M., and Sonnleitner, A. (2008). High throughput FRET screening of the plasma membrane based on TIRFM. *Cytometry A* **73**, 442–450.

Padilla-Parra, S., Auduge, N., Coppey-Moisan, M., and Tramier, M. (2008). Quantitative FRET analysis by fast acquisition time domain FLIM at high spatial resolution in living cells. *Biophys. J.* **95**, 2976–2988.

Patterson, G. H., and Lippincott-Schwartz, J. 2002. A photoactivatable GFP for selective photolabeling of proteins and cells. *Science* **297**, 1873–1877.

Periasamy, A., Wallrabe, H., Chen, Y., and Barroso, M. (2008). Chapter 22: Quantitation of protein-protein interactions: Confocal FRET microscopy. *Methods Cell Biol.* **89**, 569–598.

Piston, D. W., and Rizzo, M. A. (2008). FRET by fluorescence polarization microscopy. *Methods Cell Biol.* **85**, 415–430.

Ramalingam, T. S., Chakrabarti, A., and Edidin, M. (1997). Interaction of class I human leukocyte antigen (HLA-I) molecules with insulin receptors and its effect on the insulin-signaling cascade. *Mol. Biol. Cell* **8**, 2463–2474.

Rizzo, M. A., Springer, G., Segawa, K., Zipfel, W. R., and Piston, D. W. (2006). Optimization of pairings and detection conditions for measurement of FRET between cyan and yellow fluorescent proteins. *Microsc. Microanal.* **12**, 238–254.

Roszik, J., Szöllősi, J., and Vereb, G. (2008). AccPbFRET: An Image J plugin for semi-automatic, fully corrected analysis of acceptor photobleaching FRET images. *BMC Bioinformatics* **9**, 346.

Roszik, J., Lisboa, D., Szöllősi, J., and Vereb, G. (2009). Evaluation of intensity-based ratiometric FRET in image cytometry—Approaches and a software solution. *Cytometry A* 75A(9):761–767.

Runnels, L. W., and Scarlata, S. F. (1995). Theory and application of fluorescence homotransfer to melittin oligomerization. *Biophys. J.* **69**, 1569–1583.

Scarlata, S., and Dowal, L. (2004). The use of green fluorescent proteins to view association between phospholipase C beta and G protein subunits in cells. *Methods Mol. Biol.* **237**, 223–232.

Scholes, G. D. (2003). Long-range resonance energy transfer in molecular systems. *Annu. Rev. Phys. Chem.* **54**, 57–87.

Schmid, J. A., and Birbach, A. (2007). Fluorescent proteins and fluorescence resonance energy transfer (FRET) as tools in signaling research. *Thromb. Haemost.* **97**, 378–384.

Schreiber, A. B., Schlessinger, J., and Edidin, M. (1984). Interaction between major histocompatibility complex antigens and epidermal growth factor receptors on human cells. *J. Cell Biol.* **98**, 725–731.

Sebestyen, Z., Nagy, P., Horvath, G., Vamosi, G., Debets, R., Gratama, J. W., Alexander, D. R., and Szöllősi, J. (2002). Long wavelength fluorophores and cell-by-cell correction for autofluorescence significantly improves the accuracy of flow cytometric energy transfer measurements on a dual-laser benchtop flow cytometer. *Cytometry* **48**, 124–135.

Selvin, P. R. (2000). The renaissance of fluorescence resonance energy transfer. *Nat. Struct. Biol.* **7**, 730–734.

Shaner, N. C., Patterson, G. H., and Davidson, M. W. (2007). Advances in fluorescent protein technology. *J. Cell Sci.* **120**, 4247–4260.

Shcherbo, D., Souslova, E. A., Goedhart, J., Chepurnykh, T. V., Gaintzeva, A., Shemiakina, I. I., Gadella, T. W., Lukyanov, S., and Chudakov, D. M. (2009). Practical and reliable FRET/FLIM pair of fluorescent proteins. *BMC Biotechnol.* **9**, 24.

Shimozono, S., and Miyawaki, A. (2008). Engineering FRET constructs using CFP and YFP. *Methods Cell Biol.* **85**, 381–393.

Shubeita, G. T., Sekatskii, S. K., Dietler, G., Potapova, I., Mews, A., and Basche, T. (2003). Scanning near-field optical microscopy using semiconductor nanocrystals as a local fluorescence and fluorescence resonance energy transfer source. *J. Microsc.* **210**, 274–278.

Shyu, Y. J., Suarez, C. D., and Hu, C. D. (2008). Visualization of ternary complexes in living cells by using a BiFC-based FRET assay. *Nat. Protoc.* **3**, 1693–1702.

Simons, K., and Ikonen, E. (1997). Functional rafts in cell membranes. *Nature* **387**, 569–572.

Sliwkowski, M. X., Schaefer, G., Akita, R. W., Lofgren, J. A., Fitzpatrick, V. D., Nuijens, A., Fendly, B. M., Cerione, R. A., Vandlen, R. L., and Carraway, K. L. 3rd. (1994). Coexpression of erbB2 and erbB3 proteins reconstitutes a high affinity receptor for heregulin. *J. Biol. Chem.* **269**, 14661–14665.

Sliwkowski, M. X., Lofgren, J. A., Lewis, G. D., Hotaling, T. E., Fendly, B. M., and Fox, J. A. (1999). Nonclinical studies addressing the mechanism of action of trastuzumab (Herceptin). *Semin. Oncol.* **26**, 60–70.

Smith, P. R., Morrison, I. E., Wilson, K. M., Fernandez, N., and Cherry, R. J. (1999). Anomalous diffusion of major histocompatibility complex class I molecules on HeLa cells determined by single particle tracking. *Biophys. J.* **76**, 3331–3344.

Snyder, B., and Freire, E. (1982). Fluorescence energy transfer in two dimensions. A numeric solution for random and nonrandom distributions. *Biophys. J.* **40**, 137–148.

Song, L., Hennink, E. J., Young, I. T., and Tanke, H. J. (1995). Photobleaching kinetics of fluorescein in quantitative fluorescence microscopy. *Biophys. J.* **68**, 2588–2600.

Song, L., Jares-Erijman, E. A., and Jovin, T. M. (2002). A photochromic acceptor as a reversible lightdriven switch in fluorescence resonance energy transfer (FRET). *J. Photochem. Photobiol. A* **150**, 177–185.

Stryer, L., and Haugland, R. P. (1967). Energy transfer: A spectroscopic ruler. *Proc. Natl. Acad. Sci. USA* **58**, 719–726.

Subczynski, W. K., and Kusumi, A. (2003). Dynamics of raft molecules in the cell and artificial memberanes: Approaches by pulse EPR spin labeling and single molecule optical microscopy. *Biochim. Biophys. Acta* **1610**, 231–243.

Subramaniam, V., Hanley, Q. S., Clayton, A. H., and Jovin, T. M. (2003). Photophysics of green and red fluorescent proteins: Implications for quantitative microscopy. *Methods. Enzymol.* **360**, 178–201.

Suzuki, Y. (2000). Detection of the swings of the lever arm of a myosin motor by fluorescence resonance energy transfer of green and blue fluorescent proteins. *Methods* **22**, 355–363.

Szabó, G., Jr., Weaver, J. L., Pine, P. S., Rao, P. E., and Aszalos, A. (1995). Cross-linking of CD4 in a TCR/CD3-juxtaposed inhibitory state: A pFRET study. *Biophys. J.* **68**, 1170–1176.

Szabó, Á., Horváth, G., Szöllősi, J., and Nagy, P. (2008). Quantitative characterization of the large-scale association of ErbB1 and ErbB2 by flow cytometric homo-FRET measurements. *Biophys. J.* **95**, 2086–2096.

Szentesi, G., Horváth, G., Bori, I., Vámosi, G., Szöllősi, J., Gáspár, R., Damjanovich, S., Jenei, A., and Mátyus, L. (2004). Computer program for determining fluorescence resonance energy transfer efficiency from flow cytometric data on a cell-by-cell basis. *Comput. Methods Programs Biomed.* **75**, 201–211.

Szentesi, G., Vereb, G., Horváth, G., Bodnár, A., Fábián, A., Matkó, J., Gáspár, R., Damjanovich, S., Mátyus, L., and Jenei, A. (2005). Computer program for analyzing donor photobleaching FRET image series. *Cytometry A* **67**, 119–128.

Szöllősi, J., and Alexander, D. R. (2003). The application of fluorescence resonance energy transfer to the investigation of phosphatases. *Methods Enzymol.* **366**, 203–224.

Szöllősi, J., Tron, L., Damjanovich, S., Helliwell, S. H., Arndt-Jovin, D., and Jovin, T. M. (1984). Fluorescence energy transfer measurements on cell surfaces: A critical comparison of steady-state fluorimetric and flow cytometric methods. *Cytometry* **5**, 210–216.

Szöllősi, J., Damjanovich, S., Goldman, C. K., Fulwyler, M. J., Aszalos, A. A., Goldstein, G., Rao, P., Talle, M. A., and Waldmann, T. A. (1987). Flow cytometric resonance energy transfer measurements support the association of a 95-kDa peptide termed T27 with the 55-kDa Tac peptide. *Proc. Natl. Acad. Sci. USA* **84**, 7246–7250.

Szöllősi, J., Damjanovich, S., Balázs, M., Nagy, P., Tron, L., Fulwyler, M. J., and Brodsky, F. M. (1989). Physical association between MHC class I and class II molecules detected on the cell surface by flow cytometric energy transfer. *J. Immunol.* **143**, 208–213.

Szöllősi, J., Horejsi, V., Bene, L., Angelisova, P., and Damjanovich, S. (1996). Supramolecular complexes of MHC class I, MHC class II, CD20, and tetraspan molecules (CD53, CD81, and CD82) at the surface of a B cell line JY. *J. Immunol.* **157**, 2939–2946.

Szöllősi, J., Damjanovich, S., and Mátyus, L. (1998). Application of fluorescence resonance energy transfer in the clinical laboratory: Routine and research. *Cytometry* **34**, 159–179.

Szöllősi, J., Nagy, P., Sebestyen, Z., Damjanovich, S., Park, J. W., and Mátyus, L. (2002). Applications of fluorescence resonance energy transfer for mapping biological membranes. *J. Biotechnol.* **82**, 251–266.

Szöllősi, J., Damjanovich, S., Nagy, P., Vereb, G., and Mátyus, L. (2006). Principles of resonance energy transfer. *In* "Current Protocols in Cytometry," (Robert A. Hoffman., ed.), Wiley, Vol. Chapter 1, pp. Unit1 12, Hoboken, NJ.

Tramier, M., Piolot, T., Gautier, I., Mignotte, V., Coppey, J., Kemnitz, K., Durieux, C., and Coppey-Moisan, M. (2003). Homo-FRET versus hetero-FRET to probe homodimers in living cells. *Methods Enzymol.* **360**, 580–597.

Triantafilou, K., Triantafilou, M., Wilson, K. M., and Fernandez, N. (2000). Human major histocompatibility molecules have the intrinsic ability to form homotypic associations. *Hum. Immunol.* **61**, 585–598.

Tron, L., Szöllősi, J., Damjanovich, S., Helliwell, S. H., Arndt-Jovin, D. J., and Jovin, T. M. (1984). Flow cytometric measurement of fluorescence resonance energy transfer on cell surfaces. Quantitative evaluation of the transfer efficiency on a cell-by-cell basis. *Biophys. J.* **45**, 939–946.

Tuosto, L., Parolini, I., Schroder, S., Sargiacomo, M., Lanzavecchia, A., and Viola, A. (2001). Organization of plasma membrane functional rafts upon T cell activation. *Eur. J. Immunol.* **31**, 345–349.

Turcatti, G., Nemeth, K., Edgerton, M. D., Meseth, U., Talabot, F., Peitsch, M., Knowles, J., Vogel, H., and Chollet, A. (1996). Probing the structure and function of the tachykinin neurokinin-2 receptor through biosynthetic incorporation of fluorescent amino acids at specific sites. *J. Biol. Chem.* **271**, 19991–19998.

Tzahar, E., Waterman, H., Chen, X., Levkowitz, G., Karunagaran, D., Lavi, S., Ratzkin, B. J., and Yarden, Y. (1996). A hierarchical network of interreceptor interactions determines signal transduction by Neu differentiation factor/neuregulin and epidermal growth factor. *Mol. Cell Biol.* **16**, 5276–5287.

Van Munster, E. B., and Gadella, T. W. (2004). phiFLIM: A new method to avoid aliasing in frequency-domain fluorescence lifetime imaging microscopy. *J. Microsc.* **213**, 29–38.

van Royen, M. E., Dinant, C., Farla, P., Trapman, J., and Houtsmuller, A. B. (2009). FRAP and FRET methods to study nuclear receptors in living cells. *Methods Mol. Biol.* **505**, 69–96.

Varma, R., and Mayor, S. (1998). GPI-anchored proteins are organized in submicron domains at the cell surface. *Nature* **394**, 798–801.

Vereb, G., Meyer, C. K., and Jovin, T. M. (1997). Novel microscope-based approaches for the investigation of protein-protein interactions in signal transduction. *In* "Interacting Protein Domains, Their Role in Signal and Energy Transduction. NATO ASI Series" (L. M. G. HeilmeyerJr., ed.), **Vol. H102**, Springer-Verlag, New York.

Vereb, G., Matkó, J., Vamosi, G., Ibrahim, S. M., Magyar, E., Varga, S., Szöllősi, J., Jenei, A., Gáspár, R., Jr., Waldmann, T. A., and Damjanovich, S. (2000). Cholesterol-dependent clustering of IL-2Ralpha and its colocalization with HLA and CD48 on T lymphoma cells suggest their functional association with lipid rafts. *Proc. Natl. Acad. Sci. USA* **97**, 6013–6018.

Vereb, G., Nagy, P., Park, J. W., and Szöllősi, J. (2002). Signaling revealed by mapping molecular interactions: Implications for ErbB-targeted cancer immunotherapies. *Clin. Appl. Immunol. Rev.* **2**, 169–186.

Vereb, G., Szöllősi, J., Matkó, J., Nagy, P., Farkas, T., Vigh, L., Mátyus, L., Waldmann, T. A., and Damjanovich, S. (2003). Dynamic, yet structured: The cell membrane three decades after the Singer-Nicolson model. *Proc. Natl. Acad. Sci. USA* **100**, 8053–8058.

Vereb, G., Matkó, J., and Szöllősi, J. (2004). Cytometry of fluorescence resonance energy transfer. *Methods Cell Biol.* **75**, 105–152.

Vickery, S. A., and Dunn, R. C. (1999). Scanning near-field fluorescence resonance energy transfer microscopy. *Biophys. J.* **76**, 1812–1818.

Vogel, S. S., Thaler, C., and Koushik, S. V. (2006). Fanciful FRET. *Sci. STKE* **2006**(331), re2.

Wlodarczyk, J., Woehler, A., Kobe, F., Ponimaskin, E., Zeug, A., and Neher, E. (2008). Analysis of FRET signals in the presence of free donors and acceptors. *Biophys. J.* **94,** 986–1000.

Wolber, P. K., and Hudson, B. S. (1979). An analytic solution to the Förster energy transfer problem in two dimensions. *Biophys. J.* **28,** 197–210.

Wulfing, C., Sjaastad, M. D., and Davis, M. M. (1998). Visualizing the dynamics of T cell activation: Intracellular adhesion molecule 1 migrates rapidly to the T cell/B cell interface and acts to sustain calcium levels. *Proc. Natl. Acad. Sci. USA* **95,** 6302–6307.

Xing, Y., and Rao, I. (2008). Quantum dot bioconjugates for in vitro diagnostics & *in vivo* imaging. *Cancer. Biomark.* **4,** 307–319.

Xu, Y., Piston, D. W., and Johnson, C. H. (1999). A bioluminescence resonance energy transfer (BRET) system: Application to interacting circadian clock proteins. *Proc. Natl. Acad. Sci. USA* **96,** 151–156.

Yarden, Y., and Sliwkowski, M. X. (2001). Untangling the ErbB signalling network. *Nat. Rev. Mol. Cell Biol.* **2,** 127–137.

Yguerabide, J. (1994). Theory for establishing proximity relations in biological membranes by excitation energy transfer measurements. *Biophys. J.* **66,** 683–693.

Zal, T., Zal, M. A., and Gascoigne, N. R. (2002). Inhibition of T cell receptor-coreceptor interactions by antagonist ligands visualized by live FRET imaging of the T-hybridoma immunological synapse. *Immunity* **16,** 521–534.

Zhang, J., Campbell, R. E., Ting, A. Y., and Tsien, R. Y. (2002). Creating new fluorescent probes for cell biology. *Nat. Rev. Mol. Cell Biol.* **3,** 906–918.

Zimmermann, T., Rietdorf, J., and Pepperkok, R. (2003). Spectral imaging and its applications in live cell microscopy. *FEBS Lett.* **546,** 87–92.

CHAPTER 4

The Rainbow of Fluorescent Proteins

David W. Galbraith

Department of Plant Sciences
Institute for Biomedical Science and Biotechnology
University of Arizona
Tucson, Arizona 85721

I. Update

This chapter was published in 2004. In the 5 years, since its appearance, interest in the use of fluorescent proteins in flow and image cytometry, already significant, has notably intensified. Three areas are of particular importance: (i) the further discovery and modification of FPs, (ii) the development of novel reagents using these fluoroproteins, and (iii) the use of FPs in devising novel methods and cytometric platforms for analysis of tissues and cells. The importance of FPs in biological research was highlighted by the award in 2008 of the Nobel Prize in Chemistry to Drs Chalfie, Imamura, and Tsien, with the contribution of Doug Prasher ending up, unfortunately, as a footnote.

Discovery and modification of FPs has continued the strategies outlined in this chapter, involving identification of novel FPs from additional, mostly marine organisms, and modification of the sequences of these proteins, in a directed or undirected manner and with or without selection, to derive novel optical proper-ties. In that sense, this area has remained the same, but it has led to an ever-expanding palette of FPs suitable for different specific applications in cytometry. Particularly relevant to image cytometry has been the extended development of photomodulated FPs, whose fluorescence excitation and emission spectra are modified by illumination at specific wavelengths. This can be either to effect photoactivation (i.e., switching from a dark to a fluorescent state), or photocon-version (i.e., switching between two colors, most commonly from green to red). Recent work indicates two-photon illumination can be used to switch individual FPs in the opposite direction (i.e., from red to green fluorescent states) (Kremers *et al.*, 2009).

In terms of novel reagents, one of the most intriguing applications has involved the use of FPs to provide molecules that monitor stochastic noise in gene expres-sion within single cells, at the level of transcription and of protein expression (recently reviewed by Raj and van Oudenaarden, 2008). Implemented in both prokaryotic and eukaryotic organisms, transcriptional noise is monitored using promoter fusions to different FPs, with flow cytometry providing the read-out from individual single cells. Noise at the level of protein production is measured by fusing the FP coding sequences to those of endogenous genes, again employing flow cytometry to monitor the levels of fluorescence within single cells. Appropri-ate high-throughput methods have been developed to produce the large numbers of different FP fusions required for these kinds of experiments. Much interest is centered on determining the extent and degree of stochastic noise within single cells as related to gene function, and on establishing how cells regulate this noise, how they employ it in regulatory networks, and how it affects evolutionary fitness.

In terms of novel methods and platforms, one of the most important to emerge is that of super-resolution optical microscopy, Hell (2009) defined as optical micros-copy that breaks the rule of diffraction-based imaging proposed by Ernst Abbé, stating that light of wavelength λ cannot resolve objects that are closer together than a distance $d = \lambda/2\text{NA}$, where NA represents the numeric aperture of the lens that is used to focus the light (a distance of about 200 nm). This diffraction barrier assumes the objects to be resolved have identical optical properties. If, however, they have different properties, then the diffraction limit does not apply, and their separation can be determined with a precision representing distances much smaller than the 200 nm limitation of conventional imaging. Differences in optical proper-ties that permit super-resolution particularly involve devising ways to switch on and off the fluorescent signals on the objects of interest, such that they are detected sequentially in time. Two implementations of relevance here are termed PALM (photoactivation localization microscopy; Betzig *et al.*, 2006) and STORM (stochastic optical reconstruction microscopy; Rust *et al.*, 2006), the first being particularly significant since it employs photomodulated FPs (in this case photo-activated fluorescent protein (PA-FP)). In brief, a light pulse from a laser capable of activating inactive PA-FP molecules is applied to the specimen, to produce an optically sparse population of active FPs, whose fluorescence emission is then imaged, after excitation at a second wavelength corresponding to the excitation maximum of the PA-FP, through photon-counting using total internal reflection microscopy. This is continued until the specimen is completely photobleached. The positions of the PA-FP molecules within the captured images are then determined by statistical fitting of the centers of fluorescence emission. The activation step is repeated and the entire process then cycled many times, ultimately providing a reconstructed super-resolution image. The value of super-resolution imaging in biology cannot be overstated, and the impact of this technology will be profound. In this respect, a particularly attractive feature of PALM is the low cost of the instrument platform relative to other super-resolution imaging platforms.

A. How to Avoid Pitfalls

The primary pitfall in the use of FPs is ensuring that the scientist selects the sequence most suited for the application and organism of interest, with the main problems being a lack of expression, the observation of cellular toxicity following expression, conflicts with endogenous cellular autofluorescence and/or location, and issues emerging from the unusual stabilities of FPs within living cells. Careful preplanning of experiments is encouraged, as is contacting the leading laboratories and commercial suppliers in this area to determine the FPs optimal for the proposed use.

B. What Methods are Trusted "Winners"?

All of the methods described in the chapter are trusted and reproducible, and there is every indication the new methods and platforms described above will prove equally so.

II. The Fluorescent Proteins

Fluorescent proteins (FPs) can be defined as the class of proteins that exhibit fluorescence as a consequence of fluorophore formation by an intramolecular reaction of the amino acid side chains contained within these proteins. This class comprises approximately 30 members, all of which have been isolated from aquatic organisms. Specifically excluded from this class are all proteins whose fluorescence requires the presence of nonprotein prosthetic groups. Classifying FPs in this manner focuses attention on their unique ability to produce fluorescence after expression of the polypeptide structure without requiring additional components, including specific chaperones. Consequently, FPs comprise uniquely flexible transgenic markers capable of producing fluorescence within an essentially limitless range of species and cell types, in both prokarya and eukarya (Chalfie *et al.*, 1994; Miyawaki *et al.*, 2003; Tsien, 1998). A subset of FP proteins, termed *chromoproteins* (Labas *et al.*, 2002), display intense absorption but without subsequent fluorescence emission, but variants of these can be isolated that behave as FPs. The evolutionary relationship between chromoproteins and FPs is further discussed in Section I.F.

A. Green Fluorescent Protein

The prototype FP is the green FP (GFP) of the jellyfish *Aequorea victoria*, which was first described 40 years ago (Johnson *et al.*, 1962; Shimomura *et al.*, 1962). At least four species names exist for *Aequorea* (*A. victoria*, *A. aequorea*, *A. coerulescens*, and *A. forskalea*). This taxonomy is based on morphological features that are both plastic and developmentally regulated (for a complete discussion, see Mills, 1999). Clarification of species relationships awaits application of molecular methods.

GFP is a small ($\sim$26-kd) protein comprising 238 amino acids, organized as 11 β sheets forming a capped barrel (Ormö *et al.*, 1996; Yang *et al.*, 1996) that encloses the fluorophore and serves to exclude solvent molecules. This structure also confers unusual stability on the protein, which remains fluorescent up to 65 °C, at a pH 11, and in 1% sodium dodecylsulfate, or in 6 M guanidinium hydrochloride, and which is largely resistant to proteases. The fluorophore, a *p*-hydroxybenzylidene imidazolinone, is formed by an intramolecular rearrangement and oxidation involving ser65, tyr66, and gly67 (Tsien, 1998). Within the jellyfish, GFP serves to absorb broad-spectrum blue light produced by aequorin, a Ca^{2+}-dependent photoluminescent protein.

B. Modifications to Green Fluorescent Protein

Early work established the feasibility of altering the optical properties of GFP through modification of the protein primary sequence (Heim *et al.*, 1994; Tsien, 1998). Early modifications resulted in changes to shape of the bimodal absorbance spectrum of wild-type GFP to a unimodal peak center either in the ultraviolet (UV) or near 488 nm. Other amino acid changes, single and multiple, were found to alter the wavelengths of maximal absorption and emission of the fluorochrome. The naming conventions of mutant GFPs are confusing (Galbraith *et al.*, 1998). Some convergence has occurred though, with the following names for commonly encountered mutant GFP forms being generally recognized: cyan FP (CFP; absorbance maximum 439 nm and emission maximum 476 nm), blue FP

(BFP; absorption maximum 384 nm and emission maximum 448 nm), and yellow FP (YFP; absorption maximum 512 nm and emission maximum 529 nm). The spectral maxima given here should be regarded as approximations, because slightly different versions of these three classes of mutants have been produced via different combinations of amino acid substitutions, and these differ in the positions of their absorption and emission maxima (Tsien, 1998).

A number of desirable physiochemical and biological changes to the properties of FPs have also been achieved through alterations to the coding sequence of GFP. These include improving the translatability of the messenger RNA (mRNA) through codon optimization, eliminating cryptic introns (Haseloff et al., 1997), increasing the rate of folding of the translated products, decreasing misfolding and aggregation, and modifying observed sensitivities to inorganic anions and to changes in pH (Tsien, 1998). All of these modifications should be considered within the context of the organism and cell type of interest, as well as of the subcellular location within which the FP is to be expressed. Thus, for example, targeting to certain locations may involve encountering regions of particularly low pH levels (e.g., the plant cell surface and the lysosomal or vacuolar lumen), which can decrease FP fluorescence.

Many modifications work reasonably well across species and even kingdoms. For example, GFP codon usage optimized for mammalian expression is also compatible with expression in plants and serves to eliminate a cryptic intron, which would otherwise be spliced (Haseloff et al., 1997).

Further, modifications to the FP primary structure include topogenic alterations consequent to addition of targeting sequences. FPs can also be adapted to act as specific sensors of cellular physiology. These modifications, which do not involve changes to the core primary structure of the FP, are described in Section I.E.

C. Nonhydrozoan Fluorescent Proteins

A large number of additional sources of FPs have been found within the *Cnidaria*, particularly the anthozoa (Matz et al., 1999), and a comprehensive phylogeny has been constructed (Lubus et al., 2002; Shagin et al., 2004) (see also Section I.F). Some have GFP-like spectra, for example, that of *Renilla reniformis*. However, a particular advantage offered by the FPs of other anthozoans relates to the fact that none of the spectral variants of *Aequorea* GFP have emission maxima longer than about 529 nm (Miyawaki et al., 2003). In contrast, a number of anthozoan FPs have longer emission maxima, including the commercially available DsRed, AsRed, HcRed, and Kindling Red (see Section I.G). Fluorescence emission in the red offers some advantages with animal cells, because they generally lack autofluorescence at this wavelength. The presence of chlorophyll, which produces intense red fluorescence, reduces the utility of these FPs in photosynthetic organisms.

D. Photokinetic and Photodynamic Fluorescent Proteins

Other intriguing features of FPs include the observation of changes in excitation and emission spectra over time (Labas *et al.*, 2002; Terskikh *et al.*, 2000) and after actinic irradiation (Ando *et al.*, 2002; Chudakov *et al.*, 2003a,b; Zhang *et al.*, 2003). This behavior is particularly useful for defining promoter activities and charting subcellular dynamics of labeled proteins, because most FPs exhibit long half-lives and are, therefore, unsuited for measurement of rapid changes in protein levels or locations.

E. Modifying Fluorescent Proteins to Access Specific Cellular Locations and Specific Physiological Processes, and to Act as Specific Ligands

FPs in general readily accommodate additions to their N- and C-terminals, and linker peptide sequences are not required. The close physical proximity of the N- and C-terminus permits circular permutation of GFP (Baird *et al.*, 1999), allowing addition of sequences at equatorial locations relative to the barrel structure. A large variety of topogenic motifs have been successfully employed with FPs, with most work being done with GFP, and these allow FP targeting to most, perhaps all, subcellular compartments of eukaryotes (Lippincott-Schwartz and Patterson, 2003; Tsien, 1998). Measurement of dynamic changes in FP levels and intracellular locations is facilitated by use of photokinetic and photodynamic FPs. FP stability can also be modified by fusion to proteins of short half-life or to peptide sequences directing proteolytic degradation (Li *et al.*, 1998). A potential complication to the use of FPs is that most naturally exist as oligomers (Zhang *et al.*, 2003). Mutations have been identified that reduce the extent of intermolecular interactions, thereby decreasing or eliminating oligomer formation (Campbell *et al.*, 2002; Zhang *et al.*, 2003).

FP pairs having overlapping emission and excitation spectra are amenable to fluorescence resonance energy transfer (FRET). This has formed the basis for the family of calcium sensors termed *cameleons* pioneered by the group of Roger Tsien in San Diego (Miyawaki *et al.*, 1997, 1999) and by others (Romoser *et al.*, 1997). Cameleons couple calcium-specific modulation of calmodulin interactions to alterations in the spatial relationship of a pair of FPs synthesized within the context of a single polypeptide chain. This results in ratiometric alterations in the emission spectra of the composite molecule as a function of calcium concentration. Cameleons have been successfully employed to monitor calcium transients in nonmammalian systems, including plants (Allen *et al.*, 1999, 2000, 2001). An alternative to FRET-based measurements of calcium ions involves insertion of calmodulin in an equatorial position into the backbone of YFP, resulting in a nonratiometric calcium indicator (camgaroo-1), which increases eightfold in intensity on saturation with calcium (Baird *et al.*, 1999). An improved version of this reporter has decreased pH sensitivity and chloride interference and increased photostability and levels of productive expression at 37 °C (Griesbeck *et al.*, 2001). Camgaroos have been successfully employed in nonmammalian systems (Yu *et al.*, 2003a).

The intramolecular FRET strategy inherent to cameleons emerged from one that was initially devised for sensing protease action (Heim and Tsien, 1996) and that now has been extended for measurement of the concentrations and activities of many signaling molecules, including cyclic guanosine monophosphate (cGMP), Ras, Rap1, Ran, and protein kinases and phosphatases, as well as for measurement of membrane potential. Attaching topogenic motifs to these reporter constructions further permits localized determination of changes in the concentrations of the sensed molecules and ions. Intermolecular FRET measurements using FP pairs have also been employed for measurement of cyclic adenosine monophosphate (cAMP) levels and for detection of protein-protein interactions. For a comprehensive review of intermolecular and intramolecular FP FRET, see Zhang *et al.* (2003). Reports are also emerging of the measurement of protein-protein interactions using reconstitution of complementary fragments of GFP and YFP (Ghosh *et al.*, 2000; Hu *et al.*, 2002).

Developments of particular relevance to flow cytometry is the modification of GFP to attach specific binding sites, thereby permitting its direct use as a fluorescent ligand. This has been done through fusing of single-chain Fv proteins (Hink *et al.*, 2000) or through direct attachment of loops derived from the complementarity determining region 3 of the antibody heavy chain, followed by selection using phage display (Zeytun *et al.*, 2003).

F. Evolutionary Aspects of Fluorescent Proteins

With the identification of increasing numbers of GFP-like proteins within a variety of species, comparative sequence analyses can be used to examine evolutionary relationships and their ecological relevance (Labas *et al.*, 2002; Shagin *et al.*, 2004). Analyses of this type using all available GFP-like proteins, including the nonfluorescent chromoproteins of nonbioluminescent anthozoans, GFP-like proteins from anthozoan medusae and planktonic *Copepoda*, and colorless extracellular proteins of Bilaterians containing GFP-like protein-binding domains (Shagin *et al.*, 2004), suggest that GFP-like proteins comprise a large superfamily whose molecular function primarily revolves around pigmentation. The biological relevance of pigmentation and its evolutionary significance remain unclear. The superfamily can be divided into two lineages: one containing all FPs and the other leading to the colorless extracellular proteins. The phylogenetic analysis is consistent with evolution of FPs before the separation of *Cnidaria* and *Bilatera*. Within the *Cnidaria*, an extensive series of gene duplications then led to the appearance of a large and diverse set of FP paralogs, which underwent extensive evolutionary change during separation of Anthozoan and Hydrozoan subclasses. This gave rise, from the ancestral GFP, to the cyans, reds, and yellows, with at least 15 color diversification events being recognized (Shagin *et al.*, 2004). It is relevant for protein engineering to consider that different organisms have evolved GFP-like proteins of the same color, but that are based on different structural principles, because these different structures might provide suitable starting points for production of FPs having desired biotechnological characteristics (Shagin *et al.*, 2004).

G. Commercial Availability of Fluorescent Proteins

Three companies are sources of most of the anthozoan and hydrozoan FPs. Table I provides the optical properties of these FPs.

Table I
Optical Properties of Commercially Available Fluorescent Proteins

Name	Excitation maximum (nm)	Emission maximum (nm)	Quantum yield	Extinction coefficient (M^{-1} cm^{-1})	Comments	Reference
CopGFP	482	502	0.60	70,000		www.evrogen.com
PhyYFP	525	537	0.40	130,000		
Kindling	580	600	0.07	59,000	Activated by green illumination	
HcRed tandem	590	637	0.04	160,000		
EBFP	383	445	0.25	31,000		Patterson *et al.* (2001)
ECFP	434	477	0.40	26,000		
EGFP	489	508	0.60	55,000		
EYFP	514	527	0.61	84,000		
dsRed	558	583	0.68	72,500		
AmCyan1	458	489	0.75	39,000		www.bdbiosciences.com/clontech/products/literature/pdf/brochures/LivingColors.pdf; Clontech technical services
ZsGreen1	493	505	0.91	43,000		
ZsYellow1	529	539	0.65	20,000		
DsRed-Express	557	579	0.90	19,000		
DsRed2	563	582	NA	NA		
AsRed2	576	592	0.21	61,000		
HcRed1	588	618	0.03	20,000		
ECFP	439	476	0.40	26,000		
EGFP	484	510	0.60	55,000		
EYFP	512	529	0.60	84,000		
hrGFP	498	509	0.8	270,000		Ward and Cormier (1979)

1. *Becton Dickinson/Clontech* (www.clontech.com): A large number of vectors encoding various FPs are available from this company (www.bdbiosciences.com/clontech/gfp/index.shtml for details). These include *Aequorea* GFP and the *Aequorea* BFP, CFP, and YFP variants, and a number of anthozoan FPs. DsRed1 and DsRed2 encode the red fluorescent protein (RFP) from the Indo-Pacific sea anemone relative, *Discosoma* species. A variant of this, pTimer, exhibits the previously described time-dependent changes in fluorescent emission from green to red (Terskikh *et al.*, 2000). AmCyan encodes a variant of the wild-type *Anemonia majano* (Aiptasia anemone) CFP, AsRed encodes a variant of the *Anemonia sulcata* RFP, ZsGreen encodes a variant of the wild-type *Zoanthus* species GFP, ZsYellow encodes a variant of the wild-type *Zoanthus* species YFP. Finally, HcRed was generated by mutagenesis of a nonfluorescent chromoprotein from *Heteractis crispa*. All variants are available in codon-optimized form for expression in mammalian systems, and a variety of different constructions are available for purposes such as targeting the proteins to different subcellular locations, providing bifunctional reporters, retroviral expression, and so on. The company is also a source of various anti-FP monoclonal and polyclonal antibodies.

2. *Stratagene* (www.stratagene.com): Stratagene provides plasmids based around the coding sequence of the *R. reniformis* GFP. It has been suggested that expression of Renilla GFP in mammalian cells may be less toxic than that of *Aequorea* GFP (www.stratagene.com; however, see Kirsch *et al.*, 2003).

3. *Evrogen* (www.evrogen.com): Evrogen provides plasmids encoding HcRed and a variety of other novel cnidarian FPs. Cop-green is a GFP-like FP from the copepod *Pntelina plumata* (phylum Arthropoda, subphylum Crustacea, class Maxillopoda, subclass Copepoda, order Calanoida, family Pontellidae). Phi yellow is a GFP-like FP from the hydromedusa *Phialidium* species (class Hydrozoa, order Hydroida, suborder Leptomeduzae, family Campanulariidae). Kindling Red is derived from a mutant of the chromoprotein isolated from *A. sulcata* (Chudakov *et al.*, 2003a,b; Lukyanov *et al.*, 2000). Ase green is a fluorescent mutant of a naturally nonfluorescent GFP-like protein from *A. coerulescens* (Gurskaya *et al.*, 2003).

III. Flow Analysis Using Fluorescent Proteins

In this chapter, the discussion is only of expression of FPs within multicellular eukaryotes. Reviews have been published by Galbraith *et al.* (1998, 1999) and Hawley *et al.* (2001a).

A. Flow Analysis of Fluorescent Protein-Expressing Cells

A prerequisite to flow analysis is the ability to express sufficient quantities of FPs within the cell type of interest for the cells to be detectable using standard flow instrumentation. Various methods exist for transgenic expression in eukaryotic organisms. These can be roughly divided according to whether they confer transient or permanent expression, and further specification of the latter includes whether this involves homologous gene replacement. Expression of FPs evidently should ideally not perturb normal cellular functions or those functions and pathways under specific investigation. Reports concerning the effects of FP expression are contradictory; on the one hand, many of these describe the production of viable transgenic organisms expressing FP levels that are readily detected even by the eye, some of which have penetrated the nonscientific marketplace (www.ekac.org/gfpbunny.html; www.glofish.com/). Therefore, FP expression clearly is not grossly deleterious to living organisms (and obviously not within their source organisms). On the other hand, phenotypic effects of FP expression are sometimes noted (Gong *et al.*, 2003; Kirsch *et al.*, 2003; Liu *et al.*, 1999; Torbett, 2002). The availability of genome-wide methods for analysis of gene expression, such as microarrays, should permit detection of subtle alterations in gene expression that are FP specific. It is recommended that any studies involving microarrays include adequate replication to allow determination of statistical significance. Regarding potential variables, excessive overexpression of any protein (FP or otherwise, and especially one that has the potential to oligomerize or precipitate) can be toxic to the organism of interest, and the transgenic effects of FPs may also be a function of the ambient level of illumination of the transgenic organisms, a parameter not routinely (at least for animals) reported.

Shortly following the first reports of the expression of GFP in eukaryotic cells, methods were devised and described for flow cytometric detection of GFP in plants and animals (Anderson *et al.*, 1996; Galbraith *et al.*, 1995; Lybarger *et al.*, 1996; Ropp *et al.*, 1996; Sheen *et al.*, 1995). As with all flow cytometric procedures, single-cell suspensions must be employed. In the latter case, this involves production of wall-less cells (protoplasts) via enzymatic digestion (Galbraith *et al.*, 1999).

Most flow cytometers are equipped with single lasers producing excitation light at 488 nm, which is sufficient for detecting expression of wild-type GFP within transfected protoplasts (Galbraith *et al.*, 1995; Sheen *et al.*, 1995). Introducing mutations into the GFP coding sequence to eliminate either of its two peaks of absorbance allows selective fluorescence excitation at 409 nm (GFP containing S202F, T203I, and V163A replacements within its coding sequence) or at 488 nm (GFP containing S65T and V163A replacements) (Anderson *et al.*, 1998). The former is particularly suited for illumination using the 407-nm line produced by the krypton-ion laser or by the newer low-power violet laser diodes. Lybarger *et al.* (1996) described conditions for flow analysis of transfected 3T3 NIH cells expressing GFP, using laser excitation at 488 nm. Ropp *et al.* (1996) reported analysis of GFP- and YFP-expressing 293 cells using excitation at 407 nm. Beavis and Kalejta (1999) extended this work through evaluation of different single-laser wavelengths for discrimination of mammalian cells expressing GFP, YFP, and CFP. They found that excitation at 458 nm was suitable for any combination of these FPs, with electronic compensation adequately eliminating spectral overlap. Excitation of YFP at either 407 or 457 nm is certainly nonoptimal, based on the absorbance spectrum, and thus may be appropriate only for cells expressing high levels of this protein. Zhu *et al.* (1999) described the use of two-line excitation from an argon-ion laser (360 and 488 nm) for two-color (BFP/YFP) and three-color (BFP/GFP/YFP) flow analysis of transfected mammalian cells.

A disadvantage of excitation at 457 nm is that it requires an expensive tunable argon-ion laser. In contrast, argon-ion lasers producing single-wavelength (488-nm) light are relatively cheap. Stull *et al.* (2000) illustrated use of 488-nm illumination for the combined immunophenotypic discrimination of transfected human CD34[+] mononuclear cells expressing GFP and YFP. Hawley *et al.* (2001b) have described conditions for the analysis of three FPs using 488-nm excitation. In this work, discrimination was possible between fluorescent signals of DsRed, YFP, and GFP produced by stably transformed mammalian cell lines. Multilaser analysis of FP expression will be facilitated by the availability of solid-state lasers, which provide cost-conscious excitation at appropriate wavelengths. Telford *et al.* (2003) have demonstrated the suitability of a violet laser diode for flow analysis of cells expressing CFP. Table II provides a summary of this information.

Given the increasing availability of routine methods for flow analysis of FP expression of animal cells, numerous reports are now emerging concerning their use in specialized assays. Selected examples cover all aspects of biology, for example, including analysis of steroid-regulated gene expression (Necela and Cidlowski, 2003), of B-cell differentiation (Guglielmi *et al.*, 2003), cancer metastasis (Zhang *et al.*, 2003), RNA interference (Mousses *et al.*, 2003; Nagy *et al.*, 2003), mechanisms of viral replication (Voronin and Pathak, 2003), parental imprinting (Preis *et al.*, 2003), manipulation of Cre-Lox recombination (Van den Plas *et al.*, 2003),

Table II

Laser Wavelengths Employed for Analysis of Fluorescent Proteins Within Mammalian Cells

Laser-excitation wavelengths (nm)	Fluorescent proteins analyzed	Comments	Reference
488, 407	GFP		Anderson *et al.* (1996)
488, 407	GFP, YFP		Ropp *et al.* (1996)
488	GFP		Lybarger *et al.* (1996)
360, 488	BFP, GFP		Yang *et al.* (1998)
488	GFP, YFP		Lybarger *et al.* (1998)
360, 488	BFP, GFP, YFP		Zhu *et al.* (1999)
458, 488, 514	GFP, YFP, CFP		Beavis and Kalejta (1999)
488	GFP, YFP		Stull *et al.* (2000)
458, 488, 568	YFP, GFP, CFP, DsRed		Hawley *et al.* (2001b)
408	CFP	Violet laser diode	Telford *et al.* (2003)
413, 514	CFP, YFP	FRET	Chan *et al.* (2001)
458	CFP, YFP	FRET	He *et al.* (2003)

and measurement of the extent of syncytium formation of placental trophoblast cells (Kudo *et al.*, 2003).

Progress in higher plants has included the quantitative analysis of abscisic acid-inducible gene expression (Hagenbeek and Rock, 2001), extension of multi-parametric and clustering analyses (Bohanec *et al.*, 2002), and the global characterization of cell-type-specific gene expression (Birnbaum *et al.*, 2003). Beyond mammals and plants, FP expression in other cell types and its detection by flow cytometry has been reported for fish (Kobayashi *et al.*, 2004; Molina *et al.*, 2002; Takeuchi *et al.*, 2002) and *Drosophila* (Banks *et al.*, 2003).

For certain routine flow cytometric procedures (e.g., analysis of the cell cycle in mammalian systems), the cells are fixed before analysis. This creates problems for concurrent FP analysis because it has been reported that the fluorescence of most FPs is abolished by precipitative fixatives containing organic solvents (Tsien, 1998; Ward *et al.*, 1980). Kalejta *et al.* (1997, 1999) have described the use of membrane-associated GFP fusions for simultaneous analysis of transfected cells and the cell cycle in ethanol-fixed cells. Chu *et al.* (1999) described the use of formaldehyde fixation before ethanol fixation as a means of retaining cytoplasmic GFP fluorescence within transfected leukemic cells. It is evidently difficult to reconcile these reports with those concerning the structural instability of GFP in the presence of ethanol, although some FPs may be resistant to its denaturing effects (Yu *et al.*, 2003b).

Overall, flow cytometric analysis of FP-expressing cells is clearly now largely routine. As such, the general rules for detection of cell-associated fluorescence apply: Laser-excitation wavelengths should be matched to the excitation spectra of the FPs of interest, dichroic and barrier filters should be selected (custom-made if necessary) to optimally discriminate between the different emission FP spectra, and consideration should be given to the relative proportions of the different fluorescent proteins that are likely to be produced and their comparative brightness, to roughly balance the signals being produced and analyzed during flow cytometry.

B. Flow Analysis of FRET

Chan *et al.* (2001) established the feasibility of employing flow cytometric FRET analysis to characterize interactions between tumor necrosis factor (TNF) family members and between receptor and downstream signaling proteins. This involved fusing the interacting proteins to CFP and YFP and employing two-laser illumination at 413 and 514 nm. He *et al.* (2003) have extended this work by developing a flow cytometric method to detect FRET between CFP and YFP using the 458-nm excitation from a single tunable argon-ion laser.

C. Flow Analysis of Fluorescent Protein–Accumulating Organelles

FP-based flow analysis can also be applied to subcellular organelles, although in general reports of flow cytometric measurements of organelles are uncommon relative to those of cells. We have employed nuclear targeting of FPs to highlight plant nuclei within protoplasts and transgenic plants of tobacco (Grebenok *et al.*, 1997a,b) and have successfully used flow cytometry for the analysis of GFP accumulation within these nuclei in subcellular homogenates. In this work, we targeted the nucleoplasm using a nuclear localization signal, which required increasing the size of the GFP-fusion protein beyond the exclusion limit of the nuclear pores by further fusing it to the *β*-glucuronidase coding sequence. When applied to *Arabidopsis thaliana*, which has a nuclear genome approximately 1/13 times that of tobacco, a technical difficulty is that this composite molecule leaks from the nuclei in homogenates and this impedes flow cytometric analysis. FP fusions to proteins that are structural components of the nucleus avoid this problem (Zhang and Galbraith, 2003, unpublished observation).

Sirk *et al.* (2003) have described a novel assay for measuring the import and turnover of nuclear-encoded mitochondrial proteins in living mammalian cells. This involved production of transgenic cells within which the expression of a mitochondrially targeted form of GFP was placed under the regulation of an inducible promoter. This GFP fusion has the additional feature of being nonfluorescent before import into the mitochondria because of the presence of a signal peptide. This means that flow analysis of the extent of cellular GFP fluorescence emission, with PI gating to eliminate the contribution of dead cells, provides a means of quantifying mitochondrial accumulation of this protein within living cells.

IV. Flow Sorting Using Fluorescent Proteins

Given the ability to successfully discriminate FP-labeled cells or subcellular organelles using flow cytometry, we can then use flow sorting to selectively purify these components for further study. Sorting of mammalian cells according to the presence of detectable fluorescent signals is evidently routine, and signals from FP expression create no exceptions. Previous reviews include Galbraith *et al.* (1998, 1999) and more generally Boeck (2001).

Examples of applications of GFP sorting in mammalian systems can be found in Van Tendeloo *et al.* (2000), who described the use of GFP expression as an alternative to coselection methods for production of stably transfected hematopoietic cell lines. De Angelis *et al.* (2003) employed flow sorting of GFP-labeled cells for isolation of myogenic progenitor cells. Similar approaches have been used for isolation of neural precursors (Aubert *et al.*, 2003) and pluripotential progenitor cells (Nunes *et al.*, 2003), hepatic progenitor cells (Fujikawa *et al.*, 2003), hematopoietic stem cells (Ma *et al.*, 2002), and primary aortic endothelial cells (Magid *et al.*, 2003). Zharkikh *et al.* (2002) have reported the use of flow sorting of GFP-labeled cells for collection of different kidney cell types, and Hirai *et al.* (2003) for purification of hemogenic and nonhemogenic endothelial cells. Other applications include the generation of cell lines containing inducible promoter systems (Lai *et al.*, 2003).

In higher plants, Birnbaum *et al.* (2003) have described the use of flow cytometry and sorting for the global characterization of cell-type-specific gene expression in Arabidopsis roots. This involved use of transgenic plants in which cell-type-specific expression of GFP within the root was under the control of specific promoter or enhancer sequences. Protoplasts prepared from the different transgenic plants were sorted according to GFP fluorescence. RNA was then prepared from the sorted protoplasts and employed for hybridization to Affymetrix whole-genome microarrays. Expression data were also obtained from root segments corresponding to different distances from the root tip. Combining the results allowed creation of a road map of global gene expression of the entire root. Within this road map, it was possible to assign much of the changes in gene expression to eight characteristic local expression domains. Identification of the presence of specific genes within these domains allows hypothesis generation with respect to root development, specifically concerning the involvement of phytohormones and the roles of transcription factor networks.

Reports of flow sorting based on FP expression in other organisms have appeared. Takeuchi *et al.* (2002) and Kobayashi *et al.* (2004) have described the development of methods for isolation of primordial germ cells from rainbow trout. Finally, Makridou *et al.* (2003) described the use of sorting based on GFP and RFP expression for the production of stably transformed cell lines in *Drosophila*.

V. Conclusion

Is there a pot of gold at the end of the rainbow of fluorescent colors? Most certainly for the investigator; as scientists, we have only just begun to employ FP expression as a means of defining and exploring the properties of living cells and their subcellular components. The tools of flow cytometry and fluorescence-activated sorting provide unique ways to investigate the contributions of individuals within populations. Further work is needed to evaluate the greatly expanded numbers of FPs for their suitability in transient and transgenic expression. Continued expansion of the numbers of different FPs is, of course, anticipated, and these in turn will require evaluation. The emergence of novel flow-based assays for different cellular functions and components will continue, and improved sensitivity of end-point measurements should permit lowering the total numbers of cells or subcellular organelles required to be purified by flow sorting. Finally, further integration of these techniques is anticipated with established and emerging high-throughput methods, including genomics, proteomics, and metabolomics.

Acknowledgments

Some of the work described in this chapter has been supported by grants to D. W. G. from the National Science Foundation, primarily from the Plant Genome Program, and from the U.S. Department of Agriculture.

References

Allen, G. J., Kwak, J. M., Chu, S. P., Llopis, J., Tsien, R. Y., Harper, J. F., and Schroeder, J. I. (1999). Cameleon calcium indicator reports cytoplasmic calcium dynamics in Arabidopsis guard cells. *Plant J.* **19**, 735–747.

Allen, G. J., Chu, S. P., Schumacher, K., Shimazaki, C. T., Vafeados, D., Kemper, A., Hawke, S. D., Tallman, G., Tsien, R. Y., Harper, J. F., Chory, J., and Schroeder, J. I. (2000). Alteration of stimulus-specific guard cell calcium oscillations and stomatal closing in *Arabidopsis det3* mutant. *Science* **289**, 2338–2342.

Allen, G. J., Chu, S. P., Harrington, C. L., Schumacher, K., Hoffman, T., Tang, Y. Y., Grill, E., and Schroeder, J. I. (2001). A defined range of guard cell calcium oscillation parameters encodes stomatal movements. *Nature* **411**, 1053–1057.

Anderson, M. T., Tjioe, I. M., Lorincz, M. C., Parks, D. R., Herzenberg, L. A., Nolan, G. P., and Herzenberg, L. E. (1996). Simultaneous fluorescence-activated cell sorter analysis of two distinct transcriptional elements within a single cell using engineered green fluorescent proteins. *Proc. Natl. Acad. Sci. USA* **93**, 8508–8511.

Anderson, M. T., Baumgarth, N., Haugland, R. P., Gerstein, R. M., Tjioe, T., Herzenberg, L. A., and Herzenberg, L. A. (1998). Pairs of violet-light-excited fluorochromes for flow cytometric analysis. *Cytometry* **33**, 435–444.

Ando, R., Hama, H., Yamamoto-Hino, M., Mizuno, H., and Miyawaki, A. (2002). An optical marker based on the UV-induced green-to-red photoconversion of a fluorescent protein. *Proc. Natl. Acad. Sci. USA* **99**, 12651–12656.

Aubert, J., Stavridis, M. P., Tweedie, S., O'Reilly, M., Vierlinger, K., Li, M., Ghazal, P., Pratt, T., Mason, J. O., Roy, D., and Smith, A. (2003). Screening for mammalian neural genes via fluorescence-activated cell sorter purification of neural precursors from Sox1-gfp knock-in mice. *Proc. Natl. Acad. Sci. USA* **100**(Suppl. 1), 11836–11841.

Baird, G. S., Zacharias, D. A., and Tsien, R. Y. (1999). Circular permutation and receptor insertion within green fluorescent proteins. *Proc. Natl. Acad. Sci. USA* **96**, 11241–11246.

Banks, D. J., Hua, G., and Adang, M. (2003). Cloning of a *Heliothis virescens* 110 kDa aminopeptidase N and expression in *Drosophila* S2 cells. *Insect Biochem. Mol. Biol.* **33**, 499–508.

Beavis, A. J., and Kalejta, R. F. (1999). Simultaneous analysis of the cyan, yellow and green fluorescent proteins by flow cytometry using single-laser excitation at 458 nm. *Cytometry* **37**, 68–73.

Betzig, E., Patterson, G. H., Sougrat, R., Lindwasser, O. W., Olenych, S., Bonifacino, J. S., Davidson, M. W., Lippincott-Schwartz, J., and Hess, H. F. (2006). Imaging intracellular fluorescent proteins at nanometer resolution. *Science* **313**, 1642–1645.

Birnbaum, K., Shasha, D. E., Wang, J. Y., Jung, J. W., Lambert, G. M., Galbraith, D. W., and Benfey, P. N. (2003). A gene expression map of the *Arabidopsis* root. *Science* **302**, 1956–1960.

Boeck, G. (2001). Current status of flow cytometry in cell and molecular biology. *Int. Rev. Cytol.* **204**, 239–298.

Bohanec, B., Luthar, Z., and Rudolf, K. (2002). A protocol for quantitative analysis of green fluorescent protein-transformed plants, using multiparameter flow cytometry with cluster analysis. *Acta Biol. Crac. Ser. Bot.* **44**, 145–153.

Campbell, R. E., Tour, O., Palmer, A. E., Steinbach, P. A., Baird, G. S., Zacharias, D. A., and Tsien, R. Y. (2002). A monomeric red fluorescent protein. *Proc. Natl. Acad. Sci. USA* **99**, 7877–7882.

Chalfie, M., Tu, Y., Euskirchen, G., Ward, W. W., and Prasher, D. C. (1994). Green-fluorescent protein as a marker for gene expression. *Science* **263**, 802–805.

Chan, F. K., Siegel, R. M., Zacharias, D., Swofford, R., Holmes, K. L., Tsien, R. Y., and Lenardo, M. J. (2001). Fluorescence resonance energy transfer analysis of cell surface receptor interactions and signaling using spectral variants of the green fluorescent protein. *Cytometry* **44**, 361–368.

Chu, Y. W., Wang, R., Schmid, I., and Sakamoto, K. M. (1999). Analysis with flow cytometry of green fluorescent protein expression in leukemic cells. *Cytometry* **36**, 333–339.

Chudakov, D. M., Feofanov, A. V., Mudrik, N. N., Lukyanov, S., and Lukyanov, K. A. (2003a). Chromophore environment provides clue to "kindling fluorescent protein" riddle. *J. Biol. Chem.* **278**, 7215–7219.

Chudakov, D. M., Belousov, V. V., Zaraisky, A. G., Novoselov, V. V., Staroverov, D. B., Zorov, D. B., Lukyanov, S., and Lukyanov, K. A. (2003b). Kindling fluorescent proteins—A novel tool for precise *in vivo* photolabeling. *Nat. Biotech.* **21,** 191–194.

De Angelis, M. G. C., Balconi, G., Bernasconi, S., Zanetta, L., Boratto, R., Galli, D., Dejana, E., and Cossu, G. (2003). Skeletal myogenic progenitors in the endothelium of lung and yolk sac. *Exp. Cell Res.* **290,** 207–216.

Fujikawa, T., Hirose, T., Fujii, H., Oe, S., Yasuchika, K., Azuma, H., and Yamaoka, Y. (2003). Purification of adult hepatic progenitor cells using green fluorescent protein (GFP)-transgenic mice and fluorescence-activated cell sorting. *J. Hepatol.* **39,** 162–170.

Galbraith, D. W., Grebenok, R. J., Lambert, G. M., and Sheen, J. (1995). Flow cytometric analysis of transgene expression in higher plants: Green fluorescent protein. *Methods Cell Biol.* **50,** 3–12.

Galbraith, D. W., Anderson, M. T., and Herzenberg, L. A. (1998). Flow cytometric analysis and sorting of cells based on GFP accumulation. *Methods Cell Biol.* **58,** 315–341.

Galbraith, D. W., Herzenberg, L. A., and Anderson, M. (1999). Flow cytometric analysis of transgene expression in higher plants: Green fluorescent protein. *Methods Enzymol.* **320,** 296–315.

Ghosh, I., Hamilton, A. D., and Regan, L. (2000). Antiparallel leucine zipper-directed protein reassembly: Application to the green fluorescent protein. *J. Am. Chem. Soc.* **122,** 5658–5659.

Gong, Z., Wan, H., Tay, T. L., Wang, H., Chen, M., and Tan, Y. (2003). Development of transgenic fish for ornamental and bioreactor by strong expression of fluorescent proteins in the skeletal muscle. *Biochem. Biophys. Res. Commun.* **308,** 58–63.

Grebenok, R. J., Pierson, E. A., Lambert, G. M., Gong, F. C., Afonso, C. L., Haldeman-Cahill, R., Carrington, J. C., and Galbraith, D. W. (1997a). Green-fluorescent protein fusions for efficient characterization of nuclear localization signals. *Plant J.* **11,** 573–586.

Grebenok, R. J., Lambert, G. M., and Galbraith, D. W. (1997b). Characterization of the targeted nuclear accumulation of GFP within the cells of transgenic plants. *Plant J.* **12,** 685–696.

Griesbeck, O., Baird, G. S., Campbell, R. E., Zacharias, D. A., and Tsien, R. Y. (2001). Reducing the environmental sensitivity of yellow fluorescent protein—Mechanism and applications. *J. Biol. Chem.* **276,** 29188–29194.

Guglielmi, L., Le Bert, M., Comte, I., Dessain, M. L., Drouet, M., Ayer-Le Lievre, C., Cogne, M., and Denizot, Y. (2003). Combination of 3′ and 5′ IgH regulatory elements mimics the B-specific endogenous expression pattern of IgH genes from pro-B cells to mature B cells in a transgenic mouse model. *Biochim. Biophys. Acta Mol. Cell Res.* **1642,** 181–190.

Gurskaya, N. G., Fradkov, A. F., Pounkova, N. I., Staroverov, D. B., Bulina, M. E., Yanushevich, Y. G., Labas, Y. A., Lukyanov, S., and Lukyanov, K. A. (2003). A colorless GFP homologue from the non-fluorescent hydromedusa *Aequorea coerulescens* and its fluorescent mutants. *Biochem. J.* **373,** 403–408.

Hagenbeek, D., and Rock, C. D. (2001). Quantitative analysis by flow cytometry of abscisic acid-inducible gene expression in transiently transformed rice protoplasts. *Cytometry* **45,** 170–179.

Haseloff, J., Siemering, K. R., Prasher, D. C., and Hodge, S. (1997). Removal of a cryptic intron and subcellular localization of green fluorescent protein are required to mark transgenic *Arabidopsis* plants brightly. *Proc. Natl. Acad. Sci. USA* **94,** 2122–2127.

Hawley, T. S., Telford, W. G., and Hawley, R. G. (2001a). "Rainbow" reporters for multispectral marking and lineage analysis of hematopoietic stem cells. *Stem Cells* **19,** 18–124.

Hawley, T. S., Telford, W. G., Ramezani, A., and Hawley, R. G. (2001b). Four-color flow cytometric detection of retrovirally expressed red, yellow, green, and cyan fluorescent proteins. *Biotechniques* **30,** 1028–1034.

He, L. S., Bradrick, T. D., Karpova, T. S., Wu, X. L., Fox, M. H., Fischer, R., McNally, J. G., Knutson, J. R., Grammer, A. C., and Lipsky, P. E. (2003). Flow cytometric measurement of fluorescence (Förster) resonance energy transfer from cyan fluorescent protein to yellow fluorescent protein using single-laser excitation at 458 nm. *Cytometry* **53A,** 39–54.

Heim, R., and Tsien, R. Y. (1996). Engineering green fluorescent protein for improved brightness, longer wavelengths and fluorescence resonance energy transfer. *Curr. Biol.* **6,** 178–182.

Heim, R., Prasher, D. C., and Tsien, R. Y. (1994). Wavelength mutations and posttranslational autoxidation of Green Fluorescent Protein. *Proc. Natl. Acad. Sci. USA* **91**, 12501–12504.

Hell, S. W. (2009). Microscopy and its focal switch. *Nat. Methods* **6**, 24–32.

Hink, M. A., Griep, R. A., Borst, J. W., van Hoek, A., Eppink, M. H. M., Schots, A., and Visser, A. J. W. G. (2000). Structural dynamics of green fluorescent protein alone and fused with a single chain Fv protein. *J. Biol. Chem.* **275**, 17556–17560.

Hirai, H., Ogawa, M., Suzuki, N., Yamamoto, M., Breier, G., Mazda, O., Imanishi, J., and Nishikawa, S. (2003). Hemogenic and nonhemogenic endothelium can be distinguished by the activity of fetal liver kinase (Flk)-1 promoter/enhancer during mouse embryogenesis. *Blood* **101**, 886–893.

Hu, C. D., Chinenov, Y., and Kerppola, T. K. (2002). Visualization of interactions among bZIP and Rel family proteins in living using bimolecular fluorescence complementation. *Mol. Cells* **9**, 789–798.

Johnson, F. H., Gershman, L. C., Waters, J. R., Reynolds, G. T., Saiga, Y., and Shimomura, O. (1962). Quantum efficiency of cypridina luminescence, with a note on that of *Aequorea*. *J. Cell Comp. Physiol.* **60**, 85–104.

Kalejta, R. F., Shenk, T., and Beavis, A. J. (1997). Use of a membrane-localized green fluorescent protein allows simultaneous identification of transfected cells and cell cycle analysis by flow cytometry. *Cytometry* **29**, 286–291.

Kalejta, R. F., Brideau, A. D., Banfield, B. W., and Beavis, A. J. (1999). An integral membrane green fluorescent protein marker, Us9-GFP, is quantitatively retained in cells during propidium iodide-based cell cycle analysis by flow cytometry. *Exp. Cell Res.* **248**, 322–328.

Kirsch, P., Hafner, M., Zentgraf, H., and Schilling, L. (2003). Time course of fluorescence intensity and protein expression in HeLa cells stably transfected with hrGFP. *Mol. Cells* **15**, 341–348.

Kobayashi, T., Yoshizaki, G., Takeuchi, Y., and Takeuchi, T. (2004). Isolation of highly pure and viable primordial germ cells from rainbow trout by GFP-dependent flow cytometry. *Mol. Reprod. Dev.* **67**, 91–100.

Kremers, G. J., Hazelwood, K. L., Murphy, C. S., Davidson, M. W., and Piston, D. W. (2009). Photoconversion in orange and red fluorescent proteins. *Nat. Methods* **6**, 355–358.

Kudo, Y., Boyd, C. A. R., Kimura, H., Cook, P. R., Redman, C. W. G., and Sargent, I. L. (2003). Quantifying the syncytialisation of human placental trophoblast BeWo cells grown *in vitro*. *Biochim. Biophys. Acta Mol. Cell Res.* **1640**, 25–31.

Labas, Y. A., Gurskaya, N. G., Yanushevich, Y. G., Fradkov, A. F., Lukyanov, K. A., Lukyanov, S. A., and Matz, M. V. (2002). Diversity and evolution of the green fluorescent protein family. *Proc. Natl. Acad. Sci. USA* **99**, 4256–4261.

Lai, J. F., Juang, S. H., Hung, Y. M., Cheng, H. Y., Cheng, T. L., Mostov, K. E., and Jou, T. S. (2003). An ecdysone and tetracycline dual regulatory expression system for studies on Rac1 small GTPase-mediated signaling. *Am. J. Physiol. Cell Physiol.* **285**, C711–C719.

Lippincott-Schwartz, J., and Patterson, G. H. (2003). Development and use of Fluorescence Protein markers in living cells. *Science* **300**, 87–91.

Li, X., Zhao, X., Fang, Y., Jiang, X., Duong, T., Huang, C. C., and Kain, S. R. (1998). Generation of destabilized enhanced green fluorescent protein as a transcription reporter. *J. Biol. Chem.* **273**, 34970–34975.

Liu, H. S., Jan, M. S., Chou, C. K., Chen, P. H., and Ke, N. J. (1999). Is Green Fluorescent Protein toxic to the living cell? *Biochem. Biophys. Res. Commun.* **260**, 712–717.

Lukyanov, K. A., Fradkov, A. F., Gurskaya, N. G., Matz, M. V., Labas, Y. A., Savitsky, A. P., Markelov, M. L., Zaraisky, A. G., Zhao, X., Fang, Y., Tan, W., and Lukyanov, S. A. (2000). Natural animal coloration can be determined by a non-fluorescent GFP homolog. *J. Biol. Chem.* **275**, 25879–25882.

Lybarger, L., Dempsey, D., Franek, K. J., and Chervenak, R. (1996). Rapid generation and flow cytometric analysis of stable GFP-expressing cells. *Cytometry* **25**, 211–220.

Lybarger, L., Dempsey, D., Patterson, G. H., Piston, D. W., Kain, S. R., and Chervenak, R. (1998). Dual-color flow cytometric detection of fluorescent proteins using single-laser (488-nm) excitation. *Cytometry* **31**, 147–152.

Ma, X. Q., Robin, C., Ottersbach, K., and Dzierzak, E. (2002). The Ly-6A (Sca-1) GFP transgene is expressed in all adult mouse hematopoietic stem cells. *Stem Cells* **20**, 514–521.

Magid, R., Martinson, D., Hwang, J., Jo, H., and Galis, Z. S. (2003). Optimization of isolation and functional characterization of primary murine aortic endothelial cells. *Endothel. NY* **10,** 103–109.

Makridou, P., Burnett, C., Landy, T., and Howard, K. (2003). Hygromycin B-selected cell lines from GAL4-regulated pUAST constructs. *Genesis* **36,** 83–87.

Matz, M. V., Fradkov, A. F., Labas, Y. A., Savitsky, A. P., Zaraisky, A. G., Markelov, M. L., and Lukyanov, S. A. (1999). Fluorescent proteins from nonbioluminescent Anthozoa species. *Nat. Biotech.* **17,** 969–973.

Mills, C. E. (1999-present). Bioluminescence of *Aequorea*, a hydromedusa. Available from http://faculty.washington.edu/cemills/Aequorea.html.

Miyawaki, A., Llopis, J., Heim, R., McCaffery, J. M., Adams, J. A., Ikura, M., and Tsien, R. Y. (1997). Fluorescent indicators for Ca^{2+} based on green fluorescent protein and calmodulin. *Nature* **388,** 882–887.

Miyawaki, A., Griesbeck, O., Heim, R., and Tsien, R. Y. (1999). Dynamic and quantitative Ca^{2+} measurements using improved cameleons. *Proc. Natl. Acad. Sci. USA* **96,** 2135–2140.

Miyawaki, A., Sawano, A., and Kogure, T. (2003). Lighting up cells: Labeling proteins with fluorophores. *Nat. Cell Biol.* **5,** S1–S7.

Molina, A., Carpeaux, R., Martial, J. A., and Muller, M. (2002). A transformed fish cell line expressing a green fluorescent protein-luciferase fusion gene responding to cellular stress. *Toxicol. In Vitro* **16,** 201–207.

Mousses, S., Caplen, N. J., Cornelison, R., Weaver, D., Basik, M., Hautaniemi, S., Elkahloun, A. G., Lotufo, R. A., Choudary, A., Dougherty, E. R., Suh, E., and Kallioniemi, O. (2003). RNAi microarray analysis in cultured mammalian cells. *Genome Res.* **13,** 2341–2347.

Nagy, P., Arndt-Jovin, D. J., and Jovin, T. M. (2003). Small interfering RNAs suppress the expression of endogenous and GFP-fused epidermal growth factor receptor (erbB1) and induce apoptosis in erbB1-overexpressing cells. *Exp. Cell Res.* **285,** 39–49.

Necela, B. M., and Cidlowski, J. A. (2003). Development of a flow cytometric assay to study glucocorticoid receptor-mediated gene activation in living cells. *Steroids* **68,** 341–350.

Nunes, M. C., Roy, N. S., Keyoung, H. M., Goodman, R. R., McKhann, G., Jiang, L., Kang, J., Nedergaard, M., and Goldman, S. A. (2003). Identification and isolation of multipotential neural progenitor cells from the subcortical white matter of the adult human brain. *Nat. Med.* **9,** 439–447.

Ormö, M., Cubitt, A. B., Kallio, K., Gross, L. A., Tsien, R. Y., and Remington, S. J. (1996). Crystal structure of the *Aequorea victoria* Green Fluorescent Protein. *Science* **273,** 1392–1395.

Patterson, G., Day, R. N., and Piston, D. (2001). Fluorescent protein spectra. *J. Cell Sci.* **114,** 837–838.

Preis, J. I., Downes, M., Oates, N. A., Rasko, J. E. J., and Whitelaw, E. (2003). Sensitive flow cytometric analysis reveals a novel type of parent-of-origin effect in the mouse genome. *Curr. Biol.* **13,** 955–959.

Raj, A., and van Oudenaarden, A. (2008). Nature, nurture, or chance: Stochastic gene expression and its consequences. *Cell* **135,** 216–226.

Romoser, V. A., Hinkle, P. M., and Persechini, A. (1997). Detection in living cells of Ca^{2+}-dependent changes in the fluorescence emission of an indicator composed of two Green Fluorescent Protein variants linked by a calmodulin-binding sequence: A new class of fluorescent indicators. *J. Biol. Chem.* **272,** 13270–13274.

Ropp, J. D., Donahue, C. J., Wolfgang-Kimball, D., Hooley, J. J., Chin, J. Y. W., Cuthbertson, R. A., and Bauer, K. D. (1996). Aequorea green fluorescent protein: Simultaneous analysis of wild-type and blue-fluorescing mutant by flow cytometry. *Cytometry* **24,** 284–288.

Rust, M. J., Bates, M., and Zhuang, X. (2006). Sub-diffraction-limit imaging by stochastic optical reconstruction microscopy (STORM). *Nat. Methods* **3,** 793–796.

Shagin, D. A., Barsova, E. V., Yanushevich, Y. G., Fradkov, A. F., Lukyanov, K. A., Labas, Y. A., Ugalde, J. A., Meyer, A., Nunes, J. M., Widder, E. A., Lukyanov, S. A., and Matz, M. V. (2004). GFP-like proteins as ubiquitous metazoan superfamily: Evolution of functional features and structural complexity. *Mol. Biol. Evol.* **21,** 841–850.

Sheen, J., Hwang, S., Niwa, Y., Kobayashi, H., and Galbraith, D. W. (1995). Green Fluorescent Protein as a new vital marker in plant cells. *Plant J.* **8,** 777–784.

Shimomura, O., Johnson, F. H., and Saiga, Y. (1962). Extraction, purification and properties of Aequorin, a bioluminescent protein from luminous hydromedusan. *Aequorea. J. Cell Comp. Physiol.* **59,** 223–239.

Sirk, D. P., Zhu, Z. P., Wadia, J. S., and Mills, L. R. (2003). Flow cytometry and GFP: A novel assay for measuring the import and turnover of nuclear-encoded mitochondrial proteins in live PC12 cells. *Cytometry* **56A,** 15–22.

Stull, R. A., Hyun, W. C., and Pallavicini, M. G. (2000). Simultaneous flow cytometric analysis of enhanced green and yellow fluorescent proteins and cell surface antigens in doubly transduced immature hematopoietic cell populations. *Cytometry* **40,** 126–134.

Takeuchi, Y., Yoshizaki, G., Kobayashi, T., and Takeuchi, T. (2002). Mass isolation of primordial germ cells from transgenic rainbow trout carrying the green fluorescent protein gene driven by the vasa gene promoter. *Biol. Reprod.* **67,** 1087–1092.

Telford, W. G., Hawley, T. S., and Hawley, R. G. (2003). Analysis of violet-excited fluorochromes by flow cytometry using a violet laser diode. *Cytometry* **54A,** 48–55.

Terskikh, A., Fradkov, A., Ermakova, G., Zaraisky, A., Tan, P., Kajava, A. V., Zhao, X., Lukyanov, S., Matz, M., Kim, S., Weissman, I., and Siebert, P. (2000). "Fluorescent timer": Protein that changes color with time. *Science* **290,** 1585–1588.

Torbett, B. E. (2002). Reporter genes: Too much of a good thing? *J. Gene Med.* **4,** 478–479.

Tsien, R. Y. (1998). The Green Fluorescent Protein. *Annu. Rev. Biochem.* **67,** 509–544.

Van den Plas, D., Ponsaerts, P., Van Tendeloo, V., Van Bockstaele, D. R., Berneman, Z. N., and Merregaert, J. (2003). Efficient removal of LoxP-flanked genes by electroporation of Crerecombinase mRNA. *Biochem. Biophys. Res. Commun.* **305,** 10–15.

Van Tendeloo, V. F. I., Ponsaerts, P., Van Broeckhoven, C., Berneman, Z. N., and Van Bockstaele, D. R. (2000). Efficient generation of stably electrotransfected human hematopoietic cell lines without drug selection by consecutive FACsorting. *Cytometry* **41,** 31–35.

Voronin, Y. A., and Pathak, V. K. (2003). Frequent dual initiation of reverse transcription in murine leukemia virus-based vectors containing two primer-binding sites. *Virology* **312,** 281–294.

Ward, W. W., and Cormier, M. J. (1979). An energy transfer protein in coelenterate bioluminescence: Characterization of the *Renilla* green-fluorescent protein. *J. Biol. Chem.* **254,** 781–788.

Ward, W. W., Cody, C. W., Hart, R. C., and Cormier, M. J. (1980). Spectrophotometric identity of the energy-transfer chromophores in Renilla and Aequorea green-fluorescent proteins. *Photochem. Photobiol.* **31,** 611–615.

Yang, F., Moss, L. G., and Phillips, G. N., Jr. (1996). The molecular structure of Green Fluorescent Protein. *Nat. Biotechnol.* **14,** 1246–1251.

Yang, T. T., Sinai, P., Green, G., Kitts, P. A., Chen, Y. T., Lybarger, L., Chervenak, R., Patterson, G. H., Piston, D. W., and Kain, S. R. (1998). Improved fluorescence and dual color detection with enhanced blue and green variants of the Green Fluorescent Protein. *J. Biol. Chem.* **273,** 8212–8216.

Yu, D. H., Baird, G. S., Tsien, R. Y., and Davis, R. L. (2003a). Detection of calcium transients in *Drosophila* mushroom body neurons with camgaroo reporters. *J. Neurosci.* **23,** 64–72.

Yu, Y. A., Oberg, K., Wang, G., and Szalay, A. A. (2003b). Visualization of molecular and cellular events with green fluorescent protein in developing embryos: A review. *Luminescence* **18,** 1–18.

Zeytun, A., Jeromin, A., Scalettar, B. A., Waldo, G. S., and Bradbury, A. R. M. (2003). Fluorobodies combine GFP fluorescence with the binding characteristics of antibodies. *Nat. Biotechnol.* **21,** 1473–1479.

Zhang, J. H., Tang, J., Wang, J., Ma, W. L., Zheng, W. L., Yoneda, T., and Chen, J. (2003). Over-expression of bone sialoprotein enhances bone metastasis of human breast cancer cells in a mouse model. *Int. J. Oncol.* **23,** 1043–1048.

Zharkikh, L., Zhu, X. H., Stricklett, P. K., Kohan, D. E., Chipman, G., Breton, S., Brown, D., and Nelson, R. D. (2002). Renal principal cell-specific expression of green fluorescent protein in transgenic mice. *Am. J. Physiol. Renal Physiol.* **283,** F1351–F1364.

Zhu, J., Musco, M. L., and Grace, M. J. (1999). Three-color flow cytometry analysis of tricistronic expression of eBFDP, eGFP, and eYFP using EMCV-IRES linkages. *Cytometry* **37,** 51–59.

CHAPTER 5

Labeling Cellular Targets with Semiconductor Quantum Dot Conjugates

Kary L. Oakleaf, Michael S. Janes, Xingyong Wu,⋆ and Marcel P. Bruchez[†]

Invitrogen Corporation—Molecular Probes®
Labeling and Detection Technologies
Eugene, Oregon 97402

⋆Quantum Dot Corporation
Department of Chemistry and
Molecular Biosensor and Imaging Center
4400 Fifth Ave., Pittsburgh, PA 15213

[†]Carnegie Mellon University
Department of Chemistry and
Molecular Biosensor and Imaging Center
4400 Fifth Ave., Pittsburgh, PA 15213

I. Introduction

Fluorescence is a well-established tool for examining biomolecular expression in cells using various detection platforms including light microscopes, flow and scanning cytometers, and microplate readers. Traditionally, detection ligands such as antibodies or oligonucleotides have been conjugated to fluorochromic organic dye molecules such as fluorescein, rhodamine, or Cy® and Alexa Fluor® dyes. As a scientific tool, fluorescence methods have provided the bulk of our understanding of modern cell biology. Despite the huge successes, current fluorochromes have limitations and require elaborate equipment and expertise. Another fluorochrome technology has been developed based on the extraordinary optical properties of nanometer (10^{-9} m) scale semiconductor nanocrystalline particles, called quantum dots (QDs) (Alivisatos, 1996).

As fluorescence labels, QDs have several key advantages over traditional organic fluorochromes. First, they absorb light with efficiencies that are orders of magnitude beyond organic dye molecules. Second, if properly engineered, they have an extremely high efficiency of converting excitation energy into emitted photons, that is, quantum yields approaching 100%. These two features make QDs much brighter than any organic fluorochromes. Furthermore, their emission spectra can be engineered by changing the size and/or the composition of the nanoparticles. Hence, their emission spectra can be designed or "tuned" during synthesis and typically are narrow and symmetrical. The ability of all QDs to be excited with the same wavelengths of light, optimally in the ultraviolet (UV) to blue range, results in sufficient excitation of all QDs in a sample with subsequent emission at a color related to its size and composition (Antelman *et al.*, 2009). These extraordinary features enable multicolor labeling of cellular targets with minimal spectral overlap. Finally, the fluorescence of properly engineered QDs is extremely stable so continuous long-term imaging or repeated measurements can be performed without appreciable loss of signal, particularly when an appropriate mounting media is used. Commercially available Qdot® conjugates have allowed several research groups to exploit these benefits in various applications (Antelman *et al.*, 2009; Dahan *et al.*, 2003; Lidke *et al.*, 2004; Ness *et al.*, 2003; Perrault *et al.*, 2009; Tokumasu and Dvorak, 2003). QDs have become an important class of fluorochromes that complement the use of traditional organic fluorescent dyes and in some applications, serve to directly address challenges associated with those dyes. With this success, many researchers have quickly realized that QDs behave differently in terms of biological sample preparation as well as imaging methodologies. In this chapter, we discuss some key considerations for the effective use of QDs for labeling targets in live and fixed cells.

II. Selection of QDs and their Conjugates

A single QD is a nanometer scale crystal of semiconductor material, typically cadmium selenide (although other materials can be used), which is coated with a shell of zinc sulfide (Dabbousi *et al.*, 1997; Watson *et al.*, 2003). The nascent core-shell QD is very hydrophobic and must be coated with a layer of water-miscible material for use in biological applications. Until 2003, many coatings had been described in the literature with limited success in preserving the particle integrity and performance. However, advances through rigorous engineering have described a polymer layer (Watson *et al.*, 2003; Wu *et al.*, 2003) that affords water solubility, and allows the dots to be conjugated to biological affinity molecules. This coating ensures higher stability, brightness, and overall performance of QDs in various colors for biological applications. Organic fluorochromes, such as fluorescein, from different manufacturers have essentially the same chemical and optical properties including stability, excitation, and emission spectra. In contrast, QDs are nanoscale engineered particles, and even though they are made from the same materials and have similar emission spectra, they may display substantially different properties due to variances in core-shell synthesis procedures and surface-coating chemistries. High-quality QDs should have high extinction coefficients, high quantum yield, and narrow emission spectra expressed as full-width-at-half-maximum (FWHM). Such QDs remain stable and dispersed in aqueous buffers for months, display high fluorescence intensity, and have narrow emission spectra (FWHM is consistently as low as 27 nm for some products).

QD conjugates are available with various affinity molecules, including primary antibodies, secondary antibodies, streptavidin, biotin, and protein A. Currently, there are eight colors of QDs available with emission peaks at 525, 565, 585, 605, 625, 655, 705, and 800 nm and these probes are commercially available from Invitrogen Corporation (Carlsbad, CA, USA) as Qdot® brand conjugates and as QTracker(r) probes. QDs are named after their emission peaks, a nomenclature which has an advantage of providing users the exact emission peaks of different dots, which is important for users when selecting the appropriate color of dots for their instruments and applications. For example, the green fluorescent conjugates are named Qdot® 525 because their maximum emission is centered at 525 nm. However, it should be noted that this naming system is different from that used for the Alexa Fluor® dyes, which are named after their excitation peaks rather than emission maxima. For instance, the Alexa Fluor® 488 dye is a green light-emitting dye with an excitation maximum at approximately 488 nm but with an emission peak at around 520 nm. Because organic dyes may only be effectively excited by wavelengths within a narrow range, information about each dye's excitation maximum is important for determining the appropriate microscopy configuration. In contrast, QDs are very flexible with respect to excitation sources. While wavelengths ranging from UV to the emission peak of a given QD may be used to adequately excite these fluorophores, they have a stronger absorption coefficient in the blue and UV portions of the spectrum.

III. Labeling of Fixed Cells for Fluorescence Microscopy

A. Sample Preparation Methods for Using Quantum Dots

One major challenge in detecting intracellular antigens or oligonucleotide sequences with fluorescent probes is the ability of probes to penetrate specimens for access to the biomolecules of interest. In the case of immunofluorescence, access of the relatively large conjugated antibody to an intracellular target requires permeabilization of the cell membrane without disruption of the cell architecture and antigen distribution (Melan, 1999). This challenge is even more acute when one uses QD conjugates to label intracellular targets. Starting with a core/shell ranging from 3 to 10 nm followed by a water-miscible layer and conjugated molecular surface, the final QD can be approximately 10 nm larger than its initial core (Watson *et al.*, 2003). Monodispersed materials are substantially smaller than aggregated materials, so preparations of QDs that have significant aggregation may produce inferior labeling results. In contrast, organic fluorochrome conjugates are smaller and hence, the fixation and permeabilization for dye conjugates and QD conjugates may be different in some applications.

There are two basic types of chemical fixatives for cells: organic solvents and cross-linking reagents (Melan, 1999). While solvents such as methanol, acetone, or higher molecular weight poly-alcohols effectively permeabilize cells by delipidating membranes, they may also dehydrate proteins and other macromolecules in cells, which leads to precipitation or condensation of cellular architectures. Alternatively, cross-linking fixatives such as formaldehyde form bridges between protein molecules (Melan, 1999). Cross-linking reagents usually preserve cell structures better than solvents but may chemically destroy binding sites or render epitopes inaccessible to antibodies. Most cross-linking fixatives require some form of post-fixation permeabilization with detergents or other chemicals to ensure adequate access to intracellular targets.

For immunolabeling with QD conjugates, the fixation and permeabilization conditions vary with the cellular location of the antigen. For detection of surface antigens, fixative selection is more flexible because the epitopes are exposed on the cell surface. For example, the Her2 receptor antigens on the surface of SK-BR-3 cells were labeled equally well with both methanol and formaldehyde fixation. For labeling cytoplasmic targets such as tubulin or mitochondrial proteins, however, fixation with cold methanol at $-20\,^{\circ}$C provided satisfactory results, but fixation with 4% formaldehyde in phosphate-buffered saline (PBS) followed by 0.25% Triton(r) X-100 resulted in better preservation of these cellular structures. Labeling of nuclear targets presents a challenge in that highly variable degrees of detection are observed across cells in a given population and with different primary antibodies. Optimal labeling of nuclear targets has not been fully characterized, although cold acetone fixation at $-20\,^{\circ}$C is generally the preferred method when detecting nuclear antigens with QD conjugates. While not always feasible with more delicate monolayers of cells, tissue sections are more resistant to harsher sample treatments for antigen retrieval, and such methods may greatly aid in obtaining access to nuclear targets with QD conjugates. Hence, as with organic fluorochrome conjugates, the selection of the best fixation and permeabilization conditions for labeling biomolecules in cells and tissues with QD conjugates will vary with each new antigen to be labeled. Such an example is shown in Fig. 1.

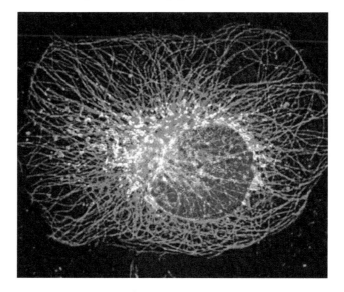

Fig. 1 High-quality staining with Qdot® conjugates at all levels of cellular structure: simultaneous detection of three cellular targets in HeLa cells fixed with 70% methanol/30% acetone using Qdot® conjugates. Microtubules (green), mitochondria (red), and nucleosomes (pseudocolored blue) were labeled with Qdot® 525 streptavidin conjugate, Qdot® 605 antimouse IgG conjugate, and Qdot® 655 antirabbit IgG conjugate, respectively, after the cells were incubated with appropriate primary antibodies. (See Plate no. 4 in the Color Plate Section.)

B. Mounting Media

A common problem in performing immunofluorescence detection with conventional fluorochromes is photobleaching. When exposed to excitation light, all fluorescent dyes will fade at some rate depending on the dye and its environment. Photobleaching is irreversible and appears to be associated with the presence of molecular oxygen (Murphy, 2001). The degree of photobleaching depends on both the intensity and the duration of illumination. To deal with this limitation of traditional fluorochromes, antifade chemicals that inhibit the generation and diffusion of reactive oxygen species have been added to mounting media (Ono *et al.*, 2001).

In contrast, high-quality QDs exhibit exceptional photostability (Bruchez *et al.*, 1998; Chan and Nie, 1998; Wu *et al.*, 2003); however, environmental sensitivity and aggregation of QDs under some conditions may result in signal stability issues (Resch-Genger *et al.*, 2008). While reasonable preservation of QD signals may be achieved in cells or tissue sections mounted in 50–90% glycerol in PBS and imaged within approximately 12 h of labeling, QD signal stability over time is far from optimal when using this formulation as a mountant. To address this limitation and provide a standardized mountant for QD-labeled cells and tissue sections, Qmount™ Qdot® mounting media was developed and is commercially available from Invitrogen Corporation (Carlsbad, CA, USA). Given the excellent long-term stability of QD conjugates in this mounting media, repeated imaging over time and archiving of samples is possible. Also, since this mounting media rapidly cures, it allows users to avoid the inconvenience of sealing coverslip edges with nail polish or other sealing reagents. For mounting samples that are labeled with both QDs and organic dyes, the self-sealing polyvinyl alcohol (PVA)-mounting medium containing 1-4-diazabicyclo[2,2,2]-octane (DABCO) from Sigma-Aldrich (St. Louis, MO, USA) may be used for adequate protection of both fluorophore types (Zhu *et al.*, 2007). However, storage of samples for longer than 2–3 days in this media is not recommended. While PVA/DABCO formulations are not optimal for dye and QD combinations, they represent the current preferred method. Traditional mounting media which have been specifically formulated for organic fluorochromes are generally incompatible with QDs and represent a significant challenge to imaging QDs.

C. Detection with a Fluorescence Microscope

QDs have a broad excitation range and have a very strong absorption coefficient in the blue and UV portions of the spectrum. A wide range of light sources such as mercury, xenon, or blue/UV lasers are suitable for QDs. Illumination of QDs with light from the UV to blue range greatly increases their fluorescence emission compared to excitation at longer wavelengths and the large Stokes shift of QDs, the distance between the peaks of excitation and emission, may improve detection sensitivity by eliminating background artifacts caused by scatter. The major fluorescence filter manufacturers provide filters specifically for detection of QD conjugates. For excitation, Semrock (Rochester, NY, USA) has a hard-coated 40 nm narrow bandpass excitation filter centered at 435 nm (FF01-435/40-25). Omega Optical (Brattleboro, VT, USA) has four excitation filters: a 100-nm broad bandpass filter centered at 415 nm (415BW100), a 45-nm narrow bandpass filter centered at 425 nm (425DF45), a 50-nm bandpass filter centered at 365 nm (365WB50), and an 80-nm bandpass filter centered at 330 nm (330WB80). Chroma Technology (Brattleboro, VT, USA) provides the 460SPUVv2, a 100-nm bandpass filter centered at 460 nm. With a mercury lamp, all excitation filters excluding the 330WB80 provide very strong excitation for high-intensity QD emission signals. The broader bandpass excitation filters may be more suitable for use with low magnification objectives (10× and 20×). With high-magnification oil objectives, these filters may induce sample heating and photo-oxidation, causing potential damage to the sample. This damage can be minimized by using either neutral density filters to reduce the intensity of light exposure or by utilizing the more narrow bandpass excitation filters. This is especially important for observing QD-labeled living cells.

The narrow and symmetrical emission spectra of QDs (Bruchez *et al.*, 1998; Watson *et al.*, 2003) make them ideal fluorochromes for multicolor fluorescence labeling. Conversely, many conventional fluorochromes have broad emission spectra and, thus, may require non-standard filters to achieve adequate spectral discrimination of fluorescence signal for multiplex analyses. Because the relatively tight emission spectra of QDs minimizes overlapping fluorescence emissions, imaging as many as four to five organic fluorochromes is possible without the requirement for more sophisticated spectral detection systems (Kosman *et al.*, 2004). In Fig. 2, cultured SK-BR-3 breast cancer cells expressing high levels of the Her2/neu receptor were first incubated with a mouse anti-Her2 antibody followed by a biotinylated goat anti-mouse IgG. Cells were then labeled with Qdot® 525, 565, 605, or 655 streptavidin conjugates, before washing and pooling the labeled cells into a single tube. Using an epifluorescence microscope equipped with a 425DF45 excitation filter and a long-pass emission filter (510ALP), all cells in the mixture were clearly visible with four distinguishable colors (Fig. 2A). When the cells in the same field were examined with individual 20 nm barrier filters centered at the emission peak of each QD color, only the cells labeled with corresponding QDs were visible (Fig. 2B-E). Because the emission spectra of the QDs are narrow, 20 nm barrier filters did not compromise the signal intensity. Furthermore, because the QDs exhibited narrow emission spectra, particularly in the visible range, there was no obvious crosstalk between the four detection channels, although the emission peaks of any two adjacent colors of QDs are

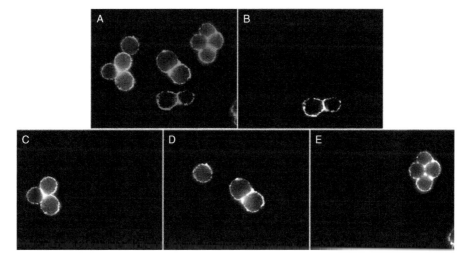

Fig. 2 Qdot® streptavidin conjugates enable effective color discrimination with simple bandpass emission filters: detection of Her2 on the surface of SK-BR-3 cells with Qdot® streptavidin conjugates. Cultured SK-BR-3 breast cancer cells expressing high levels of the Her2/neu receptor were incubated with a mouse anti-Her2 antibody followed by a biotinylated goat anti-mouse IgG. Cells were then labeled with Qdot® 525, 565, 605, or 655 streptavidin conjugates, before washing and pooling the labeled cells into a single tube. When the mixed cells were observed with a long-pass emission filter on an epifluorescence microscope, all cells in the field of view were visible with four distinguishable colors. (A) Cells were pseudocolored from gray-scale images to mimic the real colors viewed under the microscope. When the cells in the same field of view were examined with individual 20 nm bandpass filters centered at 525, 565, 605, and 655 nm (B–E), only the cells labeled with QDs having an emission peak in the detection range of the filter were visible. (See Plate no. 5 in the Color Plate Section.)

separated only by a 40- or 50-nm spectral space. Thus, multicolor fluorescence detection and discrimination can be dramatically simplified using QDs on a microscope equipped with appropriate filters.

D. Imaging QDs with a Laser Scanning Confocal Microscope

Laser scanning confocal microscopy (LSCM) is an important tool for providing optical sections and three-dimensional images of cellular structures and protein distributions. LSCM systems are typically equipped with multiple lasers that are tuned to the excitation maxima of conventional fluorochromes. Because all colors of QDs have a very broad absorbance profile, it is possible to excite multiple QDs with a single laser light in the blue or UV region. To demonstrate this, we incubated acetone-fixed human epithelial cells with a monoclonal mouse anti-nucleosome antibody, followed by biotinylated goat anti-mouse IgGs and streptavidin conjugates of Qdot® 525, 565, 585, 605, 655, or 705. The 488 nm line from an argon laser was used to excite these labeled cells. As shown in Fig. 3, all colors of QDs were effectively excited with the 488 nm laser line. These results

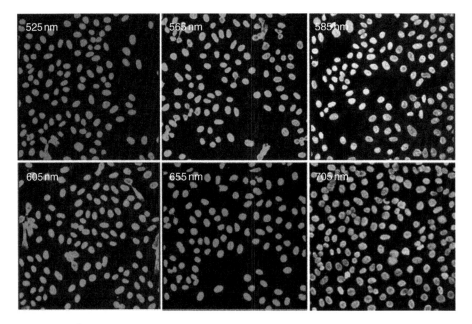

Fig. 3 Qdot® conjugates using single laser excitation source for multicolor confocal microscopy: nucleosomes in fixed human epithelial cells were labeled separately with Qdot® 525, 565, 585, 605, 655, and 705 streptavidin conjugates. The labeled cells were examined with a Leica TCS SP 2 confocal microscope. Because QDs have broad absorption profiles, all colors of the QDs were effectively excited with the 488 nm line from an argon laser. Each panel was adjusted to mimic the colors seen through a long-pass filter and a gray-scale image was used for Qdot® 705 as this color is not visible to human eyes. (See Plate no. 6 in the Color Plate Section.)

indicate that a single laser excitation source may be used for multicolor detection on a LSCM system when utilizing QDs as fluorescent labels, rendering more affordable single laser systems useful for multiplex experiments. Most microscope manufacturers currently offer LSCM systems equipped with a 405-nm excitation laser, an even more efficient excitation source for imaging QD-labeled samples.

IV. Measurements in Living Cells Using QD Conjugates

Imaging and analysis of biological responses in living cells in real time poses a special challenge for optical and fluorescence microscopy. In these systems, the events of interest may take place on time scales ranging from milliseconds to hours, and the process that an investigator is hoping to see may occur hours after applying the initial stimulus. Constitutively expressed protein labels, such as autofluorescent protein fusions (Cubitt et al., 1995), are enabling tools for microscopy that allow real time kinetic and dynamic measurements of a variety of biological phenomena within living cells. While the expression of autofluorescent proteins is an incredibly useful approach for live cell microscopy, it is limited by the fluorescence properties of these tools such as broad emission spectra and in some cases, photolability. Although advances have been made regarding the ease of this approach, particularly with viral delivery of genes, the rapid and photostable labeling made possible with QDs defines them as particularly well-suited tools for live cell imaging applications. The absorbance and quantum yield of the QDs combine to produce a material with very bright fluorescence emission that is particularly useful for imaging applications. These attributes can be harnessed in several important ways to image live cells. One persistent problem with dynamic microscopy of living cells is phototoxicity, where exposure of cells to high intensities of light results in a biological response or even in cell death (Knight et al., 2003; Zhang et al., 2006). For applications in which typical excitation conditions could be used to image fluorescent dyes, substantially lower exposure times may be used to image QD conjugates, given the much higher signal intensities typically observed with QDs. This allows the researcher to do less photodamage to the cells while collecting data over a longer time series. If acquisition of data at increased frame rates per second is desired to measure fast biological processes, the high-intensity signals from QDs effectively enable microscopy to be performed without loss of sensitivity. These properties have allowed researchers to image single QD conjugates with time resolutions of less than 0.1 s for many minutes in a single experiment (Dahan et al., 2003; Lidke et al., 2004). Additionally, the photostability of QDs represents another important property of these materials, particularly for live cell microscopy. QDs are approximately 20 times brighter than typical organic dyes and comparative photostability measurements have estimated that high-quality QDs in an appropriate buffer are as much as 100 times more photostable than common organic dyes (Bruchez et al., 1998; Walling et al., 2009; Wu et al., 2003), enabling applications such as detailed tracking of the dynamics of proteins in real time.

Techniques based on fluorescence recovery after photobleaching (FRAP) and photosensitizable fluorescent proteins (Ando *et al.*, 2002; Lippincott-Schwartz *et al.*, 2003; Tsien, 1998) have been used to study large-scale phenomena such as protein migration and trafficking but also introduce limitations. Given the intrinsic photostability limitations of autofluorescent proteins, these tools can be used only to observe large-scale changes, as tracking a single "photo-destroyed" GFP is virtually impossible and tracking photosensitized GFP molecules may be performed only for a short time. Using QD conjugates, one can track a single protein from the moment it begins to interact with the cell to its ultimate fate within the cell (Dahan *et al.*, 2003; Lidke *et al.*, 2004).

The chemical and energy-transfer properties of QD materials are in the early phases of exploration (Clapp *et al.*, 2004; Medintz *et al.*, 2003, 2004) but pose a significant opportunity for development of imaging tools for cytometry. Studies indicate that it should be possible to prepare materials that can be switched in the presence of an analyte using fluorescence resonance energy-transfer properties, in which the QDs serve as an energy-transfer donor. These could be multiplexed quite easily, and the readout would consist of a single excitation that would excite all of the QDs, and where appropriate, would transfer their energy to the acceptor, resulting in sensitized fluorescence. This would provide a very powerful set of biochemical tools, and if it were possible to deliver these materials directly into the cytoplasm, would provide an exciting opportunity for real-time cell physiology measurements. In addition, preparing the reverse materials (i.e., materials in which a cleavage event would release a signal that was previously transferred or quenched) would provide an extremely bright, selectively sensitized probe that could be used for both physiology and trafficking, as described earlier. The QDs are truly nanoscale materials, on the order of large proteins such as immunoglobulins. Even small proteins that are not directly targeted to receptors cannot easily be loaded into cells without relying upon endocytosis. However, colloidal materials have been delivered via endocytic machinery with successful release into the cytoplasmic milieu (Tkachenko *et al.*, 2003). The diffusion dynamics of QD-labeled receptors did not prove markedly different than the diffusion dynamics of receptors measured independently, although these studies have all applied to extracellular QD labeling. Recently, *in vivo* delivery of a single-wall carbon nanotube (SWNT) conjugated to a drug was reported and described effective killing of cancer cells and in effect, utilized the nanomaterial as a drug delivery mechanism (Bhirde *et al.*, 2009). It remains to be seen whether the "ball-and-chain" effect will play a role in determining the diffusion rate of QD-conjugated biological molecules when the conjugates have to be pulled through the relatively viscous environment of the cytoplasm and across intracellular membranes.

A growing number of publications demonstrate the power of QD conjugates and QTracker(r) probes for live cell imaging in a variety of cell types including embryonic stem cells (Wylie, 2007; Lin et al., 2007). Studies of the glycine receptor (GlyR) on living neurons demonstrated that the QD conjugates provided incredibly high signal to noise ratios at a single molecule level and that they could be tracked for many minutes with no degradation in the quality of the signal (Dahan *et al.*, 2003). These properties allowed a single GlyR to be tracked while diffusing the length of a neuron, sampling space within and outside of the synapse. Long-term imaging allowed careful determination that the GlyR demonstrates markedly different diffusion when it is in the synapse compared to when it is outside the synapse. These measurements would have been very difficult to generate without the optical properties of the QDs, which enabled continuous observation for many minutes (Dahan *et al.*, 2003). Similar studies using epidermal growth factor (EGF)-labeled QD particles in combination with autofluorescent proteins to track the dynamics of the erbB family of receptors revealed a new mechanism of transport along the filopodia of EGF-bound erbB2 receptors (Lidke *et al.*, 2004). In all of these studies, the use of QDs to track receptors required very low concentrations of QD conjugates, as opposed to those used to label targets by immunofluorescence at saturation. Typically, the QD conjugates are used at concentrations of less than 1 nM, as compared to 10–20 nM with antibody labeling of targets. Dynamic tracking of cell populations in chemotaxis and development has been demonstrated by two independent groups, where the QDs are delivered into the cells and then tracked as an optical probe of cell movement and lineage. The use of different colors of QDs to tag mixed cell populations of starved and healthy *Dictyostelium discoideum* showed the selective chemo-attraction of the starved cells, a process driven by chemotaxis in response to released cyclic adenosine monophosphate (Jaiswal *et al.*, 2003). In these experiments, the starved cells were seen to migrate selectively by observing motion of the QD fluorescence of one color, whereas the non-starved healthy cells remained static in culture, evident by the lack of motion in the alternative QD color. In an elegant set of experiments, phospholipid-coated QDs were prepared and introduced into selected cells in the blastocyst phase of a developing Xenopus and showed the tracking of these cells through to the developed tadpole (Dubertret *et al.*, 2002). These studies demonstrated that the QDs were stable to physiological conditions, were well retained by the cells and passed to subsequent generations, were tolerated as particles, and were non-toxic to the highly controlled and sensitive process of development. The use of QD materials to study processes in living cells is an area ripe for exploration. Many of the properties of QDs have not been fully exploited and present exciting opportunities for understanding dynamic cellular processes as they occur.

V. Conclusion

After years of intensive research and development, QDs and their conjugates are available for various uses in biological investigation. No longer simply a research subject for chemists and physicists, these materials have become powerful tools for researchers in life and biomedical sciences. Although QDs have potential advantages over conventional fluorochromes in a number of different detection modalities, microscope-based imaging represents an area wherein the properties of QDs are particularly useful. As such, fluorescence microscopy was the first application area in which scientists quickly adopted QD technology. The optical and biological properties of the QDs have enticed manufacturers of fluorescence detection instruments, including flow cytometers, to consider this new generation of fluorochromes in new instrument designs. Filter sets for microscopes designed specifically for Qdot® nanocrystals have become widely available and illumination configurations of imaging systems are emerging to allow more flexible and effective detection of QDs. Along with the influx of advanced imaging systems specifically designed for the optical properties of QDs, advances in our understanding of nanoparticle chemistry and manufacturing have and will continue to shape the utility of these important new tools and the biological applications they enable.

Acknowledgment

The authors thank Eric G. Tulsky for helpful review of the revised manuscript.

References

Alivisatos, A. P. (1996). Perspectives on the physical chemistry of semiconductor nanocrystals. *J. Phys. Chem.* **100,** 13226–13239.

Ando, R., Hama, H., Yamamoto-Hino, M., Mizuno, H., and Miyawaki, A. (2002). An optical marker based on the UV-induced green-to-red photoconversion of a fluorescent protein. *Proc. Natl. Acad. Sci. USA* **99,** 12651–12656.

Antelman, J., Wilking-Chang, C., Weiss, S., and Michalet, X. (2009). Nanometer distance measurements between multicolor quantum dots. *Nano Lett.* **9**(5), 2199–2205 (Published on Web 04/17/2009).

Bhirde, A. A., Vyomesh, P. S., Gavard, J., Zhang, G., Sousa, A. A., Masedunskas, A., Leapman, R. D., Weigert, R., Gutkind, J. S., and Rusling, J. F. (2009). Targeted killing of cancer cells *in vivo* and *in vitro* with EGF-directed carbon nanotube-based drug delivery. *ACS Nano* **3**(2), 307–316 (Published online January 13, 2009).

Bruchez, M., Jr., Moronne, M., Gin, P., Weiss, S., and Alivisatos, A. P. (1998). Semiconductor nanocrystals as fluorescent biological labels. *Science* **281,** 2013–2016.

Chan, W. C., and Nie, S. (1998). Quantum dot bioconjugates for ultrasensitive nonisotopic detection. *Science* **281,** 2016–2018.

Clapp, A. R., Medintz, I. L., Mauro, J. M., Fisher, B. R., Bawendi, M. G., and Mattoussi, H. (2004). Fluorescence resonance energy transfer between quantum dot donors and dye-labeled protein acceptors. *J. Am. Chem. Soc.* **126,** 301–310.

Cubitt, A. B., Heim, R., Adams, S. R., Boyd, A. E., Gross, L. A., and Tsien, R. Y. (1995). Understanding, improving and using green fluorescent proteins. *Trends Biochem. Sci.* **20,** 448–455.

Dabbousi, B. O., Rodriguez-Viejo, J., Mikulec, F. V., Heine, J. R., Mattoussi, H., Ober, R., Jensen, K. F., and Bawendi, M. G. (1997). (CdSe)ZnS core-shell quantum dots: Synthesis and characterization of a size series of highly luminescent nanocrystallites. *J. Phys. Chem. B* **101,** 9463–9475.

Dahan, M., Levi, S., Luccardini, C., Rostaing, P., Riveau, B., and Triller, A. (2003). Diffusion dynamics of glycine receptors revealed by single-quantum dot tracking. *Science* **302,** 442–445.

Dubertret, B., Skourides, P., Norris, D. J., Noireaux, V., Brivanlou, A. H., and Libchaber, A. (2002). *In vivo* imaging of quantum dots encapsulated in phospholipid micelles. *Science* **298,** 1759–1762.

Jaiswal, J. K., Mattoussi, H., Mauro, J. M., and Simon, S. M. (2003). Long-term multiple color imaging of live cells using quantum dot bioconjugates. *Nat. Biotechnol.* **21,** 47–51.

Knight, M. M., Roberts, S. R., Lee, D. A., and Bader, D. L. (2003). Live cell imaging using confocal microscopy induces intracellular calcium transients and cell death. *Am. J. Physiol. Cell Physiol.* **284,** C1083–C1089.

Kosman, D., Mizutani, C. M., Lemons, D., Cox, W. G., McGinnis, W., and Bier, E. (2004). Multiplex detection of RNA expression in *Drosophila* embryos. *Science* **305**(5685), 846.

Lidke, D. S., Nagy, P., Heintzmann, R., Arndt-Jovin, D. J, Post, J. N., Grecco, H. E., Jares-Erijman, E. A, and Jovin, T. M. (2004). Quantum dot ligands provide new insights into erbB/HER receptor-mediated signal transduction. *Nat. Biotechnol.* **22**, 198–203.

Lin, S., Xie, X., Patel, M. R., Yang, Y., Li, Z., Cao, F., Gheysens, O., Zhang, Y., Gambhir, S. S., Rao, J. H., Wu, J. C. (2007). Quantum dot imaging for embryonic stem cells. *BMC Biotechnology.* **7**, 67.

Lippincott-Schwartz, J., Altan-Bonnet, N., and Patterson, G. H. (2003). Photobleaching and photo-activation: Following protein dynamics in living cells. *Nat. Cell Biol.* (Suppl.), **5**, S7–S14.

Medintz, I. L., Clapp, A. R., Mattoussi, H., Goldman, E. R., Fisher, B., and Mauro, J. M. (2003). Self-assembled nanoscale biosensors based on quantum dot FRET donors. *Nat. Mater.* **2**, 630–638.

Medintz, I. L., Trammell, S. A., Mattoussi, H., and Mauro, J. M. (2004). Reversible modulation of quantum dot photoluminescence using a protein-bound photochromic fluorescence resonance energy transfer acceptor. *J. Am. Chem. Soc.* **126**, 30–31.

Melan, M. A. (1999). Overview of cell fixatives and cell membrane permeants. *In* "Method in Molecular Biology" (L. C. Javois, ed.), Vol. 115, pp. 44–55. Human Press, Totowa, NJ.

Murphy, D. B. (2001). Fluorescence microscopy. *In* "Fundamentals of Light Microscopy and Electronic Imaging" (D. B. Murphy, ed.), pp. 177–203. Wiley-Liss, New York.

Ness, J. M., Akhtar, R. S., Latham, C. B., and Roth, K. A. (2003). Combined tyramide signal amplification and quantum dots for sensitive and photostable immunofluorescence detection. *J. Histochem. Cytochem.* **51**, 981–987.

Ono, M., Murakami, T., Kudo, A., Isshiki, M., Sawada, H., and Segawa, A. (2001). Quantitative comparison of anti-fading mounting media for confocal laser scanning microscopy. *J. Histochem. Cytochem.* **49**, 305–311.

Perrault, J., Walkey, C., Jennings, T., Fischer, H., and Chan, W. C. W. (2009). Mediating tumor targeting efficiency of nanoparticles through design. *Nano Lett.* **9**(5), 1909–1915. Published online April 3.

Resch-Genger, U., Grabolle, M., Cavaliere-Jaricot, S., Nitschke, R., and Nann, T. (2008). Quantum dots versus organic dyes as fluorescent labels. *Nat. Methods* **5**(9), 763–775.

Tkachenko, A. G., Xie, H., Coleman, D., Glomm, W., Ryan, J., Anderson, M. F., Franzen, S., and Feldheim, D. L. (2003). Multifunctional gold nanoparticle-peptide complexes for nuclear targeting. *J. Am. Chem. Soc.* **125**, 4700–4701.

Tokumasu, F., and Dvorak, J. (2003). Development and application of quantum dots for immunocyto-chemistry of human erythrocytes. *J. Microsc.* **211**(Pt 3), 256–261.

Tsien, R. Y. (1998). The green fluorescent protein. *Annu. Rev. Biochem.* **67**, 509–544.

Watson, A., Wu, X., and Bruchez, M. P. (2003). Lighting up cells with quantum dots. *Biotechniques* **34**, 296–300, 302–303.

Walling, M. A., Novak, J. A., and Shepard, J. R. E. (2009). Quantum dots for live cell and *in vivo* imaging. *Int. J. Mol. Sci.* **10**, 441–491.

Wu, X., Liu, H., Liu, J., Haley, K. N., Treadway, J. A., Larson, J. P., Ge, N., Peale, F., and Bruchez, M. P. (2003). Immunofluorescent labeling of cancer marker Her2 and other cellular targets with semiconductor quantum dots. *Nat. Biotechnol.* **21**, 41–46.

Wylie, P. (2007). Multiple cell lines using quantum dots. *Methods Mol. Biol.* **374**, 113–123.

Zhu, J., Koelle, D. M., Cao, J., Vazquez, J., Huang, M. L., Hladik, F., Wald, A., and Corey, L. (2007). Virus-specific CD8+ T cells accumulate near sensory nerve endings in genital skin during subclinical HSV-2 reactivation. *J. Exp. Med.* **204**(3), 595–603.

Zhang, T., Stilwell, J., Gerion, D., Dingo, O., Cooke, P. A., Gray, J., Alivisatos, A., and Chen, F. F. (2006). Cellular effect of high doses of silica-coated quantum dot profiled with high throughput gene expression analysis and high content cellomics. *Nano Lett.* **6**(4), 800–808.

CHAPTER 6

High-Gradient Magnetic Cell Sorting

Andreas Radbruch,* **Andreas Grützkau,*** **Birgit Mechtold,**[†]
Andreas Thiel,[‡] **Stefan Miltenyi,**[§] **and Eckhard Pflüger**[§]

*Deutsches Rheuma-Forschungszentrum
10117 Berlin, Germany

[†]Institut für Genetik
Universität zu Köln, Germany

[‡]Berlin-Brandenburg Center for Regenerative Therapies
Charité 13353 Berlin, Germany

[§]Miltenyi Biotec GmbH
51429 Bergisch Gladbach, Germany

DOI: 10.1016/B978-0-12-375045-7.00006-4

I. Update

By now, 20 years hands-on experience has been made with nanosized super-paramagnetic particles and high-gradient magnetic fields which are used by the MACS® technology. The multiplicity of publications using the MACS® technology strikingly documents that high-gradient magnetic cell sorting has already made a significant impact on many research fields in cell and molecular biology, basic and clinical immunology. Although, the following chapter was already written in 1994, the protocols described herein are still up to date and applicable for almost every cell population, which can be tackled by an appropriate antibody. The success of their wide distribution is mainly achieved by the fact that MACS microbeads are almost inert and do not alter the functional status of labeled cells. Moreover, this technology is easy to establish in almost every lab interested in a particular cell type. Direct, indirect, and sequential sorting strategies are well established for all major and minor human, mouse, rat, and rhesus monkey leukocytes, stroma and tumor cells.

Over the past 15 years standard procedures using antibodies against cell-specific surface antigens were complemented by more sophisticated procedures allowing the isolations of functionally defined cell subsets according to their capacity to secrete cytokines, such as Th1 (IL-2, IFN-γ), Th2 (IL-4, IL-10), and Th17 (IL-17, IL-22) T cell subsets (Manz *et al.*, 1995). Moreover, the combined detection of cytokines, activation markers, and antigen-receptors of T lymphocytes by MHC multimer technologies nowadays allows to access antigen-specific T lymphocytes (Frentsch *et al.*, 2005).

An important development of the MACS® technology with respect to reproducibility, robustness, and effectiveness has been achieved by the introduction of automated devices, such as the AutoMACS® and the CliniMACS®. The CliniMACS® instruments are CE-marked for clinical applications in Europe, such as the isolation of stem cells (CD34) in autologous stem cell transplantations (Rosen *et al.*, 2000), the isolation of autologous haematopoietic progenitor cells (CD133) in leukemia (Koehl *et al.*, 2002), or to get access to antigen-specific T cells (CD137) for adoptive immunotherapies (Wehler *et al.*, 2008).

Recently, an instrumentation has been released which combines the use of high-gradient magnetic and flow cytometric cell sorting techniques (MACSQuant®). This device will allow an enrichment of rare cells just before analyzing them by multicolor flow cytometry and ensures the analysis of statistically significant cell numbers. Presumably, this will open the access to much more sophisticated cytometry of rare cells, for example, the detection of subpopulations.

Despite those technological developments, basic procedures of high-gradient magnetic cell sorting technology remain as described in the following hands-on article. The success of a magnetic sort regarding purity and recovery of cells of desire mainly depends on the compliance to some general principles, sometimes neglected when using the technology for the first time. Briefly, these basic rules apply to almost any cell sorting procedure and can be summarized as follows:

1. The number of cells to be labeled should be determined and adjusted the amount of MACS-conjugated antibody accordingly to achieve a maximal ratio between cell-specific and background labeling.

2. Higher proportions of dead cells and cell debris can be responsible for cell clumping and increased background labeling and should be removed prior labeling of cells, for example, by density gradient centrifugation or specific dead cell removal kits.

3. Cell suspensions should be always filtered (30 μm prewetted nylon filter) before applying them to the MACS column.

4. Buffers should be degassed in order to avoid the formation of air bubbles in the matrix of the columns.

5. Prefiltering of all buffers used for a particular sort is recommended to deplete almost any microparticles that could disturb the flow rate through the columns.

6. The flow through the MACS column should be attentively observed and using a new column is recommended when flow is slow or has stopped (check points 2–5).

7. A flow resistor (cannula) should be used if flow rate is too high, not dropwise, for example, using the large cell columns.

8. To minimize activation of cells working continuously at 4 °C is recommended, for example, in a walk-in refrigerator.

9. To ensure performance and reproducibility of sorting experiments purity, viability, and recovery of target cells should be verified by appropriate methods, for example, flow cytometry.

II. Introduction

Magnetic microparticles instead of fluorochromes can be coupled to specific ligands or antibodies against cell-surface structures. Such magnetic staining reagents can be used to isolate the labeled cells directly from complex cell mixtures. The absence of a natural magnetic background in most cell types and the ease by which extremely high magnetic forces can be produced using high-gradient magnetic technology permit the use of very small colloidal superparamagnetic particles (diameter < 100 nm) as magnetic labels. These tags do not interfere with optical analysis. Magnetically labeled cells can be stained simultaneously with fluorochromated antibodies to control and evaluate the quality of magnetic separation or to further analyze and sort the cells by flow cytometry or microscopy. Here, we describe magnetic cell separation based on staining with colloidal super-paramagnetic particles and separation of stained cells in high-gradient magnetic fields (Miltenyi et al., 1990; Molday and Molday, 1984; MACS; Fig. 1).

Although magnetism provides only one parameter for separation, magnetic cell sorting is superior to fluorescence-activated cell sorting (FACS) in several aspects

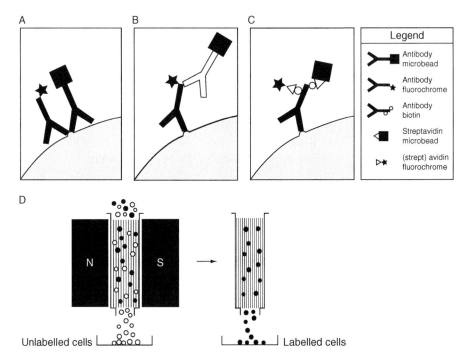

Fig. 1 Principle of high-gradient magnetic cell sorting. (A)–(C) Various principles of labeling cells with magnetic microbeads: (A) competitive binding of directly microbead-conjugated and fluorochrome-conjugated antibodies; (B) piggyback of fluorochrome-conjugated first antibody and microbead-conjugated second anti-antibody; (C) piggyback of haptenized first antibody, antihapten-conjugated microbeads, and fluorochrome-conjugated antihapten. In (D), the differential behavior of magnetic and nonmagnetic cells in a separation column inside and outside of an external magnet is illustrated.

and, in others, a valuable complementation of it. In FACS, the most powerful multiparameter sorting technology, cells are analyzed one by one and sorted one after the other. As a serial sorting device, its capacity is limited by the frequency of analysis and sorting, which is about 5000 cells per second, that is, 10^8 cells in 6 h. The capacity of MACS is only limited by the capacity of the magnetic matrix for labeled cells, which can be up to 10^9 cells. All cells are "analyzed" and sorted simultaneously allowing up to 10^{11} cells to be processed in about 30 min and giving it a leading edge in sorting of rare cells. Not only the fast processing but also the low physical forces of MACS sorting favors recovery of viable cells, compared to FACS, where the cells are submitted to considerable stress by acceleration in the nozzle. The sorting of rare cells, a combination of enrichment of the rare cells by gentle, fast, and parallel MACS followed by further enrichment by FACS, can yield enrichment rates of up to 10^8-fold in one experiment. Not many biological problems would require higher enrichment rates.

====== ## III. Application

Due to its origin in immunology, high-gradient magnetic cell sorting has been used primarily for the isolation of various cell types from human and animal hematopoietic systems (Hansel *et al.*, 1991; Irsch *et al.*, 1994; Kato and Radbruch, 1993; Schmitz and Radbruch, 1992; Vollenweider *et al.*, 1993); it has also been used for separation of plant cells and bacteria (Kronick and Gilpin, 1986). Basically, separation of organelles and chromosomes, as well as very large and fragile cells and cell aggregates, could be done by MACS, as long as immunomagnetic staining can be performed.

Designing the optimal strategy for any specific cell separation depends on

1. the availability of suitable cell-surface markers,
2. the composition of the original cell mixture,
3. frequency of wanted cells and the characteristics of the other cells,
4. specific requirements with regard to purity, recovery, and activation status of sorted cells.

Positive selection, that is, *enrichment* of wanted cells, requires one or more cell-surface markers which are specific for the cells of interest. Since the separation process distinguishes only between nonmagnetic and magnetic cells, it is important that the negative cells really remain unlabeled, that is, the staining reagent has to be titrated to eliminate background staining. Depending on the type of cell and on the surface marker used for staining, the sorted cells can be functionally influenced, for example, by staining T lymphocytes with anti-CD3 (MacDonald and Erard, 1986). This is a problem inherent to staining with antibodies which recognize and cross-link physiological receptors on the surface of the target cell and thus may induce or suppress proliferation or differentiation of the cell. As far as is known, labeling with antibody-conjugated colloidal magnetic particles has no effect other than labeling with an unconjugated cross-linking antibody. The bound microbeads will be capped off or endocytosed and degraded, depending on cell type and antigen. Similar beads have been used to treat iron deficiency in humans and animals and have been shown to be nontoxic and biodegradable (Muir and Goldberg, 1961).

Negative cell sorting, that is, labeling and *depletion* of unwanted cells from complex mixtures, will not yield homogenous, pure cell populations, unless antibodies against all unwanted cell types are available, which usually is not the case. However, the fraction of depleted, unstained cells is functionally "naive." A side effect of depletion is that dead cells, debris, and cell clumps are often removed as well from the cell suspension due to unspecific labeling.

The *isolation of rare cells* bears additional problems: First, discrimination of rare cells and cells of the major population is difficult because of the variation of immunofluorescence (%CV) which may lead to an overlap of the fluorescence distributions of frequent and rare cells, even in cases of otherwise acceptable staining. Often this may even make it difficult to identify the rare cells unambiguously for analysis and sorting. The overlap in magnetic staining is usually less significant, but is hard to access. Second, large cell numbers have to be processed in order to obtain rare cells in reasonable numbers. As a parallel enrichment method with high discriminatory power, MACS is predestined for sorting of rare cells, especially when combined with FACS (Irsch *et al.*, 1994; Kato and Radbruch, 1993; Weichel *et al.*, 1992).

IV. Materials

A. Cells and Staining Reagents

1. Single-cell suspension of good viability (Ficoll-Hypaque centrifugation removes dead cells and debris, cotton wool column aggregates, and dead cells; Esser, 1992).

2. Phosphate-buffered saline (PBS) with 1% (w/v) bovine serum albumin (BSA) or 1% (v/v) fetal calf serum (FCS) (PBS/BSA or PBS/FCS).

3. Propidium iodide (PI), 1 mg/ml in water.

4. Immunofluorescent and magnetic staining reagents in PBS/BSA with 0.05% sodium azide. For sterile sorting, reagents can be sterilized by filtration or, in cases of fluorochrome conjugates, by centrifugation, which also removes aggregates (3 min, Eppendorf centrifuge). *Titration* of staining reagents is essential to prevent background staining, that is, magnetism. Titration can be done by immunofluorescence and flow cytometric evaluation (Radbruch, 1992).

5. MACS CD14 microbeads (Miltenyi Biotec, Bergisch Gladbach, FRG; Fax: +49-2204-85197).

6. CD14-FITC (Leu M3 fluorescein conjugate; Becton-Dickinson).

7. MACS ramG1 microbeads (MACS rat antimouse IgG1 microbeads; Miltenyi Biotec).

8. CD3-FITC (Leu 4 fluorescein conjugate, mouse IgG1; Becton-Dickinson).

9. CD34 isolation kit (A1, blocking reagent; A2, modified CD34 antibody, clone QBEND-10; B, MACS microbeads; Miltenyi Biotec).

10. CD34 staining antibody (HPCA-2 phycoerythrin conjugate; Becton-Dickinson).

B. Sorting Equipment

1. Standard cell culture plasticware, such as Eppendorf tubes, sample tubes for sorter, pipettes and tips, syringes, and needles.

2. Cell filters, that is, nylon mesh with 30- to 50-μm pore size, either pieces of about 2 cm^2, glued to the bottom of a syringe (Esser, 1992), or commercially available (from Phoenix, San Diego, CA; Fax: +1-619-259-5268), autoclaved for sterility.

3. Cell counting device: Coulter counter or Neubauer chamber.

4. Centrifuge for cells and Eppendorf centrifuge.

5. MACS equipment: MACS separator and separation columns (Miltenyi *et al.*, 1990; Miltenyi Biotec, Bergisch Gladbach).

6. Ethanol (70%) in water (to sterilize and fill up the MACS reusable columns air bubble free). Alternatively MACS ready-to-use columns may be used. They become sterile and can be filled directly with buffer.

7. Flow cytometer (optional) or fluorescence microscope for separation control.

V. MACS: Staining and Sorting

A. Preparation and Staining of Cells

A single-cell suspension of good viability is a prerequisite for any good cell separation. Methods for obtaining such single-cell suspensions are described elsewhere (Esser, 1992). Since the magnetic separation process cannot distinguish between labeled single cells and aggregates containing labeled cells, it is important to avoid cell aggregates which might be composed of positive and negative cells. The cells should be resuspended carefully, for example, by finger-flicking the sediment in the tip of the tube after centrifugation, before adding new buffer. Staining should be optimized by titration of the staining reagents (Radbruch, 1992; see also Section III.A) to maximize discrimination between positive and negative cells and, even more important, to avoid background staining, which could make magnetic separation impossible. Adapting the staining to variable cell numbers should be done by adjusting the volume of the staining reagent, leaving the concentration unchanged.

1. Indirect Staining

1. Spin down all cells (usually 10^5–10^{10}).

2. Remove supernatant carefully, for example, by aspiration, and finger-flick the pellet.

3. Add a titrated amount of first staining reagent, for example, unconjugated, fluorochromated, or biotinylated antibody, specific for the marker molecule in question, to a final volume corresponding to the number of positive cells, but exceeding the total cell volume at least twofold, for example, 20 μl for 2×10^6 cells, containing 10^6 positive cells.

4. Stain for 10 min on ice.

5. Dilute the cells with PBS/BSA and spin them down (10 min, $300 \times g$).

6. Wash once: remove supernatant, resuspend the pellet by flicking, fill up with PBS/BSA, and spin down again.

7. For indirect staining, repeat steps 2–5 with second reagent, that is, stain the cells marked with the biotinylated first antibody with MACS streptavidin microbeads (diluted 1:10 from stock) or stain the unconjugated or fluorochromated first antibody with MACS anti-isotype microbeads. Stain for 15 min at 6-12 °C (in refrigerator).

8. Shake cells gently, then add a titrated amount of fluorochrome-conjugated streptavidin or, in the case of unconjugated first antibody, isotype-specific antibody or marker-specific antibody.

9. Dilute the cells with PBS/BSA and spin them down (10 min, 300 × *g*); resuspend them in PBS/BSA.

10. Check staining by microscopy or flow cytometry.

11. Store the cells on ice and in the dark, occasionally resuspending them, until they can be applied to the MACS column.

2. Direct Staining

1. Spin down all cells (usually 10^5–10^{10}).

2. Remove supernatant carefully, for example, by aspiration, and finger-flick the pellet.

3. Add the specific MACS antibody microbeads as instructed by the manufacturer.

4. Stain for 15 min at 6–12 °C (in refrigerator).

5. Add a titrated amount of the specific antibody conjugated to a fluorochrome to label the cells for sort control.

6. Stain for 10 min on ice.

7. Dilute the cells with PBS/BSA and spin them down (10 min, 300 × *g*); resuspend the cells in PBS/BSA.

8. Check staining by microscopy or flow cytometry.

9. Store the cells on ice and in the dark, occasionally resuspending them, until they can be applied to the MACS column.

B. Cell Sorting

1. Setting Up a MACS Separation Column

1. Select a column with a capacity of at least the number of positive cells you want to collect. Columns are available that can hold between 10^7 cells (type A) to 10^9 (type D) cells. Attach a three-way stopcock to the column of your choice.

2. For reusable columns: Attach a 10-ml syringe filled with 70% ethanol to the side plug of the stopcock and fill the column slowly from bottom to top with ethanol avoiding trapping air bubbles in the matrix by gently flicking the column.

Ready-to-use columns are filled with sorting buffer from the bottom and then washed with buffer once.

3. Place the column in the MACS separator, close the stopcock, remove the ethanol syringe, and replace it with a syringe filled with PBS.

4. Wash the column with several volumes of PBS and then PBS/BSA from the top of the column, and clear the side plug of the stopcock of ethanol using PBS from the syringe.

5. Immediately before separation wash the column with ice-cold PBS/BSA to cool it down, then attach a needle for flow regulation at the bottom plug of the stopcock (small needle, low flow rate; large needle, high flow rate). The whole procedure can be performed in a sterile workbench.

2. Setting Up a MACS Rare Cell Column or a MiniMACS Separation Column

For isolation of up to 10^7 labeled cells and rare cells, a MACS rare cell column or a MiniMACS column will be advantageous because it is easier to handle. Separation columns are sterile. They just have to be unpacked and filled by adding 0.5 ml of PBS/BSA or PBS/FCS. A flow resistor (needle) can be attached to slow down the flow rate for separation of dimly labeled cells. The whole procedure can be performed in a sterile workbench.

3. Magnetic Sorting

1. Cells, stained as described earlier, are passed through the MACS column placed in the MACS separator at low flow rates (0.5-2 ml/min). The flow frequency of cells can be monitored, for example, by checking drops under the microscope from time to time. When the majority of cells has passed the column, the column is washed at higher flow rates depending on the intensity of staining and following the rule:

High relative fluorescence = high flow rate,

until only few more cells come with the effluent. The cells not retained on the column in the MACS magnet are the *negative fraction*, and, if washed off at higher flow rates, the *wash fraction*. Aliquots are analyzed for depletion by microscopy or flow cytometry.

2. The column is then closed at the bottom, carefully removed from the MACS separator, and eluted outside the MACS by flushing with PBS/BSA (3–5 volumes) from the top using a syringe plunger to increase the flow rate and thus wash off the cells from the ferromagnetic matrix. If the elution volume is not too high, the cells can be processed further directly; otherwise it is necessary to spin them down (10 min, 300 × g). This fraction is the *positive fraction*. An aliquot is used for analysis of enrichment, either by microscope or by flow cytometry. For live/dead cell discrimination, propidium iodide is added to a final concentration of 0.01 μg/ml.

3. The quality of cell sorting is defined by the enrichment and depletion rates for labeled cells, the recovery of the wanted cells, and their viability.

The *enrichment rate* can be calculated as

$$f_E = \frac{\%\text{negative in original sample}}{\%\text{positive in original sample}} \times \frac{\%\text{positive in positive fraction}}{\%\text{negative in positive fraction}} \tag{1}$$

Accordingly, the *depletion rate* is

$$f_D = \frac{\%\text{positive in original sample}}{\%\text{negative in original sample}} \times \frac{\%\text{negative in positive fraction}}{\%\text{positive in positive fraction}} \tag{2}$$

The recovery is

$$\text{Recovery} = \frac{\text{absolute number of wanted cells in positive fraction}}{\text{absolute number of wanted cells in original sample}} \tag{3}$$

and the viability can be defined as

$$\text{Viability} = \frac{\%\text{live, wanted cells in positive fraction}}{\%\text{live, wanted cells in original sample}}, \tag{4}$$

although frequently it is given just as the frequency of live versus dead, wanted cells after the sort.

An essential figure in these calculations is the frequency of positive cells in the original sample. For rare cells it often simply cannot be determined with precision, making the calculation of enrichment rates difficult.

For calculation of the recovery, the cells have to be counted immediately before and after the sort. Otherwise, the sometimes considerable loss of cells during centrifugation steps will contribute to the calculated rate.

VI. Critical Aspects of the Procedure

A. Staining

As in immunofluorescence, the critical aspects of immunomagnetic staining are specificity and handling of the labeling antibody by the labeled cells. In general, short staining times improve the specificity of staining and few washing steps improve the recovery of cells throughout the staining procedure. To avoid capping of the staining antibody from the cell surface before or during the sort, one should work fast, keep the cells cold, and, whenever possible, use the reversible inhibitor of capping sodium azide (0.05% in PBS/BSA). However, antibody-conjugated microbeads require prolonged incubation times for staining on ice, so that they are used at 6–12 °C refrigerator temperature (15 min).

Problems with specificity and background usually can be improved by titration of the antibody, either in immunofluorescence and flow cytometry or in magnetic test sorts. Like any staining reagent, the magnetic microbeads can be titrated also. This can be done functionally using different concentrations of microbeads for one particular, standardized separation and determining the concentration with the best enrichment rate, that is, low background and high specific labeling. Background staining is lethal for successful separation because even slightly labeled cells will be retained in the magnetic field. Commercial preparations of magnetic microbeads, however, usually come titrated.

Fluorescent labels should be introduced after magnetic staining to avoid competition. Since the magnetic labeling will usually be carried out at nonsaturating conditions, nearly complete fluorescent labeling can be achieved.

B. Sorting

Ferromagnetic columns should be chosen according to their capacity for labeled cells. For optimal enrichment the capacity should correspond to the number of positive cells. For optimal depletion the capacity can be larger than that. Air bubbles trapped in separation columns can reduce the actual capacity of the column considerably and therefore have to be avoided. In reusable columns, they are avoided by filling the column with 70% ethanol from the bottom (see Section IV.B). Ready-to-use columns can be filled directly with buffer.

Large cell aggregates and clumps of debris may clog the column, although this is less of a problem for MACS than for FACS. Nevertheless, they should be removed by filtration of the cell suspension through a nylon mesh of defined mesh width, usually 30 or 50 μm (see Section III).

The sensitivity of the separation can be influenced by varying the flow rate of cells through the column. For depletion a slow flow rate will be chosen, whereas for enrichment a faster flow rate is preferred. In addition, magnetic labeling and optimal flow velocities depend on each other, that is, intensely labeled cells will also attach to the matrix at high flow rates, whereas dimly labeled cells will not. For individual antibodies it may be advisable to calibrate the speed of flow in test sorts for optimal sorting efficiency.

Dead cells and debris, some of which may be unspecifically labeled, in that case will attach to the MACS column and be recovered with the positive fraction. For depletion experiments, this may be a wanted effect. For enrichment of rare cells, however, one should keep in mind that unspecifically stained dead cells, even if they are rare, are also enriched and may comprise a major fraction of the positive fraction. Thus, for enrichment experiments, dead cells and debris have to be either accepted or removed before or after sorting, using Ficoll-Hypaque density gradients or cotton wool columns (Esser, 1992). For the sort control by flow cytometry, dead cells and debris can be gated out by scatter and PI fluorescence (FSC/SSC and F2/F3) (Weichel et al., 1992; Kato and Radbruch, 1993).

In general, to improve purity-enriched cells, eluted from the first column, can be passed over another column, at either the same or slightly different flow rate in order to remove unspecifically bound cells and particles from the first enrichment. An example for a double-pass column is shown in Fig. 5.

C. Sterility

The most frequent sources of contamination are staining reagents, especially if they are used by several persons over some time and stored in a refrigerator. Yeast can be eliminated by fungicides such as nystatin, but it is advisable to separate yeast from cells by Ficoll-Hypaque gradient centrifugation beforehand. Contamination of cells with bacteria can be avoided by using an antibiotic which is not normally used, such as gentamycin.

With respect to the sorting procedure, ready-to-use columns are sterile upon delivery, while multiuse columns are sterilized upon filling them with 70% ethanol (see Section IV.B). The whole MACS separation can be performed in a sterile workbench.

VII. Controls

Monitoring the magnetic sort for recovery requires a cell counting device. This could be a simple Neubauer chamber or a more sophisticated Coulter counter.

The volume of the negative fraction and the volume required for washing, that is, the time points for changing between the steps of the sorting procedure, can be either estimated, as indicated earlier (Section IV.B), or determined by collecting drops of the washing fluid from the column and judging the number of cells by phase-contrast microscopy. When no more cells leave the column, it is time to proceed to the next step. Usually, this will be after three void volumes of buffer.

With regard to specificity, the sort is monitored by immunofluorescence. In most systems tested so far, the magnetic label does not occupy all target molecules of the specific antibody on the cell surface. Sufficient target molecules are left over for labeling with the same antibody conjugated to a fluorochrome, directly or indirectly (Section III.B). Thus, the cells are labeled both magnetically and fluorescent, allowing the sorting process to be monitored by fluorescence microscopy or, preferably, by flow cytometry. Since the magnetic label does not interfere with the optical analysis, additional parameters can be included in this sort control, like propidium iodide for live/dead cell discrimination or fluorochromated antibodies specific for particular "unwanted" cells.

For cytometric analysis of rare cells, for example, in biomedical research, aliquots of the positive fraction can be directly counterstained and analyzed for the parameters in question (Irsch *et al.*, 1994; Kato and Radbruch, 1993), a unique feature of high-gradient magnetic cell sorting.

In evaluating the frequencies of contaminating cells in positive and negative fractions, which, due to the lack of better evaluation software, is usually done by setting a statistical threshold between the fluorescence distributions of labeled and unlabeled cells (Figs. 3–5), one should keep in mind that fluorescence distributions in general have a broad variation (CV). Cells from the wanted population, crossing the threshold due to this variation will be falsely classified as contaminating cells. True contaminating cell populations are those which have the same mean fluorescence as the original unwanted population.

VIII. Instruments

To magnetize the high-gradient magnetic separation columns, a strong magnet is required for generation of the magnetic fields. The design of the columns and the magnet has been described elsewhere (Miltenyi *et al.*, 1990). We have used the MACS, SuperMACS, and MiniMACS from Miltenyi Biotec (Fig. 2), together with columns from that manufacturer.

The SuperMACS can be used with columns with volumes of up to 50 ml and capacities of up to 10^9 labeled cells. It is optimal for depletion and for handling of large cell numbers. The MACS is a general purpose instrument which can be equipped with columns ranging in capacity between 10^7 and 2×10^8 cells. We have used this machine in most experiments (e.g., Section VIII.A). The MiniMACS is optimal for the efficient enrichment of up to 10^7 cells. We have used this device especially for the isolation of rare cells (Section VIII.B).

For sort control, we have used a FACScan (Becton-Dickinson) and for fluorescence-activated sorting of MACS preenriched cells a modified FACS IV (Weichel *et al.,* 1992).

IX. Results

A. Isolation and Depletion of Monocytes

A typical MACS separation using direct magnetic labeling is shown in Fig. 3. Monocytes are separated from other peripheral blood mononuclear cells (PBMC). In this experiment, 2×10^8 PBMC had been stained with 2 ml of MACS CD14 microbeads, diluted 1:5 in PBS/BSA from the purchased stock. After 15 min in refrigeration, 200 μl of CD14-FITC (50 μg/ml) was added, and the mixture incubated for another 5 min in the cold. After one wash with 15 ml of cold PBS/BSA, the cells were resuspended in 2 ml PBS/BSA. An aliquot of about 10^5 cells was analyzed in a FACScan.

The remaining cells were applied to a B2 column with a 22-guage needle, as a flow resistor, inserted into a MACS magnet. The negative cells were collected in 20-ml. An aliquot of this *negative fraction* was analyzed in the FACScan. The column with the positive cells was removed from the MACS and the positive cells were pushed off the wires from the bottom to the top of the column using a syringe attached to the side of the three-way stopcock. The column was then reinserted into the MACS magnet, and cells which had bound unspecifically in the first round were washed off. The column was then washed with 20 ml cold

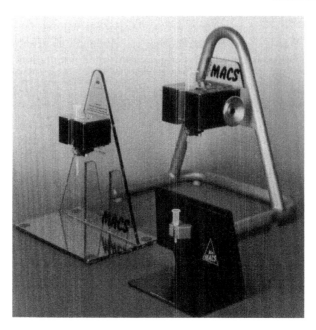

Fig. 2 Commercially available high-gradient magnetic cell sorters, the MiniMACS, the MACS, and the SuperMACS.

PBS/BSA using a 20-gauge needle. With this little intermezzo the purity of the positive fraction can be considerably increased. After the column was removed from the magnet, the flow resistor was removed from the column, a 10-ml syringe with PBS/BSA was attached to the top of the column via the adapter, and the retained cells were flushed out. An aliquot of this *positive fraction* was analyzed in the FACScan as well. Absolute cell numbers were determined in a Neubauer chamber. In this experiment, the recovery of CD14-depleted cells in the negative fraction was 89%, and the recovery of CD14-positive cells, 71%. The enrichment rate (f_E) was calculated to be 53, while the depletion rate (f_D) was 238. This MACS separation was optimized for efficient depletion. The depleted fraction could easily be used for further magnetic separations. The enrichment rate can be increased by repeating the washing cycle. We have used similar MACS separations leading to the discovery of the critical importance of murine monocytes/macrophages for the specific induction of interferon-γ expression in normal T helper lymphocytes (Schmitz and Radbruch, 1992).

B. Isolation and Depletion of T Lymphocytes

In an experiment similar to that described in Fig. 3, we separated T lymphocytes from PBMC (Fig. 4). The protocol is included to demonstrate the indirect magnetic labeling with MACS rat antimouse IgG1 microbeads. A total of 10^8 PBMC are

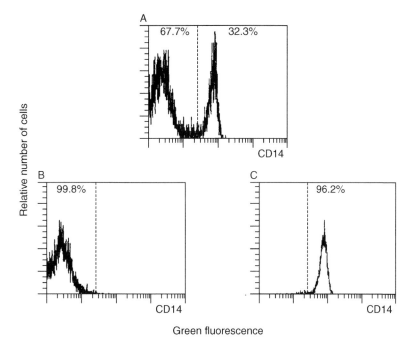

Fig. 3 Separation of CD14 monocytes from PBMC with directly conjugated MACS CD14 microbeads. For analysis the cells were stained with fluoresceine-conjugated CD14. Fluorescein-fluorescence histograms of live cells as gated by scatter and PI fluorescence. (A) Before sort: 67.7% negative cells, 32.3% positive cells; (B) negative fraction: 99.8% negative cells; (C) positive fraction after one washing cycle: 96.2%.

suspended in 350 μl PBS/BSA with CD3-FITC (50 μg/ml) and incubated for 10 min at 6–12 °C. Cells are washed with 15 ml PBS/BSA, the pellet is resuspended in 800 μl PBS/BSA, and 200 μl of MACS ramG1 microbeads are added. After 10 min at 6-12 °C (or 20 min on ice), cells are washed and resuspended in 1 ml PBS/BSA. The cell suspension is applied to a B1 column with a 23-gauge needle. The passing cells are applied onto the column again. Then, the negative cells are washed out in 15 ml PBS/BSA. After the washing step with 15 ml PBS/BSA and a 21-gauge needle, the positive fraction is collected outside of the MACS magnet with 5 ml of PBS/BSA applied from the top. All fractions are evaluated for purity, recovery, and viability as described earlier.

C. Enrichment of Rare Cells

The isolation of rare cells is a challenge for any cell sorting technology. In principle, positive selection, that is, isolation of the rare cells according to the marker molecules they express, is superior to negative selection, that is, depletion

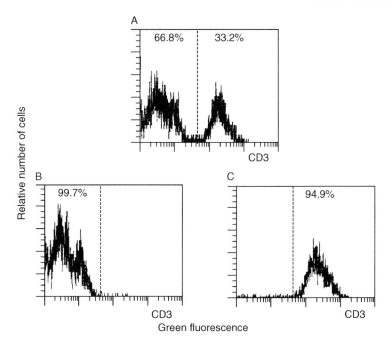

Fig. 4 Separation of CD3 T cells from PBMC with FITC-conjugated Leu 4 (IgG1) antibodies and rat antimouse IgG1 microbeads. Fluorescein-fluorescence histograms of gated live cells. (A) Before sort, (B) negative fraction f_D = 165, (C) positive fraction f_E = 37.

of all other cells according to markers they express, since usually discriminative markers are not available for all other cells, especially in the case of malignant cells. Here, we describe the purification of rare $CD34^+$ hematopoietic stem cells from peripheral blood by MACS.

For the enrichment of CD34 cells we used an improved indirect magnetic labeling technique. PBMC (5×10^8) were incubated with blocking reagent (human Ig) and haptenized CD34 antibody (clone QBEND-10) in PBS/BSA/EDTA (5 mM). Cells were washed with PBS/BSA and labeled magnetically with hapten-specific MACS microbeads (10 min at 6–12 °C) in PBS/BSA/EDTA (5 mM).

For flow cytometric analysis, phycoerythrine-conjugated HPCA-2 antibody was added. After 5–10 min incubation on ice, PBS/BSA/EDTA was added and cells were filtered through a 30-μm nylon mesh. The cells were washed, the pellet resuspended in 2 ml PBS/BSA/EDTA, and the cell suspension applied to a MiniMACS separation column which had been equilibrated with buffer before. The negative cells were washed off the column with PBS/BSA/EDTA. Retained cells were eluted from the column outside of the magnet by pipetting buffer onto the column and using the plunger supplied with the column.

An aliquot of the positive cells was used for flow cytometric analysis. The positive cells from the first column were now applied to a new separation column

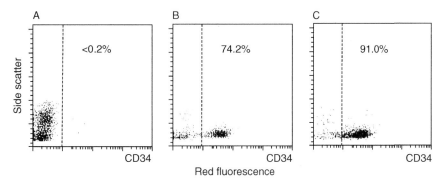

Fig. 5 Isolation of CD34 cells from PBMC with hapten-conjugated QBEND-10, hapten-specific microbeads, and physoerythrine-conjugated HPCA-2. Phycoerythrin-fluorescence versus side scatter dot plots of cells before (A) and after the first (B) and second (C) enrichment are shown. Live cells were gated according to forward scatter and PI fluorescence.

and the magnetic separation was repeated. During the second enrichment, negative cells were again washed off the column. Positive cells were eluted in PBS/BSA. Figure 5 shows flow cytometric side scatter versus phycoerythrin-fluorescence dot plots of the original fraction before separation and of the cell fractions eluted from the first and second columns. PBMCs were taken from a fresh blood sample of a normal donor with a less than 0.2% population of CD34-positive cells. The calculated enrichment rate (f_E) in this experiment is more than 4000. The typical recovery of CD34 cells is between 65% and 80% of the calculated fraction in unseparated PBMC of normal donors. MACS-enriched CD34 cells show a normal colony formation potential in semisolid short-term colony assays. Immunophenotyping revealed that peripheral CD34 cells show a different pattern of surface markers compared to bone marrow-derived CD34 cells (Kato and Radbruch, 1993; Miltenyi *et al.*, 1994).

Acknowledgment

This work was supported by the German Ministry of Science through the Genzentrum Köln.

References

Esser, C. (1992). Powerful Preselection. *In* " Flow Cytometry and Cell Sorting" (A. Radbruch, ed.), pp. 133–140. Springer, New York.

Frentsch, M., Arbach, O., Kirchhoff, D., Moewes, B., Worm, M., Rothe, M., Scheffold, A., and Thiel, A. (2005). Direct access to CD4 [+] T cells specific for defined antigens according to CD154 expression. *Nat. Med.* **11**, 1118–1124.

Hansel, T. T., DeVries, I. J., Iff, T., Rihs, S., Wandzilak, M., Betz, S., Blaser, K., and Walker, C. (1991). Induction and function of eosinophil intercellular adhesion molecule-1 and HLA-DR. *J. Immunol. Methods* **145**, 2130–2136.

Irsch, J., Irlenbusch, S., Radl, J., Burrows, P., Cooper, M., and Radbruch, A. (1994). Switch Recombination in Normal IgA1+B Lymphocytes. *Proc. Natl. Acad. Sci. USA* **91**, 1323–1327.

Kato, K., and Radbruch, A. (1993). Isolation and characterization of CD34$^+$ hematopoietic stem cells from human peripheral blood by high-gradient magnetic cell sorting. *Cytometry* **14**, 384–392.

Koehl, U., Zimmermann, S., Esser, R., Sörensen, J., Grüttner, H. P., Duchscherer, M., Seifried, E., Klingebiel, T., and Schwabe, D.(2002). Autologous transplantation of CD133 selected hematopoietic progenitor cells in a pediatric patient with relapsed leukemia. *Bone Marrow Transplant.* **29**, 927–930.

Kronick, P., and Gilpin, W. (1986). Use of superparamagnetic particles for isolation of cells. *J. Biochem. Biophys. Methods* **12**, 73–80.

MacDonald, H. R., and Erard, F. (1986). Activation requirements for resting T lymphocytes. *Curr. Top. Microbiol. Immunol.* **126**, 187–194.

Manz, R., Assenmacher, M., Pfluger, E., Miltenyi, S., and Radbruch, A. (1995). Analysis and sorting of live cells according to secreted molecules, relocated to a cell surface affinity matrix. *Proc. Natl. Acad. Sci. USA* **9264**, 1921–1925.

Miltenyi, S., Müller, W., Weichel, W., and Radbruch, A. (1990). High gradient magnetic cell separation with MACS. *Cytometry* **11**, 231–238.

Miltenyi, S., Guth, S., Radbruch, A., Pflüger, E., and Thiel, A. (1994). Isolation of CD34+ Hematopoietic Progenitor Cells by High-Gradient Magnetic Cell Sorting (MACS). *In* "The Mulhouse Manual" (E. Wunder, ed.), pp. 201–215. AlphaMed Press, Dayton, OH.

Molday, R., and Molday, L. (1984). Separation of cells labeled with immunospecific iron dextran microspheres using high gradient magnetic chromatography. *FEBS Lett.* **170**(2), 232–237.

Muir, A., and Goldberg, L. (1961). Observations on subcutaneous macrophages. Phagocytosis of iron-dextran and ferritin synthesis. *Q. J. Exp. Physiol. Cogn. Med. Sci.* **46**, 290–298.

Radbruch, A. (1992). Immunfluorescence: Basic Considerations. *In* "Flow Cytometry and Cell Sorting"(A. Radbruch, ed.), pp. 34–46. Springer, New York.

Rosen, O., Thiel, A., Massenkeil, G., Hiepe, F., Häupl, T., Radtke, H., Burmester, G. R., Gromnica-Ihle, E., Radbruch, A., and Arnold, R. (2000). Autologous stem-cell transplantation in refractory autoimmune diseases after *in vivo* immunoablation and *ex vivo* depletion of mononuclear cells. *Arthritis Res.* **2**, 327–336.

Schmitz, J., and Radbruch, A. (1992). Distinct antigen presenting cell-derived signals induce TH cell proliferation and expression of effector cytokines. *Int. Immunol.* **4**, 43–52.

Vollenweider, I., Vrbka, E., Fierz, W., and Groscurth, P. (1993). Heterogeneous binding and killing behaviour of human gamma/delta-TCR$^+$ lymphokineactivated killer cells against K562 and Daudi cells. *Cancer Immunol. Immunother.* **36**, 331–336.

Weichel, W., Irlenbusch, S., Kato, K., and Radbruch, A. (1992). Sorting of Rare Cells. *In* "Flow Cytometry and Cell Sorting" (A. Radbruch, ed.), pp. 159–167. Springer, New York.

Wehler, T. C., Karg, M., Distler, E., Konur, A., Nonn, M., Meyer, R. G., Huber, C., Hartwig, U. F., and Herr, W. (2008). Rapid identification and sorting of viable virus-reactive CD4(+) and CD8(+) T cells based on antigen-triggered CD137 expression. *J. Immunol. Methods* **339**, 23–37.

CHAPTER 7

Multiplexed Microsphere Assays (MMAs) for Protein and DNA Binding Reactions

Kathryn L. Kellar* and Kerry G. Oliver[†]

*Division of Scientific Resources
National Center for Preparedness
Detection and Control of Infectious Diseases
Centers for Disease Control and Prevention
Atlanta, Georgia 30333

[†]Radix BioSolutions Ltd.
Georgetown, Texas 78626

I. Update

The concept of multiplexing the detection of proteins and nucleic acids has become an accepted practice in research and diagnostic laboratories. Different platforms exist, but for detection of a few to 100 analytes, microsphere-based binding assays with flow cytometric analysis have been expanding as one of the most widely used formats. The instrumentation is priced to fall within the budget of an individual laboratory and the diversity and supply of commercial kits has expanded rapidly. To date, the FDA has approved as many as 58 clinical diagnostic kits incorporating multiplexed microsphere assays (MMAs). ELISAs are gradually being replaced by MMAs because they meet the same criteria for sensitivity, ease of operation, cost, and availability, with the added capacity for multiplexing. Although the commercial providers of the technology and instrumentation have remained primarily the same over the years, applications have spread to every bioresearch and clinical market, as well as the environmental arena.

Expansion of MMA technology is partially due to the product applications that have been developed on the existing platforms from the outset of scientific discovery, instead of being transitioned from other platforms. Some examples include the genotyping array for *Salmonella* subgroups (Fitzgerald *et al.*, 2007) and the gene expression profiling of microRNAs (Lu *et al.*, 2005).

The protocols and procedures for performing MMAs remain unchanged, employing common reagents and basic laboratory methods that are performed in a few hours. Magnetic microspheres, which facilitate washing the beads, are replacing polystyrene beads in the commercial market and should eliminate the often problematic wash steps performed by vacuum filtration. Instruments and software that are capable of processing greater than 100 microspheres in a single sample well are on the horizon. Additionally, instruments that perform static, nonflow, measurements of microsphere arrays have entered the market (Illumina, San Diego, CA; Christodoulides *et al.*, 2007), with more planned for the future. Several of these instruments can bring multiplexed technology both to the bedside and out into the field.

Genotyping arrays represent a large proportion of the total applications of MMAs in research and diagnostic laboratories. Although the direct hybridization format is commonly used, allele specific primer extension (ASPE), oligonucleotide ligation (OLA), and single base chain extension (SBCE) have been employed for single nucleotide polymorphism (SNP) detection. The use of "zipcode/antizipcode" or "tag/antitag" probes reduce problems with cross reactivity between targets (Bortolin *et al.*, 2004; Taylor *et al.*, 2001).

Reagent quality is the most critical issue for controlling specificity and the shelf-life of a multiplexed immunoassay. Purified, stable monoclonal antibodies provide the best products.

The quality assurance of data production can be easily managed with MMAs. Several diagnostic MMA manufacturers include additional bead sets as internal controls to monitor the proper execution of the assay steps for each sample (Zeus Scientific, Raritan, NY; Bio-Rad Laboratories, Hercules, CA); similar internal controls are available for the research market (Radix BioSolutions, Georgetown, TX). Additionally, some MMAs include a limited standard curve for analysis with each individual sample (Zeus Scientific), which has the advantage of also controlling for potential sample matrix effects that may alter the binding of analyte and antibodies.

Although large multiplexed assays take time to develop, the reward is a better understanding of the complexities of a molecular system.

II. Introduction

A flow cytometer has a limit to the size of the molecular or cellular mass that can be detected. Since DNA fragments and IgM are below that size range for all but the most sophisticated cytometers, these molecules can be detected only when bound or adsorbed onto the surface of a larger particle. This simple concept was demonstrated in the early years of flow cytometry but has seen wide application only in the last few years. Both qualitative and quantitative assays for soluble antigens, antibodies, enzymes, receptors, phosphoproteins, oligonucleotides, viral particles, peptides, and any other ligands can be performed on the surface of microspheres. The multiparametric resolving power of flow cytometry enables many combinations of fluorescent dyes to be used in these binding assays. More importantly, microspheres can be dyed with different intensities and numbers of fluorochromes or can be manufactured in a variety of sizes, so an array of microspheres can be utilized to capture multiple proteins or DNA sequences simultaneously. This capacity for multiplexing the detection of molecules within one small sample has significantly broadened the range of flow cytometric applications and has fostered the development of new instruments and computer software to automate the procedures.

In a 1977 review of flow cytometry and cell sorting, Horan and Wheeless described a potential application that could identify serum antibodies that were reactive with several antigens coated on the surface of microspheres of different sizes. For flow cytometric analysis, the microspheres were gated individually based on their light scattering properties, and the fluorescence of fluorochrome-labeled antihuman IgG or IgM was measured. In the next two decades, McHugh and his colleagues developed numerous fluorescent microsphere immunoassays for antigens as well as antibodies. Polystyrene microspheres of various sizes were used. Viral antigens were adsorbed onto the microsphere surfaces for the simultaneous detection of antibodies to cytomegalovirus and herpes simplex virus, hepatitis C core and nonstructural proteins, or four recombinant HIV proteins (McHugh et al., 1997, 1988a; Scillian et al., 1989). Antibodies to whole cell or subcellular extracts of *Candida albicans* were measured similarly (McHugh et al., 1989).

A novel assay employed human C1q-coated microspheres to capture immune complexes sequestering HIV antigens, which were identified by monoclonal antibodies to HIV proteins, followed by biotin-conjugated antimouse IgG and streptavidin-phycoerythrin (PE) (McHugh *et al.*, 1986, 1988b). These assays provided improved discrimination between infected and noninfected test subjects compared with other techniques in use at the time (McHugh and Stites, 1991). Additional applications of microspheres to binding assays for single as well as multiple analytes have been reviewed elsewhere (Kellar and Iannone, 2002; McHugh, 1994).

The number of discrete sizes of particles that can be analyzed on a standard flow cytometer is finite, so the capacity of microspheres to adsorb fluorescent dyes was important to the concept of an array of microspheres for measuring multiple analytes. Several variations of this concept have been adopted. Monosized microspheres have been dyed to different fluorescent intensities with a single fluorophore (Camilla *et al.*, 2001; Chen *et al.*, 1999; Cook *et al.*, 2001; Defoort *et al.*, 2000), as have beads of various diameters (Park *et al.*, 2000). Dyeing microspheres with two dyes has been performed to prepare first 64, and subsequently 100, spectrally distinct populations of uniformly sized microspheres for large suspension arrays (Fulton *et al.*, 1997; Kettman *et al.*, 1998). Additionally, quantum dots have been used to tag and differentiate populations of microspheres (Han *et al.*, 2001). The narrow emission bandwidth of quantum dots increases the theoretical number of microsphere populations that can be differentiated.

Commercially, these or similar microspheres are available from numerous sources, including Bangs Laboratories (Fishers, IN), Invitrogen (Molecular Probes) (Carlsbad, CA), Luminex Corporation (Austin, TX), Polysciences (Warrington, PA), Spherotech (Libertyville, IL), and Thermo Fisher Scientific (Waltham, MA).

III. Technology and Instrumentation

MMAs require the following components:

1. An array of microspheres sufficiently large enough to account for the analytes to be multiplexed.
2. Capture ligands/molecules that can be bound to the surface of the microspheres, preferably covalently.
3. Quantifiable standards for quantitative assays.
4. Directly or indirectly fluorochrome-conjugated detection ligands.
5. Any intermediate reagents between the captured ligand and final binding/ detection step.

For the majority of these assays, the detection fluorochrome can be the same for all analytes in the multiplex and serves as the "reporter," which directly correlates to the amount of analyte bound. The microsphere-embedded and reporter fluorophores must be complementary to the laser(s) excitation wavelength and the emission filters.

Multiplexed microsphere-based immunoassays have seen wide application in recent years. This assay format is an ideal substitute for ELISAs because many antigens or antibodies can be detected simultaneously from one small sample, in less time, and with a significant reduction in reagent costs (Camilla *et al.*, 2001; Carson and Vignali, 1999; Cook *et al.*, 2001; de Jager *et al.*, 2003; Kellar *et al.*, 2001; Martins *et al.*, 2002; Oliver *et al.*, 1998; Tárnok *et al.*, 2003). Quantitation of the complex network of human or mouse cytokines, chemokines and growth factors from culture supernatants, whole blood, sera, plasma, bronchial lavages, and tears has been a logical application for this technology. Several investigators have developed procedures for measuring these important modulatory proteins for studies of infectious and immune diseases (Camilla *et al.*, 2001; Carson and Vignali, 1999; Collins, 2000; Cook *et al.*, 2001; de Jager *et al.*, 2003; Hutchinson *et al.*, 2001; Kellar *et al.*, 2001; Mahanty *et al.*, 2001; Martins *et al.*, 2002; Oliver *et al.*, 1998; Prabhakar *et al.*, 2002; Tárnok *et al.*, 2003; Tripp *et al.*, 2000). Commercially available kits for the measurement of human, mouse, and rat cytokines, chemokines, growth factors, as well as an array of signaling proteins, including kinases, phosphoproteins, and cellular receptors (Bender MedSystems, Burlingame, CA; BD Biosciences, San Jose, CA; Bio-Rad Laboratories, Invitrogen; Millipore, Billerica, MA; Qiagen, Valencia, CA; R&D Systems, Minneapolis, MN) are based on the procedures demonstrated by several of the aforementioned authors and have been employed by Keyes *et al.* (2003) and Weber *et al.* (2003) in their reports of tumor-bearing nude mice and vaccine trials, respectively.

Viral and bacterial antigens (Dunbar *et al.*, 2003; Iannelli *et al.*, 1997), viral and bacterial antibodies (Bellisario *et al.*, 2001; Iannelli *et al.*, 1997; Martins, 2002; Pickering *et al.*, 2002a,b), and heterophilic antibodies in patients with disputed HIV test results (Willman *et al.*, 2001) have been detected by immunoassays on multiplexed microspheres. Competitive immunoassays have proven to be very specific and rapid for serotyping pneumococcal isolates (Park *et al.*, 2000), quantitating neutralizing antibodies to human papilloma viral genotypes (Opalka *et al.*, 2003), detecting newborn thyroxine concentrations in blood spots (Bellisario *et al.*, 2000), and screening hybridoma supernatants for monoclonal antibodies to IL-6 (Seideman and Peritt, 2002).

Another rapidly expanding area for MMAs is the detection and characterization of multiple nucleic acid sequences. Labeled PCR or gene expression products can be directly captured by hybridization to a complementary sequence-coupled microsphere population (Armstrong *et al.*, 2000; Barker *et al.*, 1994; Colinas *et al.*, 2000; Defoort *et al.*, 2000; Dunbar and Jacobson, 2000; Dunbar *et al.*, 2003; Smith *et al.*, 1998; Spiro *et al.*, 2000; Wallace *et al.*, 2003; Yang *et al.*, 2001). Alternately, oligonucleotide tags that are complementary to sequences at the 5' end of specific capture probes for PCR amplicons have been attached to microspheres and used to "zip" on biotinylated products for flow cytometric analysis (Cai *et al.*, 2000; Chen *et al.*, 2000; Iannone *et al.*, 2000; Taylor *et al.*, 2001; Ye *et al.*, 2001). These tags, or "zipcodes," can be used interchangeably in different multiplexes and allow the initial hybridization to the PCR amplicons to occur in solution.

A. Microspheres

Luminex Corporation provides the largest commercially available array of microspheres. These beads are 5.6 μm polystyrene microspheres cross-linked with divinylbenzene and contain functional carboxylic acid groups. The microspheres are dyed with different concentrations of two spectrally distinct fluorochromes resulting in a 10×10 array or 100 fluorescently distinct populations of microspheres.

Fluorescent microspheres produced by other companies resemble Luminex microspheres but may differ in the dye excitation and emission characteristics, the number of spectrally distinct dyes, and the surface functionality. For example, Bangs Laboratories market microspheres of two different sizes with five intensities of Starfire Red per size for a total of 10 distinct sets.

The introduction of quantum dots into microspheres is a realistic approach to significantly increase the number of distinct sets of microspheres available for multiplexed reactions (Han *et al.*, 2001). Practically, 10^4–10^5 microsphere populations should be obtainable using quantum dots; however, the utility of that level of multiplexing remains in question.

B. Instrumentation

Various flow instruments are compatible with MMAs. Any flow cytometer equipped with appropriate excitation sources is applicable for performing multiplexed assays. Luminex Corporation was the first to develop a benchtop cytometer designed solely for running MMAs. The Luminex 100 system utilizes a 635-nm 10-mW red diode laser and a 532-nm 10-mW yttrium aluminum garnet (YAG) laser for excitation, with four Avalanche Photodiodes (APDs) for discrimination of single microspheres and their classification dyes and a photomultiplier tube (PMT) for quantitation of the reporter fluorochrome. The system also is fitted with a 96-well XY platform that allows uninterrupted analysis of an entire microtiter plate, analogous to an ELISA plate reader. The xMAP system and the various derivations introduced by Luminex strategic partners (list available at www.luminexcorp.com) include software that classifies each microsphere set, displays the reporter median fluorescent intensity (MFIs) values, and either performs curve-fit analyses and extrapolation after acquisition or exports the data to a database or text file for analysis.

One limitation of the current Luminex 100 system is its operation as only a microsphere-based assay instrument without providing the capability for performing cell-based flow cytometric assays that is inherent in other flow cytometers. However, Janossy *et al.* (2002) have demonstrated that the Luminex 100 can be converted to an instrument capable of performing cellular analyses, and Bayer Pharmaceuticals reportedly is working on a project to provide this capability in a production instrument (Shapiro, 2003).

To meet the needs of laboratories that need an all-in-one instrument capable of performing MMAs as well as cellular analyses, BD Biosciences introduced the FACSArray Bioanalyzer system. As with the Luminex 100 system, the FACSArray is fitted with 532- and 635-nm light sources. The FACSArray has six parameter detection capabilities (two light scatter and four color measurements). This design provides the ability to detect two different microsphere classification fluorochromes and two different detection fluorochromes. The instrument was designed to perform its proprietary multiplexed Cytometric Bead Arrays (CBAs), and six-parameter four-color cellular analysis on the same instrument.

Automation of MMAs is conceptually straightforward. Several instruments have been designed specifically to perform the agitation and vacuum filtration required for assays performed on filter-bottom plates. (BioRobot 3000, Qiagen), (Janus, Perkin Elmer, Boston, MA), (AIMS, Zeus Scientific). The BioRobot 3000 is fully automated, agitates and washes filter-bottom plates, and can be linked to a twister to load the plates onto a Luminex instrument. In addition, Bio-Rad Laboratories has introduced a random access clinical analyzer, the BioPlex 2200, which has received FDA approval as a Luminex-based immunoassay MMA diagnostic device.

An alternative to flow-based systems for performing MMAs is laser scanning. A number of vendors (BD Biosciences, Palo Alto, CA; Perkin Elmer, Wellesley, MA; Acumen, Cambridge, UK) have demonstrated the capability of laser scanning systems to perform MMAs. A scanning-based system provides potential advantages such as the ability to enhance sensitivity through repeated scanning of the same sample. In addition, scanning removes the need for fluidics, which has inherent problems with sample particulates and microsphere aggregates that reduce speed. Currently, none of these systems have reduced MMAs to common practice.

C. Fluorochromes

As with traditional cellular analysis, the fluorochromes best suited for MMAs are dependent upon the excitation and emission wavelengths available with the cytometer. Many of the small flow systems, including the Luminex 100, the BD FACSArray, and the Guava system (Millipore) utilize a 532 nm YAG laser for excitation. Excitation of R-phycoerythrin (PE) at 532 nm yields a higher extinction coefficient and quantum yield compared with excitation at 488 nm or any other FL2 fluorochrome. There is also low Raman scattering, which occurs at 658 nm and, therefore, interferes minimally with PE emission at 578 nm. Thus, background signals are low and the signal-to-noise ratios are high with PE as the reporter fluorochrome for MMAs.

An important lesson learned from the original Luminex FlowMetrix™ system is the importance of the separation of emission spectra between classification dyes and quantitation dyes. The FlowMetrix system was integrated with a BD FACScan or FACSCalibur cytometer and the two classification dyes and the detection dye were excited with a 488 nm laser. Whereas spectra for classification dyes could overlap and still be properly classified, significant spectral overlap between the quantitation dye and classification dyes significantly reduced system usability due to the need for compensation. With multiple microsphere sets that contained variable amounts of classification dyes, compensation to remove the quantitation dye signal overlap into these different populations often became difficult. The current Luminex 100 system has solved this problem by integrating a two-laser system with classification and quantitation fluorochromes that do not overlap spectrally.

IV. Methods

Assay development starts with attaching capture molecules to the surface of microspheres. Capture molecules, including nucleic acids, antibodies, peptides, carbohydrates, and other ligands, can be hydrophobically adsorbed or covalently coupled to a surface functionality on the microsphere. Hydrophobic adsorption is performed in much the same manner as adsorption to ELISA plates; however, adsorption is nonspecific and does not covalently bind the capture molecule to the surface. This can be a significant pitfall when performing multiplexed assays since desorption of capture molecules from one microsphere set and resorption to another set can lead to apparent cross-reactivity in assays.

Methods for covalent coupling of capture molecules vary with the microsphere's surface functionality and the available functionality of the capture molecule. Microspheres are available with carboxyl, amine/hydrazide, maleimide, and other groups that can accept amine, carboxyl, carbohydrate, sulfhydryl, and other capture molecule linkages (Bangs Laboratories; BD Biosciences; Luminex Corp.; Polysciences; Qiagen; Radix BioSolutions, Georgetown, TX; Spherotech). There is extensive literature devoted to defining methodologies for coupling molecules of differing functionality (Hermanson, 1996; McHugh, 1994).

An additional approach to covalently attaching specific capture molecules to the surface of microspheres is through high-affinity ligand-ligate reactions such as biotin/avidin, glutathione/GST, polyhistidine/Ni chelate, or anti-immunoglobulin (Bangs Laboratories; Luminex Corp.; Invitrogen; Polysciences; Qiagen; Radix BioSolutions; Spherotech). These ligands can be inherent in the capture molecule, introduced through covalent coupling, or introduced during the molecule's synthesis. The interactions can be high affinity (e.g., biotin/avidin $K_d \sim 10e-15$) and can significantly ease the creation of coupled microspheres for the end user.

Once capture molecules are coupled to the surface of the microsphere, development of microsphere-based assays of any format follows the same regimen as assay development for other platforms. The user must decide whether the assay is to be performed as a homogeneous (one step, no wash) or heterogeneous (multiple steps with washes) assay. The format chosen will not drastically alter the development process but will dictate the amount of detection reagent used in the assay and require the user to be aware of the potential for a high-dose "hook effect" in a homogeneous assay. A high-dose "hook effect" is a condition wherein unbound analyte competes with bound analyte for limiting amounts of detection reagent. In this circumstance, a high concentration of unbound analyte reduces the signal generated from the bound analyte and thus, the signal generated will not correlate with the true analyte concentration.

The following sections describe methods for performing a capture-sandwich immunoassay, a competitive immunoassay, and a direct hybridization nucleic acid detection assay, which are the three most common formats currently being developed for MMAs.

A. Capture-Sandwich Immunoassay

The method for performing a microsphere-based capture-sandwich immunoassay is essentially the same as performing an ELISA. The recommended assay buffer, PBS-0.5% BSA-0.02% sodium azide (PBN), is also used for dilution and storage of reagents. A prewet 96-well filter-bottom microtiter plate (MultiScreen MABVN 1.2 μm, Millipore Corp., Burlington, MA) is preferred for heterogeneous assays. These assay plates are formatted similarly to an ELISA plate with blanks, standards, and samples. During the entire procedure, the fluorescent microspheres are kept in suspension and in the dark or in subdued light to minimize photobleaching. First, the capture microspheres are mixed with serial dilutions of the standards, blanks, and test samples and incubated at room temperature on a shaker until equilibrium is established, usually 1 h or less. If the assay is performed as a heterogeneous assay, then the microspheres are washed with PBS-0.05% Tween 20 by vacuum filtration after the capture phase. If tubes or paramagnetic microspheres are used, then centrifugation or magnetic separation can be performed. After the wash, the microspheres are incubated with the detection antibodies. The detection antibodies can be unlabeled, directly labeled with the detection fluorochrome, or biotin conjugated, depending on the required sensitivity, the assay format, and the sample matrix. Final detection of the assay can occur directly, with a fluorochrome-labeled anti-Ig, or fluorochrome-labeled streptavidin. If the assay is being performed as a homogeneous assay, the sample matrix will greatly influence the performance of the assay. A serum-derived sample will contain sufficient immunoglobulin to interfere with anti-Ig detection, and a tissue culture-derived sample will contain sufficient free biotin in the tissue culture media to interfere with streptavidin-based detection. All of these possible interferences must be accounted for by the use of appropriate controls and means of detection.

Initially, assays for each analyte should be developed and optimized separately. Optimization can involve titration of the amount of capture molecule coupled to the microspheres, determination of the range of the standard curves, titration of the amount of the conjugated detection antibody, or detection antibody and reporter fluorochrome, as well as the incubation times and the number of wash steps, if used. For most antibody pairs, the first optimization step can be eliminated if the concentrations (and volumes, which have also been found to be critical) that have been defined by the microsphere manufacturer, and repeatedly tested, are followed (Kellar and Douglass, 2003; Kellar et al., 2001). The range of the standard curves and the amount of the detection antibody will vary with the affinities of the antibody pairs. The most important titration is that of the detection antibody, which may need to be readjusted slightly on a regular basis. Also, fluorescent signals can be very high with the immunoassays, so MFIs should be kept within the linear range of the detectors. Once single assay standard curves are optimized, the assays are multiplexed by combining the reagents for all analytes into the same volumes used for a single analyte. The affinity of the weakest pair of antibodies will define the incubation times and the number of washes required for the multiplexed assay. A detailed protocol for performing a multiplexed assay for eight human cytokines has been reported by Kellar and Douglass (2003).

B. Competitive Immunoassay

The development of a competitive immunoassay requires a single antigen-specific antibody and a competitive antigen. Either the competitive antigen or the antigen-specific antibody can be directly coupled to the microspheres, whereas the other reagent must be labeled and act as a tracer in the presence of the samples (Bellisario et al., 2000; Park et al., 2000). The addition of the sample establishes a competition between the sample antigen and the competitive antigen for binding to the specific antibody. The assay format can be established in a true competitive format where all reagents are present at the same time. Alternatively, the sample antigen and the specific antibody can be combined, followed by the addition of the competitive antigen to measure the remaining free binding sites.

A competitive assay for thyroxine (T_4) was developed according to the procedure described by Bellisario et al. (2000). Microspheres were coupled with T_4-BSA, which was also used as the immunogen for the commercially available antibody. The microspheres (2500 in 20 μl) were added to the wells of a prewet filter plate containing 20 μl PBN, 40 μl normal human serum (diluted 1:5), 40 μl T_4 standards in dilutions ranging from 150 to 0.01 $\mu g/dl$, 20 μl of a blocking agent (mouse/rat serum 1:1), and 50 μl biotinylated anti-T_4 at 2 mg/ml. The plates were shaken for 30 min, washed three times by vacuum filtration, and then 50 μl PBN was added to each well, followed by 50 μl 8.5 $\mu g/ml$ streptavidin-PE. After a final wash, 100 μl PBN was added to each well and the microspheres were analyzed.

C. Nucleic Acid Direct Hybridization Assay

The most common multiplexed microsphere-based nucleic acid detection assay monitors the direct hybridization of a sample nucleic acid to a sequence-specific probe-conjugated microsphere. The nucleic acid sample can be labeled directly during amplification, or a number of other mechanisms, such as primer extension, oligonucleotide ligation, or rolling circle amplification, can be used to introduce a label that is dependent upon a specific hybridization event (Cai et al., 2000; Chen et al., 2000; Iannone et al., 2000; Mullinex et al., 2002). To optimize a multiplexed hybridization assay, the probe sequence and/or length can be adjusted to equalize each probe's melting temperature in order to maximize the reporter signal. Alternatively, the hybridization can be performed in the presence of one of the many chemicals demonstrated to normalize melting temperatures regardless of G/C ratios (Wood et al., 1985).

To establish a hybridization assay, prepare 10-fold dilutions from 0.1 to 100 fmol of a 5′-biotinylated complementary, synthetic oligonucleotide target to hybridize to capture oligo-coupled microspheres in order to determine the signal strength and sensitivity of the assay. The synthetic target is diluted in a final volume of 25 μl in 10 mM Tris-1 mM EDTA buffer, pH 8. Then 50 μl of microspheres (1.5×10^5 ml^{-1} of each specificity), diluted in $1.5\times$ hybridization solution (4.5 M tetramethylammonium chloride, 75 mM Tris-HCl-6 mM EDTA, pH 8, 0.15% Sarkosyl) is added to the biotinylated, complementary oligonucleotide dilution. The mixture is heated to 95 °C for 5 min and subsequently is transferred to a heat block warmed to an appropriate hybridization temperature (generally 37-55 °C). Reactions are hybridized for 20-30 min, pelleted by centrifugation, the supernatants removed, and the reactions are resuspended in 75 μl of streptavidin-PE (20 μg/ml). Reactions are incubated at room temperature for an additional 15 min and then analyzed.

Once the assay is established with synthetic target, assay performance is optimized with PCR-amplified samples. A standard PCR amplification is performed using a biotinylated reverse strand primer on known positive and known negative samples. Optimization is performed to maximize appropriate, and to minimize inappropriate, hybridization events. Optimization parameters include increasing or decreasing the volume of amplified material used in the hybridization reaction mixture and determining the optimal hybridization temperature. The optimal hybridization temperature is determined empirically by increasing or decreasing the hybridization temperature until the appropriate signals are maximized and inappropriate cross-hybridization signals are minimized.

V. Results

Quantitative multiplexed immunoassay data is generated by first gating on the size of the microspheres, selecting only single microspheres, classifying the individual microsphere subsets, and finally measuring the fluorescent intensity of the reporter fluorochrome, usually PE, per microsphere. A statistically significant number of each microsphere population is analyzed to determine the median fluorescent intensity (MFI) value. The standard curves are generated from these values and the concentrations of each analyte in each sample are extrapolated from the curves by applicable curve-fitting software. Standard curves for the chemokine IL-8 generated in a multiplexed-sandwich immunoassay are shown in Fig. 1. Figure 2 illustrates the standard curves for a MMA that combines a sandwich immunoassay for thyrotropin (TSH) with the competitive immunoassay for T_4 described in Section III.

The first immunoassays to be multiplexed on commercially available microsphere arrays were cytokine assays (Camilla *et al.*, 2001; Carson and Vignali,

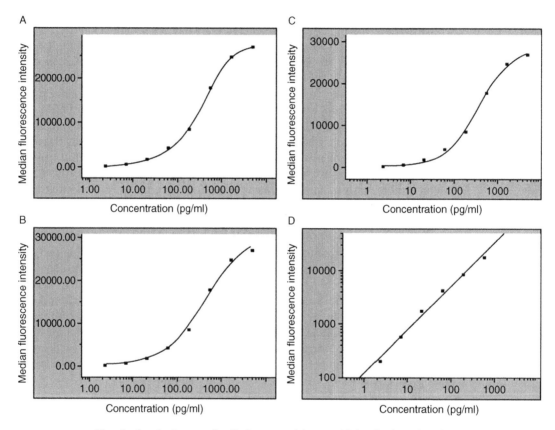

Fig. 1 Standard curves for IL-8 generated in a multiplexed microsphere-based assay. Different software packages offer different derivations of the formulas for regression analysis of standard curves. (A) An eight-point curve for IL-8 was transformed by a weighted five-parameter logistic curve-fitting model. Data analysis was performed with Bio-Plex Manager 3.0 software, Bio-Rad Laboratories, Hercules, CA. (B) The same data presented in (A) was transformed by a four-parameter logistic model with Bio-Plex Manager 3.0. (C) The same data presented in (A) was transformed by a four-parameter logistic formula with Sigma Plot 2000 software, SPSS, Chicago, IL. (D) The lower six points of the data presented in (A) were subjected to linear regression analysis of a logarithmic transformation of the values with SigmaPlot 2000 software. The lower sensitivity of the assay is visually apparent with this regression model.

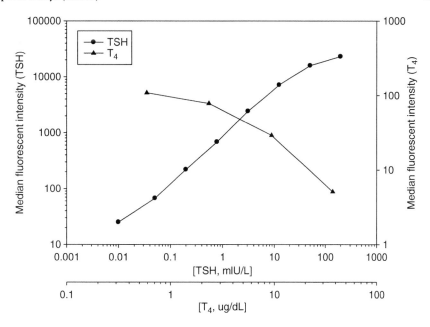

Fig. 2 A competitive immunoassay for thyroxine (T_4) and a sandwich immunoassay for thyrotropin (TSH) were combined into one multiplexed immunoassay performed as described by Bellisario *et al.* (2000). The concentration units for each analyte differ. A linear regression analysis was performed on a log transformation of the data with SigmaPlot 2000 software (Maress Lacuesta, Dr Carol Worthman, and K.L.K., unpublished data).

1999; Cook *et al.*, 2001; Hutchinson *et al.*, 2001; Kellar *et al.*, 2001; Mahanty *et al.*, 2001; Oliver *et al.*, 1998; Tripp *et al.*, 2000). The complex interactions of cytokines, chemokines, and growth factors are ideally suited to simultaneous measurements from the same sample. There is a great potential for MMAs to generate valuable baseline data on the levels of these important analytes in the serum, plasma, and other body fluids of normal subjects, as well as those with a range of infectious and immunologic diseases, but the data are emerging slowly.

Nucleic acid hybridizations are performed by using microspheres coupled with an oligonucleotide specific for the sequence of interest. Labeled complementary oligonucleotides or labeled PCR products are hybridized to these microspheres under experimentally defined conditions. Unless the nucleic acid amplification is performed under linear amplification conditions, the results will be only qualitative. A titration of complementary oligonucleotides defines only the theoretical sensitivity of the assay, since hybridization of an amplicon can be significantly affected by any secondary structure of the amplicon sequence. Figure 3 illustrates hybridization data generated both from titration of a complementary oligonucleotide and from titration of an amplicon.

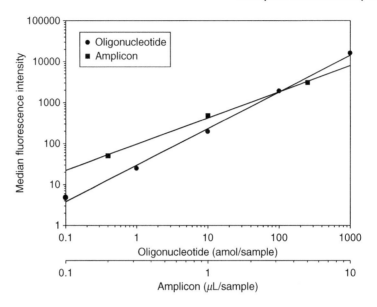

Fig. 3 Titration of oligonucleotide and PCR amplicon in a nucleic acid hybridization assay. A sequence-specific oligonucleotide containing a 5′ C12 spacer with an amino terminus was coupled covalently to Luminex 100 carboxylated microspheres. A titration of 5′ biotinylated complementary oligonucleotide or a titration of a 5′ biotinylated PCR amplicon (242 bases) was hybridized and detected as described in the text. Linear regression analysis was applied to the data that defines the sensitivity of the assay. Solid circles correspond to the sequence-specific oligonucleotide hybridization and solid squares represent the PCR amplicon hybridization.

VI. Software

For data analysis of quantitative multiplex immunoassays, the choice of software is critical. MMAs can extend over a range of 4–5 logs and require a more complex curve-fitting formulation than that used for ELISAs that extend over 2–3 logs. A five-parameter (5-P) logistic curve-fitting model that includes weighting algorithms to optimally fit the nonsymmetrical sigmoidal curves generated by these expanded standard curves is included in most software programs written for instruments for multiplexed assay analysis (Bio-Plex Manager for Bio-Plex 200, Bio-Rad; Cytometric Bead Array software for BD FACSArray, BD BioSciences; xEponent Software for Luminex 100 and 200, Luminex Corp.) and additional software packages are available directly from software specialty companies (MasterPlex CT, QT, and GT, MiraiBio, Alameda, CA; StarStation, Applied Cytometry Systems, Sheffield, UK; StatLIA, Brendan Scientific, Grosse Pointe Farms, MI). Figure 1 illustrates the transformation of the MFI values for a standard curve for IL-8 with different curve-fitting routines. The weighted 5-P formula provides a better fit than the 4-P formulas based on the Fit Probability scores (0.2819 vs. 0.1047).

Multiplexed technology yields not only many values per sample but also is applicable to high sample throughput and thus, can yield considerable data. Undoubtedly, bioinformatics will play an important role in data management as MMAs proliferate.

VII. Critical Aspects of the Methodology

A. Handling the Microspheres

The success of MMAs is often dependent on the handling of the microspheres. Sonicating, vortexing, and shaking the microspheres in tubes or on the filter plate are crucial for maintaining single-bead suspensions that facilitate robust ligand binding, rapid acquisition on the cytometer, and good assay reproducibility. However, the microspheres are susceptible to stripping of molecules from their surface if a filter-bottom microtiter plate is used with a horizontal mixer (Kellar et al., 2001).

B. Optimization and Validation of Immunoassays

The sensitivity and specificity of quantitative multiplexed immunoassays for cytokines or other antigens is directly dependent on the affinity and specificity of the individual antibodies. Monoclonal or polyclonal antibodies have been used effectively, but affinity and specificity vary from lot to lot with polyclonals. For sandwich immunoassays, the antibody pairs should be of high affinity ($K_d >$ 10e−8) and should react with different epitopes of the antigen. Some antibody pairs that work well with ELISAs may not perform as well in multiplexed immunoassays, so antibodies from different sources may need to be tested as both capture and detection antibodies to determine the best combinations. Most importantly, cross-reactivity between multiple antibodies and antigens should be carefully examined to ensure that the specificity of each antigen for its antibodies is not affected by other components of the assay (Camilla et al., 2001; de Jager et al., 2003; Kellar and Douglass, 2003).

The slopes of the standard curves for an analyte may vary with the sample matrix. As an example, if plasma is to be assayed, the standards should be assayed in normal plasma or a complex of plasma proteins that simulates plasma as much as possible. Plasma, serum, and other body fluids contain factors, such as soluble receptors, rheumatoid factor, heterophilic antibodies, and human antianimal antibodies, which will produce false positive or false negative signals and adversely affect quantitation (Hennig et al., 2000; Willman et al., 2001). Spiking of normal matrix proteins with standards and measuring the recovery of each analyte will uncover possible interference from the sample matrix. Blocking proteins may need to be included to significantly reduce or eliminate any known or potential sample matrix effects (Kellar and Iannone, 2002; Kellar et al., 2001).

Multiplexed microsphere-based immunoassays are at least as reproducible as the traditional, monoplex ELISA (Camilla *et al.*, 2001; Chen *et al.*, 1999; Kellar *et al.*, 2001). To ensure reproducibility, proper quality controls should be instituted. Each preparation of microspheres should be evaluated by preset criteria and standard curves should be given a trial run to optimize the assay reagents before samples are analyzed (Kellar and Douglass, 2003). Calibration and proper maintenance of the flow cytometer are critical. The median fluorescent intensity of 100 microspheres per analyte per sample is generally reported (Bellisario *et al.*, 2001; Carson and Vignali, 1999; Seideman and Peritt, 2002). Since each microsphere serves as the solid support for an individual assay, the MFI is highly significant. Replicates can be used to correct for pipetting errors, particularly if small sample volumes are measured.

Competitive immunoassays are relatively simple, relying on one antibody to bind the analyte captured on the microsphere unless competed off by analyte present in the sample (Park *et al.*, 2000; Opalka *et al.,* 2003). A small capture antigen may need to be complexed with a carrier molecule, such as a steroid-BSA conjugate, or linked to the microsphere via a carbon bridge to ensure that it is physically available for binding and not sterically hindered by the microsphere surface.

C. Optimization and Validation of Oligonucleotide Assays

Specificity and hybridization temperature optimums must be determined to ensure proper results. Hybridization temperature optimization is determined empirically based on the amount of signal generated with a positive control sample versus a negative control sample. The temperature is increased to maximize specificity without compromising signal. Specificity is best tested with samples positive for each individual marker. If not available, then specificity can be tested with individual complementary oligonucleotides. For the analysis of SNPs, specificity is optimal when the variant base is as close to the middle of the probe sequence as possible. To best validate cross-reactivity, a number of organizations exist that will provide samples of known sequence (Coriell Institute for Medical Research, Camden, NJ). In addition, commercial and public software packages exist that aid in probe design and elucidation of potential cross-reactivity issues (Vector NTI, Informax, Fredrick, MD; Visual OMP, DNA Software, Inc., Ann Arbor, MI; PrimerPlex, Premier Biosoft International, Palo Alto, CA; ProbeITy, Celadon Laboratories, Hyattsville, MD). For hybridization methods that "zip" the probe as opposed to the reaction product onto microspheres for analysis, Kaderali *et al.* (2003) have developed a program to assist in the design and selection of tags and antitags.

VIII. Comparison with Other Methods

ELISA remains the gold standard for immunoassay measurements of soluble analytes, and it will stay that way until a technology can equal the sensitivity, ease of operation, cost, and availability. MMAs are an attractive alternative to ELISAs because of the multiplexing capability. These assays have been demonstrated to be at least as or more sensitive, as reliable, and as easy to perform as ELISAs (Camilla *et al.*, 2001; Carson and Vignali, 1999; Chen *et al.*, 1999; Cook *et al.*, 2001; de Jager *et al.*, 2003; Kellar *et al.*, 2001; Oliver *et al.*, 1998; Prabhakar *et al.*, 2002; Tárnok *et al.*, 2003). In addition to the multiplexing capabilities, the smaller sample requirements and the generally lower cost per data point are desirable features. The major drawbacks are cost of the instrumentation, the limited palettes of commercial kits, and the inflexibility of those kits to provide only the analytes needed by the end user. As more and more kit manufacturers adopt the MMA platform, prices will decrease and the palette of assay kits will increase greatly.

A number of other multiplexed assay platforms are available to challenge the ELISA market. Thermo Fisher Scientific (Pierce) (Rockford, IL), Ray Biotech, Inc. (Norcross, GA), and others (Ekins *et al.*, 1990) have developed two-dimensional or flat arrays that involve spotting capture antibodies onto the bottom of microtiter plates, membranes, or glass slides to produce the array. The primary drawbacks of this technology are cost and lack of flexibility. First, the cost involves either the spotting equipment or purchasing prespotted arrays, as well as the cost of a sensitive scanning system capable of discriminating signal from each individual spot. Another drawback is the decreased flexibility compared with MMAs. To modify the components of a microsphere-based array requires the addition or removal of individual microsphere sets, whereas a spotted array requires reformatting and respotting of the entire array.

An additional technology that is capable of multiplexing is the multiple HPLC affinity column design introduced by Phillips (2001). This system involves generating a separate affinity column for each analyte of interest and performing the analysis in serial. Samples are labeled with Cy5 and then are detected with laser-induced fluorescence following elution from each column. The system demonstrates sensitivity and flexibility comparable to MMAs. However, the system is limited due to the complexity of developing and maintaining each column that can only be used for about 200 analyses. In addition, unlike two-dimensional spotted arrays that compensate for analyte labeling differences in different samples by comparing the ratio of two samples each labeled with Cy3 and Cy5, this system only labels proteins with Cy5. This format introduces a bias due to variation in each labeling reaction and variation in labeling due to analyte concentration differences within samples. Many of these issues are minor and can be addressed; however, until a commercial entity manufactures the system, addresses the limitation, and produces a cartridge format that is user friendly, the utility of the system will remain limited.

IX. Future Directions

Multiplexed assays hold the potential to revolutionize the way biology is studied and understood. The increased knowledge obtained from multiple bits of information about a sample can be synergistic in understanding the biologic pathways and events. MMAs hold great promise for becoming the standard format for multiplexed assays due to their flexibility, ease of use, and the compact, affordable equipment required to perform the assays. Progress toward greater sensitivity leveraged from larger signal to noise ratios as well as amplification techniques (Mullinex *et al.*, 2002) will provide an enhancement over the current technology. This is especially valuable when comparing diseased individuals with normals who may have extremely low concentrations of specific proteins in their serum. In addition, multiplexing provides a discovery tool for identifying new diagnostic markers of disease, turning these into prognostic, rather than diagnostic, disease markers.

For drug discovery, smaller, faster, and cheaper is critical and the ability to miniaturize, automate, and multiplex MMAs provides a means for more rapid, cost-efficient drug development. The small size of the microspheres used for MMAs allows for assays to be performed in very low volumes, which means less costly reagent is used. Homogeneous formats and the physical characteristics of the microspheres make MMAs amenable to robotic handling, while multiplexing generates more information on each compound within a single assay. Advances in sample acquisition, such as that outlined by Ramirez *et al.* (2003), increase throughput to approximately 1.5 s/sample, which would provide an enormous enhancement to MMA technology.

Future practical and technological advancements will greatly enhance the adoption of MMA technology. Within clinical diagnostics, a greater understanding of interfering factors in body fluids and how to block them, as well as improvements in sample preparation, will greatly increase the robustness, accuracy, and precision of MMAs. From a practical perspective, as new diagnostic panels for disease are developed, acceptance by the medical community, the FDA, and healthcare providers will be required for widespread usage. For nucleic acid applications, increases in the size of microsphere arrays, as well as enhancements in signal amplification will broaden the adoption of MMAs into the genomics world. There has been logarithmic growth in the research, development, and commercialization of MMAs since the beginning of this decade, and this trend continues.

Acknowledgment

We gratefully acknowledge the contribution of the data in Fig. 2 by collaborators, Maress Lacuesta and Dr Carol Worthman, Department of Anthropology, Emory University, Atlanta, GA. We also appreciate the support of Drs Robert Wohlhueter and Carolyn Black at the Centers for Disease Control and Prevention.

References

Armstrong, B., Stewart, M., and Mazumder, A. (2000). Suspension arrays for high throughput, multiplexed single nucleotide polymorphism genotyping. *Cytometry* **40,** 102–108.

Barker, R. L., Worth, C. A., and Peiper, S. C. (1994). Cytometric detection of DNA amplified with fluorescent primers: Applications to analysis of clonal *bcl-2* and IgH gene rearrangements in malignant lymphomas. *Blood* **83,** 1079–1085.

Bellisario, R., Colinas, R. J., and Pass, K. A. (2000). Simultaneous measurement of thyroxine and thyrotropin from newborn dried blood-spot specimens using a multiplexed fluorescent microsphere immunoassay. *Clin. Chem.* **46**, 1422–1424.

Bellisario, R., Colinas, R. J., and Pass, K. A. (2001). Simultaneous measurement of antibodies to three HIV-1 antigens in newborn dried blood-spot specimens using a multiplexed microsphere-based immunoassay. *Early Hum. Dev.* **64**, 21–25.

Bortolin, S., Black, M., Modi, H., Boszko, I., Kobler, D., Fieldhouse, D., Lopes, E., Lacroix, J. M., Grimwood, R., Wells, P., Janeczko, R., and Zastawny, R. (2004). Analytical validation of the tag-it high-throughput microsphere-based universal array genotyping platform: Application of the multiplex detection of a panel of thrombophilia-associated single-nucleotide polymorphisms. *Clin. Chem.* **50**, 2028–2036.

Cai, H., White, P. S., Torney, D., Deshpande, A., Wang, Z., Marrone, B., and Nolan, P. (2000). Flow cytometry-based minisequencing: A new platform for high-throughput single-nucleotide polymorphism scoring. *Genomics* **66**, 135–143.

Camilla, C., Mély, L., Magnan, A., Casano, B., Prato, S., Debono, S., Montero, F., Defoort, J. P., Martin, M., and Fert, V. (2001). Flow cytometric microsphere-based immunoassay: Analysis of secreted cytokines in whole-blood samples from asthmatics. *Clin. Diagn. Lab. Immunol.* **8**, 776–784.

Carson, R. T., and Vignali, D. A. A. (1999). Simultaneous quantitation of 15 cytokines using a multiplexed flow cytometric assay. *J. Immunol. Methods* **227**, 41–52.

Chen, R., Lowe, L., Wilson, J. D., Crowther, E., Tzeggai, K., Bishop, J. E., and Varro, R. (1999). Simultaneous quantification of six human cytokines in a single sample using microparticle-based flow cytometric technology. *Clin. Chem.* **45**, 1693–1694.

Chen, J., Iannone, M. A., Li, M.-S., Taylor, J. D., Rivers, P., Nelson, A. J., Slentz-Kesler, K. A., Roses, A., and Weiner, M. P. (2000). A microsphere-based assay for multiplexed single nucleotide polymorphism analysis using single base chain extension. *Genome Res.* **10**, 549–557.

Christodoulides, N., Floriano, P. N., Miller, C. S., Ebersole, J. L., Mohanty, S., Dharshan, P., Griffin, M., Lennart, A., Ballard, K. L. M., King, Jr., C. P., Langub, M. C., Kryscio, R. J., *et al.* (2007). Lab-on-a-chip methods for point-of-care measurements of salivary biomarkers of periodontitis. *Ann. N. Y. Acad. Sci.* **1098**, 411–428.

Colinas, R. J., Bellisario, R., and Pass, K. A. (2000). Multiplexed genotyping of β-globin variants from PCR-amplified newborn blood spot DNA by hybridization with allele-specific oliogodeoxynucleotides coupled to an array of fluorescent microspheres. *Clin. Chem.* **46**, 996–998.

Collins, D. P. (2000). Cytokine and cytokine receptor expression as a biological indicator of immune activation: Important considerations in the development of *in vitro* model systems. *J. Immunol. Methods* **243**, 125–145.

Cook, E. B., Stahl, J. L., Lowe, L., Chen, R., Morgan, E., Wilson, J., Varro, R., Chan, A., Graziano, F. M., and Barney, N. P. (2001). Simultaneous measurement of six cytokines in a single sample of human tears using microparticle-based flow cytometry: Allergics vs. non-allergics. *J. Immunol. Methods* **254**, 109–118.

Defoort, J. P., Martin, M., Casano, B., Prato, S., Camilla, C., and Fert V. (2000). Simultaneous detection of multiplex-amplified human immunodeficiency virus type 1 RNA, hepatitis C virus RNA, and hepatitis B virus DNA using a flow cytometer microsphere-based hybridization assay. *J. Clin. Microbiol.* **38**, 1066–1071.

De Jager, W., te Velthuis, H., Prakken, B. J., Kuis, W., and Rijkers, G. T. (2003). Simultaneous detection of 15 human cytokines in a single sample of stimulated peripheral blood mononuclear cells. *Clin. Diagn. Lab. Immunol.* **10**, 133–139.

Dunbar, S. A. and Jacobson, J. W. (2000). Application of the Luminex LabMAP in rapid screening for mutations in the cystic fibrosis transmembrane conductance regulator gene: A pilot study. *Clin. Chem.* **46**, 1498–1500.

Dunbar, S. A., Vander Zee, C. A., Oliver, K. G., Karem, K. L., and Jacobson, J. W. (2003). Quantitative, multiplexed detection of bacterial pathogens: DNA and protein applications of the Luminex LabMAP™ system. *J. Microbiol. Methods* **53**, 245–252.

Ekins, R. P., Chu, R., and Biggart, E. (1990). The development of microspot multianalyte radiometric immunoassay using dual fluorescent-labelled antibodies. *Anal. Chim. Acta* **227,** 73–96.

Fitzgerald, C., Collins, M., van Duyne, S., Mikoleit, M., Brown, T., and Fields, P. (2007). Multiplex, bead-based suspension array for molecular determination of common *Salmonella* serogroups. *J. Clin. Microbiol.* **45,** 3323–3334.

Fulton, R. J., McDade, R. L., Smith, P. L., Kienker, L. J., and Kettman, Jr., J. R. (1997). Advanced multiplexed analysis with the FlowMetrix system. *Clin. Chem.* **43,** 1749–1756.

Han, M., Gao, X., Su, J. Z., and Nie, S. (2001). Quantum-dot-tagged microbeads for multiplexed optical coding of biomolecules. *Nat. Biotech.* **19,** 631–635.

Hennig, C., Rink, L., Fagin, U., Jabs, W. J., and Kirchner, H. (2000). The influence of naturally occurring heterophilic anti-immunoglobulin antibodies on direct measurement of serum proteins using sandwich ELISAs. *J. Immunol. Methods* **235,** 71–80.

Hermanson, G. T. (1996). "Bioconjugate Techniques." Academic Press, San Diego, CA.

Horan, P. K., and Wheeless, Jr., L. L. (1977). Quantitative single cell analysis and sorting. *Science* **198,** 149–157.

Hutchinson, K. L., Villinger, F., Miranda, M. E., Ksiazek, T. G., Peters, C. J., and Rollin, P. E. (2001). Multiplex analysis of cytokines in the blood of cynomolgus macaques naturally infected with Ebola virus (Reston serotype). *J. Med. Virol.* **65,** 561–566.

Iannelli, D., D'Apice, L., Cottone, C., Viscardi, M., Scala, F., Zoina, A., Del Sorbo, G., Spigno, P., and Capparelli, R. (1997). Simultaneous detection of cucumber mosaic virus, tomato mosaic virus and potato virus Y by flow cytometry. *J. Virol. Methods* **69,** 137–145.

Iannone, M. A., Taylor, J. D., Chen, J., Li, M.-S., Rivers, P., Slentz-Kesler, K. A., and Weiner, M. P. (2000). Multiplexed single nucleotide polymorphism genotyping by oligonucleotide ligation and flow cytometry. *Cytometry* **39,** 131–140.

Janossy, G., Jani, I. V., Kahan, M., Barnett, D., Mandy, F., and Shapiro, H. (2002). Precise CD4 T-cell counting using red diode laser excitation: For richer, for poorer. *Cytometry* **50,** 78–85.

Kaderali, L., Deshpande, A., Nolan J. P., and White, P. S. (2003). Primer-design for multiplexed genotyping. *Nucleic Acid Res.* **31,** 1796–1802.

Kellar, K. L., and Douglass, J. P. (2003). Protocol. Multiplexed microsphere-based flow cytometric immunoassays for human cytokines. *J. Immunol. Methods* **279,** 277–285.

Kellar, K. L., and Iannone, M. A. (2002). Multiplexed microsphere-based flow cytometric assays. *Exp. Hematol.* **30,** 1227–1237.

Kellar, K. L., Kalwar, R. R., Dubois, K. A., Crouse, D., Chafin, W. D., and Kane, B.-E. (2001). Multiplexed fluorescent bead-based immunoassays for quantitation of human cytokines in serum and culture supernatants. *Cytometry* **45,** 27–36.

Kettman, J. R., Davies, T., Chandler, D., Oliver, K. G., and Fulton, R. J. (1998). Classification and properties of 64 multiplexed microsphere sets. *Cytometry* **33,** 234–243.

Keyes, K. A., Mann L., Cox, K., Treadway, P., Iversen, P., Chen, Y.-F., and Teicher, B. A. (2003). Circulating angiogenic growth factor levels in mice bearing human tumors using Luminex multiplex technology. *Cancer Chemother. Pharmacol.* **51,** 321–327.

Lu, J., Getz, G., Miska, E. A., Alvarez-Saavedra, E., Lamb, J., Peck, D., Sweet-Cordero, A., Ebert, B. L., Mak, R. H., Ferrando, A. A., Downing, J. R., Jacks, T., *et al.* (2005). MicroRNA expression profiles classify human cancers. *Nature* **435,** 834–838.

Mahanty, S., Bausch, D. G., Thomas, R. L., Goba, A., Bah, A., Peters, C. J., and Rollin, P. E. (2001). Low levels of interleukin-8 and interferon-inducible protein-10 in serum are associated with fatal infections in acute Lassa fever. *J. Infect. Dis.* **183,** 1713–1721.

Martins, T. B. (2002). Development of internal controls for the Luminex instrument as part of a multiplex seven-analyte viral respiratory antibody profile. *Clin. Diagn. Lab. Immunol.* **9,** 41–45.

Martins, T. B., Pasi, B. M., Pickering, J. W., Jaskowski, T. D., Litwin, C. M., and Hill, H. R. (2002). Determination of cytokine responses using a multiplexed fluorescent microsphere immunoassay. *Am. J. Clin. Pathol.* **118,** 346–353.

McHugh, T. M. (1994). Flow microsphere immunoassay for the quantitative and simultaneous detection of multiple soluble analytes. *Methods Cell Biol.* **42**, 575–595.

McHugh, T. M., and Stites, D. P. (1991). Application of bead-based assays for flow cytometry analysis. *Clin. Immunol. Newslett.* **11**, 60–64.

McHugh, T. M., Stites, D. P., Casavant, C. H., and Fulwyler, M. J. (1986). Flow cytometric detection and quantitation of immune complexes using human C1q-coated microspheres. *J. Immunol. Methods* **95**, 57–61.

McHugh, T. M., Miner, R. C., Logan, L. H., and Stites, D. P. (1988a). Simultaneous detection of antibodies to cytomegalovirus and herpes simplex virus by using flow cytometry and a microsphere-based fluorescence immunoassay. *J. Clin. Microbiol.* **26**, 1957–1961.

McHugh, T. M., Stites, D. P., Busch, M. P., Krowka, J. F., Stricker, R. B., and Hollander, H. (1988b). Relation of circulating levels of human immunodeficiency virus (HIV) antigen, antibody to p24, and HIV-containing immune complexes in HIV-infected patients. *J. Infect. Dis.* **158**, 88–1091.

McHugh, T. M., Wang, Y. J., Chong, H. O., Blackwood, L. L., and Stites, D. P. (1989). Development of a microsphere-based fluorescent immunoassay and its comparison to an enzyme immunoassay for the detection of antibodies to three antigen preparations from *Candida albicans*. *J. Immunol. Methods* **116**, 213–219.

McHugh, T. M., Viele, M. K., Chase, E. S., and Recktenwald, D. J. (1997). The sensitive detection and quantitation of antibody to HCV by using a microsphere-based immunoassay and flow cytometry. *Cytometry* **29**, 106–112.

Mullinex, M. C., Sivakamasundari, R., Feaver, W. J., Krishna, R. M., Sorette, M. P., Datta, H. J., Morosan, D. M., and Piccoli, S. P. (2002). Rolling circle amplification improves sensitivity in multiplex immunoassays on microspheres. *Clin. Chem.* **48**, 1855–1858.

Oliver, K. G., Kettman, J. R., and Fulton, R. J. (1998). Multiplexed analysis of human cytokines by use of the FlowMetrix system. *Clin. Chem.* **44**, 2057–2060.

Opalka, D., Lachman, C. E., MacMullen, S. A., Jansen, K. U., Smith, J. F., Cirmule, N., and Esser, M. T. (2003). Simultaneous quantitation of antibodies to neutralizing epitopes on virus-like particles for human papillomavirus types 6, 11, 16, and 18 by a multiplexed Luminex assay. *Clin. Diagn. Lab. Immunol.* **10**, 108–115.

Park, M. K., Briles, D. E., and Nahm, M. H. (2000). A latex bead-based flow cytometric immunoassay capable of simultaneous typing of multiple pneumococcal serotypes (multibead assay). *Clin. Diagn. Lab. Immunol.* **7**, 486–489.

Phillips, T. M. (2001). Multi-analyte analysis of biological fluids with a recycling immunoaffinity column array. *J. Biochem. Biophys. Methods* **49**, 253–262.

Pickering, J. W., Martins, T. B., Greer, R. W., Schroder, M. C., Astill, M. E., Litwin, C. M., Hildreth, S. W., and Hill, H. R. (2002a). A multiplexed fluorescent microsphere immunoassay for antibodies to pneumococcal capsular polysaccharides. *Microbiol. Infect. Dis.* **117**, 589–596.

Pickering, J. W., Martins, J. B., Schroder, M. C., and Hill, H. R. (2002b). Comparison of a multiplex flow cytometric assay with enzyme-linked immunosorbent assay for quantitaion of antibodies to tetanus, diphtheria and *Haemophilus influenzae* type b. *Clin. Diagn. Lab. Immunol.* **9**, 872–876.

Prabhakar, U., Eirikis, E., and Davis, H. M. (2002). Simultaneous quantification of proinflammatory cytokines in human plasma using the LabMAP assay. *J. Immunol. Methods* **260**, 207–218.

Ramirez, S., Aiken, C. T., Andrzejewski, B., Sklar, L. A., and Edwards, B. S. (2003). High-throughput flow cytometry: Validation in microvolume bioassays. *Cytometry* **53A**, 55–65.

Scillian, J. J., McHugh, T. M., Busch, M. P., Tam, M., Fulwyler, M. J., Chien, D. Y., and Vyas, G. N. (1989). Early detection of antibodies against rDNA-produced HIV proteins with a flow cytometric assay. *Blood* **73**, 2041–2048.

Seideman, J., and Peritt, D. (2002). A novel monoclonal antibody screening method using the Luminex-100 microsphere system. *J. Immunol. Methods* **267**, 165–171.

Shapiro, H. M. (2003). "Practical Flow Cytometry" 4th edn., Wiley, Hoboken, NJ.

Smith, P. L., WalkerPeach, C. R., Fulton, R. J., and DuBois, D. B. (1998). A rapid, sensitive, multiplexed assay for detection of viral nucleic acids using the FlowMetrix system. *Clin. Chem.* **44**, 2054–2056.

Spiro, A., Lowe, M., and Brown, D. (2000). A bead-based method for multiplexed identification and quantitation of DNA sequences using flow cytometry. *Appl. Environ. Microbiol.* **66**, 4258–4265.

Tárnok, A., Hambsch, J., Chen, R., and Varro, R. (2003). Cytometric bead array to measure six cytokines in twenty-five microliters of serum. *Clin. Chem.* **49**, 1000–1002.

Taylor, J. D., Briley, D., Nguyen, Q., Long, K., Iannone, M. A., Li, M.-S., Ye, F., Afshari, A., Lai, E., Wagner, M., Chen, J., and Weiner, M. P. (2001). Flow cytometric platform for high-throughput single nucleotide polymorphism analysis. *BioTechniques* **30**, 661–669.

Tripp, R. A., Jones, L., Anderson, L. J., and Brown, M. P. (2000). CD40 ligand (CD154) enhances the Th1 and antibody responses to respiratory syncytial virus in the BALB/c mouse. *J. Immunol.* **164**, 5913–5921.

Wallace, J., Zhou, Y., Usmani, G. N., Reardon, M., Newburger, P., Woda, B., and Pihan, G. (2003). BARCODE-ALL: Accelerated and cost-effective genetic risk stratification in acute leukemia using spectrally addressable liquid bead microarrays. *Leukemia* **17**, 1411–1413.

Weber, J., Sondak, V. K., Scotland, R., Phillip, R., Wang, F., Rubio, V., Stuge, T. B., Groshen, S. G., Gee, C., Jeffrey, G. G., Sian, S., and Lee, P. P. (2003). Granulocyte-macrophage-colony stimulating factor added to a multipeptide vaccine for resected Stage II melanoma. *Cancer* **97**, 186–200.

Willman, J. H., Hill, H. R., Martins, T. B., Jaskowski, T. D., Ashwood, E. R., and Litwin, C. M. (2001). Multiplex analysis of heterophil antibodies in patients with indeterminate HIV immunoassay results. *Am. J. Clin. Pathol.* **115**, 764–769.

Wood, W. I., Gitschier, J., Lasky, L. A., and Lawn, R. M. (1985). Base composition-independent hybridization in tetramethylammonium chloride: A method for oligonucleotide screening of highly complex gene libraries. *Proc. Natl. Acad. Sci. USA* **82**, 1585–1588.

Yang, L., Tran, D. K., and Wang, X. (2001). BADGE, BeadsArray for the Detection of Gene Expression, a high-throughput diagnostic bioassay. *Genome Res.* **11**, 1888–1898.

Ye, F., Li, M.-S., Taylor, J. D., Nguyen, Q., Colton, H. M., Casey, W. M., Wagner, M., Weiner, M. P., and Chen, J. (2001). Fluorescent microsphere-based readout technology for multiplexed human single nucleotide polymorphism analysis and bacterial identification. *Hum. Mutat.* **17**, 305–316.

CHAPTER 8

Biohazard Sorting

**Ingrid Schmid,★ Mario Roederer,† Richard A. Koup,†
David Ambrozak,† Stephen P. Perfetto,† and Kevin L. Holmes†,‡**

★Department of Hematology/Oncology
David Geffen School of Medicine
University of California at Los Angeles
Los Angeles, California 90095

†Immunology Laboratory, Vaccine Research Center
National Institute of Allergy and Infectious Diseases
National Institute of Health
Bethesda, Maryland 20892

‡Chief, Flow Cytometry Section Research Technologies Branch
NIAID, NTH, DHHS Bethesda, Maryland 20892

I. Recent Developments

Since the original publication of this chapter, "Biohazard Sorting" in 2004, the ISAC Biosafety Committee has published *standards* for sorting unfixed cells in 2007 (Cytometry, *Cytometry Part A* 71A: 414–437, 2007, see also location http://www.isac-net.org/content/category/9/135/46/ and the related chapter in Current Protocols in Cytometry, CPC Unit 3.6, 2007). The standard supersedes prior ISAC Guidelines and was driven by the following practices and procedural advances since 1997:

- Advances in cell sorter technology made high-speed cell sorting more prevalent and changed the biohazard potential of cell-sorting experiments.
- New and less expensive options for personal protection of operators have become available.
- Instrument manufacturers responded to the need for improved operator protection and have introduced instrumentation containing novel safety features.
- Newly designed safety attachments for cell sorters have become commercially available.
- With the availability of compact, easier to operate sorters many more laboratories have incorporated cell sorting into their experimentation, but often do not have dedicated operators to perform cell-sorting experiments.
- Simpler, bead-based techniques for measuring the efficiency of aerosol containment during cell sorting have been developed.
- Advances in cell biology have increased the need for live infectious cell sorting for cell culture and experiments involving molecular genetics.

The current chapter was published in 2004 and already contains many recent aspects and newer aerosol testing procedures in-place for the flow cytometry community conducting infectious cell-sorting experiments. However, it is important to note that assignment of appropriate procedures and practices for samples with variable and sometimes complex levels of biohazard risk potential such as genetically engineered cell preparations and a variety of unfixed but "pretested" samples, still remain in debate. Currently, the ISAC Biosafety Committee together with a special NIH Safety Committee and the ABSA (American Biological Safety Association) are involved in updating these standards. A primary goal for these groups is to clarify the risk assessment procedures for the assignment of biosafety practices when sorting "questionable" samples, that is, to match engineering controls with potential risk. Furthermore, these future updates will provide a foundation for modernizing the risk assessment to reflect the present knowledge and occupational safety practices as outlined in the Biosafety in Microbiological and Biomedical Laboratories (BMBL) published by the CDC. (Center of Disease Control and Prevention, 2007, 5th ed., Biosafety in Microbiological and Biomedical Laboratories, see located at http://www.cdc.gov/od/ohs/biosfty/bmbl5/bmbl5toc.htm).

A. What has Changed?

In 2009, the most common procedure for measuring aerosol containment is the Fluorescent Glo-Germ procedure, see Section III.B.3.d. This procedure has replaced the more complex bacteriophage procedures; see Sections III.B.3.b. and 3.c.

The importance of PPE, especially respirator protection is unchanged, but for laboratories utilizing PAPR devices there is a new model of PAPR helmet system, which offers novel design features of import (MaxAir Systems, BMDI, Irvine, CA).

This NIOSH-approved system consists of a HEPA filter hood with face shield attached to a lightweight helmet-based fan resulting in a hose-free PAPR system. Combined with a full body Tyvek suit, this system provides full body protection and increased mobility.

Instrument designs: The manufacturers of cell sorters have become more aware of the need to address aerosol containment in the design of their instruments. Cell sorters have been designed with engineering controls for aerosol containment and/or evacuation and they have become compact enough to be placed within Biological Safety cabinets or HEPA filtered air enclosures. Instrument manufacturers must continue to take into account the risks involved with sorting of unfixed samples and incorporate effective aerosol evacuation systems together with the ability to be housed within a Class II BSC.

B. What is the Future?

Assessment of containment in cell sorters: Although the Glo-Germ procedures offers many advantages for determination of aerosol containment, alternative methodology is being investigated which would provide more standardized and quantifiable measurements of containment. This includes the potential use of small particle counters based upon existing aerosol counting technologies.

Remote Monitoring of cell sorters: The ability to remotely monitor and operate cell sorters located in BSL-3 laboratories is very advantageous. Although the use of commercially available LAN-based remote PC operation is currently being used, the ability to utilize thin client server-based systems would locate much of the hardware (disk drives, CPU) outside of the BSL-3 laboratory, providing for better and more reliable backup procedures, as well as remote operation or monitoring using CCD cameras.

C. What Remains the Same?

While many and perhaps the majority of the unfixed organisms sorted in a flow cytometer facility are assigned to BSL-2 risk levels, it is important to realize that the possibility of aerosol generation during cell sorting raises the risk level to BSL-3. Hence, safety standards must be in place before sorting such samples with a high degree of infectious risk. Therefore, the following remain constant in these procedures:

• Sorters should be placed in a BSL-3 facility with engineering controls that are outlined by the CDC in the BMBL. Alternatively, for facilities without BSL-3 laboratory space, sorting may be performed within BSL-2 laboratories utilizing BSL-3 practices. However, careful risk assessment must be conducted by biosafety personnel according to the type of organisms present in the samples. The risk level will determine if samples can be safely sorted under these conditions.

- Operators must wear appropriate PPE, including respirator protection in the form of N-95, N-99, or N-100 respirators or PAPR's gloves, face shield, and lab coats or protective biosuit at all times. Ideally, operators should be outside of the BSL-3 laboratory.
- Sorters must be engineered with an aerosol evacuation system and this must be on throughout the sort.
- Operators must have procedures in place to maintain and test the aerosol evacuation system.
- Records must be maintained to show the aerosol evacuation system is functioning normally.

II. Introduction

Flow sorters are instruments capable of separating cell populations based on their physical properties or by exploiting differences in cell surface receptors, intracellular structures, or molecular expression. When these instruments are used for sorting unfixed samples known to harbor pathogens, the term *biohazard sorting* has been created to encompass aspects of this process pertinent to the protection of sorter operators, others involved in these experiments, and the environment. Although *biohazard sorting* is mostly associated with the processing of samples containing infectious agents (Giorgi, 1994), it also applies to cell sorting of unfixed human cell preparations or unfixed cells from other species that may carry pathogenic organisms known to infect humans (Schmid *et al.*, 1997b, 2003). Samples that have been treated with fixatives to inactivate infectious agents (Shapiro, 2003) are considered nonbiohazardous. However, a careful review of the effectiveness of a given fixation procedure against a certain pathogen is essential before classifying a sort as nonhazardous, and it is always advisable to use prudent safety practices even when sorting fixed cells (Schmid, 2000; Schmid *et al.*, 1997b, 2003).

III. Critical Aspects of the Procedure

A. Sample Handling

1. General Considerations

In the United States, all laboratories processing human blood must follow universal precautions as outlined in the Federal Code regulation: Occupational Exposure to Bloodborne Pathogens (National Committee for Clinical Laboratory Standards, 1997; United States Federal Code Regulation, 1991) and additional local and institutional regulations. Other countries have developed their own regulatory standards or adopted aspects of regulations for working with biological agents as mandated in the United States. Levels of containment (known as *biological safety level*, BSL) combine safety equipment, laboratory practices, and facility design. For specimens carrying infectious agents, safety levels are assessed according to the criteria set forth in the Centers for Disease Control and Prevention (CDC) publication *Biosafety in microbiological and biomedical laboratories*, 4th ed., available online at www.cdc.gov/od/ohs. As a general rule, pathogens encountered in a typical cell-sorting laboratory are transmitted by the parenteral route and require BSL-2 containment, for example, human immunodeficiency viruses (HIV-1, -2), hepatitis viruses (B, C, D (delta)) (Schmid *et al.*, 2003). Relevant details for the preparation of infectious samples containing HIV for flow cytometric analysis such as shipping and receiving of specimens, local sample transport, staining, and disposal have been described (Schmid *et al.*, 1999). Although cell sorting of HIV-infected specimens can be performed in a BSL-2 facility, because of the generation of aerosols during cell sorting, BSL-3 practices and personal protective equipment are mandatory for sorting HIV-positive samples to protect operators from potential exposure (Ferbas *et al.*, 1995; Giorgi, 1994; Perfetto *et al.*, 2003). Complete BSL-3 containment is required for the sorting of infectious agents causing serious or potentially lethal disease that are transmitted by the inhalation route (CDC, 1999).

2. Sample Preparation and Sort Acquisition

For proper operation of any cell sorter, samples must be prepared as single-cell suspensions. Aggregates of cells that are present in the sample to be sorted tend to accumulate at the interface of sample fluid and sheath fluid and can partially or completely clog-sorting nozzles. Any interruption of a biohazard sort enhances the risk of operator exposure to pathogens contained in the sort sample due to potential splashes and escape of sort aerosols during the manipulations required to resume sorting. Consequently, it is imperative that formation of cell aggregates during sample preparation be minimized and clumped cells be prevented from interfering with normal instrument operation. The composition of the culture medium used for sample resuspension can influence clumping. For instance, murine cells aggregate in the presence of EDTA and, therefore, should be sorted only in EDTA-free media. Passing samples through narrow-gauge syringes, sample vortexing before sorting, and sample agitation during sorting can help to disperse cell aggregates, but mixing should not be too intense or overused because it may cause cells to break apart. Furthermore, highly concentrated cell suspensions have an increased tendency to clump, so cells should be diluted to the lowest possible density for the sort speed used. Frozen cell samples that are thawed for sorting frequently contain dead cells, which may release DNA into the media. DNA then binds to the surface of live cells, and after centrifugation, these samples form solid aggregates leading to nozzle clogging problems. In these situations, adding 20 μg/ml of DNAse for 10 min at 37 °C can help. Furthermore, spinning samples at $300 \times g$ for 5–10 min is sufficient to pellet cells. Higher centrifugation speeds can damage cells and compact cells so densely that they are difficult to break up into singlets. Sort samples are often chilled to preserve cellular structures and prevent capping of antibodies bound to cell surface receptors. However, the cold can aggravate clumping, so sorting at an intermediate temperature such as 15 °C may be preferable over sorting at 4 °C. Options to remove cell aggregates include filtration through nylon mesh filters, such as using meshes with different pore sizes available from Small Parts, Inc. (Hialeah, FL), 12 × 75-mm tubes with 100 μm cell strainer caps (Becton Dickinson, Falcon), or individual 70-μm cell strainers (Becton Dickinson, Falcon). Filtering samples immediately before the sort gives cell less time to reassociate. For large cell numbers, it is advisable to distribute cell aliquots into separate tubes and filter each sample individually before placing it onto the sorter. If feasible, an in-line filter, such as from BD Biosciences (San Jose, CA), Cytek Developments (Fremont, CA), or made in the laboratory by heating the end of a clipped-off pipet tip and fusing it with nylon mesh, put on the uptake port can prevent cell aggregates from reaching the nozzle tip. Selection of a sort tip with the appropriate nozzle size for the cell size to be sorted is another essential factor. Smaller nozzle sizes provide optimal signal resolution and easy sort setup; however, to avoid sort interruptions resulting from clogged fluid lines, it is recommended that the nozzle orifice be at least four times bigger than the cell diameter (Stovel, 1977).

B. Instrumentation

1. Background

The standard jet-in-air technology used for flow sorting involves a liquid stream that contains the cells to be sorted exiting through a nozzle vibrating at high frequency. The stream is broken into individual droplets passing by high-voltage plates that electrostatically charge the droplets carrying the cells that were preselected by the operator and deflect them into the designated receptacles. Thus, jet-in-air sorters produce droplets during normal operation that can be aerosolized. The size of the sort droplets depends on the instrument operating pressure, the size of the nozzle orifice, and the vibration frequency (Ibrahim and van den Engh, 2003). Newer high-speed sorters operate with higher instrument pressures and generate large numbers of small droplets. The droplet size of an aerosol determines its movement and velocity in air and the speed of its settlement out of air due to gravitational forces (Andersen, 1958). The aerosol droplet size is also directly related to the biohazard potential because it defines particle deposition during inhalation. When inhaled, droplets larger than 5 μm remain in the upper respiratory tract while smaller ones penetrate into the lung of the exposed individual (Andersen, 1958; Sattar and Ijaz, 1987). Secondary aerosols of various droplet sizes are formed when the undeflected center stream and the side streams splash into their receptacles. During failure modes of the sorter, such as a partial nozzle block, the stream exiting the nozzle can strike a hard surface and the production of aerosols increases substantially. Again, larger amounts of secondary aerosols are produced during high-speed sorting because of the higher operating pressure compared to regular-speed low-pressure sorting. When aerosols escape from the instrument into the environment, the infectious agents they contain could be harmful if inhaled (Andersen, 1958; Musher, 2003; Schmid *et al.*, 1997a; Veccio *et al.*, 2003). Thus, for operator safety, it is important to prevent aerosol escape through instrument design or attachment of optional safety devices and verify the efficiency of aerosol containment using appropriate testing. Further critical aspects of biohazard sorting are the placement of physical barriers between operator and hazard (personal protective equipment) and mandatory operator training in the operating procedures established in the laboratory. Instruments that sort cells by a fluidic switching mechanism use an enclosed fluid system, such as FACSort (BD Biosciences), and do not produce aerosols, but cannot achieve the sort speeds needed for many research applications.

2. Instrument Design Features

Since the early 1980s when Merrill (1981) published the first paper describing the generation of aerosols by flow sorters and a method to assess their production and escape during the application of various aerosol control measures instrument manufacturers have modified cell sorter design keeping operator safety in mind. Instrument features that reduce the generation of aerosols, remove aerosols from the operator area, and place primary barriers between the operator and the potential hazard are important because they greatly reduce the risk of exposure of instrument operators and others who are present during the sort. All modern sorters have a vacuum-evacuated catch tube for the undeflected center stream and an enclosed sort compartment. Many have stream-view cameras to facilitate operator observation of sort-stream stability without the need to come close to the area of the instrument that poses the greatest hazard. Systems such as the Accudrop System (BD Biosciences), which illuminates the center stream and the deflected streams near the sort collection vials with a low-powered laser beam, allow the operator to monitor an increase in aerosol production due to a shift in stream positions and fanning. In addition, some newer model jet-in-air sorters have completely enclosed sample ports and/or auxiliary vacuum pumps that remove aerosols from the sort area as optional attachments. Useful custom modifications for improved aerosol containment on any instrument are the sealing of any openings in the sort chamber door and the sort collection area and the installation of an air evacuation system where aerosols are generated. A removable containment hood that is vented by a high-efficiency particulate air filtration (HEPA) filter/fan unit and covers the sort area and the sample introduction port (Cytek Development, Fremont, California) has become commercially available to improve containment on FACStar, FACSVantage, and FACSDiVa (BD Biosciences) cell sorters. Aerosol production is most intense during failure modes of operation; thus, sorters that shut off the sample stream when the nozzle is partially clogged offer an additional safety margin. DakoCytomation

(Fort Collins, Colorado) provides a class I biosafety cabinet attachment for their MoFlo high-speed cell sorter. Although standard cell sorters with water-cooled lasers are too large to fit into biosafety cabinets, the newest benchtop sorter, FACSAria from BD Biosciences, equipped with solid-state lasers, is small enough to be completely enclosed into a walk-in clean air and biocontainment biological safety enclosure (BioPROtect II, The Baker Co., Sanford, ME).

3. Testing the Efficiency of Aerosol Containment during Cell Sorting

a. Background

For biosafety reasons, it is important to verify that the aerosol control measures on the sorter are effective in complete prevention of aerosol escape into the environment. The first protocol described in this chapter is based on the standard gravitational force method, which uses lytic T4 bacteriophage and Petri dishes with T4-susceptible *Escherichia coli* lawns (Ferbas *et al.*, 1995; Giorgi, 1994; Merrill, 1981; Schmid *et al.*, 1997a,b). A highly concentrated suspension of T4 bacterio-phage is run through the instrument to tag aerosol droplets. *E. coli* lawn-containing Petri dishes are placed in, on, and near the flow sorter where aerosols are formed and could potentially escape. Aerosols are detected by plaque formation in the *E. coli* lawn, resulting from T4 bacteriophage landing on the Petri dishes and lysing *E. coli*. This technique measures aerosols near the aerosol source and detects droplets that rapidly settle from air, which in general constitute most of the aerosols produced during sorting. Submicrometer particles (i.e., droplet nuclei) that may contain inorganic, organic material, or infectious agents from dehydrated small (less than 5-μm) droplets can also be generated and may stay suspended in air for prolonged periods of time (Sattar and Ijaz, 1987). These droplet nuclei need to be measured with active air sampling methods (Andersen, 1958). As described in the second protocol, an Andersen Air Sampler can be used to direct room air onto Petri dishes containing T4-susceptible *E. coli* lawns. Tagging aerosol droplets with bacteriophages is an established technique that provided the phage titer is suffi-ciently high, ensures that all of the droplets generated during sorting contain T4. Because it has been established that a single phage is sufficient to generate one plaque (Merrill, 1981), the assay provides high sensitivity and the assessment of containment results by counting plaques is straight-forward (Fig. 1). However, the method requires intermediate knowledge of microbiological techniques, relies on the performance of biological materials, and even when all the materials have been pre-prepared, results take at least several hours.

A novel nonbiological method that uses melamine copolymer resin particles for rapid visualization of aerosol production and escape has been developed (Oberyszyn, 2002; Oberyszyn and Robertson, 2001). A suspension of highly fluo-rescent Glo-Germ resin particles (Glo Germ, Inc., Moab, UT) with an approxi-mate size range between 1 and 10 μm that simulates a biological sample is sorted on the instrument to be tested. Aerosol containment is measured by placing micro-scope slides around the instrument where aerosols are produced and can

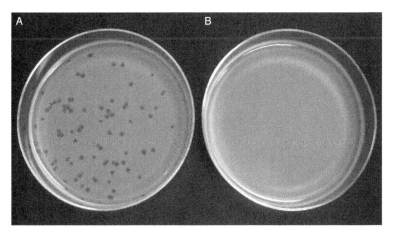

Fig. 1 Plaque formation by T4 bacteriophages on a confluent *Escherichia coli* lawn. (A) Plaques appear where T4 phages have lysed *E. coli*; (B) confluent *E. coli* lawn on a control Petri dish.

potentially escape and examining the slides for the presence of Glo-Germ particles under a fluorescent microscope. Perfetto *et al.* (2003) have increased the sensitivity and reproducibility of the original Glo-Germ method by using a viable microbial particle sampler that draws room air onto a microscope slide placed into the device. Microscopic readout of the containment test results is facilitated by concentrating the collected Glo-Germ particles onto the areas on the slide located underneath the intake ports of the particle sampler. This technique described in the third protocol in this chapter is suitable to be performed immediately before starting a biohazardous sort but for this requires ready access to a fluorescent microscope. Glo-Germ particles are highly fluorescent and, therefore, are easily detected (Fig. 2). However, meticulous cleaning and handling of the air sampler and microscope slides are essential to avoid false positives, and diligent scanning of the entire slide is needed to reliably detect escape of single particles.

b. Testing the Efficiency of Aerosol Containment During Cell Sorting Using T4 Bacteriophage Using Gravitational Deposition of Droplets

Testing is performed in regular sort mode and in failure mode to simulate conditions such as a partially blocked sort nozzle or air in the system, which result in a considerable increase of aerosol production. Aerosol containment testing should be repeated every 1–3 months and whenever the sorter was modified.

Materials and methods:

1. *E. coli,* ATCC# 11303 (*E. coli*) (The American Type Culture Collection (ATCC), Rockville, MD).

2. Bacteriophage T4 (T4 phage), ATCC# 11303-B4 (ATCC) *Prepare growth media by autoclaving the mixtures at 121 °C for 15 min.*

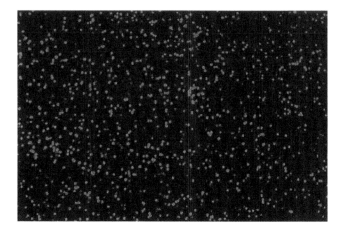

Fig. 2 Visualization of Glo-Germ particles by fluorescence microscopy using the setup and magnification as described in the text.

3. *Bottom agar*:
   - 10 g minimal agar Davis (Difco, Becton Dickinson, Franklin Lakes, NJ)
   - 13 g bacto tryptone (Difco)
   - 8 g sodium chloride
   - 2 g sodium citrate
   - 1.3 g glucose
   - distilled H_2O to 1 l
4. *Nutrient broth*:
   - 8 g bacto nutrient broth (Difco)
   - 5 g sodium chloride
   - 1 g glucose
   - distilled H_2O to 1 l
5. *Soft agar*:
   - 6.5 g minimal agar Davis (Difco)
   - 13 g of bacto tryptone (Difco)
   - 8 g sodium chloride
   - 2 g sodium citrate
   - 3 g glucose
   - distilled H_2O to 1 l
6. Chloroform
7. Sheath fluid for the flow sorter, which is compatible with maintaining T4 phage viability, such as phosphate buffered saline (PBS), Hank's balanced salt solution (HBSS). Note that sheath fluids containing detergents and/or preservatives are incompatible with T4 phage viability.

i. Preparation for Aerosol Containment Testing.
All reagents, supplies, and equipment must be sterile and proper sterile technique must be used. Although the preparation of the materials needed for the testing takes several days, the actual aerosol containment test can be completed in 1 day.

Preparation of T4 Bacteriophage Stock

1. Add 0.3–0.4 ml of nutrient broth to one vial of T4-susceptible *E. coli* and mix well. Unused *E. coli* suspension can be frozen at −20 °C for at least 6 months.

2. Initiate an overnight culture in a culture flask by inoculating approximately 100 ml of nutrient broth with 10–20 μl of reconstituted *E. coli*. Incubate the flask overnight at 37 °C in a warm room on an orbital shaker at 150 rpm or a shaking incubator at the equivalent temperature and speed.

3. Next morning, take out 2.5 ml of the *E. coli* culture and transfer it to a flask containing 50 ml of nutrient broth. For multiple cultures, set up several 50-ml flasks and incubate them at 37 °C in a warm room on an orbital shaker.

4. After 1 h, start to monitor *E. coli* growth by measuring absorbance of the culture at 600 nm in a spectrophotometer as compared to blank nutrient broth. When the culture reaches an optical density of 0.5, it has entered log-phase growth. For optimal propagation of T4 phage, phage needs to be added to the *E. coli* culture at this time. Rehydrate one vial of lyophilized T4 bacteriophage with 0.5 ml of nutrient broth and mix well. Add 0.1 ml of the T4 phage suspension to each culture flask. The remaining T4 phage suspension can be stored at 4 °C for at least 6 months.

5. Incubate the culture on the orbital shaker at 37 °C for 6–8 h until the culture starts to look more transparent.

6. Add 12 drops of chloroform to each flask, shake vigorously, transfer each culture to a 50-ml polypropylene tube, and spin at 2000 × *g* for 20 min.

7. Take off the supernatant and transfer it to screw-cap glass vials, keeping the airspace above the liquid to a minimum. Vials containing T4-phage stock can be stored indefinitely at 4 °C in the dark, but it is advisable to retitrate the stock before each aerosol containment test.

Titration of T4 Bacteriophage Stock

1. Prepare Petri dishes with bottom agar by first heating the agar to 50 °C. Using a sterile pipet put approximately 20 ml of liquified agar in each dish and let them cool with the lid ajar to avoid the accumulation of moisture. Close the lid and store dishes until needed at 4 °C. For long-term storage, wrap bottom agar dishes in plastic to prevent dehydration.

2. Initiate a new overnight *E. coli* culture by repeating step 2 from the previous section.

3. Next morning, place 0.9 ml of nutrient broth into nine 12 × 75-mm tubes. Add 0.1 ml of the T4-phage suspension to the first tube, mix well, and make a serial dilution of the stock.

4. Heat approximately 100 ml of soft agar to 50 °C. Do not exceed this temperature, because *E. coli* is heat sensitive. Add 2 ml of the *E. coli* broth culture, mix well, and transfer 4 ml of the mixture to a 15-ml culture tube. The number of tubes will depend on the number of dilutions of the T4-phage stock to be plated. Initially, it may be advisable to plate all nine dilutions of the T4-phage stock. As a general rule, when the T4-phage stock suspension has been grown to the expected titer (10^{11}–10^{12} plaque-forming units (PFU)/ml), it is sufficient to plate only the three lowest concentrations of the stock as the higher ones will produce too many plaques for accurate counting.

5. Quickly add 0.1 ml of the T4-phage dilutions to the mixture, mix gently, and pour the entire content of each of the 15-ml culture tubes onto Petri dishes containing bottom agar.

6. Incubate Petri dishes at 37 °C for plaque formation. Plaques can be seen within 4–5 h, but dishes can be incubated overnight, if this is more convenient.

7. Select a Petri dish with an intermediate number of plaques, for example, 10–100 for the calculation of PFU of T4 phage per ml. PFU/ml are calculated by multiplying the number of plaques with 100 times the inverse of the dilution factor. Note that the dilution factor is increased by 10^{-2} because only 0.1 ml of the T4-phage stock suspension is used and only 0.1 ml of the diluted stock is plated.

ii.   Testing the Efficiency of Aerosol Containment on a Flow Sorter.
Generation of an E. coli Log Growth Culture and Preparation of E. coli Lawns

1. Initiate an overnight culture in a culture flask by inoculating approximately 100 ml of nutrient broth with 10–20 μl of reconstituted *E. coli*. Incubate the flask overnight at 37 °C in a warm room on an orbital shaker at 150 rpm or a shaking incubator at the equivalent temperature and speed.

2. Heat approximately 100 ml of soft agar to 40–50 °C. Add approximately 2 ml of the *E. coli* broth culture to the soft agar and mix by swirling gently. Add 4 ml of the mixture to Petri dishes with bottom agar and incubate them for approximately 2 h at 37 °C until dishes appear slightly opaque indicating the formation of confluent *E. coli* lawns. The number of Petri dishes to be prepared depends on how many are to be used in the aerosol containment test on the sorter.

Preparation of the Instrument

1. Set up the sorter in a fast flow rate by measuring the consumption of nutrient broth using an electronic balance. Note that the flow rate varies according to the type of instrument, the differential pressure, the sort nozzle size, and the instrument pressure used. For repeated experiments, set up the sorter at a similar flow rate.

Determine the Expected Throughput of T4 Phage

1. Calculate the expected throughput of viable T4 phage in PFU/min based on the concentration of the T4 stock suspension determined by serial dilution and the measured instrument flow rate.

Measure the Actual Throughput of Viable T4 Phage Through the Sorter

1. Equip the sorter with T4 phage-compatible sheath fluid, such as PBS and HBSS, and place a culture tube containing T4-phage stock suspension onto the instrument.

2. Weigh an empty culture tube. While the phage suspension is running through the sorter, collect the center stream exiting the sort nozzle for exactly 1 min Re-weigh to determine the amount of liquid collected.

3. Mix well and immediately transfer 0.1 ml into 0.9 ml of nutrient broth. This step will minimize T4-phage activity loss through its exposure to sheath fluid.

4. Perform a titration of the T4-phage suspension by serial dilution as described in Section "Titration of T4 Bacteriophage Stock." Note that there will be a decrease in titer due to the dilution of the T4-phage suspension with sheath fluid and some loss of activity due to the stress of shearing forces. A titer more than 1×10^7 of PFU/ml is required for a successful aerosol containment test.

Testing the Instrument
Perform aerosol containment testing in mock regular sort mode

1. Distribute Petri dishes with confluent *E. coli* lawns with the lids still in place onto the instrument where aerosols are generated and where they could potentially escape. Label each Petri dish to identify its location during the test. Initially, Petri dishes may be placed at various locations within the sorter or around the room. When aerosol containment has been established, for a standard retest of a cell sorter, two Petri dishes may be placed within the sort chamber, two on top of the instrument, and two near the work area of the operator.

2. Place sufficient T4-phage stock suspension onto the sorter to perform testing of the instrument for 1–2 h. Previously published protocols have used various lengths of time for testing the instruments in mock regular sorting mode. The time chosen should take into account the settlement rate of the aerosol produced in the sorter and average sort times used in the laboratory.

3. Set up a mock sort by generating side streams and placing sort collection vials into the sort receptacles. Close the sort chamber and remove the lids from all the Petri dishes.

4. After the designated time, stop the sort, place the lids onto the Petri dishes and incubate them at 37 °C for plaque formation. Plaques can be observed already after 4–5 h of incubation, but dishes can be left overnight, if this is more convenient. Dishes that were placed outside the sort chamber should not show any plaques, if aerosol containment on the sorter was complete. If aerosol containment was incomplete, modify the sorter until satisfactory results are achieved during retesting.

Perform aerosol containment testing in failure mode

1. Repeat steps 1 and 2 from aerosol testing in regular sort mode.

2. Set up the sorter in failure mode by either directing the center stream toward the waste catcher or generating fanning of the streams. Close the sort chamber door and remove the lids from all the Petri dishes.

3. After 15 min, stop the sort, place the lids onto the Petri dishes, and incubate them at 37 °C for plaque formation. Dishes that were placed outside the sort chamber should not show any plaques if aerosol containment on the sorter was complete. If aerosol containment was incomplete, modify the sorter until satisfactory results are achieved during retesting.

c. Testing the Efficiency of Aerosol Containment During Cell Sorting Using T4 Bacteriophage by Using an Air Sampler

1. Set up the air sampler (e.g., Andersen single-stage air sampler, model N6IACFM, Grayseby-Andersen, Smyrna, GA) according to the manufacturer's instruction. The air sampler should be equipped with a two-port manifold to allow for simultaneous testing of room air in close vicinity of the sorter and further away.

2. Prepare six Petri dishes containing bottom agar with confluent *E. coli* lawn as described earlier in paragraphs 1 and 2 under "generation of an *E. coli* log growth culture and preparation of *E. coli* lawns" and label them carefully to indicate their positions and test conditions. Place one Petri dish into the sampling stage and sample the room air for 10 min before starting containment testing on the sorter. This Petri dish will serve as negative control.

3. Take two new uncovered Petri dishes with *E. coli* lawns. Place one into one sampling stage close to the sorting chamber, the other approximately 1 m away from the instrument. Start the aerosol containment testing in mock regular sort mode by running T4 phage through the sorter as described earlier and sample the room air for 10 min. Sampling times longer than 10 min can be used; however, depending on the prevalence of spores in the room air, Petri dishes may overgrow with airborne contaminants.

4. Remove the Petri dishes from the air sampler and replace their lids.

5. Take two new uncovered Petri dishes with *E. coli* lawns. Place one into one sampling stage close to the sorting chamber, the other approximately 1 m away from the instrument. Start the aerosol containment testing in mock failure mode by running T4 phage through the sorter as described earlier and sample the room air for 10 min.

6. Take one more uncovered Petri dish containing a confluent *E. coli* lawn into one sampling stage close to the sorting chamber, repeat the mock failure sort mode with the sort chamber door open, and sample the air for 10 min. This Petri dish will serve as a positive control.

7. Collect all Petri dishes and incubate at 37 °C for plaque formation.

8. Testing aerosol containment using an active air sampling method can be combined with the settle plate method or, if desired, be performed separately.

d. Testing the Efficiency of Aerosol Containment During Cell Sorting using Glo–Germ Particles and a Microbial Particle Sampler

Testing is performed before each biohazardous sort. The instrument is placed in failure mode to simulate conditions such as a partially blocked sort nozzle or air in the system, which result in a considerable increase of aerosol production. Aerosol containment results are recorded as pass/fail before each sort.

Materials and Methods

1. AeroTech 6TM viable microbial particle sampler, Cat. No. 6TM (AeroTech Laboratories, Inc., Phoenix, AZ (www.aerotechlabs.com)

2. Vacuum source, for example, in-house or a vacuum pump capable of drawing 45 l/min of air

3. Matheson flow meter: Vacuum Meter, Cat. No. 5083R60, Thomas Scientific, Swedesboro, NJ

4. Glo-Germ Particles (Glo Germ, Inc., Cat. No. GGP (www.glogerm.com)

   (a) Glo-Germ stock particles (5-μm melamine copolymer resin beads in a 5-ml volume of ethanol to yield a stock concentration ranging from 200 to 400 $\times$ 10^6 particles/ml) were washed two times in 100% ethanol by centrifugation at 900 $\times$ g for 10 min. If Glo Germ is in powder form resuspend in 5 ml of 100% ethanol before proceeding.

   (b) Resuspend in 100 ml of wash media (10% FCS + 1% Tween 20 + 0.1% (1 mg/ml) sodium azide in PBS).

   (c) Store particles in an opaque glass container at 4 °C for up to 6 years.

   (d) Particles are filtered through a 100-μm filter to create a stock suspension. This suspension is then diluted with PBS before sorting to achieve a high acquisition rate (e.g., 50,000 particles/s).

Measurement of Containment

1. Add a clean slide to a clean Petri dish without a lid and place both into the AeroTech 6TM viable microbial particle sampler. Close the top lid of the sampler and carefully secure clasps on each side. The particle sampler is kept inside a laminar flow biosafety cabinet until containment testing begins. *Note*: Slides must be very clean and can be cleaned with 70% ethanol before use. Care must be taken not to touch their surface. The microbial particle sampler can be cleaned meticulously by sonication for 5 min in a water bath followed by a wash with a mild detergent and distilled water.

2. Adjust the vacuum to the AeroTech 6TM viable microbial particle sampler to 45–55 l/min (as measured by the flow meter).

3. Place a freshly made suspension of Glo-Germ particles by adding 7–10 drops of concentrated suspension to 3 ml of PBS onto the instrument and adjust the flow rate to a high setting, 50,000 events/s. *Note*: Glo-Germ particles are small, so they require a higher forward scatter voltage setting compared to human leukocytes.

4. Place a cover over the end of the waste catcher (e.g., rubber tubing cut length-wise). When sort drawer is engaged for sorting, the center stream is deflected creating aerosols. This situation is considered the "failure mode" of operation.

5. Turn on all auxiliary vacuum sources as applicable and close all containment barriers according to the manufacturer guidelines. In some instruments (e.g., the ARIA from Becton Dickinson), the tube holder can be removed to enhance the potential loss of aerosols into the environment, therefore allowing greater test sensitivity.

6. With the instrument in "failure mode," place the AeroTech 6TM viable microbial particle sampler directly in front of the sort chamber door, for a total of 10 min.

7. Stop aerosol generation and remove the microscope slide from the Aero-Tech6TM particle sampler. *Note*: After every test, wait before opening the sort chamber until the aerosol has cleared from the sort collection chamber. The wait time required will vary and depend on the rate of air exchange within the chamber (see Section III.D). Do not allow the sample tube to back-drip to avoid contamination of the outside area with Glo-Germ particles. To remove excess Glo-Germ particles from the sample tubing, place 70% ethanol on the sample station and run it through the sample tubing for 5 min while under sort containment.

8. Examine the entire slide for the presence of particles using a fluorescent microscope with a fluorescein-exciter filter (450/490 nm) scanning with a $10\times$ objective and a fluorescein emission filter (520–640 nm).

9. All tolerances must be achieved before proceeding with a viable, potentially infectious sort.

10. Tolerance of particles outside the sort collection chamber: zero tolerance, no particles on entire slide. Any positive result must be investigated and resolved and the sorter must be retested before proceeding with sorting potentially infectious samples.

Notes and comments. As a positive control for the AeroTech 6TM viable microbial particle sampler, it is recommended to collect particles in failure mode with the sort chamber door closed and without the tube holder (ARIA specific, see number 6 above for more details) for a 5-min period. This slide should be positive for particles (>100 particles per slide) and indicates that the collection system is working correctly.

C. Laboratory Facilities

Appropriate sort facility design contributes to the protection of operators and provides a barrier to protect persons outside the room from the potential release of aerosols as a result of cell sorting potentially infectious samples. Because of the possible escape of aerosols generated during cell sorting, it is highly recommended that all biohazardous cell sorting be performed in a BSL-3 laboratory facility (CDC, 2007). BSL-3 facilities are only accessible through two self-closing doors connected through a passageway and contain washable walls, ceilings, and floors, and a ducted HEPA-filtered ventilation system. However, few older institutions have such rooms available and renovation costs are often prohibitive. When planning a new or remodeling an existing sort laboratory, facility design should take into account the BSL of future experiments and incorporate higher levels of containment, whenever possible. In the absence of a BSL-3 sort room, sorting of pathogenic samples requiring BSL-2 containment can be performed in a separate lockable room with negative air pressure equipped with easy-to-clean work surfaces and floors and a sink for hand washing (Giorgi, 1994; Schmid *et al.*, 2003). Access to the room must be limited and warning signs must be posted when experiments are in progress to prevent entry of persons not wearing the required protective clothing. Alternatively, some newer sorters are small enough to fit inside a Biocontainment Biological Safety Enclosure (The Baker Co.) and thus allow for BSL-3 containment without the cost of building a BSL-3 facility.

D. Standard Operating Procedures

All laboratories performing cell sorting on unfixed samples containing known or unknown human pathogens need to establish standard written operating procedures (SOP) to be followed diligently by all personnel involved in these experiments. Whenever there is a change in the biohazard potential of the sort samples, laboratory directors need to reassess the SOP to reflect these changes in practices. Various aspects to be addressed in detail are instrument operation, personal protection, and operator training.

1. Instrument Operation, Troubleshooting, Decontamination, and Maintenance

Instrument procedures for biohazardous sorting will include standard startup, alignment, calibration, sterilization, and shut down as established in the facility for nonhazardous sorting. In addition, it must detail the specific steps to be followed for sorting of potentially infectious specimens. Practices may involve turning on the auxiliary vacuum source at the proper setting to ensure sufficient negative airflow. The setting needs to be high enough for the removal of aerosols from the sort chamber, yet low enough to not disrupt the sort streams. Depending on the instrument, configuration and SOP setup for biohazardous sorting may include attaching the containment hood to the sorter and/or performing aerosol containment testing either before the start of each sort (Perfetto *et al.*, 2003) or at regular intervals as outlined in previous publications (Giorgi, 1994; Schmid *et al.*, 1997a,b). The waste tank needs to be filled with concentrated bleach (5.25% sodium hypochlorite solution) in sufficient quantity to achieve a final concentration of 10% bleach when the tank is full. However, bleach corrodes metal tanks, so substitution with a less corrosive disinfectant such as povidone iodine (10% final concentration) can prolong tank life. Steps to be taken during an instrument malfunction such as a nozzle blockage must be clearly defined. For example, the wait time required before opening the sort door after a clog has occurred will depend on the speed of aerosol clearance. This time can be checked with bottled smoke (Lab Safety Supply, Inc., Janesville, WI). Clearance time can also be measured by collecting Glo-Germ particles with the sort chamber door open after the sort is stopped and the sample is no longer pressurized.

Rigorous cleaning after each sort, for example, by running an alkaline solution (0.1 N sodium hydroxide) followed by an enzymatic solution, such as Coulter Clenz (Beckman-Coulter, Miami, FL), and distilled water, clears the fluid lines and sort nozzles from residual cells and cellular debris. In addition, after sorting cells of epithelial origin a hyaluronidase solution (e.g., made by dissolving 8 μg of hyaluronidase (Sigma-Aldrich, Cat. # H2126) in 25 ml of distilled water (Steve Merlin, personal communication)) can be useful for elimination of mucin-producing cells. Furthermore, utilization of a strong detergent solution, such as Contrad 70 (Decon Laboratories, Inc., Bryn Mawr, PA), can provide effective removal of organic material.

These cleaning practices improve the effectiveness of subsequent disinfection of contaminated fluid lines and sorter parts, which is an important measure to eliminate the potential for the spread of infectious agents. Per regulations outlined in the Bloodborne Pathogen Standard (U.S. Federal Code Regulation, 1991) appropriate disinfectants to be used for decontamination of equipment or work surfaces exposed to blood or other potentially infectious materials include diluted bleach, Environmental Protection Agency (EPA)-registered tuberculocides, EPA-registered sterilants or products registered to be effective against HIV or HBV as listed online at www.epa.gov/oppad001/chemregindex.html. Commonly used laboratory disinfectants (Rutala, 1996; Vesley and Lauer, 1995) and their properties are listed in Table I. Alcohols are not considered high-level disinfectants, because they are not able to inactivate bacterial spores and penetrate protein-rich materials, and isopropanol cannot kill hydrophilic viruses. In addition to disinfection of fluid lines, it is critical to decontaminate the sample uptake port in sorters with sample line backdrip. Before designing a specific sorter decontamination protocol, it is important to check with the manufacturer of a given model sorter to clarify that all instrument components that will come in contact with the disinfectant solution can tolerate exposure. Running distilled water through the sample line for at least 10 min after sorter disinfection is essential, because residual disinfectant solution may compromise cell viability in subsequent sort experiments. To this end, it is also recommended that sterile sort media be run in sufficient quantity through the sample tubing each time before starting to sort the sample of interest.

Maintaining the flow sorter in excellent working condition is critical for processing biohazardous samples because presumably the greatest operator hazard comes from sudden instrument malfunctions during a sort such as a broken fluid line, a stuck valve, a blocked waste line, or a damaged or clogged HEPA filter for air evacuation. Thus, a regular preventive maintenance schedule must be followed, either as part of an instrument service contract offered by manufacturers or performed by personnel of the sort laboratory. Nozzle tips must be cleaned

Table I

Properties of Common Disinfectants for Use in Cell Sorter Decontamination

Compound	Chlorine compounds	Ethyl alcohol	Isopropyl alcohol	Hydrogen peroxide	Iodophors
Practical requirements					
Use dilution	1/10–1/100 dilution of 0.71 M sodium hypochlorite, ~50–500 ppm[a]	70–85%[b]	70–85%[b]	3–6%	5–10%, 800–1600 ppm[a]
Contact time (min)[c]	5–30	10–30	10–30	10–30[d]	–
Inactivation by organic material[e]	Yes	No	No	Yes	Yes
Stability[f]	No[g]	Yes	Yes	Yes	Yes
Corrosive	Yes	No	No	Yes[h]	Yes
Inactivation profile					
Vegetative bacteria	Yes	Yes	Yes	Yes	Yes
Bacterial spores	Yes	No	No	Yes	Yes
Tubercle bacilli	Yes	Yes	Yes	Yes	Yes
Enveloped (lipophilic) viruses	Yes	Yes	Yes	Yes	Yes
Nonenveloped (hydrophilic) viruses	Yes	Yes[i]	No	Yes	Yes
Fungi	Yes	Yes	Yes	Yes	Yes
Protozoal parasites	Yes	No	No	No	No

Note: From Vesley and Lauer (1995) and Rutala (1996), and www.ianr.unl.edu/animaldisease/g1410.htm.

[a]Available halogen.

[b]Activity drops sharply when diluted below 50%.

[c]Optimal exposure time depends on the amount of contamination and type of contaminant.

[d]Iodophors may require prolonged contact time for inactivation of bacterial spores, tubercle bacilli, and certain fungi.

[e]Before application of the disinfectant cleaning with an enzymatic or lipophilic detergent necessary.

[f]Shelf life of solutions more than 1 week if protected from heat, light, and air.

[g]If stored at ambient room temperature concentrated bleach solutions retain their sodium hypochlorite content of 5.25% for 3 months after production. Stability of diluted bleach solutions is highly variable depending on the purity of the water used for dilution and the protection from light and air, for optimal disinfectant activity use freshly diluted solutions within 24 h.

[h]Mildly corrosive.

[i]Variable results depending on the virus.

frequently by sonication and should be replaced whenever they show wear and tear to minimize instrument trouble.

2. Personal Protection

Personal protection refers to the placement of physical primary barriers between flow sorter operators and biohazard as a means to protect the operator from exposure in the event of a breakdown of mechanical barriers (engineering devices).

Extent and types of personal protective equipment used differ among laboratories that routinely sort samples from individuals infected with HIV (Ferbas *et al.*, 1995; Giorgi, 1994; Perfetto *et al.*, 2003). Nevertheless, during biohazardous sorting, at a minimum operator's safety equipment must conform with BSL-3 recommendations (CDC, 1999) and consist of a disposable wraparound laboratory coat, gloves, safety glasses with side shields, and a respiratory mask appropriate for aerosol protection, such as N95 NIOSH-approved particulate respirators (e.g., 2300N95 Moldex respirators (Zee Medical Service, Inc., Irvine, CA)). A plastic shield may be worn over the respirator mask to provide an additional safety margin. Perfetto *et al.* (2003) described the use of a complete Depuy Bio-Hazard Respiratory System (DePuy Chesapeake Surgical, Ltd., Sterling, VA), which consists of a body suit, a helmet, and its battery-powered respiratory system with electrostatic filter media, for performing biohazardous sorting. Any personal protective equipment must always be removed whenever the operator leaves the sort room or the adjacent anteroom.

Immunization against infectious agents should be offered to personnel, if available. Prophylactic HBV vaccination is highly recommended. Postexposure prophylaxis should be available in any laboratory involved in sorting of HIV-positive specimens (Mikulich and Schriger, 2002; Schriger and Mikulich, 2002; Wang *et al.*, 2000) and should always follow the latest recommendations from CDC available online at www.cdc.gov/mmwr. Drawing a baseline serum sample from personnel before work is started, monitoring the health status of individuals, in particular compromised immunity, and periodically evaluating serum samples may be an appropriate risk assessment for laboratories that routinely sort samples containing infectious agents.

3. Operator Training and Experience

Only operators with considerable experience of frequent sorting of nonhazardous samples on the type of instrument to be used for hazardous sorting should start performing separations of samples known to carry human pathogens (Giorgi, 1994; Schmid *et al.*, 2003). The time needed for trainees to become proficient in sorter operations varies considerably, but on average 6 months to 1 year is common. Ideally, operators should also be experienced in handling samples infected with the pathogen present in the samples to be sorted (Evans *et al.*, 1990). Mandatory operator training in all the relevant safety aspects of biohazardous sorting, including the procedures involved in testing the efficiency of aerosol containment on the cell sorter is essential for reducing hazard risks to sort personnel and others involved in these experiments. Training records need to be kept on file, and retraining must be performed whenever the SOP changes because of a new sorter or cell sorting of samples with altered biohazard potential.

IV. Applications and Future Directions

Flow sorting of unfixed samples is frequently used in infectious disease studies to separate leukocyte subsets on the basis of differential cell surface expressions. Sorted subpopulations can then be examined for their response to the pathogen of interest, for cellular mechanisms of its pathogenesis, or to identify or characterize cells infected with the pathogenetic organism (Giorgi, 1994). Recent applications include investigations of gene expression patterns of cells that either carry a pathogen or have been transfected with vectors that contain genetic sequences of an infectious organism using fluorescent reporter molecules to select the cells to be isolated by sorting (Herzenberg *et al.*, 2002). Emerging applications involve preparative cell sorting of clinical samples for therapeutic interventions (Leemhuis and Adams, 2000; Lopez, 2002). Clinical cell sorting not only requires protection of instrument operators from known or unknown pathogens contained in the samples but sorting has to be performed using Good Manufacturing Practices under clean room conditions to prevent contamination of the sorted product to be reinfused into the patient (Ibrahim and van den Engh, 2003; Keane-Moore *et al.*, 2002).

Acknowledgment

This work was supported by National Institutes of Health awards CA-16042 and AI-28697.

References

Andersen, A. A. (1958). New sampler for the collection, sizing, and enumeration of viable airborne particles. *J. Bacteriol.* **76,** 471–484.

Centers for Disease Control and Prevention. (1999). Biosafety in microbiological and biomedical laboratories (BMBL). Government Printing Office, Washington, DC.

Evans, M. R., Henderson, D. K., and Bennett, J. E. (1990). Potential for laboratory exposures to biohazardous agents found in blood. *Am. J. Public Health* **80,** 423–427.

Ferbas J., Chadwick K. R., Logar A., Patterson A. E., Gilpin R. W., and Margolick J. B. (1995). Assessment of aerosol containment on the ELITE flow cytometer. *Cytometry* **22,** 45–47.

Giorgi, J. V. (1994). Cell sorting of biohazardous specimens for assay of immune function. *Methods Cell Biol.* **42,** 359–369.

Herzenberg, L. A., Parks, D., Sahaf, B., Perez, O., Roederer, M., and Herzenberg, L. A. (2002). The history and future of the fluorescence activated cell sorter and flow cytometry: A view from Stanford. *Clin. Chem.* **48,** 1819–1827.

Ibrahim, S. F., and van den Engh, G. (2003). High-speed cell sorting: Fundamentals and recent advances. *Curr. Opin. Biotechnol.* **14,** 5–12.

Keane-Moore, M., Coder, D., and Marti, G. (2002). Public meeting and workshop on safety issues pertaining to the clinical application of flow cytometry to human-derived cells. *Cytotherapy* **4,** 89–90.

Leemhuis, T., and Adams, D. (2000). Applications of high speed sorting for CD34+ hematopoietic stem cells. *In* "Emerging tools for Single-Cell Analysis" (G. Durack and J. P. Robinson, eds.), pp. 73–93. Wiley-Liss, New York.

Lopez, P. A. (2002). Basic aspects of high-speed sorting for clinical applications. *Cytotherapy* **4,** 87–88.

Merrill, J. T. (1981). Evaluation of selected aerosol-control measures on flow sorters. *Cytometry* **1,** 342–345.

Mikulich, V. J., and Schriger, D. L. (2002). Abridged version of the updated US Public Health Service guidelines for the management of occupational exposures to hepatitis B virus, hepatitis C virus, and human immunodeficiency virus and recommendations for postexposure prophylaxis. *Ann. Emerg. Med.* **39,** 321–328.

Musher, D. M. (2003). How contagious are common respiratory tract infections? *N. Engl. J. Med.* **348,** 1256–1266.

National Committee for Clinical Laboratory Standards. (1997). Protection of Laboratory Workers from Infectious Disease Transmitted by Blood, Body Fluids, and Tissue. Document M29-A.

Oberyszyn, A. S. (2002). Method for visualizing aerosol contamination in flow sorters. *In* "Current Protocols in Cytometry" (J. P. Robinson, Z. Darzynkiewicz, P. N. Dean *et al.*, eds.), Unit 3.5.1–3.5.7. Wiley, New York.

Oberyszyn, A. S., and Robertson, F. M. (2001). Novel rapid method for visualization of extent and location of aerosol contamination during high-speed sorting of potentially biohazardous samples. *Cytometry* **43**, 217–222.

Perfetto, S. P., Ambrozak, D. R., Koup, R. A., and Roederer, M. (2003). Measuring containment of viable infectious cell sorting in high-velocity cell sorters. *Cytometry* **52A**, 122–130.

Rutala, W. A. (1996). APIC guidelines for infection control practice. *Am. J. Infect. Control* **24**, 313–342.

Sattar, S. A., and Ijaz, M. K. (1987). Spread of viral infections by aerosols. *CRC Crit. Rev. Environ. Control* **17**, 89–131.

Schmid, I. (2000). Biosafety in the flow cytometry laboratory. In "In Living Color, Protocols in Flow Cytometry and Cell Sorting" (R. A. Diamond and S. DeMaggio, eds.), pp. 655–665. Springer, New York.

Schmid, I., Hultin, L. E., and Ferbas, J. (1997a). Testing the efficiency of aerosol containment during cell sorting. *In* "Current Protocols in Cytometry" (J. P. Robinson, Z. Darzynkiewicz, P. N. Dean *et al.*, eds.), Unit 3.3, Wiley, New York.

Schmid, I., Kunkl, A., and Nicholson, J. K. (1999). Biosafety considerations for flow cytometric analysis of human immunodeficiency virus-infected samples. *Cytometry* **38**, 195–200.

Schmid, I., Merlin, S., and Perfetto, S. P. (2003). Biosafety concerns for shared flow cytometry core facilities. *Cytometry* **56A,** 113–119.

Schmid, I., Nicholson, J. K. A., Giorgi, J. V., Janossy, G., Kunkl, A., Lopez, P. A., Perfetto, S., Seamer, L. C., and Dean, P. N. (1997b). Biosafety guidelines for sorting of unfixed cells. *Cytometry* **28**, 99–117.

Schriger, D. L., and Mikulich, V. J. (2002). The management of occupational exposures to blood and body fluids: Revised guidelines and new methods of implementation. *Ann. Emerg. Med.* **39**, 319–321.

Shapiro, H. M. (2003). Parameters and probes: Fixation: Why and how. *In* "Practical Flow Cytometry," 4th edn., pp. 302–306. Wiley, New York.

Stovel, R. T. (1977). The influence of particles on jet breakoff. *J. Histochem. Cytochem.* **25**, 813–820.

United States Federal Code Regulation. (1991). Occupational exposure to bloodborne pathogens. 29 CFR PART. 1910.1030.

Veccnio, D., Sasco, A. J., and Cann, C. I. (2003). Occupational risk in health care and research. *Am. J. Ind. Med.* **43**, 369–397.

Vesley, D., and Lauer, J. L. (1995). Decontamination, sterilization, disinfection, and antisepsis. *In* "Laboratory Safety" (D. O. Fleming, J. H. Richardson, J. J. Tulis, and D. Vesley, eds.), 2nd edn., pp. 219–237. American Society for Microbiology Press, Washington, DC.

Wang, S. A., Panlilio, A. L., Doi, P. A., White, A. D., Stek, M., Jr., and Saah, A. (2000). Experience of healthcare workers taking postexposure prophylaxis after occupational HIV exposures: Findings of the HIV postexposure prophylaxis registry. *Infect. Control Hosp. Epidemiol.* **21**, 780–785.

CHAPTER 9

Guidelines for the Presentation of Flow Cytometric Data

**Mario Roederer,★ Zbigniew Darzynkiewicz,†
and David R. Parks‡**

★Vaccine Research Center
National Institute of Allergy and Infectious Diseases
National Institutes of Health
Bethesda, Maryland 20892

†Braner Cancer Research Institute
New York Medical College
Hawthome, New York 10532

‡Department of Genetics
Stanford University
Stanford, California 94305

I. Introduction

In this chapter, we outline some of the concepts underlying the presentation of flow cytometric data in a consistent and informative manner. These guidelines are primarily aimed at preparation of data for publication in print but are applicable to the presentation of data in slides (e.g., during seminars). We cover some fundamental aspects of data presentation and then discuss issues particularly relevant to the presentation of data arising from flow cytometric experiments.

ESSENTIAL CYTOMETRY METHODS
Reprinted from *Methods in Cell Biology,* Volume 75 (Academic Press, 2004).

DOI: 10.1016/B978-0-12-375045-7.00009-X

We limit this discussion to the presentation of "raw" cytometric data; we will not cover more general issues of presenting derived data like charts or line plots, even though they are derived from flow cytometric analyses.

A number of excellent books and papers discuss appropriate and inappropriate graphical data presentation guidelines. One that is highly recommended is Edward Tufte's *The Visual Display of Quantitative Information* (Tufte, 2001), a highly educational and entertaining treatise on data graphing. Many of the guidelines presented in this chapter derive from suggestions that Tufte makes.

Although we touch on topics like control samples, details of good experimental design are outside the scope of this chapter. Understanding which controls are necessary to interpret data (and are, therefore, important to publish) is critical to successful publication, but our assumption is that such decisions have already been made by the author. These guidelines are to help decide *how* to present data, not *which* data to present. Most importantly, these guidelines should be used to decide what additional information needs to be present in publications to allow fully informed interpretation of the data and thus of the experiments and conclusions that are drawn.

Many of these guidelines, when implemented, may lengthen chapters. It should be noted that many journals now have mechanisms for supplying supporting data as electronic documents that can be accessed by readers over the Internet. This mechanism should be employed to provide substantiating graphs (and statistics) that will assist readers in making a full interpretation of the data.

The guidelines presented here are heavily weighted toward the presentation of immunophenotyping data. This kind of data presentation is the most common in cytometry. However, the topics covered here, to a large part, apply to other kinds of cytometric data presentation including cell cycle analysis, analysis of time-resolved events, and so on. Because these analyses have somewhat different complexities associated with their interpretation, they are the subject of a separate discourse. The topic of presentation of the data specifically related to cell cycle analysis is addressed in Chapter 11 of this volume.

Finally, we note that this represents only a beginning. The Data Presentation Standards Committee of the International Society for Analytical Cytology will strive to update, revise, and expand these guidelines and to encompass topics of a more specialized nature (e.g., cell cycle, time-resolved kinetics analysis, cell proliferation analysis, and compensation). The goal of this committee is to develop a strict set of publication guidelines that journal publishers can use to ensure consistent and accurate presentation of flow cytometric data.

Throughout this discussion, specific guidelines are set as italicized paragraphs. A summary of all guidelines (and page references of their discussion) is given in Table I.

II. General Principles of Graphical Presentation

It is important to understand the role of graphical presentation in scientific chapters. Graphs are *not* drawn to allow users to make quantitative estimates. We should never rely on a reader to look at graphs and, for example, conclude that

Table I
Summary of Guidelines

Topic	Guideline	Page
General	• Quantitative information must be presented using standard methods (numbers or charts), not by inference from univariate or bivariate distribution plots	244
	• The identity and version of the software used to generate graphs in a publication must be given	244
Graphical displays	• The number of events displayed in any graph must be indicated	244
	• To convey quantitative representation of subsets from graphical displays, a calculated frequency of gated events must be displayed. The graph itself cannot convey such information	245
	• The choice of smoothing and specific display type is up to the author. Choose whichever graph and display options most readily convey the information needed to interpret the experiments, but be consistent across all graphs within an analysis	249
	• The type of scaling on plot axes must be obvious. Numerical values for axis ticks are rarely relevant and can be eliminated except when necessary to clarify the scaling. The number of logs spanning the whole scale should be evident	251
	• A percentage value shown on a graph that indicates a proportion of gated cells refers to the fraction of the total cells shown in that graph unless otherwise indicated on the graph	252
Univariate displays	• Unless explicitly noted, the abscissa (y-axis) for a one-dimensional histogram is assumed to be linear starting at zero. In most cases, numerical axis values are irrelevant and should not be included with zero-based linear axes	245
Gating	• Whenever gated analyses are performed, an illustration of the gating process should be shown. A back-gating display is highly recommended	255
	• Unless otherwise explicitly stated, gating is assumed to have been performed subjectively	255
	• The use of control samples to set gates should be shown; the algorithm to place gates should be explicitly defined if it was not subjective	256
Statistical analyses	• When reporting intensity measurements, it is necessary to explicitly define the statistic (e.g., mean, median, a particular percentile) that is applied. All statistics should be applied to the "scaled" intensity measurement, and not to "channel" numbers	252
	• In general, statistical significance of results should be reported using standard experimental procedures (replicate analysis)	253

a particular phenotype of cells is present at a greater (or lower) frequency in one sample than in another. All quantitative information should be derived from statistical analyses that are presented in addition to the graphics—in the form of numerical annotations on the graphics or in the accompanying text (e.g., in the figure legend, results section text, or in the footnote).

> Quantitative information must be presented using standard methods (numbers or charts), not by inference from univariate or bivariate distribution plots.

The most important role of graphs is to show *patterns.* The human mind is particularly adept at pattern recognition; thus, use graphs to illustrate variations in patterns (if this is relevant). Do not show 10 nearly identical graphs on 10 samples simply to illustrate the expression patterns. Rather, show a single sample and then present the results of a statistical test to convey the variation.

Number of events displayed:	Unsmoothed			Smoothed			
	Dot plot	Density plot	Contour plot	Contour plot	Outlier contour plot	Outlier zebra plot	Outlier density plot

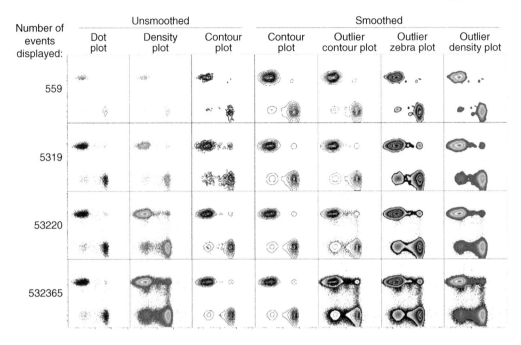

Fig. 1 Different bivariate displays of the same data. Four data sets comprising 500–500,000 events are shown in seven displays. See the discussion for advantages and disadvantages of various display methods. Note that the color scheme is not particularly relevant except to convey information to the reader; hence, the "Zebra plot" (which simply alternates black to gray transitions) is as interpretable as the density plot that employs a color scale corresponding to event density. Illustrated here is the power of advanced density estimation (smoothing); note that the smoothed graphs for 500 events look remarkably similar to the unsmoothed graphs where 1000 times as many events were analyzed. This illustrates that the smoothing algorithm employed here does not hamper the interpretation of the data distributions; in fact, it may significantly aid in visualization when low event numbers are present or when plots representing different numbers of events are to be compared.

is solely whether the graphical representation conveys the point that the author is trying to make (i.e., the distribution of expression); any color scheme that conveys this information is acceptable.

A third option for bivariate graphs is contour plots (Fig. 1). Contour plots have the desirable feature that they maintain features across a wide dynamic range of displayed events, making them particularly useful for comparing distributions from different samples. Contour plots can also reveal subtle expression patterns that the human brain does not pick out from dot plots or density plots (such as clusters of relatively less-frequent subsets or correlated expression patterns). Historically, a criticism of contour plots was that the selection of the levels at which to draw each contour line was user controlled; therefore, different analyses within the same presentation could be adjusted to look more or less alike simply by choice of levels. However, all software packages now automatically choose contour levels in an algorithmic way, so this bias no longer occurs. Different software may use unique methods to choose the levels, leading to somewhat different displays of the same data. However, as long as the same algorithm is used for all displays in a presentation, the precise algorithm for choosing the contour levels becomes unimportant.

A downside of contour plots is that events outside the last contour do not enter the display. Hence, rare populations are invisible; changes in such subsets then do not become apparent. To solve this problem, many software packages allow a combination of contour and dot plots, where any events outside the last contour plot are drawn as dots (Fig. 1). This "outlier" graph combines the advantages of contour plots (uniform density estimation over a wide dynamic range of events) with that of dot plots (revealing rare populations) and avoids the disadvantages of both methods.

Overlaying bivariate displays is often difficult. The order in which overlayed subsets are drawn significantly affects the display (i.e., one subset's events can be hidden by the other subset). Density plots cannot be overlayed, because color must be used to convey subset identity. Contour plots and dot plots can be overlayed with some success, but care must be taken that information is not hidden. In general, bivariate overlays are most useful when showing analyses such as "backgating"—showing where a subpopulation distributes among the various combinations of parameters that were used to gate that subpopulation from the main event collection (Fig. 2).

Should displays be smoothed? Smoothing of data is a topic that can be highly controversial. Many advocate that the only fair presentation is that of the "raw" data (although what constitutes raw data is itself controversial). Historically, smoothing of data displays was disparaged because the software algorithms allowed user-controlled extent of smoothing (and resmoothing). This meant that different displays within the same analysis could be smoothed to different extents, causing significantly different interpretation of the data.

At present, however, most software for flow cytometry offers more sophisticated density estimation procedures rather than generic smoothing. Such methods selectively use data from the vicinity of each point to improve the estimated density at that point (Moore and Kautz, 1986). Points with high counts use only close data for estimation, whereas in data-sparse areas a wider span of local data values are used. This process retains sharp features that are statistically justified while suppressing random noise. As shown in Fig. 1, such algorithms can be remarkably accurate, in that the smoothed displays generated from a few hundred events are quite similar to the unsmoothed displays generated from half a million events. For displays of immunofluorescence and light scatter data, density estimation is generally quite successful and is particularly helpful when data sets to be compared contain different numbers of events.

There is no reason to mandate either smoothed or unsmoothed displays. However, it is important that all displays in an analysis are created identically—using the same smoothing setting. Authors should try to avoid displaying statistical variations as though they were actual features of the distribution. In particular, unsmoothed

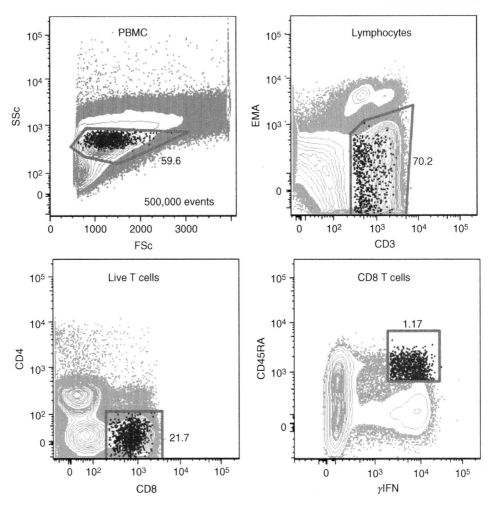

Fig. 2 Back-gating analysis. This data set is the same as that shown in Fig. 4. Instead of the typical graphical representation of the gating schema, a "back-gating" display is shown. Here, the final gated population is overlayed on each gating step as blue dots on the outlier contour plot. This allows readers to easily interpret the relative expression of the dependent markers (i.e., CD3, CD4, CD8, and EMA). (See Plate no. 7 in the Color Plate Section.)

displays for contour plots should be avoided because the contour details tend to reflect statistical "noise," leading to complex contours that are difficult to interpret.

> The choice of smoothing and specific display type is up to the author. Choose whichever graph and display options most readily convey the information needed to interpret the experiments, but be consistent across all graphs within an analysis.

There are other kinds of data display options for flow cytometric data that are far less commonly used; these are beyond the scope of this chapter. In general, however, authors should consider the likelihood that using uncommon types of displays may make data interpretation more difficult for readers.

B. Scaling Options for Display

Nearly all immunophenotyping data have been shown on a logarithmically scaled axis. The primary reason for this is that the dynamic range of the signal is so great; typically, brightly stained cells can be more than 1000 times as bright as unstained cells. To visualize both populations simultaneously, one must employ a scale that compresses the high-intensity measurements. A number of properties of the logarithmic scaling have made it the best choice.

Nonetheless, logarithmic scaling can have undesirable effects on visualization of data, particularly after fluorescence compensation. The details of the reasons for this effect are discussed extensively elsewhere (Roederer, 2001). Basically, however, the problem derives from the observation that at very low fluorescence values, the distribution of events becomes linear normal, rather than log normal, which is typical for positive immunofluorescence distributions. In particular, net fluorescence distributions for compensated cell populations often extend below zero on the dye signal axis. (There is nothing improper about measurement values below zero. Compensation involves the subtraction of correction terms, which have associated errors, so when the error in the result is greater than the mean level of fluorescence, some events inevitably end up below zero.) A log scale is particularly poor for representing low-mean, linear-normal distributions, and negative values, of course, cannot be represented on a log scale (Fig. 3).

To solve this, Parks and Moore (in preparation) have devised a new scaling algorithm called "Logicle" scaling. It is derived from the hyperbolic sine function and combines near-logarithmic compression for high-intensity values with near-linear scaling for low and negative values. As shown in Fig. 3, the Logicle scaling imparts a significantly more interpretable rendering of the data than the standard logarithmic scaling. Note that the Logicle scaling in itself does not affect statistics, gating, or other analyses; it is simply a tool for visualizing data. In some cases, use of Logicle displays should lead to more accurate selection of gating boundaries and improve statistical results by retaining very low and negative data values that would otherwise be truncated by the software.

An important feature of the Logicle scale is the presence of zero on the axis. This is not possible on a standard logarithmic axis, and its position is important. Hence, at least this tick must be clearly identified on a Logicle axis. However, the numerical values of the other markers are not critical and could be eliminated if desired. As for a logarithmic axis, the placement of the tick marks conveys the relative scaling quite well.

The type of scaling on plot axes must be obvious. Numerical values for axis ticks are rarely relevant and can be eliminated except when necessary to clarify the scaling. The number of logs spanning the whole scale should be evident.

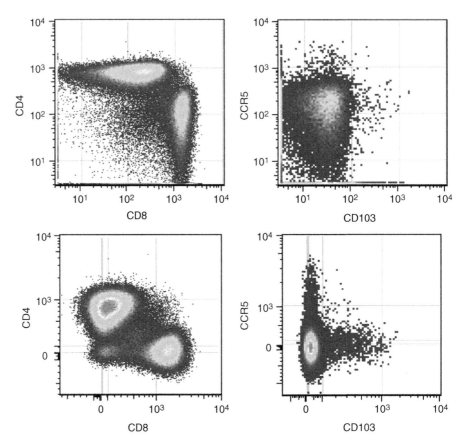

Fig. 3 Logarithmic and Logical scaled displays of two data sets. In the logarithmic CD4/CD8 plot, note the pileup of events on the axes. All of the events appearing below or left of the zero lines in the corresponding Logicle plot are contained in the axis pileup in the logarithmic plot. Also, note the distinct CD4$^-$/CD8$^-$ population near {0,0} in the Logicle display that is not apparent in the logarithmic version. In the CCR5/CD103 logarithmic plot, it looks like there is a set of CD103$^+$ events on the CCR5 baseline and another at about 200 units in the CCR5 dimension. The Logicle plot reveals these events to be a single group centered near zero in CCR5.

III. Statistics

There are two main purposes of statistical analysis in our work. One is to reduce a complex data set to a manageable number of summary values—for example, the average fluorescence intensity of a stained population. The second, which may use these summary values, is to test for statistical significance.

Flow cytometric measurements are invariably rich with quantitative information. In common experiments, half a dozen measurements are made on each of tens of thousands of cells. These cells are rarely homogeneous; thus, after identifying subsets of cells, we are left with a multitude of possible descriptors for each type of cell. Therefore, inevitably, all analyses include significant data-reduction steps in which the key features of the sample are summarized by one or a few numerical values. Although this practice is common and necessary, we should never forget that there may be considerably more information available on our data sets and should strive to ensure that our statistical reduction is appropriate for the sample.

For example, a statistic such as average fluorescence intensity does not convey the *distribution* of the expression; is it very uniform, is it broad, is it bimodal? Of course, this is where a graphical representation of the data to accompany the statistics is important: The graph conveys distribution information not retained by the selected statistics and can support the selection of a particular statistic as an appropriate representation of the sample.

Flow cytometric analyses always center on the measurement of intensity from a cell (fluorescence or scattered light). When we report the *frequency* of a cell population, what we mean is the proportion of cells that express some combination of ranges of intensities. For example, CD4 T cells may be defined as $CD4^+$ (more than background CD4 fluorescence), $CD3^+$, $CD8^-$ (no more than background CD8 fluorescence). Because all three measurements are important to defining the CD4 population, all three should be graphically illustrated to demonstrate that the gating was performed appropriately.

The simplest statistic reported is the proportion (percentage) of cells that falls within a certain combination of gates. However, even such a simple value is considerably more complex than it first appears. Consider the gating scheme shown in Fig. 4, to identify the expression of interferon-γ (IFN-γ) in stimulated CD8 T cells. The gating scheme is shown to illustrate how CD8 T cells were identified, and finally, how the cytokine expression was defined (as positive or negative). In the end, 11.2% of $CD45RA^-$ CD8 T cells were identified as positive for the cytokine. However, there are additional ways to interpret these events: the $CD8^+$ $CD45RA^-IFN_\gamma^+$ events are 82% of all CD8 T cells that secreted IFN-γ, 2.4% of all T cells, 1.7% of lymphocytes, and 1% of collected events. Which of these percentages should be reported is dependent on the particular interpretation that the presenter wants to make. However, if the percentage reported is not the fraction of events gated on that particular graph, then it must be made immediately obvious.

A percentage value shown on a graph that indicates a proportion of gated cells refers to the fraction of the total cells shown in that graph unless otherwise indicated on the graph.

Outside of statistics related to representation of a subset of cells, there is a wealth of information regarding the distribution of expression by cells (e.g., of a fluorescent marker). Often, the simplest analysis is to calculate how much of the marker is expressed on a per-cell basis. This reduction of the fluorescence measurement to a single value is an attempt to convey the central tendency of a population (e.g., the mean, median, or mode). Another common statistic is a measure of the population uniformity (such as the CV or robust CV). (*Note*: A "robust CV" is a measure of

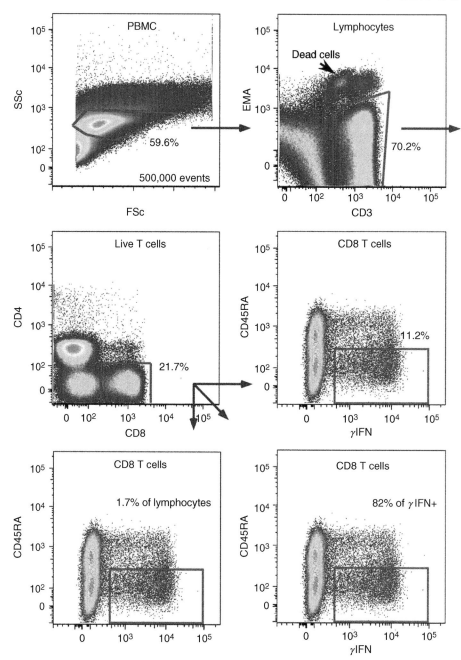

Fig. 4 Reporting subset statistics in graphical presentations. In this experiment, human peripheral blood mononuclear cells were stained with reagents to detect dead cells (EMA), as well as subsets of stimulated T cells (CD3, CD4, CD8, CD45RA, and interferon-γ [IFN-γ]). A progressive gating scheme was used to identify CD8 T cells that do not express CD45RA and do express IFN-γ, with a lymphocyte

population uniformity that is more reliable on logarithmically scaled data; discussion of its use can be found in Parks and Bigos (1997).)

Nearly all software packages calculate such statistics on gated populations of cells. It is important to understand the differences between the statistics, particularly when applied to data that cover a wide dynamic range (i.e., which is typically displayed on a logarithmic scale). It is not the intent here to cover the appropriateness of any given statistic for a distribution of data; however, it is wise to remember that with widely dispersed data, the median statistic is often a better estimate of the central tendency of a population than the mean. Statistical analyses on flow cytometric data are extensively discussed elsewhere (Parks and Bigos, 1997).

There has been some confusion (arising from older software packages) about whether it is appropriate to perform statistics on scale values (that are proportional to intensity) or "channel" values (which increase linearly across the scale, irrespective of whether the scaling is linear or logarithmic, and typically range from 0 to 256 or from 0 to 1024). All statistics reported should be on the scale values, *not* on channel values.

> When reporting intensity measurements, it is necessary to explicitly define the statistic (e.g., mean, median, a particular percentile) that is applied. All statistics should be applied to the "scaled" intensity measurement and not to "channel" numbers.

A common question regards the statistical significance associated with the measured percentage of a subset. For example, consider an analysis such as that shown in Fig. 1 done on two subjects: one subject had 11.2% cells in the $CD8^+$ $CD45RA^- IFN\text{-}\gamma^+$ gate with 500,000 events collected. For the second subject, for whom only 10,000 events were collected, this value is 8%. Are these different with statistical significance?

To answer this, we must know the precision with which each of these percentages was calculated; the precision is calculated based on the Poisson statistical

---

gate (top left), a live T-cell gate (top right), a CD8 gate (middle left), and a final subset gate (middle right). In each graphic, the fraction of cells in the gate is shown as a percentage; this percentage is a fraction of the *displayed events in that graph*. Hence, in the top right, 70.2% of lymphocytes fall in the shown gate, this is 70.2% of 59.6% or 42.2% of the entire sample. The symbol "%" was appended to this fraction to explicitly designate what the number means. Arrows between a gating region in one panel and the entire next panel clarify the gating scheme. The bottom two graphs the same data as the final gated graph, but here the emphasis was to show the representation of the $CD45RA^- IFN\text{-}\gamma^+$ cells as a fraction of lymphocytes or of $CD8^+ IFN\text{-}\gamma^+$ T cells. Hence, the percentage annotation is shown with a clarification as to what the denominator is. For these graphs, the axis values are labeled to illustrate the use of the Logicle scaling (see discussion); it would be acceptable to put a 0 under the zero tick mark; the placement of the logarithmically scaled tick above this is sufficient to convey the type of scaling used, and the values such as 10^2, 10^3, etc. could be dropped. Similarly, the numbers on the FSc (forward scatter) axis are unnecessary; the axis ticks placement conveys the linear scaling. Each graphic has been labeled with an annotation (e.g., "Lymphocytes") to illustrate that what is shown is a gated subset of the original collection; the gating scheme with percentages allows readers to estimate the number of events shown in each graph after the first (where it is explicitly shown).

distribution, from the total number of events in the *gated population,* and is represented by the square root of this number divided by the number of events. For the first subject, the number of cells counted was $500,000 \times 59.6\%$ (lymphocytes) $\times 70.2\%$ (T cells) $\times 21.7\%$ (CD8 T) $\times 11.2\%$ (cytokine) $= 5084$. Thus, the precision on the measured percentage is the square root of 5084 divided by 5084, or 0.014. The measured value is thus $11.2 \pm (11.2 \times 0.014)$ or 11.2 ± 0.16. For the second subject, we can carry through the same calculation; assuming similar fractions of cells in the lymphocyte and T-cell gates, we find that the measurement is $8\% \pm 0.9\%$. Based solely on the measurement, there is a high degree of confidence that the two values are different.

Numerical analyses of this sort are useful in deciding how much data must be acquired to ensure that purely statistical accuracy will be adequate for the intended analysis. For actual results analysis, however, biological variation is typically much greater than this. In other words, it is highly unlikely, where the first sample tested many times, that the error on this fraction would be only 1.5% ($\pm 0.16\%$). Invariably, experimental and biological variation is much greater than that afforded by flow cytometric measurements; hence, replicate analyses are the best way to estimate the significance of differences between samples. This is true for subset percentages and fluorescence intensity measurements.

> In general, statistical significance of results should be reported using standard experimental procedures (replicate analysis).

IV. Subset Analysis

Invariably, analysis and presentation of data is performed on a subset of the events that were collected. Initial subsetting is performed to eliminate debris, dead cells, and/or undesired subpopulations. Additional subsetting may be carried out to delineate subgroups within the population of interest. Because of the critical influence of this process on the interpretation of the data, an example of the gating approach should always be shown. Typically, this is done in a fashion such as that shown in Fig. 4, but a more informative (and useful) approach is a "back-gating" illustration such as that shown in Fig. 2. Here, it is immediately apparent not only how each subset was further divided, but also where within each defining gate the final population of interest resides. This is a powerful illustration that the gated subset is not an artifact of improperly positioned gates. It can also serve to illustrate more information about the underlying biology—in this example, that the gated cells, representing a stimulated population, have somewhat lower expression of CD3 and CD8 than the major CD3 or CD8 T-cell population.

> Whenever gated analyses are performed, an illustration of the gating process should be shown. A back-gating display is highly recommended.

Because gating is nearly universal in flow cytometric analysis, it is necessary to justify the gate positions. Much analysis is done visually; the scientist draws a gate around a cluster of events for further characterization. There is nothing inherently wrong with this approach; the human mind is often much better at picking out clusters of events than automated approaches (and certainly often better than blindly relying on control staining samples). Nonetheless, knowledge of this process is essential to the interpretation of the chapter.

Unless otherwise explicitly stated, gating is assumed to have been performed subjectively.

If an objective method is used to position gates (e.g., based on control samples or based on a "fluorescence minus one" [FMO] control (Kantor and Roederer, 1997; Roederer, 2001)), then this process must be explained. In addition, an example of the gates applied to the control sample should be shown. Simply stating that such controls were used to apply the gates is insufficient, because the question remains how those controls were used. Were the gates positioned so 99% of the events in the control sample were below/within the gate or 95% of the events? The percentage of cells in the control sample that is defining the gate, which is then used for comparison of the experimental sample, should always be stated. In many cases, a control sample is analyzed and then the gate is still applied visually (subjectively); this fusion of objective and subjective methodology is acceptable but should be stated.

The use of control samples to set gates should be shown; the algorithm to place gates should be explicitly defined if it was not subjective.

V. Conclusion

In this chapter, we discussed a number of guidelines; following these guidelines is a beginning step toward ensuring that presentations of flow cytometric data adhere to a consistent and understandable format. These guidelines are only a skeletal framework; future efforts by the Committee on Data Presentation Standards will provide additional (and updated) suggestions.

References

Kantor, A., and Roederer, M. (1997). FACS analysis of lymphocytes. *In* "Handbook of Experimental Immunology" (L. A. Herzenberg, D. M. Weir, and C. Blackwell, eds.), 5th edn., Vol. 2, Blackwell Science, Cambridge.

Moore, W., and Kautz, R. (1986). Data analysis in flow cytometry. *In* "Handbook of Experimental Immunology" (D. M. Weir, L. A. Herzenberg, and C. Blackwell, eds.), 5th edn., Vol. 2, Blackwell Science, Cambridge.

Parks, D. R., and Bigos, M. (1997). Collection, display, analysis of flow cytometry data. *In* "Handbook of Experimental Immunology", 5th edn., Vol. 2, Blackwell Science, Cambridge.

Roederer, M. (2001). Spectral compensation for flow cytometry: Visualization artifacts, limitations, and caveats. *Cytometry* **45**, 194–205.

Tufte, E. (2001). "The Visual Display of Quantitative Information." Graphics Press, Cheshire.

PART II

Cellular DNA Content Analysis

CHAPTER 10

Detergent and Proteolytic Enzyme–Based Techniques for Nuclear Isolation and DNA Content Analysis

Lars L. Vindeløv[*] and Ib Jarle Christensen[†]

[*]Department of Haematology
Rigshospitalet
DK-2100 Copenhagen, Denmark

[†]Finsen Laboratory
Rigshospitalet
DK-2100 Copenhagen, Denmark

Reprinted from *Methods in Cell Biology*, Volume 41 (Academic Press, 1994).
DOI: 10.1016/B978-0-12-375045-7.00010-6

I. Introduction

Performing flow cytometric DNA analysis on a suspension of nuclei has two major advantages. DNA nonspecific fluorescence from the cytoplasm is avoided, and fine-needle aspirates which contain a variable fraction of bare nuclei are well suited as starting material. The possibility of simultaneous analysis of cytoplasmic or plasma membrane-associated structures is thus sacrificed for easy and accurate determination of nuclear DNA in tissues and solid tumors.

A number of techniques have been published over the years. Some of them are listed in Table I. To choose a method for a particular purpose one should first consider the wavelengths available for excitation of the fluorochrome on the flow cytometer to be used. Second, one should assure, by testing, that high-resolution measurements can be obtained by preparing the cells in question with the method chosen. It should be remembered, in this context, that a properly aligned flow cytometer and a low flow rate are as necessary as stoichiometric staining of the DNA for obtaining good-quality histograms. The latter are characterized by a minimal amount of debris and symmetrical G_1 peaks with low (1.5–2.5%) coefficients of variation (CV).

The amount of information that can be extracted from a DNA distribution depends not only on the technical quality of the analysis but also on the methods used for standardization and statistical analysis (deconvolution) of the histogram. The data reduction obtained by statistical analysis yields the desired end-point results that will adequately and exhaustively describe the DNA distribution for most purposes. These are: (1) the number of subpopulations with different DNA content present in the sample and (2) for each subpopulation, (a) the relative size of the subpopulation, (b) the DNA index (DI), and (c) the fractions of cells in the cell-cycle phases G_1, S, and $G_2 + M$.

Table I
Some Techniques for Nuclear Isolation and DNA Analysis

References	Nuclear isolation technique	Fluorochrome/ excitation maximum
Krishan (1975, 1990)	Hypotonia	PI/536 nm
Vindeløv (1977), Vindeløv and Christensen (1990)	Nonidet-P40	PI/536 nm
Otto et al. (1979), Otto (1990)	Tween-20/citric acid	DAPI/350 nm
Taylor (1980), Palavicini et al. (1990)	Triton X-100	PI/536 nm DAPI/350 nm
Thornthwaite et al. (1980)	Nonidet-P40	PI/536 nm DAPI/350 nm
Vindeløv et al. (1983a–d), Vindeløv and Christensen (1990)	Nonidet-P40/trypsin	PI/536 nm

In the following, we describe a set of integrated methods developed in our laboratories for sample acquisition and storage, standardization, fluorochrome staining, and statistical analysis.

II. Basic Principles of the Methods

The integrated set of methods (Vindeløv and Christensen, 1990; Vindeløv *et al.*, 1983a–d) developed to solve the problems described above is outlined in Fig. 1. The analysis is performed on unfixed material. This is essential to avoid a

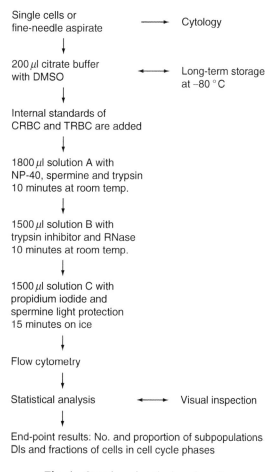

Fig. 1   Overview of methods and results.

potentially selective cell loss caused by centrifugation steps and keeps the cell requirement at a minimum. Clumping and staining artifacts caused by a fixative are avoided. Samples can be stored for long term by freezing in a citrate buffer with dimethyl sulfoxide (DMSO) (Vindeløv et al., 1983a).

The key to accurate and reproducible DI determination, as well as statistical analysis of multiple overlapping populations, is adequate internal standardization. This is achieved by adding a mixture of chicken (CRBC) and trout erythrocytes (TRBC) to the sample before staining (Vindeløv et al., 1983c). The peaks produced by the standards provide two points of reference at DIs approximately 0.30 (CRBC) and 0.80 (TRBC) and allow deconvolution of the histogram and DI determination independent of zero-point shift.

The preparation consists of three steps (Vindeløv et al., 1983b). Clean nuclei are obtained in the first step (solution A) by the combined action of the nonionic detergent Nonidet-P40 (NP-40) and trypsin. In addition, trypsinization increases the fluorescence of nuclei with dense chromatin such as granulocytes, presumably by splitting some chromosomal proteins. Spermine is essential for the stability of the unfixed nuclei during trypsinization. RNase A treatment in the second step (solution B) prevents dye binding to double-stranded RNA. Trypsin inhibitor is added, because trypsin activity after addition of propidium iodide (PI) results in unstable nuclei. In the final step (solution C), PI is added and the spermine concentration further increased for optimal stability. PI binds to double-stranded nucleic acid by intercalation (Crissman et al., 1975; Krishan, 1975).

III. Applications

Flow cytometric DNA analysis has been established as a useful research tool for a number of years. A review of applications in our laboratory may be found in Vindeløv and Christensen (1989). The methods have been used for monitoring the stability of cell lines, for sensitivity testing, and for studying the action of anticancer drugs in vitro. Likewise the methods have been used to monitor the cell-cycle perturbations produced by radiation, chemotherapy, and hormonal substances in murine and human tumors in experimental animals and in malignant tumors in patients. Since the same methods can be applied at these different levels of complexity, the results are directly comparable and may serve as a link between experimental and clinical results.

DNA analysis is currently assessed as a possible prognostic parameter in neoplastic disease (Merkel et al., 1987). In a prospective study of 249 patients with bladder cancer, ploidy and S phase size were found to have an independent prognostic value in multivariate analysis of survival and response to treatment (Vindeløv et al., 1993).

▬▬▬▬ IV. Materials

A. Citrate Buffer

Sucrose (BDH): 85.50 g (250 mM)
Trisodium citrate, 2 H_2O (Merck): 11.76 g (40 mM)
Dissolve in distilled water: ~800 ml
DMSO (Merck) is added: 50 ml
Distilled water is added to a total volume of: 1000 ml
pH is adjusted to: 7.6

B. Stock Solution

Trisodium citrate, 2H_2O (Merck): 1000 mg (3.4 mM)
Nonidet-P40 (Shell): 1000 μl (0.1%, v/v)
Spermine tetrahydrochloride (Serva, Cat. No. 35300): 522 mg (1.5 mM)
Tris (Sigma 7–9, Chemical Co., St. Louis. MO., Cat. No. T-1378): 61 mg (0.5 mM)
Distilled water is added to a total volume of: 1000 ml

C. Solution A

Stock solution: 1000 ml
Trypsin (Sigma, Cat. No. T-0134): 30 mg
pH is adjusted to: 7.6

D. Solution B

Stock solution: 1000 ml
Trypsin inhibitor (Sigma, Cat. No. T-9253): 500 mg
Ribonuclease A (Sigma, Cat. No. R-4875): 100 mg
pH is adjusted to: 7.6

E. Solution C

Stock solution: 1000 ml
Propidium iodide (Fluka Cat. No.): 416 mg
Spermine tetrahydrochloride (Serva, Cat. No. 35300): 1160 mg
pH is adjusted to: 7.6
The citrate buffer is stored at 4 °C. The staining solutions are stored in aliquots of 5 ml in capped plastic tubes at −80 °C. The tubes with solution C are wrapped in aluminum foil for light protection of the PI. Before use, the solutions are thawed in a water bath at 37 °C, but not heated to 37 °C. Solutions A and B are used at room temperature. Solution C is kept in an ice bath.

F. Internal Standards

Chicken blood can be obtained by heart puncture and collected in a tube containing 50 iu of heparin per ml of blood. The blood is subsequently diluted by citrate buffer. Rainbow trout blood can be obtained by caudal vein puncture after anesthesia with MS 222 (Sandoz, Basel, Switzerland) and should be mixed with citrate buffer immediately. The concentrations of CRBC and TRBC are determined by counting in a hemocytometer. The red cell concentrations are adjusted by dilution with citrate buffer to $CRBC = 145 \times 10^4$ cells per ml and $TRBC = 255 \times 10^4$ cells/ml. These suspensions are then mixed in equal volumes to obtain a final red blood cell concentration of 2×10^6 cells per ml and a ratio of CRBC:TRBC = 4:7, which will produce peaks of equal height in the histogram. The mixture of standards can be stored long term at $-80\,°C$ in aliquots of, for instance, 100 μl (sufficient for three to five samples), after freezing as described in Section V.B.

V. Methods

A. Sample Acquisition

Fine-needle aspiration is used for initial mechanical disagregation of normal tissues, lymphomas, and solid tumors. It is a gentle and rapid procedure causing less debris than cutting with knives or scissors. It is therefore used even when a surgical biopsy is available. The tumor is secured by pinning it with injection needles to a plate of styrofoam. In the case of very small biopsies, the tissue is wrapped in transparent plastic foil and secured during biopsy by pinning this to the plate as above. Biopsies as small as $2 \times 2 \times 2$ mm can be successfully aspirated in this way. A 0.5×25-mm (25G × 1) needle on a 20-ml disposable syringe fitted on a one-hand-operated handle (Cameco, Täby, Sweden) is adequate. Longer needles may be required for *in vivo* aspiration and thinner needles may give a better cell yield in fibrous tumors, such as some breast cancers. Suction is applied only when the point of the needle is within the tissue. The needle is moved back and forth in different directions within the tumor. The aspirate should stay within the needle during aspiration. The needle is flushed with citrate buffer (200 μl), and the cell yield is checked by counting in the hemocytometer. Aspiration is repeated until 10^6 cells have been obtained. The cells are stored by freezing (see Section V.B) or stained and analyzed on the same day. Two additional aspirates are spread on slides for cytologic examination.

B. Storage

Samples and internal standards not stained and analyzed on the same day as obtained are stored at $-80\,°C$ after freezing (Vindeløv *et al.*, 1983a). Samples have been kept in this way for up to 5 years, without change in the DNA histograms.

Samples can be frozen either as aspirates in citrate buffer or as surgical biopsies in a dry tube. It is essential that each sample is frozen and thawed only once. The cell suspension in citrate buffer is frozen in polypropylene tubes (38 × 12.5-mm test tubes with screw cap, Nunc, Roskilde, Denmark) immersed in a mixture of dry ice and ethanol (−80 °C). Before use, the samples are thawed in a water bath at 37 °C. The sample should only be thawed and not heated to 37 °C.

C. Standardization and Staining

Staining with PI (Vindeløv *et al.*, 1983b) is performed after addition of the internal standards (Vindeløv *et al.*, 1983c) to 200 µl of sample cell suspension in citrate buffer. The standards are added in an amount equal to 20% of the cells in the sample. With the concentrations chosen, 20% of the sample cell count in 0.1 µl (the volume of the hemocytometer used in our laboratory) is equal to the number of microliters of standard to be added to 200 µl of sample. Staining is performed by stepwise addition of the staining solutions. A total of 1800 µl of solution A is added to 200 µl of tumor aspirate in citrate buffer and the tube is inverted to mix the contents gently. After 10 min at room temperature, during which the tube is inverted two to three times, 1500 µl of solution B is added. The solutions are again mixed by inversion of the tube and after 10 min at room temperature 1500 µl ice-cold solution C is added. The solutions are mixed and the sample is filtered through a 25-µm nylon mesh into tubes wrapped in aluminum foil for light protection of the PI. The samples are kept in an ice-bath until analysis which should take place between 15 min and 3 h after addition of solution C (Vindeløv *et al.*, 1983b). If cells are scarce, the volumes of staining solution can be halved to increase the cell concentration.

There are few critical aspects of this simple procedure. Pure, analytic grade reagents should be used for the solutions. Accurate weighing of the reagents and accurate pipetting of the solutions are important. The samples should be handled gently throughout, since agitation will increase debris and clumping.

D. Flow Cytometric Analysis

The FACScan flow cytometer (Becton-Dickinson, Mountain View, CA) is currently used in our laboratory. Cells are analyzed at a rate of 20–40 cells/s, with a flow rate of 12 µl/s. This low rate of measurement is chosen to ensure a thin stream of sample, with the cells intersecting the laser beam in the same path. The blue 488-nm line of the argon ion laser is used for excitation. The instrument is equipped with a doublet discriminator module that allows measurement of pulse area (FL2-A), detected through a 585-nm bandpass filter.

E. Statistical Analysis

The end-point results mentioned in Section I are determined by statistical analysis of the DNA distribution (Vindeløv and Christensen, 1990). Visual inspection of the histogram plays a role in determining the quality of the analysis and the number of subpopulations present. Figure 2A shows an example of a DNA histogram. Statistical analysis or deconvolution, described briefly, involves the following steps:

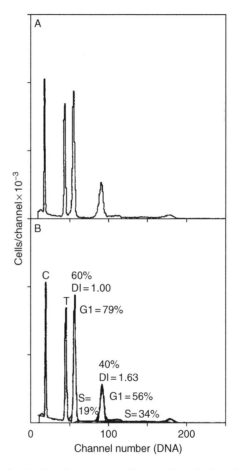

Fig. 2 (A) A DNA distribution from breast cancer. Fine-needle aspiration of a surgical biopsy stored by freezing. Visual inspection indicates a good-quality histogram with two subpopulations. (B) The histogram after statistical analysis. The peaks produced by the CRBC (C) and TRBC (T) are indicated. The first subpopulation is strictly diploid (DI = 1) and constitutes 60% of the cells. The second subpopulation is hypertriploid (DI = 1.63) and constitutes 40% of the cells. The percentages of cells in the cell-cycle phases are indicated on the figure. The CVs of the G_1 peaks are 2.3% and 2.2%.

1. Normal distributions are fitted to the peaks of the standards. Debris is subtracted by fitting a truncated exponential between the CRBC and TRBC peaks, and the zero point is corrected, based on the means of the standards.

2. A model describing the G_1 and $G_2 + M$ peaks with normal distributions, and the S phase with an exponential function of a polynomial of a given degree, is fitted to the histogram by maximum likelihood. For histograms with more than one subpopulation, a mixture of this density is fitted (Fig. 2).

3. The DIs of the G_1 peaks are calculated by comparison with known values for normal cells (Vindeløv et al., 1983d), and the areas under the curves are estimates of the fractions of cells in the cell-cycle phases G_1, S, and $G_2 + M$. Further details regarding deconvolution of DNA content frequency histograms are presented in Chapter 18 of this volume.

Figure 2B shows the deconvoluted histogram and the estimated values. Important features are the following:

1. The very accurate zero-point determination obtained by the standards.
2. A point of reference (TRBC) close to the diploid peak, which makes DI determination more accurate.
3. A slight nonlinearity of the relationship fluorescence/DNA, which has been determined experimentally and corrected for in the computer program. This allows prediction of the $G_2 + M$ mean location and thereby deconvolution of heterogeneous tumors with overlapping populations, as shown in Fig. 2.

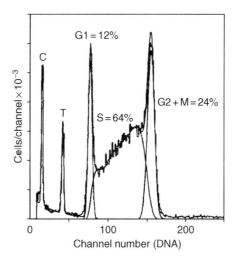

Fig. 3 An example of deconvolution of a histogram from a drug-perturbed cell population. The small cell lung cancer cell line NCI-N592 was exposed to the alkylating agent BCNU. The S phase was fitted by an eighth degree polynomial. The standards and the percentages of cells in the cell-cycle phases are indicated on the figure.

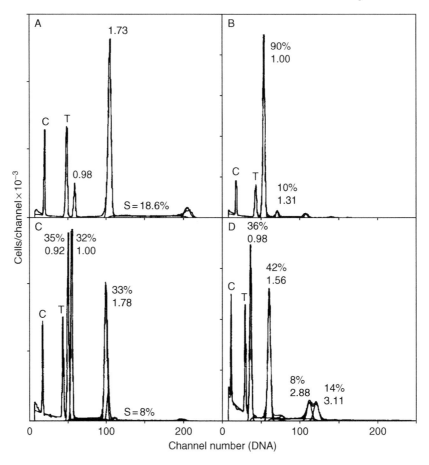

Fig. 4 (A) A DNA histogram from small cell carcinoma of the lung. Fine-needle aspiration *in vivo* of a lymph node metastasis. A single aneuploid (DI = 1.73) population is present. The peak with DI = 0.98 could represent normal cells and possibly some tumor cells. The CVs of the G_1 peaks are 1.9% and 1.7%. (B) A distribution from carcinoma of the oral cavity. Fine-needle aspiration of a small surgical biopsy. Ninety percent of the cells are strictly diploid. A small subpopulation with DI = 1.31 is present. The S phases are confounded and cannot be estimated reliably. The CV of the diploid G_1 peak is 2.2%. (C) A distribution from a non-Hodgkin lymphoma. Fine-needle aspiration *in vivo*. Three subpopulations of nearly the same size are present. The separation of the diploid (DI = 1) and the hypodiploid (DI = 0.92) peaks illustrate the need for a high resolution. The CVs range from 1.6% to 1.8%. The S phases of the diploid and hypodiploid subpopulations are confounded. An average value can be calculated. (D) A distribution from breast cancer. Fine-needle aspiration of surgical biopsy. Four subpopulations are present. Their sizes and DIs are indicated on the figure. Only one or two of the S phases can be estimated. The CVs vary from 2.1 to 2.8.

4. Histograms with a treatment-induced perturbation of the cell cycle can be deconvoluted by increasing the degree of the polynomial used for fitting the S phase. An example is shown in Fig. 3. The S phase distribution is estimated as an additional end-point result.

VI. Results

These methods were developed with emphasis on obtaining optimal results in a wide range of cells and tissues, particularly solid tumors. The staining method was developed in 1978 as a modification of an initial attempt (Vindeløv, 1977). In the past 14 years, we have analyzed approximately 20,000 samples from clinical and experimental studies. Satisfactory results have been obtained in all types of cells and tissues examined with the exception of sperm. The samples analyzed include (1) normal tissues: human lymphocytes, granulocytes and spleen and mouse lymphocytes, bone marrow, spleen, liver, kidney, and thymus; (2) human neoplasms: lung cancer, breast cancer, lymphoma, leukemia, bladder cancer, and cancer of the oral cavity; (3) human tumors in nude mice: breast cancer, lung cancer, melanoma, and colon cancer; and (4) mouse ascites tumors: JB-1, L1210, Ehrlich, and P388. Some examples are shown in Fig. 4. CVs of around 2% are obtained routinely. The theoretical and practical problems of resolving minor DNA differences as well as long-term reproducibility and comparability of results obtained by the methods have been examined in some detail (Vindeløv et al., 1983a,d).

Acknowledgments

The work was supported by grants from The Danish Cancer Society, The Danish Medical Research Council, and The Lundbeck Foundation.

References

Crissman, H. A., Mullaney, P. F., and Steinkamp, J. A. (1975). In "Methods in Cell Biology" (D. M. Prescott, ed.), Vol. 9, pp. 179–246. Academic Press, New York.

Krishan, A. (1975). J. Cell Biol. 66, 188–193.

Krishan, A. (1990). In "Methods in Cell Biology" (Z. Darzynkiewicz and H. Crissman, eds.), Vol. 33, pp. 121–125. Academic Press, San Diego.

Merkel, D. E., Dressler, L. G., and McGuire, W. L. (1987). J. Clin. Oncol. 8, 1690–1703.

Otto, F. J. (1990). In "Methods in Cell Biology" (Z. Darzynkiewicz and H. Crissman, eds.), Vol. 33, pp. 105–110. Academic Press, San Diego.

Otto, F. J., Hacker, U., Zante, J., Schumann, J., Göhde, W., and Meistrich, M. L. (1979). Histochemistry 61, 249 254.

Palavicini, M. G., Taylor, I. W., and Vindeløv, L. L. (1990). In "Flow Cytometry and Sorting" (M. R. Melamed, T. Lindmo, and M. L. Mendelsohn, eds.), 2nd edn., pp. 187–194. Wiley-Liss, New York.

Taylor, I. W. (1980). J. Histochem. Cytochem. 28, 1021–1024.

Thornthwaite, J. T., and Thomas, R. A. (1990). In "Methods in Cell Biology" (Z. Darzynkiewicz and H. Crissman, eds.), Vol. 33, pp. 111–119. Academic Press, San Diego.

Thornthwaite, J. T., Sugarbaker, E. V., and Temple, W. J. (1980). Cytometry 3, 229–237.

Vindeløv, L. L. (1977). Virchows Arch. B 24, 227–242.

Vindeløv, L. L., and Christensen, I. J., (1989). Eur. J. Haematol. Suppl. 42(48), 69–76.

Vindeløv, L. L., and Christensen, I. J. (1990). Cytometry 11, 753–770.

Vindeløv, L. L., Christensen, I. J., Keiding, N., Spang-Thomsen, M., and Nissen, N. I. (1983a). Cytometry 3, 317–322.

Vindeløv, L. L., Christensen, I. J., and Nissen, N. I. (1983b). *Cytometry* **3,** 323–327.

Vindeløv, L. L., Christensen, I. J., and Nissen, N. I. (1983c). *Cytometry* **3,** 328–331.

Vindeløv, L. L., Christensen, I. J., Jensen, G., and Nissen, N. I. (1983d). *Cytometry* **3,** 332–339.

Vindeløv, L. L., Christensen, I. J., Engelholm, S. A., Guldhammer, B. H., Højgaard, K., Sørensen, B. L., and Wolf, H. (1993). *Cytometry* (submitted for publication).

CHAPTER 11

DNA Content and Cell Cycle Analysis by the Propidium Iodide–Hypotonic Citrate Method

Awtar Krishan and Ronald Hamelik

Department of Pathology
Miller School of Medicine
University of Miami
Miami, Florida 33101

I. Introduction

Following publication of the Crissman and Steinkamp flow cytometric method (Crissman and Steinkamp, 1973) for quantitation of DNA content after staining of fixed cells with propidium iodide, we experimented with direct staining of isolated nuclei (without prior fixation and enzyme digestion) with various DNA-binding fluorochromes. The propidium iodide (PI)-citrate method (Krishan, 1975) was discovered by accident when a technician added one-tenth the amount of citrate for preparation of isotonic propidium iodide solution for identification of viable dye excluding cells. This method was particularly useful for rapid determination of cell cycle distribution in leukemic peripheral blood and bone marrow samples of patients on high-dose methotrexate rescue protocols (Krishan *et al.*, 1976).

Fried *et al.* (1976) used this method for rapid cell cycle analysis of cells in culture and Callis and Hoehn (1976) used it for detection of cells with aneuploid DNA content. Subsequently, it was used by several workers to study cell cycle phase distribution of mammalian tissue culture cells and a variety of tumor cells. Modifications have involved inclusion of RNase and a detergent in the staining solution (Vindelov, 1977; see Chapter 14, this volume). We have used this method with slight modifications for study of prostate and breast tumors (Krishan *et al.*, 2000, 2004), cells from body cavity fluids and for correlation of diagnostic marker expression with DNA aneuploidy (Krishan *et al.*, 2006) and to determine genome size of animals from the Miami zoo (Krishan *et al.*, 2005). A brief description of this simple staining method with modifications for use in formalin fixed-paraffin embedded tissues and in correlation of DNA versus protein content or diagnostic marker expression follows.

II. Application

This rapid method is especially suitable for small samples such as a drop of blood from a finger prick, tissue culture cells from multiwall plates and fine needle aspirates from solid tumors or bone marrow. For solid-tumor material, mincing of tissue with scissors or crossed scalpels and filtering through nylon cloth are essential. For formalin fixed-paraffin embedded tissue, pepsin digestion followed by antigen retrieval is used for generation of DNA distribution histograms and correlation with hormone receptor expression.

III. Materials

1. Cell pellets from tissue culture, leukemic blood, or a drop of blood from a finger prick, fine needle aspirates, biopsy samples, minced tissue or solid tumors, and body cavity fluids from peritoneal, pleural, cerebrospinal, or pericardial fluids
2. PI (Calbiochem catalog no. 537059)
3. Sodium citrate
4. Igepal-630 (Sigma-Aldrich), nonionic detergent claimed to be chemically indistinguishable from Nonidet P-40 which is no longer commercially available
5. Fine tipped Pasteur pipettes
6. Nylon cloth, 40-μm mesh (Small Parts Inc., Miami Lakes, FL)

Preparation of staining solution:

Propidium iodide: 5 mg

Sodium citrate: 100 mg

Glass-distilled water: 100 ml

Addition of a nonionic detergent in the final staining solution may help in isolation of nuclei from nonlymphoid cells such as epithelial cells in monolayer cultures.

Prepare the staining solution and store in an amber-colored bottle. We prefer a 1–l Repipet™ bottle, which allows for rapid transfer of premeasured volumes of staining solution directly into the specimen vial. The staining solution can be stored indefinitely in a cold room or at room temperature without any measurable loss of activity.

IV. Staining Procedure

1. *Peripheral blood from normal or leukemic patients:* Add a drop of the specimen directly to 1–2 ml of stain. Mix vigorously with a syringe/fine tipped pipette and run on a flow analyzer. It is important to either vigorously pipette or vortex the sample to remove the cytoplasmic fragments which may adhere to the nuclei and result in loss of resolution.

2. *Monolayer cultures:* Decant growth medium and wash cell monolayer with normal saline or balanced salt solution. Add 1–2 ml of the staining solution and scrape with a rubber "policeman." Transfer to a test tube and shake vigorously or pipette to break open cells and dislodge any cytoplasmic fragments sticking to isolated nuclei. Filter through nylon cloth and run on a flow analyzer.

3. *Surgical biopsy from a solid tumor:* Remove excess fat and necrotic tissue and wash with saline to remove blood. Cover with staining solution and dice or mince with crossed scalpels or scissors, or grate on a metal sieve. Remove large pieces and fat, and collect cell suspension in a test tube. Use a narrow-gauge needle to syringe the specimen vigorously to isolate single nuclei. Filter through a nylon cloth and run on an analyzer.

4. *Fine needle biopsy specimens:* The syringe and needle can be directly flushed into a 15-ml centrifuge tube containing 1–2 ml of staining solution. Syringe vigorously to isolate nuclei.

5. *Body cavity fluids from the peritoneal, pleural, pericardial cavities or cerebrospinal fluids:* Once the cells have been concentrated by centrifugation, an aliquot of cells can be directly mixed with the staining solution, vortexed, and analyzed.

6. *Formalin fixed/paraffin embedded 50 μm sections:* Antigen retrieval of the 50 μm sections in 0.1% sodium citrate at 90 °C for 30 min is followed by pepsin digestion to isolate the nuclei, wash to remove the enzyme and resuspension of the centrifuged pellet in the staining solution. Filter through a nylon cloth and run on an analyzer.

V. Caution

It is important that the specimen be vigorously pipetted or vortexed to dislodge cytoplasmic fragments from the nuclei.

Do not use trypsin for removal of cell monolayers or heparin as an anticoagulant for blood or bone marrow aspirates. Both of these reagents may interfere with binding of the fluorochrome and may not yield good DNA distribution histograms (Krishan *et al.*, 1978).

It is important to remember that certain DNA-intercalating agents may interfere with binding of PI to DNA. This is especially true of clinical specimens from patients on anthracycline chemotherapy. Readers are referred to Krishan *et al.* (1978) for further consideration of this artifact.

Stained specimen can be run within 5 min of staining and often can be stored as long as 24 h in an ice bucket or refrigerator at 4 °C. Longer storage leads to swelling of nuclei.

Propidium iodide may be a potential health hazard and care should be taken in handling, especially avoiding spills on hands. Proper disposal of hazardous waste is also recommended.

VI. Instrument Setup

Although peak excitation (540 nm) of PI with the 488- or 514-nm argon ion laser line is ideal, the PI-citrate method can also be used in analyzers equipped with a mercury HBO lamp light source (e.g., Quanta Flow analyzer, Beckman Coulter).

In the Coulter XL flow analyzer (argon ion 488-nm laser excitation) and the Quanta analyzer with solid state 488 excitation, we use peak emission fluorescence signal (>590 nm) to get low coefficients of variation (CV values) in DNA distribution histograms.

VII. Comments

Several modifications of the original PI-citrate method have been subsequently published. In cells with low nucleocytoplasmic ratio (e.g., lymphoid cells), the original PI-citrate method gives excellent results. In cells with large cytoplasmic mass or RNA content, use of detergents and/or RNase may be indicated.

VIII. Results

Histograms in Fig. 1 are typical of cells stained by the PI-hypotonic citrate method.

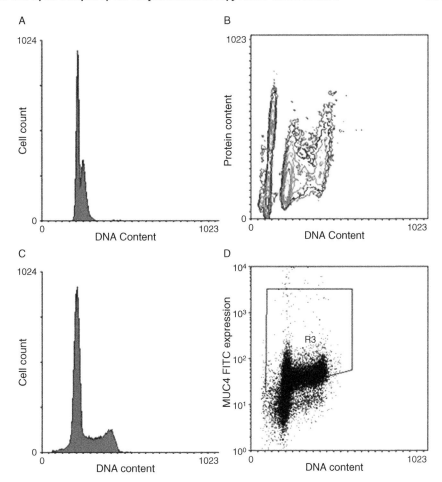

Fig. 1 (A) DNA histogram of nuclei from pleural fluid of a patient. The CV of the G_0/G_1 peak was 2.7% and a second aneuploid peak with DNA index of 1.21 is evident. (B) Contour plot of nuclei from body cavity of a patient stained with PI and fluorescein isothiocyanate to label the nuclear protein content. (C) DNA histogram of nuclei from pleural fluid of a patient. A large number of cells with S and G2/M DNA content are seen. (D) Dot plot of MUC4 FITC expression and DNA content of the cells from the sample shown in (C). Expression of MUC4 was predominantly seen in proliferating cells of this sample.

References

Callis, J., and Hoehn, H. (1976). Flow-fluorometric diagnosis of euploid and aneuploid human lymphocytes. *Am. J. Hum. Genet.* **28,** 577–584.

Crissman, H. A., and Steinkamp, J. A. (1973). Rapid, simultaneous measurement of DNA, protein, and cell volume in single cells from large mammalian cell populations. *J. Cell Biol.* **59,** 766–771.

Fried, J., Perez, A. G., and Clarkson, B. D. (1976). Rapid hypotonic method for flow cytofluorometry of monolayer cell cultures: Some pitfalls in staining and data analysis. *J. Cell Biol.* **71,** 172–181.

Krishan, A. (1975). Rapid flow cytofluorometric analysis of mammalian cell cycle by propidium iodide staining. *J. Cell Biol.* **66,** 188–193.

Krishan, A., Pitman, S. W., Tattersall, M. H. N., Paika, K. D., Smith, D. C., and Frei, III E. (1976). Flow microfluorometric patterns of human bone marrow and tumor cells in response to cancer chemotherapy. *Cancer Res.* **36,** 3813–3820.

Krishan, A., Ganapathi, R. N., and Israel, M. (1978). The effect of adriamycin and analogs on the nuclear fluorescence of propidium iodide stained cells. *Cancer Res.* **38,** 3656–3662.

Krishan, A., Oppenheimer, A., You, W., Dubbin, R., Sharma, D., and Lokeshwar, B. L. (2000). Flow cytometric analysis of androgen receptor expression in human prostate tumors and benign tissues. *Clin. Cancer Res.* **6,** 1922–1930.

Krishan, A., Arya, P., Ganjei-Azar, P., Shirley, S. E., Escoffery, C. T., and Nadji, M. (2004). Androgen and Vitamin D Receptor Expression in Archival Human Breast Tumors. *Cytometry B* **58B,** 53–60.

Krishan, A., Dandekar, P., Nathan, N., Hamelik, R., Miller, C., and Shaw, J. (2005). DNA index, genome size, and electronic nuclear volume of vertebrates from the Miami Metro Zoo. *Cytometry A* **65A,** 26–34.

Krishan, A., Ganjei-Azar, P., Jorda, M., Hamelik, R. M., Reis I. M., and Nadji, M. (2006). Detection of tumor cells in body cavity fluids by flow cytometric and immunocytochemical analysis. *Diagn. Cytopathol.* **34,** 528–541.

Vindelov, L. L. (1977). Flow microfluorometric analysis of nuclear DNA in cells from solid tumors and cell suspensions. A new method for rapid isolation and straining of nuclei. *Virchows Arch. B Cell Pathol.* **2.4,** 227–242.

CHAPTER 12

DNA Analysis from Paraffin-Embedded Blocks

David W. Hedley

Departments of Medicine and Pathology
Ontario Cancer Institute/Princess Margaret Hospital
Toronto, Ontario, Canada M4X 1K9

I. Introduction
II. Application
III. Methods
   A. Selection of Blocks
   B. Section Cutting
   C. Dewaxing
   D. Rehydration
   E. Pepsin Digestion
   F. Cell Counting
   G. Staining
   H. Data Analysis
   I. Use of Internal DNA Standards
IV. Critical Aspects of Technique
   A. Tissue Fixation
   B. Enzyme Digestion
   C. Other Factors
V. Alternative Methods for Sample Preparation
   A. Selecting Areas of Interest Within a Block
   B. Dewaxing and Rehydration
   C. Alternative Enzymatic Procedures
   D. Multiparametric Analysis
References

ESSENTIAL CYTOMETRY METHODS
Reprinted from *Methods in Cell Biology*, Volume 41 (Academic Press, 1994).

DOI: 10.1016/B978-0-12-375045-7.00012-X

I. Introduction

Routine pathological examination of tumors involves fixation, usually with formaldehyde, followed by dehydration and embedding in paraffin wax. Microtome sections can then be cut from these paraffin "blocks" for histological examination. Preserved in this way tumor tissue is remarkably durable, and because of the occasional need to cut further tissue sections for pathological review, the standard practice is to archive paraffin blocks for up to several decades. This material can be used for flow cytometric DNA analysis.

II. Application

Compared to fresh material, sample preparation from paraffin blocks is more time consuming, and the DNA histograms are generally of poorer quality in terms of CVs and debris. As originally conceived, the main advantage of the method was that it allowed retrospective analysis of material from cohorts of patients whose clinical outcome was already known, thus allowing the prognostic significance of DNA index and S phase in the various tumor types and stages to be determined. It has, however, become widely used for the prospective evaluation of individual cancer patients when fresh material is not available, or with very small tumors where the entire sample requires processing for diagnostic purposes. In these latter cases, microscopic examination of a parallel tissue section can allow selective sampling from an area of the block that contains tumor, minimizing the proportion of stromal cells in the DNA histogram.

Since its original publication (Hedley et al., 1983), many modifications of the method have been described, adapting it for use either in routine pathology laboratories or for analysis of particular tumor types (Hedley, 1989; Hitchcock and Ensley, 1993). Unfortunately, there does not appear to be one method that is ideal for all situations, and laboratories using the technique should be prepared to be creative. The purpose of this chapter is to guide users through the basic method and to review the commonly encountered problems and their possible solutions, rather than to review all of the published variations. Comprehensive and critical evaluations of the technique are discussed by Hitchcock and Ensley (1993) and by Heiden et al. (1991), and these are recommended for further reading.

III. Methods

The basic steps of the method are selection of a block that contains an adequate and representative population of tumor cells, cutting of thick sections using a microtome, dewaxing, and rehydration. This material is then subjected to enzymatic digestion, and the resulting nuclear suspension processed for DNA analysis.

A. Selection of Blocks

Although the method has been successfully applied to blocks that are several decades old (Toikkanen *et al.*, 1989), poorly preserved material is a major reason for failure to obtain an interpretable DNA histogram. This problem is further discussed in Section IV. Because %S phase is increasingly used as a prognostic determinant, and because reliable estimates require that admixture with normal stromal elements be kept to a minimum, parallel tissue sections should be examined by light microscopy and blocks selected which contain a substantial proportion of tumor cells. Methods for enriching tumor cells are discussed below.

B. Section Cutting

Sections are usually cut using a microtome and should be as thick as possible, since there is a clear relationship between the thickness of a microtome section and the extent and distribution of debris due to partially sectioned nuclei. Fifty micrometers or thicker is advised, which is greater than can be cut with some microtomes. Sections of this thickness curl up as they come off the microtome, and there is no need to uncurl them as it is more convenient to handle them in this form. Depending on tumor type and cellularity, between one and four sections are needed for DNA analysis.

C. Dewaxing

A variety of organic solvents can be used for dissolving the paraffin wax. Xylene, which was described in the original publication, is still widely used but is toxic and can be substituted for by more environmentally friendly agents such as Histoclear (National Diagnostics, Summerville, NJ). Place thick sections in a tube that is solvent compatible, add 3 ml of solvent, let stand at room temperature for 10 min, aspirate, and repeat this process, finally reaspirating the solvent.

D. Rehydration

Add 3 ml of 100% ethanol, let stand at room temperature for 10 min, and aspirate. Repeat this with 95%, 70%, and 50% ethanol for 10 min each, and finally add 100% water. During this rehydration process, the tissue will become soft and friable, and care is required to prevent aspirating pieces of tissue. If this is a problem, use either centrifugation or process tissue in nylon mesh bags. Aspirate water.

E. Pepsin Digestion

Prepare a 0.5% solution of pepsin in 0.9% saline, and adjust pH to 1.5 by adding 2 N HCl. Note that inferior grades of pepsin may have low activity or contain other digestive enzymes as contaminants. Sigma product number P7012 is of high purity and catalytic activity and gives satisfactory results. Add 1 ml to tissue, and place in a 37 °C water bath for 30 min. Agitate or briefly vortex mix periodically. During this procedure, the sample often becomes visibly turbid, and microscopic examination will show bare nuclei or intact cells resembling those seen in the parallel thin section. The enzyme digestion step is one of the most critical in the assay, and optimum conditions may differ between tumor types, or depend on the original fixation procedure. Following digestion, the nuclei should be washed once in buffered medium, filtered, and counted.

F. Cell Counting

Optimum use of equilibrium dyes such as the DNA stains requires that the ratio of dye molecules to DNA be kept within fairly narrow limits, and for this reason nuclei/cells should be counted and the final concentration adjusted. For most instruments, a concentration of 1×10^6 ml^{-1} is satisfactory. Although use of an appropriate dye: DNA ratio holds for all flow cytometric DNA analysis techniques, it is often technically more difficult to obtain reliable counts of nuclei obtained from paraffin-embedded material than with fresh tissue, due to the increased amount of debris. Counts can be made either manually with a hemocytometer or using a Coulter counter.

G. Staining

1. Dapi

The original method of Hedley *et al.* (1983) used the DNA-specific dye 4′,6′-diamidino-2-phenylindole dihydrochloride (DAPI), which binds to the minor groove of the double helix. It is obtainable from Boehringer-Mannheim and made up as a 500 μg/ml stock solution in water. This is stable at 4 °C for several months. Stain at a final concentration of 1 μg/ml at room temperature for 30 min and run on a flow cytometer, using UV excitation. DAPI has a broad blue emission spectrum, and band pass filters centered between 400 and 500 nm are suitable.

2. Propidium Iodide

Because most clinical flow cytometers use air-cooled argon lasers, propidium iodide (PI) is used more often than DAPI. Propidium iodide intercalates into the DNA double helix, and its binding is influenced by nuclear proteins to a greater extent than is DAPI. It is probably also more susceptible to the DNA denaturation that occurs during the procedure, and the CVs obtained with PI tend to be wider than those obtained with DAPI. Use a final propidium iodide concentration of

50 μg/ml for 30 min or longer, and the addition of RNase is recommended, as for propidium iodide staining of fresh material.

H. Data Analysis

For many tumor types, %S phase is a more powerful prognostic factor than DNA index, but reliable estimates of S phase are much more dependent on the extent of cell debris. Debris is a particular problem using paraffin-embedded material because of the harsh conditions involved in sample preparation and because sectioning with a microtome produces sliced nuclei. The latter show a characteristic distribution, with a disproportionate degree of debris to the left of the G_1 peak. Conventional debris subtraction overcompensates for this, giving an erroneously low (or even negative!) %S phase. Current versions of both MultiCycle and ModFit have debris subtraction that models for sliced nuclei (Fig. 1), and this has been shown to significantly improve the prognostic power of S phase (Kallioniemi et al., 1991). Use of these modeling programs is, therefore, strongly recommended for routine clinical samples.

I. Use of Internal DNA Standards

Unfortunately, there are no reliable internal standards of cellular DNA content that can be used with this method. The basic problem is that the initial fixation with formaldehyde causes cross-linkages between DNA and other macromolecules, which are then broken down during the enzyme digestion step. The degree of cross-linking varies considerably between samples, and this critically and unpredictably influences the availability of DNA for staining, rendering comparisons with internal or external standards unreliable. The least unsatisfactory standard is a piece cut from the thick section which is microscopically free of tumor; but because formalin penetrates tissue at a slow rate, the extent of fixation will vary within large blocks and may affect the stainability with DNA dyes. It is not, therefore, possible to identify with certainty hypodiploid tumors, which are rare but might carry a poorer prognosis than near-hyperdiploid tumors. By convention, the G_1 peak with the lowest DNA content is considered normal diploid, and DNA index calculated by comparing other G_1 peaks to its modal position. This gives a close correlation with the DNA index obtained for the same tumor using unfixed material, but hypodiploid tumors are erroneously assigned at a DNA index of >1.0.

IV. Critical Aspects of Technique

Compared to DNA analysis using fresh tissue, this method is technically more demanding, commonly encountered problems being failure to obtain a DNA histogram, excessive amounts of debris, or unacceptably high CVs. Very often

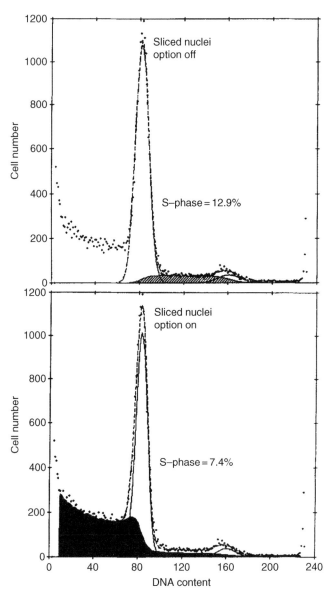

Fig. 1 Analysis of %S phase using the "sliced nuclei option" of the MultiCycle program (Phoenix Flow Systems, San Diego, CA). Note that the distribution of debris is not exponential because a major component is derived from partially sectioned nuclei. Since most cells are in the G_1 peak, debris falls off abruptly beyond this position. The program makes allowance for this during debris subtraction (right panel, shaded), and the resulting S phase estimate is a more reliable guide to prognosis (Kallioniemi *et al.*, 1991). A similar option is available in ModFit (Verity Software).

wide CVs and large amounts of debris go together. These problems can usually be traced to either the initial fixation of the tissue or to the enzyme digestion step.

A. Tissue Fixation

The quality of the histograms obtained is critically dependent on the initial fixation of the tissue. Formaldehyde-based fixatives are generally satisfactory, although the rate of penetration into tissue is slow, resulting in uneven fixation of large pieces of tissue and the possible generation of false aneuploid peaks. More exotic fixatives, such as those based on mercury or Bouin's fixative, frequently give very poor quality DNA histograms, and the pedigree of a sample should therefore be obtained if there are problems with analysis. Anecdotally, results with material fixed in Bouin's fluid are said to be improved by modifying the rehydration step to include several changes in 50% ethanol and then allowing to stand overnight in 50% ethanol.

Because the flow cytometry laboratory usually has no control over the quality of tissue fixation, it is strongly recommended that the method be worked up using material that is known to give a good quality DNA histogram, possibly obtained from another laboratory experienced with the method. It is also a good idea to include a sample of this material in each staining run, as a quality-control check.

B. Enzyme Digestion

If you are sure that the material being examined is well fixed and capable of giving a good quality DNA histogram, but are getting wide CVs or excessive debris, the problem probably lies in the enzyme digestion step. Enzyme solutions are commonly described in terms of percentage by weight of dry powder, but note that most commercial sources originate in the local abattoir, rather than the biotechnology industry and that their purity and catalytic activity varies widely. Furthermore, do not assume that the potency of an enzyme written on the label is necessarily accurate, or that preparations have an indefinite shelf life. In other words, be prepared to try other enzyme preparations and to vary the strength, pH, or incubation time.

The original method using 0.5% pepsin at pH 1.5 for 30 min at 37 °C usually gives satisfactory results, although these conditions may be too harsh for some tumor types, such as lymphomas or testis cancer, and produce nuclear fragmentation. In such cases, a reduction in pepsin concentration or incubation time would be appropriate. It should be noted that this enzymatic step was optimized for use with the minor groove binding stain DAPI and that it is not necessarily ideal for use with intercalating dyes such as propidium iodide, which are probably more susceptible to DNA degradation. There is still no completely satisfactory published review of alternative methods for enzymatic digestion, but it is possible that trypsin may give lower CVs when propidium iodide is used as the DNA stain (Tagawa *et al.*, 1993), while Heiden *et al.* (1991) have reported that the use of subtilisin Carlsberg (pronase) gives fewer cell clumps than pepsin. Details of these enzyme procedures are given in Section V.

C. Other Factors

In the dewaxing and rehydration sequence, probably the only really critical stage is ensuring complete removal of paraffin wax. Unless some unusual embedding material has been used, this should be achieved easily by using the procedure described above, but consider altering these conditions if consistently poor results are obtained despite adequate fixation and enzyme digestion. The rehydration procedure was originally taken verbatim from a local immunohistochemistry lab, and although it works there is no reason to suppose that it needs detailed adherence.

V. Alternative Methods for Sample Preparation

A. Selecting Areas of Interest Within a Block

Microscopic examination of a parallel thin section frequently shows that only a small area contains tumor cells, and in these cases detection of aneuploid peaks or estimates of %S phase will be compromised by the large admixture of stromal elements. The sample processed for flow cytometry can be enriched for tumor cells using the thin section to map out the required region of the block. Spread the thick sections flat, and trim away unwanted areas with a scalpel. Alternatively, use a scalpel blade to score around the area of interest on the surface of the block, and this will detach when the thick section is cut.

B. Dewaxing and Rehydration

Although these are readily done on a small scale using a row of test tubes, it is possible to automate the procedure using a tissue processor (Babiak and Poppema, 1991), such as that produced by Shandon-Elliot and used in routine histology laboratories. Prepare bags of 90-μm mesh nylon gauze, about 1×1 cm, and place thick sections there (Heiden *et al.*, 1991). Bags may be weighted with 1-mm diameter glass balls to make them sink. These bags are then placed in biopsy cassettes, which are loaded into a tissue processor programmed to run through a sequence of dewaxing followed by rehydration through graded ethanol to distilled water. The bags can then be placed in enzyme solution. It has been shown that centrifugation is a major cause of cell aggregation in this procedure and that this can be avoided by shaking the bags at the end of the digestion phase and adding

DNA stain, buffered to neutralize the pepsin solution, directly to the tube containing the bag (Heiden *et al.*, 1991). The supernatant is then run on the flow cytometer.

C. Alternative Enzymatic Procedures

Enzyme digestion is required to release nuclei from tissue. In addition, experiments using cells in suspension show that formalin fixation decreases stainability with DNA dyes, which is restored with pepsin treatment, usually with an improvement in the CV. The relative effects of formalin and pepsin on DNA staining vary between cell types and probably reflect differences in the formation of cross-links involving DNA and their dissolution by the enzyme and/or acidic conditions. Although the original procedure has proved to be generally applicable, there is no reason to suppose that it is the best possible method because there has been no definitive study of all the possible enzymes in all the permutations of concentration, time, tumor type, etc. A report (Tagawa *et al.*, 1993) suggests that a trypsin method originally described by Schutte *et al.* (1985) gives smaller CVs and less debris than pepsin, when used in association with propidium iodide staining, while pronase (subtilisin Carlsberg) has been reported to produce fewer aggregates (Heiden *et al.*, 1991). Methods for these enzymes, taken from the original publications, are as follows.

1. Trypsin

Dewax and rehydrate as for the pepsin method. Samples are incubated in 0.25% trypsin (Difco) in citrate buffer (3 mM trisodium citrate, 0.1% (v/v) Nonidet-P40, 1.5 mM spermine tetrachloride, 0.5 mM Tris, pH 7.6) overnight at 37 °C. Vortex, filter, and stain.

2. Pronase

Prepare a 0.1% solution of pronase (subtilisin Carlsberg, Sigma protease XXIV) in 0.01 M Tris, 0.07 M NaCl, pH 7.2. Add 1 ml to rehydrated tissue and incubate at 37 °C for 30 min to 2 h, depending on the type of tissue. At the end of the digestion the sample should be shaken, and the stain is added directly to the enzyme mix without a centrifugation step. Run on flow cytometer.

D. Multiparametric Analysis

Despite the harsh conditions used in this method, several nuclear antigens are sufficiently preserved for dual-parameter analysis of DNA content versus fluorescent antibody binding. Examples include the p105 proliferation antigen, originally described by Clevenger *et al.* (1985), and several oncogene products, including c-*myc* (Watson *et al.*, 1985). More recently, the proliferation-dependent Ki-S1 antibody has also been shown by Camplejohn *et al.* (1993) to give good staining in paraffin-embedded samples of breast cancer. Laboratories intending to develop these techniques should be prepared to vary the enzyme digestion procedure and to accept compromises between preserving sufficient nuclear protein for detection while obtaining an adequate DNA histogram.

Acknowledgment

I thank Marijka Koekebakker for her critical review of this chapter and helpful suggestions.

References

Babiak, J., and Poppema, S. (1991). *Am. J. Clin. Pathol.* **96**, 64–69.

Camplejohn, R. S., Brock, A., Barnes, D. M., Gillett, C., Raikundalia, B., Kreipe, H., and Parwaresch, M. R. (1993). *Br. J. Cancer* **67**, 657–662.

Clevenger, C. V., Bauer, K. D., and Epstein, A. L. (1985). *Cytometry* **6**, 208–214.

Hedley, D. W. (1989). *Cytometry* **10**, 229–241.

Hedley, D. W., Friedlander, M. L., Taylor, I. W., Rugg, C. A., and Musgrove, E. A. (1983). *J. Histochem. Cytochem.* **31**, 1333–1335.

Heiden, T., Wang, N., and Tribukait, B. (1991). *Cytometry* **12**, 614–621.

Hitchcock, C. L., and Ensley, J. F. (1993). *In* "Clinical Flow Cytometry: Principles and Application" (K. D. Bauer, R. E. Duque, and T. V. Shankey, eds.), pp. 93–110. Williams & Wilkins, Baltimore, MD.

Kallioniemi, O. P., Visakorpi, T., Holli, K., Heikkinen, A., Isola, J., and Koivula, T. (1991). *Cytometry* **12**, 413–421.

Schutte, B., Reynders, M. M., Bosman, F. T., and Blijham, G. H. (1985). *Cytometry* **6**, 26–30.

Tagawa, Y., Nakazaki, T., Yasutake, T., Matsuo, S., and Tomita, M. (1993). *Cytometry* **14**.

Toikkanen, S., Joensuu, H., and Klemi, P. (1989). *Br. J. Cancer* **60**, 693–700.

Watson, J. V., Sikora, K., and Evan, G. I. (1985). *J. Immunol. Methods* **83**, 179–192.

CHAPTER 13

Flow Cytometry and Sorting of Plant Protoplasts and Cells

David W. Galbraith

Department of Plant Sciences
University of Arizona
Tucson, Arizona 85721

I. Update

This chapter was published in 1994. In the 15 years since its appearance, use of flow cytometry and sorting for analysis of cells and protoplasts has found increasing applications in the basic and applied biology of plants (for recent reviews, see Galbraith, 2007a,b). Progress has occurred in a number of areas, the most dramatic and important advances and associated changes relating to (i) improvement in instrumentation, (ii) the emergence of the green fluorescent protein (GFP) and, more generally, the fluorescent proteins (FPs) as markers for flow analysis and sorting, (iii) integration of flow sorting with genomics methods, particularly for characterization of cell-specific gene expression (Birnbaum *et al.*, 2003, 2005), (iv) development of additional fluorescent markers and reporters for protoplast characterization, and (iv) applications in sorting subcellular organelles such as nuclei (Zhang *et al.*, 2008).

Improvements in flow cytometric and sorting instrumentation primarily relate to the availability of solid-state lasers, producing cheaper and more reliable illumination at an increasing number of wavelengths. This advance facilitates the use of combinations of different fluorochromes, and will be particularly important for the FPs. Secondarily, increasing use is being made of engineering data acquisition in the digital rather than analog domain, which provides increases in system accuracy, and, most recently, in dynamic range. The emergence of GFP as a versatile fluorescent marker for living cells has provided the platform for advances throughout biology, as well as three Nobel prizes! As reviewed in another chapter in this book, GFP represents but the founder member of the general class of FPs, whose membership is regularly expanding, and whose members provide a remarkable range of fluorescent properties. FP expression, in general, does not appear to perturb living organisms, either measured through visible phenotypes, or using molecular signatures determined through global gene expression analysis. More surprising, perhaps, is the limited short-term effect of protoplasting on global transcript profiles, but this means that transgenic FP expression can be used to highlight individual cells within complex tissues and organs, using cell type-specific promoters, and protoplasts prepared from these transgenic plants can then be flow sorted and their transcripts used for characterization of global gene expression within these cells (Birnbaum *et al.*, 2003, 2005). Much interest has been engendered by this approach, and a number of groups are now using these methods routinely for the analysis of specific cell types. Beyond use of the FPs, progress in plant protoplast analysis has also been associated with the use of new fluorochromes and methods developed for flow analysis of mammalian cells, one interesting example being those reagents that chart apoptosis (Galbraith, 2007a).

What has remained the same in this area relates to (i) general methods for protoplast production and purification, (ii) procedures and principles for operating flow sorters using large flow cell tips, (iii) methods for quantification of protoplast and cell size and chlorophyll contents, and (iv) methods for flow sorting, recovery, and growth of protoplasts. The caveats concerning protoplast production have also not changed; different tissues and organs, and different species require methods optimized for each situation (see, e.g., Birnbaum *et al.*, 2005; Zhang *et al.*, 2008 for descriptions of protoplast preparation from *Arabidopsis thaliana* roots and leaves). There are many examples of starting materials that are entirely recalcitrant to protoplast production. For this reason, the ability to produce protoplasts must not be considered to be "a given." The best methods for protoplast purification prior to flow analysis and sorting still involve conventional isopycnic gradient flotation using low-speed centrifugation. Since, in comparison to most mammalian cells, protoplasts are larger in size, and sometimes much larger, operation of the flow sorter requires use of large flow cell tips (diameters 100–200 μm). The general methods outlined in the chapter for large particle sorting remain the same, since the underlying physical principles are unchanged, although implementation of these methods on the different commercial flow sorters may have idiosyncrasies.

How to avoid pitfalls. Operating flow sorters with large particles is routine but not necessarily widespread. Convenient indestructible particles exist (pollen) that provide standards for setting up optimal sort settings on the instruments; it should be reiterated that this set-up should be done using particles similar in size to the fragile protoplasts that are to be sorted, and that smaller particles (such as fluorescent microspheres) are unsuitable. Protoplasts are not only large in comparison to most mammalian cells, but are also unusually fragile. This means that "crude" (i.e., unpurified) protoplast preparations comprise large amounts of subcellular debris and organelles that are light-scattering and can be fluorescent. This can lead to confusion on data acquisition in flow, since the desired protoplasts may be greatly outnumbered by the undesired debris; for this reason, protoplast purification is recommended. Protoplasts also require culture media that differ considerably in composition from the saline-based media employed for animal cells. Plants as organisms, in fact, contain very little sodium, and mechanisms that actively exclude salt from the cytoplasm frequently confer salt-tolerance in the field. Replacing the sheath fluids with protoplast compatible media evidently represents the best strategy to preserve protoplast viability, if that is an important aspect of the experimental design. It should be noted that most protoplast media contain millimolar levels of Ca^{2+} which, being incompatible with phosphate buffers, and should not be mixed. In my experience, protoplasts frequently can survive short-term exposure to saline-based media, so replacing PBS-based sheath fluids with saline represents a convenient work-around, again depending on the purpose of the experiments.

What methods are trusted "winners"? All of the methods described in the chapter are trusted and reproducible. Of methods emerging since the chapter was written, those involving FP expression are particularly robust, and well-established, and have been reproduced across many laboratories.

II. Introduction

Flow cytometry is a technique that involves measurements on populations of single cells. With the exception of special cases such as pollen, microspores, and sperm cells, higher plants are not found in the form of single cells. Instead, they are formed of complex three-dimensional tissues, comprising individual cells interconnected by means of their walls. This is a result of the way in which plant cells divide. Cytokinesis is accompanied by the production of a cell plate, which progressively extends from the mother cell wall. After cell division is complete, a continuous wall structure links the daughter cells, and the individual cells do not separate. In order to convert plant tissues into cell suspensions, so that they can be used in flow cytometry, the cell wall must be removed. Wall-less plant cells are termed *protoplasts.*

Protoplasts are prepared through the use of commercially available hydrolases (cellulases, hemicellulases, and pectinases). These enzymes solubilize the cell wall polymers, and this leads to the release of intact protoplasts as a single-cell suspension. After the removal of the cell wall, the protoplasts are usually perfectly spherical, even if derived from nonspherical cells. This makes them well suited for flow cytometric analysis, from an optical point of view. However, protoplasts (and other single plant cells such as pollen) are almost always larger than the animal cells commonly analyzed using flow cytometry; occasionally their sizes exceed the diameter of the flow cell tips normally employed in flow cytometry. For successful flow analysis of plant protoplasts and cells, flow tips larger in diameter (100–200 μm) than those employed for animal cells (60–80 μm) are therefore required.

Flow analysis has been employed for the examination of light scatter and fluorescence signals derived from different types of protoplasts, and these have provided information concerning a variety of different parameters, including cell viability, size, and chlorophyll content; protoplast cell-surface architecture; and interactions between protoplasts and microorganisms (reviewed in Galbraith, 1989, 1990). More recently, flow cytometry has been employed for analysis of protoplast membrane fluidity (Gantet *et al.*, 1990) and for the characterization of microprotoplasts (Verhoeven and Sree Ramulu, 1991) and of maize sperm cells (Zhang *et al.*, 1992).

In terms of sorting, it is clear both from theoretical considerations and empirically that the principles governing droplet formation in flow cytometry can be scaled up to accommodate large particles such as protoplasts and pollen (Harkins and Galbraith, 1987; Kachel *et al.*, 1990). This involves several elements, governed for jet-in-air sorters by equations describing the amplification due to surface tension of microscopic undulations in the diameter of the fluid jet induced by the action of a mechanoelectric transducer (Donnelly and Glaberson, 1966). Specifically, production of droplets requires that the wavelength of the imposed undulation be longer than π times the diameter of the fluid jet. This wavelength is a function of both the transducer drive frequency and the fluid velocity, which itself is a function of the system pressure and the jet diameter. Empirical observations parallel theoretical treatment and can be used to produce tables listing those drive frequencies and system pressures that allow droplet production for flow tips of different diameters (Harkins and Galbraith, 1987). In work of this type, the availability of indestructible particles of defined diameters covering the size ranges of interest is important. Conditions optimal for sorting small standard microspheres are not necessarily optimal for larger particles, due to an increased degree of interference between the particle and the process of droplet formation as the particle diameter approaches that of the fluid stream (Harkins and Galbraith, 1987). Commercial microspheres larger than about 20 μm are not readily available. However, pollen from a variety of different plant species provides standard particles spanning diameters from 20 to 95 μm that are convenient, cheap, and intrinsically autofluorescent. Pollen precursors such as microspores can also be flow analyzed and sorted based on differences in light scatter and autofluorescence (Pechan et al., 1988), and this led to populations enriched according to embryogenic potential.

Whereas microspores and pollen are mechanically extremely robust, protoplasts are very fragile and, therefore, require careful handling during flow sorting (Galbraith, 1989). Successful recovery of viable protoplasts after flow sorting was first reported using suspension culture cells and leaf mesophyll as source tissues (Galbraith et al., 1984; Harkins and Galbraith, 1984). Following development of methods for fluorescent labeling of different populations of parental protoplasts (Galbraith and Galbraith, 1979; Galbraith and Mauch, 1980), similar conditions have subsequently been used by a number of different groups for the sorting of heterokaryons formed by protoplast fusion followed by regeneration of somatic hybrid and cybrid plants (Afonso et al., 1985; Glimelius et al., 1986; Hammatt et al., 1990; Pauls and Chuong, 1987; Sjödin and Glimelius, 1989; Sundberg and Glimelius, 1991; Walters et al., 1992). A different type of electromechanical cell sorter has also been successfully employed for heterokaryon sorting (Bromova and Knopf, 1991, and references therein). Other applications of flow sorting to plant protoplasts include the purification of protoplasts representative of different leaf tissue types for analysis at the molecular level (Harkins et al., 1990).

As previously indicated, there are no theoretical limitations either on the sizes of the cells that can be analyzed and sorted using flow cytometry or on the sizes of the flow cells that can be employed. However, traditional procedures of flow cytometry and cell sorting and the designs of the instrumentation have been optimized using small biological cells and artificial microspheres (i.e., those falling in the range of 3–20 μm in diameter). When dealing with the flow analysis and sorting of larger particles, a variety of practical considerations enter the picture. This chapter, which updates methods previously described in this series (Galbraith, 1990), outlines some of the problems and their resolution in the sorting of biological cells as large as 100 μm using commercially available flow cytometric instrumentation. In principle, the procedures are applicable to all biological cells and cell aggregates within this size range, although they are illustrated for plant protoplasts, pollen, and spores of the tree-club moss Lycopodium.

III. Application

Protoplasts can be prepared from a wide variety of different plant species. The experimental conditions optimal for release of viable protoplasts vary according to the particular plant tissue under study. Details concerning protoplast preparation are beyond the scope of this chapter, and the reader is referred elsewhere (see, e.g., Lindsey, 1991). This chapter outlines the methods for the analysis and sorting of a variety of different protoplast and cell types. In principle, since all protoplasts comprise spherical structures, the methods described should be universally applicable, although some possible limitations are discussed in Section V.

IV. Materials

A. Plant Materials

Tobacco (*Nicotiana tabacum* cv. Xanthi) seed was obtained from the USDA Tobacco Research Laboratory (Oxford, NC). Seeds were germinated under sterile conditions, and the resultant plantlets maintained as vegetative plantlets in Magenta boxes on basal MS medium containing 3% (w/v) sucrose, solidified with 0.8% agar. Plants were subcultured every 14–21 days by excision of the portion of the plant containing the apical meristem and one to two expanded leaves. This process was continued for not more than three cycles before the plantlets were discarded. Pollen (*Zea mays*, *Carya illinoiensis*, paper mulberry, and ragweed) and *Lycopodium* spores were obtained from Polysciences (Warrington, PA). A cell suspension culture of *Nicotiana sylvestris*, kindly provided by Dr. Roy Jensen, was maintained in darkness under sterile conditions as 100-ml aliquots in 500-ml Erlenmeyer flasks at 25 °C, with constant orbital agitation (100 rpm). The growth medium comprised basal MS medium at pH 5.7 supplemented with 3% (w/v) sucrose and 1 mg/l 2,4-dichlorophenoxyacetic acid. Subculture was done at 5-day intervals by transfer of 50 ml of cells into 100 ml fresh medium.

B. Chemicals

Macerase and cellulysin were obtained from Calbiochem, Inc. (La Jolla, CA), cellulase from Worthington (Freehold, NJ), aniline blue from the Fisher Scientific Co. (Pittsburgh, PA), and fluorescent microspheres from Polysciences (Warrington, PA). MS (Murashige and Skoog) medium was obtained from Gibco (Grand Island, NY). All remaining chemicals were from the Sigma Chemical Co. (St. Louis, MO).

V. Procedures

A. Protoplast Preparation

1. Tobacco Leaf Tissues

Protoplasts are prepared from leaves selected from vegetative plants growing vigorously in Magenta boxes in basal MS medium containing 3% (w/v) sucrose, solidified with 0.8% agar. Leaf tissue (0.5 g) is excised under sterile conditions, sliced into 1×10-mm segments, and incubated within an 85-mm-diameter sterile plastic petridish in 10 ml of a filter-sterilized osmoticum containing 10 mM $CaCl_2$, buffered with 3 mM 2-[N-morpholino]ethane sulfonic acid (Mes), pH 5.7; the osmotic species depends on the particular application. For isolation of mesophyll protoplasts either 0.7 M mannitol or 0.35 M KCl can be employed; for applications involving epidermal protoplasts, an ionic osmoticum is essential since these protoplasts have a lower buoyant density than isotonic mannitol. The cell walls are removed by including in the osmotica 0.1% (w/v) driselase, 0.1% (w/v) macerase, and 0.1% (w/v) cellulysin. Incubation is continued for 12–15 h at room temperature. The protoplast digest is filtered through sterile cheesecloth into plastic conical centrifuge tubes and is centrifuged at $50 \times g$ for 10 min. The protoplast pellet is gently resuspended in 5 ml of a solution containing 25% (w/w) sucrose dissolved in 3 mM Mes and 10 mM $CaCl_2$, pH 5.7, and is overlaid with 5 ml of osmoticum. The protoplasts are centrifuged at $50 \times g$ for 2 min. The interface, which contains the viable protoplasts, is collected and diluted with 10 ml KCl-osmoticum, and the protoplasts are pelleted by centrifugation at $50 \times g$ for 5 min. The protoplasts are resuspended in osmoticum to a final concentration of about 10^5 ml^{-1}. The total protoplast yield is determined through hemocytometry. The proportion of viable protoplasts is found by resuspension of the protoplasts in osmoticum containing 0.1% (v/v) of a freshly prepared solution of FDA (1 mg/ml in acetone). Viable protoplasts accumulate fluorescein and appear bright green when illuminated with blue light under the fluorescence microscope (using the standard FITC filter set and mercury arc lamp illumination).

2. *N. sylvestris* Cell Suspension Cultures

Protoplasts are prepared from actively growing cell cultures 3 days after subculture. The cells contained within 150 ml of the cultures are allowed to settle under gravity for about 15 min. The supernatant is removed and the cells (about 14 ml) are resuspended in 5 volumes of an osmoticum (NTTO; Galbraith and Mauch, 1980) containing 0.5% (w/v) cellulase, 0.1% (w/v) cellulysin, 0.05% (w/v) driselase, and 0.02% (w/v) macerase. After incubation with gentle orbital shaking at 25 °C for 20 h, the protoplasts are filtered through five layers of cheesecloth into 50-ml conical tubes. The protoplasts are sedimented by centrifugation at $50 \times g$ for 5 min, resuspended in 5 ml 25% (w/v) sucrose, and are overlaid with 5 ml of osmoticum. After centrifugation at $50 \times g$, the intact protoplasts are collected from the interface (approximately 2 ml), diluted with 5 volumes of osmoticum, and collected by centrifugation at $50 \times g$ for 5 min.

B. Fluorescent Standards

Protoplasts are fixed by incubation for 60 min at 20 °C in 2% (w/v) paraformaldehyde in osmoticum. The paraformaldehyde is dissolved in 4% (w/v) by heating in water with gradual dropwise addition of 0.1 M NaOH until the solution clarifies, prior to being chilled and mixed with an equal volume of 2× concentration osmoticum. Pollen from the listed species (Table I) is autofluorescent, but fluorescence can be enhanced by staining with aniline blue. Pollen (10 mg) is stained by resuspension in 4 ml of 0.1% (w/v) aniline blue dissolved in phosphate-buffered saline, pH 9. The mixing of pollen and *Lycopodium* spores with water is facilitated by addition of 0.5% (v/v) Triton X-100.

C. Flow Cytometry

1. EPICS Series Instruments

The EPICS series flow sorters (Coulter Electronics, Hialeah, FL) are operated as detailed for the various applications.

a. Analysis of Intact Mesophyll Protoplasts

The laser is tuned to 457 nm with a power output of 100 mW. We employ two barrier filters to exclude scattered light; these have half-maximal transmittance at 510 and 515 nm (termed LP510 and LP515). A further barrier filter (LP610) screens the red channel photomultiplier. Most chlorophyll autofluorescence emission occurs above 620 nm; the first two filters are routinely present in the light path but may not be needed for this application. Integral or log integral red fluorescence, one-parameter frequency distributions are typically accumulated to a total count of 20,000.

b. Analysis of Viable Mesophyll Protoplasts

Protoplasts are stained with FDA. The laser is tuned to 457 nm with a power output of 100 mW. Fluorescence emission is detected using the green channel photomultiplier. Barrier filters LP510 and LP515 are used to eliminate scattered light. Light is reflected into the photomultiplier using a dichroic mirror (DC590)

Table I
Biological Particle Size Ranges

Particle type	Diameter (μm)
Broussonetia papyrifera pollen	13.8 ± 1.3; $N = 20$
Ambrosia elateior pollen	20.9 ± 1.2; $N = 22$
Lycopodium spores	28.6 ± 2.5; $N = 12$
Nicotiana tabacum leaf protoplasts (fixed)	33.7 ± 6.2; $N = 25$
N. tabacum leaf protoplasts	42.3 ± 6.9; $N = 20$
N. tabacum suspension culture protoplasts	41.2 ± 10.4; $N = 21$
Carya illinoensis pollen	51.1 ± 3.3; $N = 11$
Zea mays pollen	95.3 ± 4.1; $N = 11$

Note: Diameters are expressed as mean ± standards deviation (adapted from Harkins andGalbraith, 1987) that splits at a wavelength of 590 nm (for this application, a fully silvered mirror would also be appropriate; the dichroic is used for convenience in other applications). Two blue glass BG38 filters (Optical Instrument Laboratory, Houston, TX) are used to screen the photomultiplier. Integral or log integral green fluorescence, one-parameter frequency distributions are accumulated to a total count of 20,000.

c. Analysis of Viable Protoplasts from Suspension Cultures

Protoplasts are stained with FDA. The flow cytometer is operated as described in Section V.C.2 with the exception that the BG38 barrier filters can be omitted.

d. Sorting Pollen

Pollen is stained with aniline blue. The laser is tuned to 514 nm with an output of 200 mW. Barrier filters LP530 and LP540 are used to screen the green channel photomultiplier. Pulse width time-of-flight (PW-TOF) analysis is performed based on peak green fluorescence. The TOF module is set to analyze the time that the pulses of fluorescence remain above thresholds set to 50% of the peak value, although this threshold can be lowered if necessary (Galbraith et al., 1988). One-parameter frequency distributions of protoplast size result from this analysis; further one-parameter frequency distributions can be accumulated based on peak or integral fluorescence and on forward angle light scatter.

e. Sorting Fixed Mesophyll Protoplasts

This signal derives mostly from chlorophyll autofluorescence although there is some contribution to autofluorescence from paraformaldehyde fixation. The laser is tuned to 457 nm with an output of 100 mW. Barrier filters LP510 and LP515 and 590 are used to screen the red channel photomultiplier. Log integral red fluorescence signals are accumulated to give one-parameter frequency distributions. Sort windows are positioned according to the location of the peak.

f. Separation of Mesophyll and Epidermal Protoplasts

The protoplasts are prepared in ionic osmoticum. The purified protoplasts are stained with FDA, as described previously. The laser is tuned to 457 nm with a power output of 100 mW. Fluorescence emission is detected using both the red and green channel photomultipliers. Barrier filters LP510 and LP515 are used to eliminate scattered light. Light is reflected into the green photomultiplier using a dichroic mirror (DC590), with two BG38 barrier filters to eliminate entry of chlorophyll autofluorescence. Light transmitted through the dichroic mirror enters the red photomultiplier screened by barrier filter LP610. Some signal subtraction may be necessary to eliminate spectral overlap. Cell size is determined through PW-TOF based either on peak green (FDA) fluorescence (for epidermal protoplasts) or peak red fluorescence (for mesophyll protoplasts). Two-dimensional PW-TOF (FDA) versus integral red fluorescence allows the generation of contour plots permitting the sorting of epidermal from mesophyll protoplasts (Galbraith et al., 1988). Frequency distributions are typically accumulated to a total count of 20,000.

2. Coulter Elite

The Elite flow cytometer/cell sorter (Coulter Electronics, Hialeah, FL) is operated using a flow-in-quartz cell having a 100-μm-diameter orifice and a fixed sample injection assembly (Coulter part 6856762). For the described applications, the primary argon laser provides illumination at 488 nm, and the light paths leading to the four PMTs are as follows: light scattered orthogonally to the flow stream is detected by PMT1 after being reflected by a 488-nm dichroic mirror and passing through a 488-nm band-pass filter. Fluorescence emission is split according to wavelength to access the three remaining PMTs. Fluorescence transmitted by the 488-nm dichroic encounters a 488-nm laser blocking filter and then two further dichroic mirrors splitting at 550 and 625 nm. Light reflected by the 550-nm dichroic enters PMT2 after passage through a 525-nm band-pass filter. Light transmitted by the 550-nm dichroic, but reflected by the 625-nm dichroic, enters PMT3 after passage through a 575-nm band-pass filter. Light transmitted by the 625-nm dichroic enters PMT4 after passing through a 675-nm band-pass filter.

a. Analysis of Mesophyll Protoplasts Based on Chlorophyll Autofluorescence

The laser is tuned to a power output of 15 mW. The forward scatter detector is operated at 370 V and a gain of 3. PMT1 (90° light scatter) is operated at 350 V and a gain of 2. PMT4 is operated at 620 V with a gain of 3. The forward scatter discriminator is set at 100; all other discriminators are off.

b. Analysis of Lycopodium Spores

Laser power output is set to 15 mW. For measurement of forward and 90° angle light scatter (linear or log signals), the forward scatter detector is operated at 100 V with an amplification of 7.5, and PMT1 at 140 V with an amplification of 7.5. Autofluorescence is detected using PMT2 at 650 V with an amplification of 7.5. Uniparametric histograms (linear or log integral signals) are accumulated. Multiparametric histograms based on forward and 90° angle light scatter (linear or log signals) can also be accumulated. For all applications, the forward scatter discriminators are set at 100, and the remaining discriminators are turned off.

D. Cell Sorting

1. The Coulter EPICS Series Instruments

The EPICS series flow sorters can be employed for the successful sorting of protoplasts and pollen using the 100-, 155-, or 200-μm flow tips. The system pressures and transducer drive frequencies suitable for the various flow tips are listed in Harkins and Galbraith (1987). For the 200-μm flow tip, we routinely use a system pressure of 6 psi and a drive frequency of 8 kHz. The sheath fluid comprises mannitol (50.5 g/l) and glucose (68.4 g/l) buffered with 3 mM Mes, pH 5.7. We employ two 2.5-l sheath tanks connected in parallel; this provides approximately 2–3 h of sorting and analysis between refills.

Sort alignment and optimization is achieved as follows:

1. Using the sort test mode, the transducer drive amplitude is adjusted to provide a uniform and stable sorted stream.

2. The number of undulations is counted to provide an estimate of the position of droplet break off and hence the charge delay setting.

3. Analysis is initiated using standard particles that approximate the size of the protoplasts that are to be sorted (either paraformaldehyde-fixed protoplasts or aniline blue-stained pollen). Sort windows are set on the appropriate frequency distributions.

4. A sort matrix analysis is performed, to precisely define the charge delay setting that yields a sort efficiency of 100%. This can be conveniently measured using a single-cell deposition device (the Coulter Autoclone) to sort 10 particles/well in a 96-well culture plate; the actual numbers of particles recovered are then determined using an inverted light microscope.

5. Analysis and sorting of the protoplasts are initiated, using the analysis parameters described above and the sort conditions defined by use of the standard particles.

For sterile sorting, we utilize a 0.2-μm in-line filter from Pall (Ultipor Type DFA4001ARP) that can be autoclaved. The sample tube, sample pickup, and sample introduction line are autoclaved. The sheath tanks and sheath lines are thoroughly cleaned with dilute bleach and are rinsed with 70% ethanol. The sheath fluid is sterilized either by autoclaving or by passage through millipore 0.22-μm filters. Prior to sorting, the sample lines are cleared of residual ethanol by passage of sterile sheath fluid. Sorting is performed in 96-well plates prefilled with 50–100 μl of sterile growth medium. Protoplast density affects further development in culture. Either sufficient protoplasts must be sorted (about 1000 per well) or feeder cells (nonmorphogenic *N. tabacum* cell suspensions) must be included in the wells (Galbraith, 1989).

2. The Coulter Elite

Protoplasts and pollen can be sorted using the Elite with the 100-μm quartz flow tip. The system is operated at a pressure of 8 psi and a transducer drive frequency of 11.5 kHz (for pollen) and 8.9–14 kHz (for protoplasts), using the same sheath fluid as described for the EPICS series instruments.

Sort alignment and optimization are as follows:

1. Prior to instrument optimizing for sorting, the transducer is allowed to warm up completely. For the piezoelectric transducer, this involves running the drive at an amplitude of about 70% for at least 1 h. The deflection plate assembly is then moved as close as possible to the flow cell tip without blocking the laser beam (about 5 mm). The video camera is adjusted to observe the flow stream such that

one can see both the ground plate on the left edge of the screen and the laser beam intercept on the right while using only the "pan" function of the camera adjustment. The transducer drive frequency is adjusted at constant amplitude (typically about 20%) to provide as short a droplet break off point as possible. The machine is then switched to sort test mode, and the transducer amplitude is adjusted in order to give a stable sorted stream.

2. The deflection plate assembly is lowered to allow observation of two to three free droplets above the ground plane. The undulations in the fluid stream are examined using the video camera. It is important that the last attached droplet be well rounded on the left-hand side and clearly connected by a ligament to the flow stream on the right-hand side. Satellite droplets are usually seen. These can be either "fast" or "slow," depending on whether they merge with the major droplet ahead of or behind the satellite. The shape of the last attached droplet and the behavior of satellites are dependent on the amplitudes and frequencies of transducer activation. Conditions should be established that produce fast satellites, since these carry charge of the same sign as the droplets with which they shall merge and so will not affect their electrostatic deflection.

3. The cursor is moved to mark the second well-defined undulation to the right of the last attached droplet. A second point is marked between the first two free drops above the ground plate. The number of droplets between the two cursors is entered into the delay calculation program. This calculated delay setting should be very close to the optimal delay setting (as determined through sorting of particles, described below).

4. Frequency distributions are acquired, using standard particles approximating the size of the cells within the samples of interest. Sort windows are set based on these distributions.

5. A sort matrix analysis is performed, in order to empirically define the delay setting which yields a sort efficiency as close as possible to 100%. This is done by programming the sorting of batches of 25 particles onto a standard 3×1 in. glass microscope slide and counting the actual numbers recovered using microscopy. For each batch of 25, the sort delay setting is adjusted by 1-step increments to span a range of ± 5 around the calculated delay setting.

6. The phase settings and deflection plate assembly high-voltage amplitude are adjusted to obtain side streams with minimal fanning.

7. Histograms are accumulated for the cells of interest, the appropriate sort windows are defined, and the cells are sorted.

For sterile sorting, the sheath and rinse tanks are cleaned using dilute detergent and are filled with 70% ethanol. The instrument is placed in run mode for 15 min without a sample tube at the sample station; this operates the sample station on backflush. The machine is then switched to shutdown mode several times. The run/shutdown mode sequence is repeated twice, after the sheath tank is refilled first with sterile (autoclaved) water and then with sterile sheath fluid. Finally, the desired cells are sorted into a sterile collection tube.

E. Transformations of Data

1. Protoplast Size Measurement

Previous work has shown that the TOF parameter can be used for the accurate measurement of protoplast size, assuming appropriate calibration (Galbraith *et al.*, 1988). For this calibration, we require particles that approximate the size ranges over which the TOF measurements are to be made; these particles must be fluorescent and must not deform or degrade during passage through the flow cytometer. In general, artificial fluorescent microspheres with diameters similar to those of protoplasts are not readily available. We, therefore, have employed natural microspheres (pollen) stained with aniline blue for calibrating the TOF parameter (Galbraith *et al.*, 1988). Since pollen is essentially indestructible, it can also be employed for the purpose of optimizing the sort process (see Section VI). The linearity of the relationship between the TOF parameter and actual protoplast diameter can be determined directly through sorting of the protoplasts using a series of nonoverlapping sort windows spaced across the TOF frequency distribution, followed by measurement of protoplast diameters from light micrographs (Galbraith *et al.*, 1988).

2. Protoplast Chlorophyll Measurement

Correlation between the emission of red autofluorescence and protoplast chlorophyll content can be achieved in a manner analogous to that described for TOF/size measurements. Thus, different, nonoverlapping sort windows are placed on the one-dimensional frequency distributions of red autofluorescence emission and defined numbers of protoplasts are sorted. Cellular chlorophyll amounts can then be quantitated through spectrofluorometric analysis (Galbraith *et al.*, 1988).

VI. Critical Aspects of the Procedures

A. Sample Preparation

The first critical aspect relates to the biological materials under study. Familiarity with procedures of preparation and culture of protoplasts is essential for their successful use in flow analysis and sorting. Protoplasts are perhaps the most fragile type of eukaryotic cell yet subjected to flow techniques. Correspondingly, the researcher must learn to recognize features of protoplasts that indicate their viability before attempting work in flow. For this, a good light microscope equipped with differential interference contrast and epifluorescence capabilities is essential. Optical features of viable protoplasts include a turgid cytoplasm, a perfectly spherical shape, and the observation of fluorochromasia following FDA staining and of cytoplasmic streaming. Further, the organelles should be evenly dispersed through the cytoplasm and, for protoplasts prepared from cell suspension cultures, the nucleus should be well defined, centrally located, and attached to the peripheral cytoplasm by well-defined cytoplasmic strands.

In manipulation of protoplasts, care should be taken to avoid mechanical damage. Resuspension steps using Pasteur pipettes should involve minimal force. Gross changes in the osmotic potential of the various solutions encountered by the protoplasts should also be minimized, noting that different procedures of protoplast preparation often call for different levels of osmotica. One of the most critical aspects of protoplast preparation concerns the physiological status of the donor plant tissues. Variation in light intensity, day length, plant stage, and leaf number can affect protoplast yield and viability. Plantlets grown as described in sterile culture *in vitro* lack an extensive cuticle, and this appears to facilitate enzymatic degradation of the cell walls. There are reports that suggest the act of excision of leaves from these plants itself adversely affects the ability to produce protoplasts from further leaves excised at later times (Farmer and Ryan, 1992; Walker-Simmons *et al.*, 1984). Consequently, we routinely discard the plantlets after three leaf harvests and typically do not subculture them more than twice prior to starting over with new seed-derived plantlets. Finally, it is a truism to state that no one set of experimental conditions is applicable to all types of protoplasts. Differences in the composition of the various media are encountered in reports concerning the successful culture of different protoplast types. A sensible approach to flow analysis and sorting of different protoplast types is to employ optimal culture media as the sheath fluid.

B. Identification of the Population of Interest

The second critical aspect of the procedures relates to the apparently simple task of defining the objects of interest within the populations that will be subjected to flow analysis and sorting. For pollen, this is not a problem, since they usually comprise the majority of the particles within the suspension. However, protoplasts contain a very large number of internal structures and organelles which when released from broken cells can provide light scatter and autofluorescence signals during flow analysis. Use of techniques for purification of the protoplasts is recommended, such as sucrose gradient centrifugation. Protoplast purification may not always be sufficient to eliminate this problem. For example, protoplasts from mature tobacco leaves contain about 70 chloroplasts (Harkins and Galbraith, 1984). Therefore, a population of 90% intact protoplasts (which would be considered excellent according to most tissue culture standards) would produce flow cytometric frequency distributions comprising 89% free chloroplasts and 11% intact protoplasts. For most applications, therefore, the desired objects (intact protoplasts) comprise a significant minority of the particles that are being analyzed. Triggering on fluorescence rather than light scatter can be employed to eliminate the contribution of nonfluorescent, light-scattering particles. Acquisition of frequency distributions on logarithmic scales can be helpful, since it tends to highlight the presence of discrete populations. Linear frequency distributions can then be accumulated by gating on the log distributions. Finally, use of the fluorescence microscope is strongly recommended in order to determine whether appropriate staining of the protoplasts has been achieved.

C. Selection of Flow Tips

As previously noted, plant protoplasts are generally much larger than animal cells, and the sizes of those commonly employed in flow cytometric applications (30–60 μm) approach the diameters of the standard flow tips (60–75 μm). Evidently, it is important to employ flow tips larger than the cells to be analyzed, although use of flow tips much larger in diameter could lead to errors associated with the hydrodynamic process of centering the particles in the fluid stream. Larger flow tips produce larger rates of volume accumulation of sheath fluid, and this requires larger sheath tanks. The lower transducer drive actuation frequencies required by larger tips leads to the production of larger droplets, and the deflection of these droplets by the conventional defection assemblies and power supplies may not be adequate. This problem can be alleviated by increasing the length of the deflection plate assemblies or by increasing the high voltages applied to these plates.

D. Competing Pigments

A feature of higher plants is the presence of a variety of intracellular pigments, particularly those associated with photosynthesis. The presence of these pigments complicates flow analysis of protoplasts, depending on the degree of spectral overlap between these pigments and the specific fluorochromes to be used in the particular study. The researcher should be aware of these potential problems and be prepared to deal with them through appropriate selection of fluorochromes, fixation procedures, and excitation and emission wavelengths.

E. TOF Limitations

Theoretical considerations indicate that the relationship between protoplast diameter and the PW-TOF parameter will increasingly deviate from the linearity as the size of the protoplast approaches that of the beam. For the EPICS system using standard optics, deviation becomes noticeable at a protoplast diameter of about 15 μm (Galbraith *et al.*, 1988). Below this point, correct sizing can still be achieved, although it is necessary to use a series of appropriately sized standard particles in order to properly calibrate these measurements. We have not empirically established the upper limit to TOF size analysis which will be dictated by the maximal time domain that can be accommodated by the TOF circuitry. However, brief calculations suggest that this will not be a limitation: for the standard EPICS hardware, the upper limit of TOF analysis is 40 μs. For large particle sorting using the larger flow tips, the system pressure must be lowered in order to obtain a point of droplet break off above the position of the deflection assembly (Harkins and Galbraith, 1987). For the 200-μm flow tip, the highest fluid stream velocity is typically 10.8 m/s. This means that the effective upper limit for TOF measurements set by the electronics is around 400 μm, which is an order of magnitude larger than tobacco leaf protoplasts.

Biological factors might be expected to influence the linearity of the TOF analysis. For example, deformation of cellular shapes at the laser interrogation point due to hydrodynamic forces experienced within the flow cytometer might be expected to introduce errors in sizing; our data indicate that this is not the case for protoplasts (Galbraith *et al.*, 1988). On the other hand, the presence of chloroplasts within mesophyll protoplasts introduces errors in protoplast sizing when TOF measurements are performed based on FDA fluorescence, but not when they are based on chlorophyll autofluorescence (Galbraith *et al.*, 1988). This is probably due to quenching to FDA fluorescence by the highly absorbent pigments within the chloroplasts.

F. Protoplast Sorting

Critical to the successful sorting of protoplasts are the following elements: the generation of high-quality frequency distributions in which the location of the desired protoplast populations are well defined, the optimization of sort parameters in order to provide accurate and high-efficiency sorting of the desired protoplasts, the provision of sterile conditions, and the establishment of conditions for protoplast growth after sorting. Optimization of sort parameters is most easily achieved with large flow cell tips, coupled to the use of appropriate standard particles. These are detailed in further sections. Sterile conditions are achieved through the use of a combination of liquid sterilants and autoclaving. Use of a single-cell deposition device to sort protoplasts into 96-well plates also greatly facilitates maintenance of sterile conditions. Inclusion of certain antibiotics (e.g., penicillin derivatives at 50–500 mg/l) can be helpful; these are innocuous to protoplasts. In terms of optimization of culture conditions after sorting, maintenance of a minimal cell density is important. This can be achieved either by sorting large numbers of protoplasts (>1000) within the wells of the culture plates or through the use of conditioned media or feeder cells (Afonso *et al.,* 1985). Feeder cells can be provided as a liquid cell suspension or can be embedded in agarose, with or without a physical barrier. Finally, it should be noted that some types of tissue culture plates are toxic to protoplasts.

VII. Controls and Standards

We routinely employ pollen or *Lycopodium* spores for setting up and standardizing the sort process (Table I). It should be reemphasized that correct selection of sort parameters for fragile plant protoplasts requires the use of particles that approximate the size of the protoplasts, since conditions for optimal sorting of small particles may not necessarily be optimal for large cells (Harkins and Galbraith, 1987).

VIII. Instruments

The above procedures have been worked out using the Coulter EPICS and ELITE systems. Large particle sorting has also been achieved using a Becton-Dickinson FACStar Plus equipped with a 200-μm flow tip. The procedures should apply to other flow instruments operating on the same hydrodynamic principles.

IX. Results

Use of these procedures is illustrated with five examples.

A. Analysis of a Large Particle Standard: *Lycopodium*

Typical flow cytometric analyses of the optical characteristics of *Lycopodium* spores are given in Fig. 1. The Coulter ELITE was used to collect one-dimensional histograms of green autofluorescence and 90° light scatter. Unimodal, near-normal distributions are observed for both parameters.

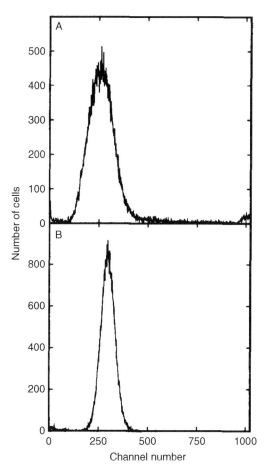

Fig. 1 Flow cytometric analysis of (A) autofluorescence (PMT2; 505–545 nm) and (B) 90° light scatter from *Lycopodium* spores. The one-dimensional frequency distributions were accumulated to a total count of 70,000. The mean and CVs (%) for the distributions were 255 (21.5%) for autofluorescence and 295 (11.1%) for light scatter.

B. Analysis of the Dimensions of Plant Protoplasts and Cells

Characterization of the TOF distributions of populations of tobacco leaf protoplasts is presented in Fig. 2A. This is also a near-normal distribution with a mode channel of 98. A correlation between the TOF parameter and actual protoplast size can be obtained by sorting protoplasts using a series of defined, nonoverlapping sort windows. Protoplast sizes are then subsequently measured using light microscopy. Figure 3 illustrates the linear relationship obtained in this manner

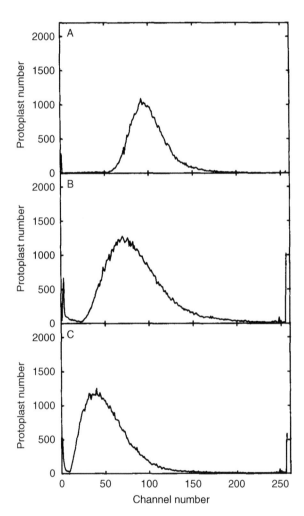

Fig. 2 Flow cytometric analysis of size according to time of flight. One-dimensional frequency distributions were accumulated to a total count of 20,000, corresponding to (A) TOF, (B) TOF^2, and (C) TOF^3 signals derived from chlorophyll autofluorescence of leaf protoplasts. The means and CVs for the distributions are 98 (18.4%), 80 (33.3%), and 48 (47.6%), respectively. From Galbraith (1990).

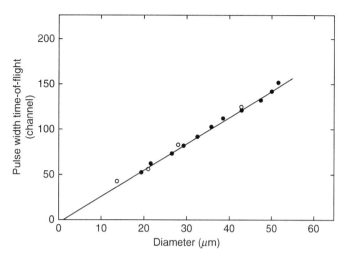

Fig. 3 Correlation between the pulse width time-of-flight parameter and protoplast or pollen/spore diameters, based on chlorophyll or aniline blue-induced fluorescence, respectively. Protoplasts, filled circles; pollen/spores, open circles. From Galbraith *et al.* (1988).

(Galbraith *et al.,* 1988). Use of standard particles allows the TOF/size relationship to be calibrated rapidly; thus, for the data presented in Fig. 2A, the true size of the protoplasts is 41 μm.

In order to produce real-time accumulations of distributions corresponding to surface area and volume, I have previously described an analog circuit that can be used to square and/or cube the TOF signal (Galbraith, 1990). This circuit operates with an accuracy that is not less than that of the EPICS MDADS. Analysis and processing of real-time distributions of TOF signal produced by protoplasts are illustrated in Fig. 2B and C. Since protoplasts are spherically symmetrical, these distributions provide a measure of protoplast surface area and volume. The major advantage of this circuit is its low cost. Future developments in this area are likely to involve manipulation of the digitized pulses, rather than operating in the analog realm, as the digital approach offers more flexibility and accuracy and less performance degradation over time.

C. Analysis of the Chlorophyll Contents of Leaf Protoplasts

Uniparametric analysis of the autofluorescence emission from a tobacco leaf protoplast suspension is presented in Fig. 4. Two peaks of fluorescence are observed on the logarithmic scale. The lower peak corresponds to chloroplasts and the upper to intact protoplasts (Harkins and Galbraith, 1984). This profile can be gated to exclude free chloroplasts and linear autofluorescence distributions obtained from the protoplasts, as shown in the inset to Fig. 4. In order to obtain a correlation between fluorescence emission and chlorophyll content, this linear distribution is divided into a series of nonoverlapping sort windows. In our

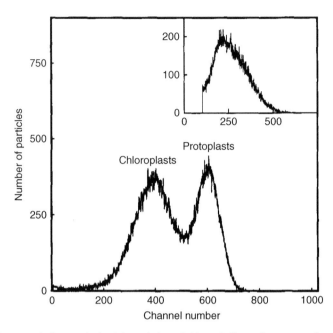

Fig. 4 Uniparametric flow analysis of the emission of chlorophyll autofluorescence from a suspension of tobacco leaf protoplasts, using a logarithmic abscissa. The lower peak corresponds to chloroplasts released from broken protoplasts. *Inset*: the logarithmic distribution is gated to include for analysis only the intact protoplasts, and the resultant distribution is displayed on a linear scale.

previous work, we selected five windows that were 15 channels wide and that were spaced by intervals 5 channels wide (Galbraith *et al.*, 1988). Specific numbers of protoplasts were sorted and their chlorophyll content was measured by spectrofluorometry. Protoplast chlorophyll content and protoplast autofluorescence yield were highly correlated (Fig. 5).

D. Combined Chlorophyll/TOF Analysis of Leaf Protoplasts

The tissues of dicotyledonous leaves comprise a variety of different cell types. Predominant are the photosynthetic tissues of the mesophyll, which contain large numbers of mature chloroplasts, and the nonphotosynthetic tissues of the epidermis and perivascular parenchyma. Biparametric flow analysis of protoplasts prepared from leaves according to TOF and chlorophyll content readily permits an identification of two populations of protoplasts (Fig. 6). As would be expected from the uniparametric analysis (Fig. 5), the mesophyll protoplasts exhibit a broad range of chlorophyll content, whereas those from the epidermis and perivascular parenchyma are essentially devoid of chlorophyll. It should be noted that for accurate sizing of the mesophyll protoplasts, TOF processing of the chlorophyll autofluorescence signal rather than the FDA fluorochromasia signal is required (Galbraith *et al.*, 1988);

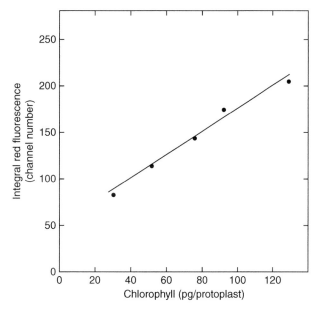

Fig. 5 Correlation between the chlorophyll autofluorescence emission and the protoplast chlorophyll content. From Galbraith *et al.* (1988).

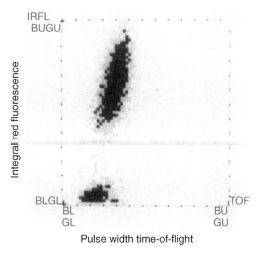

Fig. 6 Contour analysis of two-parameter measurements of the chlorophyll autofluorescence versus pulse width time-of-flight characteristics of leaf protoplasts. FDA fluorochromasia was used as the source of signals for the TOF analysis. The contour levels correspond to 5%, 15%, and 50% of the peak channel. From Galbraith *et al.* (1988).

obviously, it is not possible to analyze nonphotosynthetic protoplasts in this way. Thus for combined, accurate TOF/size analyses of these different protoplast populations, two independent TOF modules would be required.

E. Sorting and Culture of Leaf Protoplasts

Under optimal conditions, protoplasts can be sorted at efficiencies approaching 100% with no loss of viability (Afonso *et al.*, 1985; Harkins and Galbraith, 1984, 1987). The integrity of mesophyll protoplasts can be conveniently and continuously assayed by monitoring the numbers of free chloroplasts in the protoplast

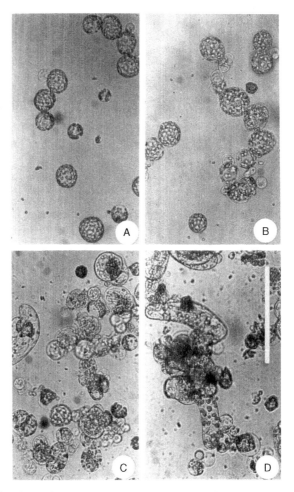

Fig. 7 Growth in culture of sorted *N. tabacum* protoplasts (A) after 1 day, (B) after 2 days, (C) after 5 days, and (D) after 12 days. (200×). Bar represents 150 μm. From Galbraith (1984).

suspension (Fig. 4). Under optimal conditions, the sorted protoplasts are metabolically active and will eventually resynthesize cell walls and enter into cell division (Fig. 7). We have successfully sorted and cultured protoplasts following induced fusion (Afonso *et al.*, 1985) and following transfection (Harkins *et al.*, 1990). Finally, we have been able to sort and culture protoplasts derived from both the photosynthetic and the nonphotosynthetic tissues of the leaf with equal efficiencies and to demonstrate cell-specific patterns of transgene expression within these sorted protoplasts (Harkins *et al.*, 1990).

Acknowledgments

I thank Georgina Lambert for valuable assistance. This work was supported by the National Science Foundation.

References

Afonso, C. L., Harkins, K. R., Thomas-Compton, M., Krejci, A. E., and Galbraith, D. W. (1985). Selection of somatic hybrid plants in *Nicotiana* through fluorescence-activated sorting of protoplasts. *Biotechnology* **3**, 811–816.

Birnbaum, K., Shasha, D. E., Wang, J. Y., Jung, J. W., Lambert, G. M., Galbraith, D. W., and Benfey, P. N. (2003). A gene expression map of the *Arabidopsis* root. *Science* **302**, 1956–1960.

Birnbaum, K., Jung, J. W., Wang, J. Y., Lambert, G. M., Hirst, J. A., Galbraith, D. W., and Benfey, P. N. (2005). Cell-type specific expression profiling in plants using fluorescent reporter lines, protoplasting, and cell sorting. *Nat. Methods* **2**, 1–5.

Bromova, M., and Knopf, U. C. (1991). High-frequency recovery of intergeneric fusion products of tobacco (*Nicotiana tabacum*) and potato (*Solanum tuberosum*) by electromechanical selection. *Plant Sci.* **74**, 127–133.

Donnelly, R. J., and Glaberson, W. (1966). Experiments on capillary instability of a liquid jet. *Proc. Roy. Soc. Lond. Ser. A* **290**, 547–556.

Farmer, E. E., and Ryan, C. A. (1992). Octadecanoid precursors of Jasmonic Acid activate the synthesis of wound-inducible proteinase inhibitors. *Plant Cell* **4**, 129–134.

Galbraith, D. W. (1984). Selection of hybrid cells by fluorescence activated cell sorting. *In* "Cell Culture and Somatic Cell Genetics of Plants" (I. K. Vasil, ed.), pp. 433–447. Academic Press, New York.

Galbraith, D. W. (1989). Flow cytometry and cell sorting: applications to higher plant systems. *Int. Rev. Cytol.* **116**, 165–227.

Galbraith, D. W. (1990). Isolation and flow cytometric characterization of plant protoplasts. *In* "Methods in Cell Biology" (Z. Darzynkiewicz and H. Crissman, eds.), Vol. 33, pp. 527–547. Academic Press, San Diego, CA.

Galbraith, D. W. (2007a). Protoplast analysis using flow cytometry and sorting. *In* "Flow Cytometry with Plant Cells" (J. Dolezel, J. Greilhuber, and J. Suda, eds.), pp. 231–250. Wiley-VCH, Weinheim.

Galbraith, D. W. (2007b). Analysis of plant gene expression using flow cytometry and sorting. *In* "Flow Cytometry with Plant Cells" (J. Dolezel, J. Greilhuber, and J. Suda, eds.), pp. 405–422. Wiley-VCH, Weinheim.

Galbraith, D. W., and Galbraith, J. E. C. (1979). A method for the identification of fusion of plant protoplasts derived from tissue cultures. *Zeitschr. Pflanzen.* **93**, 149–158.

Galbraith, D. W., and Mauch, T. J. (1980). Identification of fusion of plant protoplasts II. *Zeitschr. Pflanzen.* **98**, 129–140.

Galbraith, D. W., Afonso, C. L., and Harkins, K. R. (1984). Flow sorting and culture of protoplasts: Conditions for high frequency recovery and growth of sorted protoplasts of suspension cultures of *Nicotiana. Plant Cell Rep.* **3**, 151–155.

Galbraith, D. W., Harkins, K. R., and Jefferson, R. A. (1988). Flow cytometric characterization of the chlorophyll contents and size distributions of plant protoplasts. *Cytometry* **9**, 75–83.

Gantet, P., Hubac, C., and Brown, S. C. (1990). Flow cytometric fluorescence anisotropy of lipophilic probes in epidermal and mesophyll protoplasts from water-stressed *Lupinus-albus* L. *Plant Physiol.* **94**, 729–737.

Glimelius, K., Djupsjöbacka, M., and Fellner-Feldegg, H. (1986). Selection and enrichment of plant protoplast heterokaryons of brassicaceae by flow sorting. *Plant Sci.* **45**, 133–141.

Hammatt, N., Lister, A., Blackhall, N. W., Gartland, J., Ghose, T. K., Gilmour, D. M., Power, J. B., Davey, M. R., and Cocking, E. C. (1990). Selection of plant heterokaryons from diverse origins by flow cytometry. *Protoplasma* **154**, 34–44.

Harkins, K. R., and Galbraith, D. W. (1984). Flow sorting and culture of plant protoplasts. *Physiol. Plant* **60**, 43–52.

Harkins, K. R., and Galbraith, D. W. (1987). Factors governing the flow cytometric analysis and sorting of large biological particles. *Cytometry* **8**, 60–71.

Harkins, K. R., Jefferson, R. A., Kavanagh, T. A., Bevan, M. W., and Galbraith, D. W. (1990). Expression of photosynthesis related gene fusions is restricted by cell type in transgenic plants and in transfected protoplasts. *Proc. Natl. Acad. Sci. USA* **87**, 816–820.

Kachel, V., Fellner-Feldegg, H., and Menke, E. (1990). Hydrodynamic properties of flow cytometry instruments. *In* "Flow Cytometry and Sorting" (M. R. Melamed, T. Lindmo, and M. L. Mendelsohn, eds.), 2nd edn., pp. 27–44. Wiley-Liss, New York.

Lindsey, K. (1991). "Plant Tissue Culture Manual: Fundamentals and Applications." Kluwer, Dordrecht, The Netherlands.

Pauls, P. K., and Chuong, P. V. (1987). Flow cytometric identification of *Brassica napus* protoplast fusion products. *Can. J. Bot.* **65**, 834–838.

Pechan, P. M., Keller, W. A., Mandy, F., and Bergeron, M. (1988). Selection of *Brassica napus* L. embryogenic microspores by flow sorting. *Plant Cell Rep.* **7**, 396–398.

Sjödin, C., and Glimelius, K. (1989). *Brassica naponigra*, a somatic hybrid resistant to *Phoma lingam*. *Theor. Appl. Genet.* **77**, 651–656.

Sundberg, E., and Glimelius, K. (1991). Production of cybrid plants within Brassicaceae by fusing protoplasts and plasmolytically induced cytoplasts. *Plant Sci.* **79**, 205–216.

Verhoeven, H. A., and Sree Ramulu, K. (1991). Isolation and characterization of microprotoplasts from APM-treated suspension cells of *Nicotiana plumbaginifolia*. *Theor. Appl. Genet.* **82**, 346–352.

Walker-Simmons, M., Hollaender-Czytko, H. J., Andersen, J. K., and Ryan, C. A. (1984). Wound signals in plants - a systemic plant wound signal alters plasma-membrane integrity. *Proc. Natl. Acad. Sci. USA* **81**, 3737–3741.

Walters, T. W., Mutschler, M. A., and Earle, E. D. (1992). Protoplast fusion-derived ogura male sterile cauliflower with cold tolerance. *Plant Cell* **10**, 624–628.

Zhang, G., Campenot, M. K., McGann, L. E., and Cass, D. D. (1992). Flow cytometric characteristics of sperm cells isolated from pollen of *Zea mays* L. *Plant Physiol.* **99**, 54–59.

Zhang, C. Q., Barthelson, R. A., Lambert, G. M., and Galbraith, D. W. (2008). Characterization of cell-specific gene expression through fluorescence-activated sorting of nuclei. *Plant Physiol.* **147**, 30–40.

CHAPTER 14

DNA Content Histogram and Cell–Cycle Analysis

Peter S. Rabinovitch

Department of Pathology
University of Washington
Seattle, Washington 98195

ESSENTIAL CYTOMETRY METHODS
Reprinted from *Methods in Cell Biology*, Volume 41 (Academic Press, 1994).
Copyright © 1994 by Academic Press, Inc.,

DOI: 10.1016/B978-0-12-375045-7.00014-3

I. Introduction

The analysis of cells stained with DNA-specific fluorochromes was one of the first applications of flow cytometry, and it continues to be one of the most common uses of this technique. A primary reason for this is that this procedure can rapidly determine both DNA ploidy and cell-cycle measurements. Recent interest in the clinical application of such information has accentuated the need for greater care and accuracy in the analysis of DNA content histograms. At the same time, this interest has stimulated improvements in the computational models used to extract ploidy and cell-cycle information from DNA content histograms. This chapter reviews guidelines for DNA content histogram analysis and the principles and advances in methods of analysis.

II. DNA Content Histogram Basic Principles

The DNA content of each cell in an organism is generally highly uniform. In the resting (G_1) phase of the cell cycle there are exactly 23 chromosomes per human somatic cell, and a DNA content of approximately 7 pg/cell. This *diploid* DNA content is designated in flow cytometry by DNA index (DI) 1.0 (Hiddeman *et al.*, 1984). When diploid cells which have been stained with a dye that stoichiometrically binds to DNA are analyzed by flow cytometry, a "narrow" distribution of fluorescent intensities is obtained. This is displayed as a histogram of fluorescence intensity (*x*-axis) versus number of cells with each observed intensity. Since all G_1 cells have the same DNA content, the same fluorescence should (in theory) be detected, and only a single channel of the histogram should be filled (Fig. 1). In practice, however, instrumental errors and biological variability in DNA dye binding result in a Gaussian (normally distributed) fluorescence distribution from G_1 cells (Fig. 1). Greater variation in measurement results in broader DNA content peaks, and the term coefficient of variation (CV) is used to describe the width of the peak: CV = 100 × S.D./mean of the peak.

When beginning the process of replication, cells enter DNA synthesis, or S phase. Initially their DNA content is imperceptibly greater than the G_1 DNA content; as DNA synthesis proceeds, cellular DNA content progressively increases until, with complete DNA replication, cells enter the G_2 phase with a DNA content twice that of G_1. When DNA damage is repaired and chromosomes are organized, cells enter mitosis. Cell division returns the two daughter cells to the G_1 DNA content. In a growing cell population, the distribution of cells in various stages of S phase results in a broad distribution of DNA contents between G_1 and G_2. In the theoretical distribution, these S phase cells are easily identified between G_1 and G_2 DNA contents (Fig. 1A). In actual histograms, however, the uncertainty in measurements and Gaussian broadening of G_1, S, and G_2 phase distributions results in considerable overlap between G_1 cells and early S phase cells and G_2 cells and late S phase cells (Fig. 1B).

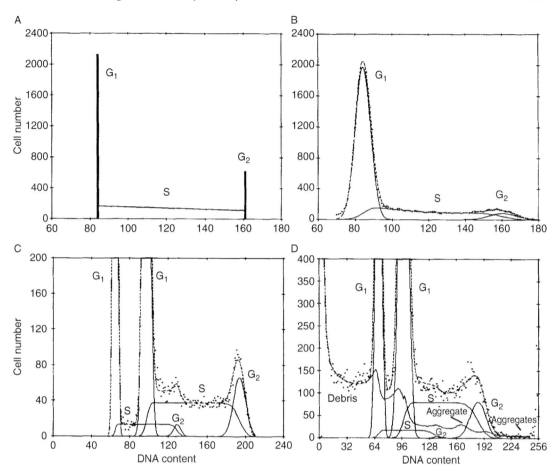

Fig. 1 The difference between a histogram from a "perfect" flow cytometer with no errors in measurement (A) and the Gaussian broadening of the histogram that is encountered in all real analyses (B). In B, actual data points are displayed as small diamonds, solid lines indicate the Gaussian G_1 and G_2 phase components and the S phase distribution, as fit with the Dean and Jett (1974) polynomial S phase model. The dashed line shows the overall fit of the model to the data. (C) The same model fitting, but to a histogram that has overlapping diploid and aneuploid cell cycles. (D) A similar histogram of diploid and aneuploid cells, but with the addition of debris resulting from extraction of nuclei from paraffin, together with aggregates of cells with cells and cells with debris. A solid line shows the combined distribution of background aggregates and debris.

When cells with an abnormal DNA content are present in tissue, a second G_1, S, and G_2 phase is present (the aneuploid cells are almost always accompanied by a component of cells with normal, diploid DNA content—for example, stromal fibroblasts, capillary endothelial cells, lymphocytes). The overlap between the diploid and aneuploid cell cycles can be variable, but adds to the complexity of the histogram analysis (Fig. 1C).

Finally, actual DNA content measurements, especially those performed from tissue specimens, frequently are complicated by the presence of debris (fragments of nuclei) and aggregation of nuclei with each other or with debris (Fig. 1D). It is the collective deviation from the theoretical simplicity of the DNA content analysis that gives importance to the computer modeling of DNA content histograms.

III. Cell–Cycle Analysis of DNA Content Histograms

DNA content histograms require mathematical analysis in order to extract the underlying G_1, S, and G_2 phase distributions; methods for this analysis have been developed and refined over the past two decades. Methods to derive cell-cycle parameters from DNA content histograms range from simple graphical approaches to more complex deconvolution methods using curve fitting. For detailed descriptions of the basic methods, the reader is referred to one of several more extensive reviews (Bagwell, 1993; Dean, 1985, 1990).

All of the simpler methods are based upon the assumption that the G_1 and G_2 phase fractions may be approximated by examining the portions of the histogram where the G_1 or G_2 phases have less overlap with S phase. There are two such approaches. The first is to calculate the area under the left half of the G_1 curve and the right half of the G_2 curve, and multiply each by two (i.e., reflecting these about the peak mean); what remains is S phase. The second approach is to use only the center-most portion of the S phase distribution and extrapolate this leftward to the G_1 mean and rightward to the G_2 mean. What remains on the left is G_1 and on the right is G_2. These methods can be reasonably accurate when one cell cycle is present and the histogram is optimal in shape. Both methods assume that the G_1 and G_2 peaks are symmetrical (DNA staining variability in tissues does not always provide this) and that the midpoint (mean) of each peak can be precisely identified. Because of the overlap of G_1 and G_2 peaks with the S phase, the mean of these peaks is not always at their maximal height (mode), especially for the G_2. If a second overlapping cell cycle is also present, then the overlap of the two cell cycles usually precludes safe use of these methods. In addition, modeling of debris and aggregates is usually not a part of these simpler graphical approaches.

The most flexible and accurate methods of cell-cycle analysis are based upon building a mathematical model of the DNA content distribution and then fitting this model to the data using curve fitting methods. The most well-established model, proposed by Dean and Jett (1974), is based upon the prediction that the cell-cycle histogram is a result of the Gaussian broadening of the theoretically perfect distribution (Fig. 1A). The underlying distribution can be recovered or "deconvoluted" by fitting the G_1 and G_2 peaks as Gaussian curves and the S phase distribution as a Gaussian-broadened distribution. As originally proposed, the shape of this broadened S phase distribution is modeled as a smooth second-order polynomial curve (a portion of a parabola). The model can be simplified by using a first-order polynomial curve (a broadened trapezoid) or a zero-order curve (a broadened rectangle). When the quality of the histogram is less than ideal, especially if G_1 or G_2 peaks are non-Gaussian (broadened bases, skewed, or having shoulders), then the simplified models may give results that are less affected by artifacts that increase the overlap of G_1, S, and G_2 peaks. This often is the case in analysis of clinical samples, as described in a subsequent section.

Some experimentally derived S phase distributions (usually from cultured cells) are more complex, and several alternative schemes have been proposed to model such distributions. The most flexible models are those of fitting S phase by the sum of Gaussians (Fried, 1976), in which the S phase is fit by a series of overlapping Gaussian curves, and the sum of broadened rectangles (Bagwell, 1979), in which the S phase is fit with a series of 5–10 broadened rectangles. In these models each of the Gaussian or broadened rectangle curves can be of any height. Therefore, the shape of the S phase is extremely flexible, and these models can fit S phase distributions that have complex shapes. This is also a primary drawback in practical use of these models, however. The very flexible S phase shape allows accurate fitting of any artifacts in the data and allows increased ambiguity in fitting the region of S near G_1 and G_2 (i.e., the areas of greatest overlap of G_1 and S and S and G_2). A generally successful compromise was suggested by Fox (1980), who added one additional Gaussian curve to Dean and Jett's polynomial S phase model. Fox's model provides a more flexible S phase shape, but still retains the smoothness of the S phase that is characteristic of the Dean and Jett model. It is especially suited to cell-cycle analysis of populations highly perturbed or synchronized by drug treatments.

Curve fitting models are almost universally fit to the histogram data by use of least square fitting. The fitting model is used to generate a mathematical expression, or function, for the predicted histogram distribution. The function has a number of parameters (usually between 7 and 22) that must be adjusted to give the optimum concordance between the fitting model and the observed data. Since the fitting function used by the model is not a simple linear equation, nonlinear least squares analysis is utilized. An excellent description of methods of nonlinear least squares analysis, and sample computer subroutines, is contained in Bevington (1969). The most commonly used technique of nonlinear least squares analysis in these applications is that described by Marquardt (1963). All of the nonlinear least square fitting techniques are *iterative:* successive approximations are made, in which the parameters in the fitting model equations are revised and the fit to the data is successively improved. When no further improvement is obtained, the fit has *converged* and is theoretically optimal. Goodness of fit is usually quantified by the χ^2 statistic:

$$\sum \frac{(y_{\mathrm{fit}_i} - y_{\mathrm{data}_i})^2}{\sigma_i^2}$$

or the reduced χ^2 statistic:

$$\chi_v^2 = \frac{\chi^2}{\mathrm{degrees \ of \ freedom}},$$

which measure the deviation of the fitting function from the data. The speed of the least square fitting is determined by the efficiency in searching for and finding the optimum combination of fitting parameter values. The Marquardt algorithm uses an optimized strategy for searching for the lowest χ^2 value along the n-dimensional "surface" defined in the space of the χ^2 versus n fitting variables.

An advantage of the least square fitting methods is that the models can be directly extended to analysis of two or even three overlapping cell cycles. The overlapping model components are mathematically deconvoluted to yield individual cell-cycle estimates. An additional advantage of curve fitting methods is that they tend to be less dependent upon the initial or "starting parameters" used to begin the fitting process. Such parameters include initial estimates of peak means and CVs, as well as the limits of the region of the histogram included in the fit. When the cell cycle and debris model is most accurate in fitting the data, the result is least dependent on starting values, and interoperator variation in results is reduced (Kallioniemi et $al.$, 1991).

It has been important to recognize that DNA content histograms from tumor tissue are often far from optimal (broad CVs, high debris, and aggregation) or complex (multiple overlapping peaks and cell cycles) and frequently contain arti-factual departures from expected shapes (e.g., skewed and non-Gaussian peak shapes). This is even more true when analyses are derived from formalin-fixed specimens. *When a skewed G_1 peak or a peak with a "tail" on the right side extends visibly into the S phase, S phase estimates should be used with extreme caution* (Shankey et $al.$, 1993a).

An important aspect of the analysis of imperfect histograms is the ability to reduce the model's complexity by using simplifying assumptions to reduce the number of model parameters being fit. This may reduce the ability of the model to fit the finer details of a histogram, but it also reduces the possibility of incorrect fitting of the data. As described above, some models may assume that a skew or broad base in G_0 or G_1 peaks is part of the S phase, which can lead to an overestimation of the true S phase. More conservative models may be more accurate in situations where CVs are wide or peaks are not well resolved, when multiple peaks are extensively over-lapping, or when background debris and aggregation are high. These situations are more fully described in later sections. The Dean and Jett algorithm may be used with a zero-order (broadened rectangle) or first-order S phase polynomial (broadened trapezoid), instead of the more flexible, but error-prone second-order polynomial. Additional constraints can be imposed to require that the CVs of the G_2 and G_1 peaks be equal (they are usually very similar), or the CVs of DNA diploid and aneuploid peaks can be made equivalent, or the G_2/G_1 ratios can be constrained to have a user-supplied value, based upon past laboratory experience.

IV. Critical Aspects of DNA Content and Cell-Cycle Analysis

This section details seven critical aspects of DNA histogram analysis:

a. Cell number

b. CV and detection of near-diploid aneuploidy

    c. Number of histogram channels, histogram range, and histogram linearity

    d. DNA content standards

    e. Debris modeling

    f. Modeling of cell or nuclear aggregation

    g. Quantitation of aggregates and debris

Guidelines for each of these aspects have been described by the DNA Cytometry Consensus Conference (Shankey *et al.*, 1993a). The following descriptions will expand upon the rationale for each of these guidelines. In this and subsequent sections, recommendations of the Consensus Conference are noted in italics.

A. Cell Number

One of the principle advantages of flow cytometry is that large numbers of cells may be analyzed in a short time. The object of acquiring larger numbers of cells is to reduce statistical fluctuations in the histogram, most apparent in areas of fewer cell numbers, and particularly in the region of S phase. Nevertheless, in order to speed the analysis, or if the tissue sample is small, there is a tendency to acquire the minimum number of events necessary. Figure 2 illustrates the effect of different cell numbers in two typical histogram types. Multiple histograms were acquired at each cell number, and the variation in the S phase measurement for each is shown. Above 10,000 events per histogram, S phase measurements are highly reproducible. Below this number, especially with fewer than 1000 events, accuracy in the S phase measurement in an individual histogram deteriorates substantially. *A minimum of 10,000 events in DNA content cell-cycle analysis has been recommended by the DNA Cytometry Consensus Conference Guidelines, although detection of DNA aneuploid populations may be possible from histograms with fewer cells or nuclei.* Note that this refers to the number of cells in the cell cycle; if a substantial proportion of events is from debris or aggregates, the total number of events acquired must be correspondingly higher in order to assure the required minimum number of intact single cells or nuclei. While ploidy information may be sometimes accurately derived from fewer cells, accurate S phase analysis may sometimes require more than this number. This is especially true if the proportion of aneuploid cells is low; again, in order to collect a minimum acceptable number of cells for the aneuploid cell-cycle analysis, the total cell number must be increased.

B. CV and Detection of Near-Diploid Aneuploidy

Distinguishing two populations with different ploidy becomes increasingly difficult as they become closer in DNA content. There is a consensus that near-diploid DNA contents cannot be reliably diagnosed in the clinical laboratory unless the histogram is bimodal, that is, there is a depression or trough between two separate peaks (Hiddeman *et al.*, 1984; Shankey *et al.*, 1993a). How close to diploid a

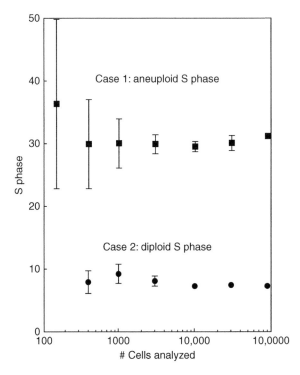

Fig. 2 Effect of cell number on accuracy of S phase analysis in DNA content histograms. Two human tumor specimens (case 1, squares, DNA aneuploid, aneuploid S phase analyzed, and case 2, circles, DNA diploid) were analyzed by acquisition of histograms with from 150 to 100,000 cells each, in triplicate. For each cell number, the mean and standard deviation of the S phase measurement was plotted.

population may be, and still be detected, depends upon both the CV of the analysis and the relative proportions of diploid and near-diploid cells. Both sample preparation and instrument performance can have a significant impact on the CV. Using the presence of a 10% dip between DNA diploid and aneuploid curves as a criterion for the detection of bimodality, the relationship between the minimum detectable DNA index, CV, and the proportion of aneuploid cells is shown in Fig. 3. Detection is optimal when the diploid and near-diploid cells are in equal proportions (Fig. 3, solid line). When the proportions of DNA diploid and aneuploid cells are unequal, then aneuploidy detection requires a lower CV or a larger ploidy difference. Note, for example, that with a CV of 6, a DI of 1.13 can be detected if the aneuploid cells are 50% of the total. If the aneuploid cells are only 5% of the total, only a DI of 1.23 can be detected with the same CV. A CV of 3.5 would be required to detect a 5% subpopulation of DI 1.13 cells. The values shown in Fig. 3 are minimum estimates, since problems with real data, such as non-Gaussian-shaped peaks or low numbers of cells in the histogram, will adversely affect peak discrimination.

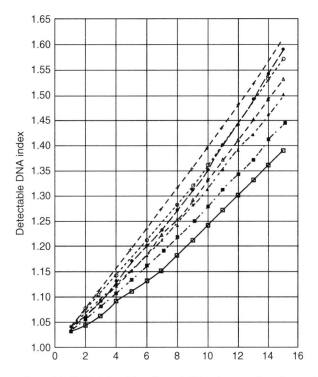

Fig. 3 Minimum detectable DNA index. The effect of CV and proportion of aneuploid cells on the ability to detect a near-diploid aneuploid peak is shown. The criterion for detectability is the presence of bimodality, with a 10% dip between diploid and aneuploid peaks. Gaussian peaks with different CV and separation were created by mathematical modeling and examined at different percentages of aneuploid cells: 5% (+), 10% (○), 25% (▲), 50% (□), 75% (■), 90% (△), and 95% (●). For any CV, as the proportion of aneuploid cells increases or decreases away from 50%, the capacity to detect near-diploid aneuploidy is diminished, and only a higher DNA index is detectable.

Note that when fitting near-diploid DNA aneuploid peaks, best results are often obtained using a software option to constrain the CVs of the diploid and aneuploid peaks to be equal. Examine this option if the fit without such constraint produces CVs for DNA diploid and aneuploid peaks that are very dissimilar.

C. Number of Histogram Channels, Histogram Range, and Histogram Linearity

The number of channels into which the DNA content histogram is digitized can have an important effect upon the data. A wide spectrum of histogram "resolutions" is in common use, ranging from 64 to 1024 channels or "bins." Too few channels may not provide the resolution needed to preserve the accuracy of the original analog signal. As an extreme example, note that if a peak were placed in channel 10, then a DI 1.05 ploidy would never be detected because the difference between channels 10 and 11 is 10%. This is more than a theoretical concern since channel numbers as low as 64 may be used in some bivariate cytograms (e.g., when one axis is DNA content and the other immunofluorescence). Conversely, if a very large number of channels is used, then unless very large numbers of cells are analyzed, statistical fluctuations in the number of events per channel will be greater. This results in a less satisfactory appearance and can produce greater uncertainty in data analysis.

Figure 4 shows practical examples of the effect of channel number on the result of histogram analysis. Variability and errors in CV, S phase fraction (SPF), and G_2/G_1 ratios (as well as calculations of DI, not shown) all increase when the mean

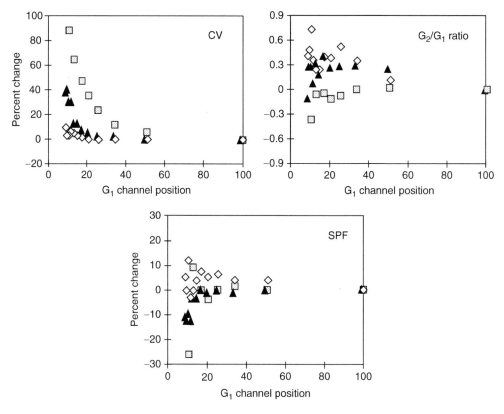

Fig. 4 The effect of reducing the number of channels used in the cell-cycle histogram. The histogram was initially analyzed with the G_1 peak in channel 100, and the G_1 channel position was subsequently reduced. The mean channel position of the G_1 peak in the histogram is shown on the abscissa versus the change in CV of the G_1 peak relative to the initial analysis (top left), the G_2/G_1 ratio (top right), and the percent S phase estimate (bottom center). The three different initial histograms had a G_1 peak CV of approximately 2 (squares, fresh tissue analysis), 4 (triangles, paraffin-embedded tissue), and 6 (diamond, paraffin-embedded tissue). Cell-cycle analysis was performed by the method of Dean and Jett using a zero-order (rectangular) S phase polynomial. Debris was analyzed with the sliced nucleus model (see text).

of the G_1 peak is placed below channel 50. CVs always rise if the G_1 peak is place in channels below 50, whereas S phase and G_2/G_1 ratio discrepancies are erratic. The magnitude of the S phase errors can be substantial, whereas the G_2/G_1 ratio variation is small (below 1%). The current recommendation is that *the lowest G_1 population should be accumulated in channels greater than 30, and probably above 50* (Shankey *et al.*, 1993a).

The range of DNA content values that is collected is another very important consideration. Valuable information can be lost if an adequate range of data to the left and right of the cell cycle(s) is not contained in the histogram. The left portion of the histogram contains most of the information describing the shape of the debris distribution, and these data are essential to proper application of debris models (see below). *Sufficient debris should be collected in the histogram to enable sound judgment of specimen quality and to allow software programs sufficient data to construct a model for debris compensation* (Shankey *et al.*, 1993a). *Setting a lower limit of data acquisition at a channel that corresponds to DI 0.1 is recommended* (Shankey *et al.*, 1993a).

Similarly, the hypertetraploid region must be examined. Hypertetraploid peaks may not be detected if these "high" data channels are discarded or are accumulated in the last "overflow" channel. In addition, data above the G_2 of the population with highest ploidy contain much of the information relating to the degree of aggregation present in the sample, and, if software aggregation modeling is applied, these data are essential to proper use of such modeling (see below). As a minimum, DNA contents up to at least 50% above the highest G_2 peak should be sampled. As an even more stringent rule, the *DNA Cytometry Consensus Guidelines recommend collecting data up to DI 6.0, or even to DI 10 if aneuploid with DI > 2.0 is present* (Shankey *et al.*, 1993a). The requirements that the diploid G_1 peak be positioned above channel 50 and that channels up to DI 6–10 be observed require that 300–500 channels are present in the histogram. Thus, while sufficient resolution is almost always achieved with 512 channels, 128-channel histograms often have insufficient resolution.

Visual observation of the position of triplets (DNA index = 3 for diploid triplets) using an expanded vertical histogram scale allows approximation of the extent of aggregation at the time of data acquisition. Dissociation or trituration of the sample to minimize aggregates is very important. The use of software aggregation modeling subsequent to data acquisition, as discussed below, can further compensate for the effects of aggregation if adequate hypertetraploid data are contained within the histogram.

Lack of histogram linearity is a common problem in many instruments. Departures from linearity can produce nonstandard G_2/G_1 ratios, altered DNA indices, and potential difficulty in computer modeling of aggregation. One common source of nonlinearity is an incorrect "zero" setting in the analog-to-digital converter—the channel in which a signal of zero intensity is placed. Problems with nonlinearity from this and other sources are most commonly manifest in the lowest and highest ends of the histogram; this can have the greatest effect on the use of a DNA content standard in the lower channels or on the evaluation of G_2 peaks and aggregates in the higher channels. *Instrument linearity should be determined on a regular basis, using standard particles and/or suitable methods* (Bagwell *et al.*, 1989), *and appropriate corrections made* (Shankey *et al.*, 1993a).

D. DNA Content Standards

Some authors have proposed that aneuploid peaks overlapping with diploid peaks might be detected by the use of external DNA content standards such as lymphocytes or nucleated red blood cells. Relative to the fluorescent standard, a shift in the position of the diploid/near-diploid composite peak away from the expected diploid position might be taken as evidence of DNA aneuploidy. However, differences in DNA staining between different cell types can result from variations in DNA dye binding (Bertuzzi *et al.*, 1990; Darzynkiewicz *et al.*, 1984; Evenson *et al.*, 1986; Heiden *et al.*, 1990; Iverson and Laerum, 1987; Klein and White, 1988; Kubbies, 1992; Wersto *et al.*, 1991; Wolley *et al.*, 1982) and this *can produce ambiguity in the correct diploid DNA content* (Shankey *et al.*, 1993a). Nonstoichiometric dye binding is also observed in dead or dying cells, apoptotic cells, or cells with DNA damage (Alanen *et al.*, 1989; Joensuu *et al.*, 1990; Kubbies, 1990; Nicoletti *et al.*, 1991; Roti Roto *et al.*, 1985; Stokke *et al.*, 1991; Telford *et al.*, 1991) and cells in different cell-cycle state and cell differentiation (Bruno *et al.*, 1991; Darzynkiewicz *et al.*, 1977). Thus, small differences in staining intensity relative to "standards" cannot be interpreted as evidence of DNA aneuploidy. Using an external reference standard to establish a range of positions or CVs to define as "diploid" results in overdiagnosis of DNA aneuploidy (Heiden *et al.*, 1990; Wersto *et al.*, 1991). *The best DNA content standard is the normal tissue component that represents the normal counterpart of the neoplastic cells* (Shankey *et al.*, 1993a). Malignant tissue almost always has at least a small component of normal diploid elements. If two peaks are present (from nonformalin-fixed tissue), but it is unclear which is aneuploid, a DNA content standard may be very useful. If human lymphocytes are added as a standard, the diploid peak position should be elevated in magnitude. Alternatively, if the standard has a DNA content that is much less than that of diploid human cells (e.g., chicken or trout red blood cells), then the standard will appear as a distinct peak at the left of the histogram that does not overlap with the diploid human cells. Software analysis of this peak will provide a ratio to diploid for evaluation relative to the expected range. Analysis of two standards simultaneously has been suggested (Vindelov *et al.*, 1983); in our experience, this rarely adds significant advantage in the clinical laboratory if histogram linearity is satisfactory (Koch *et al.*, 1984).

In the case of formalin-fixed tissue, variability in fixation and DNA dye accessibility (Becker and Mikel, 1990; Larsen *et al.*, 1986) prohibits any reproducibility in the position of the standard peak, and use of a standard is unfeasible. Even diploid cells from a paraffin block may not be a reliable standard, as there is substantial variability in DNA stainability from block to block and even within different portions of the same block (Price and Herman, 1990). The criteria that distinctly bimodal peaks be present to diagnose DNA aneuploidy must be used, and *it is recommended that the left-most peak from paraffin-embedded material be assumed to represent the DNA diploid population* (Shankey *et al.*, 1993a).

E. Debris Modeling

Damaged or fragmented cells or nuclei are almost always present in samples prepared for DNA flow cytometry; only the relative amount and origin of the debris vary. The debris produces events that are most visible on the left side of the histogram. More significantly, but often less obviously, the debris also extends into the cell-cycle region of the histogram. Since the S phase is the lowest and broadest cell-cycle compartment in the histogram, S phase calculations are most affected by the presence of debris. Thus, it is important to include modeling of the debris distribution in the histogram analysis in order to subtract the effects of the underlying debris from the cell-cycle fitting.

The earliest concept in debris fitting was that a steadily declining background debris curve could be fit by an exponential function (e^{-kx}). There are two primary reasons why a simple exponential curve does not usually provide an accurate fit. First, it is common to observe a debris component that rapidly declines with increasing DNA content combined with a portion that declines more slowly or plateaus. This latter portion has a much greater effect upon the cell-cycle fitting than would be predicted by an exponential curve. Second, debris is produced by degradation, fragmentation, or actual cutting of nuclei. Created in this manner, the fragments are always smaller than the DNA content of the nucleus from which they are derived. Thus, they are present only leftward of each DNA content position from which they are derived. In modeling this debris, each DNA content in the histogram must be considered as a separate source of debris, and, thus, the shape of the debris curve is dependent upon where the peaks in the DNA histogram are. Debris models of this kind are termed "histogram dependent."

Figure 5 illustrates the difference between the classical and histogram-dependent exponential debris. Figure 5A shows fitting of a simple exponential curve to the debris region left of the G_1 peak. This model does not take into consideration that most of the debris is created by fragmentation of G_1 nuclei; thus, the curve predicts too much background over the S and G_2 phases. Cell-cycle analysis with this model yields a 0% SPF. Application of the histogram-dependent exponential model is shown in Fig. 5B. The background debris curve drops rapidly from the left side of the G_1 peak to the right side of the G_1 peak. Fitting with this model yields an 11.2% SPF.

Figure 5C and D illustrates the shape of debris that is often observed in histograms derived from paraffin-embedded tissue. This debris is obviously not exponential in shape, as it has a much flatter distribution left of the G_1. Since the analysis of DNA histograms from cells preserved in paraffin blocks has become an increasingly important part of DNA flow cytometry, using a model which is

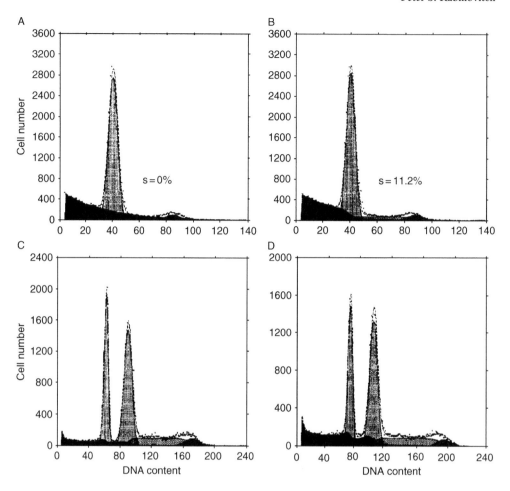

Fig. 5 Simple exponential debris model (A) and histogram-dependent exponential debris model (B) applied to a DNA diploid histogram from a fresh tissue sample containing degenerating cells. The S phase fractions resulting from the cell-cycle analysis are 0 and 11.2%, respectively. (C, D) Sliced nucleus debris modeling is applied to a DNA aneuploid tumor extracted from paraffin with microtome sectioning at 50 μm (C) or 20 μm (D). In all panels, the debris component of the fitting model is shown with horizontal hatching. Cell-cycle analyses in this and other figures were performed using the MultiCycle program written by the author (Phoenix Flow Systems, San Diego, CA).

consistent with the shape of this debris is of considerable practical importance for accurate cell-cycle analysis.

The examples in Fig. 5C and D have been fit with a model which recognizes the effect of cutting nuclei with a knife. As part of process of extraction of nuclei from paraffin blocks, sections are usually cut with a microtome at a thickness near 50 μm, and sectioning or slicing of nuclei is an unavoidable consequence. Nuclei in the path of the knife are cut randomly into two portions. If the nuclei were considered in a simple model to be identical cubes randomly cut perpendicularly to one face, then the volume of each randomly cut portion would be from near-zero to nearly full volume. In such a model, the debris distribution would be a flat plateau to the left of the whole-cell G_1 peak. This simplified model (Rabinovitch, 1988) was subsequently revised to be consistent with spherical and ellipsoidal nuclei (Bagwell *et al.*, 1991). As illustrated in Fig. 6, there tends to be a slightly greater fraction of small and large fragments produced in this model, which results from the small volume of the "crescents" produced by cutting the rounded ends of the nuclei. The distribution of sliced fragments thus exhibits a concave rather than a flat distribution. The histogram-dependent implementation of this model considers each channel of the distribution to be a discrete population of DNA contents, of which a certain proportion are cut and therefore form a flat-concave curve to the left of that channel. The probability of a nucleus being cut is proportional to its radius. The radius is proportional to the cubed-root of the volume, which is presumed to be proportional to DNA content (e.g., S and G_2 phase nuclei are larger than G_1 nuclei). The process of least squares fitting is used to determine the probability of nuclear cutting that yields the best fit to the data. Figure 5C and D illustrates the close fit of this model to histograms from paraffin-embedded cells.

As illustrated in Fig. 7, in addition to the flat-concave sliced nucleus debris, many clinical specimens from paraffin also have an additional component of more rapidly declining debris originating from degenerating cells and fragments other than those caused by cutting with the microtome. This same shape may be seen in fresh specimens that are minced with a scalpel, forced through mesh, or otherwise cut. To fit this combined effect of degradation and cutting, histogram-dependent exponential debris can be combined with sliced nucleus debris modeling. The relative contribution of each is determined by the least squares fitting. This combined model shows the greatest flexibility in modeling diverse types of histograms (Fig. 7C and F). The ability to fit the entire spectrum of the noise distribution has the additional benefit that this combined model is relatively

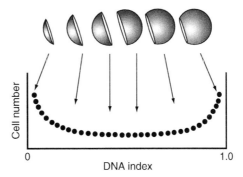

Fig. 6   The DNA content distribution of fragments of nuclei created by random sectioning. Sectioning at the "ends" of the nucleus results in larger numbers of very small and large fragments, resulting in a "smile" shaped distribution. DNA index = 1.0 corresponds to the DNA content of an intact nucleus.

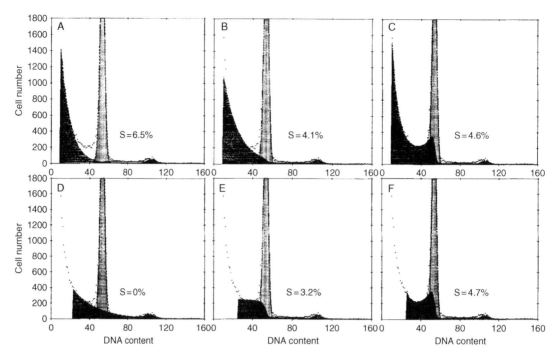

Fig. 7 Fitting of a histogram derived from paraffin-embedded diploid cells using a simple exponential background debris curve (A), histogram-dependent exponential debris (B), and the sliced nucleus debris model (C). The S phase fraction of the cell-cycle analysis is different in each case, as indicated. Simple exponential background debris applied with a left end point of the region of fitting that is closer to the G_1 peak is shown in (D), resulting in a very different S phase measurement than that in (A). The histogram-dependent exponential debris applied with the narrower fitting region is illustrated in (E), showing a 22% reduction in S phase compared to B. In contrast, the sliced nucleus model (F) is very insensitive to the change in fitting region.

insensitive to the end points chosen for the fitting region. This, in turn, reduces interoperator variation in results (Kallioniemei *et al.*, 1991). Figure 7 illustrates this behavior: the simple exponential debris shows the greatest dependence on the end points chosen for fitting (Fig. 7A and D), histogram-dependent exponential debris shows less variation (Fig. 7B and E), and the sliced nucleus model combined with histogram-dependent exponential debris (hereafter referred to simply as the sliced nucleus model) shows the least variability because it accurately fits the entire debris distribution (Fig. 7C and F).

The accuracy of the sliced nucleus modeling was evaluated by applying it to a test system comprised of PHA-stimulated diploid human lymphocytes and aneuploid HeLa cells (derived from an adenocarcinoma) and mixtures of these cell types. The cells were analyzed both fresh and after formalin fixation, paraffin embedding, and extraction from paraffin. Figure 5C and D shows examples of the mixture of lymphocytes and HeLa cells in paraffin sectioned at 50 and 20 μm, respectively. The results of cell-cycle analysis (Fig. 8) show that, for the fresh cells, the effects of the sliced nucleus debris modeling are small, except for the estimate of the lymphocyte S phase in the mixture with HeLa cells; in this case, HeLa nuclear fragments overlap the lymphocyte S phase, giving rise to a 3% overestimation of S phase without sliced nuclei debris modeling, but a satisfactory correction when using the sliced nucleus model.

For the paraffin-derived lymphocytes and HeLa cells examined independently (unmixed) there is an overestimation of S phase, which increases progressively as the section thickness decreases; this is almost completely corrected by debris modeling. When the diploid and aneuploid cells are mixed, many of the sliced HeLa nuclei overlap the lymphocyte cell-cycle distribution, resulting in an artificial elevation of the lymphocyte S phase (note in Fig. 5C and D that the majority of events between diploid and aneuploid G_1 peaks are modeled as debris). Cell-cycle fitting using the sliced nucleus model produces S phase estimates that are very close

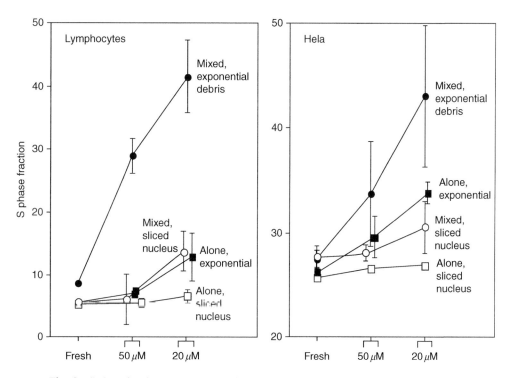

Fig. 8 S phase fraction measurements obtained with sliced nucleus debris modeling (open symbols) versus simple exponential debris (closed symbols, fit as in Fig. 7A) in cell-cycle analysis of lymphocytes alone, HeLa cells alone, and mixtures of these two cell types. The abscissa indicates results performed on fresh cells and paraffin-embedded cells sectioned at 50 or 20 μm. While each of the models yields similar results when applied to fresh cells that have little debris, when cells are mixed to simulate samples with DNA diploid and aneuploid peaks, and especially when cells are derived from paraffin, the sliced nucleus model gives S phase fractions much closer to those obtained from fresh unmixed cells.

to those of the fresh cells in 50-μm sections and nearly complete correction for 20-μm sections (the latter contain over twice the proportion of sliced nucleus debris as the former, see Fig. 5). Figure 8 also shows that the reproducibility of S phase estimates is better when sliced nucleus debris modeling is applied than when it is not. The comparison of 50 versus 20-μm sections indicates, however, that, even with an improved model, greater amounts of debris reduce the accuracy of S phase estimates. Because of this, quantitation of the proportion of debris in histograms is an important part of the assessment of cell-cycle analysis reliability.

Finally, it must be stressed that when debris modeling is used to improve the accuracy of cell-cycle measurements (as is currently recommended; see below), it is necessary that the histogram contains all of the necessary data pertaining to the debris. The requirement for an adequate range of data to the left and right of the cell cycle(s) has been described above. Similarly, it is imperative to avoid the use of light scatter gates. Although the histogram "appearance" may be improved by gating out the smaller sized fragments and debris, this gating destroys the mathematical relationships that correlate the extent of low DNA content debris with the amount of larger debris that overlaps the cell-cycle distribution.

F. Modeling of Cell or Nuclear Aggregation

Although analysis of the cell cycle(s) together with debris is sometimes sufficient to model the DNA content histogram, careful inspection of many histograms will reveal, in addition, evidence of cell aggregation. An aggregate of two G_1 cells (a doublet) will have the same DNA content as a G_2 cell and may be overlooked; however, diploid triplets will be seen at DI 3.0, quadruplets at DI 4.0, etc. In addition, S and G_2 cells and nuclear fragments (debris) also can aggregate with G_1 cells and with each other. If aneuploid cells are present, they may aggregate with diploid cells and debris, as well as with each other. The net effect upon the histogram can be complex. *Aggregates can affect detection of DNA aneuploid peaks and may cause major errors in S phase, especially for DI 2.0 tumors* (Shankey *et al.*, 1993a).

In the past, the primary approach to detection of aggregates has been to distinguish the altered pulse shape that they may produce when analyzed by a flow cytometer using a focused laser beam. Although this method may be successful, especially when examining uniformly spherical whole cells, with some cell preparations the method has less success (Rabinovitch, 1993a). If doublets of two G_1 cells pass through the laser beam aligned parallel to the laser beam, rather than perpendicular to it, then the fluorescence profile cannot be distinguished from that of a G_2 cell. In addition, an oblong G_2 cell, for example, cannot be easily distinguished from a G_1 doublet on the basis of pulse shape. Aggregates of three or more cells or nuclei may form a "spheroid" without a longer axis, and these also may not be distinguishable from a single large cell. Aggregated nuclei have no intervening cytoplasm, and their altered pulse shape may be less discernable than those of aggregated whole cells.

   The pulse shape analysis is most commonly performed by plotting fluorescence peak versus area (or less often time-of-flight or peak width vs. area) signals from each cell. A diagonal line is then drawn through the origin, making the assumption that aggregates will fall below the line (i.e., their pulse peak value will be lower than those of nonaggregates for a given pulse area). Placement of the diagonal line is subject to user interpretation. For oblong nuclei, such as from many epithelial cells, the G_1 and G_2 peak fluorescence is variable, and a diagonal that appears adequate to exclude aggregates may also exclude G_1 or G_2 cells or nuclei (Rabinovitch, 1993a).

   A DNA content histogram with aggregates may show additional peaks corresponding to doublets or triplets. In the past, software modeling of these peaks has been attempted by adding an extra peak to the cell-cycle model to fit the triplet peak position or by predicting doublets on the basis of the frequency of triplets (Beck, 1980). This approach will not, however, fit the more complicated patterns of aggregation discussed above, and aggregates may not, in fact, produce distinctly visible peaks, especially in more complex histograms with both diploid and aneuploid DNA contents. As an extension of the "histogram-dependent" modeling approach, a computer model can be applied that allows a generalized approach to the fitting of aggregation in DNA histograms (Rabinovitch, 1990, 1993a). The basis of this model is the simple assumption that any two particles, that is, elements of the histogram, have a certain probability of aggregating with each other. Thus, doublets form with a probability, p. Triplets are assumed to form by association of a doublet with a singlet; the singlet can "attach to" either of the two cells in the doublet, with a net probability of $2p^2$. Quadruplets can form in two ways: two doublets can aggregate with each other with a probability of $4p^3$ (there are four ways the two doublets can attach to each other, or $4p$ times p^2), or a triplet can combine with a singlet with a probability of $6p^3$ (there are three ways to combine the triplet with the singlet, or $3p$ times $2p^2$). The constants 2, 4, and 6 can be derived in this simplest fashion, or it is possible to modify these based on alternative models, which changes the final aggregate distribution slightly (Rabinovitch, unpublished data; Bagwell, 1993). The net aggregate distribution has a shape that is formed from the composite of all possible aggregate combinations. Expressed mathematically:

The doublet distribution : $$D(i) = p \cdot \sum_{j=1}^{I} \sum_{k=1}^{I} Y(j) \cdot Y(k)$$
$$\text{(for all } j + k = i),$$

where $Y(i)$ is the cell distribution without aggregation.

The triplet distribution : $$T(i) = 2p^2 \cdot \sum_{j=1}^{i} \sum_{k=1}^{i} D(j) \cdot Y(k)$$
$$\text{(for all } j + k = i).$$

The quadruplet distribution : $\quad Q(i) = 4p^3 \cdot \sum_{j=1}^{i} \sum_{k=1}^{i} D(j) \cdot D(k)$

$$+ 6p^3 \cdot \sum_{j=1}^{i} \sum_{k=1}^{i} T(j) \cdot Y(k) \quad \text{(for all } j + k = i).$$

The net aggregate distribution : $\quad \text{Aggregates}(i) = D(i) + T(i) + Q(i).$

Higher order aggregates can be added to the aggregate equation; in practice, however, these have little observable effect in most histograms and are usually not calculated. The aggregate distribution can be added to the cell-cycle and debris models and the combined model is fit to the observed data by using least squares fitting. This will determine the value of the variable, p; this single variable determines the amount of aggregation present in the DNA histogram, but does not change the shape of the aggregate distribution, which is determined only by $Y(i)$, the number of cells in each channel of the histogram.

An example of this fitting is shown in Figure 9. There are several aggregate peaks in the region of the DNA diploid and aneuploid cell cycles, and there are additional aggregates that are not obvious as peaks. Both the DNA diploid and DNA aneuploid S and G_2 phase measurements obtained with aggregation modeling are lower than those obtained without aggregation modeling. Comparison of S and G_2 phase measurements obtained in samples with various degrees of aggregation demonstrate that in analysis of epithelial tumors, software aggregation compensation may give more accurate S and G_2 phase measurements than hardware doublet discrimination (Rabinovitch, 1993a). Since gating from peak/area analysis affects the aggregate distribution in the histogram, it may alter the "expected" aggregate relationships. Therefore, pending future detailed study of these interactions, *use of software aggregation modeling should be restricted to data collected without hardware gating* (Shankey *et al.*, 1993a). As for debris modeling, this also means avoiding the use of light scatter gates. Although there is not yet a consensus on appropriate use of hardware gating versus software compensation for aggregate correction, it is hoped that further study will address this issue. At present, the most practical demonstration of the effectiveness of software aggregation modeling is obtained by the correlation of proliferative measurements with clinical outcome, as described subsequently.

G. Quantitation of Aggregates and Debris

The relative proportion of events analyzed by the flow cytometer that consist of cell or nuclear debris or aggregates is highly variable. The debris is generally higher in paraffin-processed tissue, due to nuclear slicing, and in degenerating or necrotic tissue, but these magnitudes are difficult to predict. To address the need for a quantitative measure of aggregates and debris, the DNA Cytometry Consensus Conference defined a parameter termed background aggregates and debris (BAD),

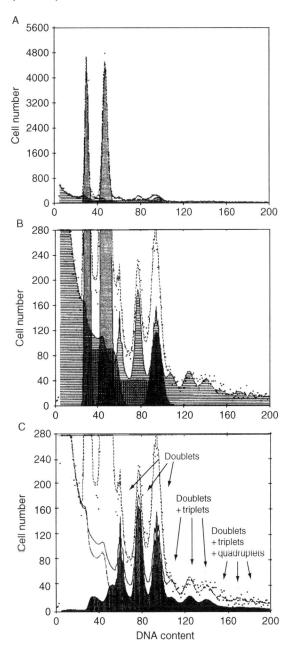

Fig. 9 Application of aggregation modeling to a DNA aneuploid histogram from a carcinoma of the breast. (A) Cell-cycle analysis without aggregation modeling. A peak is present near channel 80 that might be mistaken for a second DNA aneuploid population. (B) (20 × y-axis scale) shows the same histogram analyzed with aggregation modeling added to the sliced nucleus debris (horizontal hatching).

defined as the proportion of the histogram events between the leftmost G_1 and the rightmost G_2 that is modeled as debris or aggregates. The reason that this parameter is defined in this manner, rather than as the total percentage debris and aggregates in the entire histogram, is that left and right end points of a histogram are variable and arbitrary, depending on instrument settings. The proportion of debris in the histogram is especially sensitive to variation in the left limit of data acquisition. The BAD is unaffected by histogram end points. It is, however, very much dependent on the choice of histogram modeling. For greatest accuracy and interlaboratory comparison, it is suggested that histogram-dependent models of debris and aggregation be utilized.

V. Interpretation of DNA Content and Cell-Cycle Histograms

Choices that the operator must make, both in the cell-cycle modeling and in the interpretation of the fitted histogram, can lead to substantial variability in results. This variability has been documented in interlaboratory studies that have compared analysis and interpretation of histograms derived from replicate cells or tissues (Coon *et al.*, 1989; Hitchcock, 1991; Kallioniemi *et al.*, 1990; Wheeless *et al.*, 1989, 1991). Variation in histogram analysis may be seen both between different software programs analyzing the same data and between laboratories using the same software (Hitchcock, 1991). Some of this variability may potentially be reduced by use of fully automated "semi-intelligent" peak detection and cell-cycle analysis software (Kallioniemi *et al.*, 1994; Rabinovitch and Kallioniemi, 1991). Reproducibility in analysis within the laboratory and between laboratories is, however, most dependent on the consistent application of sensible rules for histogram interpretation.

A. Diagnosis of Aneuploidy

Both the convention on nomenclature for DNA cytometry in 1984 (Hiddeman *et al.*, 1984) and the guidelines of the DNA Cytometry Consensus (Shankey *et al.*, 1993a) indicate that *DNA aneuploidy can be reliably diagnosed only when two or more distinct (bimodal) peaks are present.* The ability to detect near-diploid DNA contents is directly related to the CV of the diploid and aneuploid peaks, as

---

Diploid and aneuploid S phases are shown by diagonal hatching, and Gaussian G_1 and G_2 peaks are shown by stippling. The total fit is indicated by the short-dashed line. (C) ($20 \times y$-axis scale) shows the individual components of the background fit. At the left of the histogram is the sliced nucleus debris (long-dashed line). The debris plus aggregates are shown as a solid line. The doublet distribution (vertical stripes) is complex in shape, reflecting the fact that all histogram components (debris, diploid G_1, S, and G_2 and aneuploid G_1, S, and G_2) aggregate with each other. The triplet distribution (diagonal stripes) has higher DNA content overall than the doublets, but there is extensive overlap. The quadruplet distribution (solid) is even higher in DNA content, but overlaps the triplet distribution to a large extent.

described previously. Detection of aneuploid populations of any DI also becomes more uncertain when the aneuploid peak is small, relative to the total number of cells in the histogram (Cusick *et al.*, 1990). *Ordinarily, for univariate DNA histogram analysis, the DNA aneuploid population should comprise at least 5–10% of the cell or nuclear events, after correction for aggregates* (Shankey *et al.*, 1993a). The number of total cells in the histogram can also be important in the identification of aneuploidy (see above). In samples with low cell number, statistical fluctuations should not be confused with an aneuploid peak; *if this is ambiguous, additional tissue samples should be used to confirm the presence or absence of DNA aneuploidy* (Shankey *et al.*, 1993a).

The requirement for the presence of bimodality in ploidy diagnosis is conservative, since in theory an aneuploid peak overlapping with the diploid peak might be detected by the use of DNA content standards. As described previously, a shift in the position of a peak with respect to standards cannot reliably be taken as evidence of DNA aneuploidy; otherwise, overdiagnosis of near-diploid aneuploidy may occur. In the past, differing criteria of diagnosis of DNA aneuploidy utilized by different investigators have complicated the interpretation and comparison of published data.

Finally, it should be noted that differentiation of DNA diploid from near-diploid cells may be more important in some tumors than others. For example, in breast cancer, aneuploid tumors with DI up to 1.3 show an improved survival compared to tumors with greater degrees of aneuploidy (Hedley *et al.*, 1993), and it has been suggested that these should be grouped together with DNA diploid cells (Toikkanen *et al.*, 1990). In contrast, the presence of near-diploid aneuploidy in some lymphomas may be clinically significant (Duque *et al.*, 1993).

B. DNA Index 2.0 Peaks: G_2 Versus DNA Tetraploidy

The difference between G_2 versus DNA tetraploid peaks cannot always be distinguished by DNA content analysis. By convention, *a definition of tetraploid peaks is DI values from 1.9 to 2.1 with proportions of cells greater than the G_2/M fraction of normal control tissue samples, after correction for aggregates* (Shankey *et al.*, 1993a). In many published studies arbitrary cutoff values, most commonly >15% total cells, have been utilized. However, some studies have correlated elevated DI 2.0 peaks with adverse clinical outcome, even when they comprise less than 15% of cells. Examples are prostatic adenocarcinoma (Jones *et al.*, 1990; Nativ *et al.*, 1989; Stephenson *et al.*, 1987) and Barrett's esophagus (Reid *et al.*, 1992). Because of this, careful assessment of the normal "control" range of G_2 values should be made in each organ system and laboratory. Hopefully, future studies will better validate the definition of tetraploidy in each organ system.

The following considerations should also be kept in mind:

1. When the diploid S phase is high, the diploid G_2 may be expected to be higher as well. In some cases of rapid proliferation a G_2 peak may be in excess of 15%.

2. The DNA index of a G_2 peak should be within the range expected for G_2 cells. Careful recording of the actual G_2/G_1 ratios for a given cell or tumor type will help to establish the laboratory confidence interval for true G_2 cells. Peaks outside this range may be aneuploid, even if they are smaller than the control G_2 level.

3. If there is evidence of S and G_2 phases for the DI 2.0 population, then there is a stronger argument that the DI 2.0 peak may be a G_1 near-tetraploid population. Note that software aggregation modeling, as described above, is often very useful in demonstrating that the G_2 of the near-tetraploid population is greater than can be accounted for by the effects of aggregation alone.

4. When a tetraploid or near-tetraploid population is present, it will, by definition, overlap with the G_2 of the "diploid" population. When the overlap is close, there is no way to accurately determine how much of the near-tetraploid peak is diploid G_2 and how much is near-tetraploid G_1. Cell-cycle analysis software should, however, be prevented from making the G_2 an unreasonably large proportion of the total near-tetraploid peak [by assuming, e.g., that the G_2 of the diploid (or lower ploidy) population is proportional to the S phase fraction of the same population].

C. Which S Phase is Most Relevant When DNA Diploid and Aneuploid Populations are Both Present?

More sophisticated cell-cycle modeling does not require that S phases be non-overlapping in order to give independent estimates of each. If at least some part of the S and G_2 phase distributions is nonoverlapping, then the relative components of each can be evaluated. Exactly how close two ploidy populations can be before their S phases can no longer be independently evaluated depends mainly upon the CV of the analysis. If there is no region of the histogram in which the two cell cycles are largely nonoverlapping, then the individual diploid and aneuploid S phases cannot be reliably established; *in this circumstance it is recommended at present that the estimate of the combined S average phase of the two populations should be reported* (Shankey *et al.*, 1993a). This recommendation will, hopefully, be subject to additional future experimental confirmation. When the ploidy value and quality of the histogram allow individual S phase estimates to be derived, the current recommendation is that *the S phase of the aneuploid population should be used for clinical interpretation* (Shankey *et al.*, 1993a). This interpretation is based, in part, upon the argument that the DNA aneuploid cells, when present, represent the malignant cells, and it is assumed that the proliferative behavior of the malignant cells is the most relevant to the biological aggressiveness of the tumor. In the past, published reports have frequently failed to specify the origin of the S phase calculation. The study of node-negative breast cancer summarized in Table I demonstrated that utilizing the DNA aneuploid S phase (when present) provided superior prognostic value compared to the use of the average of the DNA diploid and aneuploid S phase. In contrast, if only the S phase of the DNA diploid cells

Table I

The Prognostic Value of Different S Phase Estimates[a]

Method	P value[b]
Diploid S for diploid tumors, aneuploid S for aneuploid tumors, average S phase for near-diploid tumors. Classical exponential debris	0.004
As above but with sliced nuclei debris model	0.0005
Average of diploid and aneuploid S phase utilized	0.029
Diploid S phase only utilized	0.13

[a]Node-negative breast cancer. Modified after Kallioniemi *et al.* (1994).
[b]Difference (Wilcoxon-Breslow analysis) between 5-year survival with above or below median S phase.

was utilized, little prognostic information was obtained. Additional studies will hopefully also address this issue.

D. Optimal Choices of Fitting and Debris Models for Histogram Analysis

As described above, there have been a number of recent developments in cell-cycle analysis modeling, particularly in debris and aggregation compensation. It is generally assumed that these newer modeling techniques will lead to improved accuracy of S and G_2 phase measurements (Shankey *et al.*, 1993a). At the present time, however, only a limited amount of direct evidence is available to confirm this improvement in the clinical setting. Several reports indicate that the use of sliced nucleus modeling of debris, especially in samples derived from paraffin, improves the predictive strength of S phase measurements in breast (Clark *et al.*, 1992; Kallioniemi *et al.*, 1991) and prostate (Kallioniemi *et al.*, 1991) cancer. Similar results from the data set used by Kallioniemi *et al.* (1991) are shown in Table I and illustrate the improvement obtained by sliced nucleus modeling compared to classical exponential debris modeling. Based on theoretical considerations and limited published study, it is the consensus at present that *simple exponential debris modeling is not considered reliable, and histogram-dependent debris models should be used* (Shankey *et al.*, 1993a).

Comparison of sliced nucleus debris with and without software aggregation modeling is shown in Table II. The addition of aggregation modeling improves the predictive strength of S phase in this data set, supporting the use of this approach. As mentioned previously, the presence of aggregates can obscure the detection of an aneuploid peak, or an aggregate peak can be falsely interpreted as an aneuploid peak. Aggregates can especially compound the difficulty of identifying DNA "tetraploid tumors" (Shankey *et al.*, 1993a).

The use of aggregation modeling generally results in a reduction in the average values of both S and G_2 phase measurements. In diploid breast cancers, the reduction in S phase estimates averaged only 0.89 percentage points; however, aneuploid S phase estimates were reduced an average of 2.5 percentage points

Table II

Relative Risk (Top/Bottom Tertile) and Significance (P value) of S or S + G$_2$/M Phase Fraction in Cox Multivariate Analysis[a]

Model	SPF	S + G$_2$/M
Sliced nucleus debris, zero-order S	1.75 (0.01)	2 (0.001)
Sliced nucleus debris, zero-order S + aggregation	2.16 (0.0002)	2.52 (0.00001)
Sliced nucleus debris + first-order S	1.69 (0.02)	1.84 (0.002)
Sliced nucleus debris + zero-order S + CVs fixed	1.73 (0.01)	1.96 (0.002)

[a]Breast cancer, all stages, surgical treatment only. Modified after Barlow *et al.* (1994).

(Rabinovitch, 1993a). S phase estimates in bladder and prostate cancer (which are frequently tetraploid) may be reduced by 20–50% when software aggregation correction is applied (Rabinovitch, 1993a; Shankey *et al.*, 1993b). As a rule, aggregates of >5% cells can cause significant overestimation of "tetraploid" tumor S phases (Shankey *et al.*, 1993a).

Most often, S phase is the proliferative parameter evaluated in flow cytometric literature, although occasionally S + G$_2$ is utilized. Although the DNA Cytometry Consensus Guidelines recommend that *S phase, rather than S + G$_2$/M, be utilized for tumor proliferation estimates*, the relative utility of S versus S + G$_2$/M is in need of further study. Table II illustrates that, in that data set, S + G$_2$/M was of greater prognostic strength than S phase alone and that this strength was appreciably enhanced with the use of aggregation modeling. Because the G$_2$/M calculation is especially affected by the presence of doublets, it is possible that the use of improved aggregation compensation will lead to a general improvement in the prognostic strength of the S + G$_2$/M measurement.

Simplification of the cell-cycle model used in analysis of clinical specimens has been suggested in a preceding discussion, because histograms from clinical tissues often contain artifacts which are not consistent with the fitting model. An example is artifactual skewing of the G$_1$ peak which can be falsely confused with S phase. The use of a zero-order (rectangular) S phase model was compared to a first-order (trapezoidal) S phase model, as shown in Table II. The simpler zero-order S phase model provided a slightly better prognostic strength for both S and S + G$_2$. Simplifying the fitting model by constraining the CVs of diploid and aneuploid G$_1$ and G$_2$ peaks to the same value had no beneficial effect.

E. Prognostic Categories Based on Ploidy and Cell-Cycle Results

As several interlaboratory comparisons have described (Hitchcock, 1991; Kallioniemi *et al.*, 1990), the values of S and/or G$_2$/M derived by different laboratories vary considerably, due to lack of standardized techniques and use of different methods of S and G$_2$/M calculation and histogram interpretation. Thus, *use of published values for numerical cut points for survival or disease course should be avoided* (Shankey *et al.*, 1993a). At present *each laboratory must define its own range of values (high, intermediate, and low S phase for DNA diploid and DNA aneuploid tumors) based upon their own analysis of individual types of tumor* (Shankey *et al.*, 1993a). Even this method is subject to the caveat that the patient population examined locally may differ from that investigated in a published study.

Most often, clinical survival or relapse-free interval has been compared for patients with above or below mean or median S phase, although some authors have determined the S phase cutoff that yields the maximum intergroup difference in clinical outcome (Clark *et al.*, 1989). Although a single cutoff point yields a binary result—good or bad prognosis, for example—this division does not make good biological or statistical sense: cell-cycle estimates that differ by insignificant amounts may be placed in opposite prognostic categories. *A single cut point should be avoided and tertiles such as low, intermediate, and high are preferable* (Shankey *et al.*, 1993a).

At present, the DNA Cytometry Consensus Conference Guideline recommendation is that high, intermediate, and low S phase ranges for DNA diploid and DNA aneuploid tumors be calculated separately (see above). Although survival rates for diploid and aneuploid SPF ranges have occasionally been reported (Clark *et al.*, 1992), they have only rarely (Kallioniemi *et al.*, 1988) been calculated by tertiles. Use of the recommended tertiles within ploidy class may require some adjustment in current perceptions and may cause some confusion until it is more widely utilized. While the clinical hazards estimated from separate diploid and aneuploid tertile cut points (Table III) lead to improved prognostic strength, the

Table III
Estimated Hazards for a Cox Model Analysis of Breast Cancer Survival[a]

Factor	S phase range	Hazard
Size 2–5 cm		1.69
Size >5 cm		3.24
Positive nodes		2.24
Metastases		3.30
S phase tertiles, all tumors		
Low S	0–1.9	1
Intermediate S	1.9–5.4	1.54
High S	>5.4	2.10
S phase tertiles, diploid and aneuploid tumors separately		
Diploid low S	0–0.8	1
Diploid intermediate S	0.8–2.1	1.39
Diploid high S	>2.1	2.17
Aneuploid low S	0–4.1	1.50
Aneuploid intermediate S	4.1–9.1	2.35
Aneuploid high S	>9.1	2.40

[a]Breast cancer, all stages, surgical treatment only. Aneuploid S phase utilized when separately calculateable; sliced nucleus debris and aggregation modeling done with the MultiCycle program (Rabinovitch, 1993b). Modified after Barlow *et al.* (1994).

cut points are very different from the more conventional cut points derived from "all" ploidies, and the resultant hazards reflect the combined influence of ploidy and proliferative rate. Note that in the separate analysis of diploid and aneuploid cases, diploid tumor S phase values are overall much lower than aneuploid tumor S phase values. This is a consistent finding in large numbers of published reports and may be due both to intrinsically lower proliferative rates of diploid malignant cells and to the unavoidable admixture of diploid tumor cells and slower growing stromal cells in the univariate DNA histogram. Differences between the "all" versus separate prognostic categories can be appreciable in some cases. For example, diploid tumors with S phase values within the diploid-specific "intermediate" range of 0.8–1.9 or the diploid "high" range of >2.1 are associated with higher calculated risk when analyzed separately by ploidy than when all ploidies are lumped together (Table III). Similarly, aneuploid tumors with low proliferative rates of <1.9 have higher risk than the "all" low tertile would suggest (aneuploid tumors with low SPF have higher risk than diploid tumors with low SPF). Publication of analyses of additional datasets using separate diploid and aneuploid tertiles would be very valuable. As the Consensus Guidelines indicate, the specific cutoff values shown in Table III should not be directly extrapolated into other laboratories. In particular, values will be higher if sliced nucleus and aggregation modeling are not utilized.

F. Histogram Reliability and Confidence Estimation

After performing a cell-cycle analysis, how the resulting parameters are used often hinges upon the assessment of the accuracy and reliability of the DNA content and cell-cycle estimates. *Not all histograms are adequate for identification of DNA aneuploid populations, or for the estimation of S phase ... when the histogram is inadequate, due to high CVs, debris, or aggregates, it should be reported as inadequate* (Shankey *et al.*, 1993a). Criteria for the ability to detect near-diploid aneuploidy based on CV have been described previously in Fig. 3. *In general the CV of normal diploid cells in a histogram should be less than 8%* (Shankey *et al.*, 1993a). As demonstrated above, debris and aggregates can appreciably affect S phase estimation, although newer models help to compensate for these effects. Quantitating the effect of CV, debris, and aggregates on the accuracy of S phase measurement is, however, complex.

The DNA Consensus Conference Guidelines recommend that *if a sample has a high percentage of aggregates (>10% as determined by manual counting), it should be further disaggregated by mechanical means or rejected for analysis. The percentage aggregates and debris should be evaluated for each histogram ... it is recommended that >20% histogram background aggregates and debris are unsatisfactory for S phase analysis.*

The DNA Consensus Conference Guidelines further recommend that *accurate calculation of S phase generally requires that the proportion of tumor cells is greater than 15–20% of the total cells, although samples with lower proportions may be useful for DNA ploidy analysis.* When aneuploid cells are present, accurate S phase estimation requires that they should be greater than 15–20% of the total cells. This is because (1) overlap of cells from the large diploid population into the S phase region of the aneuploid population can impair the accuracy of estimation of the aneuploid S phase, and (2) even when the two populations do not overlap extensively, debris and aggregates from the more abundant population may still overlap the cell cycle of the rarer population. A common clinical setting in which low proportions of tumor cells may be found is in the presence of an abundant lymphocytic infiltrate (Eckhardt *et al.*, 1989).

Guidelines such as the above for rejection of histograms based on "poor" quality are based upon a consensus of experience, but there has been minimal publication of objective data. Existing databases of DNA content histograms and the associated clinical outcome can be used to objectively define improved reporting methods, including tests of criteria for histogram rejection. In such an analysis, "poor" histogram quality would be assumed to be associated with reduced prognostic strength of ploidy and cell-cycle results. One such analysis of predictors of poor reliability of S phase measurements is illustrated in Table IV. Consistent with the DNA Cytometry Consensus Conference Guidelines, rejection of cases with low proportions of aneuploid cells and high BAD was beneficial, although the exact rejection criteria were slightly different (more stringent for percent aneuploid cells, less stringent for BAD). A predictor of S phase reliability that is based upon a

Table IV

Predictors of S Phase Reliability: Prognostic Strength of the Remaining Cases after Rejection of Subsets of Histograms[a]

Cases removed	Percent cases removed	Hazard (relative risk)	Significance (P)
None	0	2.10	0.0005
% Aneuploid cells < 30%	14	2.54	0.00002
% Aneuploid cells < 40%	21	2.82	0.000005
BAD > 35%	10	2.40	0.00004
BAD > 28%	20	2.54	0.0002
Intramodel error			
80th percentile	20	2.30	0.0002
70th percentile	30	2.64	0.0002
Intermodel error			
90th percentile	10	2.18	0.0004
80th percentile	20	2.32	0.0002
CV > 8.6	10	2.00	0.002
CV > 7.5	20	1.94	0.006
$\chi^2 > 2.0$	10	1.93	0.005
$\chi^2 > 1.7$	20	2.05	0.003

[a]Breast cancer, all stages, surgical treatment only. Cox multivariate analysis modified after Barlow *et al.* (1994).

statistical error analysis of the least squares fitting itself, the "intramodel error" (Rabinovitch, 1993a), was also beneficial. In addition, comparison of the range of S phase values produced by alterations in the fitting model (e.g., with and without aggregation modeling, zero-order vs. first-order S phase models, with vs. without G_2/G_1 ratio constraints), the "intermodel error" (Rabinovitch, 1993a), was a moderately useful predictor of reliability. These four predictors can be combined into an overall estimate of histogram reliability (Rabinovitch, 1993a,b). In contrast, the analysis shown in Table IV failed to demonstrate that either elevated CV or higher χ^2 of the least squares fit was useful as a criterion for histogram rejection. The absence of utility of CV is surprising, but may indicate that other measures of histogram quality are more important. The χ^2 is affected by a large number of variables, not all related to goodness of the fit; these include the number of cells acquired in the histogram and the end points of the analysis region used within the histogram.

VI. Conclusion

The careful and more uniform use of guidelines for DNA content and cell-cycle analysis should lead to greater accuracy and reproducibility in performing and reporting ploidy and cell-cycle measurements. If this is done, then this simple methodology can be applied with greater confidence over a wider range of clinical settings. The future of DNA cytometry also includes greater use of multiparameter DNA analyses, especially the use of cell-type and proliferation-specific antibodies. Although the added information that is derived from the multiparameter analysis will help improve accuracy, most of these analyses will still involve cell-cycle analysis of gated histograms or bivariate histogram regions. Thus, DNA content and cell-cycle analysis will retain importance well into the future.

References

Alanen, K. A., Joensuu, H., and Klein, P. J. (1989). *Cytometry* **10,** 417–425.

Bagwell, C. B. (1979). Ph.D. Thesis, University of Miami, Coral Gables, FL.

Bagwell, C. B. (1993). *In* "Clinical Flow Cytometry: Principles and Applications" (K. D. Bauer, R. E. Duque, and T. V. Shankey, eds.), pp. 41–62. Williams & Wilkins, Baltimore, MD.

Bagwell, C. B., Baker, D., Whetstone, S., Munson, M., Hitchcox, S., Autl, K. A., and Lovett, E. J. (1989). *Cytometry* **10,** 689–694.

Bagwell, C. B., Mayo, S. W., Whetstone, S. D., Hitchcox, S. A., Baker, D. R., Herbert, D. J., Weaver, D. L., Jones, M. A., and Lovett, E. J. (1991). *Cytometry* **12,** 107–118.

Barlow, W., Kallioniemi, O.-P., Visakorpi T., Isola, J., and Rabinovitch, P. S. (1994). Submitted for publication.

Beck, H. P. (1980). *Cell Tissue Kinet.* **13,** 173–181.

Becker, R. L., Jr., and Mikel, U. V. (1990). *Anal. Quant. Cytol. Histol.* **12,** 333–341.

Bertuzzi, A., D'Agnano, I., Gandolfi, A., Graziano, A., Star, G., and Ubezio, P. (1990). *Cell Biophys.* **17,** 257–267.

Bevington, P. R. (1969). "Data Reduction and Error Analysis for the Physical Sciences," pp. 153–160. McGraw-Hill, New York.

Bruno, S., Crissman, H. A., Bauer, K. D., and Darzynkiewicz, Z. (1991). *Exp. Cell Res.* **196,** 99–106.

Clark, G. M., Dressler, L. G., Owens, M. A., Pounds, G., Oldaker, P., and McGuire, W. L. (1989). *N. Engl. J. Med.* **320,** 627–633.

Clark, G. M., Mathiew, M. C., Owens, M. A., Dressler, L. G., Eudey, L., Tormey, D. C., Osborne, C. K., Gilchrist, K. W., Mansour, E. G., Abeloff, M. D., *et al.* (1992). *J. Clin. Oncol.* **10,** 428–432.

Coon, J. S., Deitch, A. D., de Vere White, R. W., Koss, L. G., Melamed, M. R., Reeder, J. E., Weinstein, R. S., Wersto, R. P., and Wheeless, L. L. (1989). *Cancer* (*Philadelphia*) **63,** 1592–1600.

Cusick, E. L., Milton, J. I., and Ewen, S. W. B. (1990). *Anal. Cell. Pathol.* **2,** 139–148.

Darzynkiewicz, Z., Traganos, F., Sharpless, T. K., and Melamed, M. R. (1977). *Cancer Res.* **37,** 4635–4640.

Darzynkiewicz, Z., Traganos, F., Kapuscinski, J., Staino-Coico, L., and Melamed, M. R. (1984). *Cytometry* **5,** 355–363.

Dean, P. N. (1985). *In* "Flow Cytometry: Instrumentation and Data Analysis" (M. A.Van Dilla, P. N.Dean, O. D.Laerum, and M. R.Melamed, eds.), pp. 195–221. Academic Press, New York.

Dean, P. N. (1990). *In* "Flow Cytometry and Sorting" (M. R.Melamed, T.Lindmo, and M. L. Mendelsohn, eds.), 2nd edn., pp. 415–444. Wiley-Liss, New York.

Dean, P. N., and Jett, J. (1974). *J. Cell Biol.* **60,** 523.

Duque, R. E., Andreeff, M., Braylan, R. C., Diamond, L. W., and Peiper, S. C. (1993). *Cytometry* **14,** 492–496.

Eckhardt, R., Feichter, G. E., and Goerttler, K. (1989). *Anal. Quant. Cytol. Histol.* **11,** 384–390.

Evenson, D., Darzynkiewicz, Z., Jost, L., and Ballachey, B. (1986). *Cytometry* **7,** 45–53.

Fox, M. H. (1980). *Cytometry* **1,** 71.

Fried, J. (1976). *Comp. Biomed. Res.* **9,** 263–276.

Hedley, D. W., Clark, G. M., Cornelisse, C. J., Killander, D., Kute, T., and Merkel, D. (1993). *Cytometry* **14,** 482–485.

Heiden, T., Strang, P., Stendahl, U., and Tribukait, B. (1990). *Anticancer Res.* **10,** 49–54.

Hiddeman, W., Schumann, J., Andreeff, M., Barlogie, B., Herman, C. J., Leif, R. C., Mayall, B. H., Murphy, R. F., and Sandberg, A. A. (1984). *Cytometry* **5,** 445–446.

Hitchcock, C. L. (1991). *Cytometry Suppl.* **5,** 46.

Iverson, O. E., and Laerum, O. D. (1987). *Cytometry* **8,** 190–196.

Joensuu, H., Alanen, K., Klemi, P., and Aine, R. (1990). *Cytometry* **11,** 431–437.

Jones, E. C., McNeal, J., and Bruchovsky, N. (1990). *Cancer* (*Philadelphia*) **66,** 752–757.

Kallioniemi, O.-P., Blanco, G., Alavaikko, M., Hietanen, T., Mattila, J., Lauslahti, K., Lehtinen, M., and Koivula, T. (1988). *Cancer* (*Philadelphia*) **62,** 2183–2190.

Kallioniemi, O.-P., Joensuu, H., Klemi, P., and Koivula, T. (1990). *Breast Cancer Res. Treat.* **17,** 59–61.

Kallioniemi, O.-P., Visakorpi, T., Holli, K., Heikkinen, A., Isola, J., and Koivula, T. (1991). *Cytometry* **12,** 413–421.

Kallioniemi, O.-P., Visakorpi, T., Holli, K., Isola, J. J., and Rabinovitch, P. S. (1994). *Cytometry* **16,** 250–255.

Klein, F. A., and White, K. H. (1988). *J. Urol.* **139,** 275–278.

Koch, H., Bettecken, T., Kubbies, M., Salk, D., Smith, J. W., and Rabinovitch, P. S. (1984). *Cytometry* **5,** 118–123.

Kubbies, M. (1990). *Cytometry* **11,** 386–394.

Kubbies, M. (1992). *J. Pathol.* **167,** 413–419.

Larsen, J. K., Munch-Peterson, B., Christiansen, J., and Jorgensen, J. (1986). *Cytometry* **7,** 54–63.

Marquardt, F. W. (1963). *J. Soc. Ind. Appl. Math.* **11,** 431–441.

Nativ, O., Winkler, H. Z., Raz, Y., Therneau, T. M., Farrow, G. M., Myers, R. P., Zincke, H., and Lieber, M. M. (1989). *Mayo Clin. Proc.* **64,** 911–919.

Nicoletti, I., Migliorati, G., Pagliacci, M. C., Grignani, F., and Riccardi, C. (1991). *J. Immunol. Methods* **139,** 271–279.

Price, J., and Herman, C. J. (1990). *Cytometry* **11,** 845.

Rabinovitch, P. S. (1988). "MultiCycle Program." Phoenix Flow Systems, San Diego, CA.

Rabinovitch, P. S. (1990). *Cytometry Suppl.* **4,** 27.

Rabinovitch, P. S. (1993a). *In* "Clinical Flow Cytometry: Principles and Applications" (K. D. Bauer, R. E.Duque, and T. V.Shankey, eds.), pp. 117–142. Williams & Wilkins, Baltimore, MD.

Rabinovitch, P. S. (1993b). "MultiCycle Program: Advanced Version." Phoenix Flow Systems, San Diego, CA.

Rabinovitch, P. S., and Kallioniemi, O.-P. (1991). *Cytometry Suppl.* **5,** 138.

Reid, B. J., Blount, P. L., Rubin, C. E., Levine, D. S., Haggitt, R. C., and Rabinovitch, P. S. (1992). *Gastroenterology* **102,** 1212–1219.

Roti Roto, J. L., Wright, W. D., Higashikubo, R., and Dethlefsen, L. A. (1985). *Cytometry* **6,** 101–108.

Shankey, T. V., Rabinovitch, P. S., Bagwell, C. B., Bauer, K. D., Duque, R. E., Hedley, D. W., Mayall, B. H., and Wheeless, L. (1993a). *Cytometry* **14,** 472–477.

Shankey, T. V., Dougherty, S., Manion, S., and Flanigan, R. C. (1993b). *Cytometry Suppl.* **6,** 83.

Stephenson, R. A., James, B. C., Gay, H., Fair, W. R., Whitmore, W. F., and Melamed, M. R. (1987). *Cancer Res.* **47,** 2504–2509.

Stokke, T., Holte, H., Erikstein, B., Davies, C. L., Funderud, S., and Steen, H. B. (1991). *Cytometry* **12,** 172–178.

Telford, W. G., King, L. E., and Fraker, P. J. (1991). *Cell Prolif.* **24,** 447–459.

Toikkanen, S., Joensuu, H., and Klemi, P. (1990). *Am. J. Clin. Pathol.* **93**(4), 471–479.

Vindelov, I. I., Christenson, I. J., and Nissen, N. I. (1983). *Cytometry* **3,** 328–331.

Wersto, R. P., Liblit, R. A., and Koss, L. G. (1991). *Hum. Pathol.* **22,** 1085–1098.

Wheeless, L. L., Coon, J. S., Cox, C., Deitch, A. D., deVere White, R. W., Koss, L. G., Melamed, M. R., O'Connell, M. J., Reeder, J. E., Weinstein, R. S., and Wersto, R. P. (1989). *Cytometry* **10,** 731–738.

Wheeless, L. L., Coon, J. S., Cox, C., Deitch, A. D., deVere White, R. W., Fradet, Y., Koss, L. G., Melamed, M. R., O'Connell, M. J., Reeder, J. E., Weinstein, R. S., and Wersto, R. P. (1991). *Cytometry* **12,** 405–412.

Wolley, R. C., Herz, F., and Koss, L. G. (1982). *Cytometry* **2,** 370–373.

CHAPTER 15

Simultaneous Analysis of Cellular RNA and DNA Content

Zbigniew Darzynkiewicz

Cancer Research Institute
New York Medical College
Valhalla, New York 10595

I. Update

Since this chapter was published no significant changes in the methodologies, either of the assays utilizing AO or the one of PY combined with Hoechst 33342, have been reported. However, a novel and quite different approach has been developed, which can be used to differentially stain RNA and DNA as an alternative to the methods presented in the chapter. The approach is based on incorporation of BrU to RNA in live cells followed by cell fixation and immunocytochemical detection of BrU incorporated into RNA (Larsen *et al.*, 2000, see chapter 34 in this volume). DNA is counterstained with the fluorochrome of another emission color than the one used to tag the BrU Ab. The advantage of this approach is that it provides high specificity in detection of RNA. The disadvantage stems from the need to incorporate the precursor; because turnover of rRNA is relatively slow long incubation with BrU is required to label most cellular RNA. Furthermore, the potential cytotoxic effects of the incorporated BrU, including sensitization of cells to ambient light, cannot be neglected. This method, however, when combined with the time-lapse analysis of the duration of BrU incorporation pulse, can be used to analyze the rate of RNA synthesis. One application of this methodology led to the detection of segregation of RNA and DNA during apoptosis followed by their separate packaging into individual apoptotic bodies (Halicka *et al.*, 2000).

Most applications of the RNA/DNA assays described in this chapter were focused on identification of quiescent G_0 cells, characterized by RNA content several times lower compared with the cycling G_1 cells (e.g., Tanaka *et al.*, 2007; Valentin and Yang, 2008). The AO-based methodology has also been adapted to measure RNA and DNA content in cells isolated from the primary bone tumors, in which RNA content has been found to correlate with the malignant potential of these tumors (Takeshita *et al.*, 2001). The AO-based methodology provided also the means to reveal that fragmentation of DNA during apoptosis is discontinuous, reflecting sequential cleavage of various fractions of DNA differentially protected from the nucleases by chromosomal proteins (Kajstura *et al.*, 2007).

The original chapter lists numerous models of flow cytometers and sorters that were then used in conjunction with the RNA/DNA methodology. There are no restrictions at all for the models that entered market since then to be successfully used for these methods. In fact, electronic compensation is now available for most new cytometers to compensate for the overlap of the emission spectra. The use of the compensation may be particularly helpful in the AO methodology to correct for the overlap of the RNA-associated red fluorescence into the "green" channel, detecting spectrum of DNA-associated fluorescence.

II. Introduction

Acridine orange (AO) and pyronin Y (PY) are the two most common fluorochromes of RNA. Used either alone (AO) or in combination with the DNA-specific fluorochrome Hoechst 33342 (PY), these dyes have found application primarily for simultaneous, correlated (bivariate) analysis of cellular RNA and DNA content. The mechanisms of interaction with nucleic acids, and in particular the spectral changes upon binding to DNA or RNA, are very much different for each of the dyes. The binding characteristics and the specific features of the respective methods, one based on the use of AO and another utilizing PY, therefore, are described separately in this chapter.

A. Acridine Orange

AO is a unique fluorochrome in many respects. It is one of the most versatile dyes and can be used to stain a variety of different constituents of the cell (Darzynkiewicz and Kapuscinski, 1990). Because of the multitude and complexity of its interactions with different substrates, basic knowledge on the mechanisms of these interactions is required to successfully apply this dye in any staining reaction. There are three major types of application of AO in flow cytometry:

1. Supravital Cell Staining

During exposure of live cells to a low concentration of AO ($<5\ \mu$M), due to a low pH (high proton concentration) within lysosomes, this dye is specifically entrapped in these organelles. This manifests by their red luminescence. Therefore, when applied to live cells, AO can be used to estimate the efficiency of the proton pump (pH) of lysosomes (Traganos and Darzynkiewicz, 1994).

2. Differential Staining of Double-Stranded (ds) Versus Single-Stranded (ss; Denatured) DNA

This reaction is performed after cell fixation, extraction of RNA by RNase, and induction of partial DNA denaturation by acid, heat, or other means (Darzynkiewicz, 1994).

3. Differential Staining of RNA and DNA

This application is described in the present chapter. Applications 2 and 3 are based on the quite unique propensity of AO to differentially stain ds versus ss nucleic acids (Darzynkiewicz and Kapuscinski, 1990; Kapuscinski and Darzynkiewicz, 1984; Kapuscinski et al., 1982). Namely, AO shows a considerable change in its absorption and emission spectra when bound to ds, compared to ss, nucleic acids. The shift in the absorption and/or emission spectrum when the dye binds to different substrates received the name metachromasia, and, thus, AO is a metachromatic fluorochrome. The metachromatic behavior of AO is the cause for its wide applicability as a probe of the content and conformation of nucleic acids in cytochemistry and biochemistry. AO binds to ds nucleic acids by intercalation, and fluoresces green in the intercalated form when excited in blue light. The maximum absorption of AO, when bound by intercalation to DNA, is at 500–506 nm, and emission is at 520–524 nm; this is classical fluorescence emission (S_1–S_0 transitions), with a short (5 ns) lifetime (Table I).

Table I
Properties of AO and PY and Their Complexes with Natural and Synthetic Nucleic Acids

Dye	Nucleic acid	Absorption		Emission		Binding	
		λ_{max} (nm)	$E_{max} \times 10^{-4}$ (M^{-1} cm^{-1})	λ_{max} (nm)	Q_R	n	$K_1 \times 10^4$ (M^{-1})
AO (m)	None	492[a]	6.85[a]	525[a]	1[a]	–	–
AO (d)	None	466[a]	4.57[a]	–	<0.01[a]	–	–
AO	Calf thymus DNA	502[a]	5.85[a]	522[a]	2.22[a]	4[a]	5[a]
AO	Poly(dA-dT)·poly(dA-dT)	504[a]	6.23[a]	522[a]	2.13[a]	4[a]	9.2[a]
AO	Poly(dG-dC)·poly(dG-dC)	503[a]	6.64[a]	522[a]	2.06[a]	4[a]	7.9[a]
AO	Poly(dA-dC)·poly(dG-dT)	500[a]	5.17[a]	520[a]	1.83[a]	4[a]	7.4[a]
AO	rRNA	–	–	638[b,c]	[d]	1[e]	–
AO	Poly(rA)	457[e]	4.76[e]	630[b,c]	[d]	1[e]	20.6[e,f]
AO	Poly(rC)	426[e]	1.62[e]	644[b,c]	[d]	1[e]	11.5[e,f]
AO	Poly(rU)	438[e]	2.28[e]	643[b,c]	[d]	1[e]	3.6[e,f]
PY	None	547[g]	10.13[g]	565[g]	1[g]	–	–
PY	Calf thymus DNA	559[g]	7.55[g]	569[g]	0.30[g]	4–6[g]	1.74[g]
PY	Poly(dA-dT)·poly(dA-dT)	559[g]	7.23[g]	573[g]	1.26[g]	–	–
PY	Poly(dG-dC)·poly(dG-dC)	560[g]	7.68[g]	569[g]	0.22[g]	–	–
PY	Poly(dA-dC)·poly(dG-dT)	559[g]	8.18[g]	574[g]	0.36[g]	–	–
PY	rRNA (16S + 23S)	560[g]	7.04[g]	573[g]	0.29[g]	4–5	6.96[g]
PY	Poly(rA)·poly(rU)	562[g]	8.95[g]	573[g]	1.02[g]	–	–
PY	Poly(rI)·poly(rC)	563[g]	7.05[g]	574[g]	0.95[g]	–	–
PY	Poly(rG-rC·poly(rG-rC)	560[g]	7.90[g]	565[g]	0.24[g]	–	–

Note: λ_{max}, maximum wavelength; E_{max}, molar extinction coefficient at maximum; Q_R, relative quantum yield, as compared to AO monomer (Kapuscinski and Darzynkiewicz, 1987b), or in the case of PY, relative to PY free in solution (Kapuscinski and Darzynkiewicz, 1987a); n, binding site size (nucleotides), for ds nucleic acids per mole of base pairs; K_1, association constant; m, AO monomer; d, AO dimer (Kapuscinski and Darzynkiewicz, 1987b).

[a]In 0.15 N NaCl, pH 7 (Kapuscinski and Darzynkiewicz, 1987b).
[b]Suspensions in 0.1 N NaCl, pH 7 (Kapuscinski *et al.*, 1983).
[c]Noncorrected for emission monochromator and photomultiplier response.
[d]Phosphorescence.
[e]In 10 mM phosphate buffer, pH 7; 25% (v/v) ethanol as a cosolvent (Kapuscinski *et al.*, 1983).
[f]Cooperative association constant (M^{-1} nucleotides).
[g]In 10 mM NaCl, pH 7; PY concentration 5–6 μM; D:P $\sim$ 0.05 (Kapuscinski and Darzynkiewicz, 1987a).

Interaction of AO with ss nucleic acids is a complex, multistep process, which is initiated by AO intercalation between the neighboring bases, neutralization of the polymer charge by this cationic dye, and subsequent condensation and agglomeration (precipitation; solute-solid state transition) of the product (Kapuscinski *et al.*, 1982). The condensation reaction is highly cooperative. In the final product, AO molecules are interspaced with bases, forming stacks of alternating dye-base composition, which by virtue of their solid-state form are protected, to a limited degree, from interaction with oxygen or water molecules. The absorption spectrum of AO in these precipitated products is blue shifted, compared to the intercalated AO, with the maximum range between 426 and 458 nm, depending on the base composition of the nucleic acid. The emission of AO in these complexes also varies between 630 and 644 nm, depending on the base composition. The lifetime of the red emission, at room temperature, is >20 ns. These spectral properties of AO in complexes with ss nucleic acids are suggestive of the intersystem crossing (phosphorescence; T_1–S_0 transition) rather than fluorescence. The solid-state nature of the complexes, as when freezing AO in solution (Zanker, 1952), may facilitate the intersystem crossing by partially eliminating the collision quenching, which otherwise occurs due to the long lifetime of the T_1-excited state in the presence of oxygen and solvents.

In the cell, large sections of rRNA and tRNA have ds conformation. Therefore, to obtain differential staining of DNA versus RNA with AO, these sections have to be selectively denatured, under conditions in which DNA still remains double stranded. Cell treatment with AO in the presence of the chelating agent EDTA results in such selective denaturation of RNA (Darzynkiewicz *et al.*, 1975). By breaking RNA-protein interactions in ribosomes which stabilize dsRNA, EDTA promotes denaturation of dsRNA, which occurs as a result of interaction with AO. The RNA-selective denaturing properties of AO result from the fact that this ligand has a higher affinity to ssRNA than to ssDNA (Table I). The denaturation itself results from the fact that at an increased dye:phosphate ratio (D:P), binding of AO to ss nucleic acids becomes thermodynamically preferable; the weaker but more numerous 1:1 (D:P) interactions of AO with ss sections dominate thermodynamically over the stronger but fewer (1:4) sites of AO intercalation to ds regions (Kapuscinski and Darzynkiewicz, 1989). Thus, increasing D:P promotes denaturation of ds RNA sections.

Cell staining with AO is generally done in salt solutions of relatively high ionic strength (0.1–0.2 M NaCl). Because of the significant electrostatic component in binding of AO to nucleic acids, resulting in competitive interactions between the AO^+ and Na^+, it is the concentration of free AO in solution (rather than the absolute D:P calculated based on molar ratios of the dye and nucleic acid in the sample) which is of importance for selective RNA denaturation (Darzynkiewicz and Kapuscinski, 1990).

To recapitulate, AO has two distinct functions in the mechanism of metachromatic staining of RNA. Namely, it (i) denatures ds RNA and (ii) differentially stains RNA (after its conversion to ss form) versus DNA. Selective RNA

The second cause of variability in RNA content is related to cell reproduction. Dividing cells double their constituents, including the number of ribosomes, during the cell cycle. Progression through the cell cycle is thus associated with an increase in cellular RNA content, and the increase occurs throughout interphase, at a relatively constant rate, proportional to the rate of cell proliferation (Darzynkiewicz *et al.*, 1979). Cellular RNA content, therefore, is a reflection of cell maturity in the cycle, and, for example, it allows discrimination of early and late G_1 cells (Darzynkiewicz *et al.*, 1980a,b). Because cell growth in size (number of ribosomes) and the rate of proliferation are generally coupled, the RNA parameter, therefore, is also an indirect marker of cell proliferation (Darzynkiewicz, 1988).

The cells withdrawn from the cell cycle (quiescent cells, G_0, G_{1Q}) have on the average a 5- to 10-fold fewer ribosomes than their cycling counterparts (Fan and Penman, 1970; Johnson *et al.*, 1976). Thus, the difference in RNA content allows one to identify noncycling cells and can be used as a marker of their mitogenic stimulation (Darzynkiewicz *et al.*, 1976). This is the most common application of the RNA methodologies so far (Darzynkiewicz, 1990).

Measurements of cellular RNA have several potential applications in clinical oncology. RNA content of tumor cells has been shown to be a prognostic marker in several malignancies (Darzynkiewicz, 1988). This is not surprising, given that tumor progression is a consequence of uncontrolled cell proliferation and/or defective differentiation; both, as mentioned, correlate with cellular RNA content. It is also common knowledge that all markers of nucleolar activity, that is, activity related to the synthesis of preribosomal RNA, have prognostic value in tumors. Because many antitumor drugs are cell-cycle specific, cellular RNA content may also be predictive of the tumor cell sensitivity to such drugs. Furthermore, unbalanced growth of cells treated with antitumor drugs can be estimated from RNA content measurements to evaluate cell response to the drug and as a factor predictive of cell death or recovery (Traganos *et al.*, 1982). Thus, the RNA assays may be useful in customizing therapy to achieve the maximal therapeutic effect with minimal toxicity to the patient and in monitoring the treatment.

IV. Materials

A. Stock Solution of AO

Dissolve AO in distilled water (dH_2O) to obtain 1 mg/ml concentration. It is essential to have AO of high purity. Invitrogen/Molecular Probes (Eugene, OR) offers AO of high purity (Cat. No. A 1301). The stock solution may be kept in the dark (in dark or foil-wrapped bottles) at 4 °C for several months without deterioration.

B. First Step: Solution A

Triton X-100, 0.1% (v/v)

HCl, 0.08 M (final concentration)

NaCl, 0.15 M (final concentration)

Solution A may be prepared by adding 0.1 ml of Triton X-100 (Sigma Chemical Co., St. Louis, MO), 8 ml of 1 M HCl, 0.877 g NaCl, and dH$_2$O to a final volume of 100 ml. This solution may be stored at 4 °C for several months.

C. Second Step: Solution B

Acridine orange, 6 μg/ml ($\sim$20 μM)

EDTA-Na, 1 mM

NaCl, 0.15 M

Phosphate-citric acid buffer, pH 6

Solution B may be made as follows:

1. Prepare 100 ml of buffer by mixing 37 ml of 0.1 M citric acid with 63 ml of 0.2 M Na$_2$HPO$_4$.
2. Add 0.877 g NaCl, stir until dissolved.
3. Add 34 mg of EDTA disodium salt. Equivalent amounts of tetrasodium EDTA may also be used. EDTA in acid form may be used, but it requires a longer time to dissolve. Stir until dissolved.
4. Add 0.6 ml of the stock solution of AO (1 mg/ml).

Solution B is also stable and can be stored for several months at 4 °C in dark. Solutions A and B may be kept in automatic dispensing pipette bottles set at 0.4 ml for solution A and 1.2 ml for solution B; the latter in a dark bottle.

D. Nuclear Isolation Solution

Prepare a solution containing 10 mM Tris buffer (pH 7.6), 1 mM sodium citrate, 2 mM MgCl$_2$, and 0.1% (v/v) nonionic detergent Nonidet (NP-40).

E. PY and Hoechst 33342 Staining Solution

In 100 ml of Hanks' buffered saline (HBSS) (containing Ca^{2+} and Mg^{2+}), dissolve 0.2 mg of Hoechst 33342 and 0.4 mg of high-purity PY. Hoechst 33342 is available from Molecular Probes. Relatively pure PY is offered by Polysciences, Inc. (Warrington, PA; Cat No. 18614). Most batches of PY available from other sources have 30–40% impurities and should be purified by extractions with chloroform and recrystallization from methanol (Kapuscinski and Darzynkiewicz, 1987a).

≡ V. Staining Procedures Employing AO

A. Staining of Unfixed Cells

1. Transfer a 0.2-ml aliquot of the original cell suspension (not more than 2×10^5 cells) to a small tube (e.g., 2- or 5-ml volume). Chill on ice.
2. Add gently 0.4 ml of ice-cold solution A. Wait 15 s, keeping on ice.
3. Add gently 1.2 ml of ice-cold solution B. Measure cell luminescence during the next 2–10 min (equilibrium time).

The sample should be kept on ice prior to and during the measurement. Vortexing or syringing cells when immersed in solution A, especially in the absence of any serum or proteins in the original cell suspension, results in disintegration of plasma membrane and isolation of cell nuclei. RNA content of isolated nuclei, therefore, can be measured in this way. Visual inspection of the nuclei under phase-contrast or UV-light microscopy is essential to estimate the efficiency of the isolation, which can be controlled by selecting optimal time and speed of vortexing, or number of syringings.

B. Staining of Fixed Cells

1. Fix cells in suspension in 70% ethanol, on ice.
2. Centrifuge cells, remove all ethanol, rinse once, and resuspend in HBSS, at a cell density $<2 \times 10^6$ per 1 ml.
3. Withdraw 0.2 ml of cell suspension and stain, using solutions A and B, as described already for fresh, unfixed cells. In the case of fixed cells, the presence of Triton X-100 in solution A is not necessary, although it does not interfere with their staining.
4. To assess the contribution of RNA to the detected luminescence, after removal of ethanol and suspension in HBSS, the cells can be incubated with RNase A (5 Kunitz units per 1 ml) for 30 min at 37 °C, prior to being stained with AO.

The advantage of staining fixed cells is the stability of the fluorescence intensity during the measurement, due perhaps to inactivation of the endogenous nucleases by ethanol. Also, due to the possibility of control treatments with exogenous RNase or DNase, the specificity of the staining reaction can be estimated more reliably. The disadvantage is, however, lower resolution of DNA measurements reflected by higher coefficient of variation (CV) of the mean green fluorescence of $G_{0/1}$ cells and increased cell aggregation.

C. Isolation and Staining of Unfixed Nuclei from Solid Tumors

1. The tissues may be kept frozen (below −40 °C) prior to nuclear isolation. Place the freshly resected (or frozen and then thawed) tissue in the nuclear isolation solution and trim to remove the necrotic, fatty, and other undesirable portions.

2. Transfer the trimmed tissue to a new aliquot of the isolation solution and mince finely with scalpel or scissors. Mix small tissue fragments by vortexing and/or pipetting with a Pasteur pipette or syringe with a large gauge needle. Observe the release of nuclei by sampling the suspension and viewing it under the phase-contrast microscope. If the release of nuclei is inadequate, transfer the minced tissue suspended in the isolation solution into homogenizer with a glass or Teflon pestle. Homogenize by pressing the pestle several times; check the efficiency of nuclear isolation. The nuclei should be clean, unbroken, and lacking cytoplasmic tags.

Collect the nuclear suspension from the above remaining tissue fragments (allow to sediment for ~1 min) with a Pasteur pipette and filter through 40- to 60-μm-pore nylon mesh. Dilute the suspension with the isolation buffer, if necessary, so that no more than 2×10^6 nuclei per 1 ml of the isolation solution remains in the final suspension. All isolation steps should be done on ice.

3. Withdraw a 0.2-ml aliquot of nuclei suspension and stain identically as described for whole cells; that is, mix with 0.4 ml of solution A and then with 1.2 ml of solution B.

4. Control samples can be incubated with RNase A. To this end, 1 ml of final nuclear suspension is treated with 50 units of RNase A for 20 min at 24 °C, and 0.2-ml aliquots of this suspension are then processed in exactly the same way as nuclei that were not subjected to treatment with RNase.

RNA leaks from the isolated, unfixed nuclei into the buffer. It is advisable, therefore, to stain the nuclei as soon as possible after isolating them. After being stained, RNA complexed with AO is less soluble.

VI. Staining RNA and DNA with PY and Hoechst 33342

1. Fix cells in suspension in 70% ethanol on ice.
2. Centrifuge cells, remove ethanol completely, rinse once with and resuspend in HBSS containing Ca^{2+} and Mg^{2+}, at a cell density $<2 \times 10^6$ per 1 ml.
3. Admix 0.5 ml of this suspension with 0.5 ml of a solution of HBSS containing 2 μg/ml of Hoechst 33342 and 4 μg/ml of PY, prepared as described in Section III.E.
4. Measure cell fluorescence after 20 min.

VII. Critical Aspects of the Procedures

A. Acridine Orange Procedure

1. Preservation of Intact Cells

To ensure that unfixed cells are permeabilized but do not disintegrate, the presence of serum (proteins) or serum albumin in the first step is required. To this end, prior to adding solution A, all cells must be suspended in a salt solution or tissue culture medium that contains 10–20% (v/v) serum or 1% (w/v) of bovine albumin. Thus, the cells can be taken directly from tissue culture without prior centrifugation, washing, etc., and stained, as described in the procedure. Vigorous shaking, pipetting, or vortexing of cell suspensions after addition of the detergent breaks the cells and results in a release of the nuclei.

2. Critical Concentration of AO

Differential staining of DNA versus RNA requires a proper concentration of free (unbound) AO in the final staining solution and at the time of the actual act of measurement under equilibrium ($\sim$20 mM). The following problem associated with this requirement may occur:

When the cell number (density) in the original suspension exceeds 2×10^6 cells per 1 ml (or even less when cells are highly hyperdiploid and/or have excessive RNA content), the amount of bound AO is high and therefore the free dye concentration may be significantly reduced (the "mass action" law). The RNA denaturation is then incomplete and part of the RNA can stain green.

Solution: Dilute the original cell suspension to have fewer cells in the sample.

3. Diffusion of AO from Sample During Flow

With some instruments (most cell sorters) in which cell measurements take place outside the nozzle (i.e., in air), a significant diffusion of dye from the sample to the sheath fluid takes place after the stream leaves the nozzle, prior to its intersection with the laser beam. This breaks the equilibrium and lowers the actual AO concentration in the sample at the time of cell measurement. Dye diffusion is also a problem in some instruments that have a narrow sample stream and long flow channel (e.g., Cytofluorograf 50 made by Ortho Diagnostics).

Solution: Increase the AO concentration (up to 20 μg/ml) in solution B and increase sample flow rate to compensate for the diffusion. Wherever possible, use channels with favorable geometry (wider sample stream and/or shorter distance between the nozzle and intersection with the laser beam). The optimal dye concentration for a particular instrument can be established by preparing a series of solution B with different AO concentrations (e.g., from 5 to 20 μg/ml) and testing at which concentration cells in the $G_{0/1}$ cell cluster have the same green fluorescence [the lowest CV of the green fluorescence mean value, corresponding to a lack of correlation between green and red luminescence; the $G_{0/1}$ cell cluster ought to be horizontal (or vertical if axes are reversed), but never skewed (diagonal)].

4. Overlap of Emission Spectra of AO: Sensitivity of RNA Detection

One of the limitations of the AO technique is the relatively low sensitivity of RNA detection. This is primarily due to the emission spectrum overlap: the green fluorescence of AO intercalated to DNA has a long "tail" toward a higher wavelength, which is significant to even as far as 620–670 nm. Therefore, RNA measurements in cells (or cell nuclei) characterized by a high DNA:RNA ratio lack sensitivity, being obscured by the high component due to AO bound to DNA.

Solution: The following points should be considered to improve the measurement:

Use long-pass filters transmitting above 640 or 650 nm, rather than 610 or 620 nm, to measure red luminescence. This significantly reduces the DNA-associated spectral component.

Due to significant differences in absorption spectra (Table I), it is impossible to achieve maximum excitation of AO bound to both DNA and RNA simultaneously, using a single laser. The strategy for optimal excitation is described in Section VIII.

5. Contamination of the Tubing by AO

As a cationic, strongly fluorescing dye, AO is absorbed in the surface of the sample flow tubing. Binding of a similar nature is also exhibited by rhodamine 123 and some other dyes. Release of such dyes interferes with measurement of the subsequent samples, especially if the cells have low fluorescence.

Solution: Rinse the sample flow line with bleach (e.g., 10% Clorox), then 50% ethanol, and then PBS, each 10 min. Alternatively, if possible, replace the sample flow tubing after using AO, and keep this tubing for use only with AO.

B. Pyronin Y–Hoechst 33342 Procedure

In many respects, the critical points of staining with PY are similar to those of the AO methodology. The most critical aspect of the Hoechst 33342-PY procedure is the stringent requirement for the appropriate concentration of PY. Too low a concentration of the dye (or too dense a cell suspension) cannot ensure stoichiometry of the staining because of the paucity of PY in the solution (dye:binding site ratio <1). Too high a concentration of PY triggers denaturation and condensation of RNA, which quenches its fluorescence. Paradoxically, thus, by increasing the PY concentration one can completely suppress RNA fluorescence and (in the absence of Hoechst 33342) induce PY intercalation to DNA; under these conditions PY can be used as a DNA-specific fluorochrome (Portela and Stockert, 1979).

Because of these denaturing properties of PY, the RNA stainability is very sensitive to its native conformation. Under appropriate conditions, the procedure allows one to discriminate between polyribosomal RNA (which is more resistant to denaturation and shows more extensive double strandedness) and rRNA in dispersed ribosomes (Traganos *et al.*, 1988). The staining procedure, thus, can be used to measure disaggregation of polyribosomes which occurs during mitosis or hyperthermia (Fan and Penman, 1970).

It should be stressed that PY taken up by live cells is partially localized the mitochondria and lysosomes (Darzynkiewicz *et al.*, 1987). The specificity of staining of RNA with PY in live cells, therefore, is uncertain.

VIII. Controls and Standards

A. Specificity of DNA and RNA Detection

Specificity of cell staining can be assayed by measuring parallel samples, prefixed in ethanol and then suspended in PBS containing Mg^{2+} and either RNase or DNase, as follows:

1. Dissolve 5 Kunitz units of RNase A (DNase-free RNase or RNase boiled for 5 min) in 1 ml of PBS, and incubate cells in this solution for 30 min at 37 °C.
2. Dissolve 0.5 mg of DNase I in 1 ml of PBS and incubate cells in this solution for 30 min at 37 °C.

Following these incubations, the cells should be stained with AO, passing through solutions A and B, as described earlier in the chapter, or with PY-Hoechst 33342. The percentage loss of red and green luminescence as a result of treatment with RNase or DNase (AO procedure) or red and blue (PY-Hoechst 33342) is an indication of the specificity of staining of RNA or DNA, respectively.

In the case of unfixed cells, cell suspensions already treated with solutions A and B (AO procedure) may be subsequently treated with 5 Kunitz units of RNase A and incubated for 30 min at 24 °C prior to fluorescence measurements.

B. RNA Content Standards

RNA and DNA content of the measured cells can be expressed quantitatively, by comparing them with standard cells. Nonstimulated, peripheral blood lymphocytes appear to offer the best standard. The RNase-treated and untreated lymphocytes should be measured to establish the extent of RNase-specific red luminescence. The cells to be compared have to be measured under conditions identical to those used for lymphocytes. Their RNA index should be expressed as a multiplicity (or fraction) of RNA content of lymphocytes.

Lymphocytes are not uniform, and there is a minor difference in RNA content between B and T cells. A more accurate standard would, therefore, be purified populations of B or T lymphocytes. If the measured cells have a several fold higher RNA content than lymphocytes (which is often the case), or if they are aneuploid, it is convenient to use lymphocytes as an internal standard.

Note that because the extent of the spectrum overlap (higher in the case of the AO methodology) depends on the excitation wavelength and the emission filters used, it can vary from instrument to instrument, depending on minor differences in the filter specifications. Therefore, the proportion of the RNase-specific luminescence to total red luminescence varies as well. This variation does not allow one to express RNA content as a ratio of the mean intensity of red luminescence of the measured cell population to that of the lymphocytes (without an adjustment for the RNase specificity), because such an index cannot be compared between different laboratories or flow cytometers.

The lymphocytes may also serve as a standard of DNA content, to estimate the DNA index from the mean (modal) intensity of the green fluorescence of the $G_{0/1}$ population.

C. Cell Staining on Slides

Observation of cells under UV-light microscopy is often required to reveal localization and confirm specificity of cell staining. The cells may be stained with AO on microscopic slides (cytospin preparations, smears, etc.) by rinsing fixed specimens with solution A and mounting under coverslips in a drop of solution B, in equilibrium with AO. Because of AO absorption by the glass surface, the dye concentration is lowered. This can be compensated for by higher initial concentrations of the dye in solution B (~20 μg/ml). Alternatively, the slides with cells should be rinsed several times with solution B, to saturate the binding sites with AO, prior to mounting.

IX. Instruments

A. Staining with AO

Maximum absorption of AO is at ~455–490 nm. The 488-nm line of the argon ion laser is the most commonly used excitation wavelength. In the instruments illuminated by a mercury or xenon lamp, blue excitation filters can be used (BG 12, bandpass combination filters transmitting light between 460 and 500 nm). Optimal excitation can be achieved using two lasers, one tuned to 488 nm (DNA detection) and another to 457 nm (RNA detection).

The emission filters and dichroic mirror setup should discriminate green fluorescence (measured at 530 + 15 nm) and red luminescence (measured preferably above 640 or 650 nm). Electronic compensation of the spectrum overlap, while it improves the "cosmetics" of the results, should rather be avoided. If the degree of the compensation varies between the samples, the method cannot be quantitative or properly standardized.

As discussed earlier, the geometry of the flow channel affects the staining reaction (the "diffusion problem"). The author has experience with the following instruments, which can be categorized into three groups:

1. FC-4800, FC-4801 (Biophysics); FC-200, FC201, and ICP-22 (all Ortho Instruments Co.); FACS analyzer, FACScan (Becton-Dickinson); Profile and Elite (Coulter Electronics); and PAS III (PARTEC). Very good differential

stainability of RNA-DNA with AO was observed using all of the listed instruments. They have no diffusion problem and do not require an increased AO concentration compared to this protocol. Mercury lamp illumination in the ICP-22, FACS analyzer, and PARTEC instruments makes them especially well designed for measurement of cells with low RNA content.

2. FACS II, FACS III, FACStar (Becton-Dickinson); EPICS IV and C (Coulter Electronics); and 30H Cytofluorograph (Ortho) require higher (8–20 $\mu g/ml$) concentrations of AO in solution B to compensate for the dye diffusion.

3. Due to extensive dye diffusion, it is difficult to obtain good stainability using the Ortho 50H cytofluorograph with the original sorting channel. Substitution of this channel with the nonsorting one (as in the 30H cytofluorograph) improves the staining.

B. Staining with PY-Hoechst 33342

Absorption of Hoechst 33342 is in the UV-light spectrum range and optimal excitation can be achieved with a 350/356-nm wavelength line of the krypton ion laser, 351/363 argon laser, or UV filters (UG1) in the mercury lamp illumination systems (Shapiro, 1981, 1988). PY can be excited with any wavelength between 488 and 530 nm, by several lines available in either krypton or argon ion lasers (Crissman *et al.*, 1985). Differential staining of DNA and RNA by Hoechst and PY, thus, requires dual excitation, provided either by two lasers or a laser and a mercury lamp.

X. Results

Stainability of DNA and RNA with AO of HL-60 leukemic cells undergoing myeloid differentiation in the presence of dimethyl sulfoxide (DMSO) is shown in Fig. 1. A significant decrease in RNA content accompanies cell differentiation and cell arrest in the G_1 compartment. The horizontal, nonslanted position of the contour representing G_1 cells is evidence of proper staining.

Changes in cellular DNA and RNA content during mitogenic stimulation of lymphocytes are shown in Fig. 2. Based on differences in RNA content, it is possible to distinguish nonstimulated, quiescent cells from cells entering the cell cycle. Their progression through the cell cycle is paralleled by a further increase in RNA content. Correlated measurements of RNA and DNA offer a sensitive assay of lymphocyte stimulation, providing information regarding both the initial steps of stimulation (exit from G_0) and cell-cycle progression. Since stimulation of lymphocytes is a multistep process that does not always result in cell proliferation, the traditional assays based on radioactive thymidine incorporation, in contrast to the present technique, cannot detect the early steps of the stimulation process and thus are useless in such situations.

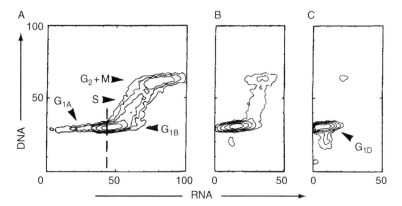

Fig. 1 The isometric contour maps representing the distribution of HL-60 cells with respect to their RNA and DNA contents after staining with AO. (A) Exponentially growing cells, (B) cells growing in the presence of 1.5% (v/v) DMSO for 3 days, and (C) cells cultured in the presence of DMSO for 5 days. In the exponentially growing population, two G_1 compartments, G_{1A} and G_{1B}, can be distinguished: Cells enter S phase from G_{1B}, whereas cells in G_{1A} are early G_1 cells. A decrease in cell proliferation observed during cell differentiation is accompanied by a decrease in RNA content. Cells in G_1 that exhibit the differentiated phenotype were denoted as G_{1D} (Darzynkiewicz et al., 1980a).

Stainability of RNA and DNA with PY and Hoechst 33342 is shown in Fig. 3. The staining reaction is very sensitive to the conformation of RNA in the cell. Therefore, as is evident, cells in mitosis (M) have lower stainability with PY than G_2 cells, despite the fact that the RNA content of M cells is relatively higher than in most G_2 cells. The hypochromicity of RNA in M cells with PY is a consequence of the denaturing properties of this dye: RNA of the polyribosomes (which are more numerous in interphase than in metaphase) is more resistant to denaturation and thus it stains more intensely compared to RNA in mitotic cells. During mitosis, polyribosomes disaggregate and RNA of individual ribosomes is more extensively denatured by PY, which leads to quenching of this dye's fluorescence (Traganos et al., 1988).

Other examples of the results and descriptions of other applications of the RNA-DNA staining techniques are presented in several reviews (Darzynkiewicz, 1988; Darzynkiewicz and Kapuscinski, 1990; Traganos, 1990).

XI. Comparison of the Methods

Different advantages and limitations characterize each of the methods and each should be considered when making the choice.

A. RNA Quantitation

Stoichiometry of RNA measurement is better assured by the AO methodology (Bauer and Dethlefsen, 1980). This is due to the fact that the total RNA content is stained by AO, and there is less variation in quantum yield resulting from differences in base composition of conformation of RNAs than in the case of PY.

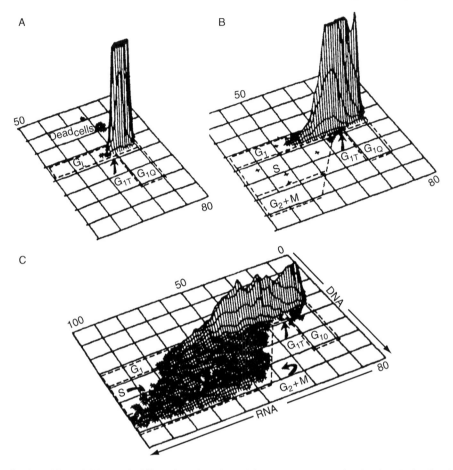

Fig. 2 Differential AO stainability of DNA and RNA in lymphocytes, unstimulated and stimulated to proliferation. Bivariate distribution (RNA versus DNA) of human peripheral blood lymphocytes, unstimulated (A) and incubated with the mitogenic agent, phytohemagglutinin for 24 h (B) and 72 h (C). Stimulated lymphocytes are distinguished from quiescent cells (G_{1Q}) by the increased RNA content, early (24 h) as cells in transition to the cycle (G_{1T}), and later (72 h) as cells progressing through the cycle, in G_1, S, and $G_2 + M$.

Resolution of the DNA measurements by AO, thus, in cells containing excessive amounts of glycosaminoglycans (primary fibroblasts, mast cells, keratinocytes) is low.

B. Analysis of RNA Conformation

Both AO and PY can be used to reveal changes in conformation of RNA. The HO342-PY technique detects changes associated with association of polyribosomes (Traganos *et al.*, 1988), whereas the AO methodology can be

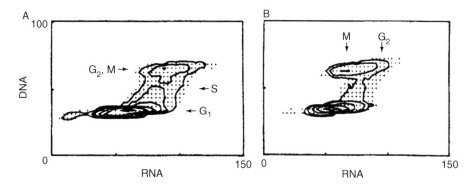

Fig. 3 Stainability of DNA and RNA with HO342-PY in CHO cells at different phases of the cell cycle. (A) Exponentially growing CHO cells and (B) cells from the colcemid-treated culture, containing approximately 20% cells in mitosis. Decreased PY stainability of mitotic cells (M) compared to G_2 is due to the fact that PY stains preferentially polyribosomal RNA and during mitosis polyribosomes dissociate (Traganos *et al.*, 1988).

applied to measure the degree of double strandedness of RNA and its sensitivity to heat denaturation (melting profile) (Darzynkiewicz *et al.*, 1975).

C. Instrument Requirements

The possibility of excitation with a single wavelength, or by use of a mercury lamp as the only source of illumination, gives the AO methodology the advantage that it can be used with simpler and less costly instruments.

Acknowledgments

This work was supported by NCI Grant ROI CA28704, the "This Close" Foundation, the Carl Inserra Fund, and the Chemotherapy Foundation. I thank Drs Frank Traganos and Jan Kapuscinski for their comments and suggestions.

References

Bauer, K. D., and Dethlefsen, L. A. (1980). Total cellular RNA content: correlation between flow cytometry and ultraviolet spectroscopy. *J. Histochem. Cytochem.* **28,** 493–498.

Brachet, J. (1940). La detection histochimique des acides pentose-nucleiques. *C. R. Seances Soc. Biol. Fil.* **133,** 88–90.

Crissman, H. A., Darzynkiewicz, Z., Tobey, R. A., and Steinkamp, J. (1985). Correlated measurements of DNA, RNA, and protein in individual cells by flow cytometry. *Science* **228,** 1321–1324.

Darzynkiewicz, Z. (1988). Cellular RNA content, a feature correlated with cell kinetics and tumor prognosis. *Leukemia* **2,** 777–787.

Darzynkiewicz, Z. (1990). Multiparameter flow cytometry in studies of the cell cycle. In "Flow Cytometry and Sorting" (M. R. Melamed, T. Lindmo, and M. L. Mendelsohn, eds.), pp. 469–501. Wiley-Liss, New York.

Darzynkiewicz, Z. (1994). Acid-induced denaturation of DNA *in situ* as a probe of chromatin structure. *Methods Cell Biol.* **41**, 527–541.

Darzynkiewicz, Z., and Kapuscinski, J. (1990). Acridine orange: A versatile probe of nucleic acids and other cell constituents. In "Flow Cytometry and Sorting" (M. R. Melamed, T. Lindmo, and M. L. Mendelsohn, eds.), pp. 291–314. Wiley-Liss, New York.

Darzynkiewicz, Z., Traganos, F., Sharpless, T., and Melamed, M. R. (1975). Conformation of RNA *in situ* as studied by acridine orange staining and automated cytofluorometry. *Exp. Cell Res.* **95**, 143–153.

Darzynkiewicz, Z., Traganos, F., Sharpless, T., and Melamed, M. R. (1976). Lymphocyte stimulation: A rapid multiparameter analysis. *Proc. Natl. Acad. Sci. USA.* **73**, 2881–2884.

Darzynkiewicz, Z., Evenson, D. P., Staiano-Coico, L., Sharpless, T., and Melamed, M. R. (1979). Correlation between cell cycle duration and RNA content. *J. Cell Physiol.* **100**, 425–438.

Darzynkiewicz, Z., Sharpless, T., Staiano-Coico, L., and Melamed, M. R. (1980a). Subcompartments of the G_1 phase of cell cycle detected by flow cytometry. *Proc. Natl. Acad. Sci. USA* **77**, 6696–6700.

Darzynkiewicz, Z., Traganos, F., and Melamed, M. R. (1980b). New cell cycle compartments identified by multiparameter flow cytometry. *Cytometry* **1**, 98–108.

Darzynkiewicz, Z., Kapuscinski, J., Traganos, F., and Crissman, H. A. (1987). Application of pyronin Y (G) in cytochemistry of nucleic acids. *Cytometry* **8**, 138–145.

Fan, H., and Penman, S. (1970). Regulation of protein synthesis in mammalian cells. II. Inhibition of protein synthesis at the level of initiation during mitosis. *J. Mol. Biol.* **50**, 655–670.

Halicka, H. D., Bedner, E., and Darzynkiewicz, Z. (2000). Segregation of RNA and separate packaging of DNA and RNA in apoptotic bodies during apoptosis. *Exp. Cell Res..* **260**, 248–255.

Hannon, G. J., Rivas, F. V., Murchison, E. P., and Steitz, J. A. (2006). The expanding universe of noncoding RNAs. *Cold Spring Harb. Quant. Biol.* **71**, 551–564.

Johnson, L. F., Levis, R., Abelson, H. T., Green, H., and Penman, S. (1976). Changes in RNA in relation to growth of the fibroblast. IV. Alterations in the production and processing of mRNA and rRNA in resting and growing cells. *J. Cell Biol.* 933–938.

Kajstura, M., Halicka, H. D., Pryjma, J., and Darzynkiewicz, Z. (2007). Discontinuous fragmentation of nuclear DNA during apoptosis revealed by discrete "sub-G_1" peaks on DNA content histograms. *Cytometry A* **71A,** 125–131.

Kapuscinski, J., and Darzynkiewicz, Z. (1984). Denaturation of nucleic acids induced by intercalating agents. Biochemical and biophysical properties of acridine orange-DNA complexes. *J. Biomol. Struct. Dyn.* **1**, 1485–1499.

Kapuscinski, J., and Darzynkiewicz, Z. (1987a). Interactions of pyronin Y (G) with nucleic acids. *Cytometry.* **8**, 129–137.

Kapuscinski, J., and Darzynkiewicz, Z. (1987b). Interactions of acridine orange with double stranded nucleic acids. Spectral and affinity studies. *J. Biomol. Struct. Dyn.* **5**, 127–143.

Kapuscinski, J., and Darzynkiewicz, Z. (1989). Structure destabilization and condensation of nucleic acids by intercalators. In "Biological Structure, Dynamics. Interactions and Stereodynamics" (R. H. Sarma and M. H. Sarma, eds.), pp. 267–281. Adenine Press. Schenectady, NY.

Kapuscinski, J., Darzynkiewicz, Z., and Melamed, M. R. (1982).Luminescence of the solid complexes of acridine orange with RNA. *Cytometry* **2**, 201–212.

Kapuscinski, J., Darzynkiewicz, Z., and Melamed, M. R. (1983). Interactions of acridine orange with nucleic acids. Properties of complexes of acridine orange with single stranded RNA. *Biochem. Pharmacol.* **32**, 3679–3694.

Larsen, J. K., Jensen, P., and Larsen, J. (2000). Flow cytometric analysis of RNA synthesis by detection of bromouridine incorporation. In "Current Protocols in Cytometry" (J. P. Robinson, Z. Darzynkiewicz, P. Dean, A. Orfao, P. Rabinovitch, and H. Tanke, eds.), pp. 7.12.1–7.12.11, Wiley, New York.

Pollack, A., Prudhomme, D. L., Greenstein, D. B., Irvin, G. L., III, Claflin, A. J., and Block, N. L. (1982). Flow cytometric analysis of RNA content in different cell populations using pyronin Y and methyl green. *Cytometry* **3**, 28–35.

Portela, R. A., and Stockert, J. C. (1979). Chromatin fluorescence by pyronin staining. *Experientia.* **35,** 1663–1665.

Shapiro, H. M. (1981). Flow cytometric estimation of DNA and RNA content in intact cells stained with Hoechst 33342 and pyronin Y. *Cytometry.* **2,** 143–150.

Shapiro, H. M. (1988). Practical Flow Cytometry, 2nd edn., Liss, New York.

Tanke, H. J., Niewenhuis, I. A. B., Koper, G. J. M., Slats, J. C. M., and Ploem, J. S. (1980). Flow cytometry of human reticulocytes based on RNA fluorescence. *Cytometry.* **1,** 313–320.

Takeshita, H., Kusuzaki, K., Kuzuhara, A., Tsui, Y., Ashihara, T., Gebhard, M. C., Mankin, H. J., Springfield, D. S., and Hirasawa, Y. (2001). Relationship between histologic grade and cytofluorometric cellular DNA and RNA content in primary bone tumors. *Anticancer Res.* **21,** 1271–1277.

Tanaka, T., Kajstura, M., Halicka, H. D., Traganos, F., and Darzynkiewicz, Z. (2007). Constitutive histone H2AX phosphorylation and ATM activation are strongly amplified during mitogenic stimulation of lymphocytes. *Cell Prolif.* **40,** 1–13.

Traganos, F. (1990). Single and multiparameter analysis of the effects of chemotherapeutic agents on cell proliferation *in vitro*. In "Flow Cytometry and Sorting" (M. R. Melamed, T. Lindmo, and M. L. Mendelsohn, eds.), pp. 773–801. Wiley-Liss, New York.

Traganos, F., and Darzynkiewicz, Z. (1994). Lysosomal proton pump activity: Supravital cell staining with acridine orange differentiates leukocyte subpopulations. *Methods Cell Biol.* **41,** 183–194.

Traganos, F., Darzynkiewicz, Z., Sharpless, T., and Melamed, M. R. (1977). Simultaneous staining of ribonucleic and deoxyribonucleic acids in unfixed cells using acridine orange in a flow cytofluorometric system. *J. Histochem. Cytochem.* **25,** 46–56.

Traganos, F., Darzynkiewicz, Z., and Melamed, M. R. (1982). The ratio of RNA to total nucleic acid content as a quantitative measure of unbalanced cell growth. *Cytometry.* **2,** 212–218.

Traganos, F., Crissman, H. A., and Darzynkiewicz, Z. (1988). Staining with pyronin Y detects changes in conformation of RNA during mitosis and hyperthermia of CHO cells. *Exp. Cell Res.* **179,** 535–544.

Valentin, M., and Yang, E. (2008). Autophagy is activated, but is not required for the G0 function of BCL-2 or BCL-xL. *Cell Cycle* **7,** 2762–2768.

Zanker, V. (1952). Quantitative absorptions- und emissionsmessungen am acridinorangekation bei normal-und tieftemperaturen im organishen lösungsmittel und ihr beitrag zur deutung des metachromatishen fluoreszenzproblemes. *Z. Phys. Chem.* **200,** 250–292.

PART III

Cell proliferation and the Death Assays

CHAPTER 16

Immunochemical Quantitation of Bromodeoxyuridine: Application to Cell–Cycle Kinetics

Frank Dolbeare★ and Jules R. Selden[†]

★Biology and Biotechnology Program
Lawrence Livermore National Laboratory
Livermore, California 94550

[†]Department of Safety Assessment
Merck Research Laboratories
West Point, Pennsylvania 19486

I. Update

The methodology described in this chapter, originally developed by Howard Gratzner and Frank Dolbeare over 25 years ago, has been widely used since then, and is still the most commonly applied approach to detect DNA replication by cytometry. Since its publication, the protocols presented in this chapter have not been significantly changed. Attempts have been made, however, to circumvent the step of partial DNA denaturation needed to make the incorporated BrdU accessible to antibody. This step requires hash conditions of cell treatment with acid or heat that often are incompatible with the concurrent immunocytochemical detection of intracellular proteins or cell-surface antigens. One of such attempts was incorporation of BrdU into DNA followed by the induction of DNA photolysis by cells exposure to UV light (SBIP; DNA Strand-Break Induction by Photolysis). The DNA strand breaks generated during photolysis were then labeled in the TUNEL reaction, utilizing BrdUTP end exogenous terminal transferase (Darzynkiewicz *et al.*, 2006). This approach has been used in many studies that required immunocytochemical detection of intracellular antigens concurrently with analysis of DNA replication (e.g., Juan *et al.*, 1997). The ABSOLUTE-T™ SBIP commercial kit based on this methodology is available from several vendors including Phoenix Flow Systems (San Diego, CA), or Molecular Probes/Invitrogen (Eugene, OR).

Salic and Mitchison (2008) have described a method to detect DNA synthesis in proliferating cells, based on the incorporation of 5-ethynyl-2′-deoxyuridine (EdU) and its subsequent detection by a fluorochrome (Alexa 568, Alexa 488, or Alexa 594) tagged azide through a Cu(I)-catalyzed [3 + 2] cycloaddition reaction ("click" chemistry). The method offers similar sensitivity to the classical BrdU incorporation protocol as described by Dolbeare and Selden but is more rapid and does not require DNA denaturation. The EdU incorporation "click" approach has been successfully used in for *in vitro* and *in vivo* cell labeling (Salic and Mitchison, 2008). A variant of the click methodology was presented by Cappella *et al.* (2008) in which the incorporated EdU was derivatized by a copper-catalyzed cycloaddition using a BrdU-tagged azide probe. The DNA-bound bromouracil was then immunocytochemically detected using commercial antibodies, also with no need for DNA denaturation step. Although prolonged cells exposure to EdU is cytotoxic to certain cell types, with some restrictions the methodology can be used for cell-cycle kinetic studies (Diermeier-Daucher *et al.*, 2009). Considering the advantages offered by the "click" approach one may predict that it will become widely used for the detection of DNA replication and cell-cycle kinetic studies in the future. Many authors, however, are still using the "classical" BrdU incorporation assay as presented in this chapter.

II. Introduction

The immunological BrdUrd techniques developed for flow cytometry and the microscope have almost replaced autoradiographic techniques with tritiated thymidine in cell-kinetic studies. The immunochemical evaluations of BrdUrd are being used largely in clinical situations to measure labeling index (LI), S phase duration (Ts), potential doubling time (Tpot), growth fraction, and drug resistance (Danova *et al.*, 1990; Duprez *et al.*, 1990; Garin *et al.*, 1991; Giaretti, 1991; Hardonk and Harms, 1990; Ito *et al.*, 1990; Raza and Preisler, 1990; Raza *et al.*, 1991a, 1992; Riccardi *et al.*, 1988; San-Galli *et al.*, 1991; Suzuki, 1988; Waldman *et al.*, 1988; Ward *et al.*, 1991; Yu *et al.*, 1992). BrdUrd can be used in the same way as [^{3}H]TdR to provide detailed cell kinetics measurements. Continuous exposure to BrdUrd provides an estimate of the fraction of noncycling cells, and pulse-chase experiments provide quantitative measures of cell progression. For use with human material, where only limited sampling is practical, the pulse-chase approach has been refined to a single time point method (Begg *et al.*, 1985). The introduction of halogen-selective antibodies (e.g., anti-BrdUrd selective and anti-IdUrd selective) (Dolbeare *et al.*, 1988; Raza *et al.*, 1991b; Shibui *et al.*, 1989) allows double-labeling experiments like those using ^{3}H- and ^{14}C-labeled nucleotides.

The incorporation of BrdUrd into cells offers a more accurate estimation of the fraction of cells in S phase than does DNA measurement alone, and the optimal method is the simultaneous measurement of both DNA content and incorporated BrdUrd in each cell (Dolbeare *et al.*, 1983). Fluorescein-conjugated anti-BrdUrd antibodies provide a green fluorescent signal that is proportional to incorporated BrdUrd when plotted on a logarithmic *y*-axis. Propidium iodide (PI) intercalation provides a red fluorescent signal that is proportional to DNA content when plotted on a linear *x*-axis. G_1 and G_2 cells are low in green fluorescence but are well resolved on the basis of red fluorescence. S phase cells have intermediate DNA contents, but are readily distinguished by their intense green fluorescence.

III. Applications

The primary applications of the technique are the quantification of cell-cycle phase fractions, phase durations, doubling time, labeling index, and growth fractions. Specific effects of agents which stimulate cell proliferation, block specific phases of the cell cycle, or slow the progression through the cell cycle can be quantified. The combination of one or more nucleoside analogs with varying amounts of time between cell exposure and sample collection can be used to determine the total cell-cycle time and the duration of the various cell-cycle compartments.

The quantitative equations for determining phase durations, for example, time duration of G_1 (Tg1), Ts, and cell-cycle time (Tc), were derived by Takahashi (1966). Although the mathematics provided by this model is beyond the scope of this chapter, a simplified mathematical equation can describe the movement of cells from compartment to compartment (Gray et al., 1990). For example, the rate of change in the number of cells in compartment i at time t, $N_i(t)$ might be described as

$$\frac{\mathrm{d}N_i(t)}{\mathrm{d}t} = l_{i-1}N_{i-1}(t) - l_i N_i(t),$$

where l_i defines the rate at which cells leave compartment $i + 1$. For compartment 1, the equation becomes

$$\frac{\mathrm{d}N_1(t)}{\mathrm{d}t} = 2l_k N_k(t) - l_1 N_1(t),$$

where k is the number of the last compartment in the cycle and the factor 2 takes account of the fact that the cell number doubles as cells move from G_2M to G_1 phase.

Nonkinetic applications of BrdUrd or IdUrd incorporation include (1) testing of drug resistance or sensitivity (Lacombe et al., 1992; Waldman et al., 1988), (2) analogs incorporated into DNA during the repair of DNA damage (unscheduled DNA synthesis) (Beisker and Hittleman, 1988; Selden et al., 1993), (3) the incorporation of BrdUrd into DNA for demonstrating sister chromatid exchanges (SCEs) by fluorescent antibody labeling (Pinkel et al., 1985), (4) Replication patterns in chromatin and nuclei during DNA synthesis (Allison et al., 1985; Nakamura et al., 1986), (5) the isolation of nascently replicated DNA using immunoaffinity columns directed against incorporated BrdUrd (Leadon, 1986), (6) BrdUrd-labeled probes for in situ hybridization (Frommer et al., 1988), and (7) analysis of proliferation during differentiation (Gratzner et al., 1985; Kaufmann and Robert-Nicoud, 1985).

IV. Materials

A. General Purpose Reagents

1. Bromodeoxyuridine or iododeoxyuridine: Generally, anti-BrdUrd antibodies are more efficient at stoichiometric binding of BrdUrd when the level of BrdUrd substitution is at about 25% of the total thymidines. At substitutions higher than this amount, steric factors limit binding, while below about 1% substitution the

possibilities of multivalent binding and the absolute signal strength are limited. In the most rigorous experiments for *in vitro* labeling of cultured cells, BrdUrd is diluted in culture media along with TdR at a defined molar ratio, and 5-fluorouracil (5-FU) is included to block endogenous thymidine synthesis. This mixture of BrdUrd + dThd = 10 μM, 5 μM 5-FU, and 5 μM dCytd ensures incorporation at a specified substitution level. Less rigorous incorporation protocols use either BrdUrd or IdUrd between 0.1 and 10 mM, assuring that at this level the analog is present in excess of intracellular dThd pools.

2. Antihalopyrimidine antibody: We describe here procedures using monoclonal antibodies against either BrdUrd or IdUrd. Animal antisera derived against one of the halopyrimidines may also be used, but historically these have lacked the needed selectivity. Table I is a list of commercially available antibromodeoxyuridine antibodies. The final sensitivity of the measurement depends on the purity, affinity, and specificity of the particular antibody. High-affinity antibodies permit quantification of low levels of BrdUrd incorporation (e.g., <0.1% substitution) (Beisker *et al.*, 1987, Vanderlaan *et al.*, 1986). Antibodies directly conjugated to fluorophores may be used, or the anti-BrdUrd antibodies may be detected using indirect immunochemical methods. For methods involving the simultaneous detection of cell-surface antigens and BrdUrd, direct conjugates are required. The monoclonal antibody pair of IU-4 and Br-3 (Caltag, South San Francisco, CA) allows double-labeling with IdUrd and BrdUrd, since Br-3 shows halogen selectivity in binding, preferring BrdUrd to IdUrd (Dolbeare *et al.*, 1988; Shibui *et al.*, 1989).

3. Phosphate-buffered saline (PBS); 0.05 M sodium phosphate, pH 7.2 (no Ca^{2+} or Mg^{2+}), and 0.15 M NaCl.

4. Antibody diluting buffer: PBS as described above containing 2× salt sodium citrate (SSC) (SSC = 0.15 M NaCl + 0.015 M sodium citrate), 0.5% Tween 20, (Sigma Chemical Company, St. Louis, MO), and a blocking protein to limit nonspecific sticking of the antibody. The blocking protein may be 1% bovine serum albumin, 1% gelatin, or 2% dry nonfat milk protein.

5. Ribonuclease A (RNase) stock solution: Ribonuclease A (Sigma Chemical Company) at 0.5 mg/ml in PBS. This solution can be stored refrigerated with 0.1 mg/ml sodium azide for up to 6 months.

6. Paraformaldehyde solution (E.M. grade, Polyscience, Inc., Warrington, PA). Stock solution is 0.25 or 1% paraformaldehyde in PBS, pH 7.2.

7. 0.1 M HCl plus 0.5% Triton X-100: Five grams of Triton X-100 (Sigma Chemical Company) in 1 l of 0.1 M HCl.

8. Wash buffer: Five grams of Tween 20 (Sigma Chemical Company) in 1 l of PBS.

9. Goat antimouse IgG-fluorescein conjugate: This antibody may be obtained from a number of sources (see "Linscott's Directory of Immunological and Biological Reagents," 7th Ed.). Depending on the particular needs of the experiment (e.g., membrane antigen labeling in addition to BrdUrd/DNA analysis) one may require an alternative fluorophore as a conjugate, such as the blue emitting

Table I
Commercial Sources of Anti–BrdUrd Antibodies[a]

Clone	Ig type	State	Source
>1 clone	IgG1	Purified	Bioclone, Australia
3D9	IgG1	Purified	Oncogen Science, Inc.
3D9	IgG1	Purified-FITC	Oncogene Science, Inc.
76-7	IgG1	Purified	Amac, Inc.
			Biodesign, Invc
			Serotec, Ltd.
			Sigma Chemical
B44	IgG1	Purified	Becton-Dickinson
B44	IgG1	Purified-FITC	Becton-Dickinson
BMC9318	IgG1	Purified	Boehringer-Mannheim
		Purified-FITC	Boehringer-Mannheim
Br3	IgG1	Purified	Caltag
Br3	IgG1	Purified biotin	Caltag
Br3		Purified-FITC	Caltag
BU5.1	IgG2a	Purified purified-FITC	Cymbus Bioscience
			IBL Research Products
			Paesel&Lorei GmbH
			Progen Biotechnik GmbH
BU1-75	IgG Rat	Ascites	Sera-Labs, Ltd.
		Purified	Accurate Chemical
		Supernatant	
Bu6-4	IgG1	Purified	Pierce Chemical
Bu20a	IgG1	Supernatant	Dako
BU-33	IgG1	Purified	Sigma
MBU	IgG1	Ascites	Medscand, USA
IU4	IgG1	Purified	Caltag
SB18	IgG1	Ascites	Accurate Chemical
			Medica, Inc.
			Sanbio BV
ZBU30	IgG1	Purified	Zymed Labs
		Purified Alkphos	Zymed Labs
		Purified biotin	Zymed Labs
		Purified-FITC	Zymed Labs
		Purified rox	Zymed Labs
		Purified phyco	Zymed Labs
Anti-BrdUrd	IgG1	Acites	Chemicon
Anti-BrdUrd	??		Janssen Biochimica
			Accurate Chemical

[a]Most of the sources are listed in Linscott's Directory of Immunological and Biological Reagents, Eighth Edition, obtainable from Linscott's Directory, Santa Rosa, CA 95404. Complete addresses of all of the above sources and their international distributors are given in the directory.

fluorophore, aminomethylcoumarin acetic acid (AMCA), or a red emitting fluorophores Texas red, Princeton red, or phycoerythrin (Molecular Probes, Eugene, Oregon).

10. PI (Sigma Chemical Company): The working solution is 10 μg/ml in PBS, pH 7.2. A stock solution of 1 mg/ml PI in 70% ethanol stored in the refrigerator is stable for at least a year. PI is a suspected carcinogen and should be handled with proper caution.

11. Exonuclease III and *Eco*RI (Bethesda Research Laboratories, Gaithersberg, MD).

12. 0.1 M Citric acid/0.5% Triton X-100.

13. *Eco*R I buffer (see recommendations of manufacturer for specific endonucleases): 0.1 M Tris-HCl, pH 7.5, containing 50 mM NaCl and 10 mM $MgCl_2$.

14. Exonuclease III buffer: 50 mM Tris-HCl, pH 8, 10 mM 2-mercaptoethanol, 5 mM $MgCl_2$.

15. 40-μm nylon mesh (Small Parts, Inc., Miami Lakes, FL).

B. Reagents for Labeling Surface Antigens Plus BrdUrd

This method follows that of Carayon and Bord (1992).

1. A biotin-conjugated antibody to cell-surface antigen, such as biotin-coupled anti-CD4.

2. Streptavidin-phycoerythrin (Becton-Dickinson, San Jose, CA).

3. 1% paraformaldehyde with 0.01% Tween 20.

4. DNase I (Sigma Chemical Company): 50 Kunitz unit/ml in PBS containing Ca^{2+} and Mg^{2+}.

C. Reagents for Staining Tissue Sections

While there are many suitable techniques for immunohistochemical staining of paraffin-embedded histologic sections, we have chosen the method of Shibuya *et al.* (1992) as illustrative of the methods. This study describes methods for dual labeling of both incorporated BrdUrd and IdUrd, using the halogen-selective monoclonal antibody pair Br-3 and IU-4 and a combination of alkaline phosphatase enzyme immunohistochemistry and immunogold staining. Simpler methods can be developed by the reader, using parts of this comprehensive method.

1. 6 μM histologic sections mounted on glass slides from tissues exposed to BrdUrd and/or IdUrd, fixed in 70% ethanol, paraffin embedded, and sectioned.

2. 4 N HCl.

3. 2.5% glutaraldehyde.

4. Gold-conjugated goat antimouse IgG and a silver enhancement kit (IntenSE M, Amersham Corp., Arlington Heights, IL).

5. 5% Acetic acid in water.

6. Tris-buffered saline (50 mM Tris, pH 7.6, and 0.15 M NaCl).

7. Alkaline phosphate-conjugated rabbit antimouse immunoglobulins (Dako Corp., Santa Barbara, CA).

8. Alkaline phosphatase substrate kit I (Vector Immunochemicals, Burlingame, CA).

9. 5% Gills No. 1 hematoxylin (Sigma).

10. HistoClear (National Diagnostics, Manville, NJ).

11. Permount (Fisher Chemical Company, Fair Lawn, NJ).

V. Procedures

A. Immunochemical Labeling Followed by Flow Cytometric Detection

1. Thermal Denaturation Method (Beisker *et al.*, 1987; Dolbeare *et al.*, 1990)

a. Use $1–5 \times 10^6$ cells previously labeled with BrdUrd or IdUrd and fixed in cold 50% ethanol or methanol/acetic acid (3/1, v/v). Centrifuge at $500 \times g$ for 2 min. (*Note*: avoid over centrifugation which can lead to serious cell clumping.) Pour off supernatant. Suspend cells by gentle vortexing.

b. Add 1.5 ml of RNase stock solution and incubate for 10 min at 37 °C.

c. Centrifuge at $500 \times g$ for 2 min, pour off supernatant, vortex to loosen pellet, and suspend cells in 2 ml of 0.25–1% paraformaldehyde solution for 30 min at room temperature.

d. Centrifuge, decant, and vortex pellet. Wash with 2 ml of PBS. Centrifuge, decant, and vortex pellet.

e. Suspend cells in 1.5 ml 0.1 M HCl/Triton X-100 for 10 min on ice.

f. Add 5 ml of PBS and centrifuge for 2 min at $500 \times g$. Drain pellets well before vortexing.

g. Suspend cells in 1.5 ml of distilled H_2O and place tubes in a water bath at 95 °C for 10 min. This is the DNA denaturing step and often must be adjusted depending on the cell type to maximize denaturation while avoiding extensive cell loss.

h. Remove samples from hot water bath and place in an ice/water mixture until the suspensions are cold. Then add 3 ml of PBS, and centrifuge $500 \times g$ for 2 min. Drain tubes and vortex gently to loosen pellet. At this point, some clumping may be observed. Disperse clumps by pipetting or syringing through a 25-gauge needle. Failure to disperse the clumps completely may prevent antibody access to BrdUrd-labeled cells. These cells may appear later as apparently unlabeled, resulting in a measurement error.

Note: Steps 2 through 8 may be deleted if HCl denaturation is used (Dolbeare *et al.*, 1983; Gratzner, 1982). Cells are incubated at room temperature with 200 μl of 1.5–4 M HCl for 20 min and then washed twice with 2.5 ml of borate or phosphate buffer to restore pH to neutral before antibody treatment.

i. Suspend cells in 100 μl of diluted anti-BrdUrd antibody for 30 min at room temperature. Antibody should be diluted in antibody buffer to 0.2–0.5 μg/ml. We have observed that staining is better at 25 °C than at 4 °C for the short incubation time.

j. Add 5 ml of wash buffer, and centrifuge at 500 $\times$ g for 2 min. Drain well, and vortex pellet.

k. Add 100 μl of diluted second antibody (goat antimouse IgG FITC conjugate, typically diluted 1:50–1:500 in antibody buffer) for 20 min at room temperature. This step may be omitted if a direct conjugate anti-BrdUrd is used.

l. Add 5 ml of wash buffer, centrifuge, drain well, and vortex pellet.

m. Suspend cells in 1.5 ml of PI working solution.

n. Filter by running sample through a 40-μm nylon mesh and analyze with a flow cytometer (optical filters for flow cytometer are described in Section VII).

2. Restriction Enzyme/Exonuclease III Method (Dolbeare and Gray, 1988)

This method avoids the use of heat for denaturation, thus preserves many other antigens, and minimizes cell loss.

a. Following treatment with RNase, wash the cells and incubate for 30 min in 1 ml 1% paraformaldehyde at room temperature.

b. Wash cells with 2 ml of PBS and incubate cells with 1 ml cold 0.1 M citric acid containing 0.5% Triton X-100 for 10 min.

c. Wash cells with 2.5 ml 0.1 M Tris-HCl, pH 7.5, and incubate in 100 μl of 0.1 M Tris-HCl containing 50 mM NaCl and 10 mM MgCl$_2$ and 10 units of EcoRI for 30 min at 37 °C.

d. Wash cells with 1 ml 0.1 M Tris-HCl, pH 7.5, and resuspend in 100 μl of exonuclease III buffer containing 30 units of exonuclease III.

e. Wash cells with 2 ml PBS and continue with thermal protocol at step i (incubation with anti-BrdUrd antibodies).

3. Double Pyrimidine Label Method (Bakker et al., 1991; Dolbeare et al., 1988; Shibui et al., 1989).

This method permits a combination of pulse label, for example, with BrdUrd, and extended labeling with, for example, IdUrd, making possible calculation of growth fraction in addition to cell kinetic parameters (phase duration, labeling index, and cell-cycle time). This method depends on halogen-selective monoclonal antibodies. A suitable pair of monoclonal antibodies is Br-3, which binds preferentially to BrdUrd over IdUrd, and IU-4, which does not show halogen selectivity.

a. Following the denaturation step, incubate cells in the presence of 100 μl Br-3 for 20 min at 25 °C.

b. Without removing the Br-3, add 100 μl of the second antibody (with affinity IdUrd > BrdUrd, e.g., IU-4) at low concentration, for example, 1:5000 dilution in antibody diluting buffer. Direct conjugate antibodies simplify the technique since both Br-3 and IU-4 are mouse monoclonal IgGl subtypes. Simultaneous double indirect antibody staining can be done if one of the antibodies is a mouse monoclonal and the second is a rat monoclonal, a rabbit polyclonal antibody, or a mouse monoclonal of a different IgG subtype.

c. For single-laser flow cytometric analysis of the double-label, FITC and phycoerythrin antibody conjugates can be used. 7-Aminoactinomycin D can also be used as the DNA stain (Bakker *et al.*, 1991).

4. Combined BrdUrd/DNA/Cell-Surface Antigen (Method of Carayon and Bord, 1992).

This method permits detection of cell-surface antigens along with proliferation status, allowing determination of proliferation in each population in a mixture of cells.

a. Wash cells with 2 ml of PBS following BrdUrd incorporation.

b. Resuspend cells in 1 ml PBS containing phycoerythrin-labeled antibody to cell-surface antigen (anti-CD4) and incubate for 30 min at 4 °C.

c. Wash cells once with 2 ml PBS and resuspend in 1 ml 1% paraformaldehyde/0.01% Tween 20 overnight at 4 °C. This fixes the first antibody onto its antigen. Centrifuge, aspirate supernatant, and loosen pellet.

d. Wash cells in 2 ml PBS to remove paraformaldehyde. Incubate in PBS containing Mg^{2+} and Ca^{2+} and 50 Kunitz units DNase I, 30 min at 37 °C.

e. Wash and resuspend cells in 150 μl of PBS containing 10% bovine serum albumin and 0.5% Tween 20 and 20 μl FITC-coupled anti-BrdUrd for 45 min at room temperature.

f. Wash cells with 2 ml PBS and resuspend in 1 ml PBS containing PI (10 μg/ml) and analyze on a flow cytometer using three-color fluorescence detection.

5. Washless Technique (Larsen, 1990)

This technique permits histochemical staining of cells without centrifugation. Reagents are added in a stepwise manner so that cell loss is minimized.

B. Immunochemical Labeling Followed by Microscopic Detection

1. Fluorescent Staining Method

a. Cells centrifuged on slides by cytospin techniques or grown on slides may be stained using the above protocols with times being reduced 50% for washing, denaturation, and incubation with antibodies.

b. After fluorescent or colorimetric staining with anti-BrdUrd, slides may be washed with PBS and then distilled water. Fluorescence or light microscopy may be used directly for determining the labeling index. Labeling index is defined here as the number of BrdUrd-labeled cells/total cells observed in the field. For bright field microscopy after color staining, counterstain cells with 100 μl 0.05% Giemsa in 0.05 M phosphate buffer (pH 6.5).

2. Histochemical Staining of Tissue Sections (Shibuya *et al.*, 1993)

This method uses a combination of colloidal gold and alkaline phosphatase enzyme immunohistochemical methods to detect both BrdUrd and IdUrd incorporated into DNA in histologic sections. Nuclei are counterstained with hematoxylin, and sections are examined by microscopy.

a. Sections are deparaffinized by immersion in xylene for 5 min followed by immersion for 2 min each in 100%, 70%, and 50% ethanol. DNA is denatured by immersion of slides in 4 N HCl for 10 min at room temperature.

b. Incubate for 30 min with 100 μl Br-3 diluted 1:20,000 in PBS with 5% normal goat serum as a blocking protein.

c. Rinse slides with 5 ml PBS.

d. Incubate slides for 30 min with 100 μl gold-conjugated goat antimouse IgG diluted 1:50 in PBS.

e. Wash slides with 5 ml PBS and fix with 100 μl 2.5% glutaraldehyde for 10 min. Rinse slides with 10 ml deionized water.

f. Silver precipitate for 20–30 min with silver enhancement kit as described in the protocol provided by the manufacturer. Result is a black staining of the nuclei containing BrdUrd.

g. Immerse slides in 5% acetic acid for at least 1 h. Shibuya *et al.* (1993) suggested that slides may be left overnight in this solution. Rinse slides with TBS following incubation.

h. Incubate 30 min with 100 μl IU-4 dilutes 1:800 in TBS with 1% normal rabbit serum as a blocking protein. Rinse slides with 5 ml TBS following incubation.

i. Incubate 30 min with 100 μl alkaline phosphatase-conjugated rabbit anti-mouse immunoglobulins diluted 1:50 in TBS. Rinse slides with 5 ml TBS following incubation.

j. Incubate with 100 μl suitable alkaline phosphatase substrate, for example, 0.2 mM naphthol ASMX phospate plus 0.2 mM Fast red TR salt in 0.05 M Tris buffer, pH 8.2, containing 5 mM Mg^{2+}. Wash with 5 ml distilled water. Result is a red staining of the nuclei containing IdUrd. BrdUrd containing nuclei also stain, but this is marginal against their previous black staining by Br-3. Counterstain nuclei with 100 μl of 5% Gills No. 1 hematoxylin, giving a light blue color to nuclei not containing either BrdUrd or IdUrd. Wash with 5 ml distilled water.

k. Air dry the slides, clear with HistoClear, and mount glass coverslip with Permount.

C. Cell Kinetics Measurements

1. Generally the window technique is used for FCM analysis (Dean *et al.*, 1984; Pallavicini *et al.*, 1985). Commercial software for these computations is available from Phoenix Flow Systems, Verity Software House, Cell Pro, and others.

An arbitrary electronic window can be selected for determining events in a particular part of the bivariate BrdUrd/DNA histogram. The number of events appearing in that window as a function of time after the BrdUrd pulse can be used to derive cell-cycle parameters, for example, G_1, S, and Tc. If cells are continuously pulsed with BrdUrd for a period equal to or greater than Tc, it is possible to evaluate the growth fraction of the tumor.

2. The alternate method proposed by Begg *et al.* (1985) permits determination of Tpot with a single sample several times following a BrdUrd pulse (see Fig. 1, the bivariate distribution at 3 h). A window may be drawn around the S phase portion and a value for relative movement, Rm, can be determined. The relationship defined by Begg *et al.* (1985) shows that Rm could be derived with the expression

$$\text{Rm} = F_\text{L} - F_{\text{G}_1}/F_\text{GM} - F_{\text{G}_1},$$

where F_L is the mean red fluorescence of the green-labeled cells and F_{G1} and F_{GM} are the mean red fluorescence of the G_1 and $G_{2+}M$ cells, respectively. Rigorous mathematical derivation of this expression was published by White and Meistrich (1986) and subsequently modified by Terry *et al.* (1991).

3. Using double halopyrimidine labels, it became possible to determine Ts and Tpot with a single biopsy sample (see Raza *et al.*, 1991a; Shibui *et al.*, 1989). One of the analogs (e.g., IdUrd) is given several hours before surgery, the second analog given just prior to surgery. Either thin-section histochemistry or flow cytometry can be used to derive quantitative terms for Ts and Tpot.

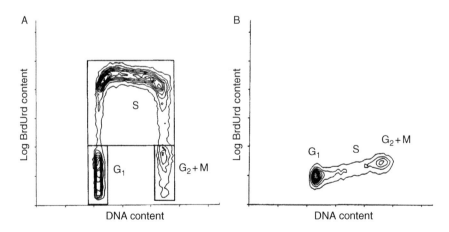

Fig. 1 A series of bivariate DNA/BrdUrd distributions taken after a 30-min pulse of 10 μM BrdUrd followed by a thymidine chase and subsequent sampling at 3-h intervals.

VI. Critical Aspects of the Procedure

A. DNA Denaturation

All antibodies to halogenated nucleosides suitable for cell kinetics studies reported to date recognize the incorporated analog only in the context of single-stranded DNA. Thus denaturation of the DNA is required before antibody staining. Extensive denaturation is neither required nor desirable, however, since denaturation leads to a loss of the ability to stain with DNA intercalating dyes, causes unacceptably broad coefficients of variation in DNA staining, and increases cell loss. As such, there must be a balance between the need to denature the DNA for antibody recognition of the BrdUrd and the need to retain intact double-stranded DNA for PI binding. Limited denaturation by either heating or with HCl or partial DNA digestion by nucleases is the preferred method. Carefully controlled denaturation may also allow resolution of tetraploid G_2 and M cells into distinct G_2 and M peaks in the bivariate histogram by exploiting the differences in denaturation of DNA in mitotic and G_2-phase cells (Nüsse et al., 1989).

B. Paraformaldehyde Fixation

Paraformaldehyde fixation will lower the sensitivity of the staining reaction probably by reducing the denaturability of the DNA and cross-linking of chromatin proteins. Even with 1% paraformaldehyde fixation, however, quantification of a 10 nM BrdUrd 30-min pulse is attainable. The paraformaldehyde treatment also helps to prevent cell loss, especially of lymphoid cells, during the staining procedure. We have also found that the fixation step improves the quality of the DNA histogram with lower CVs for the G_1 peak.

C. Nonspecific Fluorescence

Nonspecific fluorescence is due primarily to nonspecifically bound antibody (anti-BrdUrd or fluorescein-conjugated second antibody). Additional washes after antibody treatment or incorporation of 1–5% blocking protein in the wash buffer can reduce nonspecific binding of antibody. Since most nonspecific antibody binding localizes in the cytoplasm and on the cytoplasmic membrane, using nuclei rather than whole cells can also greatly reduce nonspecific fluorescence (Landberg and Roos, 1992).

D. Cell Loss

Cell loss generally results from clumping and cell adherence to the walls of the test tube being used to process the cells. Lymphoid cell loss is generally greater from this procedure but may be reduced greatly by using lower centrifuge speeds during the pelleting of cells. Limit centrifugation to between $400 \times g$ and $800 \times g$ for no more than 4 min. Clumping may depend also on the nature of the fixative used.

Frequently large clumps appear after the thermal denaturation step. Clumps may be disaggregated by a combination of mild vortexing and syringing the suspension gently through a size of 25-gauge needle. Cell adherence to the centrifuge tube can be decreased by either siliconizing the tubes or by using microfuge tubes to reduce tube surface area.

E. Double-Labeling with BrdUrd and IdUrd

When staining cells that have double halopyrimidine labels, for example, with BrdUrd pulse and continuous IdUrd label, one should be concerned with the differences in specificities and affinities of the specific antibodies. If very specific antibodies are used, there may not be a problem. As an example of halogen specificity, IU-4 has affinity for the halopyrimidines in the following order IdUrd > BrdUrd > CldUrd, whereas Br-3 has the following specificity: BrdUrd = CldUrd $\gg$ IdUrd. If some crossreactivity occurs with one of the antibodies (e.g., IU-4 will react with both IdUrd and at lower affinity with BrdUrd), then add that antibody at a much lower concentration after the specific antibody Br-3 has incubated with the cells for 20–30 min. Then continue the incubation for an additional 30 min. In this way, the Br-3 will saturate BrdUrd sites but not react with the IdUrd sites. Adding the IU-4 then will preferentially bind only to the exposed IdUrd sites. Using a lower concentration of this antibody will prevent displacement of the Br-3 from the BrdUrd sites.

F. Optical Filters for FCM

Using correct optical filters during the flow cytometric analysis will prevent the cross talk between the photomultipliers detecting the fluorescein and PI signals. PI exhibits a broad band of fluorescence ranging from 530 to 700 nm. If a 550-nm short-pass filter coupled with a 500-nm long-pass filter is used for the green fluorescence, then some PI fluorescence will be observed in the green fluorescence channels causing a skewing of the BrdUrd histogram. While commercial instruments have capabilities for compensating for some excess cross talk, they cannot correct for a large excess of one or the other fluorophore in the sample. As such, the absolute fluorescence intensity of the separate staining dyes should be approximately equal.

VII. Controls and Standards

A. Negative Controls

To some extent, the presence of G_1 cells in the cell population serve as a negative control cell population. G_1 cells should have minimal green fluorescence from antibodies to BrdUrd, since they do not contain BrdUrd. Elevated G_1 staining indicates nonspecific antibody binding or autofluorescence. The latter may result from inappropriate fixation techniques, particularly those using aldehyde fixatives. The ratio of green fluorescence intensity of the mid S phase to G_1 cells should be at least 10 and under ideal staining conditions may be several hundred. If the protocol used continuous BrdUrd labeling so unlabeled cells are not part of the sample, then a specific negative control sample where BrdUrd has been omitted should be used.

B. Cross Talk

To determine whether cross talk is occurring between fluorescence channels (i.e., whether PI fluorescence is being detected by the fluorescein detector) add PI but no antibodies after the thermal denaturation step. If the green fluorescence is above background then cross talk is present. Correct this by either reducing the concentration of PI used or by selecting more appropriate filters. We recommend a final PI concentration of about 10 μg/ml.

VIII. Instruments

A. Flow Cytometer

Any flow cytometer equipped with a single argon ion laser with two photomultiplier tubes is adequate. The instrument should also be equipped with a log amplifier to accommodate the large range of fluorescence signal generated by the anti-BrdUrd fluorescence. Both fluorescein and PI can be excited at 488 nm. Use a 514-nm band pass filter for the fluorescein fluorescence (BrdUrd content) and a 600-nm long-pass filter for the PI fluorescence (DNA content). If AMCA is used as the fluorophore for an antibody, then use an excitation of 363 and a 450-nm band pass filter to collect fluorescence. The incorporation of a doublet eliminator will prevent accumulation of doublet G_1 and early S phase cells in windows that should contain either G_2 or late S phase cells.

B. Microscope

A bright field or fluorescent microscope can be used to determine labeling indices of cells stained on slides.

IX. Results

Figure 2A shows a bivariate DNA/BrdUrd contour histogram generated by CHO cells stained according to the above protocol. This is the kind of histogram generated after a 30-min pulse of 1 mM BrdUrd. G_1 and $G_2 + M$ populations should have only background green fluorescence. S phase cells have green

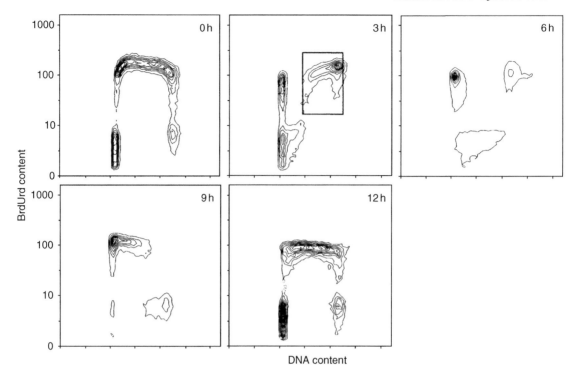

Fig. 2 (A) Bivariate histogram showing DNA content (x-axis = PI fluorescence) and BrdUrd content (y-axis = log fluorescein fluorescence) following a 30-min pulse of 1 mM BrdUrd. (B) Bivariate histogram showing DNA content and BrdUrd content of CHO not receiving BrdUrd pulse.

fluorescence and produce the horseshoe-shaped pattern with mid S phase cells having a 64-fold higher fluorescence than G_1 cells. The fraction of cells with S phase fluorescence divided by the total population will give the labeling index. Cells that have not been incubated with BrdUrd but stained by the same protocol will exhibit only background fluorescence (Fig. 2B).

The cell kinetic applicability of the method is demonstrated by Fig. 1 where a single BrdUrd pulse was given at $t = 0$ followed by a thymidine pulse chase after 30 min. Samples were taken at 30 min (0 h) and at $t = 3, 6, 9$, and 12 h, fixed in 50% ethanol, and stained by the thermal denaturation protocol. S phase cells that incorporated BrdUrd show the typical green fluorescence after 30 min. This same cohort of S phase cells progress through the cell cycle with the fluorescence appearing in $G_2 + M$ and daughter G_1 cells at 3 and 6 h and progressing further into G_1 and back into S phase at 9 h with most of the label reappearing in S phase at approximately one cell-cycle time, that is, 12 h after the initial pulse of BrdUrd.

The window method for determining fraction of cells in each compartment is shown in Fig. 3. A BrdUrd/DNA bivariate histogram was derived from an *in vivo*

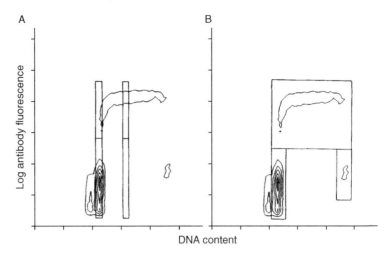

Fig. 3 Graphical presentation of windows set up to calculate the fraction of labeled cells in G_1 phase (left boxes in A) or mid S phase (right boxes in A) and the labeling index (B). Fraction of labeled cells in G_1 phase = cells in labeled G_1 box (upper G_1 window)/total cells in G_1 box (both labeled and unlabeled G_1 cells). Fraction of labeled cells in mid S phase = cells in labeled mid S box/cells in S box including both labeled and unlabeled mid S box. Labeling index = cells in labeled, high FITC fluorescence box/total cells in distribution. From Kuo *et al.* (1993).

pulse label of BrdUrd in the brown Norway rat injected 17 days previously with the myeloid leukemic tumor cells (Kuo *et al.*, 1993). The windows in Fig. 3A are drawn so that events in G_1 or mid S can be evaluated. When these counts are collected as a function of time after the BrdUrd pulse, then the kinetic parameters, T_{G_1}, T_s, and T_c can be determined. The S phase window in Fig. 3B permits the calculation of LI, that is, the fraction of cells in S phase. Figure 4A shows the quantitative analysis of the cytokinetic properties of asynchronous BNML spleen cells. Fractions of labeled cells in mid S phase (A) and in G_1 phase (B) are derived from pulse-chase bivariate distributions. T_s is the time for cells to go from one maximum or minimum in label to a second, for example, the T_s is the time for the cells to go from the first arrow in Fig. 4A to the second, that is, from the first minimum S phase label to the second. Growth fractions were obtained from bivariate distributions after continuous labeling. The labeling index increases with time until the entire growing fraction is labeled. Thus, further continuous labeling results in very little change in the fraction of labeled cells. An extrapolation of this limit (Fig. 4B) leads to the derivation of the growth fraction (0.76). Solid lines represent the best-fitted mathematical model of the experimental data points (•) derived from a computer analysis of the three sets of data. Table II is a summary of the kinetic data for BNML cells derived from the curves in Fig. 4.

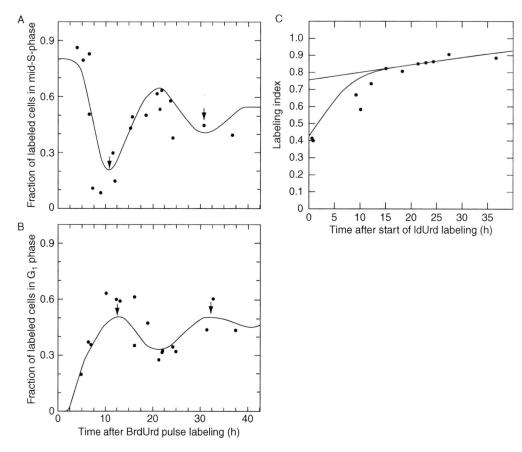

Fig. 4 Quantitative analysis of the cytokinetic properties of asynchronous BNML spleen cells. Fractions of labeled cells in mid S phase (A) and in G_1 phase (B) are derived from pulse-chase bivariate distributions. Growth fractions were derived from bivariate distributions after continuous labeling with boxes described in Fig. 3C. Solid lines represent the best-fitted mathematical model of the experimental data points (●) derived from a computer analyses of the three sets of data. From Kuo *et al.* (1993).

Table II
BNML Cytokinetic Parameters[a]

Tgl	5.4 h
Ts	12.5 h
Tg2m	2.6 h
Tc	20.5 h
GF	0.76

[a]These parameters are derived from the curve fitting, quantitative analyses shown in Fig. 3. Tgl, Ts, Tg2m, and Tc represent the time for G_1, S, G_2M phases and cell-cycling time, respectively. GF, growth fraction.

X. Summary

We have described several laboratory procedures for the immunochemical staining of the halopyrimidines, BrdUrd and IdUrd, in cell suspensions for flow cytometry and a method for staining histological sections on slides. Halogenated pyrimidine quantitation allows cell-cycle parameters, including total cell-cycle time, phase durations, and growth fraction to be determined. We have presented some flow cytometric data to demonstrate the use of these methods in determining bivariate BrdUrd/DNA histograms with CHO cells and in kinetic studies with the brown Norway rat myeloid leukemia model.

Acknowledgment

Part of this work was performed under the auspices of the U.S. Department of Energy at Lawrence Livermore National Laboratory under contract No.W-7405-Eng-48.

References

Allison, L., Arndt-Jovin, D. J., Gratzner, H., Ternynck, T., and Robert-Nicoud, M. (1985). Mapping of the pattern of DNA replication in polytene chromosome from *Chironomus thummi* using monoclonal anti-bromodeoxyuridine antibodies. *Cytometry* **6**, 584–590.

Bakker, P. J. M., Tukker, C. J., van Ofen, C. H., and Aten, J. (1991). An indirect immunofluorescence double staining procedure for the simultaneous flow cytometric measurement of iodo- and chlorodeoxyuridine incorporated into DNA. *Cytometry* **12**, 366–372.

Begg, A. C., McNally, N. J., Shrieve, D. C., and Karcher, H. A. (1985). A method to measure the duration of DNA synthesis and the potential doubling time from a single sample. *Cytometry* **6**, 620–626.

Beisker, W., and Hittelman, W. (1988). Measurement of the kinetics of DNA repair synthesis after uv irradiation using immunochemical staining of incorporated 5-bromo-2'-deoxyuridine and flow cytometry. *Exp. Cell Res.* **174**, 156–161.

Beisker, W., Dolbeare, F., and Gray, J. W. (1987). An improved immunocytochemical procedure for high-sensitivity detection of incorporated bromodeoxyuridine. *Cytometry* **8**, 235–239.

Cappella, P., Gasparri, F., Pulici, M., and Moll, J. (2008). A novel method based on click chemistry, which overcomes limitations of cell cycle analysis by classical determination of BrdU incorporation, allowing multiplex antibody staining. *Cytometry A* **73A**, 626–636.

Carayon, P., and Bord, A. (1992). Identification of DNA-replicating lymphocyte subsets using a new method to label the bromo-deoxyuridine incorporated into the DNA. *J. Immunol. Methods* **147**, 225.

Danova, M., Riccardi, A., and Mazzini, G., (1990). Cell cycle-related proteins and flow cytometry. *Haematologica*, **75**, 252–264.

Darzynkiewicz, Z., Huang, X., and Okafuji, M. (2006). Detection of DNA strand breaks by flow and laser scanning cytometry in studies of apoptosis and cell proliferation (DNA replication). *Methods Mol. Biol.* **314**, 81–94.

Dean, P. N., Dolbeare, F., Gratzner, H., Rice, G. C., and Gray, J. W. (1984). Cell-cycle analysis using a monoclonal antibody to BrdUrd. *Cell Tissue Kinet.* **17**, 427–436.

Diermeier-Daucher, S., Clarke, S. T., Hill, D., Vollmann-Zwerenz, A., Bradford, J. A., and Brockhoff, G. (2009). Cell type specific applicability of 5-ethynyl-2'-deoxyuridine (EdU) for dynamic proliferation assessment in flow cytometry. *Cytometry A* **75**, 535–546.

Dolbeare, F., Gratzner, H., Pallavicini, M., and Gray, J. W. (1983). Flow cytometric measurement of total DNA content and incorporated bromodeoxyuridine. *Proc. Natl. Acad. Sci. USA* **80**, 5573–5577.

Dolbeare, F., and Gray, J. W. (1988). Use of restriction endonucleases and exonuclease III to expose halogenated pyrimidines for immunochemical staining. *Cytometry* **9**, 631.

Dolbeare, F., Kuo, W. L., Beisker, W., Vanderlaan, M., and Gray, J. W. (1990). Using monoclonal antibodies to bromodeoxyuridine-DNA analysis. *Methods Cell Biol.* **33**, 207–216.

Dolbeare, F., Kuo, W. L., Vanderlaan, M., and Gray, J. W. (1988). Application of bromodeoxyuridine antibodies in studies of the cell cycle. *Proc. Am. Assoc. Cancer Res.* **29**, 1896.

Duprez, A., Barat, J. L., Girard, A., Hoffmann, M., and Hepner, H. (1990). Cell proliferation in human cerebral tumors. *In vivo* study of 45 cases by incorporation of 5-iododesoxyuridine. *Neurochirurgie* **36**, 157–166.

Frommer, M., Paul, C., and Vincent, P. C. (1988). Localisation of satellite DNA sequences on human metaphase chromosomes using bromodeoxyuridine-labelled probes. *Chromosoma* **97**, 11–18.

Garin, L., Barona, R., O'Connor, E., Armengot, M., and Basterra, J. (1991). Determination of the phase of synthesis by means of bromodeoxyuridine. Its clinical application in oncology of the head and neck. *An. Otorrinolaringol. Ibero Am.* **8**, 567–574.

Giaretti, W. (1991). Ploidy and proliferation evaluated by flow cytometry. An overview of techniques and impact in oncology. *Tumori* **77**, 403–419.

Gratzner, H. G. (1982). Monoclonal antibody to 5-bromo- and 5-iododeoxyuridine: A new reagent for detection of DNA replication. Science **218**, 474.

Gratzner, H. G., Ahmad, P. M., Stein, J., and Ahmad, F. (1985). Flow cytometric analysis of DNA replication during the differentiation of 3T3-L1 preadipocytes. *Cytometry* **6**, 563–569.

Gray, J. W., Dolbeare, F., and Pallavicini, M. (1990). Quantitative cell-cycle analysis. In "Flow Cytometry and Sorting" (M. R. Melamed, T. Lindmo, and M. L. Mendelsohn, eds.), 2nd edn., pp. 445–468. Wiley-Liss, New York.

Hardonk, M. J., and Harms, G. (1990). The use of 5'-bromodeoxyuridine in the study of cell proliferation. *Acta Histochem. Suppl.* **39**, 99–108.

Ito, H., Kamiryo, T., Kajiwara, K., Nishizaki, T., and Oshita, N. (1990). Use of bromodeoxyuridine in graft versus host analysis. *No Shinkei Geka* **18**, 595–599.

Juan G, Li X, and Darzynkiewicz Z. (1997). Correlation between DNA replication and expression of cyclins A and B1 in individual MOLT-4 cells. *Cancer Res.* **57**, 803–807.

Kaufmann, S. J., and Robert-Nicoud, M. (1985). DNA replication and differentiation in rat myoblasts studied with monoclonal antibodies against 5-bromodeoxyuridine, actin, and alpha 2-macroglobulin. *Cytometry* **6**, 570–577.

Kuo, W. L., Dolbeare, F., Vanderlaan, M., and Gray, J. W. (1993).Use of bromodeoxuridine in cell prolifartion studies. In "DNA Cytometric Analysis" (A. Sampedro, ed.), pp. 219–238. Oviedo University Press, Spain.

Lacombe, F., Belloc, F., Dumain, P., Puntous, M., Lopez, F., Bernard, P., Boisseau, M. R., and Reiffers, J. (1992). Quantitation of resistance of leukemic cells to cytosine arabinoside from BrdUrd/DNA bivariate histograms. *Cytometry* **13**, 730–738.

Landberg, G., and Roos, G. (1992). Flow cytometric analysis of proliferation associated nuclear antigens using washless staining of unfixed cells. *Cytometry* **13**, 230–240.

Larsen, J. (1990). Washless double staining of a nuclear antigen (Ki-67 or bromodeoxyuridine) and DNA in unfixed nuclei. *Methods Cell Biol.* **33**, 227–234.

Leadon, S. A. (1986). Differential repair of DNA damage in specific nucleotide sequences in monkey cells. *Nucleic Acids Res.* **14**, 8979–8995.

Nakamura, H., Morita, T., and Sato, C. (1986). Structural organizations of replicon domains during DNA synthetic phase in the mammalian nucleus. *Exp. Cell Res.* **165**, 291–297.

Nüsse, M., Julch, M., Geido, E., Bruno, S., DiVinci, A., Giaretti, W., and Russo, K. (1989). Flow cytometric detection of mitotic cells using the bromodeoxyuridine/DNA technique in combination with 90 degrees and forward scatter measurements. *Cytometry* **10**, 312–319.

Pallavicini, M. G., Summers, L. J., Dolbeare, F., and Gray, J. W. (1985). Cytokinetic properties of asynchronous and cytosine arabinoside perturbed murine tumors measured by simultaneous bromodeoxyuridine/DNA analyses. *Cytometry* **6**, 602–610.

Pinkel, D., Thompson, L., Gray, J., and Vanderlaan, M. (1985). Measurement of sister chromatid exchanges at very low bromodeoxyuridine substitution levels using a monoclonal antibody in Chinese hamster ovary cells. Cancer Res. **45**, 5795–5798.

Raza, A. G., Bokhari, J., Yousuf, N., Medhi, A., Mazewski, C., Khan, S., Baker, V., and Lampkin, B. (1991b). Cell cycle kinetic studies in human cancers. Development of three DNA-specific labels in three decades. *Arch. Pathol. Lab. Med.* **115**, 873–879.

Raza, A. G., Miller, M., Mazewski, C., Sheikh, Y., Lampkin, B., Sawaya, R., Crone, K., Berger, T., Reisling, J., Gray, J., Khan, S., and Preisler, H. D. (1991a). Observations regarding DNA replication sites in human cells in vivo following infusions of iododeoxyuridine and bromodeoxyuridine. *Cell Prolif.* **24**, 113–126.

Raza, A. G., and Preisler, H. D. (1990). Cellular dynamics of leukemias. *CRC Crit. Rev. Oncol.* **1**, 373–378.

Raza, A. G., Yousuf, N., Bokhari, A., Masterson, M., Lampkin, B., Yanik, G., Mazewski, G., Khan, S., and Preisler, H. D. (1992). Contribution of *in vivo* proliferation-differentiation studies toward the development of a combined functional and morphologic system of classification of neoplastic diseases. *Cancer* (Philadelphia) 69 (Suppl.), 1557–1566.

Riccardi, A., Danova, M., and Ascari, E. (1988). Bromodeoxyuridine for cell kinetic investigations in humans. *Haematologica* **73**, 423–430.

Salic, A., and Mitchison, T. J. (2008). A chemical method for fast and sensitive detection of DNA synthesis *in vivo. Proc. Natl. Acad. Sci. USA* **105**, 2415–2420.

San-Galli, F., Maire, J. P., and Guerin, J. (1991). Medulloblastoma: towards new prognostic factors. *Neurochirurgie* **37**, 3–11.

Selden, J., Dolbeare, F., Clair, J., Nichols, W., Miller, J., Kleemeyer, K., Hyland, R., and DeLuca, J. (1993). Statistical confirmation that immunofluorescent detection of DNA repair in human fibroblasts by measurement of bromodeoxyuridine incorporation is stoichiometric and sensitive. *Cytometry* **14**, 154–167.

Shibui, S., Hoshino, T., Vanderlaan, M., and Gray, J. (1989). Double labeling with iodo- and bromo-deoxyuridine for cell kinetics studies. *J. Histochem. Cytochem.* **37**, 1007–1011.

Shibuya, M., Ito, S., Davis, R. L., and Hoshino, T. (1993). Immunohistochemical double staining with immunogold-silver and alkaline phosphatase to identify nuclear markers of cellular proliferation. *Biotech. Histochem.* **68**, 17–19.

Suzuki, H. (1988). The prognostic value of flow cytometric DNA analysis in colorectal cancer patients. *Jpn. J. Surg.* **18**, 483–486.

Takahashi, M. (1966). Theoretical basis for cell cycle analysis: II. Further studies on labelled mitosis wave method. *J. Theor. Biol.* **13**, 195–202.

Terry, N. H., White, R. A., Meistrich, M. L., and Calkins, D. P. (1991). Evaluation of flow cytometric methods for determining population potential doubling times using cultured cells. *Cytometry* **12**, 234–241.

Vanderlaan, M., Watkins, B., Thomas, C., Dolbeare, F., and Stanker, L. (1986). Improved high-affinity monoclonal antibody to iododeoxyuridine. *Cytometry* **7**, 499–507.

Waldman, F., Dolbeare, F., and Gray, J. W. (1988). Clinical applications of the bromodeoxyuridine/ DNA assay. *Cytometry*, Suppl. 3, 65–72.

Ward, J. M., Wedghorst, C. M., Diwan, B. A., Konishi, N., Lubet, R. A., Henneman, J. R., and Devor, D. E. (1991). Evaluation of cell proliferation in the kidneys of rodents with bromodeoxyuridine immunohistochemistry of tritiated thymidine autoradiography after exposure to renal toxins, tumor promoters, and carcinogens. *Prog. Clin. Biol. Res.* **369**, 369–388.

White, R., and Meistrich, M. (1986). A comment on "A method to measure the duration of DNA synthesis and the potential doubling time from a single sample". *Cytometry* **7**, 486–492.

Yu, C. C., Woods, A. L., and Levison, D. A. (1992).The assessment of cellular proliferation by immunohistochemistry: a review of currently available methods and their applications. *Histochem. J.* **24**, 121–131.

CHAPTER 17

Cell-Cycle Kinetics Estimated by Analysis of Bromodeoxyuridine Incorporation

Nicholas H. A. Terry* and **R. Allen White†**

*Departments of Experimental Radiation Oncology
M. D. Anderson Cancer Center
The University of Texas
Houston, Texas 77030

†Department of Biomathematics
M. D. Anderson Cancer Center
The University of Texas
Houston, Texas 77030

ESSENTIAL CYTOMETRY METHODS
Reprinted from *Methods in Cell Biology*, Volume 63 (Academic Press, 2001).
DOI: 10.1016/B978-0-12-375045-7.00017-9

I. Introduction

Historically, the elucidation of cell-cycle kinetic parameters developed around the use of radioactively labeled thymidine, which is incorporated by cells synthesizing DNA. The fraction of labeled cells [labeling index (LI)] could, through the use of autoradiography, then be measured directly, and the percentage of labeled mitotic figures (PLM) could be counted as a function of time after labeling. The LI may be used to determine the fraction of cells synthesizing DNA, dependent on the duration of S phase, and the changing PLM curve (or its associated cousins such as the continuous labeling curve) gave dynamic information about the progression of labeled cells through the cell cycle. The tedious techniques of multiple sampling and autoradiography are now largely supplanted by the use of halogenated pyrimidines, such as bromodeoxyuridine (BrdUrd), chlorodeoxyuridine (CldUrd), and iododeoxyuridine (IdUrd) used either singly or in combination. These agents may readily be tagged using monoclonal antibodies and visualized by fluorescent probes.

The advantages of the use of monoclonal antibodies to these thymidine analogs are not only in the increased ease and speed of analysis. The more recent methodology also offers greater precision in estimating such quantities as S-phase fraction from bivariate DNA versus thymidine-analog flow-cytometric measurements, thus avoiding the problematic assumptions inherent in single-parameter DNA histogram deconvolution. Moreover, this methodology makes it possible to estimate cell-cycle kinetic parameters from measurements made at a single time after labeling, thereby making it routinely possible to evaluate clinical data concerning the relationship between cell proliferation and treatment outcome. For example, the tumor potential doubling time (T_{pot}) has been explored for utility both as a predictive assay for treatment outcome and as a selection criterion for patients for accelerated radiotherapy regimens. An understanding of the contemporary techniques of the analysis of kinetic parameters, by the use of halogenated thymidine analogs, is both essential and fundamental for current dynamic cell-kinetic studies both *in vitro* and *in vivo*.

This chapter details the methods for sample preparation and staining, flow-cytometric data acquisition, and the procedures for quantitative analysis of dynamic cell-kinetic data. Whereas the PLM technology relied on labeling cells in S phase and observing them through the window of mitosis, the two-color flow-cytometric technique described here still relies on S-phase labeling but now observes the labeled cells throughout the entire cell cycle. Importantly, it utilizes the extra information discernible from the division status of the labeled population. (Parental and filial cell generational status are distinguishable one from the other.) This technology permits the rapid evaluation of the quantities required to describe many of the growth kinetic parameters of cell populations. The basis of these techniques is that cells in S phase can be selectively labeled both *in vitro* and *in vivo* by administration of a nontoxic level of BrdUrd. The cells that incorporate

BrdUrd continue to progress through the cell cycle and may be sampled, or a biopsy or surgical specimen taken, at a known time later.

The sample, fixed in ethanol, may be processed to produce a suspension of single nuclei by enzymatic digestion with, for example, pepsin (Carlton *et al.*, 1991). The nuclei are analyzed simultaneously for BrdUrd and DNA content by flow cytometry (Dolbeare *et al.*, 1983). The BrdUrd-labeled nuclei are selectively stained by a monoclonal antibody to BrdUrd (Gratzner, 1982) using a fluorescein isothiocyanate-conjugated (FITC, green fluorescing) second antibody technique. All the nuclei are also stained with propidium iodide (PI), which fluoresces red at an intensity proportional to their DNA content, thereby simultaneously defining a reference standard for relative cell ages.

At the time of labeling, the BrdUrd-labeled cells are assumed to be completely and exclusively in the S phase, with all unlabeled cells in the G_1 and $G_2 + M$ phases of the cell cycle. In the interval between the administration of BrdUrd and sampling, the cycling cells progress unperturbed to subsequent phases of the cell cycle. In particular, the BrdUrd-labeled cells progress through S, $G_2 + M$ and, subsequently, into G_1 of the next generation. These observations were the basis for the original method for calculation of the duration of DNA synthesis (T_S), and T_{pot} from a single biopsy sample (Begg *et al.*, 1985). Subsequent modeling and experimental studies (for overviews, see Terry and White, 1996; Terry *et al.*, 1992) have considerably refined this technique.

As with all laboratory methods, sample preparation is paramount. The staining requires a balancing act between denaturing sufficient DNA for the antibodies to bind to the incorporated BrdUrd while leaving enough DNA in its normal config-uration in order to derive good quality DNA histograms with low coefficients of variation (CV) on the G_1 and $G_2 + M$ peaks. Several different techniques have been developed since its origination (Dolbeare *et al.*, 1983) to accomplish these goals. Beisker *et al.* (1987) and Dolbeare *et al.* (1990) describe a thermal denatur-ation method whose principal advantage is in increased sensitivity. Restriction enzyme/exonuclease III methods (Dolbeare and Gray, 1988) avoid the use of heat for denaturation, thus preserving many other antigens, and minimize cell loss. Other combined BrdUrd/DNA/cell-surface and other antigen methods have been described (Begg and Hofland, 1991; Carayon and Bord, 1992), which may allow determination of proliferation in identifiable populations of cells. Washless tech niques, which require no centrifugation, also help to minimize cell loss (Larsen, 1990). The kinetic analytical methods that we describe here require sample prepar-ative techniques that are both of high sensitivity, giving excellent visualization of incorporated BrdUrd, together with low CVs for the peaks in the DNA histo-grams. Such sample preparation methods generally require production of isolated nuclei, either from cells *in vitro* or disaggregated from solid tissues and tumors, by the use of digestive enzymes and then denaturation of the DNA by strong acid. The method we recommend here is developed from a Schutte *et al.* (1987) modification of the procedure originally described by Dolbeare *et al.* (1983).

II. Applications

The primary dynamic cell-kinetic applications of the technique are in the quantitation of cell-cycle-phase durations, population doubling times, cell-cycle-phase boundary transitions or stasis, together with measurement of LI and GF (growth fractions). In the clinical context, the method is being used largely to measure LI, T_S, and T_{pot} (for recent reviews, see Antognoni *et al.*, 1998; Begg, 1995; Dubray *et al.*, 1995; Terry, 1996; Terry and Peters, 1995).

BrdUrd-labeling is also useful in, for example, cell synchrony experiments (Bussink *et al.*, 1995), where it enables a more accurate estimation to be made of the proportion of cells in defined phases of the cell cycle. It is also helpful in studies where dynamic information about cell progression, or nonprogression such as following drug treatment, through the cycle may be informative (Sacks *et al.*, 1990; Terry *et al.*, 1997). Although beyond the scope of this chapter, extension of analytical methods following labeling with two halogenated pyrimidine analogs of thymidine (e.g., CldUrd and IdUrd), given either as two pulses or as a continuous infusion of one and a pulse label of the other, have allowed for refinement of measures of S-phase cells *in vivo* (Pollack *et al.*, 1993a, 1995). The use of different monoclonal antibodies allows specific, and simultaneous, visualization of the two thymidine analogs, together with total DNA content. These methods were developed for tumors and have been used to measure the GF and T_{pot} of both DNA-aneuploid tumor cells and associated DNA-diploid cells simultaneously (White *et al.*, 1994a) following continuous infusion with CldUrd. In another study, using two pulses of IdUrd and CldUrd, White *et al.* (1994b) measured three differently labeled subpopulations within S phase and the separate progression through the cell cycle of the diploid and aneuploid cells. Pollack *et al.* (1993b) used two independently visualized labels to refine calculations of T_S and T_{pot}. The proliferation kinetics of tumor cells recruited into the cycle that were previously quiescent have also been estimated (Pollack *et al.*, 1994).

III. Materials

A. Introduction

There are many ways to achieve the quality of sample labeling, tissue and tumor disaggregation, and immunochemical visualization of incorporated BrdUrd that is needed for the analytical methods. Our general approach is to work with aseptic techniques and sterile reagents. Nonconjugated anti-BrdUrd primary antibodies, followed by a fluorochrome-conjugated second antibody, allow for greater signal enhancement than do directly conjugated antibodies. The following reagents give consistent results. All solutions are prepared, filtered through a 0.22-μm filter, and most stored at 4 °C with a few exceptions:

1. Sodium borate, HCl, and Tween 20 are prepared, filtered, and stored at room temperature.

2. BrdUrd working solutions, antibodies, and PI are diluted fresh weekly and stored at 4 °C in the dark.

3. Ethanol and normal goat serum are stored at −20 °C.

B. General Laboratory Reagents

1. Calcium- and magnesium-free phosphate-buffered saline (PBS, Dulbecco's) (Gibco BRL, Life Technologies, Rockville, MD). Store the powder at 4 °C in a desiccator.

2. 5-Bromo-2′-deoxyuridine (BrdUrd) (Sigma Chemical, St. Louis, MO). Store at −20 °C in a desiccator.
   a. Stock solution for *in vitro* work is 1 mM in PBS.
   b. Working solution for injection in *in vivo* tumor/normal tissue studies is 6 mg/ml in PBS for mice and 3 mg/ml in PBS for rats.
   c. Due to the solubility limit of BrdUrd use either bromodeoxycytidine or CldUrd for continuous labeling studies using miniosmotic pumps *in vivo*. Stir to dissolve over a low heat (∼35 °C). Filter and store working solutions in the dark at 4 °C, discard after 1 week.

3. *Ethanol*. Use 200 proof, absolute, from glass or plastic containers. (The large metal drums are a source of unwanted positively charged ions which predispose for protein precipitation on fixation.) Filter and store at −20 °C.

4. 2 N hydrochloric acid (HCl).

5. Digestive enzymes:
   a. Pepsin (EM Science, Cherry Hill, NJ). Working solution is 0.04% in 0.1 N HCl.
   b. Collagenase (Sigma Chemical). Working solution is 0.1% in PBS.

6. Sodium borate decahydrate (Sigma Chemical). Working solution is 0.1 M.

7. Tween 20 (polyoxyethylene-sorbitan monolaurate) (Sigma Chemical).

8. PBT = PBS + 0.5% Tween 20.

9. Bovine serum albumin (BSA) (Sigma Chemical).

10. PBTB = PBS + 0.5% Tween 20 + 0.5% BSA.

11. Normal goat serum (Sigma Chemical). Store at −20 °C.

12. PBTG = PBTB + 1.0% normal goat serum.

13. *Anti-BrdUrd, IU-4 nonconjugated* (Caltag, South San Francisco, CA). Many other antibodies and suppliers are available. Store at −20 °C. Stock aliquots are 5 μl/tube (depending on antibody activity) in PBT, stored at −20 °C.

14. *Second antibody (goat antimouse IgG-FITC conjugate)* (Sigma Chemical and many other suppliers). Store at −20 °C aliquoted in PBTG.

15. *PI* (Sigma Chemical). Store the powder at 4 °C, desiccated, in the dark. A stock solution is 1 mg/ml PI in 70% ethanol kept at 4 °C in the dark. The working solution is 10 μg/ml in PBTB. PI is a suspected carcinogen and should be handled with care.

IV. Methods

A. Labeling and Fixation *In Vitro* and *In Vivo*

1. *In Vitro* Labeling and Fixation of Cultured Cells

  a. Incubate with a final concentration of 1 μM BrdUrd (from 1 mM stock) for 20 min at 37 °C.

  b. Aspirate off medium. Rinse twice with warmed serum-free medium (do this quickly).

  c. Refeed with fresh whole medium that should be warmed and pregassed. Return to incubator (quickly).

  d. At desired time interval trypsinize the cells as follows:

    1. Aspirate off medium, rinse with fresh warmed serum-free medium.

    2. Add 1 ml of 0.05% trypsin and incubate for 5 min at 37 °C.

    3. Tap dish gently. Add 9 ml of whole medium (containing serum). Draw up and down and transfer to a 15-ml centrifuge tube. Draw up and down several times without bubbling.

    4. Reserve an aliquot (10–50 μl) for counting in a hemocytometer (preferred) or a Coulter counter. Record the total cell yield.

    5. Centrifuge (4 min at 350 × g) remaining cells to pellet.

  e. *Fixation.* The final cell concentration should be 2 × 10^6 cells per 2 ml of solution (or any multiple of this) in 65% ethanol in PBS. Adjust the following volumes depending on the actual cell count.

    1. Aspirate off medium. Add 0.7 ml of cold PBS slowly while vortexing the pellet.

    2. Continue vortexing and trickle in 1.3 ml freezer temperature 100% ethanol. Continue to vortex for 15 s.

Leave to fix at 4 °C overnight before staining. This fixation procedure is also suitable for cells acquired by tumor fine needle aspiration (FNA) biopsy, prepared from ascites tumors or blood or bone marrow preparations.

2. *In Vivo* Labeling and Fixation of Samples from Solid Tumors and Normal Tissues

    a. Infuse BrdUrd:

        1. 100–200 mg/m^2 BrdUrd (100 mg IdUrd) for humans as a 20-min intravenous infusion. (*Note*: This is given only under the authorization of the investigational drug mechanism for approved protocols.)

        2. 60 mg/kg intraperitoneally for mouse (i.e., 0.10 ml/10 g body weight of a 6 mg/ml solution).

        3. 30 mg/kg intraperitoneally for rats.

    b. Prepare a 15-ml centrifuge tube with 65% cold ethanol (in PBS); weigh the tube containing the mixture.

    c. On tissue receipt (25 mg is an operational minimum), coarsely mince tissue with scissors and place in tube.

    d. Vortex for 15–30 s to fix the tissue; reweigh (tube + ethanol + tissue chunks), store at 4 °C for at least overnight.

B. Sample Preparation

1. Staining Procedure for Cultured Cells, Tumor Fine Needle Aspiration Biopsies, Ascites Tumors, and Blood or Bone Marrow Preparations

    The details of step c below need to be determined by microscopic observation of the pepsin digestion.

    a. Vortex fixed cells (at low speed until evenly dispersed) for 15 s and transfer 4×10^6 cells to a 15-ml centrifuge tube. Centrifuge at $350 \times g$ for 4 min at room temperature. *Note*: if whole cells, rather than nuclei, are to be prepared, proceed to step e below.

    b. Add 5 ml of pepsin (0.04% in 0.1 N HCl) per 4×10^6 cells (minimum of 5 ml of pepsin even if you have fewer cells, otherwise there is a risk of cell loss due to cells sticking to the sides of the tube).

    c. Incubate for 10 min on a rocker at room temperature. (These are appropriate conditions for many laboratory cell lines.) Incubate other cells + pepsin for 10–60 min on a rocker, either at room temperature or at 37 °C.

    Note: Optimal incubation times and temperatures vary with different cell types and tissues; therefore, a time curve should be done for each cell type. If this is not possible (i.e., one FNA from a human breast tumor), frequent observation under the microscope during pepsin digestion is strongly advised.

    d. Centrifuge the tubes containing pepsin + nuclei, aspirate off the supernatant, vortex the pellet for 5–10 s.

e. Add 3 ml of 2 N HCl (1.5 ml per 2×10^6 nuclei or cells) to each tube while vortexing at low speed and incubate stationary for 20 min at 37 °C. Shake twice during incubation.

f. Add 6 ml of 0.1 M sodium borate to each tube while vortexing, continue vortexing for 10 s, centrifuge, and aspirate off the supernatant. Vortex the pellet, then add 6 ml of PBTB while vortexing, and centrifuge.

g. Aspirate off the supernatant and add 0.2 ml (per 2×10^7 or fewer nuclei) of the previously aliquoted anti-BrdUrd monoclonal antibody in PBT at 1:100 dilution (dilution varies depending on vendor and lot number and needs to be established for each new batch). Incubate for 60 min, at room temperature, in the dark.

h. Add 3 ml of PBTB while vortexing, centrifuge, then aspirate off the supernatant.

i. While vortexing the pellet add 0.2 ml (per 2×10^7 or fewer nuclei) of second antibody (goat antimouse-FITC) in PBTG at 1:100 dilution (actual dilution depends on vendor and lot number), and incubate for 45 min in the dark at room temperature.

j. Add 3 ml of PBTB while vortexing. Save an aliquot (10–50 μl) of the suspension for counting nuclei; then centrifuge and aspirate off the supernatant.

k. While vortexing the pellet add PI (10 μg/ml in PBTB) for a final concentration of 1×10^6 nuclei per ml (based on the counts made in j above). (If there are fewer than 10^6 nuclei in total, still use a minimum PI volume of 1 ml in order to guarantee stoichiometric staining.)

l. Store overnight at 4 °C in the dark; run on the flow cytometer the next day. (We have stored stained cells/nuclei at 4 °C for 3 months with no deterioration and, if prepared aseptically, specimens older than 5 years are still evaluable, but may need reincubating with the antibodies.)

2. Staining Procedure for Solid Tumors and Tissues

This is a general procedure and will need to be adjusted where noted for specific tissues. If aseptic techniques are used throughout then fixed tissues may be stored almost indefinitely.

a. Finely mince a portion of the ethanol-fixed tumor or tissue chunks in a preweighed 60-mm dish.
   1. Air dry for approximately 5 min to evaporate surplus ethanol (the tissue should not be allowed to dry out).
   2. Reweigh (dish + tissue fragments).
   3. Transfer the tissue fragments to a 50-ml Erlenmeyer flask.

b. From this point on, all reagent volumes are calculated for approximately 0.1 g of tissue.

    1. These volumes are minima; therefore, with less than 0.1 g of tissue still use these volumes.

    2. For more than 0.1 g of tissue use appropriate multiples of each volume.

  c. Assuming a potential cell yield of $1–2 \times 10^8$ cells/g of tissue, use one of the following two dissociation solutions:

    1. For collagen-rich tissue: add 5 ml of 0.1% collagenase (in PBS) to the 50-ml Erlenmeyer flask.

      a. Incubate for 15 min in a 37 °C shaker water bath (cover the top of the flask with parafilm).

      b. Add directly to the (collagenase + tissue) slurry 5 ml of pepsin (0.04% in 0.1 N HCl) and incubate further.

Proceed to the step d below.

    2. For other solid tissues and most rodent tumors: add 5.0 ml of 0.04% pepsin (in 0.1 N HCl) to the 50-ml Erlenmeyer flask.

  c. Incubate 20–90 min either in a 37 °C shaker water bath (cover top of flask with parafilm), or at room temperature, in which case use a 15-ml centrifuge tube on a rocker.

  d. *Pepsin digestion periods.* As for *in vitro* preparations this is the only step that is routinely variable. The optimum incubation time, temperature, and extent of agitation vary widely for different tumors and normal tissues and should be checked periodically (every 10 min) by microscope to monitor for clean nuclei (with little cytoplasm attached) and to obtain the maximum nuclei yield. We usually divide fixed chunks into two flasks (or centrifuge tubes) for two separate digestion times staggered apart by 10–20 min.

    1. Head and neck squamous cell carcinomas: typically between 20 and 60 min.

    2. Colorectal adenocarcinomas: typically between 40 and 90 min.

    3. Breast and bladder tumors: 15 min collagenase + typically 20–60 min in pepsin.

    4. Normal tissues vary widely and a time course should be performed to establish optimal conditions monitored by microscopic observation of clean nuclei yield.

Note: pepsin activity reduces to zero after much longer than 60 min at 37 °C. If further tissue digestion is required add a further 3–5 ml of prewarmed 0.04% pepsin.

  e. Aspirate (pepsin + nuclei slurry) with a 10-cc syringe attached to an 18-gauge needle, remove the needle, and filter the slurry suspension through a 35-μm nylon mesh into 15-ml centrifuge tube.

  f. Save an aliquot of the suspension for counting nuclei.

    1. Store the aliquot on ice until ready to count.

    2. Record the total volume of the suspension for yield calculations below, step k.

g. Centrifuge the tubes containing (collagenase), pepsin, and nuclei at $350 \times g$ for 4 min at room temperature.

h. Aspirate off the supernatant and add 1.5 ml of 2 N HCl while vortexing, incubate stationary for 20 min at 37 °C. Gently shake the tubes twice during incubation.

i. Add 3 ml of 0.1 M sodium borate while vortexing; continue vortexing for 10 s. Centrifuge as before.

j. Aspirate supernatant and add 3 ml of PBTB while vortexing, and centrifuge.

k. Count the nuclei from the reserved aliquot (f above) and calculate the total nuclei yield. [Total yield = (the average number of nuclei in one large square of a standard hemocytometer) $\times 10^4 \times$ (volume in ml of the total suspension).]

l. Add 0.2 ml per 2×10^7 nuclei of anti-BrdUrd monoclonal antibody in PBT at 1:100 dilution and incubate for 60 min at room temperature in the dark. Use a minimum volume of 0.2 ml in order to saturate the pellet. *Note*: antibody dilution varies depending on vendor and lot; test new batches against a standard cell line, for example, CHO cells.

m. Add 3 ml of PBTB to the anti-BrdUrd/nuclei suspension, mix gently, and centrifuge as before.

n. Aspirate off the supernatant and add 0.2 ml/2×10^7 nuclei of second antibody (goat antimouse FITC in PBTG at 1:100 dilution—dilution depends on the activity of a particular batch), and incubate for 45 min at room temperature in the dark.

o. Add 3 ml of PBTB to the suspension, mix gently, and centrifuge as before.

p. Aspirate off the supernatant and add PI (10 μg/ml in PBTB) so that the final concentration is 1×10^6 nuclei/ml suspension. *Note*: skilled personnel should lose no more than 50% of the initial number of nuclei (after pepsin) due to centrifugation and aspiration.

q. Store the nuclei suspension, for at least overnight, in PI at 4 °C in the dark.

C. Flow Cytometry and Data Acquisition

While almost any flow cytometer, equipped with a single argon ion laser with two photomultiplier tubes, may be adequate, some instruments are better suited to these procedures than are others. The instrument must be equipped with log amplification to accommodate the large range of anti-BrdUrd fluorescence signal. An optimal configuration includes narrow-beam excitation optics and hardware discrimination of doublets, together with the ability to configure the optical path so that there is no cross talk between the red and green channels. The default optical configuration of most three- or more color commercial flow cytometers may readily be changed for two-color work. Most bench-top sorters, with fixed light paths and significantly defocused excitation beams are less well suited for this task.

D. Data Analysis

1. Identification of Specific Subpopulations in the Bivariate DNA Versus BrdUrd Histogram (Cytogram)

The flow-cytometric data that may be obtained following these procedures are shown in Figs. 1 and 2. Figure 1 illustrates results from a mouse mammary carcinoma, MCa-4, following a pulse-label of 60 mg/kg intraperitoneal BrdUrd. The data are presented as DNA content (x-axis = PI fluorescence) versus BrdUrd content (y-axis = log fluorescein fluorescence). Tumors were excised either shortly (20 min) after labeling (Fig. 1A), or 3 (Fig. 1B) and 6 h later (Fig. 1C). These sampling times represent periods shorter than the duration of $G_2 + M$, longer than $G_2 + M$ but shorter than the duration of S phase, and approximately the duration of S phase, respectively, for this particular model system. The boxed regions indicate the identifiable subpopulations from which the fractional quantities needed for kinetic analysis may be calculated (see Section IV.D.2). Specific calculations from the data in Fig. 1B are given in Section IV.D.2.b. The lower figures in each graph of Fig. 1 show the PI fluorescence distributions of the BrdUrd-labeled cells from which the mean fluorescence intensity of those that remain undivided at the time of sampling, $\overline{F}_L(t)$, may be computed.

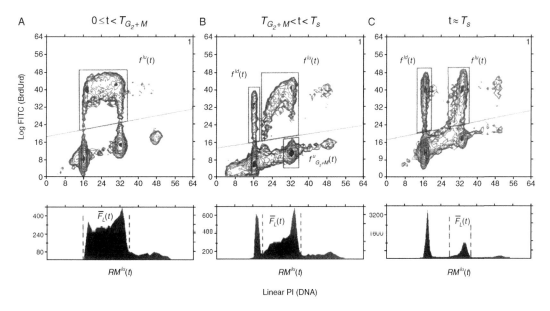

Fig. 1 Bivariate histograms of mouse mammary carcinoma MCa-4 showing DNA content (x-axis = PI fluorescence) and BrdUrd content (y-axis = log fluorescein fluorescence) following a pulse of 60 mg/kg intraperitoneal BrdUrd. Tumors were excised either shortly (20 min) after labeling (A), or 3 (B) and 6 h later (C). The boxed regions indicate the subpopulations from which the fractional quantities needed for kinetic analysis may be calculated. The lower figures in each panel show the PI fluorescence distributions of the BrdUrd-labeled cells from which their relative movements may be computed.

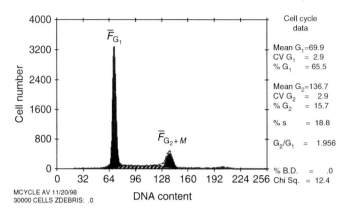

Fig. 2 The DNA content distribution (PI fluorescence) that pertains for any, and all, of the time-points shown in Fig. 1. The mean fluorescence channel numbers and positions for the G_1 and $G_2 + M$ peaks are indicated. The total number of events is calculated from a modeled fit to the data.

Figure 2 shows the DNA content distribution (PI fluorescence) that pertains for any, and all, of the sampling times shown in Fig. 1. The important quantities that need to be measured for subsequent analysis are the mean fluorescence channel numbers for the G_1 and $G_2 + M$ peaks, together with the total number of events, which may be estimated from a modeled fit to the data.

2. Estimation of Kinetic Parameters

Kinetic parameters may be estimated by analysis of the generated bivariate halogenated thymidine analog-DNA cytograms (Fig. 1), together with the total DNA distribution (Fig. 2), obtained at a time t after labeling by the following array of procedures. The particular procedure chosen depends partly on the time following labeling and partly on the number of experimental time-points available.

As shown by several authors the following measured quantities can provide specific information on kinetic parameters. It is important to observe, as seen in Table I, that these changing, dynamic quantities are informative at different times after labeling depending on the relationship between sampling times and the underlying cell-cycle parameters.

In Table I the plus signs indicate times when the measured quantity is changing and 0 and 1 are the values of the quantities when known.

The dynamic quantities listed in Table I, and selected references to their use, are

1. $f^u_{G_2 + M}(t)$, the fraction of unlabeled cells with $G_2 + M$ DNA content (White *et al.*, 1990a).

2. $f^{lu}(t)$, the fraction of labeled, undivided cells (Carlton *et al.*, 1991; Johansson *et al.*, 1998; White *et al.*, 1990a,b).

Table I

The Dynamic Quantities That May Be Measured from the Cytogram, and the Cell Cycle Periods for Which They May Be Used to Obtain Information About Kinetic Parameters

Measured quantity	Minimum $t \rightarrow$ Maximum $t \rightarrow$	0 T_{G_2+M}	T_{G_2+M} T_S	T_S $T_S + T_{G_2+M}$	$T_S + T_{G_2+M}$ $T_S + T_{G_2+M} + T_C$
$f^u_{G_2+M}(t)$		$+$	0	0	0
$f^{lu}(t)$		$+$	$+$	$+$	0
$f^{ld}(t)$		0	$+$	$+$	$+$
$RM^{lu}(t)$		$+$	$+$	1	0
$RM^l(t)$		$+$	$+$	$+$	$+$

3. $f^{ld}(t)$, the fraction of labeled, divided cells (Carlton *et al.*, 1991; White *et al.*, 1990a,b).

4. $RM^{lu}(t)$, the relative movement, between a G_1 and $G_2 + M$ DNA content, of labeled undivided cells (Begg *et al.*, 1985; Carlton *et al.*, 1991; Johansson *et al.*, 1996, 1998; Ritter *et al.*, 1992; White *et al.*, 1991).

5. $RM^l(t)$, the relative movement of all labeled cells (Johansson *et al.*, 1996; White *et al.*, 1991).

The relative movements, $RM^l(t)$ and $RM^{lu}(t)$ are defined through the equation (Begg *et al.*, 1985)

$$RM(t) = \frac{\bar{F}_L(t) - \bar{F}_{G_1}}{\bar{F}_{G_2+M} - \bar{F}_{G_1}} \tag{1}$$

and range between 0, when all the BrdUrd-labeled cells that remain undivided have a mean red fluorescence [$\bar{F}_L(t)$, see Fig. 1] equal to the mean G_1 DNA content ($\bar{F}_{G_1}$, Fig. 2), and 1 if all the undivided labeled cells have a $G_2 + M$ mean DNA content ($\bar{F}_{G_2+M}$, Fig. 2).

Evidently, an examination of the values of the measured quantities can immediately provide qualitative information on the durations of G_2 and mitosis, T_{G_2+M}, and S phase, T_S, for example, $f^{ld}(t) = 0$ implies $t \leftarrow T_{G_2+M}$ while $RM^{lu}(t) = 1$ suggests $t = T_S$.

Quantitatively, however, different circumstances will give rise to different types of measurements being made. In the clinical situation, typically only a single time-point sample is available whereas for laboratory data, either *in vitro* or *in vivo*, multiple time-point measurements might routinely be made. In what follows the methods for analysis are grouped according to both the time after labeling and the number of different sampling times from which measurements are available. It should be noted, however, that even where multiple time-point samples exist, informative measurements can be made on different quantities such that single time-point analytical methods may be appropriate. For example, single time measurements can be compared to multiple time estimates in order to gain insight into intra- versus intersample variability. In any case, careful observation of the

DNA versus BrdUrd histograms should be made prior to choosing the appropriate analytical approach.

a. Single Time-Point Measurements with $f^{ld}(t) = 0$

The procedure used here is based on methods described by Ritter *et al.* (1992). The first step computes T_S from $RM^{lu}(t)$, and the second step combines T_S and information about the fraction of labeled cells contained in $f^{lu}(t)$ and $f^u_{G_2 + M}(t)$, to determine the potential doubling time, T_{pot}. Define

$$v \equiv \ln\left[\frac{1 + f^u_{G_2 + M}(t) + f^{lu}(t)}{1 + f^u_{G_2 + M}(t)}\right], \tag{2}$$

for which it has been shown that

$$v = cT_S, \tag{3}$$

where $c = \ln(2)/T_{pot}$ and ln is the natural logarithm. Equations (2) and (3) may be derived from equations in White *et al.* (1990a). These equations are similar to those in Ritter *et al.* (1992) but are more accurate for short sampling times.

Using the value of v just obtained for cT_S, compute the initial intercept of the relative movement curve (Carlton *et al.*, 1991; White *et al.*, 1991):

$$RM^I \equiv \frac{1 - e^{-v}(1 + v)}{v(1 - e^{-v})}. \tag{4}$$

Using the relative movement of the BrdUrd-labeled undivided cells, we denote

$$RM\Delta t = RM^{lu}(t) - RM^I. \tag{5}$$

T_S can now be estimated from

$$T_S = \frac{t}{2RM\Delta t(1 + \sqrt{1 - 2RM\Delta t})} \tag{6}$$

with T_{pot} given by

$$T_{pot} = \frac{\ln(2)T_S}{v}. \tag{7}$$

Such short-time estimates should be expected to be unreliable.

b. Single Time-Point Measurements with $RM^{lu}(t) < 1$ and $f^{ld}(t) > 0$

The calculations are similar to, but are more likely to result in more precise parameter estimates than, those employed for shorter times. Define (White *et al.*, 1990a,b)

$$v \equiv ln\left[\frac{1 + f^{lu}(t)}{1 - f^{ld}(t)/2}\right]. \tag{8}$$

The initial intercept of the relative movement curve, RM^{II} under these circumstances, is given by (Carlton *et al.*, 1991)

$$RM^{II} \equiv \frac{1 - e^{-v} - ve^{-1.3v}}{v(1 - e^{-1.3v})}, \tag{9}$$

and now the changing relative movement of the BrdUrd-labeled undivided cells is

$$RM\Delta t = RM^{lu}(t) - RM^{II}. \tag{10}$$

T_S and T_{pot} are computed from

$$T_S = \frac{t}{2RM\Delta t} \tag{11}$$

$$T_{pot} = \frac{ln(2)T_S}{v}. \tag{12}$$

As an example, the data from Fig. 1B resulted in the following: $RM^{lu}(t) = 0.763$, $f^{lu}(t) = 0.166$, and $f^{ld}(t) = 0.045$, where $t = 3$ h. From these quantities, values of 8.9 h for T_S and 35 h for T_{pot} were computed (see also Section V.D).

c. Multiple Time-Point Measurements

There are several possible methods for fitting kinetic parameters from measurements made at multiple time-points after labeling (White *et al.*, 1990a,b, 1991). All of these provide, in contrast to the previously described methods, direct estimates of $T_{G_2 + M}$, T_S, and T_{pot}. The most complete approach is to fit $RM^{lu}(t)$ or $RM^{l}(t)$, $f^{lu}(t)$ and either $f^u_{G_2 + M}(t)$ or $f^{ld}(t)$, using nonlinear methods such as the Marquardt (1963) algorithm (Press *et al.*, 1992). For simplicity, however, the following linear functions may be used as a substitute.

1. Multiple time-point measurements with $f^u_{G_2 + M}(t) > 0$. Since

$$ln[1 + f^u_{G_2 + M}(t)] = cT_{G_2 + M} - ct, \tag{13}$$

standard linear regression packages may be used to compute c and T_{pot} as well as $T_{G_2 + M}$. Further T_S may be computed from Eq. (3).

2. Multiple time-point measurements with both $f^{ld}(t)$ and $f^{lu}(t) > 0$ In this case, we write

$$ln[1 + f^{lu}(t)] = c(T_S + T_{G_2 + M}) - ct \qquad (14)$$

$$ln[1 - f^{ld}(t)/2] = cT_{G_2 + M} - ct, \qquad (15)$$

and again linear regression may be used to obtain the kinetic parameters. Note that these equations hold for times up to $T_S + T_{G_2 + M}$ in contrast to single time-point methods which are limited to times shorter than T_S.

V. Critical Aspects of the Procedure

A. Labeling *In Vitro* and *In Vivo*

When labeling cells *in vitro*, it is important to minimize perturbation of the cultures. The parasynchrony induced by such insults as leaving the dishes out of the incubator for more than 1 min or so, or refeeding without using warmed, pregassed medium, will be readily discernible in the data. For *in vivo* work do not refill syringes with needles attached that have been used for injections; otherwise contamination of the stock BrdUrd solution will result.

B. Sample Preparation

Adequate sample preparation techniques have been identified as probably the most important requirement for production of the accurate flow-cytometric histograms that these studies need (Terry, 1996). This is particularly true in the case of preparations from solid tumors. Different pepsin digestion times of tumors can produce strikingly different flow-cytometric profiles despite their preparation from a homogenate of the same specimen. For example, tumors that under optimal digestion conditions would contain an aneuploid population might, if prepared inadequately, be misclassified as diploid. Because no pepsin digestion procedure can be considered "standard," even for tumors of similar histologies from similar sites, many problems may be obviated by making multiple digests, optimized for nucleus yield, from a homogenate of the same sample. In a comprehensive assessment of the sources of error in interlaboratory comparisons, Haustermans *et al.* (1995) also demonstrated that variations in the sample preparation and staining steps were the largest contributors to the overall variance.

The pepsin digestion step to give nuclei is the only part of the procedure that we routinely adjust. Depending on the sample properties, any or all of digestion time, temperature, and degree of agitation may be manipulated to optimize tissue disaggregation and nuclei yield. Frequent microscopic observation is the key to success. It should be noted that not all commercial pepsins are the same and, if bought in small quantities, may differ in their activity from batch to batch.

All centrifugation is at $350 \times g$ for 4 min at room temperature. This is a relatively gentle centrifugation and care should be taken not to disturb the pellet before aspirating off the supernatant.

Our approach to sample staining is very standardized. It is always based on knowledge of cell numbers, and the use of defined minimal volumes of reagents when counts are low. The goal is to approach a plateau of FITC staining, together with stoichiometric PI staining, 50- to 500-fold levels of FITC/background staining, and low CVs around the G_1 and G_{2+M} peaks of the DNA histogram. CVs of 2–3% about the G_1 peak are readily attainable for *in vitro* sample preparations. The 2.5–4% G_1 CVs are achievable goals for solid tumors and normal tissues. CVs in excess of 5% are generally unacceptable for subsequent quantitative analyses.

C. Flow Cytometry and Data Acquisition

High-quality flow cytometry is required for these analytical procedures. The samples must be adequately stained and, if nuclei are to be used (preferred), no undigested cells should remain in the sample. Flow should be stable as long run times may sometimes be required. The instrument should be in good optical alignment with no spectral overlap of the red and green channels. [Fluorescent signal compensation if required, but performed imprecisely, will compromise estimation of quantities such as $RM^{lu}(t)$.] For the same reason there must be good linearity of analog-to-digital converters and amplifiers. Depending on the time of sampling after labeling some of the fractional quantities required for analysis might be of low frequency. Hence, it is important to collect sufficient events, after hardware gating for doublet discrimination. For *in vitro* samples, 30,000 total nuclei usually suffice, DNA-aneuploid tumors usually require 50,000 gated events, and normal tissues, with low LI, may need 100,000 or more nuclei to be acquired.

Furthermore, Haustermans *et al.* (1995) have implicated placement of analytical regions on the flow-cytometric histograms as a significant source of potential error. Exploration of the data, by adjusting the regions of interest, usually gives sufficient feedback regarding the stability of estimates of $RM^{lu}(t)$, $f^{lu}(t)$, and $f^{ld}(t)$. There are objective criteria (White and Terry, 1992) that may be used to help distinguish BrdUrd labeled from unlabeled cells, and aid in analytical region placement, in instances when this distinction is not absolute.

D. Data Analysis

Before embarking on any numerical analysis, it is important to look carefully at the DNA versus BrdUrd histograms to ensure that the data are of sufficient quality that estimation of kinetic parameters is worthwhile. For example, in DNA-aneuploid samples multiple overlapping populations may hinder or preclude analysis. A special case is that of DNA-diploid tumors as we show in Figs. 1 and 2. In these circumstances, two completely overlapped populations (tumor parenchyma together with normal stromal and infiltrating cell populations) are present but indistinguishable, one from the other, based only on DNA contents. If tumor cells are in the majority then a minimum value for LI may be approximated and, while a reasonable value for T_S may often be obtained, the tumor T_{pot} has to be shorter than that value which results from analysis of the measured quantities. The use of whole cells, rather than nuclei, together with three-color flow cytometry, may be helpful in the case of DNA-diploid tumors if markers exist to discriminate tumor from normal cells (Begg and Hofland, 1991).

An important concept to appreciate is whether or not the time interval between BrdUrd labeling and sampling is greater, or less, than the duration of $G_2 + M (T_{G_2 + M})$. Depending on this timing, the calculation of the duration of S phase (T_S) differs by a factor of 2 (White and Meistrich, 1986). This can usually be readily deduced by inspection of the bivariate DNA versus BrdUrd flow-cytometric histogram and checking for the presence or absence of BrdUrd-labeled cells that have divided [$f^{ld}(t)$] in the period since labeling.

Although the kinetic parameters estimated by the analysis of the bivariate data are often treated as equivalent, it should be observed that there are subtle differences among them. Whereas T_S and $T_{G_2 + M}$ are direct measures of the duration of cell-cycle phases, based on the observed progression of cells, the potential doubling time is, in fact, a derived quantity depending on $RM^{lu}(t)$, $f^{lu}(t)$, and $f^{ld}(t)$. The interpretation of T_{pot} thus depends on the homogeneity of the populations making up the fractions of labeled cells. Thus, it may be necessary to fit overlapping DNA populations before computing $f^{lu}(t)$ and $f^{ld}(t)$ in order to obtain a T_{pot} value for a tumor subpopulation. Moreover, as also pointed out by Bertuzzi *et al.* (1997), changes in the pattern of cell loss can strongly influence the computed value of T_{pot}. Thus, caution is required in interpreting the relationships between T_{pot} values obtained from different tumors.

Acknowledgments

The authors thank Mrs Nalini Patel for her expert technical assistance. This work was supported by the National Institutes of Health Grant No. CA 06294 and the State of Texas Higher Education Coordinating Board Advanced Technology Program.

References

Antognoni, P., Terry, N. H. A., Richetti, A., Luraghi, R., Tordiglione, M., and Danova, M. (1998). The predictive role of flow cytometry-derived tumor potential doubling time (T_{pot}) in radiotherapy: Open questions and future perspectives (review). *Int. J. Oncol.* **12,** 245–256.

Begg, A. C. (1995). The clinical status of T_{pot} as a predictor? Or why no tempest in the T_{pot}! *Int. J. Radiat. Oncol. Biol. Phys.* **32,** 1539–1541.

Begg, A. C., and Hofland, I. (1991). Cell kinetic analysis of mixed populations using three-color fluorescence flow cytometry. *Cytometry* **12,** 445–454.

Begg, A. C., McNally, N. J., Shrieve, D. C., and Kärcher, H. (1985). A method to measure duration of DNA synthesis and the potential doubling time from a single sample. *Cytometry* **6,** 620–626.

Beisker, W., Dolbeare, F., and Gray, J. W. (1987). An improved immunocytochemical procedure for high-sensitivity detection of incorporated bromodeoxyuridine. *Cytometry* **8,** 235–239.

Bertuzzi, A., Gandolfi, A., Sinisgalli, C., Starace, G., and Ubezio, P. (1997). Cell loss and the concept of potential doubling time. *Cytometry* **29,** 34–40.

Bussink, J., Terry, N. H. A., and Brock, W. A. (1995). Cell cycle analysis of synchronized Chinese hamster cells using bromodeoxyuridine labeling and flow cytometry. *In Vitro Cell. Dev. Biol.* **31,** 547–552.

Carayon, P., and Bord, A. (1992). Identification of DNA-replicating lymphocyte subsets using a new method to label the bromo-deoxyuridine incorporated into the DNA. *J. Immunol. Methods* **147,** 225–230.

Carlton, J. C., Terry, N. H. A., and White, R. A. (1991). Measuring potential doubling times of murine tumors using flow cytometry. *Cytometry* **12,** 645–650.

Dolbeare, F., and Gray, J. W. (1988). Use of restriction endonucleases and exonuclease III to expose halogenated pyrimidines for immunochemical staining. *Cytometry* **9,** 631–635.

Dolbeare, F. A., Gratzner, H. G., Pallavicini, M. G., and Gray, J. (1983). Flow cytometric measurement of total DNA content and incorporated bromodeoxyuridine. *Proc. Natl. Acad. Sci. USA* **80,** 5573–5577.

Dolbeare, F., Kuo, W. L., Beisker, W., Vanderlaan, M., and Gray, J. W. (1990). Using monoclonal antibodies in bromodeoxyuridine-DNA analysis. *In* "Methods in Cell Biology" (Z. Darzynkiewicz and H. Crissman, eds.), Vol 33, p.207. Academic Press, San Diego, CA.

Dubray, B., Maciorowski, Z., Cosset, J.-M., and Terry N. H. A. (1995). Le point sur le temps de doublement potentiel. *Bull. Cancer/Radiother.* **82,** 331–338.

Gratzner, H. (1982). Monoclonal antibody to 5-bromo and 5-iododeoxyuridine: A new reagent for detection of DNA replication. *Science* **218,** 474–475.

Haustermans, K., Hofland, I., Pottie, G., Ramaekers, M., and Begg, A. C. (1995). Can measurements of potential doubling time (T_{pot}) be compared between laboratories? A quality control study. *Cytometry* **19,** 154–163.

Johansson, M. C., Baldetorp, B., Bendahl, P. O., Fadeel, I. A., and Oredsson, S. M. (1996). Comparison of mathematical formulas used for estimation of DNA synthesis time of bromodeoxyuridine-labelled cell populations with different proliferative characteristics. *Cell Prolif.* **29,** 525–538.

Johansson, M. C., Johansson, R., Baldetorp, B., and Oredsson, S. M. (1998). Comparison of different labelling index formulae used on bromodeoxyuridine-flow cytometry data. *Cytometry* **32,** 233–240.

Larsen, J. K. (1990). Washless double staining of a nuclear antigen (Ki-67 or bromodeoxyuridine) and DNA in unfixed nuclei. *In* "Methods in Cell Biology" (Z. Darzynkiewicz and H. Crissman, eds.), Vol. 33, p. 227. Academic Press, San Diego, CA.

Marquardt, D. W. (1963). An algorithm for least-squares estimation of nonlinear parameters. *J. Soc. Indust. Appl. Math.* **11,** 431–441.

Pollack, A., Terry, N. H. A., Van, N. T., and Meistrich, M. L. (1993a). Flow cytometric analysis of two incorporated halogenated thymidine analogues and DNA in a mouse mammary tumor grown *in vivo*. *Cytometry* **14,** 168–172.

Pollack, A., White, R. A., Cao, S., Meistrich, M. L., and Terry, N. H. A. (1993b). Calculating potential doubling time (T_{pot}) using monoclonal antibodies specific for two halogenated thymidine analogues. *Int. J. Radiat. Oncol. Biol. Phys.* **27,** 1131–1139.

Pollack, A., Terry, N. H. A., White, R. A., Cao, S., Meistrich, M. L., and Milas, L. (1994). Proliferation kinetics of recruited cells in a mouse mammary carcinoma. *Cancer Res.* **54,** 811–817.

Pollack, A., Terry, N. H. A., Wu, C. S., Wise, B. M., White, R. A., and Meistrich, M. L. (1995). Specific standing of iododeoxyuridine and bromodeoxyuridine in tumors double-labelled *in vivo*: A cell kinetic analysis. *Cytometry* **20,** 53–61.

Press, W. H., Teukolsky, S. A., Vetterling, W. T., and Flannery, B. P. (1992). "Numerical Recipes in FORTRAN. The Art of Scientific Computing," 2nd edn., Cambridge University Press, Cambridge.

Ritter, M. A., Fowler, J. F., Kim, Y., Lindstrom, M. J., and Kinsella, T. J. (1992). Single biopsy, tumor kinetic analyses: A comparison of methods and an extension to shorter sampling intervals. *Int. J. Radiat. Oncol. Biol. Phys.* **23,** 811–820.

Sacks, P. G., Oke, V., Calkins, D. P., Vasey, T., and Terry, N. H. A. (1990). Effects of β-all-trans retinoic acid on growth, proliferation and cell death in a multicellular tumor spheroid model for squamous carcinomas. *J. Cell. Physiol.* **144**, 237–243.

Schutte, B., Reynders, M. M. J., van Assche, C. L. M. V. J., Hupperets, P. S. G. J., Bosman, F. T., and Blijham, G. H. (1987). An improved method for the immunocytochemical detection of bromodeoxyuridine labeled nuclei using flow cytometry. *Cytometry* **8**, 372–376.

Terry, N. H. A. (1996). Predictive assays for radiotherapy: The role of tumor proliferation (T_{pot}) measurements. *Onkologie* **19**, 322–327.

Terry, N. H. A., and Peters, L. J. (1995). The predictive value of tumor-cell kinetic parameters in radiotherapy: Considerations regarding data production and analysis. *J. Clin. Oncol.* **13**, 1833–1836.

Terry, N. H. A., and White, R. A. (1996). Lessons from multiparameter thymidine analogue-DNA flow cytometry for one parameter DNA cytometry. *Clin. Immunol. Newslett.* **16**, 46–50.

Terry, N. H. A., White, R. A., and Meistrich, M. L. (1992). Cell kinetics: From tritiated thymidine to flow cytometry. *Br. J. Radiol. Suppl.* **24**, 153–157.

Terry, N. H. A., Milross, C. G., Patel, N., Mason, K. A., White, R. A., and Milas, L. (1997). The effect of paclitaxel on the cell cycle kinetics of a murine adenocarcinoma *in vivo*. *Breast J.* **3**, 99–105.

White, R. A., and Meistrich, M. L. (1986). A comment on "A method to measure the duration of DNA synthesis and the potential doubling time from a single sample." *Cytometry* **7**, 486–490.

White, R. A., and Terry, N. H. A. (1992). A quantitative method for evaluating bivariate flow cytometric data obtained using monoclonal antibodies to bromodeoxyuridine. *Cytometry* **13**, 490–495.

White, R. A., Terry, N. H. A., and Meistrich, M. L. (1990a). New methods for calculating kinetic *in vitro* properties using pulse labeling with bromodeoxyuridine. *Cell Tissue Kinet.* **23**, 561–573.

White, R. A., Terry, N. H. A., Meistrich, M. L., and Calkins, D. P. (1990b). Improved method for computing potential doubling time from flow cytometric data. *Cytometry* **11**, 314–317.

White, R. A., Terry N. H. A., Baggerly, K. A., and Meistrich, M. L. (1991). Measuring cell proliferation by relative movement. I. Introduction and *in vitro* studies. *Cell Prolif.* **24**, 257–270.

White, R. A., Fallon, J. F., and Savage, M. P. (1992). On the measurement of cytokinetics by continuous labeling with bromodeoxyuridine with applications to chick wing buds. *Cytometry* **13**, 553–556.

White, R. A., Pollack, A., and Terry, N. H. A. (1994a). Simultaneous cytokinetic measurement of aneuploid tumors and associated diploid cells following continuous labelling with chlorodeoxyuridine. *Cytometry* **15**, 311–319.

White, R. A., Pollack, A., Terry, N. H. A., Meistrich, M. L., and Cao, S. (1994b). Double labeling to obtain S-phase subpopulations; Application to determine cell kinetics of diploid cells in an aneuploid tumor. *Cell Prolif.* **27**, 123–127.

CHAPTER 18

Flow Cytometric Analysis of Cell Division History Using Dilution of Carboxyfluorescein Diacetate Succinimidyl Ester, a Stably Integrated Fluorescent Probe

A. Bruce Lyons, ★ **Jhagvaral Hasbold,** † **and Philip D. Hodgkin** ★,†,‡

★Discipline of Pathology
Faculty of Health Science
The University of Tasmania
Hobart, Australia

†The Centenary Institute of Cancer Medicine and Cell Biology
Sydney, Australia

†Medical Foundation
University of Sydney
Sydney, Australia

ESSENTIAL CYTOMETRY METHODS
Reprinted from *Methods in Cell Biology*, Volume 63 (Academic Press, 2001).
Copyright © 2001 by Academic Press, Inc.,
All rights of reproduction in any form reserved.

DOI: 10.1016/B978-0-12-375045-7.00018-0

I. Introduction and Background

The cells of the immune system undergo significant expansion and differentiation as a result of immune stimulation, as well as during the production of the formed hemopoietic elements. In contrast to tissues with defined microanatomy, where associations between cells and structures are relatively simple to determine, the mobile nature of lymphohemopoietic cells makes it more difficult to define lineage relationships.

Here, we review the use of a technique based on the serial dilution of a stably binding intracellular fluorochrome, carboxyfluorescein diacetate succinimidyl ester (CFSE), which allows 8–10 sequential cell divisions to be analyzed by flow cytometry (Fig. 1). When incubated with cells, the fluorescein-based CFSE crosses the cell membrane and attaches to free amine groups of cytoplasmic cell proteins. After enzymatic removal of carboxyl groups by endogenous intracellular esterases, CFSE acquires identical spectral characteristics to fluorescein, with optimal excitation by 488 nm argon laser light, emitting strongly at 519 nm, and as such is compatible with almost all single and multiple laser flow cytometers. On cell division, CFSE is distributed equally between progeny, allowing the division history of a cell population to be determined. This technique can be used to investigate the behavior of cells *in vitro*, as well as division of transferred cells *in vivo*.

Other competing techniques for monitoring cell proliferation, such as the use of tritiated thymidine incorporation, can quantify overall division behavior of a population but give no information on the division history of individual cells. Furthermore, appropriately conjugated monoclonal antibodies can be employed to identify the cells undergoing division and whether their phenotypic properties change with division number. A major advantage of the CFSE-based technique is that viable cells from defined division cycles can be recovered, allowing functional characteristics to be related to differentiation stage.

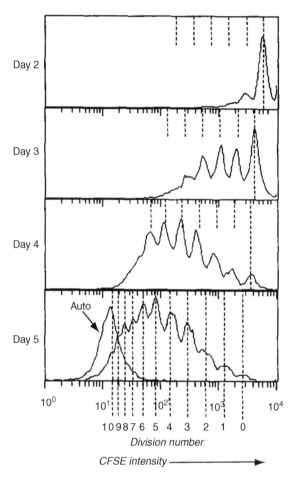

Fig. 1 Tracking division of B lymphocytes using CFSE dilution. Small dense murine B cells labeled with CFSE were cultured for up to 5 days with supernatants containing interleukin (IL)-5 (31 U/ml) and IL-4 (140 U/ml) and a source of CD40L in the form of membranes from an activated T helper cell clone (H66.61). At intervals of 2, 3, 4, and 5 days, cells were harvested and analyzed by flow cytometry. Division is characterized by sequential twofold reduction in CFSE fluorescence, resulting in equally spaced peaks on a logarithmic scale. Dashed lines indicate the division cycle number for each panel. Note the slow decay in intensity of peaks independent of division. The arrow indicates the autofluorescence level of stimulated but unstained cells. (Reproduced from Hodgkin *et al.*, 1996, by copyright permission of The Rockefeller University Press.)

Since the introduction of the CFSE-based technique in 1994, it has become the method of choice for investigating the division-related differentiation of lympho-hemopoietic cells, as well as the kinetics of cellular expansion during an immune response. Among the cell types that have been investigated using the technique are hemopoietic stem cells, T and B lymphocytes, natural killer (NK) cells, and a number of cell lines.

II. Reagents and Solutions

A. Carboxyfluorescein Diacetate Succinimidyl Ester Stock

A stock solution of 5 mM CFSE [5(and 6)-carboxyfluorescein diacetate succinimidyl ester] (MW 557, Molecular Probes, Eugene, OR) is made by dissolving CFSE in DMSO at a concentration of 2.785 mg/ml. Gentle pipetting is sometimes necessary to achieve this. The stock solution is aliquoted into convenient volumes (e.g., 50 μl) and stored frozen at -20 °C under dessicating conditions. Stock solutions can be kept for over 1 year under these conditions. To ensure reproducibility of staining intensity, aliquots are thawed a maximum of three times and then discarded.

B. Buffers and Culture Medium

Standard isotonic phosphate-buffered saline (PBS), pH 7, is usually employed in the staining procedure; however, culture medium such as RPMI with no added serum has also been successfully used. The addition of a small amount of protein such as bovine serum albumin (BSA) in the staining procedure can improve poststaining viability (see later).

For culture of cells after staining, standard culture medium such as RPMI or DMEM supplemented with 5–10% serum is routinely used.

III. Preparation and Labeling of Cells

A. Labeling of Cells with CFSE (Standard Protocol)

This standard protocol has been successfully employed to stain B and T lymphocytes of both mouse and human origin. There have been a number of "in house" modifications of the original staining procedure, including staining on ice or at room temperature, which can sometimes improve viability of sensitive cells. However, the standard protocol will be suitable for most applications.

Cells to be labeled are suspended at 5×10^7 ml in PBS/0.1% BSA, ensuring that cells are well suspended with no aggregates. The resolution of cell division is critically dependent on uniformity of staining. The inclusion of a small amount of BSA in the staining step does not markedly affect the intensity of staining achieved, but can improve the viability of cells, especially when sensitive cells such as B lymphocytes are stained. Alternatively, cells can be stained in a culture medium such as RPMI without added serum, containing 0.1% BSA which can also improve poststaining viability.

To each milliliter of cell suspension, 2 μl of CFSE stock solution is added and immediately mixed to ensure uniform staining, resulting in a final concentration of 10 μM CFSE. The cells are incubated for 10 min at 37 °C, and the labeling is quenched by adding five volumes of ice-cold RPMI/10% fetal calf serum (FCS).

Resolution of division-related fluorescence peaks is critically dependent on achieving uniform staining of the starting population, so it is essential to mix cell suspension well at the point of addition of CFSE. Another way to ensure even staining is to resuspend cells at 10^8 ml and add an equal volume of CFSE prediluted to 20 μM. Due to the labile nature of CFSE, it is important to predilute immediately before adding to the cell suspension. More detailed discussion of the methodology can be found in Lyons and Doherty (1998).

B. Determining Appropriate Levels of CFSE Staining Intensity

The standard protocol is suitable for tracking cell division from day 2 to 14 after staining, both *in vitro* (Fig. 1) and *in vivo*. For time points earlier than 2 days CFSE staining intensity may need to be reduced. This is because within the first few days after staining, there is a sharp drop in CFSE fluorescence as rapidly turned over components are catabolized. This period is then followed by a slower loss of fluorescence (Fig. 2). Note that this fluorescence loss occurs in the absence of cell division, and also occurs in divided cells, so that the relationship between fluorescence intensity between cells undergoing different numbers of cell divisions remains constant (see Fig. 1). In addition, when a CFSE-stained population of cells is both cultured *in vitro* and transferred *in vivo,* the rate of loss of CFSE in undivided cells is the same under both conditions (Lyons, 1999). As the relationship between fluorescence intensity obtained is essentially linear with respect to

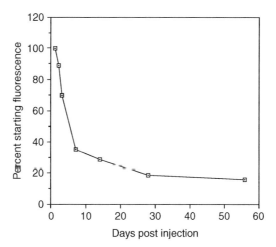

Fig. 2 Decay of CFSE fluorescence intensity of nondivided cells. Murine splenic lymphocytes 2×10^7, labeled with CFSE were injected intravenously into syngeneic hosts. At various time intervals, spleens were removed and analyzed by flow cytometry. The fluorescence intensity of nondividing cells was determined, and expressed as a percentage of starting CFSE fluorescence intensity. (From Lyons and Doherty, 1998. Copyright 1998 John Wiley & Sons. Reprinted by permission of Wiley-Liss, Inc., a subsidiary of John Wiley & Sons, Inc.)

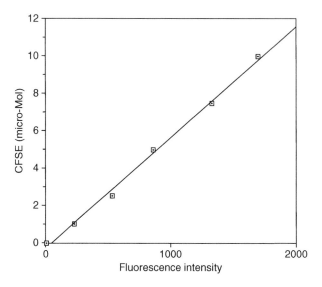

Fig. 3 Linearity of staining with respect to concentration of CFSE. Murine splenic lymphocytes were stained using the standard protocol, except differing final concentrations of CFSE were used. Fluorescence intensity (arbitrary units) was plotted against CFSE concentration, showing staining is linear with respect to CFSE concentration. (From Lyons and Doherty, 1998. Copyright 1998 John Wiley & Sons. Reprinted by permission of Wiley-Liss, Inc., a subsidiary of John Wiley & Sons, Inc.)

CFSE concentration (Fig. 3), the fluorescence intensity can be manipulated to suit the experiment. For example, lower concentrations can be used for short experiments, whereas for extremely long-term experiments, such as transfer and tracking of cells *in vivo,* a high initial staining intensity may be required. However, it is important to consider that CFSE labeling may be toxic to some cell types at high concentrations. For this reason, the experimenter will need to determine appropriate levels of staining for each application. Note that very intensely CFSE-stained cells may make flow cytometric compensation difficult or impossible due to very bright fluorescence spilling over into orange and red channels. Conversely, understaining will limit the number of cycles of division that can be resolved.

C. Improving Resolution by Presorting

If the starting population of cells being stained is heterogenous with respect to size, this will result in a broad range of starting fluorescence intensities. As a result, the separation of division-related peaks will not resolve as well as a more uniformly labeled starting population. Some researchers have overcome this limitation by presorting the stained cell population such that their fluorescence intensity is over a 40-channel interval on a 1024-channel scale (Nordon *et al.*, 1997), ensuring high resolution of division. The only caveat for this approach is to ensure that sorting for a narrow range of fluorescent intensity does not select for a distinct subset of the starting population, as this approach will tend to select cells on the basis of size.

IV. Gathering of Information Concurrent with Division

A. Cell Surface Marker Staining

The spectral characteristics of CFSE are essentially identical to fluorescein making it possible to monitor cell phenotype and division history simultaneously by the use of specific antibodies coupled with compatible fluorochromes.

Cells are labeled with CFSE by the previously mentioned standard method, and then cultured appropriately to induce proliferation and differentiation. Cells are harvested at various time points after cell culture initiation and incubated with surface marker specific antibodies for 45 min followed by secondary conjugate on ice. Cells are washed two times with ice-cold PBS/0.1% BSA/0.1% sodium azide between incubations. Stained cells are then analyzed on a flow cytometer. Appropriate fluorochromes for use in conjunction with CFSE for a single laser (488 nm) are phycoerythrin (PE) and tandem dyes such as PE/Texas Red or PE/Cy5, or alternatively peridinium chlorophyll protein (PerCP). More elaborate multilaser cytometers will support the use of other fluorochromes such as allophycocyanin (APC), allowing more complex analyses to be performed. Electronic compensation between detecting channels should be performed according to the instructions of the instrument manufacturer. However, very bright staining with CFSE may result in compensation being difficult to achieve, in which case it may be necessary to adjust labeling to suit the experiment.

It is recommended to have *in vitro* culture controls for flow cytometry analysis. The first control is use of CFSE-labeled cells either unstimulated in culture or stimulated in such a way as to maintain viability without inducing division. For example, interleukin (IL)-4 will keep resting B cells alive for 3–5 days of *in vitro* culture, without inducing proliferation. This control is important for marking the starting point for calibrating division number (i.e., division 0). The second control is use of cells that have not been CFSE labeled but stimulated to proliferate. This control is for the estimation of autofluorescence intensity, which provides information on the number of divisions that can be tracked. As the CFSE labeling can vary between experiments, in some cases, the intensity of CFSE level may interfere with the other channels in flow cytometry analysis. Therefore, it is important to use single color CFSE-labeled cells to check and appropriately adjust compensation before analysis.

In response to various stimuli immune cells proliferate and differentiate. The differentiation of activated cells is usually associated with the appearance or downregulation of cell surface markers. As the stimulated cells progress through the differentiation pathway, they display a distinct composition of surface molecules that help determine the cells functional capabilities. A number of variables can affect the differentiation profile of cultured cells such as the time of culture or the rate of proliferation. The CFSE-labeling method provides a unique opportunity to separate these two variables and examine the changes of cell surface molecule expression in relation to cell division number. An example is shown in Fig. 4, which demonstrates the division-related changes in immunoglobulin (Ig) isotype expressed on CFSE-labeled murine B cells. This method is based on small dense B cells purified from murine spleen; however, differentiation changes can be successfully followed in *in vitro* stimulated murine T cells (Gett and Hodgkin, 1998), as well as in human peripheral B and T lymphocytes (Fig. 6) (Kindler and Zubler, 1997).

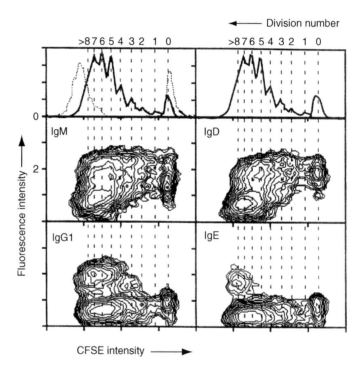

Fig. 4 Isotype switching to IgE requires more rounds of cell division than switching to IgG1. Small dense murine B lymphocytes were cultured in the presence of IL-4 (1000 U/ml) and an optimal concentration of plasma membrane of a CD40L transfected insect cell line for 5 days. The culture was harvested and stained for surface expression of different immunoglobulin isotypes. The top panels show histograms of CFSE fluorescence, with the dotted histogram on the left representing autofluorescence of unstained stimulated cells, and the dotted histogram on the right representing control unstimulated CFSE-stained cells. Dashed lines represent positions of the cell division peaks. Note that switching to IgG1 occurs after three cell divisions, but acquisition of surface IgE requires at least six rounds of cell division. (Reproduced from Hasbold *et al.*, 1998, and is used with kind permission of the publisher.)

B. Intracellular Staining

As described earlier, double staining of CFSE-labeled cells for cell surface markers provides unique information about cell division regulated differentiation. It has also proved possible to couple the CFSE method with the detection of intracellular components. Potentially, the intracellular staining method can be used to detect any cytoplasmic or nuclear molecule, depending on the sensitivity of antibody used. This method has also proved to be more sensitive than surface staining for some rare markers, especially when they are difficult to detect by a standard surface staining techniques, such as IgE.

Many different procedures have been described for fixation and permeabilization of cells for intracellular staining. The primary concern in fixing and permeabilizing CFSE-labeled cells is to retain the CFSE and, at the same time, to preserve the cell morphology and interior for antibody recognition. Most fixation and permeabilization methods described in the literature were not suitable for this purpose. However, an adaptation of a method using paraformaldehyde and Tween 20 detergent proved successful.

Appropriately stimulated CFSE-labeled cells ($\sim 10^6$/sample) are fixed with 0.25 ml 2% paraformaldehyde (PFA) for 10 min at room temperature followed by the addition of 0.25 ml PBS and 0.5 ml 0.2% Tween 20 solution. The cells are then mixed on a vortex and left overnight at 4 °C to permeabilize. Next day, cells are centrifuged and the pellets resuspended in PBS containing 0.1% BSA and 0.1% sodium azide, and left on ice for a further 30 min to block any free PFA residues. Note that after overnight incubation with PFA and Tween 20, plastic tubes usually become highly hydrophobic, and it may be difficult to remove supernatants after centrifugation. Changing tubes after overnight incubation and/or the addition of 0.05% Tween 20 to the washing buffer (PBS/BSA/0.05% Tween 20) for subsequent washes will help reduce cell losses. For intracellular staining of fixed and permeabilized cells any standard staining protocol for flow cytometry is suitable. However, if streptavidin reagents are used higher backgrounds may occur due to the nonspecific binding to cytoplasmic components.

C. Bromodeoxyuridine and DNA Staining

Commonly used methods to assess overall cell proliferation, such as tritiated thymidine incorporation and bromodeoxyuridine (BrdU) incorporation provide a comparative cumulative assessment of cell proliferation. However, BrdU staining in combination with the CFSE-labeling technique also allows the examination of the division-related proliferation rate and cell cycle profile as well as division history. The following method is designed to detect simultaneously cell division history, cell cycle position, and proliferation rate in a population of dividing cells by flow cytometry.

CFSE-labeled cells are stimulated and pulsed for 3 h before harvest with BrdU (100 μg/ml final concentration) on various days of culture. The harvested cells are fixed and permeabilized with PFA and Tween 20 as described earlier. The method of Tough and Sprent (1994) is then used to detect the level of BrdU incorporation. Cells are treated with DNase I (20 μg/ml final concentration diluted in 50 mM Tris-HCl buffer, pH 7.4, with 10 mM $MgCl_2$) for 30 min at 37 °C in a water bath. Subsequently, cells are stained with an appropriate BrdU-specific antibody conjugate. For obvious reasons, this antibody should be conjugated to fluorochromes other than fluorescein isothiocyanate (FITC). If BrdU-specific antibodies are not conjugated to fluorochromes and secondary antibody or avidin conjugate is used, the following staining controls are suggested: (1) fixed and permeabilized cells stained with secondary conjugate only and (2) cells not pulsed with BrdU but treated and stained in same way as BrdU pulsed samples. In some instances, especially when cells are proliferating at a high rate, a dramatic reduction of BrdU incorporation occurs after 72 h of *in vitro* culture due to cell overgrowth and medium depletion. Therefore, it is advisable to keep the initial starting cell density low for day 4 and 5 cultures. CFSE-labeled and BrdU-stained cells run on the flow cytometer allow an estimation of the rate of cell division and how it might differ at each division (see Fig. 5). If BrdU is being detected with a PE-conjugated antibody it is also possible to monitor simultaneously the position of each cell within the cell cycle. The DNA intercalating molecule 7-aminoactinomycin D (7-AAD) fluoresces at 647 nm and is compatible with both CFSE and PE in a single laser (488 nm) flow cytometer (Rabinovitch *et al.*, 1986). After detection of BrdU, CFSE-labeled fixed cells are incubated with 7-AAD at 1 μg/ml for 30 min on ice, before washing and analysis by flow cytometry. The 7-AAD channel should be collected in linear acquisition mode. Figure 5C indicates the simultaneous staining with CFSE and 7-AAD.

D. The Source of Division Asynchrony in Cultures

One of the most curious features of CFSE profiles is the tremendous variation in the number of divisions that otherwise identical cells will undergo at a given time point after stimulation. This level of asynchrony is affected by time and the strength of stimulation (Hodgkin *et al.*, 1997). BrdU pulsing can be used to determine whether the source of asynchrony is either due to differing rates of cell division, or variation in the time a cell responds to activation signals eliciting division. Figure 5B shows that in the case of B cells stimulated with CD40L, the rate of division is constant once cells enter their first division cycle. Therefore, at the population level, there is a difference in time taken to enter first division among apparently homogeneous cells. This procedure can also be followed *in vivo* as mice carrying transferred CFSE-labeled cells can be fed BrdU to reveal information about division asynchrony.

E. Cell Sorting on CFSE Peaks for Functional and Analytical Studies

A major advantage of the CFSE dye dilution technique is the ability to recover viable cells that have undergone defined numbers of rounds of cell division. Sorting profiles can be further refined by expression of one or more surface markers, in combination with division history.

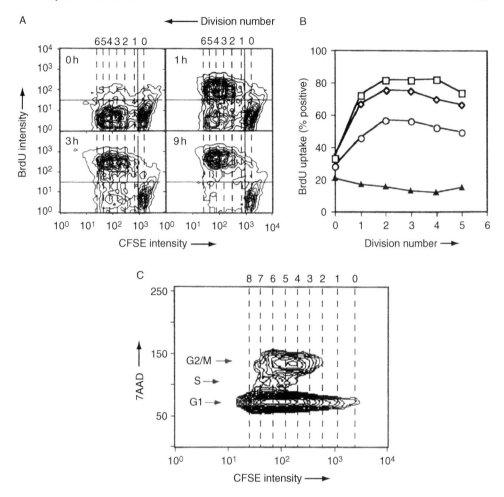

Fig. 5 Simultaneous use of bromodeoxyuridine incorporation and CFSE dilution enables rate of cell division to be compared. CFSE-stained B cells were cultured for 3 days under the same conditions as in Fig. 1. For the final 1, 3, or 9 h of incubation, cultures were pulsed with 50 μg BrdU. Cells were harvested, permeabilized, and stained with a biotinylated anti-BrdU antibody detected with streptavidin tricolor. (A) Contour plots of CFSE versus BrdU incorporation. From this data, the percentage of BrdU positive cells in each division cycle was determined, and represented graphically in (B), showing 9-h pulse (open squares), 3-h pulse (open diamonds), 1-h pulse (open circles), and 0-h pulse (background control, closed triangles). (C) A contour plot of 7-AAD stained cells. CFSE-labeled cells were stimulated with CD40L and IL-4. Four days later cells were fixed and permeabilized, and stained with 10 ng/ml 7-AAD. Dashed lines represent positions of the cell divisions. (A and B are reproduced from Hasbold *et al.*, 1998, and are used with kind permission of the publisher.)

Cells recovered in this way can be analyzed in functional assays, such as secretion of cytokines or antibodies (Gett and Hodgkin, 1998), or ability to lyse targets in the case of cytotoxic T cells. Cells can also be restained with CFSE to assess further proliferative potential. Sorted cells can be used as a source of RNA or DNA to assess levels of specific message for cytokines and other products, or to determine genomic recombination events in antigen receptors of B and T cells, or changes in DNA methylation patterns associated with differentiation.

F. Tracking the Fate of Dead Cells

Immune cells in culture exhibit varying amounts of cell death, usually by the process of apoptosis. In many instances, it is helpful to determine what contribution cell death makes to the overall proliferation behavior of cells at the population level. To investigate cell death, the membrane integrity of harvested CFSE-labeled cells can be probed with nucleic acid binding fluorochromes. Again, 7-AAD is especially useful, as it is compatible with fluorochromes such as PE, allowing surface marker expression to be simultaneously analyzed along with division and cell death. Unfixed, CFSE-stained cells are stained on ice for 30 min with 1 μg/ml 7-AAD, and washed prior to flow cytometric analysis. Alternatively, CFSE division analysis can be combined with staining for annexin V to simultaneously detect apoptotic cells (Warren and Kinnear, 1999).

On dying, CFSE-stained cells retain their fluorescence, allowing the division cycle in which cell death occurred to be determined (Hasbold *et al.*, 1999; Wells *et al.*, 1997).

===== ## V. Analysis of Data

A. Monitoring the Position of the Undivided Group

After labeling with CFSE, the small resting population runs as a log-normal distribution with respect to fluorescence intensity, as shown in Fig. 1. As mentioned previously, the mean intensity of labeled cells progressively reduces with time in culture even without any cell division (Figs. 1 and 2). Therefore, for any quantitative analysis of division history it is important to determine the mean fluorescence intensity of an undivided control group of cells. This control is usually provided by cells cultured under similar conditions but without any proliferative stimulation. A potential problem with this procedure would result if unstimulated cells lost CFSE intensity at a different rate to activated cells. Remarkably, this does not occur. Figure 6 shows fluorescence histograms and dot plots of forward scatter against CFSE fluorescence of small resting and larger activated cells at day 3 after stimulation. Clearly, the mean CFSE staining intensity of the small cells in both unstimulated and stimulated cultures is identical to that of the large activated cells, and the intensity predicted from twofold dilution of subsequent divisions is faithfully preserved. The consistent, proportional lessening of CFSE intensity across all divisions is also apparent from the time course data shown in Fig. 1. For this

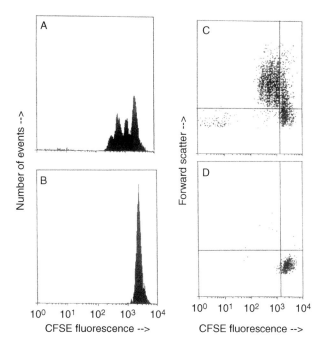

Fig. 6 Fidelity of CFSE partitioning after activation. Human mononuclear cells were activated *in vitro* with PHA and sheep red blood cells. After 3 days, control (B and D) and activated (A and C) cells were harvested and analyzed by flow cytometry. (A, B) Histograms of CFSE fluorescence intensity (C, D), CFSE fluorescence intensity versus forward scatter as an indication of size, therefore of activation status, with the lower right quadrant indicting the position of undivided, unactivated cells. This figure demonstrates that the fidelity of CFSE inheritance is not affected by blastogenesis.

reason, unstimulated control cultures can be used to indicate the position of the undivided peak at the time of harvesting an *in vitro* stimulation experiment.

B. Finding the Position of Each Peak

When a cell divides, the retained CFSE label is distributed to each daughter cell. Experience with lymphocytes indicates that this partitioning is essentially even and that the mean intensity of the cells found in division 1 is accurately half that of the undivided peak recorded at the same time. A geometric twofold reduction in intensity is faithfully followed with each successive cell division, allowing the position of the mean of cells in each division to be calculated from the starting fluorescence. When plotted as log CFSE intensity as is usually done, the peaks are evenly separated (Fig. 1). This diminution in fluorescence, however, has a natural limitation. Unlabeled but stimulated cells increase in size and autofluorescence intensity (arrowed in the lower panel of Fig. 1). This autofluorescence is essentially insignificant when compared to the CFSE intensity of cells in early divisions, but contributes increasingly to the total fluorescence as cells undergo further divisions. The mean intensity of each peak can be determined by the following equation, which requires only the experimental values for D_0 and A—the mean fluorescence intensity of the undivided cells and unlabeled cells, respectively.

$$D_i = [(D_0 - A)/2^{(i)}] + A \qquad (1)$$

where i is the division number.

C. Estimating the Number of Cells per Division

For quantitative analysis of cell proliferation it is often necessary to estimate the total cell number found in each division. As the cells in adjacent divisions always overlap to some extent, then the numbers must be estimated from the curve. This can be achieved by using an interval analysis, or a more accurate computer-based curve-fitting procedure.

1. Interval Analysis

This is the simplest procedure, and requires the use of software to prepare intervals around the predicted mean. An example is shown in Fig. 7B. Intervals are assigned around the peak channel, and the number of events in that interval is divided by the total event number (all gated live cells) to determine the proportion of cells in that division. Clearly, this procedure will become less accurate at later divisions as autofluorescence becomes significant and the overlap between divisions increases.

2. Fitting by Computer

As mentioned previously, the distribution of fluorescence intensity in the starting population is log normal. Remarkably, for T and B cells, the standard deviation of this distribution is preserved with each division, indicating that the redistribution of CFSE between daughter cells at mitosis is accurately twofold through at least seven generations. As a consequence, the standard CFSE profile of asynchronously dividing cell populations is made up of a series of log-normal distributions with the mean intensity of each predicted by Eq. (1) and with a constant standard deviation. Many computer-based curve-fitting programs are available that can be adapted to take gated flow cytometric data and fit a series of overlapping log-normal curves and return the value of the area of each. Pro Fit for Macintosh (Quantumsoft, Zurich) or Peakfit for PC (Jandel, CA) are suitable for this purpose. Figure 7D compares an interval analysis using Cellquest software (Becton Dickinson, Palo Alto, CA) with a computer-based Pro Fit analysis, and shows good correlation over the first nine divisions.

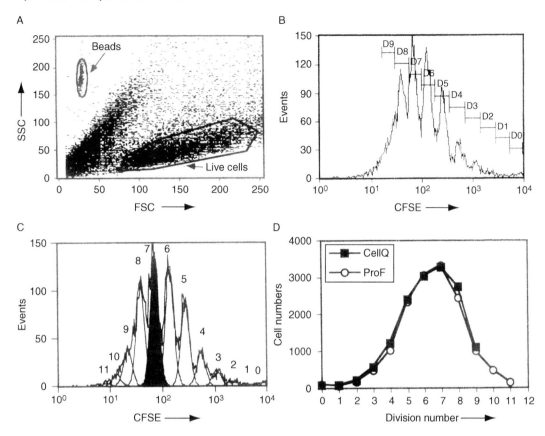

Fig. 7 Quantitative data analysis. CFSE-labeled B cells were stimulated with CD40L and IL-4 for 4 days. (A) Dot plot indicates gated live cells and the position of Calibrite beads on forward and side scatter profiles. (B) Interval analysis of the CFSE histogram using CellQuest software. Markers above the histogram represent gates that correspond to the different division numbers. (C) A series of log-normal Gaussian curves fitted to the same data by Pro Fit. Division numbers are shown above each peak. (D) This panel represents the number of live cells in each CFSE peak based on a CellQuest interval analysis (closed square) or Pro Fit curve-fitting analysis (open circles) following reference to the number of beads as described in the text. Data represents mean and SE of triplicate cultures (error bars smaller than symbol)

3. Converting to Total Cell Numbers

Both interval and curve-fitting analyses return the proportion of cells found within each division. It is often of more interest to know the absolute number of cells found within a cell culture. Estimates of total cell count can be made by reference to bead numbers run at the same time as the cells. Calibrite beads (Becton Dickinson, Palo Alto, CA) are suitable for this application. Prior to harvesting cells a known number of beads are added to the culture. Subsequently, when the culture is harvested and run on the flow cytometer the cells and beads appear with the beads easily distinguished from lymphocytes by their distinct forward and side scatter profile (Fig. 7A). A ratio of live cell events versus the number of beads acquired in the same sample is then used to determine the number of cells in culture. For example, if on acquisition 3000 live cells were counted together with 2000 beads then there were 1.5 times as many cells as beads. Thus, if 10,000 beads had been added to culture then 15,000 total cells were present. The total number of cells within each division can then be determined from the proportions calculated earlier.

D. The Method of Slicing Data from Asynchronous Cultures for Differentiation Analysis

Many lymphocyte differentiation events follow a division-based "map" that is not affected by the time in culture or time spent in a single division. Thus, CFSE can be used to separate the two variables of time and division cycle number to prepare such maps. In order to process data for a differentiation map analysis, flow cytometry data showing phenotype changes with division need to be converted to a quantitative estimate of the percentage of cells of interest in each division. Figure 8 illustrates an example of the "slicing" method. CFSE-labeled B cells were stimulated *in vitro*, and after 4 days of culture cells were stained with anti-IgG1 antibodies and analyzed by flow cytometry. The viable cell population was gated using side and forward scatter parameters. Individual peaks on the CFSE histogram profile were gated and the percentage of cells within the peak expressing surface IgG1 calculated. This proportion is then plotted against division number to reveal the division-related surface immunoglobulin expression (Fig. 8D).

VI. Application of Carboxyfluorescein Diacetate Succinimidyl Ester to *In Vitro* Culture of Lymphocytes

A. B Lymphocyte Cultures

Lymphocytes have provided a useful subject for CFSE applications, as they can be isolated in a resting state and then activated *in vitro* to undergo differentiation changes equivalent to that observed *in vivo*. B lymphocytes have proved particularly useful for exploring the effect of signals on, and the relationship between, cell division and phenotypic changes associated with differentiation mentioned previously. For these studies, it is important to begin with a homogeneous small resting population that retains differentiation potential and will label with CFSE as a tight, homogeneous population thereby giving good resolution of division peaks. Such cells can be prepared from murine spleen by teasing organs apart through a steel mesh, followed by hypotonic lysis of red cells. After removal of adherent cells by incubation of the resulting suspension in plastic tissue culture vessels, T cells are depleted using antibodies against CD4, CD8, and CD90 (Thy-1) and complement-mediated lysis. Small, dense B cells are then recovered from Percoll gradients at the

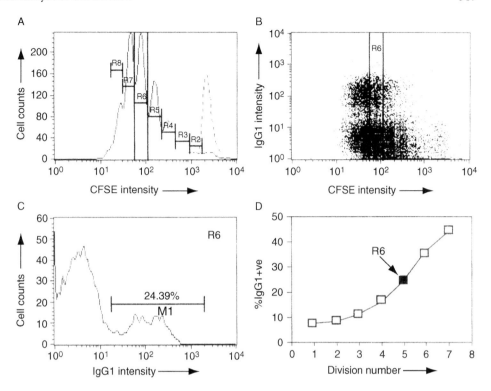

Fig. 8 Quantitative analysis of division-linked differentiation by "slicing" data from asynchronous cultures. CFSE-labeled murine B cells were stimulated for 4 days in the presence of CD40L and IL-4, then stained for surface IgG expression. (A) Peaks on the CFSE profile were gated individually. (B) A representative peak was gated as R6, with IgG1-associated fluorescence on the vertical axis. This was used to generate a histogram of IgG1 expression (C), allowing quantitation of IgG1 expression. This data was generated for division cycles 0–7, allowing changes in expression at each cycle to be represented in graphical form (D). (Reproduced from Hasbold *et al.*, 1998, and is used with kind permission of the publisher.)

65–72% interface. Detailed description of this procedure can be obtained from Hodgkin and Kehry (1995)

To induce B cell proliferation, differentiation, and isotype switching, cells are cultured with cytokines (e.g., IL-4, IL-5, or IL-6, depending on the experiment) as well as with a stimulus for proliferation (e.g., lipopolysaccharide or anti-Ig antibodies) or a source of signaling through the CD40 molecule. This latter stimulation can be delivered by an agonistic anti-CD40 antibody, purified membranes from activated T lymphocyte clones, or a CD40L transfected insect cell line.

The initial starting density of B lymphocytes should be kept low, preferably below 10^5 cells/ml, as the intense burst of cell proliferation seen in maximally stimulated cultures after around 3–5 days can result in exhaustion of medium and failure to complete the threshold number of division cycles required for differentiation to production of downstream Ig isotypes such as IgE. See Hodgkin and Kehry (1995) and Hodgkin *et al.* (1997) for a discussion of these issues. Typically B cell proliferation will begin at around 36–48 h and proceed at a rate of about one division per 8–12 h thereafter. Figures 4 and 8 demonstrate the information which can be gained using the CFSE technique with respect to division-related isotype switching to IgG1 and IgE. See Hodgkin *et al.* (1996), Kindler and Zubler (1997), and Hasbold *et al.* (1998) for more detailed discussions of B cell differentiation studies and the effects of time, stimulation dose, and division number.

B. T Lymphocyte Cultures

T lymphocytes can be purified from peripheral lymphoid tissues or blood using flow cytometry or depletion by complement-mediated lysis. Alternatively, contaminating B and other cells can be excluded from analysis by immunophenotypic staining, providing they do not interfere in a functional sense.

The CFSE division monitoring technique has been used to explore the relation between cytokine production and division number (Bird *et al.*, 1998; Gett and Hodgkin, 1998), the kinetics of T cell proliferative responses (Gett and Hodgkin, 1998; Wells *et al.*, 1997), and the acquisition of characteristics of memory cells (Lee and Pelletier, 1998). These studies also have clinical applications, such as determining antigen-specific T cell responses to *Candida* (Angulo and Fulcher, 1998). This approach could be extended to other pathogens, as well as for monitoring efficacy of vaccination.

Another application of the CFSE technique is in the determination of the frequency of responding T cells in a mixed population. Figure 9 shows the response of CFSE-labeled BALB/c splenocytes, gated for T cells, cultured with irradiated allogeneic C57/B16 splenocytes. The proportion of events in each peak can be determined, allowing the frequency of responders in the starting population to be calculated. Table I shows how the data generated from Fig. 9 can be used to calculate this frequency.

In Fig. 10, a comparison of CD4 and CD8 T cell responses to immobilized anti-CD3 and soluble anti-CD28 reveals a more vigorous response by CD8 T cells.

VII. Monitoring Lymphocyte Responses *In Vivo*

In addition to *in vitro* culture to investigate division-linked differentiation, it is possible to use animal models to gain information on the division and differentiation of lymphocytes *in vivo*. CFSE-labeled lymphocytes can be reintroduced into syngeneic hosts, and their division behavior monitored by flow cytometry of cell suspensions derived from secondary lymphoid or other tissues. The exquisite sensitivity of flow cytometry allows tracking of labeled lymphocytes in mice for up to 6 months (Lyons, 1997; Lyons and Parish, 1994).

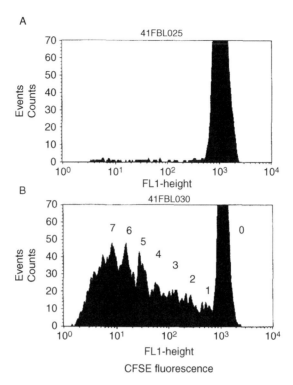

Fig. 9 *In vitro* alloresponse (mixed lymphocyte response). Murine splenic lymphocytes from BALB/c mice were stained with CFSE and cultured alone (A) or at a 16:1 ratio with C57/B16 irradiated, nonlabeled splenocytes (B). After 7 days culture, cells were stained with CD45R-PE, and T cells (CD45R-) were analyzed by flow cytometry. Note division of a proportion of T lymphocytes cultured with allogeneic stimulators, and no division when cultured alone. The number of events in each peak can be determined, allowing the frequency of responders in the starting population to be calculated, which in this case is approximately 6.5%. (From Lyons and Doherty, 1998. Copyright 1998 John Wiley & Sons. Reprinted by permission of Wiley-Liss, Inc., a subsidiary of John Wiley & Sons, Inc.)

A. Intravenous Transfer of Lymphocytes

The usual source of lymphocytes in sufficient numbers for transfer is the spleen or alternatively lymph nodes. From a single murine spleen, approximately $5-10 \times 10^7$ lymphocytes can be obtained, with roughly equal numbers of B and T cells, depending on the strain. Typically, lymph nodes yield around 10-fold fewer cells. Specific subsets of lymphocytes can be purified from cell suspensions, or alternatively, the division behavior of subsets within a mixed transferred population can be monitored using appropriately conjugated antibodies (see Section IV). The usual route for transfer to recipients is via the lateral tail vein, with a maximum volume of 0.2 ml injected. The usual injecting medium is PBS. Transfer of increasing numbers of lymphocytes shows a linear relationship to the percentage of donor cells found in the spleen, up to 5×10^7 lymphocytes transferred per

Table I
***In Vitro* Allorresponse (Mixed Lymphocyte Response)[a]**

Division number (D_n)	Mean fluorescence	Events	Divisor	Undivided cohort number
0	1124	3533	2^0	3533
1	513	158	2^1	79
2	248	281	2^2	70.25
3	116	358	2^3	44.75
4	59	320	2^4	20
5	32	436	2^5	13.62
6	15	718	2^6	11.22
7	8	628	2^7	4.90

[a]Data from Fig. 9B were used to generate this table. Note that the fluorescence intensity of peaks closely follows the predicted serial dilution with each division. The number of cells ("Events") in a given division number (D_n) are divided by 2^{D_n} to calculate the number of original undivided cells they derived from. This number is referred to as the undivided cohort number and is shown in the final column. The sum of this column is the total undivided cohort number ($= 3776$). The sum of cohorts from division 1–7 represents the number of these precursors that have been activated to proliferate within the time of the assay. In this experiment, this number is 244. Therefore, the proportion of the starting population induced into division is: 244/3776 or 6.5%, which is within published estimates of allorresponding precursor percentages obtained using limiting dilution analysis. (From Lyons and Doherty, 1998. Copyright 1998 John Wiley & Sons. Reprinted by permission of Wiley-Liss, Inc., a subsidiary of John Wiley & Sons, Inc.)

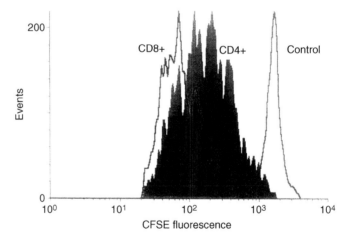

Fig. 10 *In vitro* proliferation of subsets of T cells. Murine splenic T cells purified by flow cytometry were stained with CFSE and cultured alone, or in the presence of immobilized anti-CD3 antibody and anti-CD28. After 3 days culture, cells were labeled with PE-conjugated anti-CD4 or anti-CD8 antibodies before analysis. Histograms show CFSE fluorescence of undivided control cells, CD4 T cells (shaded histogram) and CD8 T cells (open histogram). Note that CFSE staining in conjunction with immunophenotyping allows the kinetics of proliferation in different populations to be compared. In this example, the response of CD8 T cells is more vigorous than that of CD4 T cells. (From Lyons and Doherty, 1998. Copyright 1998 John Wiley & Sons. Reprinted by permission of Wiley-Liss, Inc., a subsidiary of John Wiley & Sons, Inc.)

recipient mouse, before apparent saturation occurs. At this point, donor cells will represent about 1–2% of splenic lymphocytes.

B. Preparation of Cell Suspensions and Strategies for Analysis

Cell suspensions from both lymphoid organs and other tissues can be prepared using standard techniques. In the majority of cases, where cell division occurs, it will be within secondary lymphoid organs such as spleen and nodes. It will usually be necessary to determine how many cells in total need to be analyzed to obtain sufficient events for meaningful cell division information to be obtained from the transferred cells, especially if only a subpopulation responds. Initially, it will be important to determine the percentage of CFSE-labeled cells in a given population. From this information, the total number of cells needed for meaningful analysis can be determined. Where CFSE positive events are rare, gating will be required to exclude the recipient's own cells, resulting in a data file of CFSE positive events. See Fig. 11 and Lyons and Parish (1994), Lyons (1997), and Fulcher *et al.* (1996) for examples.

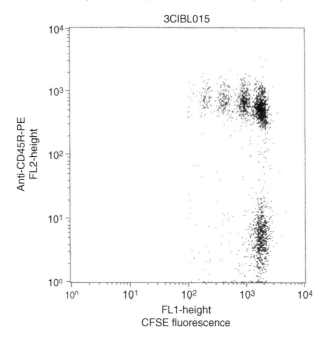

Fig. 11 *In vivo* B cell division in absence of T division after injection of splenic lymphocytes. Murine splenic lymphocytes (2×10^7) labeled with CFSE were injected intravenously. After 14 days, a cell suspension of the recipient's spleen was labeled with a CD45R antibody conjugated with PE, to enable discrimination between B and T cells. Events with green fluorescence above the autofluorescence background were collected, ensuring that only CFSE positive events were analyzed, which in this experiment represented approximately 0.5% of total events. 5000 events were collected, and anti-CD45R staining revealed cell division in the B, but not T, cell compartment. (From Lyons and Doherty, 1998. Copyright 1998 John Wiley & Sons. Reprinted by permission of Wiley-Liss, Inc., a subsidiary of John Wiley & Sons, Inc.)

VIII. Antigen Receptor Transgenic Models

One limitation in the study of immune responses at the individual cell level is the diversity of the immune repertoire, which dictates that the proportion of cells of specificity for a given epitope will be low. The development of antigen receptor transgenic animals can circumvent such limitations. Mice with B cells expressing a single immunoglobulin molecule recognizing a protein antigen such as hen egg lysozyme are available, as are mice in which defined T cell receptors are expressed, which are restricted for peptide presentation on either class I or class II major histocompatibility complex (MHC). Such mice allow the dissection of *in vivo* responses to antigen. These models are not without their shortcomings, as they are by nature nonphysiological. For example, currently available B cell receptor transgenic animals cannot be used to study immunoglobulin class switching. Also, immunization with cognate antigen can result in massive expansion of immune cells followed by equally dramatic apoptotic cell death, due to the inability of such population increases to be sustained. However, by adoptively transferring transgenic lymphocytes to recipient animals, such responses can more closely resemble physiological ones while still ensuring that the proportion of responding cells is large enough to study without exceeding the capacity of the animal to support the expanded population. Used with caution, such models combined with the CFSE cell division technique can yield much useful information on the kinetics of immune responses as well as on differentiation of lymphocytes (Fulcher *et al.*, 1996; Kurts *et al.*, 1997).

Note Added in Proof. It has recently been demonstrated that cell nuclei retain CFSE stain, allowing levels of nuclear transcription factors to be measured simultaneously with cell division (Hasbold and Hodgkin, 2000).

References

Angulo, R., and Fulcher, D. A. (1998). Measurement of *Candida*-specific blastogenesis: Comparison of carboxyfluorescein succinimidyl ester labelling of T cells, thymidine incorporation, and CD69 expression. *Cytometry* **34**, 143–151.

Bird, J. J., Brown, D. R., Mullen, A. C., Moskowitz, N. H., Mahowald, M. A., Sider, J. R., Gajewski, T. F., Wang, C.-R., and Reiner, S. L. (1998). Helper T cell differentiation is controlled by the cell cycle. *Immunity* **9**, 229–237.

Fulcher, D. A., Lyons, A. B., Korn, S., Cook, M. C., Koleda, C., Parish, C., Fazekas de St. Groth, B., and Basten, A. (1996). The fate of self-reactive B-cells depends primarily on the degree of antigen receptor engagement and availability of T-cell help. *J. Exp. Med.* **183**, 2313–2328.

Gett, A. V., and Hodgkin, P. D. (1998). Cell division regulates the T cell cytokine repertoire, revealing a mechanism underlying immune class regulation. *Proc. Natl. Acad. Sci. USA* **95**, 9488–9493.

Hasbold, J., and Hodgkin, P. D. (2000). Flow cytometric cell division tracking using nuclei. *Cytometry* **40**, 230–237.

Hasbold, J., Lyons, A. B., Kehry, M. R., and Hodgkin, P. D. (1998). Cell division number regulates IgG1 and IgE switching of B cells following CD40L and IL-4 stimulation. *Eur. J. Immunol.* **28**, 1040–1051.

Hasbold, J., Hong, J. S., Kehry, M. R., and Hodgkin, P. D. (1999). Integrating signals from IFN-gamma and IL-4 by B cells: Positive and negative effects on CD40 ligand-induced proliferation, survival, and division-linked isotype switching to IgG1, IgE, and IgG2a. *J. Immunol.* **163**, 4175–4181.

Hodgkin, P. D., and Kehry, M. R. (1995). Methods for polyclonal B lymphocyte activation and proliferation and Ig secretion *in vitro*. *In* "Handbook of Experimental Immunology" (D. M. Weir, C. Blackwell, L. A. Herzenberg, and L. A. Herzenberg, eds.) Blackwell, Oxford.

Hodgkin, P. D., Lee, J. H., and Lyons, A. B. (1996). B cell differentiation and isotype switching is related to division cycle number. *J. Exp. Med.* **184**, 277–281.

Hodgkin, P. D., Bartell, G., Mamchak, A., Doherty, K. V., Lyons, A. B., and Hasbold, J. (1997). The importance of efficacy and partial agonism in evaluating models of B lymphocyte activation. *Int. Rev. Immunol.* **15**, 101–127.

Kindler, V., and Zubler, R. H. (1997). Memory, but not naive, peripheral blood B lymphocytes differentiate into Ig-secreting cells after CD40 ligation and costimulation with IL-4 and the differentiation factors IL-2, IL-10, and IL-3. *J. Immunol.* **159**, 2085–2090.

Kurts, C., Carbone, F. R., Barnden, M., Blanas, E., Allison, J., Heath, W. R., and Miller, J. F. (1997). CD4$^+$ T cell help impairs CD8$^+$ T cell deletion induced by cross-presentation of self-antigens and favors autoimmunity. *J. Exp. Med.* **186**, 2057–2062.

Lee, W. T., and Pelletier, W. J. (1998). Visualizing memory phenotype development after *in vitro* stimulation of CD4$^+$ T cells. *Cell. Immunol.* **188**, 1–11.

Lyons, A. B. (1997). Pertussis toxin pretreatment alters the *in vivo* division behaviour and survival of B lymphocytes, after intravenous transfer. *Immunol. Cell Biol.* **75**, 7–12.

Lyons, A. B. (1999). Divided we stand: Tracking cell proliferation with carboxyfluorescein diacetate succinimidyl ester. *Immunol. Cell Biol.* **77**, 509–515.

Lyons, A. B., and Doherty, K. V. (1998). Flow cytometric analysis of cell division by dye dilution. *In* "Current Protocols in Cytometry" (J. P. Robinson, *et al.*, eds.), Unit 9.11. Wiley, New York.

Lyons, A. B., and Parish, C. R. (1994). Determination of lymphocyte division by flow cytometry. *J. Immunol. Methods* **171**, 131–137.

Nordon, R. E., Ginsberg, S. S., and Eaves, C. J. (1997). High resolution cell division tracking demonstrates the Flt3 ligand dependence of human marrow CD34$^+$CD38$^-$ cell production *in vitro*. *Br. J. Haematol.* **98**, 528–539.

Rabinovitch, P. S., Torres, R. M., and Engel, D. (1986). Simultaneous cell cycle analysis and two-color surface immunofluorescence using 7-amino-actinomycin D and single laser excitation: Applications to study of cell activation and the cell cycle of murine Ly-1 B cells. *J. Immunol.* **136**, 2769–2775.

Tough, D. F., and Sprent, J. (1994). Turnover of naive- and memory-phenotype T cells. *J. Exp. Med.* **179**, 1127–1135.

Warren, H. S., and Kinnear, B. F. (1999). Quantitative analysis of the effect of CD16 ligation on human NK cell proliferation. *J. Immunol.* **162**, 735–742.

Wells, A. D., Gudmundsdottir, H., and Turka, L. A. (1997). Following the fate of individual T cells throughout activation and clonal expansion. Signals from T cell receptor and CD28 differentially regulate the induction and duration of a proliferative response. *J. Clin. Invest.* **100**, 3173–3183.

CHAPTER 19

Antibodies Against the Ki-67 Protein: Assessment of the Growth Fraction and Tools for Cell Cycle Analysis

Elmar Endl, Christiane Hollmann, and Johannes Gerdes

Division of Molecular Immunology
Research Center Borstel
D-23845 Borstel, Germany

I. Introduction

The monoclonal antibody Ki-67 was found by intensive studies at the University of Kiel (Ki), Germany, aimed at the production of monoclonal antibodies (MAb) to Hodgkin and Sternberg Reed cells (Gerdes et al., 1983). The clone found in the 67th well of a 96-well microtiter plate (Ki-67) produced an antibody recognizing a human nuclear antigen that is exclusively expressed in proliferating cells. Detailed analysis of the cell cycle distribution revealed that this proliferation-associated

ESSENTIAL CYTOMETRY METHODS
Reprinted from *Methods in Cell Biology*, Volume 63 (Academic Press, 2001).
Copyright © 2001 by Academic Press, Inc.,
DOI: 10.1016/B978-0-12-375045-7.00019-2

antigen, later designated as the Ki-67 protein, is present in all active parts of the cell cycle, that is, G_1, S, G_2, and mitosis, but not in resting cells, that is, G_0 (Baisch and Gerdes, 1987; Endl *et al.*, 1997; Gerdes *et al.*, 1984a). Immunohistological methods based on antibodies against the Ki-67 protein provided reliable and reproducible means for the determination of the growth fraction of a given human cell population, well within the scope of a routine histological laboratory (Gerdes, 1985). Since then, Ki-67 has become one of the most cited markers for cell proliferation (Brown and Gatter, 1990; Gerdes, 1990; Scholzen and Gerdes, 2000; Schwarting, 1993).

Clinical studies prove that the Ki-67 protein is an independent prognostic marker in human neoplasms, including mammary carcinoma (Gerdes *et al.*, 1987a; Jansen *et al.*, 1998; Pavelic *et al.*, 1992), soft tissue sarcoma (Heslin *et al.*, 1998), bladder carcinoma (Popov *et al.*, 1997; Vollmer *et al.*, 1998), meningnomas (Perry *et al.*, 1998), and non-Hodgkin's lymphoma (Gerdes *et al.*, 1984b, 1987b; Grogan *et al.*, 1988; Hall *et al.*, 1988). Besides the ability to characterize the growth fraction in histopathology and cancer research, the Ki-67 protein has some additional properties that fulfill the criteria for a "robust" proliferation marker. No invalid expression of the Ki-67 protein has been reported for precancerous genetic alterations in human tumors that already lead to the unscheduled expression of cell cycle-regulating proteins like the cyclins (Jares *et al.*, 1996). Furthermore, the expression of the Ki-67 protein is not related to any DNA repair processes in cells following DNA damaging treatment such as ultraviolet (UV) exposure (Hall *et al.*, 1993), and there is also no evidence that excessive expression of the Ki-67 protein is involved in apoptosis (Coates *et al.*, 1996; Pittman *et al.*, 1994).

The characterization of the Ki-67 protein in molecular terms was thought to help to explain its role in the system of proteins regulating the cell cycle (Duchrow *et al.*, 1994; Gerdes *et al.*, 1991). The cloning of the cDNA identified two differentially spliced mRNAs encoding for the two isoforms of the Ki-67 protein with predicted molecular masses of 359 and 320 kDa (Schlüter *et al.*, 1993). Computer aided analysis revealed nuclear targeting sequences, potential phosphorylation sites, and several PEST sequences (Duchrow *et al.*, 1994) known to be present in proteins that show a high turnover and are susceptible to degradation by proteases (Rechsteiner and Rogers, 1996). There is evidence now that the Ki-67 protein is posttranslationally modified during mitosis and that this modification results from phosphorylations that are regulated by the cyclin B/CDC2 kinase pathway (Endl and Gerdes, 2000; MacCallum and Hall, 1999). Homologues to the human Ki-67 protein have been described in rodents (Gerlach *et al.*, 1997; Scholzen and Gerdes, 2000; Starborg *et al.*, 1996) and in rat kangaroo PtK2 cells (Takagi *et al.*, 1999), whereas no significant homology to other cell cycle-regulated proteins has been elucidated. The Ki-67 protein sequence therefore remains unique and its relationship to other proteins unknown. Bacterially expressed parts of the Ki-67 cDNA, however, enabled the generation of new monoclonal antibodies (Cattoretti *et al.*, 1992; Gerlach *et al.*, 1997; Key *et al.*, 1993; Scholzen and Gerdes, 2000) that are better suited for further investigations using cytometric techniques than the prototype antibody Ki-67.

Microscopy showed that the regulation of the Ki-67 protein expression is accompanied by characteristic spatial distribution patterns within the different cell cycle compartments (Braun *et al.*, 1988; Endl and Gerdes, 2000; van Dierendonck *et al.*, 1989). Expression is predominantly localized in the nucleus during interphase with an intensive staining of the nucleoli (Kill, 1996; van Dierendonck *et al.*, 1989, 1991). The Ki-67 protein is associated with the dense fibrillar components of the nucleoli, but the domains are distinct from domains containing two of the major nucleolar antigens fibrillarin and RNA polymerase I (Kill, 1996). During mitosis it is associated with the chromosomes, where it covers the chromatin (Bading *et al.*, 1989; Gerdes *et al.*, 1984a; Kill, 1996; Verheijen *et al.*, 1989a,b). Redistribution of the Ki-67 protein during postmitotic formation of the nucleus is thought to be relevant for nucleolar reorganization (Bridger *et al.*, 1998).

In addition, flow cytometric analysis revealed that the level of the Ki-67 protein expression is related to the different cell cycle compartments. It is upregulated in lymphocytes stimulated with phytohemagglutinin A just before the cells enter the very first S phase (Gerdes *et al.*, 1984a). The expression of Ki-67 protein is necessary for the transit from G_1 into S phase, as was demonstrated by a decreased incorporation of [^{3}H]thymidine in IM-9 cells incubated with antisense deoxynucleotides complementary to the deduced translation start site of the Ki-67 mRNA (Schlüter *et al.*, 1993). Staining for the Ki-67 protein increases during S phase and displays a high intensity during mitosis (Bruno and Darzynkiewicz, 1992; Endl *et al.*, 1997; Lopez *et al.*, 1991). Cells reenter the subsequent G_1 phase with a high expression of the Ki-67 protein (Bruno and Darzynkiewicz, 1992; Du Manoir *et al.*, 1991; Lopez *et al.*, 1991). Two ways of regulation are supposed for the protein expression during the following G_1 phase. Ki-67 detection can decline in cells during early G_1 with an estimated half-life of less than 1 h (Bruno and Darzynkiewicz, 1992) as calculated for HL-60 cells after vinblastine block of mitosis and pulse-chase experiments using [^{35}S]methionine in L428 cells (Heidebrecht *et al.*, 1996), whereas the level of protein is still detectable in cells that continue to grow until the cells reenter a new S phase (Du Manoir *et al.*, 1991; Lopez *et al.*, 1991).

Despite the accumulating knowledge about the structure, localization, and regulation of the Ki-67 protein, thus far there is no clue to the function of the protein. Its role in the cell cycle regulating network awaits further characterization. Considering the unique and complex character of the Ki-67 protein, a multidisciplinary and multiparameter approach to elucidate the function of the protein may therefore be not only advantageous but necessary.

II. Application

The methods to retrieve the protein in formalin-fixed paraffin-embedded tissue are standardized (Cattoretti *et al.*, 1992, 1993; Gerdes *et al.*, 1992; McCormick *et al.*, 1993a; Shi *et al.*, 1995) and can be controlled in a standardized fashion

(Ruby and McNally, 1996). However, the diversity of protocols for Ki-67 measurement in flow cytometry is sometimes confusing (Table I). The attempt of this
chapter is therefore to give a summary of methods already used in cell biology and
some examples of applications on different cell types.

Antibodies against the Ki-67 protein can be used for cell kinetic investigations
of cell cultures (Littleton *et al.*, 1991; Schwarting *et al.*, 1986; Tsurusawa and
Fujimoto, 1995), peripheral blood mononuclear cells (Campana *et al.*, 1988;
Cavanagh *et al.*, 1998; Lopez *et al.*, 1991; Palutke *et al.*, 1987), and solid tumors
(Camplejohn, 1994; Schutte *et al.*, 1995). Counterstaining with propidium iodide
enables a correlation of Ki-67 expression with the position of cells in the cell cycle,
on a single cell basis (Landberg *et al.*, 1990). This information can be used to
investigate antigens with yet unknown regulation within proliferating and resting
cells for their usefulness as proliferation markers (Endl *et al.*, 1997; Neubauer
et al., 1989; Pellicciari *et al.*, 1995).

Staining with monoclonal antibodies against the Ki-67 protein is also an easy
method to estimate the growth fraction of a given cell population. This makes it
superior to autoradiographic methods (Scott *et al.*, 1991) and detection of
incorporated nucleotide analogs like bromodeoxyuridine (Wilson *et al.*, 1996).
Additional information about the proliferation characteristics within a given cell
population can be obtained by counterstaining with specific antibodies for a
particular subpopulation of cells. Cytokeratin was used to precisely estimate the
amount of proliferating tumor cells in solid tumor preparations (Schutte *et al.*,
1995) and T cell turnover in both healthy and HIV-1 infected adults by combining
CD4 and CD8 assessment and staining with monoclonal antibodies against the
Ki-67 protein (Sachsenberg *et al.*, 1998).

Monoclonal antibodies against the Ki-67 protein in mouse and rat (MIB-5, TEC-3,
Dianova, Hamburg, Germany) have been described for proliferation assessment in
rodents (Gerlach *et al.*, 1997; Scholzen and Gerdes, 2000) and have to be evaluated for
their use in flow cytometry. Progress in the development of fluorochromes that can
easily be conjugated to the preferred antibodies for proliferation-associated antigens
may promote multiparameter analysis in proliferation studies as an alternative to the
commonly used counterstaining with DNA-specific dyes.

III. Materials and Methods

A. Preparation of Paraformaldehyde in Phosphate-Buffered Saline

Fixation and preservation of the antigen during immunochemical staining procedures are the most critical steps in the analysis of intracellular antigens (Bauer
and Jacobberger, 1994; Clevenger and Shankey, 1993). Formaldehyde is a common fixative for immunohistochemistry and has become more and more popular
for preparation of single cell suspensions in flow cytometry. It both preserves
the morphology of the cells and protects the antigen from leaking by forming

Table I
Protocols for Ki-67 Measurement in Flow Cytometry

Fixative	%	Time (min)	Temperature (°C)	Detergent	%	Time (min)	Temperature (°C)	Cells	Reference
PFA in PBS	0.5	5	4	Triton X-100	0.1	3	4	MCF7, K562	Kute and Quadri (1991)
PFA in PBS[a]	0.5	10	On ice	Triton X-100	0.1	4	On ice	HL-60	Bruno et al. (1992)
PFA in PBS	1	10	On ice	Triton X-100	0.25	5	On ice	J82	Endl et al. (1997)
Acetone	95	10	Room temp.	–	–	–	–	U937	Schwarting et al. (1986)
Acetone	100	30	–20	–	–	–	–	HeLa S3	Sasaki et al. (1987)
Acetone	80	30	Room temp.	–	–	–	–	PB-MNC	Lopez et al. (1991)
Methanol	70	30	–18	–	–	–	–	HL-60	Wersto et al. (1988)
Methanol[b]	70	10	–20	Triton X-100	0.5	15	On ice	MR65	Schutte et al. (1995)
Methanol	100	1 h	4	–	–	–	–	Molt-4	Tsurusawa and Fujimoto (1995)

[a]With varying concentrations of NaCl.
[b]First perform detergent treatment then fixation by adding pure methanol.

covalent cross-links between uncharged amino groups. Subsequent treatment with detergents enables the antibody to penetrate the cells and recognize its specific antigen.

A fresh, purified, and methanol-free solution of formaldehyde is recommended and is commercially available from vendors for electron microscopy (e.g., Polysciences, Warrington, PA). It should be kept in mind that methanol is sometimes added as a stabilizer and may therefore introduce an undesired variable in the fixation procedure.

Fresh formaldehyde/phosphate-buffered saline (PBS) solutions can be prepared from paraformaldehyde (PFA) and can be used for 1 week when stored in dark glass bottles. Stock solutions can be frozen at $-70\,^\circ$C and are suitable for several months. Avoid any inhalation during preparation. Formaldehyde is harmful if inhaled or absorbed through skin.

1. Classic Preparation of Paraformaldehyde in Phosphate–Buffered Saline

1. To prepare a solution of 2% PFA in 10 ml PBS, add 0.2 mg PFA to 1 ml distilled water and warm to 70 $^\circ$C in a water bath.
2. Add NaOH (2 M) dropwise while stirring until solution becomes clear.
3. Incubate another 30 min at 70 $^\circ$C while stirring.
4. Add 5 ml of PBS and adjust buffer to pH 7.4 by adding HCl dropwise.
5. Adjust volume to 10 ml with PBS.
6. Filter through a sterile filter (0.2 μm) to remove aggregates.

2. Rapid Preparation of Paraformaldehyde in Phosphate–Buffered Saline

1. Add 10 ml PBS to 0.2 mg PFA in a glass bottle.
2. Put the bottle in a microwave oven (600 W) and heat at medium power settings. Avoid exposure to vapor. Reduce vapor by plugging the bottle neck with cotton wool.
3. Watch the solution for bubble formation and stop heating when the first bubbles appear.
4. Remove the bottle from the microwave oven and dissolve remaining PFA by shaking.
5. If there is some undissolved PFA left, put the bottle back into the microwave oven and start the procedure again.
6. If necessary adjust to pH 7.4.
7. Filter through a sterile filter and store in dark glass bottles or freeze at $-70\,^\circ$C.

B. Paraformaldehyde Fixation

1. Spin 10^6 cells ($600 \times g$, 5 min), decant, and resuspend pellet in 500 μl ice-cold PBS.

2. Transfer the 500 μl of cell suspension into 500 μl of ice-cold PBS containing 2% PFA (final concentration 10^6 cells in 1 ml PBS containing 1% PFA). Incubate 10 min on ice.

3. Add 2 ml of ice-cold PBS containing 0.5% bovine serum albumin (BSA), spin, and decant.

4. Resuspend in 500 μl ice-cold PBS containing 0.25% Triton X-100. Incubate 5 min on ice.

5. Add 2 ml of ice-cold PBS containing 0.5% BSA, spin, and decant.

6. Resuspend in 100 μl of primary Ki-67 antibody in PBS containing 0.5% BSA. The following antibodies may be used:
   - MIB-1 for human cells, dilute to 2 μg/ml (Dianova).
   - Anti-Ki-67 polyclonal rabbit serum for human cells (DAKO, Hamburg, Germany).
   - MIB-5 for rat cells, dilute to 5 μg/ml (Dianova).
   - TEC-3 for mouse cells, dilution 1:10 (Dianova).

7. Add 2 ml of ice-cold PBS containing 0.5% BSA, spin, and decant.

8. Resuspend in 100 μl of secondary antibody. The secondary antibody may be fluorescein-conjugated goat antimouse, goat antirat, or goat antirabbit antibodies, according to the primary antibody.

9. Incubate 30 min at 4 $^\circ$C in the dark.

10. Add 2 ml of ice-cold PBS containing 0.5% BSA, spin and decant.

11. Resuspend in 500 μl PBS containing 50 μg/ml propidium iodide (PI) and 50 μg/ml RNase. Incubate 20 min at 37 $^\circ$C in the dark.

12. Analyze cells by flow cytometry.

C. Acetone Fixation of Cell Suspensions

The protocol is based on one of the first publications on the use of antibodies against the Ki-67 protein in flow cytometry (Baisch and Gerdes 1987, 1990; Gerdes and Baisch, 1984a).

1. Spin 1 × 10^6 cells (600 × g for 5 min) and decant.

2. Resuspend cells in 200 μl of 150 mM NaCl at 4 $^\circ$C. Transfer cell suspension slowly into 800 μl of pure acetone (-20 $^\circ$C) while gently shaking.

3. Incubate at -20 $^\circ$C for at least 30 min.

4. Centrifuge and decant.

5. Resuspend cells in 2 ml ice-cold PBS containing 0.5% BSA, centrifuge, and decant.

6. For antibody and DNA staining, proceed as described for paraformaldehyde fixation (Steps 6–11).

D. Paraformaldehyde/Methanol Fixation of Cell Suspensions

Methanol fixation is hypotonic and coagulant and may therefore alter the cellular morphology and solubilize nucleoproteins. An interesting variation is fixation with PFA prior to methanol treatment (Jacobberger, 1991; Schimenti and Jacobberger, 1992). This preserves the antigens of interest and may unmask additional epitopes recognized by the primary antibody, owing to the denaturing properties of alcohol fixation (Clevenger and Shankey, 1993).

1. Spin 10^6 cells ($600 \times g$, 5 min), decant, and resuspend pellet in 500 μl ice-cold PBS.
2. Transfer the 500 μl of cell suspension into 500 μl of ice-cold PBS containing 2% PFA (final concentration 10^6 cells in 1 ml PBS containing 1% PFA).
3. Incubate 10 min on ice.
4. Add 2 ml of ice-cold PBS containing 0.5% BSA, spin, and decant.
5. Resuspend in 100 μl ice-cold PBS.
6. Transfer cells to 900 μl methanol ($-20\ ^\circ$C) while gently shaking.
7. Incubate 30 min at $-20\ ^\circ$C.
8. Add 2 ml of ice-cold PBS containing 0.5% BSA, spin, and decant.
9. For antibody and DNA staining, proceed as described for paraformaldehyde fixation (Steps 6–11).

E. Cryopreservation of Cells

Experiments may require a longer storage of cells. The following procedure describes the freezing of single cell suspensions and should preserve the integrity of the Ki-67 protein for several months.

Determine the concentration and volume of the suspension, calculate the total number of available cells, and perform viability assessment. Cells are frozen at 5×10^6 cells/ml of freezing medium. Freezing medium contains 90% fetal calf serum (FCS)/0.05% azide/10% DMSO. Pellet the cells and, after removing the supernatant, gently resuspend with 1 ml volume of freezing medium. Avoid any mechanical stress. Divide into freezing vials in 1-ml aliquots.

1. Freezing

1. Wrap the vials in cotton wool and place them in a Styrofoam box.
2. Store the racks at $-20\ ^\circ$C for at least 24 h, then transfer to $-70\ ^\circ$C freezer; after 1 week transfer the vials to liquid nitrogen storage.

3. Best results are obtained by freezing rates of approximately 1 °C/h. Because only a few laboratories own apparatus for controlled freezing rates, this is a simple and easy to use method to imitate the procedure.

2. Thawing

1. Each vial thawed will require 15 ml of defrosting medium (60% RPMI and 40% FCS). Warm the defrosting medium to 37 °C before thawing the sample. Put aside 1 ml of this medium for use in final resuspension of thawed cells.

2. Remove a vial from the freezer and immediately place in a 37 °C water bath. Use constant but gently swirling until just a small ice crystal remains. Do not immerse the cap of the vial into the water bath during thawing, as the sealing ring on the vial may deform when it warms and may not be able to protect the vial from water leakage.

3. Open the vial and add 500 μl of defrosting medium. Transfer the cells immediately to the remaining defrosting medium by slowly adding the cell suspension to the medium. Avoid any mechanical stress, that is, bubble formation, by rinsing the pipette. Gently invert the tube to mix the cells.

4. Centrifuge, decant, and resuspend the pellet in the 1 ml of defrosting medium previously set aside.

F. Enzymatic Cell Isolation from Fresh Solid Tissue

The following protocol describes the disaggregation and staining of peripheral human lymphoid tissue. The different subsets of cells were characterized by the antibody CD4, by the B cell-specific antibodies CD10 and surface immunoglobulin D (IgD), and by the follicular dendritic cell-specific monoclonal antibody R4/23. Proliferative activity of the different subsets was monitored by counterstaining with the DNA binding dye Hoechst 33258 and fluorescein isothiocyanate (FITC) conjugated MIB-1 antibody. Figure 4 represents the corresponding cytograms.

1. Cut tissue into 2 mm cubes with a scalpel in a Petri dish.
2. Incubate cubes in 10 ml PBS containing 350 U/ml collagenase type IV and 500 Kunitz units/ml DNase I for 20 min at 37 °C while gently shaking.
3. Let the tissue fragments sediment, and transfer supernatant to FCS to stop enzymatic activity (final concentration 10% FCS in PBS).
4. Centrifuge, decant, and resuspend cells in cold RPMI supplemented with 10% FCS.
5. Resuspend remaining tissue fragments in PBS containing 350 U/ml collagenase type IV and 500 Kunitz units/ml DNase I for 20 min at 37 °C.
6. Repeat steps 2–4. The procedure can be repeated three times, and cells can be pooled for further proceedings.

7. Pass cells through a 53-μm filter before starting the fixation and staining procedure.

G. Simultaneous Staining of Ki-67 and Cell Surface Antigens

1. Prepare single cell suspensions from solid tissue as described in Section III.F.

2. Separate cells by centrifugation on a Ficoll-Hypaque density gradient.

3. Incubate cells with monoclonal antibody against surface antigen. The following may be used:
   - CD4, IgG1 (Leu 3a, Becton Dickinson, Franklin Lakes, NJ).
   - CD10, IgG2a (B-E3, ImmunoQuality Products, Gromingen, the Netherlands).
   - Anti-IgD, IgG1 (IgD26, DAKO).
   - CD21L, IgM (R4/23, Naiem *et al.*, 1983).

4. Wash in ice-cold PBS containing 1% FCS, spin, and decant.

5. Incubate cells with secondary antibodies conjugated with fluorochromes suitable for multiparameter analysis. The secondary antibodies may be goat antimouse IgG PE (excitation at 488 nm, emission at 578 nm) and goat antimouse IgM biotin.

6. Wash in ice-cold PBS containing 1% FCS, spin, and decant.

7. Incubate with Streptavidin-Red 670 (excitation at 488 nm, emission at 670 nm).

8. Wash in ice-cold PBS containing 1% FCS, spin, and decant.

9. Resuspend in 500 μl ice-cold PBS and transfer cell suspension to 500 μl PBS containing 2% PFA (final concentration 10^6 cells in 1 ml PBS containing 1% PFA).

10. Incubate 10 min on ice, centrifuge, and decant.

11. Resuspend cells in PBS containing 0.1% Nonidet P-40.

12. Incubate 10 min on ice, centrifuge, and decant.

13. Add FITC conjugated MIB-1 antibody (Dianova) in PBS supplemented with 1% FCS in the appropriate concentration determined by titration experiments. Incubate 30 min at 4 °C.

14. Wash in ice-cold PBS containing 1% FCS, spin, and decant.

15. If your flow cytometer is equipped with a UV excitation source, you can use Hoechst 33258 for a simultaneous assessment of DNA content. Resuspend in 500 μl PBS and add Hoechst 33258 (stock solution 1 mg in distilled water) to a final concentration of 2 μg/ml. Incubate for 15 min.

16. Analyze by flow cytometry.

IV. Critical Aspects

Cell fixation should be tested for different cell types. Antibodies should be titrated, and staining efficiency should be controlled by fluorescence microscopy (Bauer and Jacobberger, 1994; Jacobberger, 1991). Staining of the Ki-67 protein is restricted to the nucleus during interphase, with a high density of protein located in the nucleoli. This staining pattern serves as a control for proper staining procedure.

We suggest that one should try the formaldehyde fixation first, because it yields the best results with regard to staining intensity and reproducibility. Formaldehyde forms cross-links with protein end groups and is therefore well suited for preservation of the antigen, but it usually produces DNA histograms of poor quality (Clevenger and Shankey, 1993). Detergent treatment provides improved accessibility for DNA binding dyes and enhances antibody staining. Formaldehyde concentrations can be varied from 0.5% to 1%. Higher concentrations may disturb the precise measurement of DNA histograms, due to a limited binding of PI to cross-linked DNA. Triton X-100 concentrations can vary from 0.1% to 0.25%. Concentrations should be checked for loss of binding sites in an experimental setup using multiparameter analysis of membrane-bound antigens. Other modifications of the protocol include variations in time and temperature (see Table I).

The use of milder detergents like Nonidet P-40 may help in preserving membrane-bound antigens when precise measurement of cell cycle distribution by DNA intercalating dyes is not required. Because cross-linking agents like formaldehyde can mask epitopes in a way that they are no longer recognizable by the antibody, alcohol and acetone fixation represent reasonable alternatives when monoclonal antibodies against the Ki-67 protein are combined with other antibodies.

The procedure for isolation of cells from solid tissue may be modified. However, it should be kept in mind that the Ki-67 protein is unstable and can easily be destroyed by minimal quantities of trypsin or pepsin (Gerdes et al., 1991). Another alternative is the isolation of bare nuclei because the Ki-67 protein is restricted to the nucleus (Landberg and Roos, 1992; Larsen et al., 1991). A critical aspect of this procedure is that cells in mitosis will be lost during the preparation. Isolated nuclei can be fixed and processed as described for whole cells (Schutte et al., 1995). Loss of protein by detergents and buffers used during isolation procedure of cell nuclei should be controlled by other methods, for example, Western blotting.

V. Controls and Standards

Unspecific binding should always be monitored by running an appropriate isotype control. The MIB-1 antibody shows a rapid saturation curve (McCormick et al., 1993b), whereas unspecific binding sites for IgG follow different staining kinetics. The required amount of antibody should therefore be titrated to achieve a maximum signal-to-noise ratio (Bauer and Jacobberger, 1994; Jacobberger, 1991). The principal components of background fluorescence are autofluorescence of the cells (Aubin, 1979), unspecific binding of the primary and secondary antibodies, and spectral overlap from propidium iodide staining. Unspecific binding can be minimized by using monoclonal primary antibodies and $F(ab')_2$ fragments of the fluorochrome conjugated secondary antibody absorbed against species-specific immunoglobulin. Autofluorescence can be reduced by using alternative fluoro-chromes for antibody labeling such as the indodicyanine dye Cy5 (Amersham, Buckinghamshire, England). The excitation wavelength of Cy5 is in the range of 625–650 nm where autofluorescence is almost negligible (Benson *et al.*, 1979). This both increases the sensitivity of antigen detection and bypasses the problem of spectral overlap from propidium iodide staining.

An appropriate isotype control is required for an accurate determination of the percentage of cells that stain positive for the Ki-67 protein. A precise separation of positive/negative antibody staining can be achieved by normalized subtraction of the isotype control. This alternative software approach runs by subtracting the negative control fluorescence histogram at each interval from the corresponding axis interval of the positive staining distribution (Overton, 1988).

VI. Examples of Results

Figure 1 represents the typical staining pattern of the MIB-1 antibody on PFA/Triton X-100 treated HeLa cells during interphase. The staining is characterized by a strong intensity of nucleolar staining. Numerous small, discrete foci of staining in the nucleoplasm can also be recognized. The staining intensity of the nucleoplasm and the number of nucleoli and nucleoplasmic speckles may vary depending on the

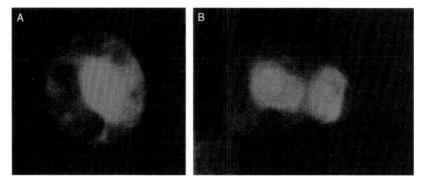

Fig. 1 Cytospin preparations of HeLa cells in interphase. Cells were fixed as single cell suspension with PFA/Triton X-100 stained with Hoechst 33258 for the representation of the nucleus (left) and MIB-1 antibody staining of the Ki-67 protein followed by FITC conjugated secondary antibody (right).

cell type and the position of the cells in the cell cycle (Braun *et al.*, 1988; Kill, 1996). The cytoplasm is completely negative.

Figure 2 represents two-parameter displays of DNA versus Ki-67 expression in cell lines derived from various species. Figure 2A and B demonstrates the variation of Ki-67 staining in human tumor cell lines with different *in vitro* growth characteristics. Figure 2A is derived from the nontumorigenic urothelial human cell line HCV29, which is sensitive to variations in the concentration of growth factors and is growth inhibited by cell-cell contact *in vitro*. Cells that are in the G_1 phase exhibit a broad variation in staining intensity that ranges from high intensity in postmitotic cells to no expression in resting cells. It should also be noted that cells enter S phase with an intermediate staining intensity. Figure 2B represents a two-parameter display DNA versus Ki-67 of HeLa cells. These cells are less sensitive to varying growth conditions *in vitro*, and all of these cells display a detectable level of Ki-67. The rat monoclonal antibody TEC-3 raised against the murine Ki-67 equivalent gives a similar profile for the mouse bladder carcinoma cell line MB49, as shown in Fig. 2C. MIB-5 staining in YB210 rat myeloma cells shows

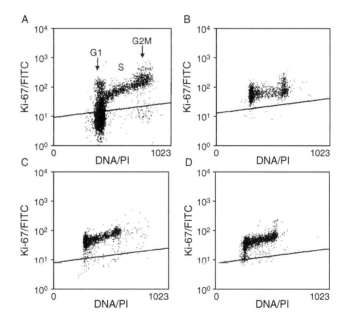

Fig. 2 Two-parameter display of Ki-67 expression versus DNA distribution for cell lines derived from different species. Cells were fixed with PFA followed by Triton X-100 permeabilization and stained for the Ki-67 protein with monoclonal antibodies MIB-1 for human cells, TEC-3 for mouse cells, and MIB-5 for rat cells. Cells were counterstained with PI to display the cell cycle distribution. Cytograms (A) and (B) represent the nontumorigenic human urothelial cell line HCV29 (A) and IM-9 cells derived from human lymphoma (B). MB49 cells isolated from mouse bladder carcinoma and the YB210 rat myeloma cell line are shown in (C) and (D), respectively. The solid line indicates the upper limit of the fluorescence intensity of the corresponding isotype control.

virtually the same staining characteristics. The solid line indicates the upper limit of the fluorescence intensity of the corresponding isotype control.

Figure 3 compares the effect of different fixation and permeabilization procedures on the detection of the Ki-67 protein using the MIB-1 antibody on HeLa cells. PFA/Triton X-100, acetone, and PFA/methanol yield the same staining pattern for HeLa cells stained with MIB-1 and goat antimouse F(ab')$_2$ fragments.

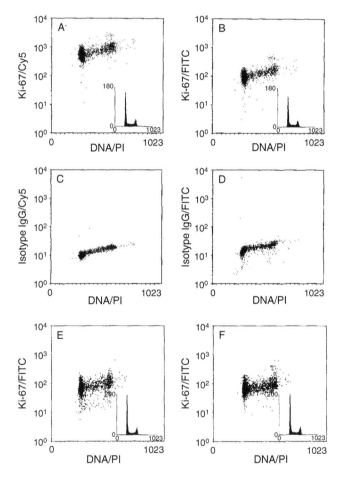

Fig. 3 Two-parameter display of Ki-67 expression versus DNA content of HeLa cells prepared with different fixation protocols and stained with different fluorochrome conjugated secondary antibodies. (A, B) HeLa cells fixed with PFA followed by Triton X-100 treatment. Cells were stained with MIB-1 antibody against the Ki-67 protein and Cy5 conjugated or FITC conjugated secondary antibodies in (A) and (B), respectively. (C, D) Corresponding cytograms for isotype IgG as primary antibody. (E, F) Display cytograms of HeLa cells stained analogously but fixed in acetone (E) and PFA/methanol (F). Histograms of the DNA distribution are included in the corresponding two-parameter display.

Sensitivity can be improved by using Cy5 (Amersham) conjugated goat antimouse F(ab')$_2$ as secondary antibodies as is shown in Fig. 3A. The coefficients of variation of the G$_1$ fraction of cells for the different fixation protocols are approximately 3% and therefore sufficient for a calculation of the cell cycle distribution for all protocols.

Figure 4 illustrates the possibility of multiparameter analysis and cell proliferation assessment using FITC conjugated MIB-1 antibody and detection of cell surface antigens that are specific for subpopulations of cells in human lymphoid

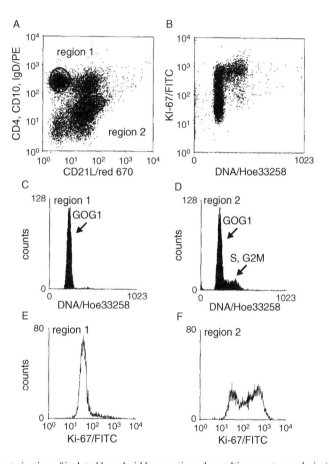

Fig. 4 Characterization of isolated lymphoid human tissue by multiparameter analysis. (A) Cells were stained with CD4, CD10, and anti-IgD followed by phycoerythrin (PE) conjugated secondary antibody and CD21L followed by Red 670 conjugated secondary antibody. (B) Corresponding simultaneous measurement of cell cycle characteristics by antibody staining of the Ki-67 protein and DNA counterstaining with Hoechst 33258 for all cells displayed in (A). (C, E) Expression of the Ki-67 protein and percentage of cells in the different cell cycle compartments for cells corresponding to region 1. Data for cells from region 2 are displayed in (D) and (F).

tissue. Several populations of cells can be identified in a two-parameter display of CD4, CD10, and anti-IgD versus CD21L (Fig. 4A). Region 1 consists of cells that show a strong reactivity to antibodies for CD4. These cells represent nonproliferating T cells, monocytes, and dendritic cells (Sprenger *et al.*, 1995). The corresponding DNA histogram confirms that nearly all of these cells are in same state of the cell cycle according to their DNA content (Fig. 4C). Assessment of the Ki-67 protein by MIB-1 antibody staining reveals that all of these cells are in G_0, since they are all negative (Fig. 4E). Cells that are positive for CD21L and stain for CD10 and IgD (region 2), however, involve both proliferating and resting cells according to their Ki-67 protein expression (Fig. 4F). More than 60% are in an active state of the cell cycle. The proliferative activity is further confirmed by the DNA staining, which reveals cells in S and G_2M phase (Fig. 4D). These cells represent germinal center B lymphocytes, and the results obtained by multiparameter analysis are in concordance with the known biology of these cells.

References

Aubin, J. E. (1979). Autofluorescence of viable cultured mammalian cells. *J. Histochem. Cytochem.* **27**, 36–43.

Bading, H., Rauterberg, E. W., and Moelling, K. (1989). Distribution of c-myc, c-myb, and Ki-67 antigens in interphase and mitotic human cells evidenced by immunofluorescence staining technique. *Exp. Cell Res.* **185**, 50–59.

Baisch, H., and Gerdes, J. (1987). Simultaneous staining of exponentially growing versus plateau phase cells with the proliferation-associated antibody Ki-67 and propidium iodide: Analysis by flow cytometry. *Cell Tissue Kinet.* **20**, 387–391.

Baisch, H., and Gerdes, J. (1990). Identification of proliferating cells by Ki-67 antibody. *In* "Methods in Cell Biology" (Z. Darzynkiewicz, J. P. Robinson, and H. A. Crissman, eds.), Vol. 41, pp. 217–226. Academic Press, New York.

Bauer, K. D., and Jacobberger, J. W. (1994). Analysis of intracellular proteins. *In* "Methods in Cell Biology" (Z. Darzynkiewicz, J. P. Robinson, and H. A. Crissman, eds.), Vol. 41, pp. 351–376. Academic Press, New York.

Benson, H. C., Meyer, R. A., and Zaruba, M. E. (1979). Cellular autofluorescence—Is it due to flavins? *J. Histochem. Cytochem.* **27**, 44–48.

Braun, N., Papadopoulos, T., and Muller-Hermelink, H. K. (1988). Cell cycle dependent distribution of the proliferation-associated Ki-67 antigen in human embryonic lung cells. *Virch. Arch. B Cell Pathol. Incl. Mol. Pathol.* **56**, 25–33.

Bridger, J. M., Kill, I. R., and Lichter, P. (1998). Association of pKi-67 with satellite DNA of the human genome in early G1 cells. *Chromosome Res.* **6**, 13–24.

Brown, D. C., and Gatter, K. C. (1990). Monoclonal antibody Ki-67: Its use in histopathology. *Histopathology* **17**, 489–503.

Bruno, S., and Darzynkiewicz, Z. (1992). Cell cycle dependent expression and stability of the nuclear protein detected by Ki-67 antibody in HL-60 cells. *Cell Prolif.* **25**, 31–40.

Bruno, S., Gorczyca, W., and Darzynkiewicz, Z. (1992). Effect of ionic strength in immunocytochemical detection of the proliferation associated nuclear antigens p120, PCNA, and the protein reacting with Ki-67 antibody. *Cytometry* **13**, 496–501.

Campana, D., Coustan-Smith, E., and Janossy, G. (1988). Double and triple staining methods for studying the proliferative activity of human B and T lymphoid cells. *J. Immunol. Methods* **107**, 79–88.

Camplejohn, R. S. (1994). The measurement of intracellular antigens and DNA by multiparametric flow cytometry. *J. Microsc.* **176**, 1–7.

Cattoretti, G., Becker, M. H., Key, G., Duchrow, M., Schluter, C., Galle, J., and Gerdes, J. (1992). Monoclonal antibodies against recombinant parts of the Ki-67 antigen (MIB 1 and MIB 3) detect proliferating cells in microwave-processed formalin-fixed paraffin sections. *J. Pathol.* **168,** 357–363.

Cattoretti, G., Pileri, S., Parravicini, C., Becker, M. H., Poggi, S., Bifulco, C., Key, G., D'Amato, L., Sabattini, E., and Feudale, E. (1993). Antigen unmasking on formalin-fixed, paraffin-embedded tissue sections. *J. Pathol.* **171,** 83–98.

Cavanagh, L. L., Saal, R. J., Grimmett, K. L., and Thomas, R. (1998). Proliferation in monocyte-derived dendritic cell cultures is caused by progenitor cells capable of myeloid differentiation. *Blood* **92,** 1598–1607.

Clevenger, C. V., and Shankey, T. V. (1993).Cytochemistry II: Immunofluorescence measurement of intracellular antigens. *In* "Clinical Flow Cytometry" (K. D. Bauer, R. E. Duque, and T. V. Shankey, eds.), pp. 157–176. Williams & Wilkins, Baltimore.

Coates, P. J., Hales, S. A., and Hall, P. A. (1996). The association between cell proliferation and apoptosis: Studies using the cell cycle-associated proteins Ki67 and DNA polymerase α. *J. Pathol.* **178,** 71–77.

Duchrow, M., Schluter, C., Wohlenberg, C., Flad, H. D., and Gerdes, J. (1994). Molecular characterization of the gene locus of the human cell proliferation-associated nuclear protein defined by monoclonal antibody Ki-67. *Cell Prolif.* **29,** 1–12.

Du Manoir, M. S., Guillaud, P., Camus, E., Seigneurin, D., and Brugal, G. (1991). Ki-67 labeling in postmitotic cells defines different Ki-67 pathways within the 2c compartment. *Cytometry* **12,** 455–463.

Endl, E., and Gerdes, J. (2000). Post-translational modifications of the Ki-67 protein coincide with two major check points during mitosis. *J. Cell Physiol.* **182,** 371–380.

Endl, E., Steinbach, P., Knuchel, R., and Hofstadter, F. (1997). Analysis of cell cycle-related Ki-67 and p120 expression by flow cytometric BrdUrd-Hoechst/7AAD and immunolabeling technique. *Cytometry* **29,** 233–241.

Gerdes, J. (1985). An immunohistological method for estimating cell growth fractions in rapid histopathological diagnosis during surgery. *Int. J. Cancer* **35,** 169–171.

Gerdes, J. (1990). Ki-67 and other proliferation markers useful for immunohistological diagnostic and prognostic evaluations in human malignancies. *Semin. Cancer Biol.* **1,** 199–206.

Gerdes, J., Schwab, U., Lemke, H., and Stein, H. (1983). Production of a mouse monoclonal antibody reactive with a human nuclear antigen associated with cell proliferation. *Int. J. Cancer* **31,** 13–20.

Gerdes, J., Lemke, H., Baisch, H., Wacker, H. H., Schwab, U., and Stein, H. (1984a). Cell cycle analysis of a cell proliferation-associated human nuclear antigen defined by the monoclonal antibody Ki-67. *J. Immunol.* **133,** 1710–1715.

Gerdes, J., Dallenbach, F., Lennert, K., Lemke, H., and Stein, H. (1984b). Growth fractions in malignant non-Hodgkin's lymphomas (NHL) as determined in situ with the monoclonal antibody Ki-67. *Hematol. Oncol.* **2,** 365–371.

Gerdes, J., Pickartz, H., Brotherton, J., Hammerstein, J., Weitzel, H., and Stein, H. (1987a). Growth fractions and estrogen receptors in human breast cancers as determined in situ with monoclonal antibodies. *Am. J. Pathol.* **129,** 486–492.

Gerdes, J., Van Baarlen, J., Pileri, S., Schwarting, R., Van Unnik, J. A., and Stein, H. (1987b). Tumor cell growth fraction in Hodgkin's disease. *Am. J. Pathol.* **128,** 390–393.

Gerdes, J., Li, L., Schlueter, C., Duchrow, M., Wohlenberg, C., Gerlach, C., Stahmer, I., Kloth, S., Brandt, E., and Flad, H. D. (1991). Immunobiochemical and molecular biologic characterization of the cell proliferation-associated nuclear antigen that is defined by monoclonal antibody Ki-67. *Am. J. Pathol.* **138,** 867–873.

Gerdes, J., Becker, M. H., Key, G., and Cattoretti, G. (1992). Immunohistological detection of tumour growth fraction (Ki-67 antigen) in formalin-fixed and routinely processed tissues. *J. Pathol.* **168,** 85–86.

Gerlach, C., Golding, M., Larue, L., Alison, M. R., and Gerdes, J. (1997). Ki-67 immunoexpression is a robust marker of proliferative cells in the rat. *Lab. Invest.* **77,** 697–698.

Grogan, T. M., Lippman, S. M., Spier, C. M., Slymen, D. J., Rybski, J. A., Rangel, C. S., Richter, L. C., and Miller, T. P. (1988). Independent prognostic significance of a nuclear proliferation antigen in diffuse large cell lymphomas as determined by the monoclonal antibody Ki-67. *Blood* **71**, 1157–1160.

Hall, P. A., Richards, M. A., Gregory, W. M., d'Ardenne, A. J., Lister, T. A., and Stansfeld, A. G. (1988). The prognostic value of Ki67 immunostaining in non-Hodgkin's lymphoma. *J. Pathol.* **154**, 223–235.

Hall, P. A., McKee, P. H., Menage, H. D., Dover, R., and Lane, D. P. (1993). High levels of p53 protein in UV-irradiated normal human skin. *Oncogene* **8**, 203–207.

Heidebrecht, H. J., Buck, F., Haas, K., Wacker, H. H., and Pawaresch, R. (1996). Monoclonal antibodies Ki-S3 and Ki-S5 yield new data on the "Ki-67" proteins. *Cell Prolif.* **29**, 413–425.

Heslin, M. J., Cordon-Cardo, C., Lewis, J. J., Woodruff, J. M., and Brennan, M. F. (1998). Ki-67 detected by MIB-1 predicts distant metastasis and tumor mortality in primary high grade extremity soft tissue sarcoma. *Cancer* **83**, 490–497.

Jacobberger, J. W. (1991). Intracellular antigen staining: Quantitative immunofluorescence. *Methods* **2**, 207–218.

Jansen, R. L., Hupperets, P. S., Arends, J. W., Joosten-Achjanie, S. R., Volovics, A., Schouten, H. C., and Hillen, H. F. (1998). MIB-1 labelling index is an independent prognostic marker in primary breast cancer. *Br. J. Cancer* **78**, 460–465.

Jares, P., Campo, E., Pinyol, M., Bosch, F., Miquel, R., Fernandez, P. L., Sanchez-Beato, M., Soler, F., Perez-Losada, A., Nayach, I., Mallofre, C., Piris, M. A., *et al.* (1996). Expression of retinoblastoma gene product (pRb) in mantle cell lymphomas. Correlation with cyclin D1 (PRAD1/CCND1) mRNA levels and proliferative activity. *Am. J. Pathol.* **148**, 1591–1600.

Key, G., Becker, M. H. G., Baron, B., Duchrow, M., Schlueter, C., Flad, H. D., and Gerdes, J. (1993). New Ki-67 equivilant murine monoclonal antibodies (MIB 1-3) generated against bacterially expressed parts of the Ki-67 cDNA containing three 66bp repetitive elements encoding for the Ki-67 epitope. *Lab. Invest.* **68**, 629–635.

Kill, I. R. (1996). Localisation of the Ki-67 antigen within the nucleolus. Evidence for a fibrillarin-deficient region of the dense fibrillar component. *J. Cell Sci.* **109**, 1253–1263.

Kute, T. E., and Quadri, Y. (1991). Measurement of proliferation nuclear and membrane markers in tumor cells by flow cytometry. *J. Histochem. Cytochem.* **39**, 1125–1130.

Landberg, G., and Roos, G. (1992). Flow cytometric analysis of proliferation associated nuclear antigens using washless staining of unfixed cells. *Cytometry* **13**, 230–240.

Landberg, G., Tan, E. M., and Roos, G. (1990). Flow cytometric multiparameter analysis of proliferating cell nuclear antigen/cyclin and Ki-67 antigen: A new view of the cell cycle. *Exp. Cell Res.* **187**, 111–118.

Larsen, J. K., Christensen, I. J., Christiansen, J., and Mortensen, B. J. (1991). Washless double staining of unfixed nuclei for flow cytometric analysis of DNA and nuclear antigen (Ki-67 or bromodeoxyuridine). *Cytometry* **12**, 429–437.

Littleton, R. J., Baker, G. M., Soomro, I. N., Adams, R. L., and Whimster, W. F. (1991). Kinetic aspects of Ki-67 antigen expression in a normal cell line. *Virch. Arch. B Cell Pathol. Incl. Mol. Pathol.* **60**, 15–19.

Lopez, F., Belloc, F., Lacombe, F., Dumain, P., Reiffers, J., Bernard, P., and Boisseau, M. R. (1991). Modalities of synthesis of Ki-67 antigen during the stimulation of lymphocytes. *Cytometry* **12**, 42–49.

MacCallum, D. E., and Hall, P. A. (1999). Biochemical characterization of pKi67 with the identification of a mitotic-specific form associated with hyperphosphorylation and altered DNA binding. *Exp. Cell Res.* **252**, 186–198.

McCormick, D., Chong, H., Hobbs, C., Datta, C., and Hall, P. A. (1993a). Detection of the Ki-67 antigen in fixed and wax-embedded sections with the monoclonal antibody MIB1. *Histopathology* **22**, 355–360.

McCormick, D., Yu, C., Hobbs, C., and Hall, P. A. (1993b). The relevance of antibody concentration to the immunohistological quantification of cell proliferation-associated antigens. *Histopathology* **22**, 543–547.

Naiem, M., Gerdes, J., Abdulaziz, Z., Stein, H., and Mason, D. Y. (1983). Production of a monoclonal antibody reactive with human dendritic reticulum cells and analysis of human lymphoid tissue. *J. Clin. Pathol.* **36,** 167–175.

Neubauer, A., Serke, S., Siegert, W., Kroll, W., Musch, R., and Huhn, D. (1989). A flow cytometric assay for the determination of cell proliferation with a monoclonal antibody directed against DNA-methyltransferase. *Br. J. Haematol.* **72,** 492–496.

Overton, W. (1988). Modified histogram subtraction technique for analysis of flow cytometry data. *Cytometry* **9,** 619–626.

Palutke, M., KuKuruga, D., and Tabaczka, P. (1987). A flow cytometric method for measuring lymphocyte proliferation directly from tissue culture plates using Ki-67 and propidium iodide. *J. Immunol. Methods* **105,** 97–105.

Pavelic, Z. P., Pavelic, L., Lower, E. E., Gapany, M., Gapany, S., Barker, E. A., and Preisler, H. D. (1992). c-myc, c-erbB-2, and Ki-67 expression in normal breast tissue and in invasive and noninvasive breast carcinoma. *Cancer Res.* **52,** 2597–2602.

Pellicciari, C., Mangiarotti, R., Bottone, M. G., Danova, M., and Wang, E. (1995). Identification of resting cells by dual-parameter flow cytometry of statin expression and DNA content. *Cytometry* **21,** 329–337.

Perry, A., Stafford, S. L., Scheithauer, B. W., Suman, V. J., and Lohse, C. M. (1998). The prognostic significance of MIB-1, p53, and DNA flow cytometry in completely resected primary meningiomas. *Cancer* **82,** 2262–2269.

Pittman, S. M., Strickland, D., and Ireland, C. M. (1994). Polymerization of tubulin in apoptotic cells is not cell cycle dependent. *Exp. Cell Res.* **215,** 263–272.

Popov, Z., Hoznek, A., Colombel, M., Bastuji-Garin, S., Lefrere-Belda, M. A., Bellot, J., Abboh, C. C., Mazerolles, C., and Chopin, D. K. (1997). The prognostic value of p53 nuclear overexpression and MIB-1 as a proliferative marker in transitional cell carcinoma of the bladder. *Cancer* **80,** 1472–1481.

Rechsteiner, M., and Rogers, S. W. (1996). PEST sequences and regulation by proteolysis. *Trends Biochem. Sci.* **21,** 267–271.

Ruby, S. G., and McNally, A. C. (1996). Quality control of proliferation marker (MIB-1) in image analysis systems utilizing cell culture based control materials. *Am. J. Clin. Pathol.* **106,** 634–639.

Sachsenberg, N., Perelson, A. S., Yerly, S., Schockmel, G. A., Leduc, D., Hirschel, B., and Perrin, L. (1998). Turnover of $CD4^+$ and $CD8^+$ T lymphocytes in HIV-1 infection as measured by Ki-67 antigen. *J. Exp. Med.* **187,** 1295–1303.

Sasaki, K., Murakami, T., Kawasaki, M., and Takahashi, M. (1987). The cell cycle associated change of the Ki-67 reactive nuclear antigen expression. *J. Cell Physiol.* **133,** 579–584.

Schimenti, K. J., and Jacobberger, J. W. (1992). Fixation of mammalian cells for flow cytometric evaluation of DNA content and nuclear immunofluorescence. *Cytometry* **13,** 48–59.

Schlüter, C., Duchrow, M., Wohlenberg, C., Becker, M. H., Key, G., Flad, H. D., and Gerdes, J. (1993). The cell proliferation-associated antigen of antibody Ki-67: A very large, ubiquitous nuclear protein with numerous repeated elements, representing a new kind of cell cycle-maintaining proteins. *J. Cell Biol.* **123,** 513–522.

Scholzen, T., and Gerdes, J. (2000). The Ki-67 protein: From the known and the unknown. *J. Cell Physiol.* **182,** 311–322.

Schutte, B., Tinnemans, M. M., Pijpers, G. F., Lenders, M. H., and Ramaekers, F. C. (1995). Three parameter flow cytometric analysis for simultaneous detection of cytokeratin, proliferation associated antigens and DNA content. *Cytometry* **21,** 177–186.

Schwarting, R. (1993). Little missed markers and Ki-67. *Lab. Invest.* **68,** 597–599.

Schwarting, R., Gerdes, J., Niehus, J., Jaeschke, L., and Stein, H. (1986). Determination of the growth fraction in cell suspensions by flow cytometry using the monoclonal antibody Ki-67. *J. Immunol. Methods* **90,** 65–70.

Scott, R. J., Hall, P. A., Haldane, J. S., van Noorden, S., Price, Y., Lane, D. P., and Wright, N. A. (1991). A comparison of immunohistochemical markers of cell proliferation with experimentally determined growth fraction. *J. Pathol.* **165,** 173–178.

I. Update

Since this chapter appeared in 2004, the number of publications concerning the phosphorylated form of histone H2AX (the gene is now officially designated H2AFX) has increased from about 100 to almost 1000. More is known concerning H2AX phosphorylation in relation to chromatin structure, the nature of the DNA damage that can stimulate phosphorylation, and the processes involved in removal of γH2AX after DNA breaks are repaired. The majority of papers continue to examine γH2AX formation in processes related to DNA damage signaling or its use as an indicator of response to DNA damaging agents. Phosphorylation of H2AX is also studied in relation to chromatin dynamics, apoptosis, meiosis, tumorigenesis, viral activation, telomere dysfunction, and genomic instability. The potential for using γH2AX in the analysis of the response of individual tumors to treatment or for identifying genotoxic agents is gaining momentum. On the technical side, several papers have described quantitative methods for the measurement of γH2AX by flow or image cytometry. Mouse monoclonal and rabbit polyclonal antibodies are now available commercially from several sources. As these are not all equally sensitive or specific, routine testing of antibodies (including testing different lot numbers from the same supplier) has become more important.

A. Pitfalls to Avoid

- Failing to confirm specificity of γH2AX antibody staining.
- Equating all γH2AX foci to the presence of DNA double-strand breaks (the signal can remain long after the break is rejoined).
- Equating increases in γH2AX intensity averaged over a cell with increases in the number of DNA double-strand breaks (foci may be larger, not more numerous).
- Not realizing that an excessive γH2AX signal may represent apoptotic DNA fragmentation rather than drug-induced DNA breaks.
- Failing to appreciate the importance and variability in background expression of γH2AX when attempting to detect small amounts of induced DNA damage (cells for study should be chosen carefully).
- Assuming that the kinetics of γH2AX formation and loss will be the same as the kinetics for other molecules that form foci in association with double-strand breaks (colocalization may only occur early or late in the process of repair).
- Failing to appreciate the role of blocked replication forks in stimulating formation of γH2AX (i.e., detecting γH2AX foci may require cells to proliferate).

B. Recommended Approaches

- Western blotting to confirm antibody specificity and dose response.
- Flow cytometry analysis for measuring heterogeneity in γH2AX expression within a population, for multiparameter analysis, and for measuring the rate of formation and loss of γH2AX.

- Image cytometry for greatest sensitivity, for measurement of γH2AX foci size/intensity, for analysis of patterns of expression in tissues and tumors, and for examining colocalization between γH2AX and other proteins involved in DNA damage processing.

II. Introduction

Histone proteins are essential components of a dynamic nucleosomal structure that undergoes various chemical modifications associated with chromatin packaging, transcription, replication, and repair. Rogakou *et al.* (1998, 1999) reported that rapid phosphorylation of a specific serine at the C-terminal end of a minor nucleosomal histone protein, H2AX, occurred only at sites surrounding DNA double-stranded breaks. This very early event preceded the actions of most repair enzymes involved in homologous recombination and nonhomologous end-joining of these breaks. Subsequently, serine-139-phosphorylated H2AX, designated γH2AX, was found to colocalize with certain repair enzymes at sites of DNA damage (Celeste *et al.*, 2002, 2003b; Paull *et al.*, 2000). Mice lacking H2AX were radiation sensitive, immune deficient, and growth retarded (Celeste *et al.*, 2002). Cells from these mice were also radiosensitive and chromosomally unstable (Bassing *et al.*, 2002). In mice deficient in the p53 tumor suppressor, loss of even a single H2AX allele decreased latency of development of lymphomas and other tumor types (Celeste *et al.*, 2003a).

Mechanistically, members of the PI3 kinase family, primarily ATM and ATR, have been implicated in phosphorylation of H2AX (Burma *et al.*, 2001; Paull *et al.*, 2000; Redon *et al.*, 2002). Immediately after irradiation, microscopically visible spots, or "foci," begin to form, and within an hour, each focus contains thousands of γH2AX molecules covering about 2 Mb of DNA flanking the break. What is perhaps most exciting is that an individual DNA double-stranded break can now be visualized using antibody staining and fluorescence microscopy (Rothkamm and Lobrich, 2003; Sedelnikova *et al.*, 2002). In comparison, previous methods used to detect DNA double-stranded breaks, such as neutral filter elution, pulsed field gel electrophoresis, or neutral comet assay, had a detection limit of about 40–100 breaks/cell.

A model proposed by Redon *et al.* (2002) suggested that PI3 kinases like ATM are attracted to the DNA double-stranded break and phosphorylate H2AX molecules as they progress away from the break. The change in chromatin structure appears to activate ATM that can then phosphorylate H2AX (Bakkenist and Kastan, 2003). Then γH2AX or perhaps ATM itself (Andegeko *et al.*, 2001) attracts DNA repair enzymes, and once the break is rejoined, the kinase dissociated and phosphatases removed the phosphate group(s) of H2AX. Loss of γH2AX has been coupled to the rejoining of double-stranded breaks and the actions of protein phosphatase 1 (Nazarov *et al.*, 2003; Siino *et al.*, 2002).

A modification to this model suggested that γH2AX may alter the efficiency or accuracy of repair, perhaps through interactions with intermediary proteins in the repair process and/or through chromatin structure modifications (Downs and Jackson, 2003). Greater sensitivity to killing by ionizing radiation has been associated with a slower rate of loss of γH2AX and a higher proportion of cells tumor retaining γH2AX foci (MacPhail et al., 2003a; Rothkamm and Lobrich, 2003; Taneja et al., 2004).

III. Considerations in the Use of γH2AX as a Measure of Double-Stranded Breaks

The number of γH2AX foci formed in a cell after exposure to ionizing radiation appears to be directly related to the number of DNA double-stranded breaks (Sedelnikova et al., 2002). Linear dose-response relationships can be generated by counting individual foci or by measuring γH2AX intensity using flow cytometry (MacPhail et al., 2003a; Rothkamm and Lobrich, 2003). However, the level of the substrate H2AX is variable between different cell types and is dependent on DNA content (MacPhail et al., 2003b; Rogakou et al., 1998). In addition, the amount of γH2AX that is formed is dependent on the kinase and phosphatase activity of the cell (Paull et al., 2000). Therefore, it is important to appreciate that the signal one sees using *flow cytometry* cannot be directly converted to the number of double-stranded breaks per cell. Moreover, the background level of γH2AX is cell-type-dependent and cell-phase-dependent; levels of γH2AX in G_o/G_1 phase cells are about three times lower than cells in S and G_2 phases (MacPhail et al., 2003b). Conversely, damaged mitotic cells and some apoptotic cells express very high levels of this molecule (Rogakou et al., 1999, 2000). Although γH2AX foci are described as covering about Mb surrounding a break, chromatin conformation has also been shown to influence foci sizes. Mitotic cells or cells treated with agents that alter chromatin structure (e.g., hypertonic saline or histone deacetylase inhibitors) develop much larger foci (Banuelos et al., 2007; Reitsema et al., 2004).

The sensitivity for detecting DNA double-stranded breaks using γH2AX is greatest using microscopic analysis of individual foci in intact nuclei. In theory, one double-stranded break can be detected in a single cell or even fractions of breaks within populations of damaged cells (Rothkamm and Lobrich, 2003). However, because S-phase cells can also demonstrate foci at sites of "background" damage, the greatest sensitivity for detecting radiation- or drug-induced damage is achieved using nonproliferating cells. Mouse embryonic stem cells and many transformed cell types, such as M059J human glioma cells, also show high background levels of foci, which reduces sensitivity for detecting induced double-stranded breaks (Banáth et al., 2009; Paull et al., 2000).

Table I compares H2AX expression with a single-cell gel electrophoresis method, the comet assay, a method that can also detect DNA damage in individual cells (Olive, 2002). Although the comet assay is more versatile because it can be

Table I

Comparison Between γH2AX Antibody Binding and Comet Assay for Detection of DNA Damage

Endpoint	γH2AX assay	Comet assay
Types of DNA damage detected	DNA double-stranded breaks, blocks to replication leading to double-stranded breaks, apoptosis (early)	Base damage, single- and double-stranded breaks, protein and interstrand cross-links, apoptosis (all stages)
Sensitivity for detecting double-stranded breaks	1–5 per cell	40–100 per cell
Single-cell detection	Yes	Yes
Independent of factors other than DNA damage or cell cycle position	No	Yes
Methods of detection	Flow and image cytometry immunoblotting (single cells and tissue sections)	Image cytometry (single cells only)
Influence of cell cycle position on double-stranded breaks		
Background	Three times lower in G_1 phase	Three times lower in S phase
Double stranded break induced	No effect	Three times lower in S phase
Ease of detection	Simple (using flow cytometry)	Time consuming (using image analysis)

modified to detect many more types of DNA damage, it is relatively insensitive for detecting double-stranded breaks (Olive *et al.*, 1991). It is, however, a more direct measure of double-stranded breaks that is not influenced by kinase and phosphatase activity or by variations in the amount of substrate available for phosphorylation, so damage can be directly compared between different cell types. The comet assay is, therefore, useful in conjunction with γH2AX detection.

Phosphorylation of γH2AX has been identified by binding of antibodies to phosphorylated serine-139 at the C-terminus of histone H2AX. Both monoclonal and polyclonal antibodies are commercially available, and detection is possible via immunoblotting, immunocytochemistry, immunohistochemistry, or flow cytometry. Immunoblotting confirm the specificity of the antibody for serine 139 histone H2AX, and it has been used to show that accessibility of chromatin to the antibody is not the cause for the cell cycle variation in γH2AX detected by flow cytometry (MacPhail *et al.*, 2003a,b). Immunocytochemistry is the most sensitive method for detecting DNA damage because individual foci can be recognized and counted. Information on foci size, number, and intensity is also possible with appropriate software and good-quality confocal (or deconvolution) images (MacPhail *et al.*, 2003a). However, the accuracy of measuring discrete foci in whole nuclei is compromised for doses higher than about 2 Gy (50 or so breaks per diploid cell). Examining individual optical sections can extend this limit higher, but with loss of accurate information on the total number of foci per cell.

Flow cytometry is useful for bivariate analysis of cell cycle responses for ease of measurement of the kinetics of development and loss of γH2AX after DNA damage, and for rapid analysis of heterogeneity in response. For both flow and image analysis, samples can be counterstained with DNA binding dyes such as 4′ 6-diamidino-2-phenylindole dihydrochloride hydrate (DAPI). For γH2AX detection, primary antibody staining is followed by staining with a secondary antibody that is often fluorescein isothiocyanate (FITC) conjugated. The intensity of γH2AX staining increases linearly with staining up to about 10–20 Gy (250–500 breaks/diploid cell) (MacPhail et al., 2003a,b). Using flow cytometry, saturation begins to occur after DNA damage exceeds about 500 breaks because there is a limit to the amount of H2AX that can be phosphorylated and neighboring 2 Mb foci will begin to overlap.

Maximum development of γH2AX foci size after exposure to ionizing irradiation requires about 30–60 min after treatment. After or perhaps even during this time, there is a progressive loss of γH2AX foci as damage is repaired. The rate of loss is cell-type- and species-dependent, with typical half-times of 2–3 h for radiation-resistant rodent cells but 6 h or more for radiosensitive human tumor cells. Interestingly, the foci that remain hours after irradiation often appear larger, and daughter cells appear to inherit foci patterns. Rates of development and loss are similar for treatment with drugs that cause DNA double-stranded breaks with the caveat that if long drug treatments are used, foci numbers will likely represent the steady-state level of γH2AX formation and loss. Many agents like ultraviolet (UV) radiation and cisplatin do not cause direct double-stranded breaks but produce increasing levels of γH2AX as a function of time after exposure. Expression of γH2AX after treatment with these agents appears to be a result of γH2AX foci formed at stalled and/or damaged replication forks (Limoli et al., 2002; Olive and Banath, 2009).

There are some unresolved issues about the nature of the damage that stimulates γH2AX focus formation and exactly how these foci are able to promote DNA repair. However, the ability to accurately predict cell killing based only on γH2AX expression measured an hour after treatment with various drugs is remarkable (Banath and Olive, 2003). In this study, V79 fibroblasts were exposed for 30 min to six drugs known to produce double-stranded breaks and then allowed to form foci for 1 h. The percentage of cells that showed γH2AX antibody-binding levels greater than the untreated sample was found to be correlated with the percentage of cells that were killed by the treatment.

IV. Methods of Analysis

A. Reagents

1. Tris buffered saline (TBS), pH 7.4.
2. Tris, serum and detergent permeabilizing buffer (TST) for flow analysis: TBS containing 4% fetal bovine serum and 0.1% Triton X-100.

3. TTN permeabilizing buffer for slides: TBS containing 1% bovine serum albumin and 0.2% Tween-20.

4. Primary antibody: Phosphoserine-139 H2AX mouse monoclonal antibody, polyclonal antibodies and b Biotin and FITC-conjugated antibodies are also available.

5. Secondary antibody: Alexa 488 goat antimouse IgG (H + L)F(ab′)2 fragment conjugate or Alexa 594 goat antimouse immunoglobulin G (IgG) (H + L) F(ab′)2 fragment conjugate (Molecular Probes).

6. Stock solution of DAPI (Sigma).

7. Vectashield antifade solution (Vector Laboratories).

B. Cell Preparation Procedures for Flow Cytometry Analysis of γH2AX

1. After treatment and incubation for repair, centrifuge $2-5 \times 10^5$ single cells at $600-800 \times g$ and resuspend the cell pellet in 0.3-ml phosphate buffered saline TBS in a 4.5-ml tube. Then add 0.7 ml 99.9% ethanol to the tube while vortexing. Fixed samples can be maintained at $-20\ ^{\circ}C$ for at least 2 weeks before analysis.

2. On the day of analysis, add 1 ml of cold TBS to the cells in ethanol. Centrifuge cells at $800 \times g$ and then resuspend the pellet in 1 ml of cold TST for 10 min to permeabilize and rehydrate the cells.

3. Centrifuge samples and resuspend the pellet in 200 μl mouse monoclonal anti-γH2AX antibody diluted 1:500 in TST. Place tubes on a shaker platform and incubate, covered, for 2 h at room temperature.

4. Wash cells by centrifuging and resuspending the pellet in TST. Repeat this procedure.

5. Resuspend pellet in 200 μl of secondary antibody (e.g., Alexa 488-conjugated antimouse IgG, diluted 1:200 in TST), and return to the shaker platform, covered, for incubation for 1 h at room temperature. During shaking, take care that cells remain in solution and do not form a meniscus above the liquid.

6. Wash cells by centrifuging and resuspending pellet in TBS with 4% fetal bovine serum.

7. Centrifuge and resuspend pellet in TBS containing 1 $\mu g/ml$ of DAPI.

8. Analyze samples on a dual-laser flow cytometer using UV excitation for DAPI and 488-nm excitation for γH2AX. Perform gating on forward scatter and time of flight to eliminate signals from debris and doublets.

C. Cell Preparation Procedures for Image Cytometry Analysis of γH2AX

1. Grow cells on washed and sterile glass slides by first pipetting 10-μl cells (10^4 cells in serum-free medium) onto a localized region of the slide and allowing cells to attach for 30 min before adding complete medium to cover the coverslips.

Cells are usually incubated overnight to maximize cell spreading and reduce nuclear thickness.

2. After treatment and allowing time for foci to develop, rinse coverslips in TBS. Then place slides for 15–30 min in freshly prepared 2% paraformaldehyde in PBS. Frozen PBS sections of tissues are prepared after embedding tissue in OCT embedding compound. After allowing cut sections to dry for 1 min, place slides in 2% paraformaldehyde for 15–30 min. After fixation, the following steps can be performed for cells and tumor sections attached to slides.

3. Rinse slides in PBS and permeabilize cells by submersing in −20 °C methanol for 1 min.

4. Rinse slides in TBS and submerse in TTN for 20 min.

5. Deposit 25 μl anti-γH2AX diluted 1:500 in TTN on 1-cm^2 parafilm squares. Invert these squares over the cell or tissue region on the coverslips/slides. Place in a humidified chamber for 2 h at room temperature or 4 °C overnight.

6. Rinse coverslips/slides with TBS and incubate with 25 μl fluorescent secondary antibody diluted 1:200 in TTN for 1 h at room temperature in the dark.

7. Rinse coverslips in TBS several times for 5 min each time in the dark and immerse in 0.05 μg/ml of DAPI for 5 min followed by a 5-min rinse in TBS. A final dip in fixative can be used.

8. Drain coverslips, add 10 μl of antifade mounting medium next to cells before inverting coverslip over the slide. Seal coverslip.

9. Slides are viewed using a fluorescent microscope with motorized focus capacity that allows deconvolution of about 50-nm slices through the cells. Our cell images were acquired using a 100× Neofluor objective and a Q-Imaging 1350 EX digital camera. Images are analyzed using Northern Eclipse 6.0 software (Empix Imaging). A macro was designed to identify foci positions and associated intensity and size after application of a thresholding algorithm and a mask to remove residual background staining (MacPhail et al., 2003a). Alternatively, simply counting foci by eye is quite acceptable using coded slides.

D. Special Considerations

Using flow cytometry, γH2AX intensity can be expressed relative to the intensity of the control untreated population. These samples should be collected, stained, and analyzed at the same time as the test cells. When using asynchronous cells, radiation and drugs can block cells in various phases of the cell cycle. Because the control population will typically display a DNA histogram and γH2AX intensity consistent with asynchronous, unperturbed cells, simple comparisons of γH2AX intensity in treated populations with the mean fluorescence of controls may not be valid. In this situation, it may be necessary to limit analysis to cells in a specific phase of the cell cycle (e.g., based on DAPI staining intensity).

Foci in irradiated mitotic cells typically stain about five times more intensely than foci of cells in other phases of the cell cycle, and these cells can appear as a separate population above the G_2 phase cells. Apoptotic cells in which the membrane is "intact" also stain about 5–10 times more intensely with γH2AX antibodies than normal cells. In both cases, foci size appears to be increased. Both mitotic and apoptotic cells can be distinguished from the remaining cells within the population based on differential γH2AX staining and, if appropriate, omitted from analysis.

Some studies may be more easily interpreted if synchronized cells are used for analysis. Twofold differences in foci number will occur through the cell cycle simply as a (probabilistic) result of the increase in DNA at risk for damage. The situation is even more complex as cells enter S phase, because the initiation of DNA synthesis can result in a twofold to threefold higher background levels of γH2AX. Even with bivariate analysis of DAPI-stained cells, it is not possible to distinguish G_1 from early S-phase cells based on DNA content alone.

Colocalization of γH2AX with repair complexes may be a useful way to determine the kinetics of repair enzyme recruitment to sites of DNA damage. However, it is important to consider whether foci of γH2AX that colocalize with foci composed of various repair factors (e.g., BRCA1, Mre11, Rad51) might be an indication that repair of those particular breaks has not been successful. In this case, a higher percentage of colocalized foci could indicate poorer repair. The probability of (false) colocalization will increase as γH2AX foci size and number increase, so statistical approaches will be required. Application of optical sectioning methods will reduce the possibility that two foci in different planes can appear colocalized.

V. Typical Results

A. Cell Cycle Variation in γH2AX Expression Detected Using Flow Cytometry

Bivariate analysis of γH2AX intensity versus DNA content indicates that untreated S- and G_2-M-phase cells have a higher level of expression of γH2AX than cells in G_1 phase (Fig. 1). Visually, the individual foci that are present in untreated S-phase cells typically appear much smaller than those in irradiated cells (Fig. 2). After drug or radiation treatment, two general patterns of γH2AX staining have been observed. After exposure of asynchronously growing cultures to X-rays, bleomycin, doxorubicin, or etoposide, γH2AX expression is largely independent of cell cycle position. For these drugs, the relative increase in γH2AX intensity can be used as an indication of cell killing. Alternatively, one can determine the percentage of cells that show more γH2AX than the untreated controls (i.e., cells are outside a gate placed around the control population). The percentage of cells that fall into this gate has been shown to correlate with the (clonogenic) fraction of cells that survive treatment (Banáth and Olive, 2003). However, this

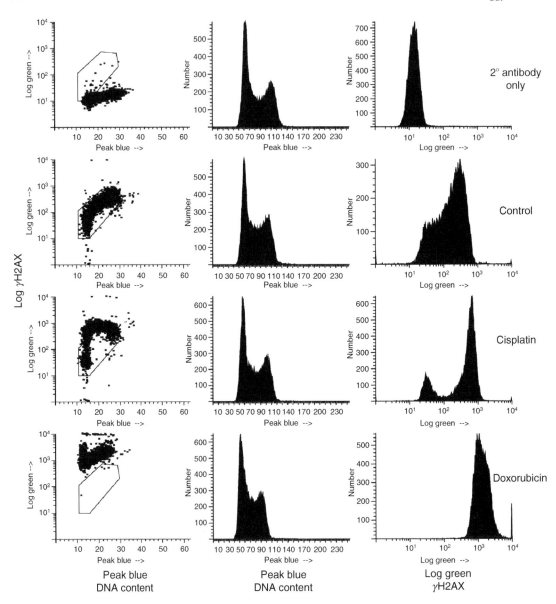

Fig. 1 Bivariate analysis of γH2AX and DNA content in hamster V79 cells. The control sample shows the typical increased expression of γH2AX in S/G₂ phase cells relative to G₁ phase cells. One hour after 2 h of exposure to cisplatin, S-phase cells show increased expression of γH2AX. One hour after a 30-min exposure to doxorubicin, all cells show increased expression of γH2AX.

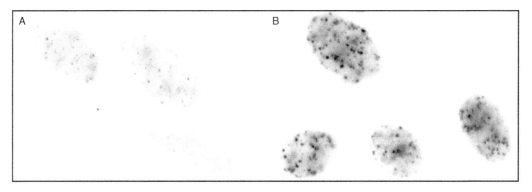

Fig. 2 Appearance of γH2AX foci in 4′,6-diamidino-2-phenylindole dihydrochloride hydrate (DAPI)-stained nuclei of human lung fibroblasts (HFL-1 cells) exposed to 0 Gy (A) or 1 h after 2 Gy (B). S-phase cells in unirradiated samples show many smaller foci (two upper cells in panel A).

approach needs to be established independently for each drug and is likely to depend on both treatment time and recovery time. For example, cells treated with cisplatin, tirapazamine, 4-nitroquinoline-N-oxide, or high-dose hydrogen peroxide show damage that is greatest in S-phase cells (but not excluded from G_1 cells). S-phase-dependent damage is likely to be a result of γH2AX produced at stalled replication forks, and foci development may be initiated as treated cells enter S phase. This makes timing an important consideration for determining whether cells have responded to some drugs by forming foci. In this case, the percentage of cells that retain foci measured after all cells have had an opportunity to enter S phase may be more informative for cell response than the relative increase in γH2AX measured soon after exposure.

B. Rate of Loss of γH2AX and Residual γH2AX as Measures of Cell Response to Radiation

The rate of loss of γH2AX after exposure to X-rays has been shown to correlate with radiation sensitivity for a series of human tumor cell lines—generally, tumor cells that are more radiation sensitive show a slower loss of γH2AX than radiation-resistant cells (MacPhail *et al.*, 2003a). Figure 3A shows the kinetics of development and loss of γH2AX in SiHa cervical carcinoma cells as a function of time after exposure to various radiation doses. In this cell line as in most others, γH2AX loss rate after irradiation is dose independent. Using flow cytometry, background levels of γH2AX through the cell cycle and the radiation-induced G_2 block can complicate measurements of loss rate after doses below 2 Gy. Figure 3B indicates that the maximum γH2AX signal has been reduced to 50% by about 5 h after irradiation of this cell line, independent of dose. By 24 h, those cells that retain foci can be counted and compared to the percentage of cells that survive irradiation (Fig. 3C). The percentage of cells that lack foci may be useful in providing a rough estimate of sensitivity to ionizing radiation and other agents. The appearance of selected SiHa cells examined 24 h after exposure to 6 Gy is shown in Fig. 4. Note that foci

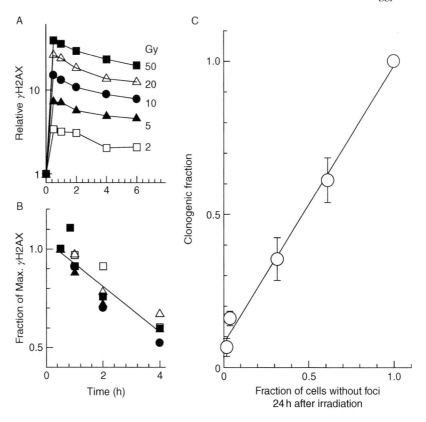

Fig. 3 Kinetics of development and loss of γH2AX measured by flow cytometry in exponentially growing SiHa cervical carcinoma cells examined as a function of time after exposure to 0–50 Gy. (A) Relative γH2AX intensity was calculated by dividing the mean fluorescence of the treated sample by the mean fluorescence of the control sample. (B) γH2AX intensity was divided by the maximum intensity of γH2AX achieved for each dose; the half-time for loss of γH2AX is independent of radiation dose. (C) The percentage of SiHa cells that lack foci 24 h after irradiation was compared to the percentage of cells that survived to form colonies.

can be quite large and variable in number, and occasionally they appear in micronuclei.

C. Double-Stranded Breaks in Human Lymphocytes

The excellent sensitivity of γH2AX for detecting DNA double-stranded breaks is illustrated with human lymphocytes exposed to low doses of ionizing radiation (Fig. 5). Damage by 5 cGy can be detected using flow cytometry (e.g., about one double-stranded break per cell). Blood was collected by venipuncture directly into gradient tubes (Becton Dickinson CPT tubes) for lymphocyte separation by centrifugation. After irradiation, 1 h for focus formation, fixation, and antibody

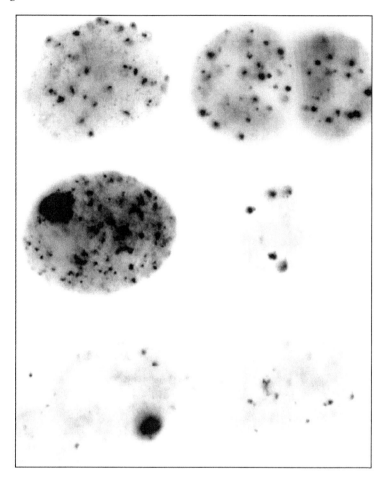

Fig. 4 Appearance of representative SiHa 4′,6-diamidino-2-phenylindole dihydrochloride hydrate (DAPI)-stained nuclei 24 h after exposure to 6-Gy X-rays (only 10% of cells will remain clorogenic). The majority of nuclei exhibit γH2AX foci that are variable in both number and size. The percentage of cells that lacked foci correlated with the percentage that survived to form colonies (Fig. 3C).

staining, samples were analyzed for γH2AX. To obtain the dose-response relationship, the mean fluorescence intensity of the irradiated samples was divided by the mean fluorescence of the control cells (note that the sensitivity for detecting double-stranded breaks is greatly increased in this case because of the absence of proliferating cells). In comparison, the limit of detection for double-stranded breaks in proliferating, asynchronous hamster V79 cells is about 20 cGy (MacPhail *et al.*, 2003a). A dose-response curve for white blood cells analyzed using the neutral comet assay is shown for comparison (Banath *et al.*, 1998); this is the only other method available for the measurement of initial DNA double-stranded breaks in individual cells and is almost two orders of magnitude less sensitive.

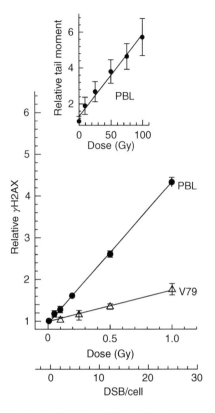

Fig. 5 Sensitivity for detecting double-stranded breaks in human peripheral blood lymphocytes (PBLs) exposed to X-rays. The response of PBL was measured 1 h after exposure to X-rays. The decrease in sensitivity of hamster V79 cells (MacPhail *et al.*, 2003a) is largely a result of the higher background expression of γH2AX in S-phase cells, which reduces sensitivity for detecting radiation-induced damage. Shown for comparison purposes, (inset) the response of PBL in the neutral comet assay measured immediately after irradiation (Banáth *et al.*, 1998); from a comparison of slopes, the comet is 75 times less sensitive for detecting double-stranded breaks. Error bars are standard deviations for three experiments.

D. Detection of Apoptosis and Influence of Loss of Membrane Integrity

As previously mentioned, apoptotic cells have been found to stain intensely with γH2AX antibodies, which is not surprising because apoptosis results in extensive DNA fragmentation. Interestingly, TK6 cells that stained positive for antibodies against activated caspase-3 showed high levels of γH2AX when examined 4 h after irradiation but near-normal levels 24 h later despite that most of the cells had then undergone apoptosis (MacPhail *et al.*, 2003b). It is possible that at later times, apoptotic cells may no longer stain intensely because γH2AX is lost at the time when membrane integrity is lost. To examine this possibility, asynchronous TK6 human lymphoblasts were exposed to 6-Gy X-rays and fixed 12 h later. By this time, 24% of the (apoptotic) cells had become permeable to propidium iodide (PI), an indication that membrane integrity had been lost in those cells. Cells were sorted into two populations: those that were negative for PI staining and those that were positive. Cells that were negative for PI showed only 12% apoptotic cells as measured using the neutral comet assay, and this value agreed well with the population of cells (15%) that expressed high levels of γH2AX (Fig. 6). However, PI-permeable cells that were 65% apoptotic/necrotic based on the comet assay showed only 12% of cells with high levels of γH2AX. Therefore, the presence of high levels of γH2AX in a cell can be an indication of apoptosis, but caution must be exercised if membrane integrity is lost. Lack of γH2AX expression in a cell that shows other evidence of apoptosis (e.g., positive for activated caspase-3 or positive in the comet assay) can be an indication that apoptosis has progressed to necrosis. This result does not rule out the possibility that γH2AX is still present but no longer readily detectable once cells have lost membrane integrity.

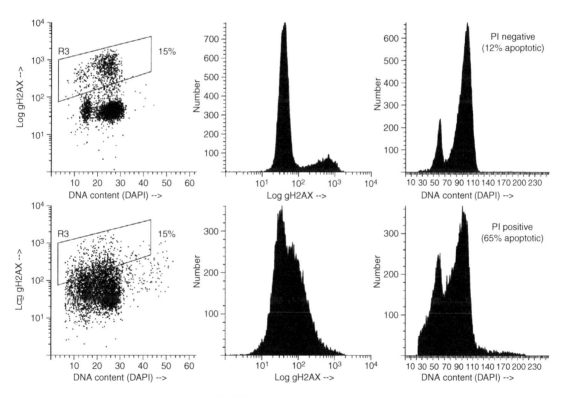

Fig. 6 Expression of γH2AX in apoptotic cells that retain membrane integrity. TK6 cells were exposed to 6 Gy and 12 h later, cells were incubated with the membrane impermeant dye, propidium iodide (PI). PI-positive and PI-negative cells were sorted, and sorted cells were then analyzed for the percentage of apoptotic cells using the comet assay (Olive *et al.*, 1993) or fixed and stained for γH2AX expression.

E. Analysis of DNA Damage in Cells from Tissues and Tumors

Both flow and image cytometry can be used to examine double-stranded breaks produced in cells from normal tissues and solid tumors exposed to DNA double-stranded breaking agents. Flow cytometry was used to analyze γH2AX in spleen cells and SCCVII tumor cells removed from C3H mice 1 h after exposure to 60 mg of etoposide per kilogram of body weight (Fig. 7). This treatment killed about 30% of the tumor cells when measured using a clonogenic assay. The percentage of cells that showed γH2AX levels greater than the control level was consistent with this value. Interestingly, spleen cells removed from the same treated mouse showed about the same enhancement of γH2AX staining relative to the control spleen cells, suggesting that the therapeutic ratio, at least by this criterion, may not be particularly favorable for proliferating normal tissues.

Fluorescent images of γH2AX tumor sections can also be prepared from tumors or normal tissues exposed to DNA damaging agents. In Fig. 8, C3H mice were exposed to X-rays, and 1 h later, the DNA binding dye Hoechst 33342 was administered intravenously to stain nuclei of cells surrounding blood vessels. Frozen sections prepared from kidneys were fixed in paraformaldehyde and stained for γH2AX. Virtually, all nuclei show γH2AX antibody binding that is radiation dose dependent. Fluorescence detection of gH2AX in formalin-fixed, paraffin-embedded sections is also possible.

VI. Possible Applications

Applications of this method might include but are not limited to the following:

1. Identification of sites of double-stranded breaks for studies on the kinetics of recruitment of DNA repair enzymes to these sites (Fernandez-Capetillo et al., 2003; Paull et al., 2000).
2. Sensitive detection of DNA double-stranded breaks by ionizing radiation (applications in radiation biodosimetry) (Redon et al., 2009).
3. Prediction of tumor cell killing by ionizing radiation (radiosensitivity) based on γH2AX loss rate or residual damage after treatment (MacPhail et al., 2003a; Taneja et al., 2004).
4. Prediction of tumor cell killing by drugs (chemosensitivity) based on level of γH2AX expression after treatment (Banath and Olive, 2003; Halicka et al., 2009).
5. Detection of cells in early phases of apoptosis (Huang et al., 2003; MacPhail et al., 2003b; Rogakou et al., 2000).
6. Examination of the physical distribution of DNA damage within tumor sections using γH2AX immunohistochemistry (e.g., location of damaged cells relative to necrosis, blood vessels, and stroma).
7. Comparison between the amount of DNA damage produced in tumors in relation to normal tissues (therapeutic ratio).

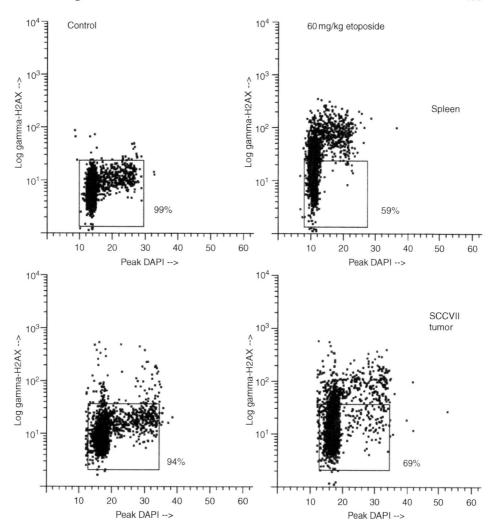

Fig. 7 Flow cytometry analysis of γH2AX in SCCVII tumor cells and spleen cells from C3H mice. Tissues were removed and fixed 1 h after drug administration. Etoposide creates DNA double-stranded breaks in proliferating cells that contain its target enzyme topoisomerase II. The percentage of SiHa tumor cells that survived this treatment was 71% when measured using a clonogenic assay. This value correlates well with the percentage of cells that show control levels of γH2AX (69%). Note that spleen cells responded similarly to tumor cells from the same mouse.

8. Use as a tool for examining radiation bystander effects or adaptive responses that may involve DNA repair pathways and damage signaling (Ballarini *et al.*, 2002).

9. Identification of radiation-resistant hypoxic tumor cells (in combination with damage by ionizing radiation) (Olive and Banáth, 2004).

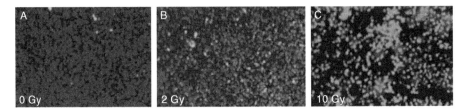

Fig. 8 Identification of γH2AX in frozen sections of kidneys from C3H mice exposed to ionizing radiation. Images were obtained using a 20× neofluor objective (individual foci are not visible at this magnification). Blue is Hoechst 33342 nuclear staining; red is γH2AX antibody staining. (See Plate no. 8 in the Color Plate Section.)

10. Use as a tool for examining chromosomal break points or telomere abnormalities (Takai *et al.*, 2003).

11. Characterization of genomic instability (Bassing *et al.*, 2002; Ivanov *et al.*, 2003).

12. Tool in studies of V(D)J recombination and meiotic recombination (Hamer *et al.*, 2003).

13. Use in examining the importance of higher order chromatin organization and dynamics in DNA damage and repair (Mailand *et al.*, 2007).

14. Use as a biomarker for H2AX targeted drugs in cancer treatment (Taneja *et al.*, 2004).

VII. Conclusion

The formation of γH2AX foci at sites of double-stranded breaks is an important event in the recognition and signaling of the presence of the break. Exactly how these foci promote repair is not yet known, although both chromatin changes and repair enzyme retention are probably involved. Although there are unresolved questions concerning the nature of the physical damage that provokes γH2AX focus formation, this event offers an opportunity to measure, with exceptional sensitivity, a DNA lesion underlying chromosome damage and cell killing that is produced, directly or indirectly, by various agents. Colocalization studies with residual γH2AX provide an ideal way to visualize repair enzyme activity at sites of unrepaired breaks.

Acknowledgments

This work was supported by the Canadian Cancer Society. The expert collaboration of Dr Judit P. Banath and Susan H. MacPhail is gratefully acknowledged.

References

Andegeko, Y., Moyal, L., Mittelman, L., Tsarfaty, I., Shiloh, Y., and Rotman, G. (2001). Nuclear retention of ATM at sites of DNA double strand breaks. *J. Biol. Chem.* **276**, 38224–38230.

Bakkenist, C. J., and Kastan, M. B. (2003). DNA damage activates ATM through intermolecular autophosphorylation and dimer dissociation. *Nature* **421**, 499–506.

Ballarini, F., Biaggi, M., Ottolenghi, A., and Sapora, O. (2002). Cellular communication and bystander effects: A critical review for modelling low-dose radiation action. *Mutat. Res.* **501**, 1–12.

Banáth, J. P., and Olive, P. L. (2003). Expression of phosphorylated histone H2AX as a surrogate of cell killing by drugs that create DNA double-strand breaks. *Cancer Res.* **63**, 4347–4350.

Banáth, J. P., Fushiki, M., and Olive, P. L. (1998). Rejoining of DNA single- and double-strand breaks in human white blood cells exposed to ionizing radiation. *Int. J. Radiat. Biol.* **73**, 649–660.

Banáth, J. P., Banuelos, C. A., Klokov, D., Macphail, S. M., Lansdorp, P. M., and Olive, P. L. (2009). Explanation for excessive DNA single-strand breaks and endogenous repair foci in pluripotent mouse embryonic stem cells. *Exp. Cell Res.* **315**(8), 1505–1520.

Banuelos, C. A., Banáth, J. P., MacPhail, S. H., Zhao, J., Reitsema, T., and Olive, P. L. (2007). Radiosensitization by the histone deacetylase inhibitor PCI-24781. *Clin. Cancer Res.* **13**, 6816–6826.

Bassing, C. H., Chua, K. F., Sekiguchi, J., Suh, H., Whitlow, S. R., Fleming, J. C., Monroe, B. C., Ciccone, D. N., Yan, C., Vlasakova, K., Livingston, D. M., Ferguson, D. O., *et al* (2002). Increased ionizing radiation sensitivity and genomic instability in the absence of histone H2AX. *Proc. Natl. Acad. Sci. USA* **99**, 8173–8178.

Burma, S., Chen, B. P., Murphy, M., Kurimasa, A., and Chen, D. J. (2001). ATM phosphorylates histone H2AX in response to DNA double-strand breaks. *J. Biol. Chem.* **276**, 42462–42467.

Celeste, A., Petersen, S., Romanienko, P. J., Fernandez-Capetillo, O., Chen, H. T., Sedelnikova, O. A., Reina-San-Martin, B., Coppola, V., Meffre, E., Difilippantonio, M. J., Redon, C., Pilch, D. R., *et al* (2002). Genomic instability in mice lacking histone H2AX. *Science* **296**, 922–927.

Celeste, A., Difilippantonio, S., Difilippantonio, M. J., Fernandez-Capetillo, O., Pilch, D. R., Sedelnikova, O. A., Eckhaus, M., Ried, T., Bonner, W. M., and Nussenzweig, A. (2003a). H2AX haploinsufficiency modifies genomic stability and tumor susceptibility. *Cell* **114**, 371–383.

Celeste, A., Fernandez-Capetillo, O., Kruhlak, M. J., Pilch, D. R., Staudt, D. W., Lee, A., Bonner, R. F., Bonner, W. M., and Nussenzweig, A. (2003b). Histone H2AX phosphorylation is dispensable for the initial recognition of DNA breaks. *Nat. Cell Biol.* **5**, 675–679.

Downs, J. A., and Jackson, S. P. (2003). Cancer: Protective packaging for DNA. *Nature* **424**, 732–734.

Fernandez-Capetillo, O., Celeste, A., and Nussenzweig, A. (2003). Focusing on foci: H2AX and the recruitment of DNA-damage response factors. *Cell Cycle* **2**, 426–427.

Hamer, G., Roepers-Gajadien, H. L., van Duyn-Goedhart, A., Gademan, I. S., Kal, H. B., van Buul, P. P., and de Rooij, D. G. (2003). DNA double-strand breaks and gamma-H2AX signaling in the testis. *Biol. Reprod.* **68**, 628–634.

Halicka, H. D., Ozkaynak, M. F., Levendoglu-Tugal, O., Sandoval, C., Seiter, K., Kajstura, M., Traganos, F., Jayabose, S. and Darzynkiewicz, Z. (2009). DNA damage response as a biomarker in treatment of leukemias. *Cell Cycle* **8**, 1720–1724.

Huang, X., Traganos, F., and Darzynkiewicz, Z. (2003). DNA damage induced by DNA topoisomerase I and topoisomerase II inhibitors detected by histone H2AX phosphorylation in relation to the cell cycle phase and apoptosis. *Cell Cycle* **2**, 614–619.

Ivanov, A., Cragg, M. S., Erenpreisa, J., Emzinsh, D., Lukman, H., and Illidge, T. M. (2003). Endopolyploid cells produced after severe genotoxic damage have the potential to repair DNA double strand breaks. *J. Cell Sci.* **116**, 4095–4106.

Limoli, C. L., Giedzinski, E., Bonner, W. M., and Cleaver, J. E. (2002). UV-induced replication arrest in the xeroderma pigmentosum variant leads to DNA double-strand breaks, gamma-H2AX formation, and Mre11 relocalization. *Proc. Natl. Acad. Sci. USA* **99**, 233–238.

MacPhail, S. H., Banath, J. P., Yu, T. Y., Chu, E. H., Lambur, H., and Olive, P. L. (2003a). Expression of phosphorylated histone H2AX in cultured cell lines following exposure to X-rays. *Int. J. Radiat. Biol.* **79**, 351–358.

MacPhail, S. H., Banath, J. P., Yu, T. Y., Chu, E., and Olive, P. L. (2003b). Cell cycle-dependent expression of phosphorylated histone H2AX: Reduced expression in unirradiated but not x-irradiated G(1)-phase cells. *Radiat. Res.* **159**, 759–767.

Mailand, N., Bekker-Jensen, S., Faustrup, H., Melander, F., Bartek, J., Lukas, C., and Lukas, J. (2007). RNF8 ubiquitylates histones at DNA double-strand breaks and promotes assembly of repair proteins. *Cell* **131**, 887–900.

Nazarov, I. B., Smirnova, A. N., Krutilina, R. I., Svetlova, M. P., Solovjeva, L. V., Nikiforov, A. A., Oei, S. L., Zalenskaya, I. A., Yau, P. M., Bradbury, E. M., and Tomilin, N. V. (2003). Dephosphorylation of histone gamma-H2AX during repair of DNA double-strand breaks in mammalian cells and its inhibition by calyculin A. *Radiat. Res.* **160**, 309–317.

Olive, P. L. (2002). The comet assay. An overview of techniques. *Methods Mol. Biol.* **203**, 179–194.

Olive, P. L., and Banáth, J. P. (2004). Phosphorylation of histone H2AX as a measure of radiosensitivity. *Int. J. Radiat. Oncol. Biol. Phys.* **58**, 331–335.

Olive, P. L., and Banáth, J. P. (2009). Kinetics of H2AX phosphorylation after exposure to cisplatin. *Cytometry B Clin. Cytom.* **76**, 76–90.

Olive, P. L., Wlodek, D., and Banáth, J. P. (1991). DNA double-strand breaks measured in individual cells subjected to gel electrophoresis. *Cancer Res.* **51**, 4671–4676.

Olive, P. L., Frazer, G., and Banáth, J. P. (1993). Radiation-induced apoptosis measured in TK6 human B lymphoblast cells using the comet assay. *Radiat. Res.* **136**, 130–136.

Paull, T. T., Rogakou, E. P., Yamazaki, V., Kirchgessner, C. U., Gellert, M., and Bonner, W. M. (2000). A critical role for histone H2AX in recruitment of repair factors to nuclear foci after DNA damage. *Curr. Biol.* **10**, 886–895.

Redon, C., Pilch, D., Rogakou, E., Sedelnikova, O., Newrock, K., and Bonner, W. (2002). Histone H2A variants H2AX and H2AZ. *Curr. Opin. Genet. Dev.* **12**, 162–169.

Redon, C. E., Dickey, J. S., Bonner, W. M., and Sedelnikova, O. A. (2009). γ-H2AX as a biomarker of DNA damage induced by ionizing radiation in human peripheral blood lymphocytes and artifical skin. *Adv. Space Res.* (in press).

Reitsema, T., Banath, J. P., MacPhail, S. H., and Olive, P. L. (2004). Chromatin condensation by hypertonic saline and expression of phosphorylated histone H2AX. *Radiat. Res.* **161**, 402–408.

Rogakou, E. P., Pilch, D. R., Orr, A. H., Ivanova, V. S., and Bonner, W. M. (1998). DNA double-stranded breaks induce histone H2AX phosphorylation on serine 139. *J. Biol. Chem.* **273**, 5858–5868.

Rogakou, E. P., Boon, C., Redon, C., and Bonner, W. M. (1999). Megabase chromatin domains involved in DNA double-strand breaks *in vivo*. *J. Cell Biol.* **146**, 905–916.

Rogakou, E. P., Nieves-Neira, W., Boon, C., Pommier, Y., and Bonner, W. M. (2000). Initiation of DNA fragmentation during apoptosis induces phosphorylation of H2AX histone at serine 139. *J. Biol. Chem.* **275**, 9390–9395.

Rothkamm, K., and Lobrich, M. (2003). Evidence for a lack of DNA double-strand break repair in human cells exposed to very low X-ray doses. *Proc. Natl. Acad. Sci. USA* **100**, 5057–5062.

Sedelnikova, O. A., Rogakou, E. P., Panyutin, I. G., and Bonner, W. M. (2002). Quantitative detection of (125)IdU-induced DNA double-strand breaks with gamma-H2AX antibody. *Radiat. Res.* **158**, 486–492.

Siino, J. S., Nazarov, I. B., Svetlova, M. P., Solovjeva, L. V., Adamson, R. H., Zalenskaya, I. A., Yau, P. M., Bradbury, E. M., and Tomilin, N. V. (2002). Photobleaching of GFP-labeled H2AX in chromatin: H2AX has low diffusional mobility in the nucleus. *Biochem. Biophys. Res. Commun.* **297**, 1318–1323.

Takai, H., Smogorzewska, A., and de Lange, T. (2003). DNA damage foci at dysfunctional telomeres. *Curr. Biol.* **13**, 1549–1556.

Taneja, N., Davis, M., Choy, J. S., Beckett, M. A., Singh, R., Kron, S. J., and Weichselbaum, R. R. (2004). Histone H2AX phosphorylation as a predictor of radiosensitivity and target for radiotherapy. *J. Biol. Chem.* **279**, 2273–2280.

CHAPTER 21

Assays of Cell Viability: Discrimination of Cells Dying by Apoptosis

Zbigniew Darzynkiewicz, Xun Li, and Jianping Gong

Cancer Research Institute
New York Medical College
Valhalla, New York 10595

DOI: 10.1016/B978-0-12-375045-7.00021-0

I. Update

Of all the chapters published in the cytometry-devoted MCB volumes (Vols. 33, 41, 42, 63, 64, and 75) this chapter has been most widely cited (over 240 citations) and the frequency of its citations still remains high. It appears that the principles of the differential staining of live *versus* dying *versus* dead cells described in the chapter, that provide foundation for variety of methodologies used in cell necrobiology, are still of interest to wide audience of readers. However, since its publication in 1994 numerous new methods, particularly the assays designed for detecting apoptotic cells, have been developed (reviews in Darzynkiewicz *et al.*, 2004; Kaufmann *et al.*, 2008).

Of these newer approaches perhaps the most widely used is the annexin V affinity assay. This methodology is based on the use of fluorochrome tagged anticoagulant protein annexin V which in the presence of Ca^{2+} binds with high affinity to phosphatidylserine, one of the plasma membrane phospholipids. Phosphatidylserine in live cells is located on the inner leaflet of plasma membrane but during apoptosis flips to the external surface (van Engeland *et al.*, 1998). Binding of annexin V is usually combined with cells exposure to PI or 7-aminoactinomycin D (7-AAD), the charged fluorochromes that are excluded from live and early apoptotic cells but penetrate plasma membrane and bind to nucleic acids of late apoptotic and necrotic cells. Combined analysis of annexin V binding and dye exclusion test allows one to identify live (annexin V negative) *versus* early apoptotic (annexin V positive, PI or 7-AAD negative) *versus* late apoptotic plus necrotic (PI or 7-AAD positive) cells. The virtue of annexin V affinity assay is its simplicity.

A large group of assays relies on the detection of caspases activation (review in Kaufmann *et al.*, 2008). Activation of each of the caspases can be detected by (i) immunocytochemical approach, using commercially available polyclonal or monoclonal Abs to the individual-cleaved caspases, or Abs to the specific caspase-cleaved substrates such as poly(ADP-ribose)polymerase (PARP); (ii) using peptide substrates with a caspase-cleavage site and having a fluorochrome on one end and fluorochrome quencher on another end of the molecule; the peptide cleavage separates quencher from the fluorochrome leading to enhancement of fluorescence; (iii) using the affinity ligands such as fluorochrome-labeled inhibitors of caspases (FLICA). Of these three approaches the most specific with respect to identification of individual caspases are the immunocytochemical assays. The positive features of the FLICA methodology are low cost of the reagents and simplicity of the labeling protocol. Applicability of the FLICA to detect apoptosis *in vivo*, in animal tissues, is another asset of these probes (Darzynkiewicz *et al.*, 2009). Their specificity to identify activated caspases, however, is in question (Pozarowski *et al.*, 2003).

The DNA fragmentation (TUNEL) assay described in this chapter is still widely used as a marker of apoptotic cells. The methodology of this assay, however, had advanced. The new approach uses BrdUTP rather than biotin- or digoxygenin-conjugated nucleotide to label DNA strand breaks, and the presence of the incorporated BrdU is detected immunocytochemically with BrdU Ab (Li and Darzynkiewicz, 1995, Darzynkiewicz *et al.*, 2008). This improvement led to increase in sensitivity of the assay and made it simpler. The widely used APO-BRDU TUNEL assay kit (Phoenix Flow Systems, San Diego, CA) which provides the easy to use prepared reagents and a lucid protocol is based on this methodology.

The most recent approaches to measure apoptosis rely on the use of supravital fluorochromes of the SYTO family (review in Wlodkowic *et al.*, 2008). Upon SYTO fluorochrome penetration into the cell significant degree of attenuation of its fluorescence is seen in apoptotic cells. While the mechanism responsible for this phenomenon is still not fully understood, the approach to use SYTO probes as a specific marker of apoptotic cells appears very attractive. The attractiveness stems from low cost (per assay) of the probes, their wide choice with respect to different excitation/emission spectra and the possibility to use them supravitally, with no need for cell fixation. The combination of the green-fluorescing SYTO 16 with PI, similar as annexin V/PI assay, makes it possible to identify live *versus* early apoptotic *versus* late apoptotic and necrotic cells. The SYTO probes appear to be ideal candidates to be used in combination with the surface immunophenotyping to concurrently assess, by multivariate analysis, frequency of apoptosis in cells of different subtypes.

Recent progress in field of necrobiology led to expansion of our knowledge on the differences in the mode of cell death depending on nature of the death inducer, cell type, cell cycle phase, cell environment, etc. It became apparent that during the cell demise molecular mechanisms and morphological changes may significantly vary and the caspase-independent mechanisms can cooperate or even substitute caspases. Often, the features of dying cells do not allow one to simply classify them as live, apoptotic, or necrotic cells. The recently proposed classification lists the submodes of cell death termed "autophagy," "entosis," mitotic catastrophe," necrosis," necroptosis," and pyroptosis"(Kroemer *et al.*, 2009). It is expected that combinations of cytometric probes will be developed to characterize many of the above subtypes of cell demise.

II. Introduction

A. The Modes of Cell Death: Apoptosis and Necrosis

Two distinct modes of cell death, apoptosis and necrosis, can be recognized based on differences in the morphological, biochemical, and molecular changes of the dying cell. The assays of cell viability presented in this chapter are discussed in light of their applicability to differentiate between these two mechanisms.

1. Apoptosis

The terms "apoptosis," "active cell death," "cell suicide," and "shrinkage necrosis" are being used, often interchangeably, to define a particular mode of cell death characterized by a specific pattern of changes in nucleus and cytoplasm (Arends *et al.*, 1990; Compton, 1992; Wyllie, 1985, 1992). Because this mode of cell death plays a role during the programmed cell death, as originally described in embryology, the term "programmed cell death" is also being used synonymously (albeit incorrectly, in the context of denoting the mode of cell death) with apoptosis.

The role of apoptosis in embryology, endocrinology, and immunology is the subject of several reviews (e.g., Tomei and Cope, 1991). The wide interest in apoptosis in oncology, so apparent in recent years, stems from the observations that this mode of cell death is triggered by a variety of antitumor drugs, radiation, or hyperthermia and that the intrinsic propensity of tumor cells to respond by apoptosis is modulated by expression of several oncogenes such as *bcl*-2, c-*myc*, or tumor suppressor gene p53 and may be prognostic of treatment (Wyllie, 1992; Sachs and Lotem, 1993; Schwartzman and Cidlowski, 1993).

Extensive research is underway in many laboratories to understand the mechanism of apoptosis. Knowledge of the molecular events of this process may be the basis for new antitumor strategies. Apoptosis affecting CD4$^+$ lymphocytes of HIV-infected patients also appears to play a pivotal role in pathogenesis of AIDS (Meyaard *et al.*, 1992).

The most common feature of apoptosis is active participation of the cell in its demise. The cell mobilizes a cascade of events that lead to its disintegration and the formation of the "apoptotic bodies" which are subsequently engulfed by the neighboring cells without invoking inflammation (Arends *et al.*, 1990; Compton, 1992; Tomei and Cope, 1991; Wyllie, 1985, 1992). Increased cytoplasmic Ca^{2+} concentration, cell dehydration, increased lipid peroxidation, chromatin condensation originating at the nuclear periphery, activation of endonuclease which has preference to DNA at the internucleosomal (linker) sections, proteolysis, fragmentation of the nucleus, and fragmentation of the cell are the most characteristic events of apoptosis. On the other hand, even during advanced stages of apoptosis, the structural integrity and the transport function of the plasma membrane are preserved. Also preserved and functionally active are the mitochondria and lysosomes.

Thus, regardless of cell type, or the nature of event which triggers apoptosis, this mode of cell death has many features in common. Some of these features can be analyzed by image or flow cytometry, and several methods have been described to identify apoptotic cells (Darzynkiewicz *et al.*, 1992).

Mitotic death, also termed delayed reproductive death, shows some features of apoptosis and thus may represent delayed apoptosis (e.g., Chang and Little, 1991; Tounekti *et al.*, 1993); it occurs as a result of cell exposure to relatively low doses of drugs or radiation, which induce irreparable damage, but allow cells to complete at least one round of division.

2. Necrosis

Necrosis is an alternative to the apoptotic mechanism of cell death. Most often it is induced by an overdose of cytotoxic agents and is a cell response to a gross injury. However, certain cell types do respond even to pharmacological concentrations of some drugs or moderate doses of physical agents by necrosis rather than apoptosis and the reason for the difference in response is not entirely clear. While apoptosis requires active participation of the involved cell, often even in terms of initiation of the *de novo* protein synthesis, necrosis is a passive and degenerative process. *In vivo*, necrosis triggers the inflammatory response in the tissue, due to a release of cytoplasmic constituents to intercellular space, often resulting in scar formation. In contrast, remains of apoptotic cells are phagocytized not only by the "professional" macrophages, but also by other neighboring cells, without evoking any inflammatory reaction. The early event of necrosis is swelling of cell mitochondria, followed by rupture of the plasma membrane, and release of the cytoplasmic content (reviews in Tomei and Cope, 1991).

B. Types of Cell Viability Assays

1. Assays of Plasma Membrane Integrity

It has been recognized during the past several decades that one of the major features discriminating dead from live cells is loss of the transport function and physical integrity of the plasma membrane. A plethora of assays of cell viability has been developed based on this phenomenon. Thus, for example, because the intact membrane of live cells excludes a variety of charged dyes such as trypan blue or propidium iodide (PI), incubation with these dyes results in selective labeling of dead cells, while live cells show no, or minimal, dye uptake (Horan and Kappler, 1977). By virtue of its simplicity, the PI exclusion assay appears to be the most popular in flow cytometry. A short ($\sim$5 min) incubation with 10–20 μg/ml of this dye, in isotonic media, labels the cells that have impaired transport function of the plasma membrane. Such cells cannot exclude this charged dye, which, after crossing the plasma membrane, binds to DNA and dsRNA and fluoresces intensively.

An assay based on a similar principle makes use of the nonfluorescent substrate of esterases, fluorescein diacetate, which is taken up by live cells, and, upon hydrolysis, the product (fluorescein) is detected due to its strong green fluorescence (Hamori *et al.*, 1980). Incubation of cells in the presence of both PI and fluorescein diacetate labels live cells green (fluorescein) and dead cells red (PI). This is a convenient assay, widely used in flow cytometry. Another assay combines the use of PI and Hoechst 33342 (HO342). Both are described later in this chapter.

Ethidium monoazide, as PI, is also excluded from live cells but enters cells that have damaged plasma membrane (Riedy *et al.*, 1991). This dye, upon binding to nucleic acids, can be covalently attached to the latter in the photochemical reaction. Therefore, exposure of cells to this dye, followed by their illumination with white light, irreversibly labels dead cells. The cells can then be fixed and counterstained with another dye of different color: the distinction between the dead and live cells at the time of their exposure to the dye remains preserved after subsequent fixation, washings, and counterstaining. This method, which is often used to exclude dead cells during immunophenotyping, is described by Stewart and Stewart (1994).

Integrity of the plasma membrane can also be probed by cell resistance to trypsin and DNase: while live cells remain intact during incubation with these enzymes, dead cells are nearly totally digested and removed from the analysis (Darzynkiewicz *et al.*, 1984a).

2. Function of Cell Organelles

There are several assays of cell viability based on the functional tests of cell organelles. Thus, for example, the cationic dye rhodamine 123 (Rh123) accumulates in mitochondria of live cells due to the mitochondrial transmembrane potential (Johnson *et al.*, 1980). Cell incubation with 1–10 μg/ml of Rh123 results in labeling of live cells while dead (necrotic) cells show no Rh123 uptake. As in the case of fluorescein diacetate and PI, the combination of Rh123 and PI labels live cells green (Rh 123) and dead cells red (Darzynkiewicz *et al.*, 1982).

Another functional assay involves lysosomes. Incubation of cells in the presence of 1–2 μg/ml of the metachromatic dye acridine orange (AO) results in uptake of this fluorochrome by lysosomes of live cells which luminesce in red wavelength. Dead (necrotic) cells exhibit weak green and no red fluorescence at that low an AO concentration. This assay is described elsewhere (Traganos and Darzynkiewicz, 1994) as a test of the lysosome active transport (proton pump).

As discussed earlier, cells which undergo apoptosis have a preserved plasma membrane and several cell functions remain little changed, relative to live cells. Therefore, this simple discrimination of live and dead cells by dye exclusion, or based on functional assays of some organelles, is inadequate to identify cells that die by apoptosis, especially at early stages of cell death. Other assays, therefore, have to be used in such situations.

3. Identification of Apoptotic Cells

Two distinct features of apoptotic cells, extensive DNA cleavage and preservation of the cell membrane, provide the basis for the development of most flow cytometric assays to discriminate between apoptosis and necrosis. DNA cleavage can be detected by extraction of the degraded, low MW DNA from the ethanol-prefixed or detergent-permeabilized cells and subsequent cell staining with DNA fluorochromes. Apoptotic cells then show reduced DNA stainability (content of high MW DNA) (Darzynkiewicz *et al.*, 1992). Alternatively, the numerous DNA strand breaks in apoptotic cells can be labeled with biotinylated or digoxigenin-conjugated nucleosides in the reaction employing exogenous terminal deoxynucleotidyl transferase (TdT) or DNA polymerase (nick translation) (Gorczyca *et al.*, 1992, 1993b,c).

Another marker of apoptosis is chromatin condensation, which can be assayed by DNA *in situ* sensitivity to denaturation (Hotz *et al.*, 1992, 1994). Changes in plasma membrane permeability to HO342 resulting in differential staining of DNA with this dye can also discriminate between live and apoptotic cells (Dive *et al.*, 1992; Ormerod *et al.*, 1992, 1993).

Practical aspects of several cell viability assays, their specificity in terms of discrimination between apoptosis and necrosis, and applicability in different cell systems are described in the following sections.

III. Changes in Light Scatter During Cell Death

Cell death is accompanied by a change in its property to scatter light and thus light scatter measurement is one of the simplest assays of cell viability. At early stages, apoptosis is characterized by markedly reduced cell ability to scatter light in forward direction and either by an increase (Swat *et al.*, 1991) or no change in the 90° angle light scatter (Darzynkiewicz *et al.*, 1992). At later stages, both the forward and right angle light scatter signal are decreased. These changes are a reflection of chromatin condensation, nuclear fragmentation, cell shrinkage, and shedding of apoptotic bodies. The major advantage of the assay based on light scatter measurement is its simplicity and that it can be combined with analysis of the surface immunofluorescence, for example, to identify the phenotype of the dying cell. It can also be combined with functional assays such as of the mitochondrial potential, lysosomal proton pump, exclusion of PI, or plasma membrane permeability to such dyes as Hoechst 33342.

Reduced forward light scatter, however, is not a very specific marker of apoptosis. Mechanically broken cells, isolated cell nuclei, and necrotic cells also have low light scatter properties. Furthermore, a loss of cell-surface antigens may accompany apoptosis, especially at later stages of cell death, and this may complicate the bivariate light scatter/surface immunofluorescence analysis. This approach, therefore, requires several controls and should be accompanied by another, more specific assay, or at least by confirmation of apoptosis by microscopy.

Because different models of flow cytometers have somewhat different positions and characteristics of their light scatter detectors, the degree of light scatter changes in the same cell systems may vary from instrument to instrument and be responsible for the lack of reproducibility of this assay between laboratories.

IV. Cell Sensitivity to Trypsin and DNase

A short exposure of live cells simultaneously to trypsin and DNase has little effect on their morphology, function, or viability. Conversely, cells with damaged plasma membranes are digested by these enzymes to such a degree that, for all practical purposes, they can be gated out during measurement, for example, by raising the triggering threshold of the light scatter, DNA, or protein fluorescence signals (Darzynkiewicz *et al.*, 1984a).

A. Reagents

1. DNase I, 200 μg/ml in HBSS (with Mg^{2+} and Ca^{2+}). Aliquots of 100 μl of DNase in microfuge tubes may be stored at $-40\,°C$.
2. Trypsin, 0.5%, in HBSS. Aliquots of 10 μl of trypsin may be stored at $-40\,°C$ in microfuge tubes.

B. Procedure

1. Centrifuge cells (e.g., dispersed mechanically from the solid tumor). Suspend the pellet (approximately 10^6 cells) in 0.2 ml of HBSS.
2. Add a 100-μl aliquot of DNase I, incubate 15 min at 37 °C, then add 100 μl of trypsin, and incubate for an additional 30 min. Add 5 ml of HBSS with 10% calf serum or HBSS containing the soybean trypsin inhibitor, to inactivate trypsin. Centrifuge.
3. Rinse cells once in HBSS, suspend in HBSS, fix cells in suspension, or stain with a desired fluorochrome (e.g., after permeabilization with 0.1% Triton X-100 in the presence of 1% albumin) shortly after this procedure.

C. Results

This procedure removes all cells with damaged plasma membrane (dead cells) from the suspension and thus all remaining cells exclude PI. However, because of the use of trypsin, which digests the cell coat, epitopes of many cell-surface antigens of live cells may not be preserved. Incubation of so-treated cells in culture medium at 37 °C for several hours restores most antigens.

Since elimination of cells by digestion with these enzymes is selective to cells with damaged plasma membrane (exclusion of trypsin and DNase), this method can be used to remove necrotic or very late apoptotic cells, as well as those cells which have been mechanically damaged (e.g., by sonication or syringing), but not early apoptotic cells. Also digested are isolated nuclei. Somewhat similar results can be obtained by separation of live cells by the Ficoll-Hypaque gradient centrifugation.

V. Fluorescein Diacetate (FDA) Hydrolysis and PI Exclusion

FDA is a substrate for esterases, the ubiquitous enzymes that are present in all types of cells. It can penetrate into live cells. The product of hydrolysis, fluorescein, is highly fluorescent and charged. It becomes, therefore, entrapped in the cell. This assay, which combines counterstaining of dead cells with PI, has been widely used in classical cytochemistry and in flow cytometry (Hamori *et al.*, 1980).

A. Reagents

1. FDA, dissolved in acetone, 1 mg/ml (stock solution).
2. PI, dissolved in distilled water, 1 mg/ml (stock solution).

B. Procedure

1. Suspend approximately 10^6 cells in 1 ml of HBSS.
2. Add 2 μl of the FDA stock solution.
3. Incubate cells at 37 °C for 15 min.
4. Add 20 μl of the PI stock solution.
5. Incubate 5 min at room temperature, and analyze by flow cytometry.

Both dyes are excited in blue light (488-nm line of the argon ion laser). Measure green fluorescence at 530 ± 20 nm and red fluorescence at >620 nm. Live cells show green, dead cells, and red fluorescence.

This method, which is based on analysis of the integrity of the plasma membrane, as other methods based on the same principle, discriminates between necrotic cells (or cells with damaged membrane) and live cells, but not between live and apoptotic cells.

VI. Rh123 Uptake and PI Exclusion

Uptake of the cationic fluorochrome Rh123 is specific to functionally active mitochondria: the dye accumulates in the mitochondria due to the transmembrane potential of these organelles (Johnson *et al.*, 1980). Thus, live cells with intact plasma membrane and charged mitochondria concentrate this dye and exhibit strong green fluorescence. In contrast, dead cells fail to stain significantly with this dye. A dual-staining with Rh123 and PI was proposed as the cell viability test (Darzynkiewicz *et al.*, 1982). This is also an assay of the mitochondrial transmembrane potential.

A. Reagents

1. Rh123 (available from Molecular Probes), dissolved in distilled water, 1 mg/ml (stock solution).
2. PI, dissolved in distilled water, 1 mg/ml (stock solution).

B. Procedure

1. Add 5 μl of Rh123 stock solution to approximately 10^6 cells suspended in 1 ml of tissue culture medium (or HBSS). Incubate 5 min at 37 °C.
2. Add 20 μl of the PI stock solution. Keep 5 min at room temperature. Analyze by flow cytometry.

Both dyes are excited with blue light (e.g., 488-nm laser line). Rh123 fluoresces green (530 $\pm$ 20 nm), while PI, as described above, in red.

C. Results

Live cells stain with Rh123 green and exclude PI. Dead cells, which have damaged mitochondria, have minimal green fluorescence but stain with PI (Fig. 1) (Darzynkiewicz *et al.*, 1982).

VII. PI Exclusion Followed by Counterstaining with Hoechst 33342

As mentioned earlier, the charged dye PI is excluded by live cells. On the other hand, the bisbenzimidazole dye HO342 penetrates through the plasma membrane and can stain DNA in live cells. The method combining these dyes to discriminate between live and dead cells was introduced by Stöhr and Vogt-Schaden (1980) and modified by several authors (e.g., Pollack and Ciancio, 1990; Wallen *et al.*, 1983). In this method, the cells are first exposed to PI and subsequently stained with HO342. Compared with live cells, HO342 fluorescence is suppressed in dead cells.

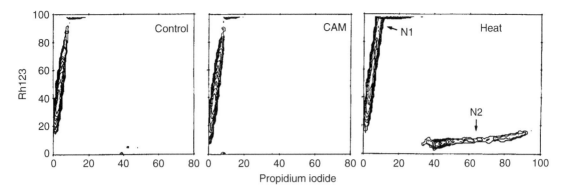

Fig. 1 Stainability of live, apoptotic, and necrotic cells with Rh123 and PI. The mitochondrial probe Rh123 (green fluorescence) is taken up by live cells from control HL-60 cultures (control); live cells exclude PI. Induction of apoptosis in approximately 40% cells by incubation with 0.15 μM camptothecin for 4 h (CAM) neither changes Rh123 uptake nor affects exclusion of PI. Cell heating (45 °C, 2 h) causes necrosis (heat). Early necrotic (N1) cells have elevated Rh123 uptake but exclude PI. Late necrotic cells (N2) have minimal Rh123 fluorescence and stain intensely with PI. (reprinted with permission from Darzynkiewicz *et al.*, 1992).

The latter, however, stain more intensely with PI. This method, in modification of Pollack and Ciancio (1990), is described in the following sections.

A. Reagents

1. *Stock solution of HO342*: 1 mM HO342 in distilled water.
2. *Staining solution of PI*: 20 μg/ml of PI in PBS.
3. *Staining solution of HO342*: dilute the HO342 stock solution 1:4 in PBS (Ca^{2+} and Mg^{2+} free).
4. *Fixative*: 25% ethanol in PBS.

B. Staining Procedure

1. Centrifuge 3×10^5–10^6 cells, decant medium, and vortex the pellet.
2. Add 100 μl of PI staining solution, vortex, and keep on ice for 30 min.
3. Add 1.9 ml of fixative (reagent No. 4) and vortex.
4. Add 50 μl of HO342 staining solution, vortex, and keep on ice for at least 30 min. Samples are stable for up to 3 days in this solution, when kept at 0–4 °C.

C. Instrumentation

Excitation of HO342 is in UV light, with maximum at 340 nm. PI can be excited with blue or green light, but it has also an absorption band in the UV light spectrum. Excitation of both PI and HO342, thus, can be achieved using, for example, the 351-nm line of the argon ion laser or, in the case of illumination with a high pressure mercury lamp, using the UG1 filter.

Fluorescence of HO342 is in blue wavelength and a combination of optical filters and dichroic mirrors is required to obtain maximum transmission at 480 ± 20 nm. Fluorescence of PI is measured with long pass filter >620 nm.

D. Results

Figure 2 illustrates positions of live, apoptotic, and necrotic cells following cell staining with PI and HO342. As is evident, the early apoptotic cells (Apl) have lowered stainability with HO342 compared to live cells from the control culture. The loss of HO342 stainability is likely a result of DNA degradation and extraction of low MW DNA from the cells following their permeabilization with ethanol. PI fluorescence of Ap cells is also low. The loss of plasma membrane function late in apoptosis results in increased stainability with PI (Ap2). Necrotic cells, or cells with destroyed plasma membrane function (e.g., by repeated freezing and thawing) have high PI and low HO342 fluorescence.

VIII. Hoechst 33342 Active Uptake and PI Exclusion

The PI-HO342 method presented in Fig. 2, based on cell staining with PI followed by their exposure to HO342 after permeabilization (fixation) with 25% ethanol, is very much different from the method which utilizes the same dyes, but

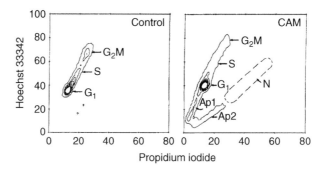

Fig. 2 Stainability of apoptotic and necrotic cells with PI and HO342 under conditions when HO342 is applied after cell exposure to PI and following their permeabilization. HL-60 cells, untreated (control) and treated with 0.15 μM camptothecin (CAM) for 6 h to induce apoptosis (Del Bino *et al.*, 1991) were incubated with 10 μg/ml of PI, then permeabilized with 25% ethanol and stained with HO342, as originally described by Pollack and Ciancio (1990). In this assay, the cells with undamaged plasma membrane exclude PI and stain predominantly blue with HO342. The cells that cannot exclude PI have a more intense PI fluorescence and proportionally lower HO342 fluorescence. The live cells from control culture stain strongly with HO342, in proportion to their DNA content. The early apoptotic cells from the CAM-treated cultures show diminished HO342 fluorescence (Ap1). Late apoptotic cells (Ap2) cannot exclude PI and stain more intensely with PI. The position of necrotic cells (N), which also cannot exclude PI but have higher DNA content than Ap2, is indicated with a broken outline. (reprinted with permission from Del Bino *et al.*, 1991).

detects apoptotic cells based on their increased uptake of HO342, under supravital conditions in isotonic media, as described by Dive *et al.* (1992) and Ormerod *et al.* (1992, 1993). These authors observed that during a short exposure to HO342 apoptotic cells have stronger blue fluorescence, perhaps as a result of the increased permeability to this dye, compared to nonapoptotic cells. The combination of supravital HO342 uptake and PI exclusion, together with analysis of the cells' light scatter properties, provides an attractive assay of apoptosis.

A. Reagents

1. HO342, dissolved in distilled water 0.1 mg/ml, stock solution.
2. PI, dissolved in distilled water, 1 mg/ml, stock solution.

B. Procedure

1. Suspend approximately 10^6 cells in 1 ml of culture medium, with 10% serum. Add 10 μl of stock solution of HO342.
2. Incubate cells for 7 min.
3. Cool the sample on ice and centrifuge.
4. Resuspend cell pellet in 1 ml PBS. Add 5 μl of the PI stock solution. Analyze.

C. Results

As described earlier, excitation of HO342 is in UV light. Although optimally excited at green or blue wavelength, PI can also be excited in UV. Apoptotic cells have increased blue fluorescence of HO342, compared to live, nonapoptotic cells (Dive *et al.*, 1992; Ormerod *et al.*, 1993). The intensity of the cells' blue fluorescence changes with time of incubation with HO342, and the optimal time should be determined for best discrimination of apoptotic cells. The decrease in light scatter, combined with HO342 uptake, is also helpful to identify apoptotic cells. Necrotic cells are counterstained with PI.

IX. Controlled Extraction of Low MW DNA from Apoptotic Cells

A. Degradation of DNA in Apoptotic Cells

One of the early events of apoptosis is the activation of endonuclease which nicks DNA preferentially at the internucleosomal (linker) sections (Arends *et al.*, 1990; Compton, 1992). Although exceptions have been reported (Cohen *et al.*, 1992; Collins *et al.*, 1992), DNA degradation is a very specific event of apoptosis and the electrophoretic pattern of DNA indicating the presence of DNA sections the size of mono- and oligonucleosomes is considered a trademark of apoptosis.

Fixation of cells in ethanol is inadequate to preserve the degraded, low MW DNA inside apoptotic cells: this portion of DNA leaks out during the subsequent rinse and staining procedure, and therefore less DNA in these cells stains with any DNA fluorochrome (Darzynkiewicz *et al.*, 1992). Thus, the appearance of cells with low DNA stainability, lower than that of G_1 cells ("sub-G_1 peaks", "A_0"cells) in cultures treated with various cytotoxic agents, has been considered to be a marker of cell death by apoptosis (Nicoletti *et al.*, 1991; Telford *et al.*, 1991; Umansky *et al.*, 1981).

Because the degree of DNA degradation varies depending on the stage of apoptosis, cell type, and often the nature of the apoptosis-inducing agent, the degree of extraction of low MW DNA during the staining procedure (and thus separation of apoptotic from live cells) is not always reproducible. It has been observed, however, that addition of phosphate-citric acid buffer at pH 7.8 to the rinsing fluid enhances extraction of the degraded DNA (Gong *et al.*, 1993; Gorczyca *et al.*, 1993c). This approach can be used to modulate the extent of DNA extraction from apoptotic cells to the desired level.

B. Reagents

1. *Phosphate-citric acid buffer*: prepare solutions of 0.2 M Na_2HPO_4 and 0.1 M citric acid. Mix 192 ml of 0.2 M Na_2HPO_4 with 8 ml of 0.1 M citric acid; pH should be approximately 7.8.
2. *Hanks' buffered salt solution (HBSS)*.
3. *Fixative*: 70% ethanol.

C. Procedure

1. Fix cells in suspension in 70% ethanol (i.e., add 1 ml of cells suspended in HBSS containing approximately 10^6 cells into 9 ml of 70% ethanol in tube) on ice. Cells in fixative (ethanol) can be stored at $-20\,^\circ$C for up to 1 week.

2. Centrifuge cells, decant ethanol, suspend cells in 10 ml of HBSS, and centrifuge.

3. Suspend cells in 1 ml of HBSS, into which you may add 0.2–1.0 ml of the phosphate-citric acid buffer (reagent No. 1). Add less (e.g., 0.2 ml or none) buffer if DNA degradation in Ap cells is extensive and DNA extraction effective, so the Ap cells are well separated from G_1 cells. Add more of the buffer (up to 1.0 ml) if DNA is not markedly degraded and there are problems with separating Ap cells (Ap cells overlap with G_1 cells). Incubate at room temperature for 5 min.

4. Centrifuge cells and add 1 ml of HBSS containing 20 μg/ml of PI and 5 Kunitz units of the DNase-free RNase A. (Boil RNase for 5 min before use if it is not DNase free.) Incubate cells for 30 min at room temperature. Use the laser

excitation line at 488 nm or blue light (BG12 filter) and measure cell red fluorescence (>620 nm) and forward light scatter. Ap cells should have a diminished forward light scatter signal and decreased PI fluorescence compared to the cells in the main peak (G_1).

5. Cellular DNA may be stained with other fluorochromes instead of PI, and other cell constituents may be counterstained as well. The following is the procedure to simultaneously stain DNA and protein with DAPI and sulforhodamine 101, respectively

• After point No. 3 of this procedure suspend cell pellet in 1 ml of a staining solution that contains

Triton X-100, 0.1%

$MgCl_2$, 2 mM

NaCl, 0.1 M

Pipes buffer, 10 mM; final pH 6.8

DAPI, 1 μg/ml

Sulforhodamine 101, 20 μg/ml.

(*N. B.*: This solution is stable and can be stored, in a dark bottle, at 0–4 °C for several weeks). Cells are analyzed while suspended in this solution.

• Use excitation with UV light (e.g., 351-nm argon ion line or UG1 filter for mercury lamp illumination). Measure the blue fluorescence of DAPI (460-500 nm) and red fluorescence of sulforhodamine 101 (>600 nm).

D. Results

Results are shown in Fig. 3. As is evident, apoptotic cells are well separated from live cells based on differences in DNA content. They also have diminished protein content. As mentioned, the degree of separation may be modified by varying the concentration of the phosphate-citric acid buffer (reagent No. 1).

This buffer can also be added to the staining medium (e.g., in a 1:10 to 1:5 proportion) when unfixed cells are stained with the DNA-specific dyes, following cell permeabilization with detergents, to extract low MW DNA from apoptotic cells.

Because the cell-cycle position of the nonapoptotic cells can be estimated (Fig. 2), this method offers the opportunity to investigate the cell-cycle specificity of apoptosis. Another advantage of this method is its simplicity and applicability to any DNA fluorochrome (Telford *et al.*, 1991) or instrument.

This method, however, is not very specific in terms of detection of apoptotic cells. In addition to apoptotic cells, the "sub G_1" peak can also represent mechanically damaged cells, cells with lower DNA content (e.g., in a sample containing cell populations with a different DNA index), or cells with different chromatin structure (e.g., cells undergoing erythroid differentiation) in which accessibility of DNA to the fluorochrome is diminished (Darzynkiewicz *et al.*, 1984b).

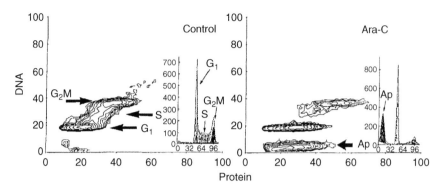

Fig. 3 Bivariate analysis of DNA and protein content distribution of control HL-60 cells and cells treated with 25 μM 1 β-arabinofuranosylcytosine (Ara-C) for 4 h. DNA content frequency histogram of these cells is shown in the inset. The presence of apoptotic cells (Ap) in Ara-C-treated culture correlates with disappearance of S phase cells. A clear distinction between Ap and nonapoptotic cells was obtained by rinsing the ethanol-prefixed cells with phosphate-citric acid buffer (pH 7.8), as described in the procedure.

It should be stressed that this method can only be used on cells that are fixed in ethanol or permeabilized with detergents. Cell fixation in formaldehyde or glutar-aldehyde cross-links low MW DNA to other constituents and precludes its extraction. Only very late apoptotic cells, in which DNA degradation is significantly advanced, show diminished DNA content following fixation with the cross-linking agents. See Chapter 22 for the description of possible pitfalls and problems associated with use of this methodology.

X. Sensitivity of DNA *In Situ* to Denaturation

Chromatin condensation is one of the early events of apoptosis. It has been observed that sensitivity of DNA *in situ* to denaturation is increased during chromatin condensation, for example, as during mitosis or in G_0 (Darzynkiewicz *et al.*, 1987), and that chromatin condensation in apoptotic cells is also accompanied by increased DNA denaturability (Hotz *et al.*, 1992). This method, and its application to identify apoptotic cells, is described elsewhere (Darzynkiewicz, 1994).

XI. Detection of DNA Strand Breaks in Apoptotic Cells

As mentioned, extensive DNA cleavage is a characteristic event of apoptosis. The presence of DNA strand breaks in apoptotic cells can be detected by labeling the 3'OH termini in DNA breaks with biotin- or digoxigenin-conjugated nucleotides, in the enzymatic reaction catalyzed by exogenous TdT (Gorczyca *et al.*, 1992, 1993a,b), or DNA polymerase (Gold *et al.*, 1993). The method for detection of apoptotic cells by labeling DNA strand breaks with biotinylated dUTP (b-dUTP) by TdT is as follows:

A. Reagents

1. *Fixatives*: 1% methanol-free formaldehyde (paraformaldehyde; available from Polysciences, Inc., Warrington, PA) in PBS, pH 7.4; 70% ethanol.
2. *Reaction buffer for TdT (5× concentrated)*:
   - Potassium (or sodium) cacodylate, 1 M
   - Tris-HCl, 125 mM, pH 6.6 at 4 °C
   - Bovine serum albumin (BSA), 1.25 mg/ml
   - Cobalt chloride, 10 mM.
   (The complete buffer is available from Boehringer-Mannheim Biochemicals, Indianapolis, IN; Cat. No. 220 582.)
3. TdT in storage buffer
   - Potassium cacodylate, 0.2 M
   - EDTA, 1 mM
   - 2-Mercaptoethanol, 4 mM
   - Glycerol, 50% (v/v), pH 6.6 at 4 °C
   - TdT, 25 units per 1 μl.
   (Available from Boehringer-Mannheim; Cat. No. as above.)
4. Biotin-16-dUTP, 50 nmol in 50 μl. (Available from Boehringer-Mannheim; Cat. No. 1093 070.)
5. Saline-citrate buffer. Dilute 20× concentrated saline-sodium citrate buffer (SSC) fivefold with distilled water to obtain 4× concentrated solution (0.6 M NaCl, 0.06 M Na citrate), and add:
   - Fluoresceinated avidin, 2.5 μg/ml (final concentration)
   - Triton X-100, 0.1% (v/v)
   - Dry milk, 5% (w/v) final concentration.
6. Rinsing buffer. Dissolve in HBSS:
   - Triton X-100, 0.1% (v/v)
   - BSA, 0.5%.
7. PI buffer. Dissolve in HBSS:
   - PI, 5 μg/ml
   - 0.1% DNase-free RNase A.

B. Procedure

1. Fix cells in suspension in 1% formaldehyde for 15 min on ice.
2. Sediment cells, resuspend pellet in 5 ml HBSS, and centrifuge.
3. Resuspend cells in 70% ethanol on ice. The cells can be stored in ethanol, at -20 °C for up to 3 weeks.
4. Rinse cells in HBSS and resuspend the pellet ($<10^6$ cells) in 1.5 ml microfuge tube in 50 μl of the solution which contains:
   - Reaction buffer (reagent No. 2), 10 μl
   - Biotin-16-dUTP (reagent No. 4), 1 μl (1 μg)
   - TdT (reagent No. 3), 0.2 μl (5 units)
   - Distilled H_2O, 38.8 μl.
   Proportions of the reagents are scaled up proportionally to the number of samples examined at a given time, and the solution is then subdivided into 50-μl aliquots per sample.
5. Incubate cells in this solution for 30 min at 37 °C, then add 1.3 ml of the rinsing buffer (reagent No. 6), centrifuge, and resuspend the pellet in 100 μl of the saline-citrate buffer containing fluoresceinated avidin (reagent No. 5). Incubate at room temperature for 30 min.
6. Add 1.3 ml of the rinsing buffer and centrifuge. Rinse again with rinsing buffer. Resuspend the cell pellet in 1 ml of PI solution (reagent No. 7). Incubate 30 min at room temperature. Measure cell green (fluorescinated avidin) and red (PI) fluorescence after illumination with blue light (488 nm).

C. Commercially Available Kits for Labeling DNA Strand Breaks

ONCOR Inc. (Gaithersburg, MD) provides a kit (Apopt Tag kit, Cat. No. S7100-kit) specifically designed to label DNA strand breaks in apoptotic cells, which utilizes digoxigenin-conjugated dUTP and TdT. Description of the method is included with the kit. The results obtained with use of this kit compare favorably with the data shown in Fig. 4 (Li *et al.*, 1996).

D. Results

Simultaneous detection of DNA strand breaks and analysis of DNA content makes it possible to identify the cell-cycle position of both cells in apoptotic population as well as the cell with undegraded DNA (Fig. 4).

Identification of apoptotic cells based on the detection of DNA strand breaks is the most specific assay of apoptosis, by flow cytometry. Namely, the cells with DNA strand breaks which accompany necrosis, or the cells with primary breaks induced by X-irradiation, up to a dose of 25 Gy, have markedly lower

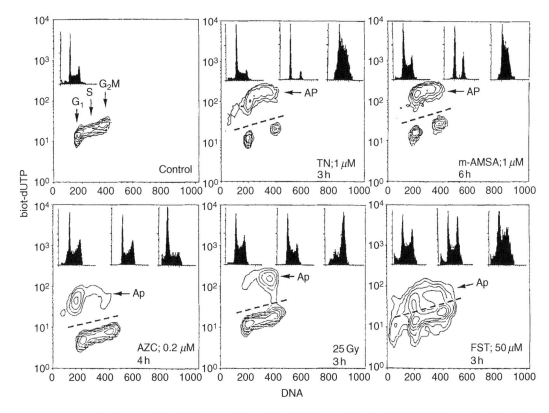

Fig. 4 Detection of the apoptosis-associated DNA strand breaks in HL-60 cells treated with various drugs and γ radiation. DNA strand breaks in apoptotic cells (Ap) are labeled with biotinylated dUTP (b-dUTP) in the reaction catalyzed by terminal deoxynucleotidyl transferase (TdT). Simultaneous counterstaining of DNA allows correlation of the presence of DNA strand breaks with cell position in the cell cycle. By gating analysis of the b-dUTP-labeled and b-dUTP-unlabeled cell populations, the cell-cycle distribution of nonapoptotic and apoptotic cells, respectively, can be estimated. In contrast to the staining of DNA of the ethanol-fixed cells (Fig. 3), only a minor amount of DNA is extracted from apoptotic cells in this method. Note the S phase specificity of apoptosis after treatment with teniposide (TN) and amsacrine (*m*-AMSA), G_1 specificity after treatment with azacytidine (AZC), and G_2M specificity after γ radiation (25 Gy). Apoptosis induced by Fostriecin (FST) is not cell cycle specific. Insets show the DNA content frequency histograms of the total (left), nonapoptotic (middle), and apoptotic (right) cell populations; the frequency scales are arbitrary (scaled to maximum frequency). Dashes, the gating thresholds. (reprinted with permission from Gorczyca *et al.*, 1993c).

incorporation of b-dUTP compared to that of apoptotic cells (Gorczyca *et al.*, 1993a). This method is applicable to clinical material, to measure apoptosis of blast cells in peripheral blood or bone marrow induced during chemotherapy of leukemias (Gorczyca *et al.*, 1993a) or in solid tumors, from needle biopsy specimens (Gorczyca *et al.*, 1994).

XII. A Selective Procedure for DNA Extraction from Apoptotic Cells Applicable to Gel Electrophoresis and Flow Cytometry

A. Recovery of the Degraded DNA from Apoptotic Cells and Its Analysis by Gel Electrophoresis

As mentioned earlier, one of the most characteristic features of apoptosis is activation of an endonuclease which has preference to the linker DNA sections; products of this enzyme are discontinuous DNA sections of a size equivalent to mono- and oligonucleosomose, and they form a typical "ladder" during gel electrophoresis (Arends *et al.*, 1990). Such an electrophoretic pattern is considered to be a hallmark of apoptosis, and although exceptions have been observed (e.g., Cohen *et al.*, 1992), analysis of DNA size on agarose gels is a widely used procedure to reveal apoptosis. Because such analysis should often be used to confirm the data by flow cytometry, a simple procedure of DNA electrophoresis in agarose gels is included in this chapter.

One of the methods to identify apoptotic cells, presented in this chapter, is based on cell fixation in 70% ethanol followed by cell hydration and extraction of low MW DNA with phosphate-citric acid buffer at pH 7.8 (see Section IX.A). This method can be combined with the analysis of DNA extracted with the buffer by gel electrophoresis (Gong *et al.*, 1994), as described in the following sections.

B. Reagents

1. *Phosphate-citric acid buffer*: Mix 192 ml of 0.2 M Na_2HPO_4 with 8 ml of 0.1 M citric acid; the final pH is 7.8.
2. *Nonidet NP-40*: Dissolve 0.25 ml of Nonidet NP-40 in 100 ml of distilled water.
3. *RNase A*: Dissolve 1 mg of DNase-free RNase A in 1 ml of distilled water.
4. *Proteinase K*: Dissolve 1 mg of proteinase K in 1 ml of distilled water.
5. *Loading buffer*: Dissolve 0.25 g of bromophenol blue and 0.25 g of xylene cyanol FF in 70 ml of distilled water. Add 30 ml of glycerol (dyes are available, e.g., from Bio-Rad Laboratories, Richmond, CA).
6. *DNA molecular weight standards*: Use the standards that provide DNA between 100 and 1000 bp size (e.g., from Integrated Separation Systems, Natick, MA).
7. *Electrophoresis buffer (TBE, 10 × concentrated)*: Dissolve 54 g Tris base and 27.5 g boric acid in 980 ml of distilled water; add 20 ml of 0.5 M EDTA (pH 8.0).
8. *Agarose gel (0.8%)*: Dissolve 1.6 g of agarose in 200 ml of TBE.
9. *Ethidium bromide (EB)*: Stock solution, dissolve 1 mg of EB in 1 ml of distilled water; for staining gels (working solutions), add 100 μl of the stock solution to 200 ml TBE.

C. Procedure

1. Fix 10^6–10^7 cells in suspension in 10 ml of 70% ethanol on ice. Cells may be subjected to DNA extraction and analysis after 4 h fixation in ethanol, or stored in fixative, preferably at $-20\,°C$, for up to several weeks.

2. Centrifuge cells at $1000 \times g$ for 5 min. *Thoroughly remove ethanol.* Resuspend cell pellet in 40 μl of phosphate-citric acid buffer (reagent No. 1) and transfer to 0.5-ml volume Eppendorf tubes. Keep at room temperature for at least 30 min, occasionally shaking.

3. Centrifuge at $1500 \times g$ for 5 min. Transfer the supernatant to new Eppendorf tubes and concentrate by vacuum, for example, in SpeedVac concentrator (Savant Instruments, Inc.; Farmingdale, NY) for 15 min.

4. Add 3 μl of 0.25% Nonidet NP-40 and 3 μl of RNase A solution. Close the tube to prevent evaporation and incubate at $37\,°C$ for 30 min.

5. Add 3 μl of proteinase K solution and incubate for an additional 30 min at $37\,°C$.

6. Add 12 μl of the loading buffer, transfer entire contents of the tube to 0.8% agarose horizontal gel.

7. Load a sample of DNA standards on the MW standard lane of the gel.

8. Run electrophoresis at 2 V/cm for 16–20 h.

9. To visualize the bands, stain the gel with 5 μg/ml of ethidium bromide and illuminate with UV light.

10. Cells remaining in the pellet (after step 3 of this procedure) can be counterstained with PI (in the presence of RNase A) or DAPI, as described in Section VIII, for analysis by flow cytometry.

Further details of the procedure of gel preparation and electrophoresis can be found in Sambrook *et al.* (1989).

D. Results

This simple procedure allows one to analyze the MW of DNA extracted from apoptotic cells by gel electrophoresis and to detect apoptotic cells, from the very same preparations, by flow cytometry (Gong *et al.*, 1994). Figure 5 illustrates both DNA analysis by gel electrophoresis and DNA content of the cells from which DNA was extracted, measured by flow cytometry.

E. Advantages of the Procedure: Comparison with Other Methods of DNA Extraction

The procedure of DNA analysis presented above offers several advantages over other methods used to extract DNA from apoptotic cells (e.g., Arends *et al.*, 1990; Compton, 1992):

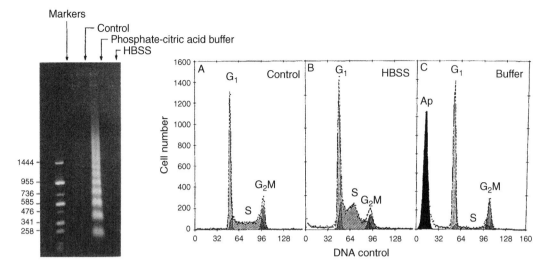

Fig. 5 DNA gel electrophoresis and corresponding DNA content frequency distribution histograms of HL-60 cells, untreated (Control, panel A) and treated with 0.15 μM camptothecin for 3 h (B and C). As shown by us before (Del Bino *et al.*, 1991) such treatment induces apoptosis of S phase cells. The cells shown in panel B were rinsed with Hanks' buffered salt solution (HBSS) while those in panels A and C were rinsed with phosphate-citric acid (PC) buffer. DNA extracted with HBSS and PC were analyzed by gel electrophoresis as shown in the respective lanes. Markers: MW markers, as indicated (base pairs).

1. The cells may be fixed in ethanol and stored indefinitely, a feature which is attractive for use with clinical material, which may be collected at different times, transported, and analyzed later.

2. Fixation in ethanol inactivates endogenous enzymes, preventing autolysis after sample collection.

3. Ethanol also inactivates HIV and other viral or bacterial pathogens, making the sample safer to handle.

4. The method is rapid and uses less toxic reagents (no phenol, chloroform, etc.).

5. The low MW DNA is selectively extracted and is not mixed with high MW DNA in the sample. Bands of uniform intensity, therefore, can be more easily obtained on gel electrophoretograms, when equal amounts of DNA (e.g., estimated by spectrophotometry; absorption at 260 nm) are applied per well.

6. The ratio of high to low MW DNA (extractable and nonextractable from the ethanol-fixed cells) may serve as an index of apoptosis in cell populations. High MW DNA can be extracted by standard biochemical procedures.

7. The cells remaining after extraction of low MW DNA can be subjected to flow cytometry and, thus, the percentage of apoptotic cells can be estimated in literally the same samples from which DNA was used for electrophoresis.

▆▆▆▆▆▆▆▆▆ ## XIII. Comparison of the Methods: Confirmation of the Apoptotic Mode of Cell Death

Each of the presented methods has its advantages and limitations. Some of them have already been discussed. In general, the methods based on analysis of integrity and transport function of the plasma membrane (exclusion of PI, FDA hydrolysis, trypsin, and DNase sensitivity), although rather simple and inexpensive to use, fail to identify apoptotic cells. They can be used to identify necrotic cells, mechanically damaged cells, or very late stages of apoptosis.

Identification of apoptotic cells is more complex. None of the methods described in the literature, when used alone, can provide total assurance of detection of apoptosis. The most specific appears to be the method based on the detection of DNA strand breaks by labeling them with biotin- or digoxigenin-conjugated nucleotides. The number of DNA strand breaks in apoptotic cells is so large that the degree of cell labeling in this assay appears to be an adequate discriminator between apoptotic and necrotic cells. This assay also allows one to discriminate between apoptotic cells and cells with primary DNA strand breaks caused by high doses of radiation (Gorczyca *et al.*, 1992). However, one has to keep in mind that there may be situations when apoptosis is induced in the absence of DNA degradation (e.g., Cohen *et al.*, 1992) and, vice versa, extensive DNA degradation even selective to internucleosomal DNA may accompany necrosis (Collins *et al.*, 1992).

A higher degree of assurance of identification of apoptosis can be obtained by using more than one viability assay on the same sample. It is advisable, therefore, to simultaneously probe DNA cleavage (e.g., by labeling DNA strand breaks) and integrity of the plasma membrane (e.g., PI exclusion). Preservation of the plasma membrane integrity and extensive DNA breakage is a more specific marker of apoptosis than DNA strand breakage alone. Likewise, the status of mitochondria, probed by Rh123, which remains little changed in apoptotic cells, may be used as an additional marker of the mode of cell death.

In situations when apoptosis is not paralleled by DNA degradation (Cohen *et al.*, 1992; Collins *et al.*, 1992), analysis of DNA sensitivity to denaturation Darzynkiewicz (1994) appears to be the method of choice. The increased supravital cell stainability with HO342 combined with a decrease in forward light scatter may also be applied in such cases.

It should be stressed, however, that regardless of the flow cytometric method(s) used to identify apoptosis, this mechanism of cell death should be confirmed by inspection of cells under light or electron microscopy. Morphological changes during apoptosis have a very specific pattern and analysis of cell morphology should be the deciding factor in situations when there is any ambiguity regarding the mechanism of cell death.

Acknowledgments

Supported by NCI Grant ROI CA28704, the "This Close" Foundation, the Carl Inserra Fund, the Chemotherapy Foundation, I thank Dr Frank Traganos for his comments and suggestions.

References

Arends, M. J., Morris, R. G., and Wyllie, A. H. (1990). Apoptosis. The role of endonuclease. *Am. J. Pathol.* **136**, 593–608.

Chang, W. P., and Little, J. B. (1991). Delayed reproductive death in X-irradiated Chinese hamster ovary cells. *Int. J. Radiat. Biol.* **60**, 483–496.

Cohen, G. M., Sun, X.-M., Snowden, R. T., Dinsdale, D., and Skilleter, D. N. (1992). Key morphological features of apoptosis may occur in the absence of internucleosomal DNA fragmentation. *Biochem. J.* **286**, 331–334.

Collins, R. J., Harmon, B. V., Gobe, G. C., and Kerr, J. F. R. (1992). Internucleosomal DNA cleavage should not be the sole criterion for identifying apoptosis. *Int. J. Radiat. Biol.* **61**, 451–453.

Compton, M. M. (1992). A biochemical hallmark of apoptosis: internucleosomal degradation of the genome. *Cancer Metastasis Rev.* **11**, 105–119.

Darzynkiewicz, Z. (1994). Acid-induced denaturation of DNA in situ as a probe of chromatin structure. *Methods Cell Biol.* **41**, 527–541.

Darzynkiewicz, Z., Traganos, F., Staiano-Coico, L., Kapuscinski, J., and Melamed, M. R. (1982). Interactions of rhodamine 123 with living cells studied by flow cytometry. *Cancer Res.* **42**, 799–806.

Darzynkiewicz, Z., Williamson, B., Carswell, E. A., and Old, L. J. (1984a). The cell cycle specific effects of tumor necrosis factor. *Cancer Res.* **44**, 83–90.

Darzynkiewicz, Z., Traganos, F., Kapuscinski, J., Staiano-Coico, L., and Melamed, M. R. (1984b). Accessibility of DNA *in situ* to various fluorochromes: Relationship to chromatin changes during erythroid differentiation of Friend leukemia cells. *Cytometry.* **5**, 355–363.

Darzynkiewicz, Z., Traganos, F., Carter, S., and Higgins, P. J. (1987). *In situ* factors affecting stability of the DNA helix in interphase nuclei and metaphase chromosome. *Exp. Cell Res.* **172**, 168–179.

Darzynkiewicz, Z., Bruno, S., Del Bino, G., Gorczyca, W., Hotz, M. A., Lassota, P., and Traganos, F. (1992). Features of apoptotic cells measured by flow cytometry. *Cytometry.* **13**, 795–808.

Darzynkiewicz, Z., Huang, X., Okafuji, M., and King, M. A. (2004). Cytometric methods to detect apoptosis. *Methods Cell Biol.* **75**, 307–342.

Darzynkiewicz, Z., Galkowski, D., and Zhao, H. (2008). Analysis of apoptosis by cytometry using TUNEL assay. *Methods.* **44**, 250–254.

Darzynkiewicz, Z., Pozarowski, P., Lee, B. W., and Johnson, G. L. (2009). Fluorochrome-labeled inhibitors of caspases (FLICA): Convenient *in vitro* and *in vivo* markers of apoptotic cells for cytometric analysis. *Methods Mol. Biol.* (in press).

Del Bino, G., Lassota, P., and Darzynkiewicz, Z. (1991). The S-phase cytotoxicity of camptothecin. *Exp. Cell Res.* **193**, 27–35.

Dive, C., Gregory, C. D., Phipps, D. J., Evans, D. L., Milner, A. E., and Wyllie, A. H. (1992). Analysis and discrimination of necrosis and apoptosis (programmed cell death) by multiparameter flow cytometry. *Biochim.* Biophys. Acta **1133**, 275–282.

Gold, R., Schmied, M., Rothe, G., Zischler, H., Breitschopt, H., Wekerle, H., and Lassman, H. (1993). Detection of DNA fragmentation in apoptosis: application of *in situ* nick translation to cell culture systems and tissue sections. *J. Histochem. Cytochem.* **41**, 1023–1030.

Gong, J., Li, X., and Darzynkiewicz, Z. (1993). Different patterns of apoptosis of HL-60 cells induced by cycloheximide and camptothecin. *J. Cell. Physiol.* **157**, 263–270.

Gong, J., Traganos, F., and Darzynkiewicz, Z. (1994). A selective procedure for DNA extraction from apoptotic cells applicable for gel electrophoresis and flow cytometry. *Anal. Biochem.* **218**, 314–319.

Gorczyca, W., Bruno, S., Darzynkiewicz, R. J., Gong, J., and Darzynkiewicz, Z. (1992). DNA strand breaks occurring during apoptosis: Their early *in situ* detection by the terminal deoxynucleotidyl transferase and nick translation assays and prevention by serine protease inhibitors. *Int. J. Oncol.* **1**, 639–648.

Gorczyca, W., Bigman, K., Mittelman, A., Ahmed, T., Gong, J., Melamed, M. R., and Darzynkiewicz, Z. (1993a). Induction of DNA strand breaks associated with apoptosis during treatment of leukemias. *Leukemia.* **7,** 659–670.

Gorczyca, W., Gong, J., and Darzynkiewicz, Z. (1993b). Detection of DNA strand breaks in individual apoptotic cells by the *in situ* terminal deoxynucleotidyl transferase and nick translation assays. *Cancer Res.* **52,** 1945–1951.

Gorczyca, W., Gong, J., Ardelt, B., Traganos, F., and Darzynkiewicz, Z. (1993c). The cell cycle related differences in susceptibility of HL-60 cells to apoptosis induced by various antitumor agents. *Cancer Res.* **53,** 3186–3192.

Gorczyca, W., Tuziak, T., Kram, A., Melamed, M. R., and Darzynkiewicz, Z. (1994). Detection of apoptosis in fine-needle aspiration biopsies by *in situ* end-labeling of fragmented DNA. *Cytometry.* **15,** 169–175.

Hamori, E., Arndt-Jovin, D. J., Grimwade, B. G., and Jovin, T. M. (1980). Selection of viable cells with known DNA content. Cytometry **1,** 132–135.

Horan, P. K., and Kappler, J. W. (1977). Automated fluorescent analysis for cytotoxicity assays. *J. Immunol. Methods* **18,** 309–316.

Hotz, M. A., Traganos, F., and Darzynkiewicz, Z. (1992). Changes in nuclear chromatin related to apoptosis or necrosis induced by the DNA topoisomerase II inhibitor fostriecin in MOLT-4 and HL-60 cells are revealed by altered DNA sensitivity to denaturation. *Exp. Cell Res.* **201,** 184–191.

Hotz, M. A., Gong, J. P., Traganos, F., and Darzynkiewicz Z. (1994). Flow cytometric detection of apoptosis. Comparison of the assays of *in situ* DNA degradation and chromatin changes. *Cytometry.* **15,** 237–244

Johnson, L. U., Walsh, M. L., and Chen, L. B. (1980). Localization of mitochondria in living cells with rhodamine 123. *Proc. Natl. Acad. Sci.* USA **77,** 990–994.

Kaufmann, S. H., Lee, S. H., Meng, X. W., Loegering, D. A., Kottke, T. J., Henzing, A. J., Ruchaud, S., Samejima, K., and Earnshaw, W. C. (2008). Apoptosis-associated caspase activation assays. *Methods.* **44,** 262–272.

Kroemer, G., Galluzzi, L., Vandenabeele, P., Abrams, J., Alnemri, E. S., Baehrecke, E. H., Blagosklonny, M. V., El-Deiry, W. S., Golstein, P., Green, D. G., Hengartner, M., Knight, R. A. (2009). Classification of cell death: Recommendations of the nomenclature committee on cell death 2009. *Cell Death Differ.* **16,** 3–11.

Li, X., and Darzynkiewicz Z. (1995). Labeling DNA strand breaks with BrdUTP. Detection of apoptosis and cell proliferation. *Cell Prolif;* **28,** 571–579.

Meyaard, L., Otto, S. A., Jonker, R. R., Mijnster, M. J., Keet, R. P. M., and Miedema, F. (1992). Programmed death of T cells in HIV-1 infection. *Science.* **257,** 217–219.

Nicoletti, I., Migliorati, G., Pagliacci, M. C., Grignani, F., and Riccardi, C. (1991). A rapid and simple method for measuring thymocyte apoptosis by propidium iodide staining and flow cytometry. *J. Immunol. Methods* **139,** 271–279.

Ormerod, M. G., Collins, M. K. L., Rodriguez-Tarduchy, G., and Robertson, D. (1992). Apoptosis in interleukin-3-dependent haemopoietic cells. Quantification by two flow cytometric methods. *J. Immunol. Methods* **153,** 57–63.

Ormerod, M. G., Sun, X.-M., Snowden, R. D., Davies, R., Fearnhead, H., and Cohen, G. M. (1993). Increased membrane permeability of apoptotic thymocytes: a flow cytometric study. *Cytometry.* **14,** 595–602.

Pollack, A., and Ciancio, G. (1990). Cell cycle phase-specific analysis of cell viability using Hoechst 33342 and propidium iodide after ethanol preservation. *Methods Cell Biol.* **33,** 19–24.

Pozarowski, P., Huang, X., Halicka, D. H., Lee, B., Johnson, G., and Darzynkiewicz, Z. (2003). Interactions of fluorochrome-labeled caspase inhibitors with apoptotic cells. A caution in data interpretation. *Cytometry A* 55A, 50–60.

Riedy, M. C., Muirhead, K. A., Jensen, C. B., and Stewart, C. C. (1991).Use of a photolabeling technique to identify nonviable cells in fixed homologous or heterologous cell populations. *Cytometry.* **12,** 133–139.

Sachs, L., and Lotem, J. (1993). Control of programmed cell death in normal and leukemic cells: new implications for therapy. *Blood.* **82,** 15–21.

Sambrook, J., Fritsch, E. F., and Maniatis, T. (1989). Molecular Cloning: A Laboratory Manual, 2nd edn., Cold Spring Harbor Laboratory., Cold Spring Harbor, NY.

Schwartzman, R. A., and Cidlowski, J. A. (1993). Apoptosis: the biochemistry and molecular biology of programmed cell death. *Endocr. Rev.* **14,** 133–155.

Stewart, C. C., and Stewart, S. J. (1994). Cell preparation for the identification of leukocytes. *Methods Cell Biol.* **41,** 40–60.

Stöhr, M., and Vogt-Schaden, M. (1980). A new dual staining technique for simultaneous flow cytometric DNA analysis of live and dead cells. In "Flow Cytometry IV" (O. D. Laerum, T. Lindmo, and E. Thorud, eds.), pp. 96–99. Universitetforslaget, Bergen, Norway.

Swat, W., Ignatowicz, L., and Kisielow, P. (1991). Detection of apoptosis of immature CD4+8+ thymocytes by flow cytometry. *J. Immunol. Methods* **137,** 79–87.

Telford, W. G., King, L. E., and Fraker, P. J. (1991). Evaluation of glucocorticoid- induced DNA fragmentation in mouse thymocytes by flow cytometry. *Cell Prolif.* **24,** 447–459.

Tomei L. D., and Cope F. O., eds. (1991). "Apoptosis: The Molecular Basis of Cell Death." Cold Spring Harbor Laboratory, Plainview, NY.

Tounekti, O., Pron, G., Belehradek, J., and Mir, L. M. (1993). Bleomycin, an apoptosis- mimetic drug that induces two types of cell death depending on the number of molecules internalized. *Cancer Res.* **53,** 5462–5469.

Traganos, F., and Darzynkiewicz, Z. (1994). Lysosomal proton pump activity: Supravital cell staining with acridine orange differentiates leukocyte subpopulations. *Methods Cell Biol.* **41,** 183–194.

Umansky, S. R., Korol', B. R., and Nelipovich, P. A. (1981). *In vivo* DNA degradation in thymocytes of gamma-irradiated or hydrocortisone-treated rats. *Biochim. Biophys. Acta* **655,** 281–290.

van Engeland, M., Nieland, L. J., Ramaekers, F. C., Schutte, B., and Reutelingsperger, C. P. (1998). Annexin V-affinity assay: A review on an apoptosis detection system based on phosphatidylserine exposure. Cytometry **31,** 1–9.

Wallen, C. A., Higashikubu, R., and Roti Roti, J. L.(1983) Comparison of the cell kill measured by the Hoechst-propidium iodide flow cytometric assay and the colony formation assay. *Cell Tissue Kinet.* **16,** 357–365.

Wlodkowic, D., Skommer, J., and Darzynkiewicz, Z. (2008). SYTO probes in cytometry of tumor cell death. *Cytometry A* 73A, 496–507.

Wyllie, A. H. (1985). The biology of cell death in tumours. *Anticancer Res.* **5,** 131–142.

Wyllie, A. H. (1992). Apoptosis and the regulation of cell numbers in normal and neoplastic tissues: an overview. *Cancer Metastasis Rev.* **11,** 95–103.

CHAPTER 22

Difficulties and Pitfalls in Analysis of Apoptosis

Zbigniew Darzynkiewicz,[*] **Elżbieta Bedner,**[†] **and Frank Traganos**[*]

[*]Brander Cancer Research Institute
New York Medical College
Valhalla, New York 10595

[†]Department of Pathology
Pomeranian School of Medicine
Szczecin, Poland

I. Update

Among the chapters from the MCB volumes dedicated to cytometry (Vols. 33, 41, 42, 63, 64, and 75), the chapters addressing methodological aspects of analysis of apoptosis were most widely cited. This is a reflection that cytometry has become

DOI: 10.1016/B978-0-12-375045-7.00022-2

detailed protocols and instructions for use. However, most of these chapters and protocols fail to adequately address certain problems and difficulties that are often encountered in the analysis of apoptosis. The common pitfalls and inappropriate uses of the methodology are apparent from reviewing the literature. Some of these problems are generic to most of the methods: they generally pertain to data interpretation, in particular how the frequency of apoptotic cells (AI) relates to the incidence of cell death in cultures or in tissue. Other problems are specific to particular methods or cell systems. Certain issues associated with the inappropriate use of flow cytometry in the analysis of apoptosis were described previously (Darzynkiewicz *et al.*, 1998). The most common errors in measurement of apoptosis as well as the frequent mistakes in the analysis and interpretation of the data have been updated and are discussed in this chapter.

III. AI may not be Correlated with Incidence of Cell Death

It is often assumed that the frequency of apoptotic cells (AI), *in vivo* or in cultures, is a reflection of "how many" cells underwent apoptosis, for example, as a result of the treatment or over a given time interval. Thus, for example, when a particular drug is administered to a culture and several hours later the percentage of apoptotic cells is higher in this culture than in a culture treated with another drug, an assumption is often made that the first drug was more effective in *inducing* apoptosis or in *cell kill*. Likewise, the increased AI *in vivo*, in the tissue, is often interpreted as reflecting the increased *cell death incidence* in this tissue. This is not always a correct assumption because apoptosis is a kinetic event. The entire apoptotic process, from the initiation to the total disintegration of the cell, is of short and variable duration. The time window during which individual apoptotic cells demonstrate their characteristic features (markers) that allow them to be recognizable varies depending on (1) the method used, (2) the cell type, and/or (3) the nature of the inducer of apoptosis. Thus, for example, variable estimates of the AI in the same cell population are expected when different methods, differing in the width of the time window through which they recognize apoptosis, are used. Furthermore, some inducers may slow down or accelerate the apoptotic process. This may occur if the rate of either formation and/or shedding of apoptotic bodies, endonucleolysis, or proteolysis is affected by the inducer or by the growth conditions (e.g., temperature, pH). Thus, when the duration of apoptosis is shortened, the frequency of apoptotic cells (AI) is diminished even if the incidence of cell death remains the same. Conversely, prolongation of apoptosis in absence of any change in incidence of cell death manifests by the increase in AI. Protease inhibitors, for example, including inhibitors of serine proteases, delay nuclear fragmentation and prolong the process of apoptosis (Hara *et al.*, 1996). Induction of apoptosis in their presence, therefore, is expected to be reflected by the increased AI compared to parallel cultures with the same incidence of cell death but where apoptosis is of shorter duration.

The duration of apoptosis is also different in different cell types and tissues, as well as *in vivo* and *in vitro*. Cells of hematopoietic tumor lines (e.g., such as HL-60 cells) *in vitro*, when triggered by DNA damage, for example, by DNA topoisomerase inhibitors, progress through the entire process of apoptosis rapidly; the cells disintegrate totally within 4–6 h after the treatment. The same treatment of MCF-7 breast carcinoma cells triggers apoptosis after a 24-h delay and leads to an apoptotic process that is of much longer duration (Del Bino *et al.*, 1999). Apoptosis *in vivo*, within tissues, appears to be rapid, with the remains of apoptotic cells completely removed from the tissue in a short time. This is evidenced by the fact that under conditions of homeostasis the AI is often similar to the mitotic index (e.g., 1–2%). Because the duration of mitosis is approximately 1 h, the duration of apoptosis must be similarly short. It is quite possible, however, that the rate of removal of the remains of apoptotic cells by neighboring cells and by macrophages also varies depending on the tissue type.

In conclusion, an observed increase in AI, for example, as a result of a particular treatment, may indicate that indeed the incidence of cells dying by apoptosis was increased by the treatment. As discussed, however, it may also indicate that the same number of cells were dying but that the duration of apoptosis was prolonged. A combination of both, namely, increased incidence of apoptosis and prolongation of this process, could occur as well. Unfortunately, no methods yet exist to arrest cells in apoptosis and therefore to obtain a cumulative estimate of the rate of cell entrance to apoptosis. Such an approach is available, for example, for mitosis, which can be arrested by microtubule poisons, and the rate of cell entrance to mitosis can be calculated to obtain a quantitative estimate of the *cell birth rate* (Darzynkiewicz *et al.*, 1986). Perhaps the apoptotic process can be arrested by some inhibitors of caspases, which would then allow one to measure the rate of cell entrance to apoptosis or the *cell death rate*. Nevertheless, the percentage of apoptotic cells in a cell population as presently estimated by any given method (AI) is not a measure of the incidence of cells dying by apoptosis.

To estimate the incidence of cell death, for example, as a result of treatment with chemical or physical agent, the absolute number (not the percentage) of live cells should be measured in the treated culture and compared with the appropriate untreated control. A correction should also be made to account for the rate of cell proliferation. The latter may be obtained from the classic cell growth curves, when the number of live cells is plotted against the time in culture. Alternatively, it may be obtained from rate of cell entrance to mitosis in a stathmokinetic experiment by arresting cells in mitosis (Darzynkiewicz *et al.*, 1986). The observed deficit in the actual number of live cells from the expected number of live cells estimated based on the rate of cell birth provides an estimate of the cumulative cell loss (death) during the measured time interval. Indirectly, the cell proliferation rate can be inferred from the percentage of cells incorporating bromodeoxyuridine (BrdU) or from the mitotic index, under the assumption that the treatment which induces apoptosis does not affect the duration of any particular phase of the cell cycle. The estimate of apoptosis (by detection of cells with DNA strand breaks) may be combined with analysis of BrdU incorporation and cell cycle position, by flow or laser scanning cytometry (Li *et al.*, 1996). This methodology, which reveals AI and the fraction of cells replicating DNA in the same sample, may be particularly useful in evaluating the proliferative potential of tumors.

IV. Difficulties in Estimating Frequency of Apoptosis by Analysis of DNA Fragmentation

A common misconception in analysis of apoptosis is that the amount of fragmented (low MW, "extractable") DNA detected in cultures, tissue, or cell extracts, etc., is proportional to the frequency of apoptosis. Many methods were developed to estimate the amount of fragmented DNA, and numerous reagent kits are being sold for that purpose. They include direct quantitative colorimetric analysis of "soluble" DNA, densitometry of "DNA ladders" on gels, and immunochemical assessment of nucleosomes. Some of these approaches are advertised by the vendors as quantitative, in the sense that they are intended to provide information regarding the frequency of apoptosis in cell populations. Such claims are grossly incorrect. Namely, the amount of fragmented (low MW) DNA that can be extracted from a single apoptotic cell varies over a wide range depending on the stage of apoptosis. Although early during apoptosis only a small fraction of DNA is degraded, when apoptosis is more advanced nearly all DNA is fragmented. Thus, the amount of low MW DNA that is extracted from a single apoptotic cell varies manyfold depending on the stage of apoptosis. As a result, the total content of low MW DNA extracted from the cell population, or the ratio of the low to high MW fraction, does not provide information about the frequency of apoptotic cells (AI), even in relative terms, for example, for comparison of cell populations. For this reason, biochemical methods based on analysis of fragmented DNA cannot be used to quantitatively estimate the frequency of apoptosis.

DNA "laddering" observed during electrophoresis provides evidence of internucleosomal DNA cleavage that is considered one of the hallmarks of apoptosis (Arends *et al.*, 1990). Analysis of DNA fragmentation by gel electrophoresis to detect such laddering is thus a valuable method to demonstrate the apoptotic mode of cell death. It should not be used, however, as a means to quantitate the frequency of apoptosis.

In some cell systems, apoptosis may occur without internucleosomal DNA cleavage; the products of DNA fragmentation are large DNA sections that cannot be easily extracted from the cell (Oberhammer *et al.*, 1993). Obviously, in these systems, apoptosis cannot be revealed by the presence of DNA laddering on gels or by analysis of low MW products. These instances are discussed in the next section of this chapter.

V. The Lack of Evidence is not Evidence for the Lack of Apoptosis

There are numerous publications describing cell death that resembles apoptosis which lacks, however, one or more characteristic apoptotic features ("atypical apoptosis") (e.g., Cohen *et al.*, 1992; Collins *et al.*, 1992; Ormerod *et al.*, 1994;

Zakeri *et al.*, 1993; Zamai *et al.*, 1996). Thus, for example, apoptosis-associated DNA endonucleolysis frequently terminates after generating 50- to 300-kb breaks and does not proceed to generate internucleosomal-sized DNA fragments (Oberhammer *et al.*, 1993). Such cells contain relatively few *in situ* DNA strand breaks compared with classic apoptosis. Methods based on detection of DNA laddering on gels will fail to identify apoptosis in such situations. [It should be noted, however, that the 50- to 300-kb fragments can be detected by pulsed field electrophoresis (Oberhammer *et al.*, 1993).] Because of the paucity of DNA strand breaks under such circumstances, it is also difficult to identify such cells by the DNA strand break TdT-mediated dUTP-biotin nick end labeling (TUNEL) assay (Del Bino *et al.*, 1999).

Apoptosis is often induced by agents that are inhibitors of a particular enzyme or metabolic pathway that is associated with apoptosis. Identification of apoptotic cells based on activity of this enzyme or analysis of the pathway involved will be unsuccessful. For example, apoptosis can be induced by certain protease inhibitors (Hara *et al.*, 1996). Because these inhibitors (perhaps by inhibiting proteolysis of nuclear lamin) prevent nuclear fragmentation, this feature (nuclear fragmentation) cannot be used as a marker distinguishing apoptotic cells.

Application of more than one method, each based on a different principle (i.e., detecting a different cellular feature of apoptosis), offers a better chance of detecting atypical apoptosis than does any single method. As mentioned, if DNA in apoptotic cells is fragmented to 50- to 300-kb sections it is not extractable, and such cells cannot be identified as apoptotic either by the method based on analysis of DNA content or DNA laddering during electrophoresis. It is likely, however, that such apoptotic cells can be recognized based on their reduced F-actin stainability with fluorescein isothiocyanate (FITC)-phalloidin (Endersen *et al.*, 1995), by their reactivity with a fluorochromed annexin V (Koopman *et al.*, 1994), by the drop in mitochondrial transmembrane potential detected by transmembrane potential-sensing flourochrome probes (Cossarizza *et al.*, 1994; Zamzani *et al.*, 1998), or by other markers.

VI. Misclassification of Apoptotic Bodies or Nuclear Fragments as Single Apoptotic Cells

Identification of apoptotic cells often relies on cellular DNA content measurements by flow cytometry. It was initially observed that following cell fixation in ethanol and staining of their DNA, apoptotic cells were recognized by virtue of their lower stainability compared to G_1 cells (Umansky *et al.*, 1981). On the DNA content frequency histograms they occupied a position between the origin of the DNA coordinate and the G_1 peak ("sub-G_1" cells). Their decreased DNA stainability was explained as partially due to the extraction of fragmented, low MW DNA during the staining procedure (Darzynkiewicz *et al.*, 1992; Gong *et al.*, 1994). To optimize the distinction between intact G_1 and apoptotic cells, the extent of

DNA extraction can be enhanced by using buffers of higher molar strength prior to fixation (Gong *et al.*, 1994). The apoptotic (sub-G_1) cells can thus be distinguished based on their fractional DNA content, which may partly be due to DNA extraction during the staining procedure but may also reflect DNA loss due to shedding of apoptotic bodies, or even to changes in chromatin structure (condensation) that make DNA less accessible to the fluorochrome.

The major problem with this methodology stems from the fact that, commonly, the analysis is performed on cells that were subjected to treatment with a detergent or hypotonic solution instead of fixation. Such treatment lyses the plasma membrane and gives rise to the following artifacts. (1) Because the nucleus of an apoptotic cell is fragmented, numerous individual chromatin fragments are present in a single cell. On cell lysis, each individual chromatin fragment, having a fractional DNA content, is separately released and, when measured, is erroneously identified as a single apoptotic cell. Therefore, the percentage of objects represented by the sub-G_1 peak significantly overestimates the actual AI. (2) Similar problems arise when, for instance, chromosomes are released from lysed mitotic cells. Both individual chromosomes as well as chromosome aggregates having a fractional DNA content may mistakenly be identified as apoptotic cells. This problem is exacerbated when apoptosis is induced by agents that increase the proportion of mitotic cells, for example, taxol or other microtubule poisons. (3) Often, following cell irradiation or treatment of cells with DNA damaging drugs, micronuclei are formed. Their number depends on the degree of DNA damage and duration of treatment. Having a fractional DNA content, micronuclei may also be erroneously identified as apoptotic cells.

A gentle permeabilization of the cell with a detergent but in the presence of exogenous proteins such as serum or serum albumin prevents lysis of the plasma membrane. It was shown that the presence of 1% (w/v) albumin or 10% (v/v) serum protects cells from lysis (e.g., induced by 0.1% Triton X-100) without affecting their permeabilization by detergent. In fact, this method is used for simultaneous analysis of DNA and RNA as well as for detection of apoptotic cells characterized by reduced DNA content (Darzynkiewicz, 1994). However, apoptotic or nonapoptotic cells suspended in saline containing detergent and serum proteins remain very fragile, and pipetting, vortexing, or even shaking the tube containing the suspension causes their lysis and release of the cell constituents into solution.

Logarithmic amplification of the fluorescence signal is frequently used to measure and display cellular DNA content in the methods that employ detergents to quantitate apoptosis. A logarithmic scale allows one to measure and record events with 1 or even 0.1% of the DNA content of intact, nonapoptotic cells. The majority of such objects cannot be individual apoptotic cells. In the case of cell lysis by detergents, as discussed earlier, these objects represent nuclear fragments, individual apoptotic bodies, individual chromosomes, chromosome aggregates, micronuclei, or contaminating bacteria.

To exclude objects with a minimal DNA content that may not be apoptotic cells from analysis, it is advisable during fluorescence measurement to set the threshold of DNA detection at a constant level, for example, at 1/10th or 1/20th the fluorescence value of intact G_1 cells. This would eliminate all particles with a DNA content less than 10% or 5% of that of G_1 cells from the analysis. Although the AI may then be underestimated, the underestimate is constant and introduces less error than would occur if all objects with a fractional DNA content were counted. In essence, there is little reason to use a logarithmic scale to measure DNA content because a linear scale provides better assurance that objects with a minimal DNA content are not included in the analysis.

Difficulties in identification of apoptotic cells by this methodology may also occur when G_2/M or late S phase cells undergo apoptosis, and the extraction of low MW DNA from them is insufficient to shift them to a position below the G_1 peak. On DNA frequency histograms, such cells may overlap with the nonapoptotic G_1 cells. Likewise, when cells grow at two DNA ploidy levels, the apoptotic cells of a higher DNA ploidy despite the loss of DNA may still overlap in DNA content with the nonapoptotic cells of lower DNA ploidy. As mentioned, a more extensive DNA extraction, which is provided using high-molarity buffers and improves separation of apoptotic cells on DNA frequency histograms (Gong et al., 1994), may be useful in such situations.

The method of identification of apoptotic cells on the basis of their fractional DNA content, even if it is based on analysis of fixed cells (i.e., is devoid of artifacts of cell lysis discussed earlier), is not very specific. Mechanically damaged cells, in particular cell fragments that may remain in suspension, for example, after isolation of cells from the tissue, will have fractional DNA content and be indistinguishable from apoptotic cells. Likewise, late necrotic cells are also characterized by loss of DNA and therefore may have similar DNA stainability as apoptotic cells on DNA histograms.

VII. Apoptosis Versus Necrosis Versus "Necrotic Stage" of Apoptosis

There are many differences between typical apoptotic and necrotic cells (Arends et al., 1990; Kerr et al., 1972; Majno and Joris, 1995), and they have provided a basis for development of numerous markers and methods that can discriminate between these two modes of cell death (Darzynkiewicz et al., 1992, 1997a,b). The major difference stems from the early loss of integrity of the plasma membrane during necrosis. This event results in a loss of the ability of the cell to exclude many fluorochromes. In contrast, the plasma membrane and membrane transport functions remain, to a large extent, preserved during the early stages of apoptosis. The permeability of a cell to PI, or its ability to retain some fluorescent probes such as products of enzyme activity, is the most common marker distinguishing apoptosis from necrosis. Thus, for example, a combination of fluorochrome-conjugated annexin V with PI was proposed to distinguish live cells (unstainable with both dyes) from apoptotic cells (stainable with annexin V but unstainable with PI) from

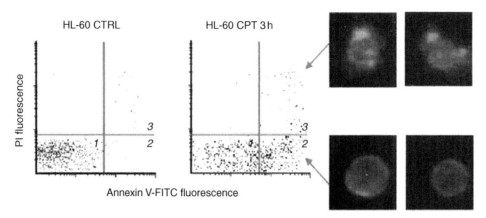

Fig. 1 Discrimination between nonapoptotic, early apoptotic, and late apoptotic ("necrotic stage" of apoptosis) or necrotic cells following staining with annexin V-FITC conjugate and PI and fluorescence measurement by laser scanning cytometry (LSC). The cells in quadrant 1 are live; they exclude PI and do not bind annexin V-FITC. They predominate in HL-60 untreated (control) cultures. The cells in quadrant 2 are early apoptotic; they bind annexin V-FITC but still exclude PI. The cells in quadrant 3 are late apoptotic (or necrotic) and they bind both PI and annexin V-FITC. The quadrant 2 and 3 cells are more frequent in the camptothecin-treated sample (CPT). The possibility of observing the morphology of the cells selected based on their fluorescence, as offered by LSC, helps to confirm the identity of apoptotic cells. (See Plate no. 9 in the Color Plate Section.)

necrotic cells (stainable with both dyes) (Koopman *et al.*, 1994) (Fig. 1). The fluorochrome-conjugated annexin V binds to phosphatidylserine, the phospholipid of the plasma membrane that is inaccessible to this conjugate in live cells but becomes exposed on the outer leaflet of the plasma membrane and, therefore, accessible during apoptosis (Fadok *et al.*, 1992; Koopman *et al.*, 1994). Although this approach works well in many instances, it has limitations and possible pitfalls, as follows:

1. Late stage apoptotic cells resemble necrotic cells to such an extent that, to define them, the term "apoptotic necrosis" was proposed (Majno and Joris, 1995). This is a consequence of the fact that the integrity of the plasma membrane of late apoptotic cells is compromised, which makes the membrane leaky and permeable to charged cationic dyes such as PI. Thus, since the ability of such cells to exclude these dyes is lost, the discrimination between late apoptosis and necrosis cannot be accomplished by methods based on the use of plasma membrane permeability probes or annexin V.

2. The permeability and asymmetry of plasma membrane phospholipids (accessibility of phosphatidylserine) may change, for example, as a result of prolonged treatment with proteolytic enzymes (trypsinization), mechanical damage (e.g., cell removal from flasks by a rubber policeman, cell isolation from solid tumors, or even repeated centrifugations), electroporation, or treatment with some drugs.

3. Many flow cytometric methods designed to quantify the frequency of apoptotic or necrotic cells are based on the differences between live versus apoptotic versus necrotic cells in the permeability of their plasma membrane to different fluorochromes such as PI, 7-aminoactinomycin D (7-AAD), or Hoechst dyes. It should be stressed, however, that plasma membrane permeability may vary depending on the cell type and on many other factors, unrelated to apoptosis or necrosis. The assumption, therefore, that live cells maximally exclude a particular dye, early apoptotic cells are somewhat leaky, while late apoptotic or necrotic cells are totally permeable to the dye, and that these differences are large enough to identify these cells, is not universally applicable.

It is particularly difficult to discriminate between apoptotic and necrotic cells in suspensions from solid tumors. Necrotic areas form in tumors as a result of massive local cell death, for example, due to poor accessibility to oxygen and growth factors when tumors grow in size and their local vascularization becomes inadequate. Needle aspirate samples or cell suspensions from the resected tumors may contain many cells from the necrotic areas. Such cells are indistinguishable from late apoptotic cells by many markers. Because the AI in solid tumors, representing spontaneous- or treatment-induced apoptosis, should not include cells from the necrotic areas, one has to carefully eliminate such cells from analysis. Because incubation of cells with trypsin and DNase I selectively and totally digests all cells whose plasma membrane integrity is compromised, that is, primarily necrotic cells (Darzynkiewicz *et al.*, 1994), such a procedure may be used to remove necrotic cells from suspensions. It should be noted, however, that late apoptotic cells have somewhat permeable plasma membranes and are expected to also be sensitive to this treatment.

VIII. Selective Loss of Apoptotic Cells During Sample Preparation

Relatively early during apoptosis, cells will detach from the surface of culture flasks and float in the medium. Thus, the standard procedure of discarding the medium, followed by trypsinization or EDTA treatment of the attached cells and their collection, results in selective loss of apoptotic cells that are discarded with the medium. Such loss may vary from flask to flask depending on how the culture is handled, for example, the degree of mixing or shaking, efficiency in discarding the old medium. Surprisingly, cell trypsinization and discarding the medium is still occasionally reported by some authors. Needless to say, such an approach cannot be used for quantitative analysis of apoptosis. To estimate the frequency of apoptotic cells in adherent cultures, it is essential to collect floating cells, pool them with the trypsinized ones, and measure them as a single sample. It should be stressed that trypsinization, especially if prolonged, results in digestion of cells with a compromised plasma membrane. Thus, collection of cells from cultures by trypsinization is expected to cause selective loss of late apoptotic and necrotic cells.

Similarly, density gradient separation of cells (e.g., using Ficoll-Hypaque or Percoll solutions) may result in selective loss of dying and dead cells. This is due to the fact that early during apoptosis the cells become dehydrated, have condensed nuclei and cytoplasm, and therefore have a higher density compared to nonapoptotic cells. Knowledge of any selective loss of dead cells in cell populations purified by such an approach is essential when one is studying apoptosis.

Repeated centrifugations lead to cell loss by at least two mechanisms. One involves electrostatic cell attachment to the tubes and may be selective to a particular cell type. Thus, for example, preferential loss of monocytes and granulocytes was observed during repeated centrifugation of white blood cells, while lymphocytes remained in suspension (Bedner *et al.*, 1997). Cell loss is of particular concern when hypocellular samples ($<5 \times 10^4$ cells) are processed. In such a situation, carrier cells in excess (e.g., chick erythrocytes) may be added to preclude disappearance of the cells of interest through centrifugations. The second mechanism of cell loss involves preferential disintegration of fragile cells. Because apoptotic cells, especially at late stages of apoptosis, are very fragile, they may selectively be lost from samples that require centrifugation or are repeatedly vortexed, pipetted, etc. Addition of serum or bovine serum albumin to cell suspensions, shortened centrifugation time, and decreased gravity force all may have a protective effect against cell breakage by mechanical factors. Apoptotic cells may also preferentially disintegrate in biomass cultures that require constant cell mixing.

IX. Live Cells Engulfing Apoptotic Bodies Masquerade as Apoptotic Cells

Exposure of phosphatidylserine on the outer leaflet of the plasma membrane that occurs during apoptosis (Fadok *et al.*, 1992) makes such cells and their fragments (apoptotic bodies) attractive to neighboring cells, which phagocytize them. In addition to professional phagocytes, cells of fibroblast or epithelial lineage also have the ability to engulf apoptotic bodies. It is frequently observed, especially in solid tumors, that the cytoplasm of both nontumor as well as tumor cells located in the neighborhood of apoptotic cells contains inclusions typical of apoptotic bodies. The remains of apoptotic cells engulfed by neighboring cells contain altered plasma membranes, fragmented DNA, and other constituents with attributes characteristic of apoptosis. Thus, if the distinction is based on any of these attributes, the live, nonapoptotic cells that phagocytized apoptotic bodies cannot be distinguished from genuine apoptotic cells by flow cytometry (Fig. 2).

X. The Problems with Commercial Kits and Reagents

A large number of commercial kits designed to detect apoptosis have become available, and reagent companies are racing to introduce new kits, often advertising them as "unique apoptosis detection kits." Some of these kits have solid

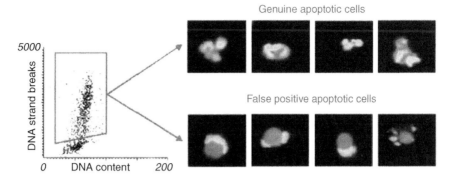

Fig. 2 Discrimination between genuine and false positive apoptotic cells by LSC. Peripheral blood mononuclear cells were obtained from a leukemia patient during chemotherapy (Gorczyca *et al.*, 1993a). DNA strand breaks were labeled with fluoresceinated dUTP using exogenous terminal deoxynucleotidyltransferase as described (Gorczyca *et al.*, 1992, 1993a). Genuine apoptotic cells are characterized by morphology typical of apoptosis. However, the nonapoptotic cells resembling monocytes, which contain cytoplasmic inclusions stainable with PI and having DNA strand breaks, also are present in this sample. These cells (false positive) most likely represent monocytes that phagocytized apoptotic bodies. By flow cytometry, which does not allow morphological identification, such cells are mistakenly classified as apoptotic. (See Plate no. 10 in the Color Plate Section.)

experimental foundations and have been repeatedly tested on a variety of cell systems. Other kits, however, especially those advertised by vendors who do not fully explain the principle of detection of apoptosis on which the kit is based, and do not list its chemical composition, may not be universally applicable. It is not uncommon for a kit to be introduced after being tested on one or two cell lines using a single agent to trigger apoptosis (generally, classic apoptosis of either a T cell leukemia line treated with the Fas ligand or HL-60 cells treated with a DNA topoisomerase inhibitor).

Before application of any new kit, it is advisable to confirm that at least three to four independent laboratories have already successfully used it on different cell types. Furthermore, it is good practice to initially use the new kit in parallel with a well-established methodology, in a few experiments. This would allow one, by comparison of the apoptotic indices, to estimate the time window of detection of apoptosis by the new method and its sensitivity, compared to the one that is already established and accepted in the field.

One potential pitfall in using commercial kits has been described by Bedner *et al.* (1999a). The problem pertains an erroneous identification of live nonapoptotic eosinophils as apoptotic cells. This misclassification was due to the fact that there are trace amounts of unconjugated FITC in most commercially available reagents and kits (e.g., FITC-conjugated dUTP, FITC-conjugated avidin, FITC-conjugated primary or secondary antibodies). This FITC reacts avidly with granule proteins of eosinophils, labeling them very strongly. Thus, all the methods of identification of apoptotic cells that rely on the use of FITC-conjugated reagents and permeabilized cells are expected to nonspecifically label eosinophils, which then can be erroneously classified as apoptotic cells by flow cytometry (Bedner *et al.*, 1999a).

XI. Cell Morphology is still the Gold Standard for Identification of Apoptotic Cells

Apoptosis was originally defined as a specific mode of cell death based on characteristic changes in cell morphology (Kerr *et al.*, 1972) (Fig. 3). The characteristic morphological features of apoptosis and necrosis are listed in Table I. Although individual features of apoptosis may serve as markers for detection and analysis of the proportion of apoptotic cells in the cell populations studied by flow cytometry or other quantitative methods, the mode of cell death should *always* be identified by inspection of cells by light or electron microscopy. Therefore, when quantitative analysis is done by flow cytometry, it is essential to *confirm* the mode of cell death on the basis of morphological criteria. Furthermore, if there is any ambiguity regarding the mechanism of cell death, the morphological changes should be the deciding attribute in resolving the uncertainty.

It should be stressed that optimal preparations for light microscopy require cytospining of live cells following by their fixation and staining on slides. The cells are then flat and their morphology is easy to assess. On the other hand, when the cells are initially fixed and stained in suspension, then transferred to slides and analyzed under the microscope, their morphology is obscured by the unfavorable geometry: the cells are spherical and thick and require confocal microscopy to reveal details such as early signs of apoptotic chromatin condensation.

Differential staining of cellular DNA and protein of cells on slides with DAPI and sulforhodamine 101, respectively, is rapid and simple and provides very good morphological resolution of apoptosis and necrosis (Darzynkiewicz *et al.*, 1997a). A combined cell illumination with ultraviolet (UV) light (to excite the DNA fluorochrome, e.g., DAPI) and light transmission microscopy utilizing interference contrast (Nomarski illumination) is our favorite method of cell visualization to identify apoptotic cells (Fig. 3). Other DNA fluorochromes, such as PI, 7-AAD, or acridine orange (AO), can be used as well.

XII. Laser Scanning Cytometry: Have your Cake and Eat it too

As mentioned earlier, characteristic changes in cell morphology provided the deciding criteria for identification of apoptotic cells. Quantitation of AI by microscopy or by classic image analysis techniques, however, was cumbersome and slow, and selection of cells for visual inspection was biased. Flow cytometry provided rapid and unbiased analyses but did not allow for morphological identification of the measured cell. Although the cells could be electronically sorted for their

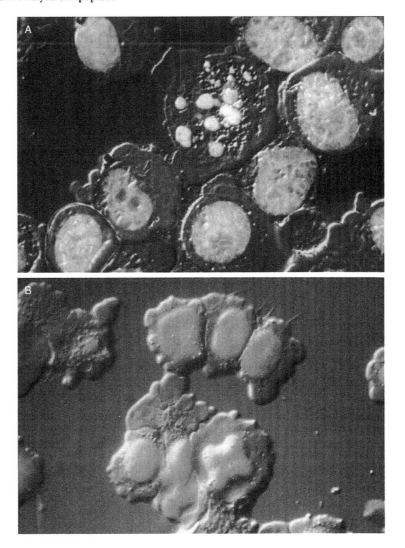

Fig. 3 Morphology of apoptotic cells from U937 cultures treated with tumor necrosis factor α (TNF-α), cytocentrifuged, fixed in formaldehyde, and stained with 4,6-diamidino-2-phenylindole (DAPI) (A) or 7-AAD (B) as revealed by their examination by fluorescence microscopy combined with interference contrast (Nomarski) illumination. Photographs were taken with a Nikon Microphot-FXA, with a 40× objective. (See Plate no. 11 in the Color Plate Section.)

identification, cell sorters are expensive, the procedure is time-consuming and cumbersome, and there are many technical problems involved in recovering apoptotic cells for morphological analysis following sorting.

A new instrument, the laser scanning cytometer (LSC), satisfies both requirements for analysis of apoptosis: rapidity of the measurements and the possibility of

Table I

Morphological Criteria for Identification of Apoptosis or Necrosis

Apoptosis	Necrosis
Reduced cell size, convoluted cell shape	Cell and nuclear swelling
Plasma membrane undulations (blebbing/budding)	Patchy chromatin condensation
	Swelling of mitochondria
Chromatin condensation (DNA hyperchromicity)	Vacuolization in cytoplasm
	Plasma membrane rupture (ghost-like cells)
Loss of the structural features of the nucleus (smooth appearance)	Dissolution of nuclear chromatin (karyolysis)
Nuclear fragmentation (karyorrhexis)	Attraction of inflammatory cells
Presence of apoptotic bodies	
Dilatation of the endoplasmic reticulum	
Relatively unchanged cell organelles	
Shedding of apoptotic bodies	
Phagocytosis of the cell remnants	
Cell detachment from tissue culture flasks	

morphological examination of the selected cells. The LSC is a microscope-based cytofluorometer that combines advantages of flow and image cytometry (Kamentsky and Kamentsky, 1991; Pozarowski *et al.*, 2006). The fluorescence of individual cells can be measured rapidly, with sensitivity and accuracy comparable to those of flow cytometry. Cell staining and measurement on slides eliminates cell loss, which inevitably occurs during the repeated centrifugations necessary for sample preparation for flow cytometry. Since the spatial *x-y* coordinates of each cell on the slide are recorded, these cells can be relocated after the initial LSC measurement, for example, for visual microscopy after staining with another dye to carry out additional image analysis. Furthermore, because the geometry of the cells cytocentrifuged or smeared on the slide is more favorable for morphometric analysis than is the case for cells in suspension, more information on cell morphology can be obtained by laser scanning than by flow cytometry. Advantages and limitations of the LSC for analysis of apoptosis have been reviewed (Bedner *et al.*, 1999b; Darzynkiewicz *et al.*, 1999; Li and Darzynkiewicz, 1999).

Two attributes of LSC make it an instrument of choice for analysis of apoptosis. As mentioned, the first attribute is the possibility of morphological examination of the cells of interest. Thus, several thousand cells can be measured per sample, with rates approaching 100 cells/s, to quantify the frequency of apoptotic cells identified on the basis of a particular marker. Morphology of the presumed apoptotic cells can then be discerned following their relocation and microscopic examination, as shown in Fig. 4. This attribute of the LSC made it possible, for example, to identify the false positive apoptotic cells in bone marrow of leukemia patients undergoing chemotherapy. The latter, showing the presence of DNA strand breaks (TUNEL positive), were actually nonapoptotic monocytes and macrophages that engulfed

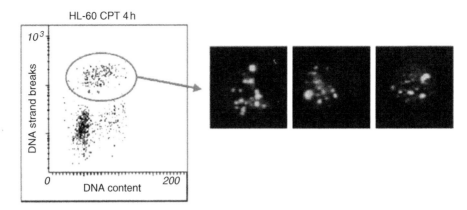

Fig. 4 Morphological identification of apoptotic cells detected based on the presence of DNA strand breaks, after relocation, by LSC. HL-60 cells treated with 0.15 μM CPT for 4 h were subjected to DNA strand break labeling with FITC-conjugated digoxigenin antibody (Gorczyca *et al.*, 1992) and counterstained with PI. Note nuclear fragmentation and yellow fluorescence (due to colocalization and spectral overlap of FITC and PI) of the nuclear fragments. (See Plate no. 12 in the Color Plate Section.)

apoptotic bodies, the products of disintegration of apoptotic cells (Bedner *et al.*, 1998, 1999b) (Fig. 2). Flow cytometry, of course, was unable to discriminate between the genuine apoptotic cells and the false positive ones. The LSC thus allows one to quantify the frequency of apoptotic cells at rates approaching those of flow cytometry, and it also allows one to confirm the accuracy of classification based on the gold standard of apoptosis, morphology.

The second attribute that is unique to the LSC and of great value in analysis of apoptosis relates to the possibility of repeated measurements of the same set of cells and integration (merging) of the results of all sequential measurements into a single file (Kamentsky and Kamentsky, 1991; Pozarowski *et al.*, 2006). Multivariate analysis of the integrated data reveals correlations between measured cell features. This capability of the LSC was employed to combine analysis of functional features of live cells with cell attributes that can be probed only after fixation (Li and Darzynkiewicz, 1999; Li *et al.*, 2000). Specifically, the functional changes that occur during apoptosis, dissipation of the mitochondrial transmembrane potential ($\Delta\Psi_m$; Petit *et al.*, 1995; Zamzani *et al.*, 1998), and oxidative stress (increase in ROIs; Hedley and McCulloch, 1996; Sheng-Tanner *et al.*, 1998) were correlated with the attributes measured in fixed cells, the cell cycle position, and the presence of DNA strand breaks. The cells were first measured when alive to assess their $\Delta\Psi_m$ or ROIs, and then the cells were fixed and subjected to analysis of DNA content and/or presence of DNA strand breaks. The results of both measurements were integrated into a single file for multivariate analysis. It was possible, therefore, to reveal the status of DNA fragmentation or cell cycle position of the same cell whose $\Delta\Psi_m$ or ROI level was measured. This approach appears to be of particular utility in mapping sequences of intracellular events that include both functional and structural changes. It is currently used in our laboratory to map the sequences and correlate the following: the decrease in intracellular pH; rise in Ca^{2+}; ROIs; changes in reduced glutathione; exposure of phosphatidylserine on the outer leaflet of the plasma membrane; activation of caspases; dissipation of $\Delta\Psi_m$; translocation of Bax to, and leakage of cytochrome *c* from, mitochondria; nuclear translocation of nuclear factor (NF-κB); chromatin condensation; and DNA fragmentation, events that occur during apoptosis (Li *et al.*, 2000).

Acknowledgments

This work was supported by the National Cancer Institute (Grant ROI CA 28 704), by the Chemotherapy Foundation, and by "This Close" for Cancer Research Foundation.

References

Arends, M. J., Morris, R. G., and Wyllie, A. H. (1990). Apoptosis: The role of endonuclease. *Am. J. Pathol.* **136**, 593–608.

Bedner, E., Burfeind, P., Gorczyca, W., Melamed, M. R., and Darzynkiewicz, Z. (1997). Laser scanning cytometry distinguishes lymphocytes, monocytes and granulocytes by differences in their chromatin structure. *Cytometry* **29**, 191–196.

Bedner, A., Burfeind, P., Hsich, T.-C., Wu, J. M., Augero-Rosenfeld, M. E., Melamed, M. R., Horowitz, H. W., Wormser, G. P., and Darzynkiewicz, Z. (1998). Cell cycle effects and induction of apoptosis caused by infection with human granulocytic Ehrlichiosis pathogen measured by laser scanning cytometry. *Cytometry* **33**, 47–55.

Bedner, E., Halicka, H. D., Cheng, W., Salomon, T., Deptala, A., Gorczyca, W., Mclamed, M. R., and Darzynkiewicz, Z. (1999a). High affinity of binding of fluorescein isothiocynate to eosinophils detected by laser scanning cytometry: A potential source of error in analysis of blood samples utilizing fluorescein-conjugated reagents in flow cytometry. *Cytometry* **36**, 77–82.

Bedner, E., Li, X., Gorczyca, W., Melamed, M. R., and Darzynkiewicz, Z. (1999b). Analysis of apoptosis by laser scanning cytometry. *Cytometry* **35**, 181–195.

Cohen, G. M., Su, X.-M., Snowden, R. T., Dinsdale, D., and Skilleter, D. N. (1992). Key morphological features of apoptosis may occur in the absence of internucleosomal DNA fragmentation. *Biochem. J.* **286**, 331–334.

Collins, R. J., Harmon, B. V., Gobe, G. C., and Kerr, J. F. R. (1992). Internucleosomal DNA cleavage should not be the sole criterion for identifying apoptosis. *Int. J. Radiat. Biol.* **61**, 451–453.

Cossarizza, A., Kalashnikova, G., Grassilli, E., Chiappelli, F., Salvioli, S., Capri, M., Barbieri, D., Troiano, L., Monti, D., and Franceschi, C. (1994). Mitochondrial modifications during rat thymocyte apoptosis: A study at a single cell level. *Exp. Cell Res.* **214**, 323–330.

Darzynkiewicz, Z. (1994). Simultaneous analysis of cellular RNA and DNA content. *In* "Methods in Cell Biology" (Z. Darzynkiewicz, J. P. Robinson, and H. A. Crissman, eds.), Vol. 41, pp. 401–420. Academic Press, San Diego.

Darzynkiewicz, Z., and Li, X. (1996). Measurements of cell death by flow cytometry. *In* "Techniques in Apoptosis. A User's Guide" (T. G. Cotter and S. J. Martin, eds.) pp. 71–106. Portland Press, London.

Darzynkiewicz, Z., and Pozarowski, P. (2007). All that glitters is not gold: All that binds FLICA is not caspase. A caution in data interpretation—and new opportunities. *Cytometry A* **71A**, 536–537.

Darzynkiewicz, Z., Traganos, F., and Kimmel, M. (1986). Assay of cell cycle kinetics by multivariate flow cytometry. *In* "Techniques of Cell Cycle Analysis" (J. W. Gray and Z. Darzynkiewicz, eds.), pp. 291–332. Humana Press, Clifton, NJ.

Darzynkiewicz, Z., Bruno, S., Del Bino, G., Gorczyca, W., Hotz, M. A., Lassota, P., and Traganos, F. (1992). Features of apoptotic cells measured by flow cytometry. *Cytometry* **13**, 795–808.

Darzynkiewicz, Z., Li, X., and Gong, J. (1994). Assays of cell viability. Discrimination of cells dying by apoptosis. *In* "Methods in Cell Biology" (Z. Darzynkiewicz, J. P. Robinson, and H. A. Crissman, eds.), Vol. 41, pp. 16–39. Academic Press, San Diego.

Darzynkiewicz, Z., Juan, G., Li, X., Murakami, T., and Traganos, F. (1997a). Cytometry in cell necrobiology: Analysis of apoptosis and accidental cell death (necrosis). *Cytometry* 27, 1–20.

Darzynkiewicz, Z., Li, X., Gong, J., and Traganos, F. (1997b). Methods for analysis of apoptosis by flow cytometry. *In* "Manual of Clinical Laboratory Immunology" (N. R. Rose, E. C. de Macario, J. D. Folds, H. C. Lane, and R. Nakamura, eds.), pp. 334–344. ASM Press, Washington, DC.

Darzynkiewicz, Z., Bedner, E., Traganos, F., and Murakami, T. (1998). Critical aspects in the analysis of apoptosis. *Hum. Cell* 11, 3–12.

Darzynkiewicz, Z., Bedner, E., Li, X., Gorczyca, W., and Melamed, M. R (1999). Laser scanning cytometry. A new instrumentation with many applications. *Exp. Cell Res.* 249, 1–12.

Del Bino, G., Darzynkiewicz, Z., Degraef, C., Mosselmans, R., Fokan, D., and Galand, P. (1999). Comparison of methods based on annexin V binding, DNA content or TUNEL for evaluating cell death in HL-60 and adherent MCF-7 cells. *Cell Prolif.* 32, 25–37.

Dragovich, T., Rudin, C. M., and Thompson, C. B. (1998). Signal transduction pathways that regulate cell survival and cell death. *Oncogene* 17, 3207–3213.

Endersen, P. C., Prytz, P. S., and Aarbakke, J. (1995). A new flow cytometric method for discrimination of apoptotic cells and detection of their cell cycle specificity through staining of F-actin and DNA. *Cytometry* 20, 162–171.

Fadok, V. A., Voelker, D. R., Cammpbell, P. A., Cohen, J. J., Bratton, D. L., and Henson, P. M. (1992). Exposure of phosphatidylserine on the surface of apoptotic lymphocytes triggers specific recognition and removal by macrophages. *J. Immunol.* 148, 2207–2216.

Gong, J., Traganos, F., and Darzynkiewicz, Z. (1994). A selective procedure for DNA extraction from apoptotic cells applicable for gel electrophoresis and flow cytometry. *Anal. Biochem.* 218, 314–319.

Gorczyca, W., Bruno, S., Darzynkiewicz, R., Gong, J., and Darzynkiewicz, Z. (1992). DNA strand breaks occurring during apoptosis: Their early *in situ* detection by the terminal deoxynucleotidyl transferase and nick translation assays and prevention by serine protease inhibitors. *Int. J. Oncol.* 1, 639–648.

Gorczyca, W., Bigman, K., Mittelman, A., Ahmed, T., Gong, J., Melamed, M. R., and Darzynkiewicz, Z. (1993a). Induction of DNA strand breaks associated with apoptosis during treatment of leukemias. *Leukemia* 7, 659–670.

Gorczyca, W., Gong, J., and Darzynkiewicz, Z. (1993b). Detection of DNA strand breaks in individual apoptotic cells by the *in situ* terminal deoxynucleotidyl transferase and nick translation assays. *Cancer Res.* 52, 1945–1951.

Hara, S., Halicka, H. D., Bruno, S., Gong, J., Traganos, F., and Darzynkiewicz, Z. (1996). Effect of protease inhibitors on early events of apoptosis. *Exp.Cell Res.* 232, 372–384.

Hedley, D. W., and McCulloch, E. A. (1996). Generation of oxygen intermediates after treatment of blasts of acute myeloblastic leukemia with cytosine arabinoside: Role of bcl-2. *Leukemia* 10, 1143–1149.

Hotz, M. A., Traganos, F., and Darzynkiewicz, Z. (1992). Changes in nuclear chromatin related to apoptosis or necrosis induced by the DNA topoisomcrase II inhibitor fostriecin in MOLT-4 and HL-60 cells are revealed by altered DNA sensitivity to denaturation. *Exp. Cell Res.* 201, 184–191.

Kamentsky, L. A., and Kamentsky, L. D. (1991). Microscope-based multiparameter laser scanning cytometer yielding data comparable to flow cytometry. *Cytometry* 12, 381–387.

Kaufmann, S. H., Lee, S. H., Meng, X. W., Loegering, D. A., Kottke, T. J., Henzing, A. J., Ruchaud, S., Samejima, K., and Earnshaw, W. C. (2008). Apoptosis-associated caspase activation assays. *Methods* 44, 262–272.

Kerr, J. F. R., Wyllie, A. H., and Curie, A. R. (1972). Apoptosis: A basic biological phenomenon with wide-ranging implications in tissue kinetics. *Br. J. Cancer* 26, 239–257.

Kerr, J. F. R., Winterford, C. M., and Harmon, V. (1994). Apoptosis, its significance in cancer and cancer therapy. *Cancer* 73, 2013–2026.

Koester, S. K., Schlossman, S. F., Zhang C., Decker, S. J., and Bolton, W. E. (1998). APO2.7 defines a shared apoptotic-necrotic pathway in a breast tumor hypoxia model. *Cytometry* 33, 324–332.

Koopman, G., Reutelingsperger, C. P. M., Kuijten, G. A. M., Keehnen, R. M. J., Pals, S. T., and van Oers, M. H. J. (1994). Annexin V for flow cytometric detection of phosphatidylserine expression of B cells undergoing apoptosis. *Blood* 84, 1415–1420.

Kroemer, G. (1998). The mitochondrion as an integrator/coordinator of cell death pathways. *Cell Death Differ.* 5, 547–548.

Kroemer, G., Galluzzi, L., Vandenabeele, P., Abrams, J., Alnemri, E. S., Baehrecke, E. H., Blagosklonny, M. V., El-Deiry, W. S., Golstein, P., Green, D. G., Hengartner, M., Knight, R. A., *et al.* (2009). Classification of cell death: Recommendations of the nomenclature committee on cell death 2009. *Cell Death Differ.* 16, 3–11.

Li, X., and Darzynkiewicz, Z. (1995). Labeling DNA strand breaks with BdrUTP. Detection of apoptosis and cell proliferation. *Cell Prolif.* 28, 571–579.

Li, X., and Darzynkiewicz, Z. (1999). The Schrödinger's cat quandary in cell biology: Integration of live cell funtional assays with measurements of fixed cells in analysis of apoptosis. *Exp. Cell Res.* 249, 404–412.

Li, X., Melamed, M. R., and Darzynkiewicz, Z. (1996). Detection of apoptosis and DNA replication by differential labeling of DNA strand breaks with fluorochromes of different color. *Exp. Cell Res.* 222, 28–37.

Li, X., Du, L., and Darzynkiewicz, Z. (2000). During apoptosis of HL-60 and U-937 cells caspases are activated independently of dissipation of mitochondrial electrochemical potential. *Exp. Cell Res.* 257, 290–297.

Majno, G., and Joris, I. (1995). Apoptosis, oncosis, and necrosis. An overview of cell death. *Am. J. Pathol.* 146, 3–16.

Meier, P., and Evan, G. (1998). Dying like flies. *Cell* 95, 295–298.

Nicoletti, I., Migliorati, G., Pagliacci, M. C., Grignani, F., and Riccardi, C. (1991). A rapid and simple method for measuring thymocyte apoptosis by propidium iodide staining and flow cytometry. *J. Immunol. Methods* 139, 271–280.

Nuñez, G., Benerdict, M. A., Hu, Y., and Inohara, N. (1998). Caspases: The proteases of the apoptotic pathway. *Oncogene* 17, 3237–3245.

Oberhammer, F., Wilson, J. M., Dive, C., Morris, I. D., Hickman, J. A., Wakeling, A. E., Walker, P. R., and Sikorska, M. (1993). Apoptotic death in epithelial cells: Cleavage of DNA to 300 and/or 50 kb fragments prior to or in the absence of internucleosomal fragmentation. *EMBO J.* 12, 3679–3684.

Ormerod, M. G. (1998). The study of apoptotic cells by flow cytometry. *Leukemia* 12, 1013–1025.

Ormerod, M. G., O'Neill, C. F., Robertson, D., and Harrap, K. R. (1994). Cisplatin induced apoptosis in a human ovarian carcinoma cell line without a concomitant internucleosomal degradation of DNA. *Exp. Cell Res.* 211, 231–237.

Ormerod, M. G., Cheetham, F. P. M., and Sun, X.-M. (1995). Discrimination of apoptotic thymocytes by forward light scatter. *Cytometry* 21, 300–304.

Petit, P. X., LeCoeur, H., Zorn, E., Dauguet, C., Mignotte, B., and Gougeon, M. L. (1995). Alterations of mitochondrial structure and function are early events of dexamethasone-induced thymocyte apoptosis. *J. Cell Biol.* 130, 157–165.

Pozarowski, P., Holden, E., and Darzynkiewicz, Z. (2006) Laser scanning cytometry: Principles and applications. *Meth. Cell Biol.* 319, 165–192.

Reed, J. C. (1998). Bcl-2 proteins. *Oncogene* 17, 3225–3236.

Sheng-Tanner, X., Bump, E. A., and Hedley, D. A. (1998). An oxidative stress-mediated death pathway in irradiated human leukemia cells mapped using multilaser flow cytometry. *Radiat. Res.* 150, 636–647.

Smolewski, P., Bedner, E., Du, L., Hsieh, T.-C., Wu, J. M., Phelps, D. J., and Darzynkiewicz, Z. (2001). Detection of caspases activation by fluorochrome-labeled inhibitors: Multiparameter analysis by laser scanning cytometry. *Cytometry* 44, 73–82.

Umansky, S. R., Korol', B. R., and Nelipovich, P. A. (1981). *In vivo* DNA degradation in the thymocytes of gamma-irradiated or hydrocortisone-treated rats. *Biochim. Biophys. Acta* **655**, 281–290.

Vaux, D. L., and Korsmeyer, S. J. (1999). Cell death in development. *Cell* **96**, 245–254.

Zakeri, Z. F., Quaglino, D., Latham, T., and Lockshin, R. A. (1993). Delayed internucleosomal DNA fragmentation in programmed cell death. *FASEB J.* **7**, 470–478.

Zamai, L., Falcieri, E., Marhefka, G., and Vitale, M. (1996). Supravital exposure to propidium iodide identifies apoptotic cells in the absence of nucleosomal DNA fragmentation. *Cytometry* **23**, 303–311.

Zamzani, N., Brenner, C., Marzo, I., Susin, S. A., and Kroemer, G. (1998). Subcellular and submito-chondrial mode of action of Bcl-2-like oncoproteins. *Oncogene* **16**, 2265–2282.

PART IV

Cell Surface Immunophenotyping

CHAPTER 23

Cell Preparation for the Identification of Leukocytes

Carleton C. Stewart and Sigrid J. Stewart

Laboratory of Flow Cytometry
Roswell Park Cancer Institute
Buffalo, New York 14263

ESSENTIAL CYTOMETRY METHODS
Reprinted from *Methods in Cell Biology*, Volume 63 (Academic Press, 2001).
Copyright © 2001 by Academic Press, Inc.,
DOI: 10.1016/B978-0-12-375045-7.00023-4

I. Introduction

Immunophenotyping is the term applied to the identification of cells using antibodies to antigens expressed by these cells. These antigens are actually functional membrane proteins involved in cell communication, adhesion, or metabolism. Immunophenotyping using flow cytometry has become the method of choice in identifying and sorting cells within complex populations. Applications of this technology have occurred in both basic research and clinical laboratories. The National Committee for Clinical Laboratory Standards (now NCCLS) has prepared guidelines for flow cytometry that describe in detail the current recommendations for processing clinical samples (Borowitz, 1993; Landay, 1992).

Advances in flow cytometry instrumentation design and the availability of new fluorochromes and staining strategies have led to methods for immunophenotyping cells with two or more antibodies simultaneously (Stewart, 1994, 1996; Stewart and Stewart, 1994a,b, 1995, 1997a,b, 1999). With this progress has come the realization that the use of a single antibody is insufficient for identification of a unique cell population. Cells within one population may have proteins in common with cells of another population. By evaluating the unique repertoire of proteins using several antibodies together, each coupled with a different fluorochrome, any given cell population can be identified and their frequency in a specimen determined. The actual function of many proteins to which specific monoclonal antibodies bind are now known so that we not only know the identity of the cell but we can also determine what it is doing. For example, using antibodies to cytokines and their receptors it is now possible to identify those cells making them (Picker *et al.*, 1995; Prussin, 1997; Romagnani, 1997) and those cells responding to them (Debili *et al.*, 1995; Falini *et al.*, 1995; Olsson, *et al.*, 1992; Wognum *et al.*, 1993).

As the number of antibodies used for phenotyping increases so does the complexity caused by the overlapping spectra of the fluorochromes. Controls must also be evaluated along side the experimental samples to insure the data is collected and interpreted correctly. These process controls must be simple and inexpensive to prepare and faithfully used every time the instrument acquires data so that problems can be recognized and addressed when they occur.

Several strategies for staining cells with antibodies labeled with different colored fluorochromes are the focus of this chapter. This chapter has been divided into topics or sections to describe why the recommended procedures are performed as they are described. The behavior of antibodies toward cells is illustrated and described in the first section. Understanding the ways antibodies bind to cells using the law of mass action and the artifacts that can ensue provides a rational basis for the procedures described in the remaining sections. In Section III, we consider the tandem complexes, which are two fluorochromes chemically combined to provide unique excitation/emission spectra. These tandem fluorochromes represent a unique set of problems that must be understood to appropriately solve them. In Section IV, methods for cell preparation and staining are described in detail, and we consider dead cells, their evaluation, and their effect on data interpretation. In Section V, we describe the method for evaluating antibody quality by titering. Finally, in Section VI, the formulation of all solutions is described.

II. Antibodies

A brief review of the characteristics of antibodies germane to their use in immunophenotyping is appropriate. Antibodies bind to three-dimensional molecular structures on antigens called epitopes, and each antigen contains hundreds of different epitopes. A monoclonal antibody is specific for a single epitope, whereas polyclonal antibodies are actually the natural pool of hundreds of monoclonal antibodies produced within the animal, each one binding to its unique epitope. It is most common to use monoclonal antibodies to determine the immunophenotype of cells.

A. Antibody Structure

The basic unit of all antibodies comprises a heavy and light chain, the former defining the antibody isotype. As depicted in Fig. 1, the minimum functional antibody molecule consists of two heavy and two light chains linked together with disulfide bonds. The part of the molecule that contains only heavy chains is known as the Fc [the fragment (F) of the antibody molecule that exhibits a nonvariable or constant (c) amino acid sequence within a given isotype and subclass] portion, whereas the part that contains both heavy and light chains is the $F(ab')_2$ portion. The Fc receptor binding domain and the complement binding and activation domain (Fig. 1) are in the Fc portion. Variations in the structure of the heavy chains lead to "isotypes" and there are "subclasses" within some isotypes. These variations are of practical importance because they define the repertoire of cells to which the Fc portion of the antibody will bind.

In the $F(ab')_2$ portion, the light chains are either κ or λ, and they are associated with the heavy chains by disulfide bonding; the designation ab is "antigen binding," and the numeral 2 refers to the fact that there are two antigen binding sites in the basic functional subunit. These two can be chemically separated to produce an F(ab) using papain instead of pepsin for digestion. It is noteworthy that murine and rat antibodies almost always have κ light chains and that most *in vitro* hybridoma-produced monoclonal antibodies are of the γ [immunoglobulin G (IgG)] isotype. Table I shows some important properties of immunoglobulins.

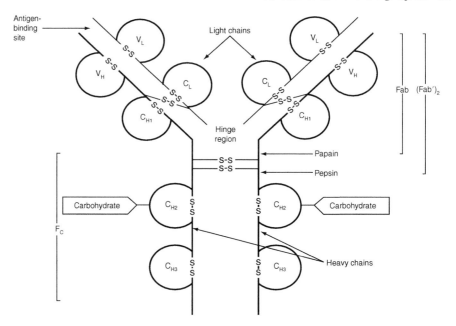

Fig. 1 Schematic representation of an IgG molecule. The Fab portion consists of the light chain and a fragment of the heavy chain. These two chains are held together by disulfide bonds. For any given IgG molecule there is a region of variable amino acid sequence (V_L) on the light chain and the heavy chain (V_H) that produces the epitope-binding site. Papain digestion produces two Fab fragments for each molecule. There are also regions where the amino acid sequence is constant for IgG molecules of similar isotype and subclass. There is one constant region for light chains (C_L) and three for heavy chains (C_{H1}, C_{H2}, and C_{H3}). The heavy chains are held together by disulfide bonds. If pepsin is used to digest the antibody, a fragment [$F(ab')_2$] is produced containing two epitope-binding sites. The Fc fragment is that portion of the Ab posterior to the hinge region. From Stewart and Stewart (1994a).

B. Antibody Binding Kinetics

The single most important factor in immunophenotyping is antibody quality. This is even more important for intracellular than for membrane immunophenotyping. To better understand the basis for evaluating antibody quality, a brief review of the law of mass action may be helpful. We will illustrate the law using monoclonal antibodies (Ab) that bind to a single epitope (E) on an antigen. A more detailed description of the law for immunophenotyping cells is available (Stewart and Mayers, 2000). In Eq. (1), a single monoclonal antibody binding site rapidly binds to a single epitope on the antigen with a forward rate constant k_f. As the concentration of AE increases, there is a tendency for the complex to come apart with a reverse rate constant of k_r. The affinity constant K_a is equal to the ratio of the forward and reverse rate constants. The quality of an antibody is almost solely the result of the reverse reaction

Table I
Human Immunoglobulins[a]

		Immunoglobulin properties			Binding to Fc receptor class on leukocytes					
Isotype	Subclass	Serum mg/ml	Complement fixing	Protein A reaction	NK cells	B cell	Monocyte	Neutrophil	Eosinophil	Basophil
IgG	–	–	–	–	III	II	I, II	II, III	II	II, III
	IgG1	9	Yes	Yes	III	II	I, II	II, III	II	II, III
	IgG2	3	Yes	Yes	–	II	II	II	II	II
	IgG3	1	Yes	No	III	II	I, II	II, III	II	II, III
	IgG4	0.5	No	Yes	–	II	I, II	II	II	II
IgM	–	1.5	Yes	No	–	–	–	–	–	–
IgA	IgA1	3	No	No	–	–	II, III	–	–	–
	IgA2	0.5	No	No	–	–	–	II, III	–	–
IgD	–	0.3	No	No	–	–	–	–	–	–
IgE	–	Nil	No	No	–	–	–	–	II	I

[a]Modified and expanded from Roitt *et al.* (1989) and Stewart and Mayers (2000). FcRI is CD64, FcRII is CD32, and FcRIII is CD16.

$$Ab + E \underset{k_r}{\overset{k_f}{\rightleftharpoons}} AbE. \tag{1}$$

At equilibrium, the change in concentration of each component is zero so that

$$K_a = \frac{k_f}{k_r} = \frac{[AbE]}{[Ab][E]}. \tag{2}$$

Antibodies are immunoglobulins (Ig), and they are naturally sticky molecules. This can be easily demonstrated by centrifuging the contents of a new vial of antibody from your favorite supplier at $15,000 \times g$ for 5 min and observing the pellet of IgG aggregates.

Rule: Always centrifuge antibodies at $15,000 \times g$ prior to immunophenotyping to remove aggregates.

Immunoglobulin also binds nonspecifically because of its high electrostatic attraction to proteins. This nonspecific binding to proteins also obeys the laws of mass action, but occurs at a much lower affinity ($K_a = 10^4$ l/M). All IgGs bind nonspecifically in about the same concentration range at or above 30 μg/ml/10^7 cells.

Figure 2 compares the nonspecific binding of IgG ($K_a = 5 \times 10^4$) and the specific binding of an antibody with a K_a of 5×10^8 and with a K_a of 5×10^9 for a membrane epitope and an intracellular epitope. In Fig. 2A, for membrane antigen, the nonspecific binding curve is well separated for both antibodies, so it is easy to resolve epitope positive cells from nonspecific binding. In Fig. 2B, for intracellular

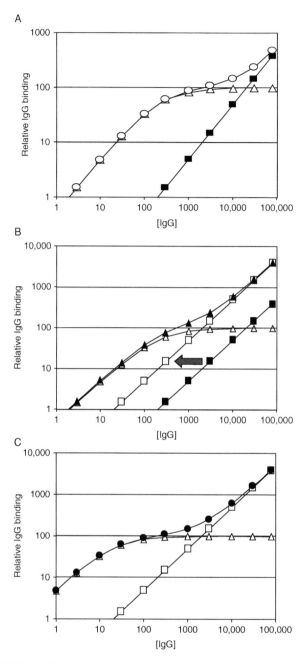

Fig. 2 Antibody binding kinetics. Antibodies are actually immunoglobulins (Ig) that can bind to cells via their antigen (epitope) binding sites or by nonspecific binding. (A) For binding to membrane antigens on viable cells, the affinity constant for the antibody describes the kinetics of specific binding to its epitope. What we measure in the flow cytometer is the sum (○) of specific (△) antibody and

antigens, the nonspecific binding curve has shifted to the left an order of magnitude because of the increased concentration of protein targets inside versus on the membrane of cells, that is, [E], which in this case is protein, has increased in Eq. (2). Because the concentration of protein targets has increased, there is increased nonspecific binding (Ig-protein) at equilibrium. Since nonspecific binding is so high, specific binding of the lower affinity antibody can no longer be resolved as there is no definable plateau region. This demonstrates the importance of the affinity constant in describing antibody quality.

Rule: K_a is the most important descriptor of antibody quality.

Unfortunately, the value of this single most important parameter is almost never available from your favorite supplier of antibodies. We can still determine antibody quality without knowing its K_a. In Fig. 2C, increasing the K_a of the antibody to 5×10^9 results in a lower concentration to saturate the desired epitopes so that they can be resolved from nonspecific binding. Antibody binding can readily be appreciated by plotting the amount of free antibody as a function of the affinity constant as shown in Fig. 3. For a low-affinity antibody ($K_a \approx 10^7$–10^8), 99% of it is free, that is, it is not bound to any epitope at the concentration required for epitope saturation. All of this free antibody is available to bind nonspecifically to extra- or intracellular proteins causing data misinterpretation. If, however, good antibodies are used (10^9–10^{10}), the concentration for epitope saturation is much lower and so is the amount of free antibody.

Figure 4 demonstrates the effect that low-affinity antibodies have on immunophenotyping and its interpretation. In Fig. 4, a preparation with 85% viable cells is stained with two antibodies having a similar low K_a of $\sim 5 \times 10^8$. As shown in Fig. 4A, Ab1 stains a mutually exclusive population from Ab2, but there is a population on the 45° line that appears to coexpress both Ab1 and Ab2. These are the permeable dead cells, which allow entry of low-affinity Ab1 and Ab2 into the high concentration of intracellular protein to which they bind nonspecifically. In Fig. 4B, Ab3 has an order of magnitude higher K_a so the concentration of Ig required for epitope saturation is much lower and the concentration of free antibody to nonspecifically bind at epitope saturation is much lower. However, Ab1 still binds nonspecifically producing an erroneous frequency and a heterogeneous expression as both the live and dead cells bind it. Note that in Fig. 4A the Ab1$^+$ cells were a circular cluster compared to the ellipsoid cluster in Fig. 4B.

Rule: Dead cells should always be determined.

The presence of dead cells should always be considered in data interpretation.

---

nonspecific (■) immunoglobulin binding. (B) If the same antibody is used for intracellular staining, the nonspecific binding increases markedly (□) due to the increased intracellular protein concentration, causing the curve for nonspecific binding (■) to shift to the left (arrow). There is no change in the specific binding curve so it is the same one shown in (A), but the sum (▲) measured by the flow cytometer produces a curve in which specific binding (△) can no longer be resolved. (C) The only way to correct this problem is to use a better, higher affinity antibody (●).

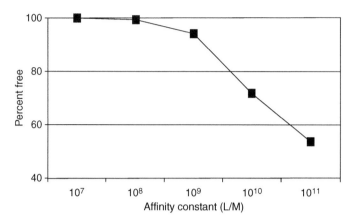

Fig. 3 Effect of affinity constant on antibody binding. This figure shows the relationship between the amount of free antibody at 50% epitope saturation as a function of the affinity constant. If the affinity constant is 10^7, 99.93% of antibody is free, but if it is 10^{11}, virtually all of it binds to the epitope.

C. Influence of Epitope, Fluorochrome, and Fixation

Besides antibody quality, there are two other factors to consider when choosing antibodies for multiparameter immunophenotyping, the epitope on the antigen and the fluorochrome conjugated to the antibody. Figure 5 shows an acute B-lineage lymphocytic leukemia stained with CD19 monoclonal antibody from two different clones. Because this particular leukemia exhibits aberrant CD19 expression, it is totally absent when stained with CD19, clone SJ25C1, but was found to be heterogeneously expressed when the CD19, derived from clone J4.119, was used. This is not a fluorochrome intensity effect because on normal B cells the staining intensities for both CD19 clones are virtually identical.

Rule: Antibodies to lineage antigens from at least two clones should be used for immunophenotyping malignant cells.

The fluorochrome can also make a profound difference on epitope detection. As shown in Fig. 6, CD69, when conjugated with phycoerythrin (PE), clearly resolves both activated T cells and the higher epitope expressing micromegakaryocytes. When the same antibody is used, conjugated with APC, a 10-fold reduction in fluorescence intensity occurs; thus, only the brighter micromegakaryocytes are resolved, and the T cells are completely negative.

Rule: Always select antibodies conjugated with the brightest fluorochromes for immunophenotyping cells that express antigens at low density.

Choose the fluorochromes in order of brightness: PE, phycoerythrin-CY5 tandem complex (PECY5), phycoerythrin-Texas Red tandem complex (PETR), allophycocyanin (APC), fluorescein isothiocyanate (FITC), and peridinin chlorophyll protein (PerCP).

Fixation can also dramatically affect epitope detection, especially when intracellular staining is performed. Epitopes on fixed and permeabilized cells may no

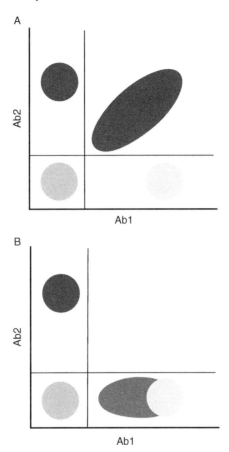

Fig. 4 Effect of dead cells and antibody affinity on immunophenotyping. In (A), Ab1 and Ab2 have a K_a of 5×10^7 l/M, and 15% of cells are dead. The dead cells are permeable to both antibodies and bind nonspecifically to the higher concentration of intracellular protein. This produces a cluster of double positive events on a $45°$ angle whose frequency is equal to the dead cells in the suspension. If there were actually viable cells that are positive for both antibodies, the dead cells would obscure their frequency and the data misinterpreted. In (B), Ab3 has an affinity of 10^9, so it has no appreciable nonspecific binding, but Ab1 still binds nonspecifically as well as specifically, so the cluster of positive cells now exhibits the same heterogeneous expression as dead cells, but the viable CD3$^+$ cells cannot be resolved. This would lead to an erroneous evaluation of Ab1$^+$ cells and data misinterpretation.

longer bind to their specific antibody because the epitope has been changed. As shown in Fig. 7, CD19$^+$ and CD3$^+$ cells are clearly resolved when viable cells are stained prior to fixation with formaldehyde. If, however, the cells are fixed first and then stained, CD3 fluorescence is not different, but CD19 fluorescence is reduced.

Rule: Always test the effect of fixation on epitope expression.

The epitope for the CD19 used is sufficiently altered by formaldehyde fixation that it is no longer recognized by this clone (SJ25C1) of CD19. This might be

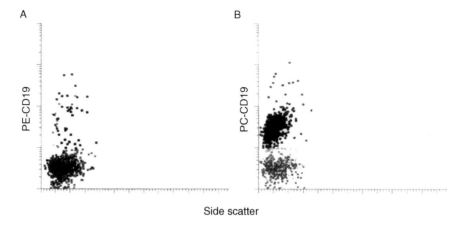

Fig. 5 Differing monoclonal antibody epitope binding. An acute B-lineage lymphocytic leukemia was stained with two different clones of CD19. In (A), the CD19 antibody produced by clone SJ25C1 was used. None of the leukemic cells are positive. Leukemic cells are positive when the antibody produced by clone J4.119 was used (B).

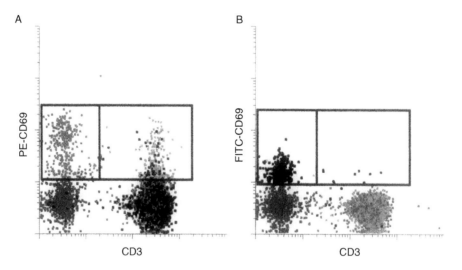

Fig. 6 Importance of fluorochrome intensity. Blood was stained with either PE-CD69 (A) or FITC-CD69 (B) and APC-CD3. In (A), CD3 negative, CD69 positive population of micromegakaryocytes exhibits a mean channel fluorescence (MCF) of 77. The CD3 positive T cell population exhibits a MCF of 44 at a resolution of 128. When APC-CD69 (B) was used, the MCF for micromegakaryocytes is only 35, and no CD69 positive T cells are found.

corrected by using a different clone of CD19, by using a different fixation procedure, or both. The correct interpretation of data may depend on knowledge of the behavior of the antibodies used for immunophenotyping in different settings.

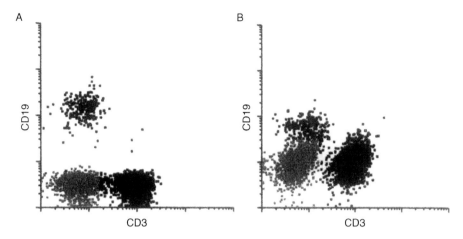

Fig. 7 Effect of fixation on antibody binding. Human blood cells were prepared by lysing the erythrocytes with lysing solution. In (A), they were first stained with the antibody combination FITC-CD3 and PE-CD19 and then fixed in 2% Ultrapure formaldehyde. In (B), cells were first fixed in 2% Ultrapure formaldehyde and then stained with the same antibody combination. CD19, but not CD3, fluorescence is significantly reduced by prefixation.

D. The Effect of Drugs on Expression

Drugs can also have a profound influence on protein expression. It is customary to use the drug Brefeldin A when measuring intracellular cytokines. CD69 is an activation protein found on T cells (Testi *et al.*, 1989). Since CD69 expression is induced by stimulation, its membrane expression, shown in Fig. 8, may be completely blocked if the activation does not occur in the absence of the drug. This is caused by the inability to transport the CD69 protein to the surface membrane if it is inhibited by Brefeldin A.

Rule: Always test the effect of a drug on antigen expression.

A similar phenomenon can occur for concomitantly expressed membrane proteins that have a high turnover rate.

E. Primary and Secondary Antibody

For immunophenotyping using a single color, it is customary to first stain cells with a primary antibody, usually a monoclonal antibody, directed to the epitope of interest. A secondary polyclonal antibody that is specific to epitopes on the primary antibody (the primary antibody, therefore, acting as an antigen in this reaction) and that is covalently coupled with a fluorochrome is then used to bind to the primary antibody, effectively coloring the cell to which the primary antibody has bound.

This method has the advantage that a single second antibody can be used for staining many different primary antibodies. Unconjugated antibodies are also less expensive than conjugated primary antibodies, and the latter may be unavailable.

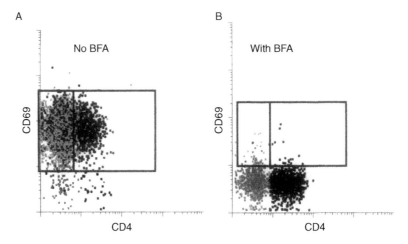

Fig. 8 Effect of the drug Brefeldin A on membrane CD69 expression. CD69 is an activation antigen on T cells that is upregulated by specific antigens and mitogens. Human blood cells were prepared by culturing them 4 h with phorbol myristatic acid (25 ng/ml) and ionomycin (1 μg/ml) at 37 °C in the absence (A) or presence (B) of Brefeldin A. The cells were put on ice and stained with the antibody combination FITC-CD4, PE-CD69, and PerCP-CD3. After gating each bivariate histogram on CD3, the CD69 expression on the CD3$^+$ T cells is shown. In addition to the complete absence of CD69 expression in the Brefeldin A treated cells (B), the CD4 expression is considerably less in the non-Brefeldin A (A) compared to the Brefeldin A (B) treated cells. Thus, drugs can affect protein expression.

Using a second antibody also provides an amplification of fluorescence that may be useful for resolving cells that have a low frequency of epitopes. This advantage, however, can be negated by increased background.

The F(ab′)$_2$ fragment of an affinity purified second antibody should always be used. It is of course implicit in this scheme that the second antibody does not react with the cellular epitope. Second antibodies are available with every fluorochrome conjugated to them.

Rule: Always use F(ab) or F(ab′)$_2$ fragments for second antibodies.

A typical second antibody might be labeled "FITC conjugated goat F(ab′)$_2$ antimouse IgG antibody (heavy and light chain specific, purified by affinity chromatography)." This label contains considerable information in abbreviated form. "Goat antimouse IgG" means the antibody was prepared by immunizing goats with mouse IgG. "Purified by affinity chromatography" means the goat serum was passed over an affinity column (usually Sepharose beads) to which mouse IgG was bound and, after elution from the column, the F(ab′)$_2$ fragments were prepared and conjugated with fluorescein isothiocyanate ("FITC conjugated").

Since this goat polyclonal second antibody was prepared using a mouse IgG affinity column, this preparation contains antibodies that are specific for the heavy chain of mouse IgG. Because all isotypes found in serum, that is, IgG, IgM, and IgA, have light chains (Table I), they will also bind this second antibody because it is also "light chain specific." Therefore, this polyclonal second reagent is not specific for mouse IgG at all. In order for a second antibody to be specific for IgG it must have no light chain activity. If the antibody was only heavy chain specific the label would read "γ specific" or "γ2b specific," etc. A simple way to determine if a second antibody to murine IgG has light chain activity is to stain murine spleen cells with it; no cells should be positive. If there are positive cells, the reagent is defective and should not be used in applications where heavy chain specificity is required.

Any polyclonal second antibody contains all the isotypes found in the serum of the animal used to produce it (e.g., IgM, IgG, IgA). Because IgM is a pentomer and IgAs are dimers, it is very important that they all be reduced to the basic structure shown in Fig. 1. Furthermore, the FcR on polyclonal antibodies will bind to FcR on cells. These properties will increase the background binding of second antibodies, which can be effectively eliminated by using Fab fragments.

F. Blocking

Most immunophenotyping experience has been derived using hematopoietic cells in general and lymphocytes in particular. We have already described the nonspecific antibody binding, which is least for lymphoid cells and most for myeloid cells. Fc binding to Fc receptors can also lead to data misinterpretation, because each leukocyte lineage expresses a unique repertoire (Table I). This binding is not nonspecific, as so often referred to in publications, it is a saturable ligand-receptor interaction that is highly specific. Blocking the receptors with IgG prior to staining with a primary antibody can reduce this potential problem.

In the hypothetical mixture of cells shown in Fig. 9, some may have both the desired epitope and Fc receptors, FcR (cell A), some cells may have only FcR (cell B), other cells may have only epitopes (cell C), and some cells may have neither (cell D). Only the cells with the desired epitope (A and C) should be identified by the mouse monoclonal antibody (MAb) and the second antibody conjugated with a fluorochrome [fluoresceinated goat antimouse Ig (FGAM)]. When the MAb is added to the cells it can bind to the desired epitope via the $F(ab')_2$ portion (A and C), or to the cell via the Fc portion (A and B). If the sample contains an equal portion of each cell type, three-quarters of the cells present will bind the MAb. Since the second antibody is a fluorochrome-conjugated $F(ab')_2$ goat antimouse IgG, it binds to the murine MAb (but not to any FcR). Since one-quarter of the cells have only FcR and no epitopes, these cells are inappropriately labeled and counted as epitope positive, thereby overestimating the percentage of epitope positive cells.

To account for this problem, the common practice is to use an "isotype" control. This control consists of the same cells, incubated with a myeloma protein having no epitope specificity, but having the same isotype and subclass as the specific antibody. As shown in Fig. 9A, this myeloma protein is presumed to bind to cells having FcR (cells A and B). The second antibody is added, as before, to stain cells that have bound the isotype protein. The percentage of positive cells revealed by

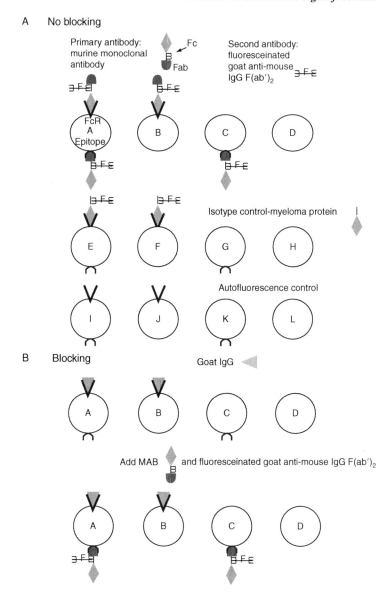

Fig. 9 Indirect immunofluorescence staining. (A) The primary antibody is shown as a stick symbol that binds to cells with epitopes shown by the half circle. The Fc portion binds to its receptor, shown by the V. The second antibody binds to the primary antibody at the m (A–D). The stick symbol for the isotype control (a myeloma protein) has the same shape for the FcR but no antibody binding site because there are no epitopes on the cells (E–H). And cells that have not been stained serve as an autofluorescence control (I–L). Epitope positive and FcR positive cells are resolved. (B) To block Fc binding, cells are incubated for 10 min with goat Ig at a concentration that will saturate FcR (200 μg/ml/ 10^7 cells), shown as a triangle. The murine primary MAb is then added followed after washing by the fluorochrome-conjugated second antibody. The cells are washed, fixed, and analyzed. Only epitope positive cells are resolved.

the isotype protein is then subtracted from the percentage obtained using the specific antibody. This procedure leads to an underestimate of the epitope positive cells because one-quarter of the cells express both FcR and epitopes, and this group is subtracted from the total. Thus, isotype controls may actually lead to an erroneous conclusion.

Rule: Never subtract positive cells in a control sample from positive cells in an experimental sample.

The autofluorescence control should always be analyzed. It contains no antibodies and is otherwise processed the same way as the other samples. This control provides a baseline to determine the minimum fluorescence above which positive cells are identified. The isotype control and autofluorescence control should give identical results if a properly titered high-quality antibody is used at exactly the same IgG concentration as the isotype control. To the extent that they differ is indicative of an improperly titered antibody, a poor-quality antibody, or both.

Treating the cells with normal immunoglobulin (blocking) prior to staining them with specific antibodies can reduce the Fc binding of antibodies to cells. As shown in Fig. 9, a murine monoclonal antibody was used. Because the second antibody is derived from the goat, the blocking immunoglobulin is also derived from the goat. This is to prevent binding of the second antibody to the block. The cells are incubated for 10 min with an excess of goat IgG where the IgG binds the FcR (cells A and B). Without washing the cells, the primary murine MAb is added wherein it binds only to its epitopes (cells A and C). Because the labeled second antibody was made against murine IgG in the goat, it binds to the desired MAb but not the blocking goat IgG. To be sure that the block is effective, the "isotype control" antibody can be used in place of the primary antibody. Even if it did bind to unblocked cells, it should not bind to the blocked ones. Thus, the cells treated with an isotype control, which must be adjusted in concentration to equal that for each antibody used or adjusted to the concentration to equal the concentration of the highest antibody used, should look identical to the untreated cells that exhibit only autofluorescence.

Rule: For indirect staining, block with the immunoglobulin fraction from the same species from which the second antibody was made.

In Table I, the approximate amounts of Ig isotypes and IgG subclasses in the serum of most mammals is shown. The FcR blocking reaction is concentration dependent; there is not a high enough concentration of IgG in serum to effectively block all FcR; therefore, only purified IgG at a high enough concentration should be used for blocking.

As shown earlier in Section II.B, nonspecific binding cannot be blocked. We add 10 μl of normal IgG at 1 mg/ml to 50 μl of cells at 20×10^6 ml^{-1}. This concentration provides enough IgG of each subclass to effectively block all FcR binding. The effect of blocking is shown in Fig. 10. A population of both small cells and large cells are positive in the histograms on the top of each panel. The antibody staining for the cells is shown in the histogram along the side of each panel. By projecting

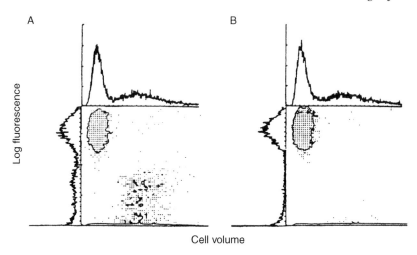

Fig. 10 Effect of blocking on MAb binding to mononuclear cells (MNC). Human cells were isolated from blood using neutrophil isolation medium (NIM, Cardinal Associates, Santa Fe, NM). Cells were adjusted to 20×10^6 ml^{-1} in PBS and 50 μl were used for labeling. (A) Cells were incubated with fluoresceinated CD3 for 15 min. (B) Cells were first incubated for 10 min with 10 μg mouse IgG, and then fluoresceinated CD3 was added for 15 min. After washing, the samples were analyzed. FcR binding to monocytes of CD3 is eliminated by blocking.

each histogram event into the bivariate display, the correlation of cell size with antibody staining can be visualized. In this example, directly fluoresceinated anti CD3 (FL-CD3) was used to label mononuclear cells (MNC). When cells are incubated with FL-CD3, many large cells are dimly stained (Fig. 10A). When the cells are incubated for 10 min with mouse Ig (mIg) prior to the addition of the FL-CD3, no large cells are dimly stained (Fig. 10B). Thus, blocking the cells with mIg prior to staining them had a profound effect on the type of cells that were stained by the antibody. Mouse, rather than goat IgG was used for blocking because the specific antibody was directly conjugated.

G. Directly Conjugated Antibodies

Primary antibodies are currently available conjugated with fluorescein (F), PE, PETR, or PECY5, APC, and biotin (B). Because there is no second antibody to mouse Ig, it can be used for blocking. Using a directly conjugated specific antibody, the cells can be preincubated with murine IgG as the block prior to staining with a labeled murine MAb. Rat IgG would be used if the labeled specific MAb were of rat origin. When using a biotinylated antibody, labeled avidin is used as the second reagent. Avidins can be purchased conjugated with FITC, PE, PETR, PECY5, PerCP, or APC.

H. Wash Versus No Wash

There is currently an ongoing discussion as to whether one should wash specimens or run them in lysing reagent without washing. Simply adding 50 μl of blood or bone marrow to a cocktail of antibodies, incubating 5 min, adding a lysing reagent for 5 min, and fixing the sample would mean that the specimen is completely ready for analysis in 10 min. For intracellular markers the incubation time is increased to 45 min to allow for permeabilization and diffusion of the antibody. Because this homogeneous assay is so simple, it is most attractive to suppliers and to users. The advantage is reduced processing time and a perception that cells are not lost. So why wash?

There are several reasons to consider washing the specimen. First, when measuring light chains on B cells, the specimen must be washed to remove immunoglobulins that will otherwise bind the antibodies to them instead of the cells. While healthy individuals may not have soluble free shed membrane antigens, patients with disease often have them and they will bind the antibodies in suspension thereby reducing their concentration for binding to cells. This can and does significantly bias the cellular immunophenotyping result often unbeknownst to the laboratorian. Finally, when multiple antibodies with their fluorochromes are used, the soluble fluorochrome spectra will be present in the cell stream and detected by all photomultiplier tubes (PMTs). This will cause a baseline offset to the steady-state light conditions of the cell stream, reducing proportionately the signal (positive cells)-to-noise ratio directly affecting instrument sensitivity. Populations of cells exhibiting dim fluorescence in a washed specimen may completely disappear in an unwashed one. How often does this occur? Every time you do not wash a specimen, but you will never know if you do not wash.

Using the CAP survey results for 1998, the most commonly used lysing reagents were BDLyse (41%), Q-Prep (38%), ammonium chloride (7.6%), OrthoLyse (4.3%), and all others (9.1%). Our studies as well as several others show that ammonium chloride lysis is not different than the commercially prepared lysing reagents (Bossuyt *et al.*, 1997; Carter *et al.*, 1992; Tamul *et al.*, 1994). The advantage of using an ammonium chloride solution is that it can be prepared from basic chemicals, which are an order of magnitude less expensive than the commercially obtained preparations.

III. Tandem Fluorochromes

Several tandem fluorochromes are available to provide for a third and fourth color excitable by a single or dual laser flow cytometer. The properties of these reagents have been previously described (Stewart and Stewart, 1993) and are summarized in Table II. Not included in this discussion is the new tandem fluorochrome, APC-CY7 (Beavis, 1996), which may be useful for combining with those considered here.

Table II

Comparison of Reagents for Three and Four Color Phenotyping

Parameter	PerCP[a]	PETR[a]	PECY5[a]	APC[b]
Relative intensity	Dim	Medium	Bright	Medium
Compensation: amount	None	High	Low	High
Batch variability	None	Some	Significant	None
Light sensitivity	Stable	Stable	Unstable	Stable
Nonspecific binding				
Lymphocytes	None	None	None	None
Monocytes	None	None	High	None
Granulocytes	None	None	Low	None
Availability of direct conjugates	Fair	Poor	Good	Good
Emission (nm)	673	613	670	670

[a]Excitation at 488 nm.
[b]Excitation at 635 nm.

The PECY5 fluorochrome is the brightest tandem currently available because the CY5 absorption bandwidth is optimal for PE emission. This is reflected in the low amount of (FL2 – %FL3) compensation required (usually <3% and for some constructs, <0.5%). However, there is a great deal of variability among batches and suppliers. PerCP requires no compensation, but its emission energy is less than that found for fluorescein. It should be used for antibodies to highly expressed cellular epitopes. PerCP also rapidly degrades at high laser power density.

The PETR tandem is intermediate in its fluorescence emission intensity because the TR absorption bandwidth is not as ideally matched to PE emission as CY5. This energy mismatch results in a high degree of PE emission requiring considerable compensation (20-30%) similar to the overlap found for fluorescein in the PE channel. As shown in Fig. 11, PECY5 tandems can exhibit unacceptable variation in their compensation requirements between different suppliers of directly conjugated antibodies (Fig. 11B and D) or on different batches of the same antibody (Fig. 11C and D). A high amount of compensation is required when these reagents are combined with APC, and the amount depends on the construct and the antibody used (Stewart and Stewart, 1999).

Although otherwise an ideal fluorochrome because it is so bright, the PECY5 tandem exhibits two other problems not found for PETR, PerCP, or APC reagents. This tandem is exquisitely sensitive to light, and the CY5 molecule will become irreversibly degraded on short-term exposure (<1 h) to ambient light. This problem can be easily recognized as an increase in PE signal like that shown in Fig. 12, requiring increasingly more compensation and a concomitant decrease in CY5 signal with light exposure. To prevent this problem, all staining should be performed in subdued light and samples stored in the dark.

Rule: Keep PECY5 reagents in the dark.

The third problem is caused by the ability of the PECY5 fluorochrome to specifically bind to monocytic cells (Fig. 13). This problem is exacerbated by the tendency

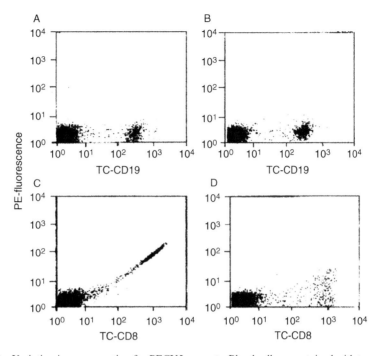

Fig. 11 Variation in compensation for PECY5 reagents. Blood cells were stained with two separate batches of PECY5-CD19 or PECY5-CD8 and compensation for FL2 – %FL3 is shown using 488 nm excitation. In (A), the PECY5-CD19 stained cells were properly compensated. When stained with a second batch (B), the same compensation is acceptable. In contrast, when the settings for PECY5-CD19 are used for PECY5-CD8, they are unacceptable (C) and new compensation settings must be found (D). If these new settings were used for PECY5-CD19, this reagent would be markedly overcompensated.

for some suppliers to overconjugate antibodies with this tandem. This causes an increase in both nonspecific binding to all cells and specific binding to monocytic lineage cells. A PECY5 isotype control is essential, as blocking is ineffective in eliminating this fluorescence. This control is useless if it is not conjugated in a similar manner and used at exactly the same concentration as the antibody. Although it has been suggested that CD64 is the receptor for the CY5 binding on monocytes (van Vugt *et al.*, 1996), our data do not support this claim because CD64 negative monocytic leukemias and some B myeloid lineage leukemias that are CD64 negative also bind PECY5 conjugated antibodies very strongly.

IV. Cell Preparation and Staining Procedures

The preparation of cells depends on their source. Cells contaminated with very high frequencies of erythrocytes must be lysed. Cells in tissues must be disaggregated, while cells adherent to the surface of culture containers must be removed.

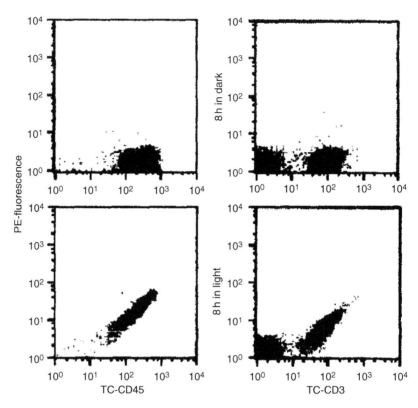

Fig. 12 Effect of light exposure on PECY5 tandem fluorescence. Blood was stained with either PECY5-CD45 or PECY5-CD3 in two sets of tubes, and one was stored 8 h in the dark while the other was left exposed to light. Light causes the degradation of the PECY5 linkage so that efficient energy transfer is reduced, resulting in increased PE photon emission.

Cells must be uniformly monodispersed in suspension. Several chapters in this series provide procedures for dispersing cells to meet this requirement. Each of these conditions has their own set of problems, and there is no consensus on the best approach to take. The procedures described here represent 20 years of experience in immunophenotyping cells of all types. The problems associated with any procedure are cell selection, cell death, and loss of epitopes. Thus, whatever procedure is selected, all three should be determined so that the quality of the final cell suspension for immunophenotyping is known.

We recommend a whole blood (or bone marrow) lysing method to remove erythrocytes, and there are several reagents commercially available for this purpose. Although all of them work with blood, some are better than others for lysing bone marrow and fetal erythroid cells. The cost may also be a factor. Lysing erythrocytes is not recommended before staining because the platelets and

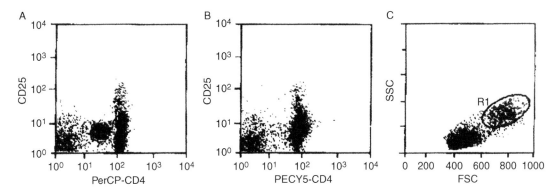

Fig. 13 PECY5 binding to monocytes. In (A), blood is stained with PerCP-CD4 and PE-CD25. Monocytes, light gray, resolved by R1 in (C) and exhibiting dim CD4 fluorescence are clearly resolved from the brighter staining T cells. In (B), blood is stained with PECY5-CD4. Because the PECY5 fluorochrome and CD4 both bind to monocytes they are much brighter and no longer can be resolved from the T cells. In (C), the forward scatter (FSC) versus side scatter (SSC) is shown.

erythrocyte debris are concentrated along with the leukocytes during the subsequent washing steps, causing difficulty in the analysis of the data. Therefore, postlabeling lysis is favored.

For dispersing tissue, we recommend multiple needle sticks with a 21-gauge needle attached to a 1-ml syringe. The biopsy material is then aspirated and expelled five times into 300 μl of PBS. Several sticks throughout the tissue results in an average yield of about 5000 cells per stick (Stomper *et al.*, 1997). Our yield is in excess of 10^8 cells/g and represents the highest we have ever experienced by any method. Although it may be more tedious to perform, one must weigh the scientific benefit of high yield and less bias using this method with ease of preparation. Mechanical tissue dispersal methods produce high amounts of debris (generated by the mechanically killed cells), and enzymes may destroy the desired epitopes for immunophenotyping and kill cells.

When using cultured cells, the usual procedure for removing them is to use a surface membrane protein protease such as trypsin. This kind of treatment can also destroy the desired epitopes for phenotyping. Mechanical dispersal, such as scraping with a rubber policeman can cause severe aggregation and cell death as they are torn from the plate. We recommend treating cultures with 10 mM cold EDTA for 10 min followed by vigorous pipetting with the tip of a Pasteur pipette close to the surface of the dish. The cells are usually less adherent and quickly disperse into the rapidly flowing medium.

In the following procedures a "wash" means to add 3.5-ml phosphate buffered saline to the tube, centrifuge the cells at 1500 × *g* for 3 min at 4 °C, and decant the supernatant by inverting the tubes and blotting the lip by touching its rim to an absorbent towel. Thoroughly resuspend the cells in the residual buffer that has drained back to the bottom of the tube (not to exceed 100 μl). Note the centrifugation speed is three to five times that found in most publications, that is why we do not lose cells by washing. All labeling procedures are carried out on ice. Room temperature may be fine for lymphocytes but can be problematic for other hematopoietic cells due to internalization of the antibody.

Each antibody should be titered prior to use so that the correct amount to optimally stain the cells is used (see Section V). Our general practice is to have the optimal amount of antibody in 5 or 10 μl, but no more. Commercially available antibodies that have already been titered may be diluted differently by the supplier so that more or less than 10 μl is required. In these cases, use the suppliers recommended amount (unless it is found to be incorrect by titering according to Section III).

For all the following procedures, put 50 μl of a cell suspension (do not exceed a concentration of 20×10^6 cells/ml blood, or bone marrow) into 12×75-mm tubes containing 10 μg of blocking immunoglobulin. Do not block when measuring Fc receptors unless you know the antibody you are using is not blocked by receptor occupancy.

A. Staining Cells with One Antibody

Indirect labeling with one antibody and a second antibody:

1. Add 10 μl of goat IgG (1 mg/ml) for 10 min
2. Add primary mouse antibody for 15 min
3. Wash
4. Add fluoresceinated goat antimouse IgG F(ab')$_2$ (FGAM) second antibody for 15 min
5. Wash

For the control tube, add the isotype control antibody instead of the first antibody in step 2. Also prepare a tube containing unstained cells; this will be used to measure cellular autofluorescence.

Rule: For indirect labeling with a second antibody, always block with IgG from the same species as the second antibody.

Indirect labeling with a biotinylated antibody and avidin:

1. Add 10 μl of x IgG (1 mg/ml) for 10 min
2. Add B-Ab for 15 min
3. Wash
4. Add F-avidin for 15 min
5. Wash

Here, x is the species of IgG that is the same as the antibody. For murine antibodies block with normal mouse IgG; for rat antibodies block with normal rat IgG.

Direct labeling with one antibody:

1. Add 10 μl of x IgG (1 mg/ml) for 10 min.
2. Add F-Ab for 15 min.
3. Wash.

Rule: For indirect labeling with biotinylated antibodies or for direct labeling, block with the immunoglobulin fraction from the same species as the antibody. Never block if the epitope being measured is the Fc receptor.

For example, CD16 is the FcR III receptor on granulocytes and natural killer (NK) cells. Some CD16 antibodies will not bind to this receptor if it is blocked.

B. Staining Cells with Two Antibodies

One directly conjugated antibody with one unconjugated antibody:

1. Add 10 μl of goat IgG (1 mg/ml) for 10 min
2. Add primary mouse antibody for 15 min
3. Wash
4. Add phycoerythrinated goat antimouse Ig (PEGAM) secondary antibody for 15 min
5. Wash
6. Add 10 μl of x IgG (1 mg/ml) for 10 min
7. Add F-Ab for 15 min
8. Wash

Rule: Always perform the indirect second antibody step first and block after this step with IgG from the same species as the primary antibody.

This is because the second antibody will have free binding sites that must be blocked so it will not bind the directly conjugated second antibody when it is added subsequently in step 7. For example, suppose that both the unconjugated and conjugated antibodies are murine MAbs. After step 5, some cells have bound MAb1 and PEGAM to them. Because free binding sites remain on the PEGAM, the F-MAb may bind to them when it is added. To prevent this binding, these free sites must be blocked by the addition of mouse IgG shown in step 6.

If a rat MAb had been used then rat IgG would be used for the block. If one MAb is rat and the other mouse then the appropriate second antibody and block would be used: Unlabeled rat, use phycoerythrinated goat antirat Ig (PEGAR) and block with rat IgG; unlabeled mouse, use PEGAM and block with mouse IgG. FGAM or fluoresceinated goat antirat Ig (FGAR) could also be used in combination with PE-labeled antibodies. In summary, always perform the indirect labeling step first and block with normal IgG of the first antibody species before adding the directly conjugated antibody.

Two directly conjugated antibodies:

1. Add 10 μl of x IgG (1 mg/ml) for 10 min
2. Add F-Ab and PE-Ab for 15 min
3. Wash

One antibody conjugated with fluorescein, one with biotin:

1. Add 10 μl of x IgG (1 mg/ml) for 10 min
2. Add F-Ab and B-Ab for 15 min
3. Wash
4. Add PE-avidin for 15 min
5. Wash

Here, x IgG is derived from the species in which the conjugated antibodies were produced.

C. Staining Cells with Three Antibodies

There are several strategies that can be used for staining cells with three antibodies. For the reagents, TC (third color) refers to either PerCP, PETR, or PECY5.

Two fluorochrome-conjugated antibodies and one unconjugated antibody:

1. Add 10 μl of goat IgG (1 mg/ml) for 10 min
2. Add unlabeled mouse antibody for 15 min
3. Wash
4. Add B-GAM for 15 min
5. Wash
6. Add 10 μl of x IgG (1 mg/ml) for 10 min
7. Add F-Ab, PE-Ab, and TC-avidin for 15 min
8. Wash

Here, x IgG is derived from the same species as the conjugated antibody. In this procedure, a biotinylated second antibody was used followed by a TC-avidin to illustrate one of the many strategies that can be used to provide for the desired fluorochrome combination. A fluorochrome-conjugated second antibody could also have been used instead of the biotinylated one.

Three antibodies—two directly conjugated and one biotinylated antibody:

1. Block with 10 μl of x IgG (1 mg/ml) for 10 min
2. Add FL-Ab, PE-Ab, and B-Ab
3. Wash
4. Add TC-avidin
5. Wash

Again, x IgG is derived from the same species as the three directly conjugated antibodies.

Three directly conjugated antibodies:

1. Block with 10 μl of x IgG (1 mg/ml) for 10 min
2. Add F-Ab, PE-Ab, and TC-Ab for 15 min
3. Wash

D. Staining Cells with Four Antibodies

It is now possible to excite the four fluorochromes F, PE, PETR, and PECY5 or PerCP with a single laser or with two lasers where APC is used instead of PETR. This provides for the opportunity to use four antibodies simultaneously.

Two directly conjugated, one biotinylated, and one unconjugated antibody:

1. Add 10 μl of goat IgG (1 mg/ml) for 10 min
2. Add primary mouse Ab for 15 min
3. Wash
4. Add phycoerythrin Texas Red tandem complex conjugated to goat anti-mouse Ig (PETRGAM)
5. Wash
6. Block with x IgG (1 mg/ml) for 10 min
7. Add F-Ab, PE-Ab, and B-Ab for 15 min
8. Wash
9. Add PECY5-avidin or PerCP-avidin for 15 min
10. Wash

Here, x IgG is derived from the same species as the unlabeled antibody.
Three directly conjugated and one biotinylated antibody:

1. Add 10 μl of x IgG (1 mg/ml) for 10 min
2. Add F-Ab, PE-Ab, PECY5 (or PerCP)-Ab, and B-Ab for 15 min
3. Wash
4. Add PETR-avidin for 15 min
5. Wash

Other combinations for the third color are also possible. For example, an antibody directly conjugated with PETR and a PECY5 or PerCP-avidin could have been used.
Four directly conjugated antibodies:

1. Add 10 μl of x IgG (1 mg/ml) for 10 min
2. Add F-Ab, PE-Ab, PETR-Ab, and PECY5 (or PerCP)-Ab for 15 min
3. Wash

The four-color panel for immunophenotyping human cells offers the advantage of combining three antibodies for subset identification with CD45 as the fourth antibody used for resolving leukocytes from debris and other cells in general and for resolving lymphocytes in particular (Mandy *et al.*, 1992; Stelzer *et al.*, 1993). The data analysis strategy for this application will be discussed in the chapter on analysis.

E. Intracellular Staining

Unless no surface membrane staining is desired, intracellular staining should always be done after all surface membrane staining has been performed and after the cells have been fixed. Formaldehyde fixation can result in the cross-linking of epitopes already discussed in Section IV.C. There are several permeabilization reagents commercially available. They usually contain Triton X-100, Nonidet P-40 (NP-40), or saponin. Because saponin permeabilization is reversible, the wash buffer, if performed, should also contain saponin. Only antibodies with directly conjugated fluorochromes or those with biotin can be used if the cells have been stained with antibodies to membrane-bound antigens, because a second antibody will also bind to the membrane antibodies. The general procedure is as follows.

To cells in 50 μl of buffer:

1. Add the amount of permeabilizing reagent recommended by the supplier for 10 min
2. Add the conjugated antibody(ies) to the suspension and incubate 30 min
3. Add 3.5 ml of **PBS**
4. Centrifuge cells at 1500 $\times$ *g* for 3 min, decant supernatant, and resuspend cells in 500 μl of formaldehyde

It is also wise to use an isotype control adjusted to the exact concentration as the antibody to the intracellular antigens. Nonspecific binding occurs at a concentration 10–50 times less than for surface phenotyping because of the increased intracellular protein concentration as discussed earlier in Section II.B.

F. Cell Fixation

Cells are often fixed after staining so data acquisition can be performed at a later time.

1. Thoroughly resuspend pelleted cells in residual PBS after staining
2. Add 3 ml of lysing reagent and agitate cells for 3 min (rock, roll, or tumble). This step may be omitted if no erythrocytes are present
3. Centrifuge cells at 1500 $\times$ *g* for 3 min. Decant supernatant and resuspend pellet
4. Wash
5. Add 300 μl of 2% Ultrapure formaldehyde

Paraformaldehyde is a powder that when in aqueous solution becomes Ultra-pure formaldehyde. This can be purchased already prepared from PolySciences (Malvern, PA). We resuspend cells in only 300 μl of this fixative to produce a more concentrated suspension so acquisition of data will be faster. If desired, cells may be centrifuged and resuspended in less formaldehyde to have them more concentrated for data acquisition. The formaldehyde concentration should not be less than 1% for reproducible forward scatter light (FSC) versus side scatter light (SSC) characteristics. Use only Ultrapure formaldehyde because other formulations will produce increased autofluorescence. Prolonged storage of samples results in increased autofluorescence, which can compromise the resolution of dimly stained cells. Therefore, data should always be acquired within 5 days of preparation.

G. Measuring Cell Viability

As discussed in Section II, dead cells can produce errors that lead to data misinterpretation. This can be avoided by evaluating their number in the cell suspension.

1. Unfixed cell suspensions:
   a. Resuspend cell pellet in 1 ml of PBS containing 10 μl (200 μg/ml) of propidium iodide per ml
   b. Analyze samples after 5 min of incubation
2. Fixed cell suspensions (Riedy *et al.*, 1991):
   a. Resuspend cell pellet (prior to fixation) in residual PBS and add enough ethidium monoazide for a final concentration of 5 μg/ml. (Some batches require less EMA to reduce nonspecific fluorescence due to its binding to any protein.)
   b. Put samples 18 cm from a 40 W fluorescent light for 10 min
   c. Wash and fix cells as described in Section IV

V. Titering Antibodies

For trouble-free immunophenotyping it is essential to determine antibody quality and titer. This should be done for every new antibody and is recommended for new batches of previously tested antibodies. Generally, antibodies to mammals other than human cells are of poorer quality than those to human cells because of the lack of regulatory rules that govern their production. This is especially true for antibodies to intracellular antigens. By properly titering an antibody, its quality can be established according to the law of mass action. Since 3 μg/100 μl/10^6 cells exhibit the initial concentration where the nonspecific binding component for all IgGs becomes detectable, this represents a convenient starting concentration for titering.

Antibodies such as CD3 or CD19, and many more, generally stain a cluster of cells whose membrane antigen is expressed in a fairly uniform manner. This results in a distinct cell cluster. Antibodies such as CD25 or CD38, and several others, generally stain cells in a heterogeneous manner that produces a continuum of negative to positive cells. Such a continuum is often exactly like that observed for nonspecific binding, making discrimination between it and specific binding more subjective. Finally, intracellular staining requires longer incubation and wash times because antibody diffusion is restricted by the infrastructure of the cell.

A high-affinity antibody at the proper antibody titer is most important if good immunophenotyping data is to be obtained. Because MAb are myeloma proteins whose antibody specificity is known, they behave both like myeloma proteins (binding by Fc and nonspecifically) and like specific antibodies (binding to epitope). For any given MAb, there may be a different degree of Fc and nonspecific binding. Proper titering can reduce nonspecific binding as well as provide the desired specific epitope binding, whereas IgG blocking eliminates the problem of FcR binding. To optimize specific binding, the mean channel fluorescence is the only parameter that is measured.

A. Processing Cells

To titer an antibody, target cells that express the epitope, target cells that do not express the epitope, the specific antibody, and an isotype control exactly matched in IgG concentration to the specific antibody are required. A target cell suspension containing a mixture of epitope positive and negative cells such as blood is best. An epitope positive and negative cell line mixed together is also appropriate.

1. Adjust the specific antibody and isotype control to 30 μg/ml in PBS and prepare five serial 1/3 dilutions of each. Also prepare a tube containing only cells (autofluorescence).
2. Make a mixture of epitope positive and epitope negative target cells if necessary, and adjust them to 2×10^7 per ml in PBS.
3. Add the antibody dilutions to five appropriately labeled 12×75-mm tubes and the isotype control dilutions to another five tubes. Prepare a tube in which only PBS is added. Then add 50 μl of cells to all tubes.
4. For extracellular epitopes, follow the staining procedure described in Section IV.A. For intracellular epitopes, follow the staining procedure described in Section IV.E.
5. Acquire data ungated.

B. Analysis of Results

Refer to Chapter 45 in Methods in Cell Biology, vol. 64 for data acquisition and analysis.

1. For Antibodies That Resolve Discrete Populations

       a. Create a bivariate histogram. If a gated population is desired, create a bivariate plot of FSC versus SSC and establish the appropriate region for gating the univariate histogram.

       b. Establish a marker to distinguish between positive and negative cells using unstained cells (autofluorescence).

       c. Analyze all the files and record the mean channel fluorescence (MCF) of the positive and negative cells (see Fig. 14A).

       d. For each antibody dilution, compute the ratio of the positive MCF by the negative MCF and plot these values as a function of the dilution (see Fig. 14B). Do this for both the specific antibody (Fig. 14B) and the isotype control.

When unblocked cells are stained with too high an antibody concentration (Fig. 14A), there is a uniform shift to the right of the negative cells so some of them are above the marker established for positive cells. As the antibody is diluted, nonspecific binding decreases so the epitope positive cells are clearly resolved. Finally, the epitope positive cells decrease in mean channel fluorescence as antibody is no longer in excess and the epitopes are no longer saturated. The kinetics can be visualized by plotting the computed ratios as shown in Fig. 14B. The optimal antibody titer is the dilution that produces the maximum signal-to-noise ratio, and it is often not the concentration that produces the highest percentage of positive cells.

2. For Antibodies That Resolve Nondiscrete Populations or for Intracellular Epitopes

If a discrete positive population is not found, a slightly different approach is used.

       a. Create a univariate histogram, but do not establish a marker for positive versus negative cells.

       b. Analyze all files and record the mean channel fluorescence of the antibody stained and isotype control stained cells.

       c. For each dilution, compute the ratio of the MCF of the antibody stained to the isotype control stained cells and plot these values as a function of the dilution as shown in Fig. 15.

Many antibodies bind heterogeneously to cells and exhibit a "ski slope" appearance of nonspecific binding. To properly titer them it is necessary to use the specific antibody and the isotype control at exactly the same concentrations. No marker is used, and the MCF of the isotype control and antibody are determined. The ratios of MCF values for the antibody and isotype control are computed, and plotted as shown in Fig. 15. The titer is the concentration that produces the maximum signal-to-noise ratio, and this ratio must be greater than 3.

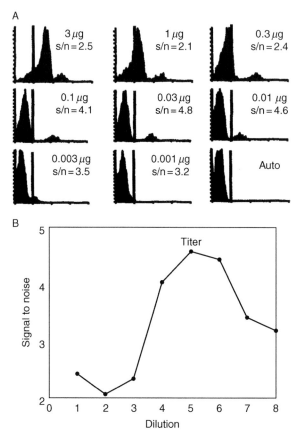

Fig. 14 Titering antibodies to epitopes with discrete expression. (A) To determine the titer, target cells are stained with various dilutions of antibody and the mean channel fluorescence of positive and negative cells determined. The marker may be set using an isotype control or unstained cells (auto). The ratio of the MCF of the positive cells to negative cells is calculated. We call this the signal-to-noise ratio (S/N). Because each antibody, when properly titered, is its own isotype control, there is no need to use a separate one. (B) By plotting S/N as a function of the concentration an inverted parabolic shaped curve is produced whose zenith is the titer. This concentration produces the best distinction between positive and negative cells. The left side of the curve represents rapid changes in nonspecific binding and very slow changes in specific binding. The right side reflects decreased saturation of epitopes by antibody.

C. Verification of Specific Antibody Binding

Contrary to well-established dogma, the hallmark of nonspecific binding is that it cannot be blocked. No matter how much Ig is added, no matter how much other protein is added, the cells will always bind more of it. The cellular protein excess relative to the specific epitope concentration causes this unlimited binding capacity in the working range of immunophenotyping. To discriminate between specific and

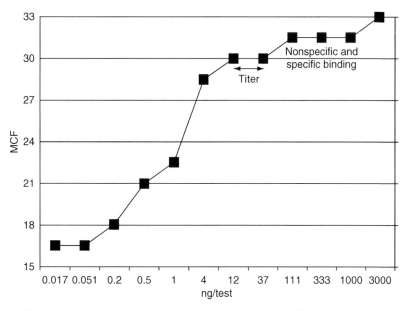

Fig. 15 Titering antibodies to epitopes with heterogeneous expression. Blood leukocytes were stained with serial 1/3 dilutions of FITC-CD38 starting with 3000 ng in a final volume of 100 μl. The mean channel fluorescence of lymphocytes was determined for the CD38 and plotted as a function of antibody concentration. Above 30 ng/test, the nonspecific binding component begins to appear. Below 10 ng/test epitope saturation no longer occurs and there is a steep slope to the titration curve. The titer is determined as the shoulder area where epitope saturation has occurred and nonspecific binding is low. If no shoulder is found, the antibody is no good. For this study, the mean channel fluorescence (MCF) for cellular autofluorescence was 5. Note that the titer of the CD38 antibody used here is 12 ng/test in a final volume of 100 μl.

nonspecific binding an unconjugated specific antibody is used to block the measured specific binding of a conjugated antibody, but this block will have no effect on nonspecific binding. Only two tubes with target cells are necessary to verify that a properly titered antibody is specifically binding to cellular epitopes. If this verification fails, specific binding is not being measured. This procedure requires an appropriate target cell expressing the desired epitope, an unconjugated and a fluorochrome-conjugated or biotinylated antibody.

1. Pipette into two tubes 10^6 target cells in 50 μl.
2. Having estimated the titer of the antibody, add an unconjugated antibody at a concentration three times the titer into the first tube. Incubate 15 min.
3. Now add to both tubes a directly conjugated antibody (with fluorochrome or biotin) at the estimated titer. Incubate 15 min for surface or 45 min for intracellular staining.
4. Acquire the data using the flow cytometer.

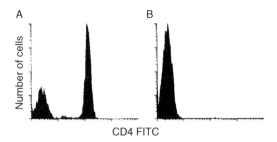

Fig. 16 Verification of antibody specificity. Human blood was incubated with either CD4-FITC (A) for 15 min or CD4 at three times titer, for 10 min, followed by CD4-FITC (B) for 15 min at the correct titer.

Figure 16 shows the expected result if the antibody is specifically binding. The binding of the fluorescent antibody is completely inhibited by the blocking unconjugated antibody (Fig. 16B). If the antibody is nonspecifically binding there is little or no effect of the block as the fluorescence is no different from the blocked and unblocked specimen (not shown). This test should always be performed when evaluating intracellular antigens.

VI. Solutions and Reagents

A. PBS (with Sodium Azide)

Phosphate-buffered saline containing no calcium or magnesium and supplemented with 0.1% sodium azide is used for all dilutions and washes. The pH should be adjusted to 7.2. While not necessary, PBS may be supplemented with protease and nuclease free 0.5% bovine serum albumin (PAB) (Sigma Chemical, St. Louis, MO). Be sure to check osmolality of the final solution (290 ± 5 mOsm for human cells). *Note*: Handle sodium azide with extreme caution.

B. Two Percent Ultrapure Formaldehyde

A 10% solution of Ultrapure formaldehyde can be obtained from PolySciences (Malvern, PA). This solution is diluted 1/5 with PBS to create the working solution. Do not use impure formaldehyde or formalin as these solutions will markedly increase autofluorescence.

C. Lysing Reagent

1. 1.6520 g ammonium chloride
2. 0.2000 g potassium bicarbonate
3. 0.0074 g EDTA (tetra)

Make up to 200 ml with distilled water; use at room temperature.

This reagent must be prepared daily because HCO_3 combines with NH_4Cl in this strong acid weak base molecule to form CO_2, thereby rendering the solution ineffective in lysing erythrocytes. We recommend weighing the reagents and storing them as packets. The dry reagents are dissolved in water when required.

D. Propidium Iodide Stock Solution

Note: Handle propidium iodide with extreme caution.

Propidium iodide (MW 668), 7.4×10^{-5} M (20 mg/100 ml) (Sigma Chemical), is prepared in PBS. Store in a dark, foil-wrapped container to protect from light.

E. Ethidium Monoazide Stock Solution

Ethidium monoazide (EMA) (Molecular Probes, Eugene, OR) is very light sensitive and must be stored at $-20\,^\circ$C in a dark vial or foil-wrapped container. The stock solution is prepared in PBS at 5 mg/ml. This can then be dispersed into small aliquots of 50-100 μg/ml and stored at $-20\,^\circ$C. These small aliquots can then be thawed one at a time as needed to do the assays. Discard remaining EMA after thawing.

References

Beavis, A. J., and Pennline, K. J. (1996). ALLO-7: A new fluorescent tandem dye for use in flow cytometry. *Cytometry* **24**, 390–394.

Borowitz, M. (1993). Clinical applications of flow cytometry: Immunophenotyping of leukemic cells. *NCCLS* **13**, 1–107.

Bossuyt, X., Marti, G. E., and Fleisher, T. A. (1997). Comparative analysis of whole blood lysis methods for flow cytometry. *Cytometry* **30**, 124–133.

Carter, P. H., Resto-Ruiz, S., Washington, G. C., Ethridge, S., Paline, A., Vogt, R., Waxdal, M., Fleisher, T., Noguchi, P. D., and Marti, G. E. (1992). Flow cytometric analysis of whole blood lysis, three anticoagulants and 5 cell preparations. *Cytometry* **13**, 66–74.

Debili, N., Wendling, F., Cosman, D., Titeux, M., Florinso, C., Dusanter-Fourt, I., Schooley, K., Methia, N., Charon, M., Nador, R., Bettaieb, A., and Vainchenker, W. (1995). The Mpl receptor is expressed in the megakaryocytic lineage from late progenitors to platelets. *Blood* **2**, 329–401.

Falini, B., Pileri, S., Pizzolo, G., Durkop, H., Flenghi, L., Stirpe, F., Martelli, M. F., and Stein, H. (1995). CD30 (Ki-1) molecule: A new cytokine receptor of the tumor necrosis factor receptor superfamily as a tool for diagnosis and immunotherapy. *Blood* **85**, 1–14.

Landay, A. L. (1992). Clinical applications of flow cytometry: Quality assurance and immunophenotyping of peripheral blood lymphocytes. *NCCLS* **12**, 1–76.

Mandy, F. F., Bergeron, M., Recktenwald, D., and Izaguime, C. A. (1992). A simultaneous three color T-cell subsets analysis with single laser flow cytometers using T cell gating protocol. *J. Immunol. Methods* **156**, 151–162.

Olsson, I., Gullberg, U., Lantz, M., and Richter, J. (1992). The receptors for regulatory molecules of hematopoiesis. *Eur. J. Haematol.* **48**, 1–9.

Picker, L. J., Singh, M. K., Zdraveski, Z., Treer, J. R., Waldrop, S. L., Bergstresser, P. R., and Maino, V. C. (1995). Direct demonstration of cytokine synthesis heterogeneity among human memory/effector cells by flow cytometry. *Blood* **86**, 1408–1419.

Prussin, C. (1997). Cytokine flow cytometry: Understanding cytokine biology at the single-cell level. *J. Clin. Immunol.* **17,** 195–204.

Riedy, M. C., Muirhead, K. A., Jensen, C. P., and Stewart, C. C. (1991). The use of a photolabeling technique to identify nonviable cells in fixed homologous or heterologous cell populations. *Cytometry* **12,** 133–139.

Roitt, I., Brostoff, J., and Male, D. (1989). "Immunology," Chap. 5, pp. 5.1–5.11. Mosby, St. Louis, MO.

Romagnani, S. (1997). The Th1/Th2 paradigm. *Immunol. Today* **18,** 263–266.

Stelzer, G. T., Shults, K. E., and Loken, M. R. (1993). CD45 gating for routine flow cytometric analysis of human bone marrow specimens. *In* "Clinical Flow Cytometry" (A. Landay, K. Ault, K. Bauer, and P. Rabinovitch, eds.), pp. 265–280.

Stewart, C. C. (1994). Multiparameter flow cytometry. *In* "Immunochemistry" (C. J. van Oss, ed.), Chap. 32, pp. 849–866. Dekker, New York.

Stewart, C. C. (1996). Clinical applications of multiparameter flow cytometry. *In* "Haematology 1996, Education Programme of the 26th Congress of the International Society of Haematology" (J. R. McArthur, S. H. Lee, J. E. Wong, and Y. W. Ong, eds.) The Int'l Society of Haematology, Singapore.

Stewart, C. C., and Mayers, G. L. (2000). Kinetics of antibody binding to cells. *In* "Immunophenotyping" (C. C. Stewart and J. K. A. Nicholson, eds.), Wiley, New York.

Stewart, C. C., and Stewart, S. J. (1993). Immunological monitoring utilizing novel probes, clinical flow cytometry. *Ann. N.Y. Acad. Sci.* **677,** 94–112.

Stewart, C. C., and Stewart, S. J. (1994a). Cell preparation for the identification of leukocytes. *In* "Methods in Cell Biology" (Z. Darzynkiewicz, J. Robinson, and H. Crissman, eds.), Vol. **41,** pp. 39–60. Academic Press, New York.

Stewart, C. C., and Stewart, S. J. (1994b). Multiparameter analysis of leukocytes by flow cytometry. *In* "Methods in Cell Biology" (Z. Darzynkiewicz, J. Robinson, and H. Crissman, eds.), Vol. **41,** pp. 61–79. Academic Press, New York.

Stewart, C. C., and Stewart, S. J. (1995). The use of directly and indirectly labeled monoclonal antibodies in flow cytometry. *In* "Methods in Molecular Biology" (W. C. Davis, ed.), Vol. **45,** pp. 129–147. Humana Press, Totowa, NJ.

Stewart, C. C., and Stewart, S. J. (1997a). Titering antibodies. *In* "Current Protocols in Cytometry" (J. P. Robinson, Z. Darzynkiewicz, P. Dean, L. Dressler, P. Rabinovitch, C. Stewart, H. Tanke, and L. Wheeless, eds.), pp. 4.1.1–4.1.13. Wiley, New York.

Stewart, C. C., and Stewart, S. J. (1997b). Immunophenotyping. *In* "Current Protocols in Cytometry" (J. P. Robinson, Z. Darzynkiewicz, P. Dean, L. Dressler, P. Rabinovitch, C. Stewart, H. Tanke, and L. Wheeless, eds.), pp. 6.2.1–6.2.15. Wiley, New York.

Stewart, C. C., and Stewart, S. J. (1999). Four-color compensation. *Cytometry* **38,** 161–175.

Stomper, P. C., Nava, M. E. R., Budnick, R. M., and Stewart, C. C. (1997). Specimen mammography-guided fine-needle aspirates of clinically occult benign and malignant lesions. Analysis of cell number and type. *Invest. Radiol.* **32,** 277–281.

Tamul, K. R., O'Gorman, M. R. G., Donovan, M., Schmitz, J. L., and Folds, J. D. (1994). Comparison of a lysed whole blood method to purified cells preparations for lymphocyte immunophenotyping: Differences between healthy controls and HIV positive specimens. *J. Immunol. Methods* **167,** 237–243.

Testi, R., Philips, J. H., and Lanier, L. L. (1989). Leu-23 induction as an early marker for functional CD3/T cell antigen receptor triggering: Requirement of receptor cross-linking, prolonged elevation of intracellular (Ca++) and stimulation of protein kinase C. *J. Immunol.* **142,** 1854.

Van Vugt, M. J., van den Herik-Oudijk, I. E., and van de Winkel, J. G. J. (1996). Binding of PE-CY5 conjugates to the human high-affinity receptor for IgG (CD64). *Blood* **88,** 2358–2360.

Wognum, A. W., van Gils, F. C. J. M., and Wagemaker, G. (1993). Flow cytometric detection of receptors for interleukin-6 on bone marrow and peripheral blood cells of humans and rhesus monkeys. *Blood* **81,** 2036–2043.

CHAPTER 24

Multicolor Immunophenotyping: Human Immune System Hematopoiesis

Brent Wood

Department of Laboratory Medicine
University of Washington
Seattle, Washington 98195

I. Update

Since the initial publication of this chapter, instrumentation suitable for higher level (polychromatic) flow cytometry has become more common in the research environment and has begun to appear in clinical laboratories. With this expansion, the lasers used on these instruments have become smaller, more stable, and more powerful, and many of the fluorochromes that were relatively exotic at the time are now commonly available as direct antibody conjugates from a number of manufacturers. Consequently, there are many more choices for configuring reagents into

ESSENTIAL CYTOMETRY METHODS
DOI: 10.1016/B978-0-12-375045-7.00024-6

panels emphasizing normal and abnormal maturation patterns, and those presented in this chapter should be used only as a guide to what is possible and not a prescription for what is ideal. Nevertheless, the basic levels of expression for antigens at specific maturational stages within a particular lineage remain unchanged and can be used as a starting point for the design and validation of improved reagent combinations (see, e.g., Wood, 2006). Additionally, the immunophenotypic demonstration of deviation from normal patterns of maturation as a marker for hematopoietic neoplasia is even more convincingly demonstrated in the literature and is increasingly being incorporated in routine clinical practice.

A. Methodological Points

1. The basic method of specimen preparation utilizing ammonium chloride lysis in combination with a small amount of formaldehyde continues to be a robust and inexpensive method of specimen preparation for surface immunophenotyping.

2. For assays where antigens of interest are present in plasma and plasma removal is required to allow cell staining, for example, immunoglobulin light chains, three washes of a small aliquot of blood, for example, 100 μl, with phosphate buffered saline (PBS) prior to antibody labeling followed by the standard ammonium chloride lysing procedure produces excellent results. This is a suitable alternative to prelysing of the sample, which may result in some cell activation as well as some increased cell loss with washing, particularly when done in large volumes.

3. When combining multiple antibodies for use in whole blood lysis methods on human material, that is, in the presence of plasma, care must be taken to ensure that antibody interactions mediated by complement component C1q do not occur (see Wood and Levin, 2006). This is particularly a problem for murine IgG2 class antibodies, and preferential use of IgG1 class reagents will minimize this problem. If interactions are encountered and such reagents must be used, removal of plasma by washing prior to labeling will greatly minimize this artifact.

II. Introduction

Hematopoiesis begins with a quiescent stem cell that proliferates under environmental influences giving rise to progeny capable of differentiation along multiple lineages. In humans, this process occurs largely in the bone marrow, the one exception being T-cell differentiation, which occurs principally in the thymus. Differentiation along each lineage occurs in a relatively linear fashion through a sequential series of stages, resulting in fully functional mature forms that may undergo further activation and differentiation as a consequence of their interaction with different microenvironments and stimuli. For instance, a B cell differentiates from stem cells in the bone marrow of humans and matures through a series of stages producing the functional mature B cell (McKenna *et al.*, 2001), whereupon it exits the bone marrow and populates lymphoid tissue throughout the body. Additional maturation and processing of these naive B cells occurs in the germinal center of lymph nodes and other lymphoid tissue after antigen exposure, resulting in subsequent maturation to plasma cells or memory B cells (MacLennan, 1994).

The maturation of hematopoietic cells is a result of the tightly regulated and sequential expression of genes and gene products (Payne and Crooks, 2002). As a consequence, the derived protein products exhibit predictable and reproducible patterns of expression with maturation that correlate with morphological or functional stages. Cell-surface proteins represent a subset of these proteins with functions as diverse as extracellular enzymes, growth factor receptors, signal transduction molecules, and adhesion molecules. The functions for many of these molecules are relatively poorly understood, many being identified simply by the presence of antigenic reactivity. Some cell-surface antigens exhibit expression restricted to a particular cell lineage, and others are more widely expressed but with differing levels depending on maturational stage or activation state. Many of these antigens can be used to aid in the identification of maturational stages and lineage assignment once their patterns of expression are known. In addition to defining normal maturational states, disease states often exhibit alterations in antigenic expression reflecting abnormalities in maturation or function, and these alterations can be used in the diagnosis and monitoring of disease states.

Flow cytometry is an excellent technique for the examination of patterns of protein expression, particularly on the cell surface. Using fluorescent-labeled antibodies that bind with high avidity and specificity to selected antigens, a wealth of information regarding hematopoietic maturation has been generated over the past two decades. This chapter describes the identification of normal maturational stages of hematopoietic cells in human bone marrow by multiparametric flow cytometry using seven, eight, and nine simultaneous antibodies directed against antigens commonly evaluated in a clinical setting.

III. Methodology

Anticoagulated human bone marrow was obtained from the hematopathology laboratory at the University of Washington Medical Center as residual material sent for clinical evaluation. This material was used in accordance with a Human Subjects Protocol approved by our Institutional Review Board.

First, 100 μl of normal bone marrow containing roughly 10^6 cells was incubated at room temperature for 30 min, with 100 μl of a pretitered cocktail of monoclonal antibodies appropriate for the cell lineage being evaluated. The sample was then treated with 1.5 ml of buffered NH_4Cl containing 0.25% formaldehyde for 10 min at room temperature followed by a single wash in PBS using centrifugation at

550 × *g* for 5 min. The resulting pellet was resuspended in 100 μl of PBS and roughly 1,000,000 events acquired on a modified LSRII (Becton-Dickinson Immunocytometry Systems, San Jose, CA) capable of up to 10-color analysis. The instrument contains four spatially separated lasers (22 mW 407 nm diode, 100 mW variable 488 nm diode, 8 mW 594 nm HeNe, and 25 mW 638 nm diode) with independent fiberoptic-coupled detector arrays for each laser. The data were compensated and analyzed after acquisition using software developed in our laboratory. For demonstration of B-cell maturation, serum light chain was removed by prelysing 100 μl of bone marrow with 1.5 ml of buffered NH_4Cl for 10 min at room temperature followed by a single wash with PBS; the sample was then prepared as described earlier, with the lyse step omitted. To demonstrate erythroid maturation, 1 ml of bone marrow was subjected to density-gradient centrifugation (Ficoll) and the resulting mononuclear layer was washed twice with PBS before being used in the assay; the subsequent lysis step was omitted.

Fluorochromes used for these experiments included Pacific blue, fluorescein isothiocyanate (FITC), phycoerythrin (PE), PE-Cy5.5, PerCp-Cy5.5, PE-Cy7, Alexa 594, allophycocyanin (APC), Alexa 700, and APC-Cy7. Zenon reagents (Molecular Probes, Eugene, Oregon) were used for CD11b-Alexa 594, and all other antibodies were directly conjugated and obtained from either Beckman-Coulter or Becton-Dickenson.

IV. Normal Immunophenotypic Patterns of Maturation

A. CD45 and Side Scatter Gating

CD45 is a transmembrane tyrosine phosphatase ubiquitously expressed by white blood cells (WBCs) and is important in modulating signals derived from integrin and cytokine receptors (Hermiston *et al.*, 2003). On WBCs, the level of CD45 is differentially expressed throughout maturation, being lower on blasts and immature forms and highest on mature myelomonocytic and lymphoid cells. In contrast, with erythroid differentiation, the low level of CD45 seen on blasts rapidly decreases and mature erythrocytes consistently show a lack of CD45 expression.

Orthogonal light scatter in large part reflects the presence of subcellular components capable of scattering light, in particular the presence of granules or vacuoles. It also correlated with overall cell size or volume and thus shows a mild increase as cells enlarge with maturation. Consequently, the early acquisition of large cytoplasmic primary granules in neutrophilic lineage cells is accompanied by a marked increase in side scatter that allows their ready separation from progenitor cells. Further maturation along the neutrophilic lineage is marked by acquisition of smaller secondary granules and loss of primary granules with a concomitant decrease in side scatter. Monocytic differentiation is marked by a moderate increase in cell size, including increased amounts of cytoplasm containing few vacuoles, resulting in a mild increase in side scatter from that seen in blasts, though not to the degree seen in neutrophilic differentiation. In contrast, lymphoid differentiation is accompanied by a mild decrease in cell size with the small amount of associated cytoplasm containing essentially no granules or vacuoles. Consequently, lymphoid cells show a moderate decrease in side scatter from that seen in blasts. Erythroid differentiation is accompanied by a marked increase in cell size in early forms with a gradual decrease in size to the erythrocyte stage, with all stages having only modest amounts of cytoplasm without significant cytoplasmic structures. This results in an early increase in side scatter with a gradual decline to a final level similar to that seen in lymphocytes.

When combined, CD45 expression and orthogonal light scatter provide a powerful, simple, and reproducible method for distinguishing major cell lineages in bone marrow (Borowitz *et al.*, 1993). As a result, CD45-side scatter gating has become one of the most common techniques used for immunophenotyping in the clinical laboratory. The characteristic patterns seen using this method are illustrated in Fig. 1.

B. Stem Cells and Blasts

The earliest easily identifiable hematopoietic cell in the marrow is a quiescent cell having the ability to differentiate along multiple lineages depending on the environmental influences to which it is exposed (i.e., a hematopoietic stem cell).

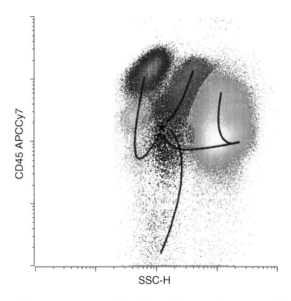

Fig. 1   CD45 versus side scatter representation of normal human bone marrow. Each of the major cell lineages differentiates from the early blast/stem cell population (yellow) (*asterisk*). The major more mature blast population (red) is easily identified by this display. Subsequent maturation toward the neutrophil (green), monocyte (purple), B lymphocyte (blue), and erythrocyte (gray) populations are indicated by the appropriate black line. (See Plate no. 13 in the Color Plate Section.)

The frequency of this population in normal bone marrow is low but can be routinely identified if a sufficiently large number of cells are evaluated (Bender *et al.*, 1991; D'Arena *et al.*, 1998; Gaipa *et al.*, 2002; Macedo *et al.*, 1995; McGuckin, 2003; Terstappen *et al.*, 1992; Tjonnfjord *et al.*, 1996). This population shows the expression of low forward and side scatter, intermediate CD45, intermediate human leukocyte antigen-DR (HLA-DR), bright CD34, low to absent CD38, intermediate CD133, low to intermediate CD13, low CD117, variable CD90, low CD123 and CD135, and very low CD33 without the expression of antigens characteristic of a particular lineage or later stages of differentiation (Fig. 2). As the cells mature, they begin to proliferate with a gain in the expression of CD38 and CD117, mildly decreased expression of CD34 and CD133, and decreased CD90. Further maturation results in each cell demonstrating a commitment to differentiation along a single lineage with resulting fragmentation of this relatively uniform-appearing population into different components, such as myelomonocytic, erythroid, lymphoid, and others. This is accompanied by the acquisition of lineage-associated antigens and patterns of antigen expression characteristic of that lineage, with all lineages ultimately showing a loss of CD34, CD133, CD117, and CD38. The earliest and most lineage-specific antigen expressed during myelomonocytic differentiation is CD64, with an associated mild increase in CD13, increased CD33, HLA-DR, and CD15, and increased side scatter. Early B cells show a loss of CD33 and CD13, a decrease in CD45 and side scatter, and the acquisition of CD10 and CD19 with retention of HLA-DR. Differentiation along the erythroid lineage is characterized by the acquisition of bright CD71 intermediate CD36 and a decrease in CD45, HLA-DR, CD13, and CD33 while showing a mild initial increase in side scatter.

C. Neutrophilic Maturation

Neutrophils mature from blasts through a linear process, historically divided into morphological stages termed *promyelocytes*, *myelocytes*, *metamyelocytes*, *bands*, and *neutrophils*. It is not until the band/neutrophilic stage that the cells are fully functional and capable of bactericidal action. Although these stages have associated antigenic changes, from an immunophenotypic perspective they are not so clearly defined and largely appear as a single continuum (Elghetany, 2002; Terstappen and Loken, 1990; Terstappen *et al.*, 1990, 1992). A selection of informative and useful antigens and their expression patterns with maturation are presented in Table I. Of particular interest are antigens that exhibit dramatic increases in expression at defined stages of maturation, such as acquisition of CD15 and CD66b by promyelocytes, acquisition of CD11b, CD11c, CD24, and CD66a by myelocytes, gain of CD55 by metamyelocytes, gain of CD35 and CD87 by bands, and gain of CD10 by granulocytes. Another useful group of antigens including CD13, CD44, and CD55 is expressed early in maturation and exhibits a decrease at an intermediate stage of maturation only to increase with terminal differentiation. A less common pattern of antigen expression, exemplified by

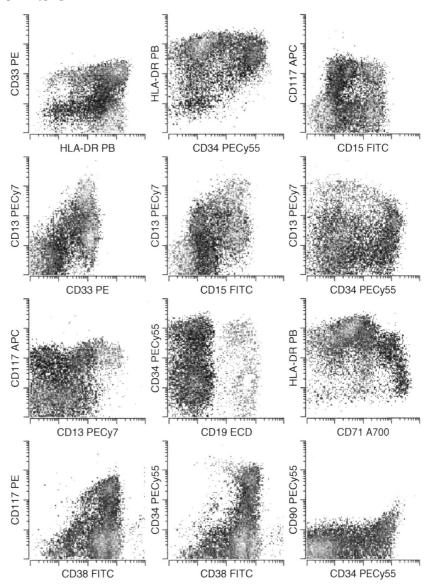

Fig. 2 Maturation and lineage commitment of blasts in human bone marrow. Normal human bone marrow was evaluated with two combinations of antibodies, one containing nine simultaneous fluorochromes (top three rows) and one containing eight simultaneous fluorochromes (bottom row) as labeled. The blast population is displayed following CD45/side scatter gating with stem cell (yellow), blast (red), neutrophil (green), monocyte (lavender), B cell (light blue), and erythroid (blue) populations colored. Note the complex patterns of maturation present in this subset of views that allow the unique identification of each of the major cell lineages. These views represent less than 20% of the total information contained in these antibody combinations. (See Plate no. 14 in the Color Plate Section.)

Table I
Surface Antigen Expression During Neutrophil Maturation

	Blast	Promyelocyte	Myelocyte	Metamyelocyte	Band	Neutrophil
CD10	−	−	−	−	−	++
CD11b	−	−	++	++	++	++
CD11c	−	−	++	++	++	++
CD13	++/+++	+++/++	+	+/++	++	+++
CD14	−	−	−	−	−	+
CD15	−	++	+++	+++	+++	+++
CD16	−	−	−	+	++	+++
CD18	++	+	+++	++	++	++
CD24	−	−	+	++	++	++
CD33	++	+++	++	+	+	+
CD35	−	−	−	−	++	++
CD44	+++	++	+	+	++	+++
CD45	++	+	+	+	++	++
CD55	+++	+	+	+++	+++	+++
CD64	−	+	++	++	−	−
CD65	−	++	++	++	+++	+++
CD66a	−	−	++	++	++	++
CD66b	−	+++	+++	++	++	++
CD87	−	−	−	−	++	++
CD117	++	+	−	−	−	−

CD64, is characterized by an absence of expression on blasts, an increase at intermediate stages of maturation, and a loss with terminal differentiation. A final group of antigens (e.g., CD33) is expressed at relatively high levels on early myeloid forms and declines with maturation. The combined use of antigens from each of these groups in multiparametric combinations serves a particularly useful function in more precisely defining maturational stages and makes the appreciation of maturational abnormalities in disease states more readily apparent. Examples of these expression patterns are displayed in Fig. 3.

D. Monocytic Maturation

Monocytes mature from blasts as a continuum through one intermediate stage termed the *promonocyte* (Terstappen and Loken, 1990; Terstappen *et al.*, 1992). The expression levels of individual antigens are listed in Table II and illustrated in Fig. 4. With maturation toward promonocytes, blasts gain the expression of CD64, CD33, HLA-DR, CD36, and CD15, with an initial mild decrease in CD13 and an increase in CD45. Further maturation toward mature monocytes shows a progressive increase in CD14, CD11b, CD13, CD36, and CD45, with a mild decrease in HLA-DR and CD15. Unlike with maturing myeloid forms, the expression of CD64, HLA-DR, and CD33 are retained on maturing monocytes at relatively high levels and serve as useful monocytic markers. Mature monocytes

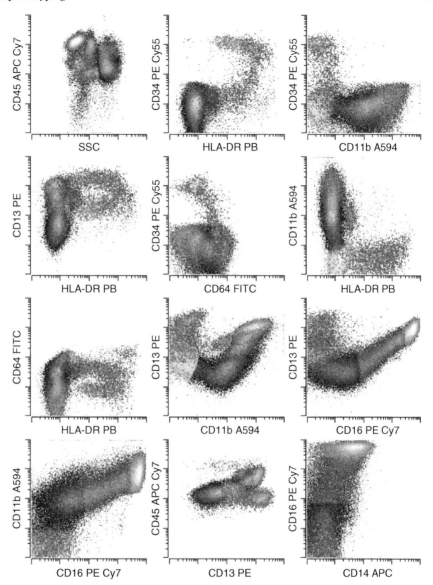

Fig. 3 Maturation of cells during neutrophilic differentiation. Normal human bone marrow was evaluated with a single combination of nine simultaneous antibodies. Maturation from early (yellow) and late (green) blasts proceeds through promyelocyte (light blue), myelocyte (blue), metamyelocyte/band (lavender), and neutrophil (pink) stages. Eosinophils can also be noted by their increased auto-fluorescence (orange). Though recognizable, these stages are not discretely defined and the subdivisions are somewhat arbitrary. The blasts have been emphasized and other populations largely excluded to improve visibility of the maturational relationships. These views represent less than 20% of the total information contained in these antibody combinations. (See Plate no. 15 in the Color Plate Section.)

Table II
Surface Antigen Expression During Monocyte Maturation

	Blast	Promonocyte	Monocyte
CD4	−/+	+	+
CD11b	−	++	+++
CD13	++	+/++	++/+++
CD14	−	+/++	+++
CD15	−	++	+
CD16	−	−	−/+
CD33	+++	+++	+++
CD36	−	++	+++
CD45	+	++	+++
CD64	−	++	+++
HLA-DR	++	+++	++/+++

show expression of bright CD14, relatively bright CD33, variably bright CD13, bright CD36 and CD64, and low CD15.

E. Erythroid Maturation

Examination of erythroid maturation requires a different method of specimen preparation than the red blood cell lysis methods commonly used to process bone marrow. Commonly used lysing reagents such as NH_4Cl with subsequent washing steps generally leave only very early erythroid precursors (proerythroblasts) to evaluate or at a minimum noticeably compromise more mature erythroid forms as detected by a prominent loss in forward scatter. Additionally, mature red blood cells must also be significantly reduced in number without undue loss of immature forms if one wants to examine antigens highly expressed on mature red blood cells (e.g., CD235a). Density-gradient centrifugation with either Percoll or Ficoll is an acceptable alternative for this purpose.

Erythroid forms have historically been divided into maturational stages based on their morphological appearance with the nucleated forms termed *proerythro-blasts*, *basophilic erythroblasts*, *polychromatophilic erythroblasts*, and *orthochroma-tophilic erythroblasts*, and the anucleate forms termed *polychromatophilic erythrocytes* and *erythrocytes*. These morphological stages correlate reasonably well with observed immunophenotypic changes (De Jong *et al.*, 1995; Loken *et al.*, 1987; Rogers *et al.*, 1996; Scicchitano *et al.*, 2003), although the differences between polychromatophilic and orthochromatophilic stages are ill-defined. The earliest erythroid forms to arise from blasts are identified by their acquisition of bright CD71, intermediate CD36 and expression of CD117 with intermediate CD45, and slightly higher forward and side scatter. Glycophorin A (CD235a) is expressed at a low level at this stage. Maturation to the basophilic erythroblast is accompanied by a decrease in CD45 and acquisition of bright CD235a expression to the level seen throughout the remainder of erythroid maturation, including

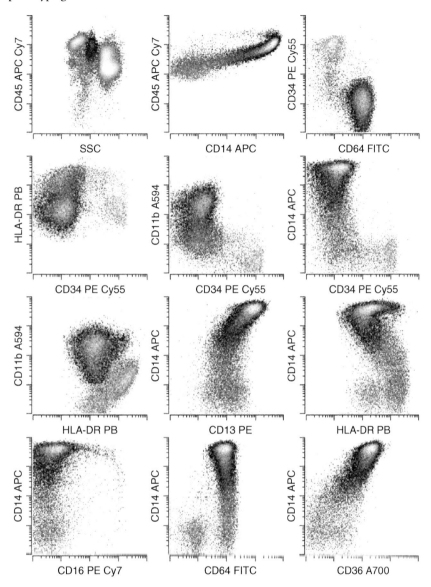

Fig. 4 Maturation of cells during monocytic differentiation. Normal human bone marrow was evaluated with a single combination of nine simultaneous antibodies. Maturation from early (yellow) and late (green) blasts proceeds through promonocyte (light blue) and monocyte (lavender) stages. Though recognizable, these stages are not discretely defined and the subdivisions are somewhat arbitrary. The blasts have been emphasized and other populations largely excluded to improve visibility of the maturational relationships. These views represent less than 20% of the total information contained in these antibody combinations. (See Plate no. 16 in the Color Plate Section.)

mature erythrocytes. Transition to polychromatophilic/orthochromatophilic erythroblasts shows a further loss of CD45 and HLA-DR and a mild decrease in CD36. After these stages, the cells rapidly lose CD71 and CD36 after becoming anucleate. These changes are described in Table III and illustrated in Fig. 5.

F. B–Cell Maturation

In humans, B-cell maturation occurs from the hematopoietic stem cell in the bone marrow and proceeds through relatively discrete well-defined stages unlike the continuum seen in myelomonocytic and erythroid maturation (Davis *et al.*, 1994; Dworzak *et al.*, 1997, 1998; Loken *et al.*, 1987; Longacre *et al.*, 1989; Lucio *et al.*, 1999; McKenna *et al.*, 2001). Upon reaching a functional stage characterized by expression of immunoglobulin on the cell surface, B cells exit the bone marrow and undergo further maturation in the presence of antigen in peripheral tissues such as lymph node. A subset of these B cells will undergo terminal differentiation to plasma cells and few will return to the bone marrow. Consequently, bone marrow B cells typically consist of a mixture of maturing B cells, a few mature but naive B cells, and a small number of plasma cells.

The immunophenotypic stages of B-cell maturation present in normal bone marrow are detailed in Table IV and examples of the observed patterns of antigen expression are shown in Fig. 6. The antigens expressed by the earliest identifiable B cells include TdT, CD79a, and low CD10 without CD19, but the size of this population is very small and is generally not observed. The next stage (early) is the first that can be readily identified and shows the acquisition of CD19, an antigen that serves as a useful marker of B cells through terminal plasma cell differentiation. At this stage, the B cells express the immature antigens CD34 and nuclear TdT with bright CD10 and low CD45 without CD20 or surface immunoglobulin. Transition to the intermediate stage of maturation results in an abrupt decrease in CD10, a gain in CD45, gradual acquisition of CD20, and a loss of CD34 and nuclear TdT. The last immature stage (late) is generally small in number but is characterized by the brightest level of CD20, a further decrease in CD10,

Table III
Surface Antigen Expression During Erythrocyte Maturation

	Blast	*Proerythroblast*	*Basophilic*	*Poly/Ortho*	*Retic*	*Mature*
CD34	++	−/+	−	−	−	−
CD36	−	++	+++	++	−/+	−
CD38	++	+	−/+	−	−	−
CD45	++	+	−/+	−	−	−
CD71	−	+++	+++	+++	+/++	−
CD117	++	++	−	−	−	−
CD235a	−	+/++	+++	+++	+++	+++
HLA-DR	++	++	+	−	−	−

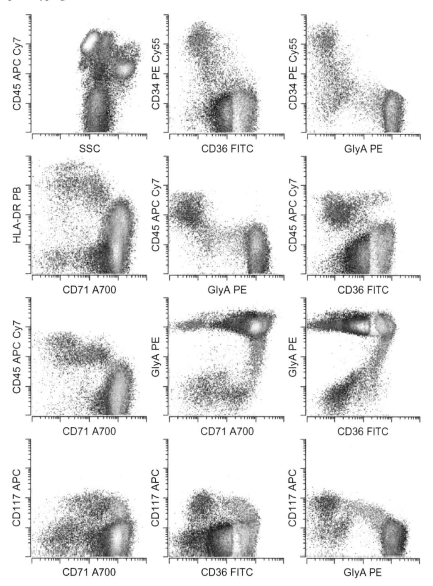

Fig. 5 Maturation of cells during erythroid differentiation. Normal human bone marrow was evaluated with a single combination of seven simultaneous antibodies. Maturation from early (yellow) and late (red) blasts proceeds through proerythroblast (green), basophilic erythroblast (yellow), poly-chromatophilic and orthochromatophilic (light blue and blue), and erythrocyte (lavender) stages. Though recognizable, these stages are not discretely defined and the subdivisions are somewhat arbitrary. The blasts have been emphasized and other populations largely excluded to improve visibility of the maturational relationships. These views represent less than 20% of the total information contained in these antibody combinations. (See Plate no. 17 in the Color Plate Section.)

Table IV
Surface Antigen Expression During B-Cell Maturation

	Early	Intermediate	Late	Mature	Plasma cell
CD10	+++	++	+	−	−
CD19	++	+++	++	++	++
CD20	−	+/++	+++	++	−
CD21	−	−	−	++	+
CD22	+	+	++	++	−
CD24	+++	+++	+++	++	−
CD34	++	−	−	−	−
CD38	++	+++	++	+/++	++++
CD45	+	++	+++	+++	++
HLA-DR	++	+++	++	++	+/−
cIgM	−	++	++	++	−
Kappa	−	−	+	++	+
Lambda	−	−	+	++	+
TdT	+++	−	−	−	−

increased CD45, and acquisition of surface immunoglobulin. Mature B cells show a loss of CD10, bright CD20, and the brightest CD45 with expression of surface immunoglobulin. Plasma cells are readily identified by their extremely bright expression of CD38 or CD138 without CD20. The relative proportions of each of these maturational stages are dependent on the overall regenerative state of the bone marrow and the presence of ongoing inflammatory reactions.

G. Miscellaneous

A number of other cell populations mature or are present in normal bone marrow and include eosinophils, basophils, dendritic cells, natural killer cells, and megakaryocytes. For most of these populations, the patterns of maturation are either not well defined or difficult to visualize using routine methods of analysis.

Of these populations, eosinophil maturation is the easiest to demonstrate and is characterized by one immature stage (eosinophilic myelocyte), showing expression of high side scatter, intermediate CD45 at a level slightly higher than neutrophilic myelocytes, low to intermediate CD11b, intermediate CD13, and low CD33 with bright CD66b without CD16. Maturation to the mature eosinophil is accompanied by an increased level of CD45, a mild decrease in side scatter, and an increase in CD11b with a decrease in CD33. Eosinophilic maturation is illustrated in Fig. 7.

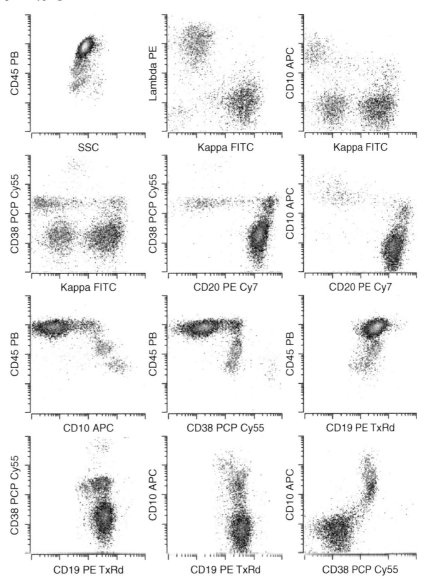

Fig. 6 Maturation of cells during B-cell differentiation. Normal human bone marrow was evaluated with a single combination of eight simultaneous antibodies. Maturation from early (yellow) B cells proceeds through an intermediate stage (light blue) before acquiring surface light-chain expression (blue and red: kappa and lambda, respectively) as late immature B cells (bright CD20) before reaching maturity (bright CD45 without CD10). Rare plasma cells are also present (green). Note the relatively discrete nature of the maturational stages. These views represent less than 20% of the total information contained in these antibody combinations. (See Plate no. 18 in the Color Plate Section.)

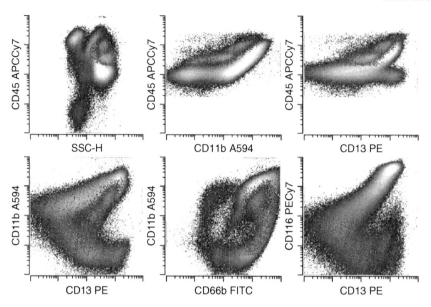

Fig. 7 Maturation of cells during eosinophilic differentiation. Normal human bone marrow was evaluated with a single combination of nine simultaneous antibodies. Maturation from blasts proceeds through the eosinophilic myelocyte stage (yellow) to the mature eosinophil (red). Note the high side scatter present throughout maturation, as well as the increased CD45 and lower CD11b relative to neutrophilic precursors. These views represent less than 20% of the total information contained in these antibody combinations. (See Plate no. 19 in the Color Plate Section.)

V. Abnormal Immunophenotypic Patterns of Maturation

The normal patterns of antigenic expression detailed earlier reflect the well-orchestrated expression of genes characteristic of normal differentiation. In hematopoietic neoplasms, an increasing variety of specific genetic abnormalities have been described that are either directly or indirectly capable of perturbing these normal patterns of protein expression. Stem cell disorders such as myelodysplastic syndromes serve as an informative model of maturational dysregulation and illustrate the association between genetic and immunophenotypic abnormalities (Kussick and Wood, 2003a; Wells *et al.*, 2003). Figure 8 shows an example of the type of abnormalities seen in myelodysplasia. Immunophenotypic abnormalities can be identified in most hematopoietic neoplasms including acute myeloid and lymphoid leukemia (Weir and Borowitz, 2001), myeloproliferative disorders (Kussick and Wood, 2003b), and B- and T-cell lymphoma (Stetler-Stevenson and Braylan, 2001). These abnormalities can be used to diagnose, classify, and monitor disease following therapy and techniques capable of detecting these abnormalities, such as flow cytometry, play an increasingly important role in clinical medicine.

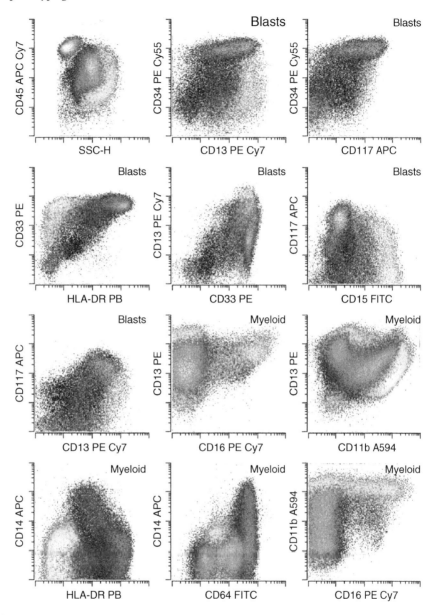

Fig. 8 Myelodysplasia. Human bone marrow was evaluated with a single combination of nine simultaneous antibodies. Blasts are increased in number (red) and show abnormal expression of CD117, CD34, CD13, and CD45. In addition, the neutrophilic lineage shows hypogranularity (decreased side scatter) with abnormal maturational expression of CD13, CD11b, and CD16 with a dyssynchronous pattern of CD13 and CD16 expression. Monocytes also show a relative increase in immature forms with abnormal CD45 and CD14 expression. These findings are characteristic of myelodysplasia. (See Plate no. 20 in the Color Plate Section.)

References

Bender, J. G., Unverzagt, K. L., Walker, D. E., Lee, W., van Epps, D. E., Smith, D. H., Stewart, C. C., and To, L. B. (1991). Identification and comparison of CD34+ cells and their subpopulations from normal peripheral blood and bone marrow using multicolor flow cytometry. *Blood* **77**, 2591–2596.

Borowitz, M. J., Guenther, K. L., Shults, K. E., and Stelzer, G. T. (1993). Immunophenotyping of acute leukemia by flow cytometric analysis. Use of CD45 and right-angle light scatter to gate on leukemic blasts in three-color analysis. *Am. J. Clin. Pathol.* **100**, 534–540.

D'Arena, G., Musto, P., Cascavilla, N., and Carotenuto, M. (1998). Thy-1 (CDw90) and c-Kit receptor (CD117) expression on CD34+ hematopoietic progenitor cells: A five dimensional flow cytometric study. *Haematologica* **83**, 587–592.

Davis, R. E., Longacre, T. A., and Cornbleet, P. J. (1994). Hematogones in the bone marrow of adults. Immunophenotypic features, clinical settings, and differential diagnosis. *Am. J. Clin. Pathol.* **102**, 202–211.

De Jong, M. O., Wagemaker, G., and Wognum, A. W. (1995). Separation of myeloid and erythroid progenitors based on expression of CD34 and c-kit. *Blood* **86**, 4076–4085.

Dworzak, M. N., Fritsch, G., Fleischer, C., Printz, D., Froschl, G., Buchinger, P., Mann, G., and Gadner, H. (1997). Multiparameter phenotype mapping of normal and post-chemotherapy B lymphopoiesis in pediatric bone marrow. *Leukemia* **11**, 1266–1273.

Dworzak, M. N., Fritsch, G., Froschl, G., Printz, D., and Gadner, H. (1998). Four-color flow cytometric investigation of terminal deoxynucleotidyl transferase-positive lymphoid precursors in pediatric bone marrow: CD79a expression precedes CD19 in early B-cell ontogeny. *Blood* **92**, 3203–3209.

Elghetany, M. T. (2002). Surface antigen changes during normal neutrophilic development: A critical review. *Blood Cells Mol. Dis.* **28**, 260–274.

Gaipa, G., Coustan-Smith, E., Todisco, E., Maglia, O., Biondi, A., and Campana, D. (2002). Characterization of CD34+, CD13+, CD33-cells, a rare subset of immature human hematopoietic cells. *Haematologica* **87**, 347–356.

Hermiston, M. L., Xu, Z., and Weiss, A. (2003). CD45: A critical regulator of signaling thresholds in immune cells. *Annu. Rev. Immunol.* **21**, 107–137.

Kussick, S. J., and Wood, B. L. (2003a). Using 4-color flow cytometry to identify abnormal myeloid populations. *Arch. Pathol. Lab. Med.* **127**, 1140–1147.

Kussick, S. J., and Wood, B. L. (2003b). Four-color flow cytometry identifies virtually all cytogenetically abnormal bone marrow samples in the workup of non-CML myeloproliferative disorders. *Am. J. Clin. Pathol.* **120**, 854–865.

Loken, M. R., Shah, V. O., Dattilio, K. L., and Civin, C. I. (1987). Flow cytometric analysis of human bone marrow: I. Normal erythroid development. *Blood* **69**, 255–263.

Longacre, T. A., Foucar, K., Crago, S., Chen, I. M., Griffith, B., Dressler, L., McConnell, T. S., Duncan, M., and Gribble, J. (1989). Hematogones: A multiparameter analysis of bone marrow precursor cells. *Blood* **73**, 543–552.

Lucio, P., Parreira, A., van den Beemd, M. W., van Lochem, E. G., van Wering, E. R., Baars, E., Porwit-MacDonald, A., Bjorklund, E., Gaipa, G., Biondi, A., Orfao, A., Janossy, G., *et al.* (1999). Flow cytometric analysis of normal B cell differentiation: A frame of reference for the detection of minimal residual disease in precursor-B-ALL. *Leukemia* **13**, 419–427.

Macedo, A., Orfao, A., Ciudad, J., Gonzalez, M., Vidriales, B., Lopez-Berges, M. C., Martinez, A., Landolfi, C., Canizo, C., and San Miguel, J. F. (1995). Phenotypic analysis of CD34 subpopulations in normal human bone marrow and its application for the detection of minimal residual disease. *Leukemia* **9**, 1896–1901.

MacLennan, I. C. (1994). Germinal centers. *Annu. Rev. Immunol.* **12**, 117–139.

McGuckin, C. P., Pearce, D., Forraz, N., Tooze, J. A., Watt, S. M., and Pettengell, R. (2003). Multiparametric analysis of immature cell populations in umbilical cord blood and bone marrow. *Eur. J. Haematol.* **71**, 341–350.

McKenna, R. W., Washington, L. T., Aquino, D. B., Picker, L. J., and Kroft, S. H. (2001). Immuno-phenotypic analysis of hematogones (B-lymphocyte precursors) in 662 consecutive bone marrow specimens by 4-color flow cytometry. *Blood* **98**, 2498–2507.

Payne, K. J., and Crooks, G. M. (2002). Human hematopoietic lineage commitment. *Immunol. Rev.* **187**, 48–64.

Rogers, C. E., Bradley, M. S., Palsson, B. O., and Koller, M. R. (1996). Flow cytometric analysis of human bone marrow perfusion cultures: Erythroid development and relationship with burst-forming units-erythroid. *Exp. Hematol.* **24**, 597–604.

Scicchitano, M. S., McFarland, D. C., Tierney, L. A., Narayanan, P. K., and Schwartz, L. W. (2003). *In vitro* expansion of human cord blood CD36+ erythroid progenitors: Temporal changes in gene and protein expression. *Exp. Hematol.* **31**, 760–769.

Stetler-Stevenson, M., and Braylan, R. C. (2001). Flow cytometric analysis of lymphomas and lym-phoproliferative disorders. *Semin. Hematol.* **38**, 111–123.

Terstappen, L. W., and Loken, M. R. (1990). Myeloid cell differentiation in normal bone marrow and acute myeloid leukemia assessed by multi-dimensional flow cytometry. *Anal. Cell Pathol.* **2**, 229–240.

Terstappen, L. W., Safford, M., and Loken, M. R. (1990). Flow cytometric analysis of human bone marrow. III. Neutrophil maturation. *Leukemia* **4**, 657–663.

Terstappen, L. W., Buescher, S., Nguyen, M., and Reading, C. (1992). Differentiation and maturation of growth factor expanded human hematopoietic progenitors assessed by multidimensional flow cytometry. *Leukemia* **6**, 1001–1010.

Tjonnfjord, G. E., Steen, R., Veiby, O. P., and Egeland, T. (1996). Lineage commitment of CD34+ human hematopoietic progenitor cells. *Exp. Hematol.* **24**, 875–882.

Weir, E. G., and Borowitz, M. J. (2001). Flow cytometry in the diagnosis of acute leukemia. *Semin. Hematol.* **38**, 124–138.

Wells, D. A., Benesch, M., Loken, M. R., Vallejo, C., Myerson, D., Leisenring, W. M., and Deeg, H. J. (2003). Myeloid and monocytic dyspoiesis as determined by flow cytometric scoring in myelodys-plastic syndrome correlates with the IPSS and with outcome after hematopoietic stem cell transplan-tation. *Blood* **102**, 394–403.

Wood (2006). *Arch. Path. Lab. Med.* **130**, 680–690.

Wood and Levin (2006). *Cytometry* **70**, 321–328.

CHAPTER 25

Differential Diagnosis of T-Cell Lymphoproliferative Disorders by Flow Cytometry Immunophenotyping. Correlation with Morphology

Wojciech Gorczyca and Sorina Tugulea

Genzyme Genetics (New York Laboratory)
New York, NY 10019

DOI: 10.1016/B978-0-12-375045-7.00025-8

I. Introduction

T-cell lymphoproliferative disorders are a heterogeneous and relatively uncommon group of lymphoid tumors that can present as adenopathy, hepatosplenomegaly, skin lesions, a mass involving various other organs, neutropenia (or other cytopenia), or leukemia/lymphocytosis (Al-Hakeem *et al.*, 2007; Arrowsmith *et al.*, 2003; Ascani *et al.*, 1997; Attygalle *et al.*, 2007; Chan, 1999; Chan *et al.*, 1999; de Bruin *et al.*, 1997; Falcao *et al.*, 2007; Farcet *et al.*, 1990; Ferry, 2002; Gorczyca, 2006; Gorczyca *et al.*, 2003, 2004; Greenland *et al.*, 2001; Harris *et al.*, 2000; Jaffe, 2006; Jaffe *et al.*, 1984; Karube *et al.*, 2008; Kinney and Kadin, 1999; Lamy and Loughran, 2003; Loughran, 1998a,b; Martinez-Delgado, 2006; Mioduszewska, 1979; Oshtory *et al.*, 2007; Pileri *et al.*, 1998; Savage, 2007; Stein *et al.*, 2000; Swerdlow *et al.*, 2008; ten Berge *et al.*, 2003; Vega *et al.*, 2007; Weisberger *et al.*, 2000, 2003; Willemze *et al.*, 2007). Predominantly, nodal distribution is characteristic for angioimmunoblastic T-cell lymphoma (AITL), peripheral T-cell lymphoma, unspecified (PTLU), and anaplastic large cell lymphoma (ALCL). Extranodal location is seen in mycosis fungoides (MF), cutaneous ALCL, extranodal NK/T-cell lymphoma, nasal type (ENKTCL), enteropathy-type T-cell lymphoma (ETTL), hepatosplenic $\gamma\delta$ T-cell lymphoma (HSTL), and subcutaneous panniculitis-like T-cell lymphoma. Leukemic blood involvement is typical for T-cell prolymphocytic leukemia (T-PLL), adult T-cell lymphoma/leukemia (ATLL), and Sézary's syndrome (SS). The diagnosis of T-cell lymphoproliferations often requires multitechnology approach including morphology, flow cytometry (FC), immunohistochemistry, cytogenetics/fluorescence *in situ* hybridization (FISH), viral studies and molecular testing (e.g., polymerase chain reaction, PCR) (Ashton-Key *et al.*, 1997; Bakels *et al.*, 1993; Ballester *et al.*, 2006; Baumgartner *et al.*, 2003; Bruggemann *et al.*, 2007; Chiaretti *et al.*, 2004; Coiffier *et al.*, 1988; Dearden and Foss, 2003; Deleeuw *et al.*, 2007; Dupuis *et al.*, 2006a,b; Falcao *et al.*, 2007; Gaulard *et al.*, 1991; Gorczyca *et al.*, 2002; Graux *et al.*, 2006; Grogg *et al.*, 2005; Jamal *et al.*, 2001; Lakkala-Paranko *et al.*, 1987; Lepretre *et al.*, 2000; Man *et al.*, 2002; Martinez-Delgado, 2006; Speleman *et al.*, 2005). Although there is no single phenotypic marker specific for T-cell lymphoma/leukemia, FC plays an important role in the diagnosis and subclassification of both mature (peripheral/postthymic) and immature T-cell lymphoproliferations. We present the flow cytometric immunophenotypic data from 275 peripheral (mature/postthymic) T-cell disorders and 79 precursor tumors (T-lymphoblastic leukemia/lymphoma; T-ALL/LBL).

II. Materials

Flow cytometric samples from Genzyme (New York Laboratory) containing abnormal monocytic populations were submitted for this study. FC data were reanalyzed and correlated with cytomorphology, cytochemistry and bone marrow studies. The neoplasms were classified according to the World Health Organization (WHO) classification of hematopoietic tumors (Swerdlow *et al.*, 2008). There were 42 cases of T-PLL, 60 cases of T-LGL, 12 cases of ATLL, 11 cases of HSTL, 22 cases of AITL, 12 cases of SS, 4 cases of ETTL, 87 cases of PTLU, 25 cases of ALCL, and 79 cases of T-ALL/LBL.

III. Methods

We used heparinized bone marrow (BM) aspirate, blood and fresh tissue specimens for FC analysis and processed the specimens within 24 h of collection. We obtained a leukocyte cell suspension from blood and BM specimens after red blood cell (RBC) lysis with ammonium chloride lysing solution, followed by 5 min of centrifugation. The cell pellet was suspended with an appropriate amount of RPMI 1640 (GIBCO, NY). Fresh tissue samples were disaggregated with a sterile blade, followed by passage through a mesh filter ($<100 \ \mu$m). The cells were then washed in RPMI media and centrifuged at 1500 rpm for 5 min. To minimize nonspecific binding of antibodies, we incubated the cells in RPMI media supplemented with 10% heat-inactivated fetal bovine serum (FBS) at 37 °C water bath for 30 min. The samples were wash with 0.1% sodium azide/10% FBS phosphate-buffered saline (PBS) buffer and assessed viability using both trypan blue and 7-aminoactinomycin D (Sigma Chemical Co., St. Louis, MI) exclusion assays. Immunophenotypic analysis was performed on Becton Dickinson Immunocytometry System FACS Canto instruments (San Jose, CA) using panels of six-color directly labeled combinations of antibodies used at a saturating concentration. Internal negative controls within each tube and isotype controls for immunoglobulin G1 (IgG1), IgG2a, and IgG2b were used as negative controls. We monitored the instrument fluorescence detector's settings and calibration according to manufacturer's recommendation, using FACS 7-Color Set up Beads (BD Biosystem) and lot specific batch match control for antibodies conjugated with tandem dyes PerCP Cy5.5 and PerCP Cy7(Becton Dickinson). The system linearity was evaluated using Sphero Rainbow Calibration Particles (Spherotech, Inc. Cat. #RCP-30-5A).

IV. Identification of Abnormal T-Cell Population by Flow Cytometry

Abnormal T-cell population may be identified by FC based on the following criteria:

1. loss or aberrant (dim or variable) expression of the pan-T antigen(s) (CD2, CD3, CD5, and/or CD7);
2. aberrant expression of CD4 or CD8 (subset restriction, dual-positive CD4/CD8 expression, or lack of both markers);
3. loss of CD45 expression (or aberrantly dim expression);
4. increased forward scatter (FSS);
5. aberrant expression of T-cell receptor (TCR), for example, dim expression of TCR$\alpha\beta$, positive TCR$\gamma\delta$, and lack of both TCRs;
6. presence of blastic markers (TdT, CD34, CD1a, and/or CD117);
7. expression of NK-cell associated markers (CD16, CD56, and/or CD57);
8. positive CD30;
9. presence of additional markers, such as HLA-DR, CD10, CD11c, CD13, CD19, CD20, CD25, CD33, or CD103.

A. Loss or Aberrant (dim) Expression of the Pan-T Antigen(s)

Loss of CD2, CD3, CD5, and/or CD7 or their aberrant expression (i.e., dim or variable expression) is often present in T-cell neoplasms. Loss of at least one pan-T marker was seen in ∼68% of cases in this series (354 tumors including 79 T-ALL/LBL and 275 peripheral T-cell tumors). One pan T-cell markers was missing in ∼42%, two antigens in ∼19%, three in ∼6%, and all four antigens in <2%. In mature (postthymic) tumors, CD7 was most often lost, and among precursor tumors CD3 was most often absent (see Tables I–V for details). Positive expression of all pan-T antigens does not exclude T-cell lymphoma. Positive expression of all four antigens (including expression on subset and dim/variable expression) was identified in ∼29% of peripheral (mature) tumors and in ∼23% of precursor lesions. Normal (moderate to bright) expression of all four antigens was present in 13.6% of all tumors, including 8/87 cases (∼9%) of PTLU, 2/46 cases (∼4%) of T-LGL leukemia, 2/25 (∼8%) cases of ALCL, 3/22 cases (∼14%) of AITL, 27/42 cases (∼64%) of T-PLL, and 6/79 cases (∼7.5%) if T-ALL/LBL.

Diminished expression of pan-T antigens is less specific for T-cell neoplasms than complete lack of the expression. Diminished expression of one or even two markers is often observed in reactive processes, such as viral infections (especially mononucleosis, where CD7 is often abnormally expressed), medication-associated changes, or in the lymph nodes involved by non-T-cell disorders (e.g., diffuse large B-cell lymphoma and non-Hodgkin lymphoma). CD7 shows most often aberrant

Table I

Immunophenotypic Profile of Precursor T-Lymphoblastic Leukemia (n = 79; Male = 62; Female = 17; Age: 11–79 Years (Average 33.7 Years; <20 Years Old = 22; ≥20 Years Old = 57))

Marker	Frequency (%)	Comments
CD1a$^+$	26.5	
CD2$^+$	68.3	Includes 1 case positive on subset; 7 cases had dim expression
CD3$^+$	32.9[a]	Includes 5 cases positive on subset; 10 cases had dim expression
CD5$^+$	89.9	27 cases had dim expression
CD7$^+$	97.5	2 cases had dim expression
All pan-T antigens+	22.8	Includes 5 cases with CD3 positive on subset and 6 cases (7.6%) with normal expression of all pan-T markers
CD4$^+$	11.4	
CD8$^+$	6.3	
Dual CD4/8$^+$	36.6	
Dual CD4/8$^-$	45.5	
CD10$^+$	32.9	Includes 5 cases positive on subset
CD11b$^+$	2.5	
CD11c$^+$	0	
CD13$^+$	8.9	
CD33$^+$	13.9	
CD34$^+$	35.4	
CD45$^{+/bright}$	22.8	
CD45$^{+moderate}$	60.7	
CD45^{+dim}	15.2	
CD45$^-$	1.3	
CD56$^+$	15.2	
CD57$^+$	2.5	
CD117$^+$	10.1	Includes 2 cases positive on subset
HLA-DR$^+$	7.6	
TCR$\alpha\beta^+$	19.7	
TCR$\gamma\delta^+$	10.6	
TCR$^-$	69.7	
TdT$^+$	87.3	
TdT$^-$/CD34$^-$	5.0	Two of these cases were CD1a+

[a]Surface CD3.

dim expression in non-neoplastic lesions, followed by CD5. In many normal samples, CD5 expression may be dim on a subset of benign T cells. None of the non-T processes showed aberrant (dim) expression of more than two markers. Therefore, presence of dim expression for three or four pan-T antigens is considered suggestive of T-cell lymphoma/leukemia. Diminished expression of one or more antigens is suggestive of malignancy, only if accompanied by other abnormalities (e.g., increased forward scatter). Fig. 1 presents aberrant expression of pan-T antigens in peripheral T-cell lymphoma.

Table II

Comparison between the Phenotype of Mature and Precursor T-Cell Neoplasms

Antigen	Peripheral T-cell disorders[a] (%)	Precursor T-lymphoblastic lymphoma/leukemia (%)
CD45		
Bright[+]	90	23
Moderate[+]	6	61
Dim[+]	1	15
Negative	3	1
Pan T-cell antigens		
CD2[+]	90	68
CD3[+]	68	33
CD5[+]	81	90
CD7[+]	52	97
All pan-T antigens positive	29	23
CD4/CD8		
CD4[+]	64	11
CD8[+]	12	6
CD4/CD8[+]	6	37
CD4/CD8[−]	17	45
TCR		
TCR$\alpha\beta^+$	65	20
TCR$\gamma\delta^+$	8	10
TCR[−]	27	70
CD1a[+]	0	26
CD10[+]	8	33
CD13[+]	0	9
CD33[+]	0	14
CD34[+]	0	35
CD117[+]	4	10
HLA-DR[+]	3	8
TdT[+]	0	87
TdT[−]/CD34[−]	100	5

[a]PTLU+PLL+SS+ALCL+AITL+ATLL+ALCL+hepatosplenic lymphoma.

B. Aberrant Expression of CD4 or CD8 (Subset Restriction, Dual–Positive CD4/CD8 Expression or Lack of both Markers)

Restricted CD4 or CD8 expression, dual CD4/CD8 expression, or lack of both CD4 and CD8 raises the possibility of T-cell neoplasm. Some non-T-cell disorders, such as Hodgkin lymphoma, may show marked predominance of CD4[+] (or rarely CD8[+]) cells. Increased CD4:CD8 ratio may also be observed in granulomatous lymphadenitis (sarcoidosis) and reactive pleural effusions. In some viral infections, the CD4/CD8 ration is reversed. Dual expression of CD4/CD8 is typical for thymocytes and therefore is not diagnostic for T-cell malignancy in the lesions obtained from mediastinum. Angioimmunoblastic T-cell lymphoma (AITL), adult T-cell lymphoma/leukemia (ATLL), and Sézary's syndrome/mycosis fungoides

Table III

Immunophenotypic Profile of T-Cell Prolymphocytic Leukemia (T-PLL) ($n = 42$ cases)

Marker	Frequency (%)
CD2$^+$	98
CD3$^+$	88[a]
CD5$^+$	96
CD7$^+$	91
All pan-T antigens normally expressed	64[b] ⎫ 100%
One or more pan-T antigen abnormal (negative and/or dim)	35 ⎭
One pan-T antigen negative	18
Two pan-T antigens negative	4
Three pan-T antigens negative	0
Four pan-T antigens negative	0
CD4$^+$	57
CD8$^+$	22
CD4/CD8$^+$	14
CD4/CD8$^-$	6
TCR$\alpha\delta^+$	91
TCR$\gamma\delta^+$	0
TCR$^-$	9
CD10$^+$	0
CD11c$^+$	2
CD13/CD33	0
CD16$^+$	0
CD25$^+$	6
CD45$^+$	93
CD45$^-$	7
CD56$^+$	6
CD57$^+$	0
CD117$^+$	17[c]
HLA-DR$^+$	2

[a]Surface CD3.

[b]Among 27 cases (64%) with normal expression of all 4 pan-T antigens, 1 case was CD45$^-$, 7 cases were CD117$^+$, 4 cases were dual CD4/CD8$^-$, and 1 case was dual CD4/CD8$^+$.

[c]Expression of CD117 was restricted to CD8$^+$ cases.

(SS/MF) are CD4$^+$. Majority of T-PLL ($\sim$57%), PTLU ($\sim$60%), and ALCL ($\sim$64%) express CD4, but they may be also CD8$^+$, dual CD4/CD8$^+$ or dual CD4/CD8$^-$. Majority of T-LGL leukemias and enteropathy-type T-cell lymphomas are CD8$^+$, and majority of hepatosplenic T-cell lymphoma and LGL leukemias with NK-cell phenotype are dual CD4/CD8$^-$. Occasional T-cell disorders (e.g., AITL) may show normal pattern of CD4/CD8 expression due to mixture of neoplastic and reactive cells and therefore may be difficult to be identified by FC. Dual CD4/CD8 expression is observed in 14% of T-PLL, 4.6% of PTLU, and $\sim$11% of T-LGL leukemia. Lack of expression of CD4/CD8 is often seen in

Table IV

Immunophenotypic Profile of T-Cell Large Granular Lymphocyte (LGL) Leukemia ($n = 46$ Cases) and NK-LGL Leukemia ($n = 14$ Cases)

Marker	T-LGL leukemia (%)	NK-LGL leukemia (%)
CD2[+]	100[a]	100
CD3[+]	100[b]	0
CD5[+]	78.3[c]	0[g]
CD7[+]	89.1[d]	100
All pan-T antigens positive	69.5[e]	0
One pan-T antigen negative	28.3	0
Two pan-T antigens negative	2.2	100
Three pan-T antigens negative	0	0
Four pan-T antigens negative	0	0
CD4[+]	0	0
CD8[+]	84.8	28.6
CD4/CD8[+]	10.9	0
CD4/CD8[−]	4.3	71.4
TCR$\alpha\delta$[+]	76.1	7.1
TCR$\gamma\delta$[+]	23.9	0
TCR[−]	0	92.9
CD11b[+]	21.7	50.0
CD11c[+]	52.2	71.4
CD16[+]	34.8[f]	71.4[h]
CD25[+]	0	n/a
CD45[+]	100	100
CD56[+]	26.1[f]	71.4[h]
CD57[+]	95.6[f]	64.3[h]
HLA-DR[+]	10.9	0

[a]Includes 5 cases with dim expression.
[b]Includes 2 cases with dim expression.
[c]Includes 23 cases with dim expression.
[d]Includes 1 case positive on subset and 14 cases with dim expression.
[e]Only two of these cases (4.3% of all cases) had normal expression of all pan-T markers.
[f]None of the cases coexpressed all three markers. None of the cases was dual CD16/CD56[+] or only CD16[+]. Two cases (4.3%) were only CD56[+] and 18 cases (39.1%) were only CD57[+].
[g]One case showed small subset of cells positive.
[h]One case was CD16[+] only, one case was CD56[+] only, and two cases were CD57[+] only. Five cases showed coexpression of all three markers.

precursor and less often in mature T-cell disorders. Dual CD4/CD8[−] phenotype is seen in 100% of NK-cell LGL leukemia, 91% of hepatosplenic T-cell lymphoma, 25% of PTLU, 8% of ALCL, and 6% of T-PLL.

C. Loss of CD45 Expression (or Aberrantly Dim Expression)

Lack of CD45 is rarely observed in peripheral T-cell disorders (~3%). When present this is highly suspicious for malignancy. In contrast to precursor B-lymphoblastic lymphoma/leukemia (B-ALL), precursor T-cell neoplasms only

Table V

Immunophenotypic Profile of Peripheral T-Cell Lymphoma, Unspecified (PTLU) ($n = 87$ Cases)

Marker	Frequency (%)	Comments
CD2$^+$	90.8	Includes 3 cases with dim expression
CD3$^+$	60.9[a]	Includes 9 cases with dim expression
CD5$^+$	85.0	Includes 13 cases with dim expression
CD7$^+$	49.4	Includes 10 cases with dim expression
All pan-T antigens normally expressed	9.2	
All pan-T antigens positive, but at least one is abnormal (dim or variable)	14.9	100%
One pan-T antigen negative	46.0	
Two pan-T antigens negative	21.8	
Three pan-T antigens negative	8.0	
Four pan-T antigens negative	0	
CD4$^+$	59.8	
CD8$^+$	10.3	
CD4/CD8$^+$	4.6	
CD4/CD8$^-$	25.3	
TCR$\alpha\delta^+$	71.9	
TCR$\gamma\delta^+$	4.7	
TCR$^-$	23.4	
CD10$^+$	8.0	
CD11c$^+$	14.9	
CD13/CD33	0	
CD16$^+$	0	
CD25$^+$	16.1	
CD45$^+$	97.7	Includes 8 cases (9.2%) with dimmer than normal expression and 77 cases (88.5%) with normal (bright) expression
CD45$^-$	2.3	
CD56$^+$	10.3	
CD57$^+$	5.7	
CD117$^+$	0	
HLA-DR$^+$	5.7	

[a]Surface CD3.

rarely show loss of CD45 (1%). Among mature (postthymic) T-cell disorders, loss of CD45 is most often seen in T-PLL ($\sim$7%), followed by PTLU ($\sim$2%). Benign (reactive) conditions do not show loss or aberrantly dim expression of CD45.

D. Increased Forward Scatter (FSS)

Increased forward scatter (FSC) may be observed in subset of T-cell lymphoproliferations characterized by predominance of large lymphocytes (e.g., ALCL, large cell type of PTLU). Occasional cases of T-LGL leukemia and other lymphomas may display mildly increased side scatter (SSC).

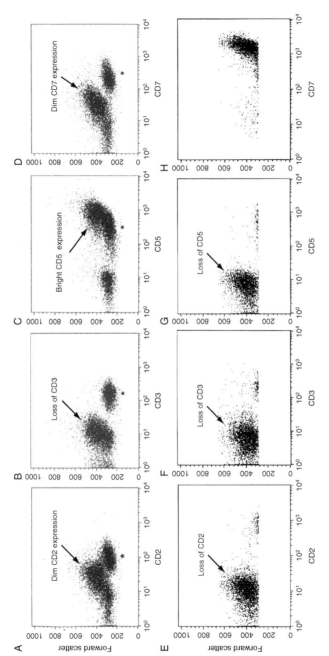

Fig. 1 Identification of abnormal T-cell by flow cytometry. Dim expression of pan-T-antigens (A–D) and loss of pan-T antigens (E–H). In the upper panels, neoplastic T-cells (arrow) show increased forward scatter and dim expression of CD2 (A), lack of CD3 (B), bright expression of CD5 (C), and dim CD7 (D). Residual benign T-cells (*) have low forward scatter and moderate expression of all pan-T antigens. In the lower panels, neoplastic T-cells show loss of CD2 (E), CD3 (F), and CD5 (G) expression.

E. Aberrant Expression of T-Cell Receptor (TCR), for example, Dim Expression of TCRαβ, Positive TCRγδ and Lack of both TCRs

Dim expression of TCRαβ, lack of both TCR-associated antigens or positive expression of TCRγδ on a significant proportion of T-cells indicates a malignant process.

F. Presence of Blastic Markers (TdT, CD34, CD1a, and/or CD117)

CD117, marker associated with acute myeloid leukemia and gastrointestinal stroma tumors can be rarely expressed in both peripheral (mature) and precursor T-cell lymphomas/leukemias (4% and 10%, respectively). Interestingly, among mature (peripheral) T-cell disorders CD117 expression is restricted to CD8$^+$ tumors. The presence of TdT, CD34, and/or CD1a indicates immature T-cell population.

G. Expression of NK-Cell Associated Markers (CD16, CD56, and/or CD57)

NK-cell markers (CD16, CD56, and CD57) are typically expressed by T-LGL leukemia. CD56 may be expressed in T-cell lymphomas including PTLU, ALCL, and hepatosplenic γδ T-cell lymphoma. Activated T-cells in reactive conditions may show dim CD56 expression on subset.

H. Positive CD30

CD30 is expressed by ALCL and a subset of other T-cell disorders, including PTLU and enteropathy-type T-cell lymphoma. Subset of cells in AITL may be CD30$^+$.

I. Presence of Additional Markers, Such as HLA-DR, CD10, CD11c, CD13, CD19, CD20, CD25, CD33, or CD103

Benign T-cells are negative for HLA-DR. T-cells in AITL often express CD10 (>60%). CD10 is positive on a subset of benign thymocytes (thymic tissue). CD25 can be present in different T-cell disorders and is typically positive in ATLL. CD103, a marker expressed by hairy cell leukemia is often positive in enteropathy-type T-cell lymphoma. Activated T-cells in reactive processes may show expression of HLA-DR, CD11c, and CD38 on subset. Subset of T-ALL/LBL may display aberrant expression of pan-myeloid markers, CD13, and/or CD33. Rare cases of PTLU coexpress B-cell associated markers (CD19 and/or CD20).

V. Precursor T-Lymphoblastic Lymphoma/Leukemia (T-ALL)

Precursor T-lymphoblastic leukemia/lymphoma (T-ALL/LBL) is a neoplasm of T-lymphoblasts which accounts for approximately 20–25% of patients with ALL (15% of childhood and 25% of adult ALL) (Pui *et al.*, 2004). T-ALL has clinical, immunologic, cytogenetic, and molecular features that are distinct from those with B-ALL (Chiaretti *et al.*, 2004, 2005; Garand and Bene, 1994; Gassmann *et al.*, 1997; Schneider *et al.*, 2000; Yeoh *et al.*, 2002;). T-ALL occurs more often in male than female patients (at ratio ∼6:1) and involves bone marrow and blood, mediastinum and less commonly lymph nodes, skin, gonads, and central nervous system (Amadori *et al.*, 1983; Baccarani *et al.*, 1982; Ferrando and Look, 2003; Jaffe, 2001; Khalidi *et al.*, 1999; Ross *et al.*, 2003 Shuster *et al.*, 1990;). Mediastinal involvement and adenopathy are more common in younger patients than in patients older than 60 years.

T-lymphoblasts are positive for TdT, one or more of pan-T antigens (CD2, CD3, CD5, and CD7) and CD45, and may be positive for CD34 and CD1a (see Table I for the frequency of antigen expression in T-ALL/LBL). CD45 expression is often dim to moderate, but may be bright similarly to mature T-cell lymphomas. Almost all cases of T-ALL/LBL display aberrant expression of pan-T markers (loss or diminished expression). Among pan-T antigens, CD7 is most often expressed and surface CD3 is often absent (staining for cytoplasmic CD3 is positive, however). Most of the cases are dual CD4/CD8$^-$ or dual CD4/CD8$^+$, and only rare cases are either CD4$^+$ or CD8$^+$. T-ALL/LBLs most often lack TCR expression, about 1/3 of cases are TCR$\alpha\beta^+$, and only rare cases are TCR$\gamma\delta^+$. A subset of T-ALL/LBL is positive for CD10, CD56, and pan-myeloid antigens. Figure 2 presents an example of FC analysis of T-ALL/LBL.

VI. Peripheral (Mature/Postthymic) Lymphoma Versus T-ALL

Table II presents a comparison between the Peripheral (mature) and precursor T-cell neoplasms. Presence of blastic markers (CD34, TdT) and CD1a indicates T-ALL/LBL (lack of both TdT and CD34 was observed in only 5% T-ALL/LBL cases). Aberrant expression of pan-myeloid antigens, CD13 and/or CD33 is seen in subset of T-ALL/LBL, but not in mature T-cell tumors. T-ALL/LBLs usually have dim or moderate CD45 expression, whereas peripheral T-cell tumors typically display bright CD45. As far as pan-T antigens are concerned, immature tumor more often are CD2 and/or CD3 negative, and CD7+, whereas peripheral tumors very often show loss of CD7. The frequency of CD5 expression is comparable in both categories. Evaluation of CD4 and CD8 expression is very helpful in differential diagnosis: dual positive or dual negative CD4/CD8 expression is more often seen in T-ALL/LBL, whereas CD4 expression is more typical for mature neoplasms. Lack of TCR, positive CD10 or positive CD117 is seen more frequently in T-ALL/LBL.

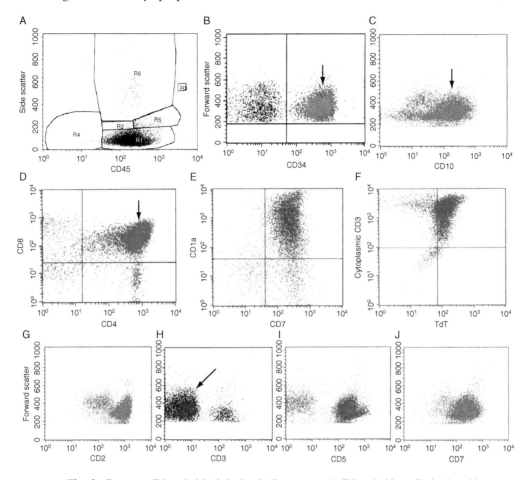

Fig. 2 Precursor T-lymphoblastic leukemia (bone marrow). T-lymphoblasts display low side scatter and moderate expression of CD45 (A) and are positive for CD34 (B), CD10 (C), CD4/CD8 (D), CD1a (E), TdT (F), CD2 (G), CD5 (I), and CD7 (J). Surface CD3 is negative (H) but cytoplasmic CD3 is positive (F).

VII. Thymocytes from Thymic Hyperplasia/Thymoma Versus T-ALL

Dual positive expression of CD4/CD8 is rarely observed in peripheral (mature/postthymic) T-cell lymphoproliferative disorders and, therefore, when present is suspicious for an immature T-cell population (T-ALL/LBL shows coexpression of CD4 and CD8 in ~40% of cases). Thymocytes from either hyperplastic thymus or thymoma are always CD4/CD8 positive and therefore the diagnosis of mediastinal T-ALL/LBL cannot be based on the presence of CD4/CD8 coexpression. Regardless whether from thymoma or benign thymic hyperplasia, benign immature

T-cells (thymocytes) display characteristic variable expression of surface CD3 antigen. Majority of cells (small T-cells) show positive but variable (smeared-like) CD3 expression, whereas larger T-cells are CD3 negative (Fig. 3). This smeared pattern is never observed in surface CD3$^+$ T-ALL/LBL. Small more mature T-cells from thymoma/thymic hyperplasia are CD10$^-$ and larger immature T-cells (surface CD3$^-$) shows positive expression of CD10.

VIII. Mature T-Cell Lymphoproliferative Disorders

A. T-Cell Prolymphocytic Leukemia (T-PLL)

T-prolymphocytic leukemia (T-PLL) is a rare mature (postthymic) T-cell lymphoproliferative disorder which affects adults, occurs more frequently in men and is characterized by aggressive clinical course and poor outcome (Bartlett and

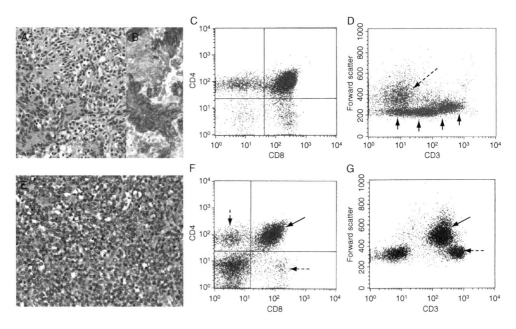

Fig. 3 Comparison between flow cytometric features of thymocytes (A–D) and T-lymphoblasts (E–G). (A, B) Histologic section of thymoma. Tumor cells are positive for cytokeratin (AE1/AE3). Flow cytometric analysis reveals immature T-cells with dual expression of CD4/CD8 (C) and variable (smeared) expression of surface CD3 ((D); arrows). There is additional population of larger T-cells with negative expression of CD3 (dashed arrow). (E) T-lymphoblastic lymphoma (mediastinum). Lymphoblasts are dual positive for CD4/CD8 (F). There are distinct populations of benign T-cells expressing CD4 or CD8 (dashed arrows). In contrast, thymocytes (C) shows gradual transition from CD4$^+$/CD8$^-$ cells to CD4$^-$/CD8$^+$ cells with majority of cells positive for both antigens. T-lymphoblasts (arrow) show moderate expression of surface CD3 (G) without typical for thymocytes smeared patter (compare with D). Normal (benign) T-cells (dashed arrow, G) show brighter CD3 expression and lower forward scatter, when compared to lymphoblasts. (See Plate no. 21 in the Color Plate Section.)

Longo, 1999; Brito-Babapulle *et al.*, 1997; Cao and Coutre, 2003; Dearden, 2006a,b; Matutes, 1998, 2002; Matutes *et al.*, 1988; Pawson *et al.*, 1997). T-PLL is composed of small to medium-sized lymphocytes with prominent nucleoli. The principal disease characteristics are organomegaly (especially splenomegaly and lymphadenopathy), anemia, thrombocytopenia, skin lesions, and prominent (often rapidly increasing) lymphocytosis in blood (most often $>100 \times 10^9 \, l^{-1}$). Skin, liver, and other organs may be also involved (Harris *et al.*, 1999; Mallett *et al.*, 1995; Magro *et al.*, 2006). A subset of patients experiences an initial (median 33 months, range: 6–103 months) indolent clinical course with stable moderate leukocytosis, with subsequent progression to aggressive stage (Garand *et al.*, 1998). There is no association with human T-cell lymphotropic viruses (HTLV-I/II) (Pawson *et al.*, 1998c).

Immunophenotypic analysis by flow cytometry shows expression of CD45 and pan-T antigens (CD2, CD3, CD5, CD7) and lack of expression of B-cell markers, HLA-DR, CD1a, TdT, and CD34 (Table III presents phenotypic characteristics of T-PLL). Majority of cases shows moderate/bright expression of CD45. Rare cases of T-PLL may be CD45 negative. Normal expression of pan-T antigens is present in ~65% of cases; feature rarely seen in other T-cell lymphoproliferations. In 57% of cases the T-PLL are CD4$^+$, in 22% they are CD8$^+$, in 14% they coexpress CD4/CD8 and in the remaining 6% they are CD4/CD8$^-$. Significant subset of CD8$^+$ cases showed coexpression of CD117. Figures 4 and 5 present examples of T-PLL.

B. T-Large Granular Lymphocyte Leukemia (T-LGL)

Large granular lymphocyte leukemia (LGL leukemia) is characterized by lymphoid cells with abundant cytoplasm and azurophilic (MPO-negative) granules. Two main variants of LGL proliferations can be recognized: T-cell LGL leukemia and NK-cell LGL leukemia (Agnarsson *et al.*, 1989; Chan *et al.*, 1986, 1992; Greer *et al.*, 2001; Kwong and Wong, 1998; Kwong *et al.*, 1997; Lamy and Loughran, 1998, 1999, 2003; Loughran, 1993, 1998a,b, 1999; Loughran and Starkebaum, 1987a,b; Loughran *et al.*, 1985; Matutes and Catovsky, 2003; Osuji *et al.*, 2006; Pandolfi *et al.*, 1990; Rose and Berliner, 2004; Sandberg *et al.*, 2006; Sokol and Loughran, 2003). T-LGL leukemia involves blood, bone marrow, liver, and spleen. It has indolent clinical course and may be associated with neutropenia, red cell aplasia, hypergammaglobulinemia, and rheumatoid arthritis.

Clonal LGL proliferations are assumed to be derived from normal LGL cells, which comprise 10–15% of peripheral blood mononuclear cells. Based on the phenotype, LGL are classified into two major lineages: CD3$^+$ T-LGL leukemia with cytotoxic T-cell phenotype and CD3$^-$ T-LGL leukemia with NK-cell phenotype. On FC analysis, T-LGL leukemia has a mature T-cell phenotype with CD3, CD8, and TCR$\alpha\beta$ expression. Rare cases may be CD4$^+$/CD8$^+$, CD4$^+$/CD8$^-$, or CD4$^-$/CD8$^-$. CD11b, CD11c, and NK-cell associated markers are variably expressed (Table IV). Of the NK-cell associated markers, CD57 is most frequently expressed (~96%), whereas CD16 or CD56 is present in approximately 1/3 cases

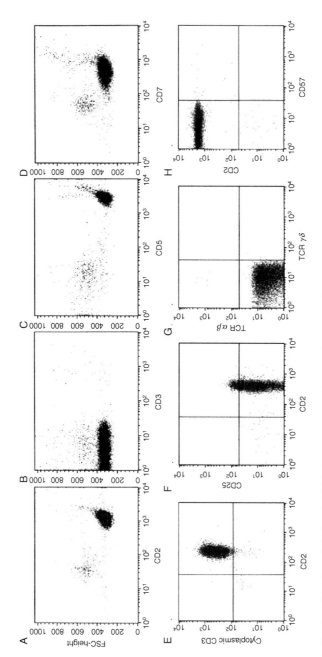

Fig. 4 T-cell prolymphocytic leukemia (T-PLL). Lymphocytes display bright expression of CD2 (A), CD5 (B), and CD7 (D). Surface CD3 is absent (C) but cytoplasmic CD3 is present (E). Neoplastic lymphocytes do not coexpress CD25 (F) or CD57 (H). They show aberrant lack of TCRα/β and TCRγ/δ (G).

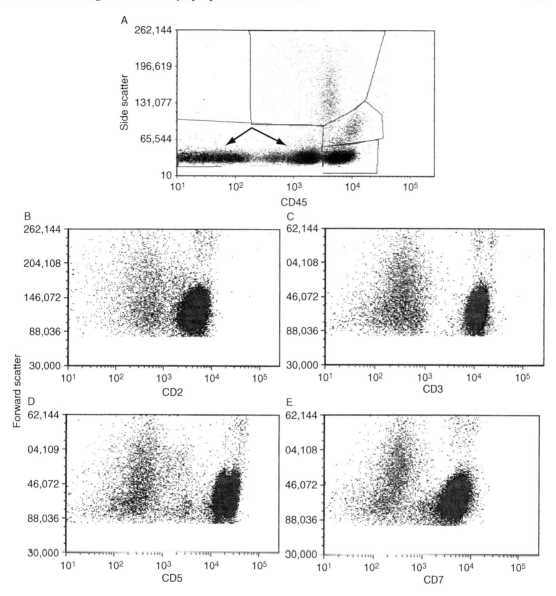

Fig. 5 T-PLL with aberrant expression of CD45 (negative on subset and dim to moderate on remaining cells (A), arrows). The expression of all pan-T markers is normal (B–E).

(CD16 in ~35% and CD56 in ~26%). T-LGL leukemias generally lack dual CD16/CD56 coexpression or coexpression of all NK-cell associated markers (CD16/CD56/CD57), a feature more typical for NK-cell proliferations. In our series of 46 cases, there were no cases with CD16 expression only, whereas 2 cases (~4%)

had only CD56 expression and 18 cases (~39%) were only CD57$^+$ (CD16 and CD56 were negative). Among pan-T markers, CD5 is most often aberrantly expressed (either dim or absent). CD2 and CD3 were positive in all 46 cases. Thirty-five cases (~76%) had TCR$\alpha\beta$ phenotype and the remaining 11 cases (~24%) were TCR $\gamma\delta^+$. Figure 6 presents T-LGL with $\gamma\delta$ phenotype.

LGL leukemia with NK cell phenotype (NK-LGL leukemia) is less common and is usually classified with NK-cell disorders. It usually has a chronic, nonprogressive clinical course. Patients are either asymptomatic or have a gradual increase in circulating large granular lymphocytes. It differs clinically from T-LGL leukemia by lack of neutropenia, anemia, or rheumatoid arthritis. The phenotypic hallmark of NK-LGL leukemia is lack of CD3 and CD5 and positive expression of CD2 and CD7. In our series, CD45, CD2, and CD7 were positive in all 14 cases, whereas CD3 and CD5 were always negative. In contrast to T-LGL leukemias, LGL with NK-cell phenotype more often shows expression of CD16 (~71%) and CD56 (~71%), with 35% of cases coexpressing those two antigens (see Table IV for details). CD57 was present in ~64% cases. Majority of cases (~71%) were CD4/CD8$^-$, and only subset of cases showed CD8 expression (~29%). Table IV summarizes the phenotype of T-LGL leukemias.

C. Adult T-Cell Lymphoma/Leukemia (ATLL)

Adult T-cell lymphoma/leukemia (ATLL) is a peripheral (mature/postthymic) T-cell lymphoproliferative disorder associated with a retrovirus, human T-cell leukemia virus type 1 (HTLV-1), and is characterized by highly pleomorphic lymphoid cells (Blattner *et al.*, 1982; Chiaretti *et al.*, 2004; Jaffe *et al.*, 1985; Levine *et al.*, 1994; Loughran, 1996; Oshiro *et al.*, 2006; Pawson, 1999; Pawson *et al.*, 1998a,b; Poiesz *et al.*, 1980; Takatsuki *et al.*, 1985; Vitale *et al.*, 2006; Yasunami *et al.*, 2007). Tumor cells have prominent nuclear irregularities, often referred to as *cloverleaf* or *flower* cells (Fig. 7). ATLL affects adults with leukemic or subleukemic presentation, cutaneous involvement, lymphadenopathy, and/or organomegaly. Hypercalcemia is present in up to 50% of patients and may results in renal failure.

In our series of 12 cases, leukemic cells were positive for CD4 (100%), CD2 (~92%), CD3 (75%), CD5 (100%), TCR$\alpha\beta$ (~83%), and CD25 (100%). CD7 was expressed only in ~8%. Both TCR$\alpha\beta$ and TCR$\gamma\delta$ were negative in ~16%. NK-cell associated markers (CD16, CD56, and CD57) were negative.

D. Anaplastic Large Cell Lymphoma (ALCL)

Anaplastic large cell lymphoma (ALCL) is a T-cell lymphoma composed of large pleomorphic cells, which have irregular kidney-shaped nuclei ("hallmark cells") and are positive for CD30, and in majority of cases ALK (anaplastic lymphoma kinase protein) (Amin and Lai, 2007; Cataldo *et al.*, 1999; Chadburn *et al.*, 1993; Chan, 1998; Chan *et al.*, 1989; Delsol *et al.*, 1988; Falini, 2001;

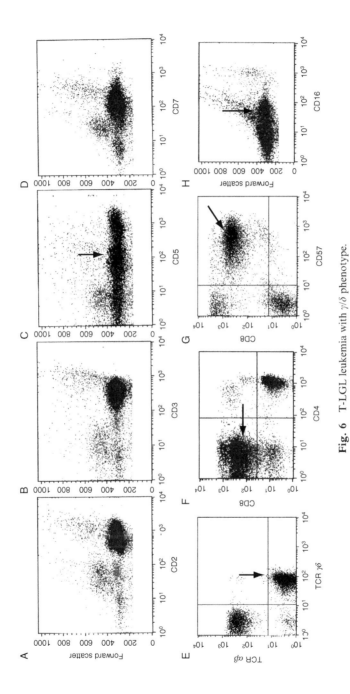

Fig. 6 T-LGL leukemia with γ/δ phenotype.

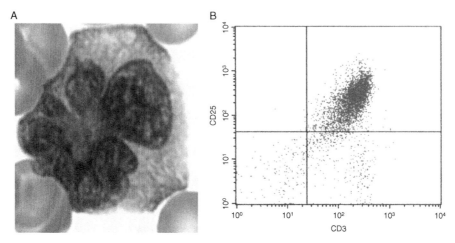

Fig. 7 Adult T-cell lymphoma/leukemia (ATLL). Tumor cells have irregular nuclei (A) and are positive for CD25 (B).

Greer *et al.*, 1991; Inghirami *et al.*, 1994; Jaffe, 2001; Kalinova *et al.*, 2007; Kinney and Kadin, 1999; Kinney *et al.*, 1993; Lamant *et al.*, 1996, 2007; Leoncini *et al.*, 2000; Liang *et al.*, 2004; Nakamura *et al.*, 1997; Onciu *et al.*, 2003; Pileri *et al.*, 1990, 1995, 1998; Shiota and Mori, 1997; Stein *et al.*, 1990, 2000; Szomor *et al.*, 2003; ten Berge *et al.*, 2000, 2002, 2003; Thompson *et al.*, 2005; Tort *et al.*, 2001; Vecchi *et al.*, 1993; Weisenburger *et al.*, 2001; Zinzani *et al.*, 1996). Subset of ALCL harbors t(2;5)(p23;q35) leading to fusion between *ALK* and *NPM* (Stein, Foss *et al.*, 2000). Other chromosomal aberrations involving *ALK*, which lead to expression and constitutive activation of ALK include $t(1;2)(q21;p23)^{TPM3/ALK}$, $inv2(p23q23)^{ATIC/ALK}$, $t(2;3)(p23;q21)^{TFG/ALK}$, $t(2;17)(p23;q23)^{CLTC/ALK}$, $t(2;19)$ $(p23;q13.1)^{TPM4/ALK}$, and $t(2;X)(p23;q11–12)^{MSN/ALK}$ (Amin and Lai, 2007; Benharroch *et al.*, 1998; Duyster *et al.*, 2001; Falini *et al.*, 1998, 1999a,b; Greenland *et al.*, 2001; Kinney and Kadin, 1999; Lamant *et al.*, 1999; Liang *et al.*, 2004; Ma *et al.*, 2000). ALCL involves lymph nodes and extranodal sites, including skin, soft tissues, lung, and bone and less often gastrointestinal tract. ALCL occurs at any age including children.

Flow cytometry may show cluster of cells with increased forward scatter (Fig. 8) as well as increased side scatter, which put tumor cells in "monocytic" or "granulocytic" regions on CD45 versus side scatter display (as with other high grade lymphomas, FC may underestimate the number of tumor cells). In our series of 25 cases, CD45 was positive in 24 cases (96%), CD2 in 16 (64%), CD3 in 8 (32%), CD5 in 11 (44%), and CD7 in 7 (28%) cases. All four pan-T antigens were positive only in 2 cases (8%), whereas one antigen was negative in 5 cases (20%), two antigens were negative in 6 cases (24%), three antigens were negative in 7 cases (28%), and all four antigens were negative in 5 cases (20%). All tumors were positive for CD30. Sixteen cases expressed CD4 (64%), 6 cases expressed CD8 (24%), one case was

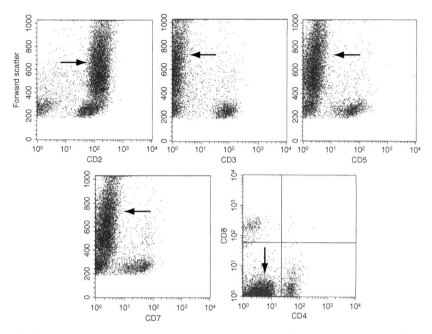

Fig. 8 Anaplastic large cell lymphoma. Large tumor cells (arrow) have markedly increased forward scatter. They are positive for CD2 and lack CD3, CD5, CD7, and both CD4 and CD8.

CD4/CD8$^+$ (4%), and two cases were dual CD4/CD8$^-$ (8%). In a FC series of 19 cases reported by Juco *et al.* the neoplastic cells expressed CD45, HLA-DR, and CD30 in all cases, CD2 in 71%, CD3 in 32%, CD4 in 63%, CD5 in 26%, CD7 in 32%, CD8 in 21%, and CD25 in 88% (Juco *et al.*, 2003). In a series reported by Kesler *et al.* (29 cases), CD4 was expressed most commonly (80%), followed by CD2 (72%), CD3 (40%), and CD5 and CD7 (32% each); CD45 was expressed in 23 of 25 cases and CD13 in 7 of 9 cases (Kesler *et al.*, 2007). Flow cytometry is helpful in identifying cases with leukemic blood involvement, body cavities (effusions), or CSF. In blood or bone marrow analysis, the sensitivity of FC approaches that of molecular testing for *NPM/ALK* (PCR) (Damm-Welk *et al.*, 2007).

E. Peripheral T-Cell Lymphoma, Unspecified (PTLU)

Peripheral T-cell lymphoma, unspecified (PTLU) is a heterogeneous group of tumors with variable clinical features, histology, genetic alteration, response to treatment, and prognosis. It is a mature T-cell lymphoma which does not fulfill morphologic, phenotypic, or genetic criteria for any distinctive mature T-cell lymphoma category, hence the designation "unspecified" (Arrowsmith *et al.*, 2003; Ascani *et al.*, 1997; Attygalle *et al.*, 2007; Ballester *et al.*, 2006; Bekkenk *et al.*, 2003; Campo *et al.*, 1998; Chan, 1999; de Bruin *et al.*, 1994; Dearden and

Foss, 2003; Falini *et al.*, 1990; Gaulard *et al.*, 1990a,b, 1991; Haioun *et al.*, 1992; Harris *et al.*, 1999; Hastrup *et al.*, 1991; Jaffe, 2006; Jaffe *et al.*, 1999; Kim *et al.*, 2002; Lakkala-Paranko *et al.*, 1987; Lepretre *et al.*, 2000; Lopez-Guillermo *et al.*, 1998; Martinez-Delgado, 2006; Mioduszewska and Kulczycka, 1988; Mioduszewska, 1979, 1985; Mioduszewska and Porwit-Ksiazek, 1984; Pileri *et al.*, 1998; Rudiger *et al.*, 2002; ten Berge *et al.*, 2003; Thorns *et al.*, 2007; Zettl *et al.*, 2004). The distinction from other (specific) peripheral T-cell disorders is based on clinical, morphologic, phenotypic, and genetic data, but is not always clear, especially at the time of initial diagnosis. The main entities from which PTLU has to be differentiated include nodal involvement by T-PLL, lymphomatous variant of ATLL, ALCL, AITL, and reactive T-cell infiltrates in benign processes (such as Kikuchi lymphadenitis) or accompanying B-cell lymphomas (e.g., T-cell/histiocyte-rich large B-cell lymphoma, T/HRBCL), classical HL or nonhematopoietic tumors. Leukemic blood involvement by PTLU needs to be differentiated from T-PLL, SS, ALCL, T-LGL leukemia, ATLL, and reactive processes.

The expression of CD45 is normal (i.e., bright) in majority of cases ($\sim$89%). Dimmer than usual expression of CD45 is seen in $\sim$9%, and negative CD45 in $\sim$2%. Although normal (moderate/bright) expression of all four pan-T antigens is occasionally seen (8/87 cases; 9.1%), most cases show aberrant expression of one or more of the pan-T antigens (loss of the antigen or aberrantly dim or variable expression; Fig. 9). Twenty-one cases ($\sim$24%) displayed positive expression of all pan-T antigens, including 8 cases (9.2%) with normal pattern and 13 cases ($\sim$15%) with aberrantly dim (or variable) expression of at least one of the pan-T markers. Remaining 66 cases ($\sim$76%) showed loss of one (40 cases; $\sim$46%), two (19 cases; $\sim$22%), or three (7 cases; $\sim$8%) pan-T antigens. Among pan-T antigens, CD7 was most frequently lost ($\sim$50%), whereas lack of CD2, CD3, and CD5 was observed in $\sim$9% (8/87), 39% (34/87), and $\sim$15% (13/87), respectively. Additionally, dim expression of CD2, CD3, CD5, and CD7 was noted in 3.4%, 10.3%, 14.9%, and 11.5% cases, respectively. In summary, $\sim$91% of cases (79/87) showed aberrant expression of at least one pan-T antigen (loss or dim expression). Although CD4$^+$ lymphomas predominated ($\sim$60%), dual negative expression of CD4/CD8 was observed on significant proportions of tumors (22/87; $\sim$25%). The remaining tumors were CD8$^+$ ($\sim$10%) and dual CD4/CD8$^+$ ($\sim$4.6%). Of additional markers, CD11c, CD15, CD25, CD30, CD56, CD57, CD117, EMA, and HLA-DR were occasionally expressed (CD10 expression is more typical for AITL but may rarely be seen in PTLU). Table V presents the immunophenotypic profile of PTLU.

F. Angioimmunoblastic T-Cell Lymphoma (AITL)

Angioimmunoblastic T-cell lymphoma (AITL), an uncommon but aggressive nodal peripheral (mature) T-cell lymphoma is characterized by systemic symptoms, (reactive) polyclonal hypergammaglobulinemia, and generalized lymphadenopathy with a polymorphous lymphoid infiltrate and increased vascularity

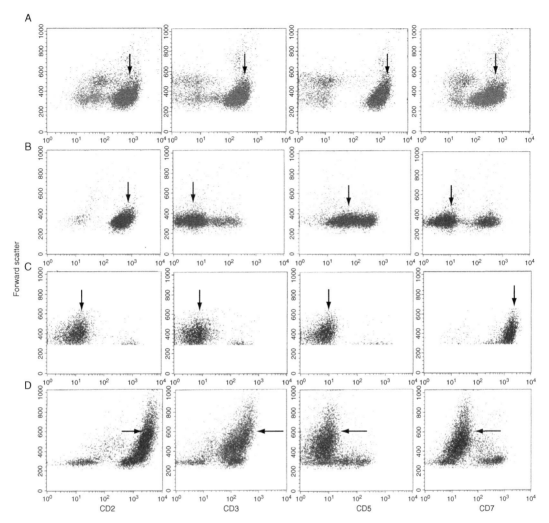

Fig. 9 PTLU—flow cytometry. Four different examples of PTLU, analyzed by flow cytometry. (A) PTLU with normal expression of all four pan T markers (no loss or aberrant dim expression of CD2, CD3, CD5, and CD7). (B) PTLU with loss of CD3 and dim expression of CD5. (C) PTLU with loss of all pan-T antigens except CD7 (arrow). (D) PTLU with increased forward scatter (arrow), aberrant bright expression of CD2, normal expression of CD3, and aberrant loss of both CD5 and CD7.

(Attygalle *et al.*, 2002, 2007a,b; Dogan *et al.*, 2003; Dunleavy *et al.*, 2007; Ferry, 2002; Frizzera *et al.*, 1989; Jaffe, 1995; Kaneko *et al.*, 1988; Lachenal *et al.*, 2007; Lorenzen *et al.*, 1994; Pro and McLaughlin, 2007; Steinberg *et al.*, 1988; Tan *et al.*, 2006; Weiss *et al.*, 1986). The expression of the chemokine CXCL13 by the neoplastic cells in conjunction with gene-expression profile suggest that AITL mostly likely derived from follicular helper T-cells, a finding that explains many of its pathological and clinical features (Dunleavy *et al.*, 2007; Dupuis *et al.*, 2006; Grogg *et al.*, 2005; Jaffe, 2006). Common clinical symptoms include skin rash, arthritis, and edema with pleural and/or peritoneal effusions.

Immunophenotyping reveals predominance of T-cells. Both $CD4^+$ and $CD8^+$ T-lymphocytes are present, the former usually predominate (Fig. 10). Clusters of atypical T-cells usually display aberrant expression of pan-T markers, most often loss of CD3 and/or CD7. The expression of T-markers may be much dimmer than in residual small (benign) T-cells. Majority of cases of AITL display coexpression of CD10 by neoplastic T-cells. The number of $CD10^+$ T-cells in individual tumor may vary, however. Attygalle *et al.* reported CD10 expression in 89% of AITL (Attygalle, Chuang *et al.*, 2007). Cook *et al.* reported the presence of rare benign $CD10^+$ T-cells in reactive lymph nodes and B-cell disorders and therefore the identification of minute population of $CD10^+$ T-cells should not be considered an indication of AITL (Cook, Craig *et al.*, 2003). Bcl-6 is expressed by T-cells in approximately one-third of cases.

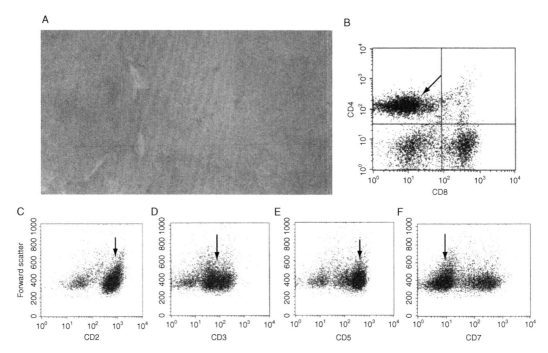

Fig. 10 AITL—flow cytometry. Histology (low power) shows effacement of the architecture by a polymorphous lymphovascular infiltrate (A). Flow cytometry shows mixed population of $CD4^+$ and $CD8^+$ T-cells (B). Neoplastic $CD4^+$ T-cells express CD2 (C), CD3 (D), and CD5 (E). CD7 is aberrantly absent (F). The expression of CD3 is dim (D).

G. Hepatosplenic T-Cell Lymphoma

Hepatosplenic T-cell lymphoma is a systemic disease preferentially affecting young males (15–30 years of age) derived from functionally immature innate effector cells, most often of $\gamma\delta$ T-cell origin, characterized by pancytopenia and prominent hepatosplenomegaly without adenopathy (Alonsozana *et al.*, 1997; Belhadj *et al.*, 2003; Chang and Arber, 1998; Cooke *et al.*, 1996; Farcet *et al.*, 1990; Jaffe, 2006; Jaffe *et al.*, 2003; Przybylski *et al.*, 2000; Vega *et al.*, 2007; Wong *et al.*, 1995). It is an aggressive tumor involving the liver, spleen, and bone marrow with characteristic intrasinusoidal distribution of tumor cells.

The tumor cells are positive for CD3, CD45, and TCR γ/δ (Belhadj *et al.*, 2003; Cooke *et al.*, 1996; Farcet *et al.*, 1990; Gorczyca *et al.*, 2002; Gorczyca, 2006; Vega *et al.*, 2007; Wong *et al.*, 1995) but rare cases of hepatosplenic T-cell lymphoma TCR$\alpha\beta^+$ have been reported (Kumar *et al.*, 2001; Lai *et al.*, 2000; Macon *et al.*, 2001; Suarez *et al.*, 2000). Majority of cases are dual CD4/CD8$^-$ and are negative for CD5 (rare cases display dim expression of CD5). Most cases are positive for CD2, CD7, TIA-1 and are negative for both CD4 and CD8 (only rare tumors may be CD8$^+$). NK-cell associated antigens are variably expressed: CD16 is positive in approximately 1/4 of cases, CD56 is positive in majority of cases and CD57 is negative (Gorczyca, 2004, 2006; Gorczyca *et al.*, 2002). Granzyme B and perforin are usually negative.

H. Sézary's Syndrome

Sézary's syndrome (SS) is an aggressive eryhtrodermic cutaneous T-cell lymphoma with poor prognosis. It is defined historically by the triad of erythroderma, generalized lymphadenopathy, and the presence of neoplastic T-cells (Sézary's cells) in skin, lymph nodes, and peripheral blood (Diamandidou *et al.*, 1996; Hoppe *et al.*, 1990; Kim and Hoppe, 1999; Kohler *et al.*, 1997; Marti *et al.*, 2003; Wieselthier and Koh, 1990; Willemze *et al.*, 2005). In a recent report of the International Society for Cutaneous Lymphomas (ISCL), criteria recommended for the diagnosis of SS include one or more of the following: an absolute Sézary's cell count of at least 1000 cells/mm^3; the demonstration of immunophenotypical abnormalities (an expanded CD4$^+$ T-cell population resulting in CD4/CD8 ratio more than 10, loss of any or all of the pan-T cell antigens, or both); or the demonstration of a T-cell clone in the peripheral blood by molecular or cytogenetic methods.

Sézary's syndrome is characterized by the presence of atypical lymphocytes with irregular convoluted nuclei. Flow cytometry immunophenotyping demonstrates predominance of CD4$^+$ T-cells with aberrant loss of CD7 in a majority of cases. In our series of 12 cases, CD2, CD3, CD5, and CD7 were positive in 11 (91.7%), 12 (100%), 11 (91.7%), and 2 (16.7%) cases, respectively. All four pan-T antigens were positive in 1 case, 1 antigen was negative in 10 cases (83.3%), and 2 antigens were negative in 1 case. All tumors were TCR$\alpha\beta^+$. One case was CD8$^+$ and the remaining cases were CD4$^+$. CD8$^+$ case showed unusual coexpression of CD117. In the skin biopsy, the infiltrate is similar to MF, but epidermotropism may be absent and cellular composition is often more monotonous (Sentis *et al.*, 1986; Trotter *et al.*, 1997; Willemze *et al.*, 2005). The bone marrow (BM) is only rarely involved by SS, and when positive, it is usually subtle with interstitial accumulation of small T-cells (Sibaud *et al.*, 2003).

I. Enteropathy Type T-Cell Lymphoma (ETTL)

Enteropathy-type T-cell lymphoma is a large cell lymphoma with cytotoxic T-cell phenotype occurring most commonly in the small intestine, with tendency to spread along the mucosa to large intestinum and/or duodenum and stomach (Ashton-Key *et al.*, 1997; Baumgartner *et al.*, 2003; Chott *et al.*, 1998, 1999; de Bruin *et al.*, 1994, 1997; Isaacson, 1995, 2000; Jaffe *et al.*, 1999, 2003; Murray *et al.*, 1995; Pricolo *et al.*, 1998). Patients have either history of celiac disease or present with the symptoms of celiac disease at the time of diagnosis (Zettl *et al.*, 2007).

The neoplastic cells have the following phenotype: CD45$^+$, CD2$^{+/-}$, CD3$^+$, CD5$^-$, CD7$^+$, CD11c$^+$, CD43$^+$, CD25$^+$, CD30$^+$, CD56$^{+/-}$, and CD103$^+$. The expression of CD103 is a characteristic flow cytometric immunophenotypic feature. Most cases are CD8$^+$, but a subset of tumors is negative for both CD4 and CD8.

IX. Conclusions

FC plays an important role in diagnosis, classification, and monitoring of the T-cell lymphoproliferative disorders. Aberrant expression of pan-T cell antigens, TCR, and CD45, abnormal CD4/CD8 ratio (including dual expression or dual negativity), and presence of additional phenotypic markers (e.g., TdT, CD34, CD117, CD1a, HLA-DR, CD30, CD103, CD10) help identify abnormal T-cell populations and their further subclassification. Once the aberrant immunophenotype has been detected by FC, cytomorphological correlation is essential for diagnosis. In difficult cases, additional methodology, such as cytogenetics/FISH and/or PCR may be required to confirm the malignant nature of the process.

References

Agnarsson, B. A., Loughran, T. P., Jr. *et al.* (1989). The pathology of large granular lymphocyte leukemia. *Hum. Pathol.* **20**(7), 643–651.

Al-Hakeem, D. A., Fedele, S., *et al.* (2007). Extranodal NK/T-cell lymphoma, nasal type. *Oral Oncol.* **43**(1), 4–14.

Alonsozana, E. L., Stamberg, J., *et al.* (1997). Isochromosome 7q: The primary cytogenetic abnormality in hepatosplenic gammadelta T cell lymphoma. *Leukemia* **11**(8), 1367–1372.

Amadori, S., Meloni, G., *et al.* (1983). Long-term survival in adolescent and adult acute lymphoblastic leukemia. *Cancer* **52**(1), 30–34.

Amin, H. M., and Lai, R. (2007). Pathobiology of ALK + anaplastic large-cell lymphoma. *Blood* **110**(7), 2259–2267.

Arrowsmith, E. R., Macon, W. R., *et al.* (2003). Peripheral T-cell lymphomas: Clinical features and prognostic factors of 92 cases defined by the revised European American lymphoma classification. *Leuk. Lymphoma.* **44**(2), 241–249.

Ascani, S., Zinzani, P. L., *et al.* (1997). Peripheral T-cell lymphomas. Clinico-pathologic study of 168 cases diagnosed according to the R.E.A.L. Classification. *Ann. Oncol.* **8**(6), 583–592.

Ashton-Key, M., Diss, T. C., *et al.* (1997). Molecular analysis of T-cell clonality in ulcerative jejunitis and enteropathy-associated T-cell lymphoma. *Am. J. Pathol.* **151**(2), 493–498.

Attygalle, A., Al-Jehani, R., *et al.* (2002). Neoplastic T cells in angioimmunoblastic T-cell lymphoma express CD10. *Blood* **99**(2), 627–633.

Attygalle, A. D., Chuang, S. S., *et al.* (2007a). Distinguishing angioimmunoblastic T-cell lymphoma from peripheral T-cell lymphoma, unspecified, using morphology, immunophenotype and molecular genetics. *Histopathology* **50**(4), 498–508.

Attygalle, A. D., Kyriakou, C., *et al.* (2007b). Histologic evolution of angioimmunoblastic T-cell lymphoma in consecutive biopsies: Clinical correlation and insights into natural history and disease progression. *Am. J. Surg. Pathol.* **31**(7), 1077–1088.

Baccarani, M., Corbelli, G., *et al.* (1982). Adolescent and adult acute lymphoblastic leukemia: Prognostic features and outcome of therapy. A study of 293 patients. *Blood* **60**(3), 677–684.

Bakels, V., Van Oostveen, J. W., *et al.* (1993). Diagnostic and prognostic significance of clonal T-cell receptor beta gene rearrangements in lymph nodes of patients with mycosis fungoides. *J. Pathol.* **170**(3), 249–255.

Ballester, B., Ramuz, O., *et al.* (2006). Gene expression profiling identifies molecular subgroups among nodal peripheral T-cell lymphomas. *Oncogene* **25**(10), 1560–1570.

Bartlett, N. L., and Longo, D. L. (1999). T-small lymphocyte disorders. *Semin. Hematol.* **36**(2), 164–170.

Baumgartner, A. K., Zettl, A., *et al.* (2003). High frequency of genetic aberrations in enteropathy-type T-cell lymphoma. *Lab. Invest.* **83**(10), 1509–1516.

Bekkenk, M. W., Vermeer, M. H., *et al.* (2003). Peripheral T-cell lymphomas unspecified presenting in the skin: Analysis of prognostic factors in a group of 82 patients. *Blood* **102**(6), 2213–2219.

Belhadj, K., Reyes, F., *et al.* (2003). Hepatosplenic gammadelta T-cell lymphoma is a rare clinicopathologic entity with poor outcome: Report on a series of 21 patients. *Blood* **102**(13), 4261–4269.

Benharroch, D., Meguerian-Bedoyan, Z., *et al.* (1998). ALK-positive lymphoma: A single disease with a broad spectrum of morphology. *Blood* **91**(6), 2076–2084.

Blattner, W. A., Kalyanaraman, V. S., *et al.* (1982). The human type-C retrovirus, HTLV, in Blacks from the Caribbean region, and relationship to adult T-cell leukemia/lymphoma. *Int. J. Cancer.* **30**(3), 257–264.

Brito-Babapulle, V., Maljaie, S. H., *et al.* (1997). Relationship of T leukaemias with cerebriform nuclei to T-prolymphocytic leukaemia: A cytogenetic analysis with in situ hybridization. *Br. J. Haematol.* **96**(4), 724–732.

Bruggemann, M., White, H., *et al.* (2007). Powerful strategy for polymerase chain reaction based clonality assessment in T-cell malignancies report of the BIOMED-2 concerted action BHM4 CT98-3936. *Leukemia* **21**(2), 215–221.

Campo, E., Gaulard, P., *et al.* (1998). Report of the European task force on lymphomas: Workshop on peripheral T-cell lymphomas. *Ann. Oncol.* **9**(8), 835–843.

Cao, T. M., and Coutre, S. E. (2003). T-cell prolymphocytic leukemia: Update and focus on alemtuzumab (Campath-1H). *Hematology* **8**(1), 1–6.

Cataldo, K. A., Jalal, S. M., *et al.* (1999). Detection of t(2;5) in anaplastic large cell lymphoma: Comparison of immunohistochemical studies, FISH, and RT-PCR in paraffin-embedded tissue. *Am. J. Surg. Pathol.* **23**(11), 1386–1392.

Chadburn, A., Cesarman, E., *et al.* (1993). CD30 (Ki-1) positive anaplastic large cell lymphomas in individuals infected with the human immunodeficiency virus. *Cancer* **72**(10), 3078–3090.

Chan, J. K. (1998). Anaplastic large cell lymphoma: Redefining its morphologic spectrum and importance of recognition of the ALK-positive subset. *Adv. Anat. Pathol.* **5**(5), 281–313.

Chan, J. K. (1999). Peripheral T-cell and NK-cell neoplasms: An integrated approach to diagnosis. *Mod. Pathol.* **12**(2), 177–199.

Chan, W. C., Link, S., *et al.* (1986). Heterogeneity of large granular lymphocyte proliferations: Delineation of two major subtypes. *Blood* **68**(5), 1142–1153.

Chan, J. K., Ng, C. S., *et al.* (1989). Anaplastic large cell Ki-1 lymphoma. Delineation of two morphological types. *Histopathology* **15**(1), 11–34.

Chan, W. C., Gu, L. B., *et al.* (1992). Large granular lymphocyte proliferation with the natural killer-cell phenotype. *Am. J. Clin. Pathol.* **97**(3), 353–358.

Chan, A. C., Ho, J. W., *et al.* (1999). Phenotypic and cytotoxic characteristics of peripheral T-cell and NK-cell lymphomas in relation to Epstein-Barr virus association. *Histopathology* **34**(1), 16–24.

Chang, K. L. and Arber, D. A. (1998). Hepatosplenic gamma delta T-cell lymphoma—Not just alphabet soup. *Adv. Anat. Pathol.* **5**(1), 21–29.

Chiaretti, S., Li, X., *et al.* (2004). Gene expression profile of adult T-cell acute lymphocytic leukemia identifies distinct subsets of patients with different response to therapy and survival. *Blood* **103**(7), 2771–2778.

Chiaretti, S., Li, X., *et al.* (2005). Gene expression profiles of B-lineage adult acute lymphocytic leukemia reveal genetic patterns that identify lineage derivation and distinct mechanisms of transformation. *Clin. Cancer Res.* **11**(20), 7209–7219.

Chott, A., Haedicke, W., *et al.* (1998). Most CD56+ intestinal lymphomas are CD8+CD5-T-cell lymphomas of monomorphic small to medium size histology. *Am. J. Pathol.* **153**(5), 1483–1490.

Chott, A., Vesely, M., *et al.* (1999). Classification of intestinal T-cell neoplasms and their differential diagnosis. *Am. J. Clin. Pathol.* **111**(1 Suppl. 1), S68–S74.

Coiffier, B., Berger, F., *et al.* (1988). T-cell lymphomas: Immunologic, histologic, clinical, and therapeutic analysis of 63 cases. *J. Clin. Oncol.* **6**(10), 1584–1589.

Cook, J. R., Craig, F. E., *et al.* (2003). Benign CD10-positive T cells in reactive lymphoid proliferations and B-cell lymphomas. *Mod. Pathol.* **16**(9), 879–885.

Cooke, C. B., Krenacs, L., *et al.* (1996). Hepatosplenic T-cell lymphoma: A distinct clinicopathologic entity of cytotoxic gamma delta T-cell origin. *Blood* **88**(11), 4265–4274.

Damm-Welk, C., Schieferstein, J., *et al.* (2007). Flow cytometric detection of circulating tumour cells in nucleophosmin/anaplastic lymphoma kinase-positive anaplastic large cell lymphoma: Comparison with quantitative polymerase chain reaction. *Br. J. Haematol.* **138**(4), 459–466.

de Bruin, P. C., Kummer, J. A., *et al.* (1994). Granzyme B-expressing peripheral T-cell lymphomas: Neoplastic equivalents of activated cytotoxic T cells with preference for mucosa-associated lymphoid tissue localization. *Blood* **84**(11), 3785–3791.

de Bruin, P. C., Connolly, C. E., *et al.* (1997). Enteropathy-associated T-cell lymphomas have a cytotoxic T-cell phenotype. *Histopathology* **31**(4), 313–317.

Dearden, C. (2006a). The role of alemtuzumab in the management of T-cell malignancies. *Semin. Oncol.* **33**(2 Suppl. 5), S44–52.

Dearden, C. E. (2006b). T-cell prolymphocytic leukemia. *Med. Oncol.* **23**(1), 17–22.

Dearden, C. E., and Foss, F. M. (2003). Peripheral T-cell lymphomas: Diagnosis and management. *Hematol. Oncol. Clin. North Am.* **17**(6), 1351–1366.

Deleeuw, R. J., Zettl, A., *et al.* (2007). Whole-genome analysis and HLA genotyping of enteropathy-type T-cell lymphoma reveals 2 distinct lymphoma subtypes. *Gastroenterology* **132**(5), 1902–1911.

Delsol, G., Al Saati, T., *et al.* (1988). Coexpression of epithelial membrane antigen (EMA), Ki-1, and interleukin-2 receptor by anaplastic large cell lymphomas. Diagnostic value in so-called malignant histiocytosis. *Am. J. Pathol.* **130**(1), 59–70.

Diamandidou, E., Cohen, P. R., *et al.* (1996). Mycosis fungoides and Sezary syndrome. *Blood* **88**(7), 2385–2409.

Dogan, A., Attygalle, A. D., *et al.* (2003). Angioimmunoblastic T-cell lymphoma. *Br. J. Haematol.* **121**(5), 681–691.

Dunleavy, K., Wilson, W. H., *et al.* (2007). Angioimmunoblastic T cell lymphoma: Pathobiological insights and clinical implications. *Curr. Opin. Hematol.* **14**(4), 348–353.

Dupuis, J., Boye, K., *et al.* (2006a). Expression of CXCL13 by neoplastic cells in angioimmunoblastic T-cell lymphoma (AITL): A new diagnostic marker providing evidence that AITL derives from follicular helper T cells. *Am. J. Surg. Pathol.* **30**(4), 490–494.

Dupuis, J., Emile, J. F., *et al.* (2006b). Prognostic significance of Epstein-Barr virus in nodal peripheral T-cell lymphoma, unspecified: A Groupe d'Etude des Lymphomes de l'Adulte (GELA) study. *Blood* **108**(13), 4163–4169.

Duyster, J., Bai, R. Y., *et al.* (2001). Translocations involving anaplastic lymphoma kinase (ALK). *Oncogene* **20**(40), 5623–5637.

Falcao, R. P., Rizzatti, E. G., *et al.* (2007). Flow cytometry characterization of leukemic phase of nasal NK/T-cell lymphoma in tumor biopsies and peripheral blood. *Haematologica* **92**(2), e24–5.

Falini, B. (2001). Anaplastic large cell lymphoma: Pathological, molecular and clinical features. *Br. J. Haematol.* **114**(4), 741–760.

Falini, B., Pileri, S., *et al.* (1990). Peripheral T-cell lymphoma associated with hemophagocytic syndrome. *Blood* **75**(2), 434–444.

Falini, B., Bigerna, B., *et al.* (1998). ALK expression defines a distinct group of T/null lymphomas (ALK lymphomas) with a wide morphological spectrum. *Am. J. Pathol.* **153**(3), 875–886.

Falini, B., Pileri, S., *et al.* (1999a). ALK+ lymphoma: Clinico-pathological findings and outcome. *Blood* **93**(8), 2697–2706.

Falini, B., Pulford, K., *et al.* (1999b). Lymphomas expressing ALK fusion protein(s) other than NPM-ALK. *Blood* **94**(10), 3509–3515.

Farcet, J. P., Gaulard, P., *et al.* (1990). Hepatosplenic T-cell lymphoma: Sinusal/sinusoidal localization of malignant cells expressing the T-cell receptor gamma delta. *Blood* **75**(11), 2213–2219.

Ferrando, A. A. and Look, A. T. (2003). Gene expression profiling in T-cell acute lymphoblastic leukemia. *Semin. Hematol.* **40**(4), 274–280.

Ferry, J. A. (2002). Angioimmunoblastic T-cell lymphoma. *Adv. Anat. Pathol.* **9**(5), 273–279.

Frizzera, G., Kaneko, Y., *et al.* (1989). Angioimmunoblastic lymphadenopathy and related disorders: A retrospective look in search of definitions. *Leukemia* **3**(1), 1–5.

Garand, R., and Bene, M. C. (1994). A new approach of acute lymphoblastic leukemia immunophenotypic classification: 1984–1994 the GEIL experience. Groupe d'Etude Immunologique des Leucemies. *Leuk. Lymphoma.* **13**(Suppl. 1), 1–5.

Garand, R., Goasguen, J., *et al.* (1998). Indolent course as a relatively frequent presentation in T-prolymphocytic leukaemia. Groupe Francais d'Hematologie Cellulaire. *Br. J. Haematol.* **103**(2), 488–494.

Gassmann, W., Loffler, H., *et al.* (1997). Morphological and cytochemical findings in 150 cases of T-lineage acute lymphoblastic leukaemia in adults. German Multicentre ALL Study Group (GMALL). *Br. J. Haematol.* **97**(2), 372–382.

Gaulard, P., Bourquelot, P., *et al.* (1990a). Expression of the alpha beta and gamma delta T-cell receptors in peripheral T-cell lymphomas. *Nouv. Rev. Fr. Hematol.* **32**(1), 39–41.

Gaulard, P., Bourquelot, P., *et al.* (1990b). Expression of the alpha/beta and gamma/delta T cell receptors in 57 cases of peripheral T-cell lymphomas. Identification of a subset of gamma/delta T-cell lymphomas. *Am. J. Pathol.* **137**(3), 617–628.

Gaulard, P., Kanavaros, P., *et al.* (1991). Bone marrow histologic and immunohistochemical findings in peripheral. T-cell lymphoma: A study of 38 cases. *Hum. Pathol.* **22**(4), 331–338.

Gorczyca, W. (2004). Differential diagnosis of T-cell lymphoproliferative disorders by flow cytometry multicolor immunophenotyping. Correlation with morphology. *Methods Cell Biol.* **75**, 595–621.

Gorczyca, W. (2006). "Flow Cytometry in Neoplastic Hematopathology." Taylor and Francis, London.

Gorczyca, W., Weisberger, J., *et al.* (2002). An approach to diagnosis of T-cell lymphoproliferative disorders by flow cytometry. *Cytometry* **50**(3), 177–190.

Gorczyca, W., Tsang, P., *et al.* (2003). CD30-positive T-cell lymphomas co-expressing CD15: An immunohistochemical analysis. *Int. J. Oncol.* **22**(2), 319–324.

Gorczyca, W., Tugulea, S., *et al.* (2004). Flow cytometry in the diagnosis of mediastinal tumors with emphasis on differentiating thymocytes from precursor T-lymphoblastic lymphoma/leukemia. *Leuk. Lymphoma.* **45**(3), 529–538.

Graux, C., Cools, J., *et al.* (2006). Cytogenetics and molecular genetics of T-cell acute lymphoblastic leukemia: From thymocyte to lymphoblast. *Leukemia* **20**(9), 1496–1510.

Greenland, C., Dastugue, N., *et al.* (2001). Anaplastic large cell lymphoma with the t(2;5)(p23;q35) NPM/ALK chromosomal translocation and duplication of the short arm of the non-translocated chromosome 2 involving the full length of the ALK gene. *J. Clin. Pathol.* **54**(2), 152–154.

Greer, J. P., Kinney, M. C., *et al.* (1991). Clinical features of 31 patients with Ki-1 anaplastic large-cell lymphoma. *J. Clin. Oncol.* **9**(4), 539–547.

Greer, J. P., Kinney, M. C., *et al.* (2001). T cell and NK cell lymphoproliferative disorders. *Hematology (Am. Soc. Hematol. Educ. Program)* 259–281.

Grogg, K. L., Attygalle, A. D., *et al.* (2005). Angioimmunoblastic T-cell lymphoma: A neoplasm of germinal-center T-helper cells? *Blood* **106**(4), 1501–1502.

Haioun, C., Gaulard, P., *et al.* (1992). Clinical and biological analysis of peripheral T-cell lymphomas: A single institution study. *Leuk. Lymphoma.* **7**(5–6), 449–455.

Harris, N. L., Jaffe, E. S., *et al.* (1999). World Health Organization classification of neoplastic diseases of the hematopoietic and lymphoid tissues: Report of the Clinical Advisory Committee Meeting-Airlie House, Virginia, November 1997. *J. Clin. Oncol.* **17**(12), 3835–3849.

Harris, N. L., Jaffe, E. S., *et al.* (2000). The World Health Organization classification of neoplastic diseases of the haematopoietic and lymphoid tissues: Report of the Clinical Advisory Committee Meeting, Airlie House, Virginia, November 1997. *Histopathology* **36**(1), 69–86.

Hastrup, N., Hamilton-Dutoit, S., *et al.* (1991). Peripheral T-cell lymphomas: An evaluation of reproducibility of the updated Kiel classification. *Histopathology* **18**(2), 99–105.

Hoppe, R. T., Wood, G. S., *et al.* (1990). Mycosis fungoides and the Sezary syndrome: Pathology, staging, and treatment. *Curr. Probl. Cancer.* **14**(6), 293–371.

Inghirami, G., Macri, L., *et al.* (1994). Molecular characterization of CD30+ anaplastic large-cell lymphoma: High frequency of c-myc proto-oncogene activation. *Blood* **83**(12), 3581–3590.

Isaacson, P. G. (1995). Intestinal lymphoma and enteropathy. *J. Pathol.* **177**(2), 111–113.

Isaacson, P. G. (2000). Relation between cryptic intestinal lymphoma and refractory sprue. *Lancet* **356**(9225), 178–179.

Jaffe, E. S. (1995). Angioimmunoblastic T-cell lymphoma: New insights, but the clinical challenge remains. *Ann. Oncol.* **6**(7), 631–632.

Jaffe, E. S. (2001). Anaplastic large cell lymphoma: The shifting sands of diagnostic hematopathology. *Mod. Pathol.* **14**(3), 219–228.

Jaffe, E. S. (2006). Pathobiology of peripheral T-cell lymphomas. *Hematology (Am. Soc. Hematol. Educ. Program)* 317–322.

Jaffe, E. S., Blattner, W. A., *et al.* (1984). The pathologic spectrum of adult T-cell leukemia/lymphoma in the United States. Human T-cell leukemia/lymphoma virus-associated lymphoid malignancies. *Am. J. Surg. Pathol.* **8**(4), 263–275.

Jaffe, E. S., Clark, J., *et al.* (1985). Lymph node pathology of HTLV and HTLV-associated neoplasms. *Cancer Res.* **45**(Suppl. 9), 4662s–4664s.

Jaffe, E. S., Krenacs, L., *et al.* (1999). Extranodal peripheral T-cell and NK-cell neoplasms. *Am. J. Clin. Pathol.* **111**(1 Suppl. 1), S46–S55.

Jaffe, E. S., Krenacs, L., *et al.* (2003). Classification of cytotoxic T-cell and natural killer cell lymphomas. *Semin. Hematol.* **40**(3), 175–184.

Jamal, S., Picker, L. J., *et al.* (2001). Immunophenotypic analysis of peripheral T-cell neoplasms. A multiparameter flow cytometric approach. *Am. J. Clin. Pathol.* **116**(4), 512–526.

Juco, J., Holden, J. T., *et al.* (2003). Immunophenotypic analysis of anaplastic large cell lymphoma by flow cytometry. *Am. J. Clin. Pathol.* **119**, 205–212.

Kalinova, M., Krskova, L., *et al.* (2007). Quantitative PCR detection of NPM/ALK fusion gene and CD30 gene expression in patients with anaplastic large cell lymphoma-residual disease monitoring and a correlation with the disease status. *Leuk. Res.* **32**(1), 25–32.

Kaneko, Y., Maseki, N., *et al.* (1988). Characteristic karyotypic pattern in T-cell lymphoproliferative disorders with reactive angioimmunoblastic lymphadenopathy with dysproteinemia-type features. *Blood* **72**(2), 413–421.

Karube, K., Aoki, R., *et al.* (2008). Usefulness of flow cytometry for differential diagnosis of precursor and peripheral T-cell and NK-cell lymphomas: Analysis of 490 cases. *Pathol. Int.* **58**(2), 89–97.

Kesler, M. V., Paranjape, G. S., *et al.* (2007). Anaplastic large cell lymphoma: A flow cytometric analysis of 29 cases. *Am. J. Clin. Pathol.* **128**(2), 314–322.

Khalidi, H. S., Chang, K. L., *et al.* (1999). Acute lymphoblastic leukemia. Survey of immunophenotype, French-American-British classification, frequency of myeloid antigen expression, and karyotypic abnormalities in 210 pediatric and adult cases. *Am. J. Clin. Pathol.* **111**(4), 467–476.

Kim, Y. H. and Hoppe, R. T. (1999). Mycosis fungoides and the Sezary syndrome. *Semin. Oncol.* **26**(3), 276–289.

Kim, K., Kim, W. S., *et al.* (2002). Clinical features of peripheral T-cell lymphomas in 78 patients diagnosed according to the Revised European-American lymphoma (REAL) classification. *Eur. J. Cancer.* **38**(1), 75–81.

Kinney, M. C., and Kadin, M. E. (1999). The pathologic and clinical spectrum of anaplastic large cell lymphoma and correlation with ALK gene dysregulation. *Am. J. Clin. Pathol.* **111**(1 Suppl. 1), S56–S67.

Kinney, M. C., Collins, R. D., *et al.* (1993). A small-cell-predominant variant of primary Ki-1 (CD30)$^{+}$ T-cell lymphoma. *Am. J. Surg. Pathol.* **17**(9), 859–868.

Kohler, S., Kim, Y. H., *et al.* (1997). Histologic criteria for the diagnosis of erythrodermic mycosis fungoides and Sezary syndrome: A critical reappraisal. *J. Cutan. Pathol.* **24**(5), 292–297.

Kumar, S., Lawlor, C., *et al.* (2001). Hepatosplenic T-cell lymphoma of alphabeta lineage. *Am. J. Surg. Pathol.* **25**(7), 970–971.

Kwong, Y. L., Chan, A. C., *et al.* (1997). Natural killer cell lymphoma/leukemia: Pathology and treatment. *Hematol. Oncol.* **15**(2), 71–79.

Kwong, Y. L., and Wong, K. F. (1998). Association of pure red cell aplasia with T large granular lymphocyte leukaemia. *J. Clin. Pathol.* **51**(9), 672–675.

Lachenal, F., Berger, F., *et al.* (2007). Angioimmunoblastic T-cell lymphoma: Clinical and laboratory features at diagnosis in 77 patients. *Medicine (Baltimore)* **86**(5), 282–292.

Lai, R., Larratt, L. M., *et al.* (2000). Hepatosplenic T-cell lymphoma of alphabeta lineage in a 16-year-old boy presenting with hemolytic anemia and thrombocytopenia. *Am. J. Surg. Pathol.* **24**(3), 459–463.

Lakkala-Paranko, T., Franssila, K., *et al.* (1987). Chromosome abnormalities in peripheral T-cell lymphoma. *Br. J. Haematol.* **66**(4), 451–460.

Lamant, L., Dastugue, N., *et al.* (1999). A new fusion gene TPM3-ALK in anaplastic large cell lymphoma created by a (1;2)(q25;p23) translocation. *Blood* **93**(9), 3088–3095.

Lamant, L., de Reynies, A., *et al.* (2007). Gene expression profiling of systemic anaplastic large-cell lymphoma reveals differences based on ALK status and two distinct morphologic ALK+ subtypes. *Blood* **109**(5), 2156–2164.

Lamant, L., Meggetto, F., *et al.* (1996). High incidence of the t(2;5)(p23;q35) translocation in anaplastic large cell lymphoma and its lack of detection in Hodgkin's disease. Comparison of cytogenetic analysis, reverse transcriptase-polymerase chain reaction, and P-80 immunostaining. *Blood* **87**(1), 284–291.

Lamy, T., and Loughran, T. P. (1998). Large granular lymphocyte leukemia. *Cancer Control* **5**(1), 25–33.

Lamy, T. and Loughran, T. P., Jr. (1999). Current concepts: Large granular lymphocyte leukemia. *Blood Rev.* **13**(4), 230–240.

Lamy, T., and Loughran, T. P., Jr. (2003). Clinical features of large granular lymphocyte leukemia. *Semin. Hematol.* **40**(3), 185–195.

Leoncini, L., Lazzi, S., et al. (2000). Expression of the ALK protein by anaplastic large-cell lymphomas correlates with high proliferative activity. *Int. J. Cancer* **86**(6), 777–781.

Lepretre, S., Buchonnet, G., et al. (2000). Chromosome abnormalities in peripheral T-cell lymphoma. *Cancer Genet. Cytogenet.* **117**(1), 71–79.

Levine, P. H., Cleghorn, F., et al. (1994). Adult T-cell leukemia/lymphoma: A working point-score classification for epidemiological studies. *Int. J. Cancer* **59**(4), 491–493.

Liang, X., Meech, S. J., et al. (2004). Assessment of t(2;5)(p23;q35) translocation and variants in pediatric ALK+ anaplastic large cell lymphoma. *Am. J. Clin. Pathol.* **121**(4), 496–506.

Lopez-Guillermo, A., Cid, J., et al. (1998). Peripheral T-cell lymphomas: Initial features, natural history, and prognostic factors in a series of 174 patients diagnosed according to the R.E.A.L. Classification. *Ann. Oncol.* **9**(8), 849–855.

Lorenzen, J., Li, G., et al. (1994). Angioimmunoblastic lymphadenopathy type of T-cell lymphoma and angioimmunoblastic lymphadenopathy: A clinicopathological and molecular biological study of 13 Chinese patients using polymerase chain reaction and paraffin-embedded tissues. *Virchows. Arch.* **424**(6), 593–600.

Loughran, T. P., Jr. (1993). Clonal diseases of large granular lymphocytes. *Blood* **82**(1), 1–14.

Loughran, T. P. (1996). HTLV infection and hematologic malignancies. *Leuk. Res.* **20**(6), 457–458.

Loughran, T. P. (1998a). Large granular lymphocytic leukemia: An overview. *Hosp. Pract. (Off. Ed.)* **33**(5), 133–138.

Loughran, T. P., Jr. (1998b). Chronic T-cell leukemia/lymphoma. *Cancer Control* **5**(1), 8–9.

Loughran, T. P., Jr. (1999). CD56+ hematologic malignancies. *Leuk. Res.* **23**(7), 675–676.

Loughran, T. P., Jr., and Starkebaum, G. (1987a). Clinical features in large granular lymphocytic leukemia. *Blood* **69**(6), 1786.

Loughran, T. P., Jr., and Starkebaum, G. (1987b). Large granular lymphocyte leukemia. Report of 38 cases and review of the literature. *Medicine (Baltimore)* **66**(5), 397–405.

Loughran, T. P., Jr., Kadin, M. E., et al. (1985). Leukemia of large granular lymphocytes: Association with clonal chromosomal abnormalities and autoimmune neutropenia, thrombocytopenia, and hemolytic anemia. *Ann. Intern. Med.* **102**(2), 169–175.

Ma, Z., Cools, J., et al. (2000). Inv(2)(p23q35) in anaplastic large-cell lymphoma induces constitutive anaplastic lymphoma kinase (ALK) tyrosine kinase activation by fusion to ATIC, an enzyme involved in purine nucleotide biosynthesis. *Blood* **95**(6), 2144–2149.

Macon, W. R., Levy, N. B., et al. (2001). Hepatosplenic alphabeta T-cell lymphomas: A report of 14 cases and comparison with hepatosplenic gammadelta T-cell lymphomas. *Am. J. Surg. Pathol.* **25**(3), 285–296.

Magro, C. M., Morrison, C. D., et al. (2006). T-cell prolymphocytic leukemia: An aggressive T cell malignancy with frequent cutaneous tropism. *J. Am. Acad. Dermatol.* **55**(3), 467–477.

Mallett, R. B., Matutes, E., et al. (1995). Cutaneous infiltration in T-cell prolymphocytic leukaemia. *Br. J. Dermatol.* **132**(2), 263–266.

Man, C., Au, W. Y., et al. (2002). Deletion 6q as a recurrent chromosomal aberration in T-cell large granular lymphocyte leukemia. *Cancer Genet. Cytogenet.* **139**(1), 71–74.

Marti, R. M., Pujol, R. M., et al. (2003). Sezary syndrome and related variants of classic cutaneous T-cell lymphoma. A descriptive and prognostic clinicopathologic study of 29 cases. *Leuk. Lymphoma.* **44**(1), 59–69.

Martinez-Delgado, B. (2006). Peripheral T-cell lymphoma gene expression profiles. *Hematol. Oncol.* **24**(3), 113–119.

Matutes, E. (1998). T-cell prolymphocytic leukemia. *Cancer Control* **5**(1), 19–24.

Matutes, E. (2002). Chronic T-cell lymphoproliferative disorders. *Rev. Clin. Exp. Hematol.* **6**(4), 401–420; discussion 449–450.

Matutes, E., and Catovsky, D. (2003). Classification of mature T-cell leukemias. *Leukemia* **17**(8), 1682–1683; author reply 1683.

Matutes, E., Brito-Babapulle, V., et al. (1988). T-cell chronic lymphocytic leukaemia: The spectrum of mature T-cell disorders. *Nouv. Rev. Fr. Hematol.* **30**(5–6), 347–351.

Mioduszewska, O. (1979). T-cell lymphomas. *Arch. Geschwulstforsch.* **49**(8), 685–693.

Mioduszewska, O. (1985). T cell leukemia/lymphoma: Histogenesis and etiology. *Patol. Pol.* **36**(2), 121–129.

Mioduszewska, O., and Porwit-Ksiazek, A. (1984). Significance of immunological, cytochemical and immunohistochemical methods in the diagnosis of malignant lymphomas. *Mater. Med. Pol.* **16**(1), 44–49.

Mioduszewska, O., and Kulczycka, E. (1988). Immunopathological characteristics and course of various peripheral T-cell lymphomas. *Nowotwory* **38**(1), 39–53.

Murray, A., Cuevas, E. C., *et al.* (1995). Study of the immunohistochemistry and T cell clonality of enteropathy-associated T cell lymphoma. *Am. J. Pathol.* **146**(2), 509–519.

Nakamura, S., Shiota, M., *et al.* (1997). Anaplastic large cell lymphoma: A distinct molecular pathologic entity: A reappraisal with special reference to p80(NPM/ALK) expression. *Am. J. Surg. Pathol.* **21**(12), 1420–1432.

Onciu, M., Behm, F. G., *et al.* (2003). ALK-positive anaplastic large cell lymphoma with leukemic peripheral blood involvement is a clinicopathologic entity with an unfavorable prognosis. Report of three cases and review of the literature. *Am. J. Clin. Pathol.* **120**(4), 617–625.

Oshiro, A., Tagawa, H., *et al.* (2006). Identification of subtype-specific genomic alterations in aggressive adult T-cell leukemia/lymphoma. *Blood* **107**(11), 4500–4507.

Oshtory, S., Apisarnthanarax, N., *et al.* (2007). Usefulness of flow cytometry in the diagnosis of mycosis fungoides. *J. Am. Acad. Dermatol.* **57**(3), 454–462.

Osuji, N., Matutes, E., *et al.* (2006). T-cell large granular lymphocyte leukemia: A report on the treatment of 29 patients and a review of the literature. *Cancer* **107**(3), 570–578.

Pandolfi, F., Loughran, T. P., *et al.* (1990). Clinical course and prognosis of the lymphoporliferative disease of granular lymphocytes. A multicenter study. *Cancer* **65**, 341–348.

Pawson, R. (1999). Malignancy: Human T-cell lymphotropic virus type I and adult T-cell leukaemia/lymphoma. *Hematology* **4**(1), 11–27.

Pawson, R., Dyer, M. J., *et al.* (1997). Treatment of T-cell prolymphocytic leukemia with human CD52 antibody. *J. Clin. Oncol.* **15**(7), 2667–2672.

Pawson, R., Mufti, G. J., *et al.* (1998a). Management of adult T-cell leukaemia/lymphoma. *Br. J. Haematol.* **100**(3), 453–458.

Pawson, R., Richardson, D. S., *et al.* (1998b). Adult T-cell leukemia/lymphoma in London: Clinical experience of 21 cases. *Leuk. Lymphoma.* **31**(1–2), 177–185.

Pawson, R., Schulz, T., *et al.* (1998c). Absence of HTLV-I/II in T-prolymphocytic leukaemia. *Br. J. Haematol.* **102**(3), 872–873.

Pileri, S., Falini, B., *et al.* (1990). Lymphohistiocytic T-cell lymphoma (anaplastic large cell lymphoma CD30+/Ki-1+ with a high content of reactive histiocytes). *Histopathology* **16**(4), 383–391.

Pileri, S. A., Piccaluga, A., *et al.* (1995). Anaplastic large cell lymphoma: Update of findings. *Leuk. Lymphoma.* **18**(1–2), 17–25.

Pileri, S. A., Ascani, S., *et al.* (1998a). Peripheral T-cell lymphoma: A developing concept. *Ann. Oncol.* **9**(8), 797–801.

Pileri, S. A., Milani, M., *et al.* (1998b). Anaplastic large cell lymphoma: A concept reviewed. *Adv. Clin. Path.* **2**(4), 285–296.

Poiesz, B. J., Ruscetti, F. W., *et al.* (1980). Detection and isolation of type C retrovirus particles from fresh and cultured lymphocytes of a patient with cutaneous T-cell lymphoma. *Proc. Natl. Acad. Sci. USA* **77**(12), 7415–7419.

Pricolo, V. E., Mangi, A. A., *et al.* (1998). Gastrointestinal malignancies in patients with celiac sprue. *Am. J. Surg.* **176**(4), 344–347.

Pro, B., and McLaughlin, P. (2007). Angioimmunoblastic T-cell lymphoma: Still a dismal prognosis with current treatment approaches. *Leuk. Lymphoma.* **48**(4), 645–646.

Przybylski, G. K., Wu, H., *et al.* (2000). Hepatosplenic and subcutaneous panniculitis-like gamma/delta T cell lymphomas are derived from different Vdelta subsets of gamma/delta T lymphocytes. *J. Mol. Diagn.* **2**(1), 11–19.

Pui, C. H., Relling, M. V., *et al.* (2004). Acute lymphoblastic leukemia. *N. Engl. J. Med.* **350**(15), 1535–1548.

Rose, M. G., and Berliner, N. (2004). T-cell large granular lymphocyte leukemia and related disorders. *Oncologist* **9**(3), 247–258.

Ross, M. E., Zhou, X., *et al.* (2003). Classification of pediatric acute lymphoblastic leukemia by gene expression profiling. *Blood* **102**(8), 2951–2959.

Rudiger, T., Weisenburger, D. D., *et al.* (2002). Peripheral T-cell lymphoma (excluding anaplastic large-cell lymphoma): Results from the Non-Hodgkin's Lymphoma Classification Project. *Ann. Oncol.* **13**(1), 140–149.

Sandberg, Y., Almeida, J., *et al.* (2006). TCRgammadelta + large granular lymphocyte leukemias reflect the spectrum of normal antigen-selected TCRgammadelta + T-cells. *Leukemia* **20**(3), 505–513.

Savage, K. J. (2007). Peripheral T-cell lymphomas. *Blood Rev.* **21**(4), 201–216.

Schneider, N. R., Carroll, A. J., *et al.* (2000). New recurring cytogenetic abnormalities and association of blast cell karyotypes with prognosis in childhood T-cell acute lymphoblastic leukemia: A pediatric oncology group report of 343 cases. *Blood* **96**(7), 2543–2549.

Sentis, H. J., Willemze, R., *et al.* (1986). Histopathologic studies in Sezary syndrome and erythrodermic mycosis fungoides: A comparison with benign forms of erythroderma. *J. Am. Acad. Dermatol.* **15**(6), 1217–1226.

Shiota, M., and Mori, S. (1997). Anaplastic large cell lymphomas expressing the novel chimeric protein p80NPM/ALK: A distinct clinicopathologic entity. *Leukemia* **11**(Suppl. 3), 538–540.

Shuster, J. J., Falletta, J. M., *et al.* (1990). Prognostic factors in childhood T-cell acute lymphoblastic leukemia: A Pediatric Oncology Group study. *Blood* **75**(1), 166–173.

Sibaud, V., Beylot-Barry, M., *et al.* (2003). Bone marrow histopathologic and molecular staging in epidermotropic T-cell lymphomas. *Am. J. Clin. Pathol.* **119**(3), 414–423.

Sokol, L. and Loughran, T. P., Jr. (2003). Large granular lymphocyte leukemia and natural killer cell leukemia/lymphomas. *Curr. Treat. Options Oncol.* **4**(4), 289–296.

Speleman, F., Cauwelier, B., *et al.* (2005). A new recurrent inversion, inv(7)(p15q34), leads to transcriptional activation of HOXA10 and HOXA11 in a subset of T-cell acute lymphoblastic leukemias. *Leukemia* **19**(3), 358–366.

Stein, R. S., Greer, J. P., *et al.* (1990). Large-cell lymphomas: Clinical and prognostic features. *J. Clin. Oncol.* **8**(8), 1370–1379.

Stein, H., Foss, H. D., *et al.* (2000). CD30(+) anaplastic large cell lymphoma: A review of its histopathologic, genetic, and clinical features. *Blood* **96**(12), 3681–3695.

Steinberg, A. D., Seldin, M. F., *et al.* (1988). NIH conference. Angioimmunoblastic lymphadenopathy with dysproteinemia. *Ann. Intern. Med.* **108**(4), 575–584.

Suarez, F., Wlodarska, I., *et al.* (2000). Hepatosplenic alphabeta T-cell lymphoma: An unusual case with clinical, histologic, and cytogenetic features of gammadelta hepatosplenic T-cell lymphoma. *Am. J. Surg. Pathol.* **24**(7), 1027–1032.

Swerdlow, S. H., Campo, E., Harris, N. L., Jaffe, N. L., Pileri, S. A., Stein, H., Thiele, J., and Vardiman, J. W. (2008). "WHO Classification of Tumors of Hematopoietic and Lymphoid Tissues." IARC Press, Lyon.

Szomor, A., Zenou, P., *et al.* (2003). Genotypic analysis in primary systemic anaplastic large cell lymphoma. *Pathol. Oncol. Res.* **9**(2), 104–106.

Takatsuki, K., Yamaguchi, K., *et al.* (1985). Clinical diversity in adult T-cell leukemia-lymphoma. *Cancer Res.* **45**(9 Suppl.), 4644s–4645s.

Tan, B. T., Warnke, R. A., *et al.* (2006). The frequency of B- and T-cell gene rearrangements and epstein-barr virus in T-cell lymphomas: A comparison between angioimmunoblastic T-cell lymphoma and peripheral T-cell lymphoma, unspecified with and without associated B-cell proliferations. *J. Mol. Diagn.* **8**(4), 466–475; quiz 527.

ten Berge, R. L., Oudejans, J. J., *et al.* (2000). ALK expression in extranodal anaplastic large cell lymphoma favours systemic disease with (primary) nodal involvement and a good prognosis and occurs before dissemination. *J. Clin. Pathol.* **53**(6), 445–450.

ten Berge, R. L., Meijer, C. J., *et al.* (2002). Expression levels of apoptosis-related proteins predict clinical outcome in anaplastic large cell lymphoma. *Blood* **99**(12), 4540–4546.

ten Berge, R. L., de Bruin, P. C., *et al.* (2003). ALK-negative anaplastic large-cell lymphoma demonstrates similar poor prognosis to peripheral T-cell lymphoma, unspecified. *Histopathology* **43**(5), 462–469.

Thompson, M. A., Stumph, J., *et al.* (2005). Differential gene expression in anaplastic lymphoma kinase-positive and anaplastic lymphoma kinase-negative anaplastic large cell lymphomas. *Hum. Pathol.* **36**(5), 494–504.

Thorns, C., Bastian, B., *et al.* (2007). Chromosomal aberrations in angioimmunoblastic T-cell lymphoma and peripheral T-cell lymphoma unspecified: A matrix-based CGH approach. *Genes Chromosomes Cancer* **46**(1), 37–44.

Tort, F., Pinyol, M., *et al.* (2001). Molecular characterization of a new ALK translocation involving moesin (MSN-ALK) in anaplastic large cell lymphoma. *Lab. Invest.* **81**(3), 419–426.

Trotter, M. J., Whittaker, S. J., *et al.* (1997). Cutaneous histopathology of Sezary syndrome: A study of 41 cases with a proven circulating T-cell clone. *J. Cutan. Pathol.* **24**(5), 286–291.

Vecchi, V., Burnelli, R., *et al.* (1993). Anaplastic large cell lymphoma (Ki-1+/CD30+) in childhood. *Med. Pediatr. Oncol.* **21**(6), 402–410.

Vega, F., Medeiros, L. J., *et al.* (2007). Hepatosplenic and other gammadelta T-cell lymphomas. *Am. J. Clin. Pathol.* **127**(6), 869–880.

Vitale, A., Guarini, A., *et al.* (2006). Adult T-cell acute lymphoblastic leukemia: Biologic profile at presentation and correlation with response to induction treatment in patients enrolled in the GIMEMA LAL 0496 protocol. *Blood* **107**(2), 473–479.

Weisberger, J., Wu, C. D., *et al.* (2000). Differential diagnosis of malignant lymphomas and related disorders by specific pattern of expression of immunophenotypic markers revealed by multiparameter flow cytometry (Review). *Int. J. Oncol.* **17**(6), 1165–1177.

Weisberger, J., Cornfield, D., *et al.* (2003). Down-regulation of pan-T-cell antigens, particularly CD7, in acute infectious mononucleosis. *Am. J. Clin. Pathol.* **120**(1), 49–55.

Weisenburger, D. D., Anderson, J. R., *et al.* (2001). Systemic anaplastic large-cell lymphoma: Results from the non-Hodgkin's lymphoma classification project. *Am. J. Hematol.* **67**(3), 172–178.

Weiss, L. M., Strickler, J. G., *et al.* (1986). Clonal T-cell populations in angioimmunoblastic lymphadenopathy and angioimmunoblastic lymphadenopathy-like lymphoma. *Am. J. Pathol.* **122**(3), 392–397.

Wieselthier, J. S., and Koh, H. K. (1990). Sezary syndrome: Diagnosis, prognosis, and critical review of treatment options. *J. Am. Acad. Dermatol.* **22**(3), 381–401.

Willemze, R., Jaffe, E. S., *et al.* (2005). WHO-EORTC classification for cutaneous lymphomas. *Blood* **105**(10), 3768–3785.

Willemze, R., Jansen, P. M., *et al.* (2007). Subcutaneous panniculitis-like T-cell lymphoma: Definition, classification and prognostic factors. An EORTC Cutaneous Lymphoma Group study of 83 cases. *Blood.*

Wong, K. F., Chan, J. K., *et al.* (1995). Hepatosplenic gamma delta T-cell lymphoma. A distinctive aggressive lymphoma type. *Am. J. Surg. Pathol.* **19**(6), 718–726.

Yasunami, T., Wang, Y. H., *et al.* (2007). Multidrug resistance protein expression of adult T-cell leukemia/lymphoma. *Leuk. Res.* **31**(4), 465–470.

Yeoh, E. J., Ross, M. E., *et al.* (2002). Classification, subtype discovery, and prediction of outcome in pediatric acute lymphoblastic leukemia by gene expression profiling. *Cancer Cell* **1**(2), 133–143.

Zettl, A., Rudiger, T., *et al.* (2004). Genomic profiling of peripheral T cell lymphoma, unspecified, and anaplastic large T-cell lymphoma delineates novel recurrent chromosomal alterations. *Am. J. Pathol.* **164**(5), 1837–1848.

Zettl, A., deLeeuw, R., *et al.* (2007). Enteropathy-type T-cell lymphoma. *Am. J. Clin. Pathol.* **127**(5), 701–706.

Zinzani, P. L., Bendandi, M., *et al.* (1996). Anaplastic large-cell lymphoma: Clinical and prognostic evaluation of 90 adult patients. *J. Clin. Oncol.* **14**(3), 955–962.

CHAPTER 26

B Cell Immunophenotyping

Nicole Baumgarth

Center for Comparative Medicine
University of California, Davis
California 95616

I. Update

This chapter on B cell immunophenotyping aims to provide information on the identification and quantification of (mainly murine) B cells, their developmental stages, and their subsets in steady state and following antigen challenge.

DOI: 10.1016/B978-0-12-375045-7.00026-X

The biological context of these events is provided and with it a description of the phenotypic changes and the differences that help to identify B cell subsets by multicolor flow cytometry.

Since publication of the original chapter, none of the fundamentals regarding the development and subset distribution of B cells have had to be revised. However, progress has been made in a couple of areas that I will briefly discuss here. For those areas, brief descriptions and updated references are also provided in the chapter itself.

A. Immunophenotyping of Bone Marrow Cells

While the bone marrow is not regarded as a ready source for the small subset of B-1 cells, two recent studies identified a small subset of cells found in the adult bone marrow and the fetal liver, as precursors for B-1 but not B-2 cells. These cells were identified as $Lin^- CD19^+ B220^{lo/-} AA4.1^+$ (Esplin et al., 2009; Montecino-Rodriguez et al., 2006).

More comprehensive information exists now also on the mature B cell populations in the bone marrow, the site where most B cell development takes place. Three pieces of new evidence are relevant for immunophenotyping. First, mature $CD19^+ B220^+ IgM^{lo} IgD^{hi}$ B cells exist in specialized extravascular areas of the bone marrow from which they freely circulate. These cells were shown to respond to blood-borne pathogens and upregulate lectin peanut agglutinin (PNA) binding in the bone marrow (Cariappa et al., 2005). PNA-binding should thus no longer be regarded as characteristic exclusively of germinal center B cells but also of B cells activated via T-independent mechanisms (Manser, 2004). Second, recent studies provided evidence that CD93, an antigen recognized by the mAb493, is expressed not only on early B cell developmental stages but is reexpressed by long-lived bone marrow plasma cells (Chevrier et al., 2009). Third, we recently identified a novel population of mature B-1 cells in the bone marrow with the following phenotype: $IgM^+ IgD^{lo/-} CD5^{+/-} CD23^- CD43^+$. These cells spontaneously produce IgM (Choi and Baumgarth, submitted for publication).

B. Peripheral B Cell Subsets

Increasing information exists regarding the distinction of transitional B cell subsets in the spleen. Initial studies suggested the presence of three transitional B cell stages in the spleen: T1 ($CD93^+ IgM^{hi} CD23^-$), T2 ($CD93^+ IgM^{hi} CD23^+$), and T3 ($CD93^+ IgM^{lo} CD23^+$), (Allman et al., 2001; Loder et al., 1999). However, more recent data suggest that the "T3" population does not give rise to mature B cells and might instead represent an anergic B cell stage (Teague et al., 2007), thus the T2 population represents the last immature B cell population in the spleen.

C. Genetic Reporter Constructs as Means to Identify B Cell Subsets

The constantly rising number of genetically engineered mice now increasingly includes mice that express fluorescence probes driven by specific genes that signal certain developmental stages of B cells. Most widely used are constructs driving GFP (EGFP or YFP) expression. These fluorescent probes can be excited with many commercial flow cytometers and can be combined with the usual set of cell surface markers to provide powerful means of identifying cells by their gene expression profile. This has the advantages to providing genetic and functional evidence for a certain developmental stage/activation status in addition to the surface phenotype. Some examples include (i) mice with reporter constructs driven by expression of RAG1/2 (Kuwata *et al.*, 1999), that is, signaling RAG locus activity (necessary for VDJ recombination); (ii) a transgenic V(D)J recombination substrate that signals the presence and accessibility of the recombinase complex during early B cell development (Borghesi and Gerstein, 2004); and (iii) detection of BLIMP-1 gene activity, a transcription factor that drives terminal B cell differentiation, with an YFP construct (Ohinata *et al.*, 2005). It is likely that these and other mice will find increasing use for the tracking of B cell developmental and activation events, if obstacles regarding their distribution and accessibility can be overcome or at least minimized.

Tips for successful immunophenotyping:

• Standardize staining procedures and never use antibody conjugates without their prior titration.

• Standardize each flow cytometer using fluorescent beads prior to each run.

• Include lineage markers that stain cells other than B cells for exclusion of "sticky" cells (i.e., CD3, CD4, CD8, GR-1, F4/80, DX-5 all can be on the same color "dump channel"), and one that identifies B cells (best: CD19), particularly for staining of B cells in the bone marrow and other sites where they are a minor leukocyte subset.

• Always include a live/dead marker as dead cells are "sticky."

• Compare staining pattern of a new reagent to an appropriate unstained control (i.e., one that contains all stains, except the one of interest).

• Be sure to collect an adequate number of events for the size of the population you are interested in. Aim for a minimum of 50–100 events per population, more if possible.

II. Introduction

B lymphocytes provide the humoral, that is, antibody-mediated arm of the adaptive immune system. The term "B cell" was coined in the late 1950s from the organ of their first identification in chickens, the bursa of Fabricius. In most mammals, including humans and mice, B cell development after birth occurs in the bone marrow. In the mouse, hematopoietic stem cells can also be found throughout life in the spleen. Whether these cells contribute significantly to adult murine B lymphopoiesis has not been studied in detail. In addition to the steady-state B cell populations that are generated by the normal B cell developmental processes, functionally distinct B cell subsets appear following infection and/or immunization (Fig. 1). As discussed below, their phenotypes are at least in part delineated and multicolor flow cytometry can be used for their identification. The aim of this chapter is to provide information on the phenotypes of the different B cell developmental stages in the bone marrow and on mature B cell subsets in the periphery (Fig. 1). Information provided here is given mainly for mice, although it appears that mouse and human B cell development are remarkably similar

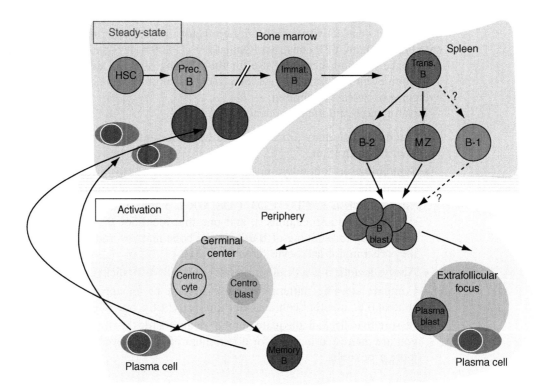

Fig. 1 Steady state and activation induced immunophenotypically distinct B cell subsets of mice. Outline of the different developmental and differentiation stages resulting in functionally and phenotypically distinct B cell subsets and their location. Bone marrow: HSC, hematopoietic stem cells; Prec. B, committed B cell precursor; Immat. B, immature B cell; Spleen: Trans. B, transitional B cell; B-2, follicular recirculating mature B cell; MZ, marginal zone B cell; B-1, B-1 (CD5+, B220^{lo} igM^{hi}) cell; Periphery, all secondary lymphoid tissues including spleen, lymph node, and tissue-associated lymphoid aggregates. Note that in mice marginal zone B cells are found only in the spleen.

(Ghia *et al.*, 1998). Fetal B cell development differs in a number of important aspects from adult development in both mice and humans. Phenotypic characteristics of fetal B cells and B cell precursors, as well as B cells from other species are less well studied and will not be reviewed in this chapter. Because of the spatial separation of B cell development on the one hand and the generation of mature B cell subsets and B cell activation on the other, I will distinguish between B cell immunophenotyping of cells from bone marrow cells and peripheral lymphoid tissues.

III. Immunophenotyping of B Cell Developmental Stages in the Bone Marrow

Like all leukocytes, B cells develop from hematopoietic stem cells in the bone marrow. A common lymphoid progenitor has been identified in mouse bone marrow that gives rise to T and B cells but not to cells of the myeloid or erythroid lineage (Kondo *et al.*, 1997). The phenotype of the committed B cell precursor is still a subject of debate. B cell development from such committed B cell precursors is characterized by an ordered process of immunoglobulin (Ig) gene rearrangements resulting in the expression of the B cell receptor (BCR). Although the genetic rearrangement profile and the developmental potential of a B cell is the ultimate proof of its developmental stage, each gene rearrangement step is characterized by a defined set of alterations in cell surface immunophenotype. Flow cytometry in conjunction with *in vitro* culture systems that recapitulate B cell development from the earliest committed precursor to the immature B cells stage were instrumental in identifying these phenotypic alterations. B cell development in the bone marrow concludes when a fully rearranged functional BCR is expressed on the cell surface and the BCR does not strongly bind to bone marrow-expressed self-antigens.

A number of different marker combinations have been used to classify the developmental stages of B cells in mice and humans. The most frequently used are those developed by three groups: (1) Hardy *et al.* (Allman *et al.*, 1999; Hardy *et al.*, 1991; Li *et al.*, 1996), (2) Melchers and colleagues (Melchers *et al.*, 1995; Rolink *et al.*, 1994; Rolink *et al.*, 1996), and (3) Osmond (Lu *et al.*, 1998; Osmond, 1990). Some confusion and discrepancies exist in the literature because each group used different markers and different nomenclatures to identify the various stages of development. It is hoped that with the availability of multicolor flow cytometry and the simultaneous use of larger numbers of markers many of these discrepancies can be resolved. Table I summarizes the different nomenclatures developed by these groups, their phenotypes and how they relate to the genotype (Ig gene rearrangement profile) of the developing B cells. Figure 2 shows the analysis of a C57BL/6 wild-type mouse bone marrow using multicolor staining combinations to differentiate the Hardy Fractions A–F.

Table I
Nomenclature and Phenotype of B-Cell Developmental Stages in Mouse Bone Marrow

	Pre-Pro B	Pro-B	Late Pro-B	Early Pre-B	Late Pre-B	Immature B	Mature B
Hardy et al. Fractions:	$A_1 A_2$ $B220^{lo}$ $AA4.1^+$ $CD24^-$ $CD4^{+/-}$	B $B220^{lo}$ $CD43^+$ $CD24^+$	C $B220^{lo}$ $CD43^+$ $CD24^{hi}$ $BP-1^+$	C' $B220^+$ $CD43^+$ $CD24^{hi}$ $BP-1^+$	D $B220^+$ $CD43^-$ IgM^- IgD^-	E $B220^+$ $CD43^-$ IgM^+ $IgD^{-/lo}$	F $B220^{hi}$ $CD43^-$ IgM^+ IgD^+
Melchers et al.	Pro-B $B220^+$ $CD19^+$ $c\text{-}kit^+$ $CD25^-$	Pre B-I $B220^+$ $CD19^+$ $c\text{-}kit^+$ $CD25^-$		Large Pre-B-II $B220^+$ $CD19^+$ $c\text{-}kit^-$ $CD25^+$ $c\mu^+$ large cycling	Small Pre-B-II $B220^+$ $CD19^+$ $c\text{-}kit^-$ $CD25^+$ small noncycling	Immature B $B220^+$ $CD19^+$ IgM^+	Mature B $CD19^+$ IgM^+ IgD^+
Osmond et al.	Early Pro-B TdT^+ $B220^-$	Intermediate Pro-B TdT^+ $B220^+$	Late Pro-B TdT^- $B220^+$	Large Pre-B TdT^- $B220^+$ $c\mu^+$ large, cycling	Small Pre-B TdT^- small $B220^+$ cm^+	Immature B TdT^- $B220^+$ $sIgM^+$	Mature B TdT^- $B220^+$ $sIgM^+$ IgD^+
Ig heavy-chain rearrangement	Germline (μ sterile transcripts)	$D_H{\to}J_H D_H J_H$	$D_H J_H{\to}V_H$	$V_H D_H J_H$	$V_H D_H J_H$	$V_H D_H J_H$	$V_H D_H J_H$
Ig light-chain rearrangement	Germline	Germline	Germline	Germline	$V_L \to J_L$	$V_L J_L$ (kappa or kappa + lambda)	$V_L J_L$

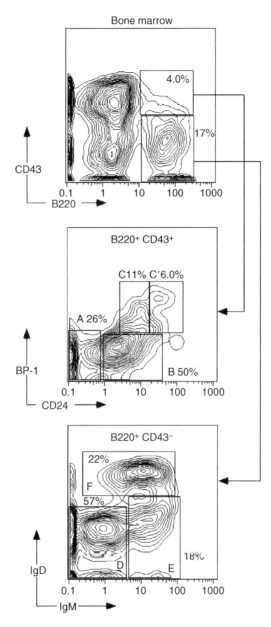

Fig. 2 B cell development is accompanied by distinct changes in cell surface immunophenotype. Staining was done according to Hardy *et al.* (1991). Shown are 5% contour plots of bone marrow cells from a 3-month-old C57BL/6 mouse gated for exclusion of propidium iodide (live) and a lymphocyte FSC/SSC. Early B cell developmental stages are distinguished from later stages of development by the expression of CD43 on B220+ cells (top panel). Note that CD43 expression is very low and requires a bright conjugate (best with either PE or APC). Those CD43+ B220+ B cells (middle panel) are further

A. The Earliest Committed B Cell Precursor

Many differences exist in the literature with regard to the phenotype of the earliest committed B cell progenitor. According to the Basel Group, expression of the markers CD19 and c-kit unequivocally identifies these cells in the bone marrow of adult mice (Rolink et al., 1996) (Table I). In contrast, Hardy and colleagues found B cell lineage-restricted cells among CD19$^-$ c-kit$^-$ B220^{dull} cells (Allman et al., 1999). Both groups have provided evidence that CD19 is expressed on the cell surface after B220 (CD45R) is expressed (Allman et al., 1999; Nikolic et al., 2002; Rolink et al., 1996). The CD19$^+$ c-kit$^+$ cells, identified as earliest committed B cell precursor by Melchers et al., appear to constitute a later developmental stage than the Fraction A$_1$ cells (AA4.1$^+$, B220^{dull} CD4$^{+/-}$ CD11b^{lo} CD19$^-$ CD24$^-$ CD43$^-$ BP1$^-$ c-kit$^-$ Sca1^{lo} IL7R$^-$ (Allman et al., 1999; Li et al., 1996; Rolink et al., 1996)) identified by Allman et al. (1999) (see Table I). Thus, the main discrepancies between these two groups seem to regard the definition of commitment to the B cell lineage, rather than differences in the phenotypic identification of the cells.

In contrast, Montecino-Rodriguez et al. (2001) recently described a AA4.1$^+$ B220$^-$ CD24$^+$ CD43$^+$ CD19$^+$ population in the bone marrow of adult BALB/c mice that represents 0.1-0.5% of nucleated cells. These cells appeared to have bipotential B cell/macrophage progenitor capability, suggesting that a small subset of not yet fully committed B cell precursors might express CD19 prior to acquisition of B220. Alternatively, B220 expression might under certain conditions be downmodulated after CD19 acquisition. This population seems to show many of the characteristics described for Fraction B cells of Hardy et al. (1991) (with the exception of B220), including the findings of D$_H$J$_H$ rearrangement (Montecino-Rodriguez et al., 2001). Thus, it is also possible that low expression of B220 that characterizes early B cell developmental stages according to Hardy et al. was characterized by Monetecino-Rodriguez et al. as B220 negative. Nonetheless, a small fraction (around 5%) of this population seems to have retained its ability to differentiate into macrophages, but not dendritic cells (Montecino-Rodriguez et al., 2001); a finding not easily reconciled with the findings of Hardy et al. and Melchers et al., but consistent with data from neonatal and adult B cell lines (Davidson et al., 1992; Martin et al., 1993) and in vitro cocultures of B cells with splenic fibroblasts (Borrello and Phipps, 1995, 1999).

Based on the unusual properties of the Lin$^-$ AA4.1$^+$ B220$^{lo/-}$ CD19$^+$ bone marrow precursor population and the fact that bipotential macrophage/B cell precursors have been associated with the development of a small B cell subset, termed B-1, Montecino-Rodriguez et al. (2006) further characterized these cells and more recently demonstrated that they efficiently give rise to B-1a and B-1b

divided into Fractions A-C' as shown. Bottom panel shows the separation of late pre-B from immature and mature B cells (Fractions D-F) according to their levels of expression of surface IgM and IgD. See Table I for further explanation. Note that Fraction A was found to be heterogeneous and only some of the cells within this fraction are committed to become B cells (see text).

cells but not to conventional B-2 cells following adoptive transfer. They also showed this precursor population to be present in the fetal liver of mice, thought to be a major source for the small and distinct B-1 cell subset (see discussion under Section III). Studies by Esplin et al. (2009) have confirmed these findings. They also suggested that these cells can be derived from adult lymphoid progenitors, that is, the lymphoid progenitors common to all lymphoid cells in the adult. In contrast to the original characterization of these cells (Montecino-Rodriguez et al., 2001), this study did not find any evidence for the Lin^{-} AA4.1^{+} B220$^{lo/-}$ CD19^{+} cells to be affiliated with myeloid cells.

B. Ig Gene Rearrangement and B Cell Immunophenotype

Terminal commitment to the B cell lineage appears to already have occurred when Ig gene rearrangement is initiated (Allman et al., 1999). Although it has been argued that the presence of macrophages with at least partial Ig heavy-chain rearrangement indicate that a "lineage switch" from B cell to myeloid cell might be possible (Davidson et al., 1992; Martin et al., 1993; Montecino-Rodriguez et al., 2001).

The signals that initiate this complex Ig rearrangement event are not fully understood. Variable region gene recombination (VDJ-recombination) occurs first on the Ig heavy chain in the order of D→J, V→DJ (Table I). Expression of RAG 1 and RAG 2 genes are essential for rearrangement to occur (Schatz et al., 1989; Shinkai et al., 1992). A successfully rearranged heavy-chain variable region is then paired with the Igμ heavy-chain constant region and expressed on the cell surface in conjunction with the surrogate light chains lamda5 and V-preB and the signaling chains Igα and Igβ (CD79a/CD79b) as "pre-B cell receptor" (Karasuyama et al., 1994). Unfortunately, pre-B cell receptor expression appears very low and/or transient, and this unique receptor cannot be used to identify pre-B cells immediately ex vivo. They can, however, be visualized by flow cytometry via specific mAb-staining if overexpressed (Kawano et al., 2006).

In general, heavy-chain rearrangement precedes light-chain rearrangement. Light chains are encoded by V and J genes only and use one of two constant regions: kappa and lamda. Lamda rearrangement occurs only when kappa rearrangement results in nonproductive rearrangement, or when the BCR recognizes self-antigens (re-rearrangement). Successful gene rearrangement on one of the two heavy and four light-chain alleles shuts down further rearrangement on the other allele(s) (allelic exclusion). This ensures that every B cell expresses mature BCR with exactly one specificity. IgM^{+} IgD^{-} B cells leave the bone marrow at the immature stages and migrate toward the spleen.

Phenotypic changes that characterize each stage in B cell development have been described (Table I). The appearance of cytoplasmic Igμ signals successful rearrangement of the Ig heavy chain. On the cell surface, this seems to coincide with expression of CD25 and loss of expression of c-kit (Rolink et al., 1994). At this stage, pre B cells undergo strong IL-7-dependent clonal expansion and the cells are relatively large and blast-like (Melchers *et al.*, 1995). Reexpression of RAG occurs following this expansion phase for the induction of light-chain rearrangement. These cells become smaller, lose CD43 (as analyzed with the mAb S7), express increased levels of CD24 and are BP-1$^+$ (Hardy *et al.*, 1991; Li *et al.*, 1996; Osmond *et al.*, 1998). The expression of surface IgM without IgD characterizes the final step of development in the bone marrow. These cells are termed immature B cells. Depending on the specificity of the receptor, further rearrangement of the light chain might occur at this developmental stage in the bone marrow (Melchers *et al.*, 1995). The phenotype of B cells that alter their specificity by further rearrangement of their light chains, a process termed re-rearrangement, has not been defined. However, eventually many of those cells will express lamda- rather than kappa-light chains (Meffre *et al.*, 2000).

Apart from the various B cell precursors outlined earlier, fully rearranged, mature, recirculating B220$^+$ CD23$^+$ IgM^{lo} IgD^{hi} B cells (Fig. 2) are found in the extravascular spaces of the bone marrow of mice, organized into discrete areas around vascular sinusoids. At that location they can respond rapidly to blood-borne pathogens with activation, visualized by their increased ability to bind to lectin peanut agglutinin (PNA) (Cariappa *et al.*, 2005). In addition, we recently identified a novel population of CD19^{hi} B220^{lo} CD23$^-$ CD43$^+$ CD5$^{+/-}$ B-1 cells in the bone marrow of adult mice (Choi and Baumgarth, submitted for publication). Nearly all of these cells spontaneously secrete IgM and together with B-1 cells in the spleen appear as the major natural IgM-secreting cell populations in nonmanipulated mice. In addition, both plasma cells and memory B cells, induced following antigen exposure, can home to the bone marrow. Their phenotypes are discussed below.

In conclusion, the bone marrow is a tissue in which B cells of all developmental stages can be isolated. Although controversy surrounds the phenotype of the earliest committed B cell, good marker combinations for the identification of most of the later developmental stages of B cells are available. As shown in Fig. 2, a single 6–8 color staining combination allows the near-complete delineation of B cell development from the pro-B cell stage to the fully mature B cells.

IV. Peripheral B Cell Populations

Immature B cells travel from the bone marrow to the spleen. In the spleen, these cells are referred to as transitional B cells. In addition to these recent bone marrow emigrants, three functionally and phenotypically distinct mature B cell subsets can be separated in the spleen: Follicular B cells, the main recirculating mature B cell population; marginal zone B cells; and B-1 cells (Fig. 3). In mice, resting lymph nodes do not usually harbor either transitional B cells or marginal zone B cells, but instead are comprised of mainly follicular B cells and a very small frequency (0.2%) of B-1 cells.

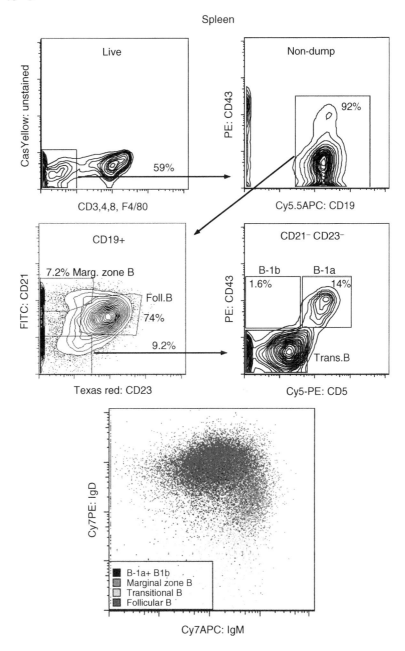

Fig. 3 Identification of major B cell subsets in murine spleen. Shown are 5% contour plots from the spleen of a 3-month-old female C57BL/6 mouse gated on a lymphocyte FSC/SSC profile and exclusion of propidium iodide. B cells are identified by their lack of expression of CD3, 4, 8 and F4/80 (top left panel) and their expression of CD19 (top right). Those cells are then subdivided into marginal zone B cells (Marg. Zone B), follicular B cells (Foll. B), and transitional B cells (trans. B) B 1a and B 1b cells. The bottom overlay dot plot shows the various levels of IgM/IgD expression levels of the above identified cell subsets.

Infection and/or immunization-induced B cell activation results in the early accumulation of B cells and plasmablasts that form extrafollicular foci at the borders to T cell areas. Somewhat later, anatomical structures termed "germinal centers" (GC) develop within the primary follicles of spleen and lymph nodes. They contain large numbers of rapidly cycling B cells that undergo immunoglobulin isotype switching and affinity maturation. Germinal center reactions lead to the development of high affinity, Ig-isotype-switched memory B cells. Both germinal center reaction and the extrafollicular foci generate antibody-secreting plasma blasts and terminally differentiated plasma cells. There is currently no marker that can distinguish the origin of these end-stage B cell effector populations.

A. Transitional B Cells

$B220^+$ IgM^{hi} $IgD^{-/lo}$ $CD21^{-/lo}$ B cells from the bone marrow undergo a final selection process in the spleen. It has been calculated that only about 20% of the bone marrow emigrants will be selected into the mature B cell pool. The mechanisms that govern this selection process are unknown. Explained by their high levels of expression of CD95 (Fas), transitional B cells are exquisitely sensitive to antigen-induced apoptosis, presumably as a safeguard mechanism to ensure that strongly self-reactive B cells are eliminated from the B cell pool. Identification of transitional B cells in the spleen can be made based on the surface markers B220 or CD19, CD21, CD23, CD5, IgM and IgD (Fig. 3). Carsetti and colleagues (Loder et al., 1999) showed that two discrete subsets of immature B cells can be identified: Transitional 1 and Transitional 2 B cells. T1 cells lack expression of CD5, CD21, and CD23 and express either no or very little IgD. T2 cells have gained CD23, CD21 and express higher levels of IgD, but in contrast to mature B cells still express high levels of IgM. Neither of these subtypes is found in peripheral tissues other than blood and spleen.

A third population of $CD5^-$ $CD21^+$ $CD23^+$ B cells was identified by Allman et al. and labeled "T3" as a further differentiation step from transitional to mature B cells (Allman et al., 2001). These cells are distinguished from T2 cells by their low expression of IgM and from mature follicular B cells by their expression of CD93 (identified with the mAb 493), a marker expressed on B cell precursors and on terminally differentiated plasma cells in the bone marrow (Chevrier et al., 2009) as well as on transitional B cells in the spleen (Rolink et al., 1998), but not on mature B cells. However, recent studies suggest that these cells are distinct from other transitional B cells and do not give rise to mature B cells upon adoptive transfer. They have a distinct V_H gene repertoire and might harbor increasing numbers of autoreactive B cells (Teague et al., 2007).

B. Follicular B Cells

Immature B cells develop to become "mature" B cells. They are also called follicular B cells. Once selected into the mature B cell pool, that is, recruited into the B cell follicles, B cells reduce their expression of IgM and instead further gain expression of IgD. The functional consequences of this change in the ratio of these Ig isotypes on the cell surface have not been fully elucidated. These cells are CD23$^+$ CD21^{int} and CD5$^-$. Follicular B cells make about 70-80% of the splenic B cell population (Fig. 3) and are the main B cell population in most peripheral tissues, with the exception of the peritoneal and pleural cavities. Because of their relative abundance, their expression of CD23 and high levels of expression for IgD, this population is fairly easily identified by flow cytometry (Fig. 3). Follicular B cells are remarkably homogeneous with regard to their surface phenotype. One exception is expression of the nonclassical MHC molecule CD1. This molecule is expressed at high levels on marginal zone B cells, but also on a small not yet functionally characterized subset of B cells that otherwise resemble follicular B cells (Amano *et al.*, 1998).

C. Marginal Zone B Cells

Like follicular B cells, marginal zone B cells develop from transitional B cells in the spleen via an unknown developmental pathway. They make about 5-10% of splenic B cells (Fig. 3). Cells with marginal zone phenotype are not found in the lymph nodes. They recirculate in the blood of humans but not in mice. The term marginal zone B cell was developed from the fact that these cells reside in the marginal zone of the spleen. It is here that blood-borne antigens are filtered through the marginal zone synoids of the spleen. Consistent with this observation is that marginal zone B cells are responding rapidly to antigen exposure (Martin and Kearney, 2002). They are also particularly sensitive to LPS stimulation due to the fact that they express high levels of toll-like receptor molecule CD180 (unpublished) that facilitates responsiveness to LPS (Nagai *et al.*, 2002). Furthermore, they express a number of other receptors on their cell surface that facilitate interaction with the innate immune system, including the nonclassical MHC molecule CD1 (Amano *et al.*, 1998; Roark *et al.*, 1998) and the complement receptor CD21 (Takahashi *et al.*, 1997). The latter is known as a potent costimulatory molecule for B cells. Thus, marginal zone B cells are characterized phenotypically by their high levels of expression of IgM, CD1, CD21, their lack of expression of CD23, and low expression of IgD. At least some CD19$^+$ CD23$^-$ CD21^{hi} marginal zone B cells express considerable levels of IgD (overlay profile Fig. 3). Further studies are required to determine whether these cells constitute a different developmental or activation stage.

It is important to note that not all B cells that reside in the marginal zone have the phenotypic characteristics of marginal zone B cells. Other B cell populations that reside in the marginal zone are some plasma blasts and plasma cells as well as B-1 cells. Hence, the pathologist's definition of marginal zone B cells is broader than the term defined by their phenotypic characteristics by flow cytometry.

D. B-1 Cells

B-1 cells constitute a very small proportion of B cells in the spleen (roughly 1%) but they produce the majority of the circulating natural antibodies, particularly IgM (Baumgarth et al., 1999; Hayakawa and Hardy, 2000). They are very rare in resting lymph nodes and bone marrow ($<0.2\%$), and make about 1% of B cells in the spleen (Fig. 3), but they are the majority B cell population in the peritoneal and pleural cavities of mice (Fig. 4). Depending on the strain, age, and sex of the mice, up to 70% of B cells in the peritoneal cavity are B-1 cells. Peritoneal and pleural cavity B-1 cells are IgM^{hi}, IgD^{lo} $CD23^{negative}$, $CD11b^+$ and $CD43^+$ (Fig. 4). (Wells et al., 1994). A further distinction is made as to their expression of CD5. Roughly 2/3 to 3/4 of these cells express the otherwise T cell-restricted marker CD5 (Fig. 4). $CD5^+$ B-1 cells are termed B-1a cells and $CD5^-$ B-1 cells as B-1b. It is important to point out that expression of CD5 can only be seen when bright CD5-conjugates are used, such as PE and APC reagents. Because staining for CD5 on B-1 cells is substantially lower than that observed on T cells (data not shown), CD5 conjugate evaluation for the identification of B-1 cells must be done on those cells. A good source for B-1 cells is the peritoneal cavity (Fig. 4).

Importantly and in contrast to the peritoneal cavity, B-1 cells in the spleen or other secondary lymphoid tissues do not express CD11b. Due to their small numbers and the lack of other unique identifiers, this makes particularly the analysis of the splenic $CD5^-$ B-1b cell population challenging. In addition, it has been shown that CD5 expression can be induced in follicular B cells in vitro under certain stimulation conditions (Cong et al., 1991) and that certain self-reactive/anergic B cells might express CD5 (Hippen et al., 2000). It is presently unclear whether a distinction has to be made between follicular B cells expressing CD5 and distinct B cell precursors that develop into $CD5^+$ B-1 cells, as suggested by the identification of distinct B-1 cell precursors in the bone marrow (Esplin et al., 2009; Montecino-Rodriguez et al., 2006) and the unique regulation of B-1 cell responses (Alugupalli et al., 2004; Choi and Baumgarth, 2008). No marker has been described to date that could distinguish the potential different origins of such cells.

V. Summary

Taken together, at least six distinct B cell subsets can be identified in the periphery that develop in the absence of antigen challenge: T1 and T2 transitional B cells, follicular B cells, marginal zone B cells, and B-1a and B-1b cells. Combinations of markers can be designed that differentiate all these B cell populations. Figure 3 shows flow cytometric analysis of mouse splenic B cell populations using a 9-color staining combinations in just one antibody cocktail. Such an approach provides the greatest accuracy, since all populations are quantified in one stain. The challenge in detecting nonantigen-induced splenic/peripheral B cell subsets is the distinction between the nonfollicular B cells, all of which are IgM^{hi} $IgD^{lo/neg}$

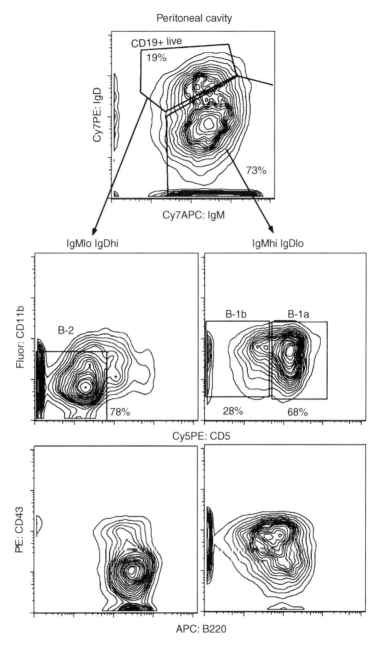

Fig. 4   The peritoneal cavity contains large numbers of B-1 cells. Shown are cells harvested from the peritoneal cavity of a 3-month-old female C57BL/6 mouse. Shown are 5% contour plots of cells gated on exclusion of propidium iodide and a lymphocytic FSC/SSC profile. B-1 and B-2 cells are separated according to their differential expression of IgM and IgD (top panel). The large differences between B-1 and B-2 cells in their expression of CD11b, CD5, CD43, and B220 are indicated in the middle and bottom panels. Note the separation of B-1 cells into B-1a and B-1b according to their expression of CD5. In contrast to B-2 cells, both B-1a and B-1b cells express CD11b, CD43, and low levels of B220.

and CD23$^-$. Flow cytometric evaluations of these relatively small populations should be done by acquiring at least 200,000 events. This ensures high enough event counts for each of the cell populations to be identified. However, if only one particular B cell population of interest is to be examined, marker combinations with smaller number of antibodies can be designed.

VI. Antigen-Induced B Cell Subsets

As shown in Fig. 1, exposure to antigen induces a number of different B cell activation events. Mature B cells respond to antigen encounter and appropriate costimulatory signals with proliferation (clonal expansion) and differentiation to antibody-secreting B cells or plasma cells. Immunohistologically, the cells might form "extrafollicular foci" at the edges of the T and B cell zones in secondary lymphoid tissues at early times after infection/immunization, or they form GC within B cell follicles. The exact molecular events that determine whether a B cell will form an extrafollicular foci or GC are not clear, but might be due at least in part to the affinity of the BCR for antigen (Phan *et al.*, 2006) and environmental cues (Kim *et al.*, 2008). Furthermore, it has been demonstrated that SAP gene-expressing CD4 T cells are required for GC B cell differentiation into long-lived plasma cells and memory B cells, but not for the generation of short-lived plasma blasts that develop from extrafollicular foci (Crotty *et al.*, 2003).

A. Germinal Center B Cells

By immunohistochemistry, B cells in GC can be divided into centrocytes and centroblasts. As the name suggests centroblasts are relatively large cells that are rapidly proliferating. These cells acquire hypermutations and might undergo iso-type switching. Centroblasts give rise to centrocytes, which express a mutated Ig gene and express surface IgG, IgA, or IgE instead of IgM. These cells undergo rapid Fas-mediated apoptosis, unless rescued by signaling through antigen-binding via BCR and costimulation via CD40 (MacLennan, 1994). It is believed that this selection process provides the basis for the increased affinity of antibody responses during the course of an infection/antigen challenge.

Identification of centrocytes and centroblasts by immunophenotyping has not been achieved in the mouse. All GC B cells are B220$^+$ CD19$^+$, CD21/35$^+$ CD22 and CD40$^+$, like most other B cell subsets. They differ immunophenotypically from other subsets by their increased ability to bind the PNA (Butcher *et al.*, 1982) and expression of an antigen identified by the mAb GL7 (Han *et al.*, 1996; Shinall *et al.*, 2000). They also express higher levels of CD24 (HSA) compared to follicular B cells. GC B cells do not express CD43 and CD138 (syndecan-1), consistent with their inability to spontaneously secrete antibodies. Also, GC B cells downregulate CD23 and in the mouse CD38. The latter is in contrast to human GC B cells which express CD38 at high levels. Thus, CD38 cannot be used in the mouse to identify

GC B cells (Oliver *et al.*, 1997; Ridderstad and Tarlinton, 1998). GC B cells appear heterogeneous with regard to their expression of surface Ig. Although IgD is rapidly downregulated during the GC reaction, some IgD$^+$ IgM$^+$ as well as IgD$^-$ IgM$^+$ and isotype-switched surface Ig-G, -A and -E$^+$ cells can be found at varying degrees throughout the GC reaction (Shinall *et al.*, 2000). Heterogeneity of PNA$^+$ B220$^+$ GC B cells has been reported also for a number of other surface markers. How these differences relate to the immunohistochemically identified cell populations of the GC has not been elucidated (Han *et al.*, 1997; Shinall *et al.*, 2000).

GC cells can give rise to either memory B cells, or to long and short-lived (affinity-matured) antibody-secreting cells/plasma cells (MacLennan, 1994).

B. Plasma Cells

Following antigen-encounter a subset of B cells returns to the bone marrow as terminal differentiated long-lived plasma cells (Benner *et al.*, 1981). Cells of similar phenotype are also located in various mucosal tissues and in the splenic red pulp. The plasma cell migration event to bone marrow is dependent on the chemokine CXCL12 (SDF-1) and on expression of CXCR4 (Hargreaves *et al.*, 2001).

Identification of plasma cells can be tricky. These cells, although cytoplasmic Ig positive may be surface Ig negative and many have either lost or strongly downregulated the pan-B cell markers B220 and CD19 as well as MHC II (Underhill *et al.*, 2003). Reexpression of CD93 might identify bone marrow plasma cells (Chevrier *et al.*, 2009). Syndecan-1 (CD138) has been widely used as a marker to identify plasma cells. However, although anti-CD138 specifically stains plasma cells in the periphery, not all plasma cells are syndecan positive (Underhill *et al.*, 2003, unpublished data). CD43 (staining with either antibody clone S7 or S11) seems to be expressed on most plasma cells, as is CD27 (Jung *et al.*, 2000). Since CD43 is also expressed on all T cells, some APC as well as early B cell precursors, a staining combination that includes a pan T cell marker and markers to exclude macrophages, granulocytes, and monocytes in conjunction with CD43, syndecan, staining for cytoplasmic Ig and CD24 (to exclude precursor B cell populations) seems currently to be the best combination to enable enumeration of plasma cells in the bone marrow. CD9 might also help to further differentiate plasma cells from other cell populations, since its expression otherwise seems limited to splenic marginal zone B cells and to B-1 cells (Won and Kearney, 2002).

C. Memory B Cells

The bone marrow has also been identified as a source for long-lived memory B cells (Gray, 1993). Memory B cells are defined as cells that can respond to repeat antigen exposure with faster and stronger antibody responses. Moreover, these cells are able to survive long term after adoptive transfer. The phenotype of the memory B cell pool has not been fully elucidated. Most but not all cells have lost expression of IgM and IgD and express a downstream Ig isotype, mostly IgG (Gray, 1993; McHeyzer-Williams *et al.*, 1991; Schittek and Rajewsky, 1990). However, more recently, the presence of IgM-expressing memory B cells has been reported (Klein *et al.*, 1999), consistent with the fact that antigen-specific IgM can be found even at late points after antigen exposure. In human, CD27 has been identified as a marker for memory B cells (Klein *et al.*, 1999), although some evidence suggests that IgM + CD27 + B cells comprise a population of antigen nonexperienced B cells (Weller *et al.*, 2008). CD27 upregulation was noted also on post-GC B cells in the mouse (Jung *et al.*, 2000). Furthermore, ligation of CD27 has been shown to inhibit differentiation of B cells to the plasma cell stage in mice, suggesting expression of this marker might be necessary for the maintenance or generation of the memory B cell pool (Raman *et al.*, 2000).

Controversy has arisen as to whether a subset of memory B cells exists in spleen and bone marrow that has lost surface expression of B220 and expresses CD11b, a marker most commonly observed on myeloid cells (Bell and Gray, 2003; Driver *et al.*, 2001; McHeyzer-Williams *et al.*, 2000), but also on B-1 cells. Identification of this presumed B220$^-$ CD11b$^+$ memory B cell subset has been done by staining for antigen-binding and surface Ig expression (Driver *et al.*, 2001; McHeyzer-Williams *et al.*, 2000). However, others suggested that this population consists of "antigen-capture cells" of myeloid origin that carry antigen-binding IgG on their cell surface via Fcg receptors (Bell and Gray, 2003), consistent with early adoptive transfer experiments that had not shown any evidence for memory cells within the B220 negative fraction of bone marrow and spleen (Manz *et al.*, 1997). This recent controversy demonstrates the need for the identification of additional markers that help to unequivocally identify memory B cell subsets. Much of our understanding on the regulation and development of the immune system has arisen from the ability to isolate and characterize discrete subsets. Multicolor flow cytometry allows us to isolate increasingly smaller subpopulations that previously have resisted analysis. However, without the availability of discrete cell markers these efforts would be futile.

D. Cells Within Extrafollicular Foci

Extrafollicular foci give rise mainly, if not only, to short-lived antibody-secreting cells. It is not clear whether the phenotype of the B cells seeding this reaction is different than those from the mature B cells that initiate the GC reaction. Although marginal zone B cells are reported to quickly respond to antigen encounter with antibody secretion, thus suggesting that there main differentiation pathway is that of the extrafollicular response (Martin *et al.*, 2001), it was recently shown that they can give rise to both extrafollicular foci as well as GC (Song and Cerny, 2003). B cells in the extrafollicular foci rapidly expand and thus have a blast-like FSC profile. Isotype-switching occurs in these foci, although initially IgM antibodies are being secreting by plasma blasts. Using an antigen-specific detection system, IgG1$^+$ NP-specific plasma blasts from extrafollicular foci were identified by the following phenotype: sydecan-1$^+$ B220$^-$ PNA^{lo} (Smith *et al.*, 1996). This differentiated them from the IgG1$^+$ NP-specific syndecan$^-$ B220$^+$ PNA^{hi} GC B cells (Smith *et al.*, 1996). Whether the phenotype of IgM$^+$ plasma blasts is similar to those that have undergone isotype switching has not been reported. Also, it is not clear whether B cells that leave the GC and develop into antibody-secreting plasma blasts and plasma cells differ in phenotype from those that have arisen via the extrafollicular route. It appears that expression of syndecan-1 and CD43 is a shared property of antibody-secreting B cells independent of their differentiation site. But as discussed earlier not all antibody-secreting plasma cells are syndecan-1$^+$, thus leaving open the possibility that GC-derived plasma cells might not necessarily express this surface receptor.

VII. Antigen–Specific B Cells

B cells recognize native unprocessed antigen via their antigen-specific receptor. This can be exploited for the identification of antigen-specific B cells by flow cytometry. Hayakawa *et al.* were first to demonstrate this using the fluorochrome phycoerythrin as an immunogen (Hayakawa *et al.*, 1987). Using cell suspensions from mice immunized with phycoerythrin, they were able to identify the specific cells by their ability to bind to phycoerythrin by flow cytometry. Lalor and colleagues used a similar strategy (Lalor *et al.*, 1992). They conjugated the hapten NP to biotin, or to a fluorochrome. NP-specific B cells in NP-KLH immunized mice were identified by binding to NP and expression of surface IgG1.

This strategy is not restricted to haptens or fluorochromes. In principle, any molecule that can be conjugated to a fluorochrome or biotin should be usable to stain antigen-specific B cells. Our studies outlined in Fig. 5 demonstrated that the hemagglutinin-molecule of influenza virus can be biotinylated and used to identify influenza virus-specific B cells (Doucett *et al.*, 2005). Thus, identification of antigen-specific B cells is considerably more straightforward than identification of antigen-specific T cells, which require a haplotype-matched MHC-tetramer reagent that contains a certain peptide of interest. Also, because the native intact antigen is used, B cells with multiple different epitope specificities can be identified with one reagent. The only restriction is the necessity for BCR expression by the cell. Thus, fully differentiated plasma cells might not express sufficient surface Ig to be labeled. However, intracytoplasmic staining with antigen-specific reagents can overcome this potential limitation (unpublished observation).

In conclusion, large numbers of markers are available that can be used in conjunction with multicolor flow cytometry to evaluate B cell developmental stages in the bone marrow and mature B cell subset in peripheral tissues of mice and similarly of humans. Although currently not fully exploited, the ability to easily develop staining reagents for antigen-specific B cells is a very powerful tool for the evaluation of B cell responses to infection or immunization alike. The challenge lies in developing good reagent combinations that work on the particular

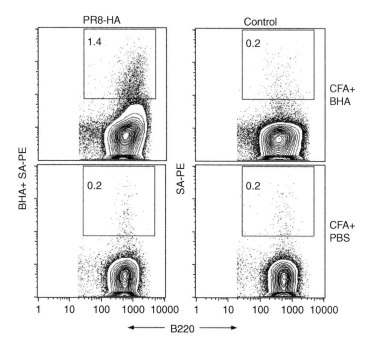

Fig. 5 Identification of influenza virus-specific cells with biotinylated hemagglutinin. Hemagglutinin from influenza virus A/PR8 was purified and biotinylated as outlined elsewhere (Doucett *et al.*, 2005). Shown are 5% contour plots with outliers of draining inguinal lymph node cells from mice immunized at the base of tail for 15 days with Complete Freund's adjuvant (CFA) plus hemagglutinin (BHA), top panel, or PBS, bottom panel. Cells were gated for exclusion of propidium iodide and expression of B220. Left panels show staining with biotinylated hemagglutinin plus streptavidin (SA)-PE. Right panels show staining with SA-PE only.

instruments of the investigator to provide clear results. The marker combinations used in the examples provided here might be a good starting point for those interested in comprehensively analyzing B cell subsets in mice. For a discussion on the choice of fluorochromes and fluorochrome combinations in multicolor flow cytometry, the reader is referred to see Baumgarth and Roederer (2000).

References

Allman, D., Li, J., and Hardy, R. R. (1999). Commitment to the B lymphoid lineage occurs before DH-JH recombination. *J. Exp. Med.* **189,** 735–740.

Allman, D., Lindsley, R. C., DeMuth, W., Rudd, K., Shinton, S. A., and Hardy, R. R. (2001). Resolution of three nonproliferative immature splenic B cell subsets reveals multiple selection points during peripheral B cell maturation. *J. Immunol.* **167,** 6834–6840.

Alugupalli, K. R., Leong, J. M., Woodland, R. T., Muramatsu, M., Honjo, T., and Gerstein, R. M. (2004). B1b lymphocytes confer T cell-independent long-lasting immunity. *Immunity* **21,** 379–390.

Amano, M., Baumgarth, N., Dick, M. D., Brossay, L., Kronenberg, M., Herzenberg, L. A., and Strober, S. (1998). CD1 expression defines subsets of follicular and marginal zone B cells in the spleen: Beta 2-microglobulin-dependent and independent forms. *J. Immunol.* **161,** 1710–1717.

Baumgarth, N., and Roederer, M. (2000). A practical approach to multicolor flow cytometry for immunophenotyping. *J. Immunol. Methods* **243**, 77–97.

Baumgarth, N., Herman, O. C., Jager, G. C., Brown, L., and Herzenberg, L. A. (1999). Innate and acquired humoral immunities to influenza virus are mediated by distinct arms of the immune system. *Proc. Natl. Acad. Sci. USA* **96**, 2250–2255.

Bell, J., and Gray, D. (2003). Antigen-capturing cells can masquerade as memory B cells. *J. Exp. Med.* **197**, 1233–1244.

Benner, R., Hijmans, W., and Haaijman, J. J. (1981). The bone marrow: The major source of serum immunoglobulins, but still a neglected site of antibody formation. *Clin. Exp. Immunol.* **46**, 1–8.

Borghesi, L., and Gerstein, R. M. (2004). Developmental separation of V(D)J recombinase expression and initiation of IgH recombination in B lineage progenitors *in vivo*. *J. Exp. Med.* **199**, 483–489.

Borrello, M. A., and Phipps, R. P. (1995). Fibroblasts support outgrowth of splenocytes simultaneously expressing B lymphocyte and macrophage characteristics. *J. Immunol.* **155**, 4155–4161.

Borrello, M. A., and Phipps, R. P. (1999). Fibroblast-secreted macrophage colony-stimulating factor is responsible for generation of biphenotypic B/macrophage cells from a subset of mouse B lymphocytes. *J. Immunol.* **163**, 3605–3611.

Butcher, E. C., Rouse, R. V., Coffman, R. L., Nottenburg, C. N., Hardy, R. R., and Weissman, I. L. (1982). Surface phenotype of Peyer's patch germinal center cells: Implications for the role of germinal centers in B cell differentiation. *J. Immunol.* **129**, 2698–2707.

Cariappa, A., Mazo, I. B., Chase, C., Shi, H. N., Liu, H., Li, Q., Rose, H., Leung, H., Cherayil, B. J., Russell, P., von Andrian, U., and Pillai, S. (2005). Perisinusoidal B cells in the bone marrow participate in T-independent responses to blood-borne microbes. *Immunity* **23**, 397–407.

Chevrier, S., Genton, C., Kallies, A., Karnowski, A., Otten, L. A., Malissen, B., Malissen, M., Botto, M., Corcoran, L. M., Nutt, S. L., and Acha-Orbea, H. (2009). CD93 is required for maintenance of antibody secretion and persistence of plasma cells in the bone marrow niche. *Proc. Natl. Acad. Sci. USA* **106**, 3895–3900.

Choi, Y. S., and Baumgarth, N. (2008). Dual role for B-1a cells in immunity to influenza virus infection. *J. Exp. Med.* **205**, 3053–3064.

Cong, Y. Z., Rabin, E., and Wortis, H. H. (1991). Treatment of murine CD5-B cells with anti-Ig, but not LPS, induces surface CD5: Two B-cell activation pathways. *Int. Immunol.* **3**, 467–476.

Crotty, S., Kersh, E. N., Cannons, J., Schwartzberg, P. L., and Ahmed, R. (2003). SAP is required for generating long-term humoral immunity. *Nature* **421**, 282–287.

Davidson, W. F., Pierce, J. H., and Holmes, K. L. (1992). Evidence for a developmental relationship between CD5 + B-lineage cells and macrophages. *Ann. N. Y. Acad. Sci.* **651**, 112–129.

Doucett, V. P., Gerhard, W., Owler, K., Curry, D., Brown, L., and Baumgarth, N. (2005). Enumeration and characterization of virus-specific B cells by multicolor flow cytometry. *J. Immunol. Methods* **303**, 40–52.

Driver, D. J., McHeyzer-Williams, L. J., Cool, M., Stetson, D. B., and McHeyzer-Williams, M. G. (2001). Development and maintenance of a B220- memory B cell compartment. *J. Immunol.* **167**, 1393–1405.

Esplin, B. L., Welner, R. S., Zhang, Q., Borghesi, L. A., and Kincade, P. W. (2009). A differentiation pathway for B1 cells in adult bone marrow. *Proc. Natl. Acad. Sci. USA* **106**, 5773–5778.

Ghia, P., Boekel, E. t., Rolink, A. G., and Melchers, F. (1998). B-cell develoment: A comparison between mouse and man. *Immunol. Today* **19**, 480–485.

Gray, D. (1993). Immunological memory. *Annu. Rev. Immunol.* **11**, 49–77.

Han, S., Zheng, B., Schatz, D. G., Spanopoulou, E., and Kelsoe, G. (1996). Neoteny in lymphocytes: Rag1 and Rag2 expression in germinal center B cells. *Science* **274**, 2094–2097.

Han, S., Zheng, B., Takahashi, Y., and Kelsoe, G. (1997). Distinctive characteristics of germinal center B cells. *Semin. Immunol.* **9**, 255–260.

Hardy, R. R., Carmack, C. E., Shinton, S. A., Kemp, J. D., and Hayakawa, K. (1991). Resolution and characterization of pro-B and pre-pro-B cell stages in normal mouse bone marrow. *J. Exp. Med.* **173**, 1213–1225.

Hargreaves, D. C., Hyman, P. L., Lu, T. T., Ngo, V. N., Bidgol, A., Suzuki, G., Zou, Y. R., Littman, D. R., and Cyster, J. G. (2001). A coordinated change in chemokine responsiveness guides plasma cell movements. *J. Exp. Med.* **194,** 45–56.

Hayakawa, K., and Hardy, R. R. (2000). Development and function of B-1 cells. *Curr. Opin. Immunol.* **12,** 346–353.

Hayakawa, K., Ishii, R., Yamasaki, K., Kishimoto, T., and Hardy, R. R. (1987). Isolation of high-affinity memory B cells: Phycoerythrin as a probe for antigen-binding cells. *Proc. Natl. Acad. Sci. USA* **84,** 1379–1383.

Hippen, K. L., Tze, L. E., and Behrens, T. W. (2000). CD5 maintains tolerance in anergic B cells. *J. Exp. Med.* **191,** 883–890.

Jung, J., Choe, J., Li, L., and Choi, Y. S. (2000). Regulation of CD27 expression in the course of germinal center B cell differentiation: The pivotal role of IL-10. *Eur. J. Immunol.* **30,** 2437–2443.

Karasuyama, H., Rolink, A., Shinkai, Y., Young, F., Alt, F. W., and Melchers, F. (1994). The expression of Vpre-B/lambda 5 surrogate light chain in early bone marrow precursor B cells of normal and B cell-deficient mutant mice. *Cell* **77,** 133–143.

Kawano, Y., Yoshikawa, S., Minegishi, Y., and Karasuyama, H. (2006). Pre-B cell receptor assesses the quality of IgH chains and tunes the pre-B cell repertoire by delivering differential signals. *J. Immunol.* **177,** 2242–2249.

Kim, S. J., Caton, M., Wang, C., Khalil, M., Zhou, Z. J., Hardin, J., and Diamond, B. (2008). Increased IL-12 inhibits B cells' differentiation to germinal center cells and promotes differentiation to short-lived plasmablasts. *J. Exp. Med.* **205,** 2437–2448.

Klein, U., Rajewsky, K., and Kuppers, R. (1999). Phenotypic and molecular characterization of human peripheral blood B-cell subsets with special reference to N-region addition and J kappa-usage in V kappa J kappa-joints and kappa/lambda-ratios in naive versus memory B-cell subsets to identify traces of receptor editing processes. *Curr. Top. Microbiol. Immunol.* **246,** 141–146; discussion 147.

Kondo, M., Weissman, I. L., and Akashi, K. (1997). Identification of clonogenic common lymphoid progenitors in mouse bone marrow. *Cell* **91,** 661–672.

Kuwata, N., Igarashi, H., Ohmura, T., Aizawa, S., and Sakaguchi, N. (1999). Cutting edge: Absence of expression of RAG1 in peritoneal B-1 cells detected by knocking into RAG1 locus with green fluorescent protein gene. *J. Immunol.* **163,** 6355–6359.

Lalor, P. A., Nossal, G. J., Sanderson, R. D., and McHeyzer-Williams, M. G. (1992). Functional and molecular characterization of single, (4-hydroxy-3-nitrophenyl)acetyl (NP)-specific, IgG1 + B cells from antibody-secreting and memory B cell pathways in the C57BL/6 immune response to NP. *Eur. J. Immunol.* **22,** 3001–3011.

Li, Y. S., Wasserman, R., Hayakawa, K., and Hardy, R. R. (1996). Identification of the earliest B lineage stage in mouse bone marrow. *Immunity* **5,** 527–535.

Loder, F., Mutschler, B., Ray, R. J., Paige, C. J., Sideras, P., Torres, R., Lamers, M. C., and Carsetti, R. (1999). B cell development in the spleen takes place in discrete steps and is determined by the quality of B cell receptor-derived signals. *J. Exp. Med.* **190,** 75–89.

Lu, L., Smithson, G., Kincade, P. W., and Osmond, D. G. (1998). Two models of murine B lymphopoiesis: A correlation. *Eur. J. Immunol.* **28,** 1755–1761.

MacLennan, I. C. (1994). Germinal centers. *Annu. Rev. Immunol.* **12,** 117–139.

Manser, T. (2004). Textbook germinal centers? *J. Immunol.* **172,** 3369–3375.

Manz, R. A., Thiel, A., and Radbruch, A. (1997). Lifetime of plasma cells in the bone marrow. *Nature* **388,** 133–134.

Martin, F., and Kearney, J. F. (2002). Marginal-zone B cells. *Nat. Rev. Immunol.* **2,** 323–335.

Martin, M., Strasser, A., Baumgarth, N., Cicuttini, F. M., Welch, K., Salvaris, E., and Boyd, A. W. (1993). A novel cellular model (SPGM 1) of switching between the pre-B cell and myelomonocytic lineages. *J. Immunol.* **150,** 4395–4406.

Martin, F., Oliver, A. M., and Kearney, J. F. (2001). Marginal zone and B1 B cells unite in the early response against T-independent blood-borne particulate antigens. *Immunity* **14,** 617–629.

McHeyzer-Williams, M. G., Nossal, G. J., and Lalor, P. A. (1991). Molecular characterization of single memory B cells. *Nature* **350**, 502–505.

McHeyzer-Williams, L. J., Cool, M., and McHeyzer-Williams, M. G. (2000). Antigen-specific B cell memory: Expression and replenishment of a novel b220(-) memory b cell compartment. *J. Exp. Med.* **191**, 1149–1166.

Meffre, E., Casellas, R., and Nussenzweig, M. C. (2000). Antibody regulation of B cell development. *Nat. Immunol.* **1**, 379–385.

Melchers, F., Rolink, A., Grawunder, U., Winkler, T. H., Karasuyama, H., Ghia, P., and Andersson, J. (1995). Positive and negative selection events during B lymphopoiesis. *Curr. Opin. Immunol.* **7**, 214–227.

Montecino-Rodriguez, E., Leathers, H., and Dorshkind, K. (2001). Bipotential B-macrophage progenitors are present in adult bone marrow. *Nat. Immunol.* **2**, 83–88.

Montecino-Rodriguez, E., Leathers, H., and Dorshkind, K. (2006). Identification of a B-1 B cell-specified progenitor. *Nat. Immunol.* **7**, 293–301.

Nagai, Y., Shimazu, R., Ogata, H., Akashi, S., Sudo, K., Yamasaki, H., Hayashi, S., Iwakura, Y., Kimoto, M., and Miyake, K. (2002). Requirement for MD-1 in cell surface expression of RP105/CD180 and B-cell responsiveness to lipopolysaccharide. *Blood* **99**, 1699–1705.

Nikolic, T., Dingjan, G. M., Leenen, P. J., and Hendriks, R. W. (2002). A subfraction of B220(+) cells in murine bone marrow and spleen does not belong to the B cell lineage but has dendritic cell characteristics. *Eur. J. Immunol.* **32**, 686–692.

Ohinata, Y., Payer, B., O'Carroll, D., Ancelin, K., Ono, Y., Sano, M., Barton, S. C., Obukhanych, T., Nussenzweig, M., Tarakhovsky, A., Saitou, M., and Surani, M. A. (2005). Blimp1 is a critical determinant of the germ cell lineage in mice. *Nature* **436**, 207–213.

Oliver, A. M., Martin, F., and Kearney, J. F. (1997). Mouse CD38 is down-regulated on germinal center B cells and mature plasma cells. *J. Immunol.* **158**, 1108–1115.

Osmond, D. G. (1990). B cell development in the bone marrow. *Semin. Immunol.* **2**, 173–180.

Osmond, D. G., Rolink, A., and Melchers, F. (1998). Murine B lymphopoiesis: Towards a unified model. *Immunol. Today* **19**, 65–68.

Phan, T. G., Paus, D., Chan, T. D., Turner, M. L., Nutt, S. L., Basten, A., and Brink, R. (2006). High affinity germinal center B cells are actively selected into the plasma cell compartment. *J. Exp. Med.* **203**, 2419–2424.

Raman, V. S., Bal, V., Rath, S., and George, A. (2000). Ligation of CD27 on murine B cells responding to T-dependent and T-independent stimuli inhibits the generation of plasma cells. *J. Immunol.* **165**, 6809–6815.

Ridderstad, A., and Tarlinton, D. M. (1998). Kinetics of establishing the memory B cell population as revealed by CD38 expression. *J. Immunol.* **160**, 4688–4695.

Roark, J. H., Park, S. H., Jayawardena, J., Kavita, U., Shannon, M., and Bendelac, A. (1998). CD1.1 expression by mouse antigen-presenting cells and marginal zone B cells. *J. Immunol.* **160**, 3121–3127.

Rolink, A., Grawunder, U., Winkler, T. H., Karasuyama, H., and Melchers, F. (1994). IL-2 receptor alpha chain (CD25, TAC) expression defines a crucial stage in pre-B cell development. *Int. Immunol.* **6**, 1257–1264.

Rolink, A., ten Boekel, E., Melchers, F., Fearon, D. T., Krop, I., and Andersson, J. (1996). A subpopulation of B220 + cells in murine bone marrow does not express CD19 and contains natural killer cell progenitors. *J. Exp. Med.* **183**, 187–194.

Rolink, A. G., Andersson, J., and Melchers, F. (1998). Characterization of immature B cells by a novel monoclonal antibody, by turnover and by mitogen reactivity. *Eur. J. Immunol.* **28**, 3738–3748.

Schatz, D. G., Oettinger, M. A., and Baltimore, D. (1989). The V(D)J recombination activating gene, RAG-1. *Cell* **59**, 1035–1048

Schittek, B., and Rajewsky, K. (1990). Maintenance of B-cell memory by long-lived cells generated from proliferating precursors. *Nature* **346**, 749–751.

Shinall, S. M., Gonzalez-Fernandez, M., Noelle, R. J., and Waldschmidt, T. J. (2000). Identification of murine germinal center B cell subsets defined by the expression of surface isotypes and differentiation antigens. *J. Immunol.* **164,** 5729–5738.

Shinkai, Y., Rathbun, G., Lam, K. P., Oltz, E. M., Stewart, V., Mendelsohn, M., Charron, J., Datta, M., Young, F., Stall, A. M., and Alt, F. W. (1992). RAG-2-deficient mice lack mature lymphocytes owing to inability to initiate V(D)J rearrangement. *Cell* **68,** 855–867.

Smith, K. G., Hewitson, T. D., Nossal, G. J., and Tarlinton, D. M. (1996). The phenotype and fate of the antibody-forming cells of the splenic foci. *Eur. J. Immunol.* **26,** 444–448.

Song, H., and Cerny, J. (2003). Functional heterogeneity of marginal zone B cells revealed by their ability to generate both early antibody-forming cells and germinal centers with hypermutation and memory in response to a T-dependent antigen. *J. Exp. Med.* **198,** 1923–1935.

Takahashi, K., Kozono, Y., Waldschmidt, T. J., Berthiaume, D., Quigg, R. J., Baron, A., and Holers, V. M. (1997). Mouse complement receptors type 1 (CR1;CD35) and type 2 (CR2;CD21): Expression on normal B cell subpopulations and decreased levels during the development of autoimmunity in MRL/lpr mice. *J. Immunol.* **159,** 1557–1569.

Teague, B. N., Pan, Y., Mudd, P. A., Nakken, B., Zhang, Q., Szodoray, P., Kim-Howard, X., Wilson, P. C., and Farris, A. D. (2007). Cutting edge: Transitional T3 B cells do not give rise to mature B cells, have undergone selection, and are reduced in murine lupus. *J. Immunol.* **178,** 7511–7515.

Underhill, G. H., Kolli, K. P., and Kansas, G. S. (2003). Complexity within the plasma cell compartment of mice deficient in both E- and P-selectin: Implications for plasma cell differentiation. *Blood* **102,** 4076–4083.

Weller, S., Mamani-Matsuda, M., Picard, C., Cordier, C., Lecoeuche, D., Gauthier, F., Weill, J. C., and Reynaud, C. A. (2008). Somatic diversification in the absence of antigen-driven responses is the hallmark of the IgM + IgD+ CD27+ B cell repertoire in infants. *J. Exp. Med.* **205,** 1331–1342.

Wells, S. M., Kantor, A. B., and Stall, A. M. (1994). CD43 (S7) expression identifies peripheral B cell subsets. *J. Immunol.* **153,** 5503–5515.

Won, W. J., and Kearney, J. F. (2002). CD9 is a unique marker for marginal zone B cells, B1 cells, and plasma cells in mice. *J. Immunol.* **168,** 5605–5611.

PART V

Cytogenetics/Chromatin Structure

CHAPTER 27

Telomere Length Measurements Using Fluorescence *In Situ* Hybridization and Flow Cytometry

Gabriela M. Baerlocher★,† and Peter M. Lansdorp★,‡

★Terry Fox Laboratory
British Columbia Cancer Agency
Vancouver, British Columbia
Canada V5Z 1L3

†Department of Hematology and Department of Clinical Research
University Hospital Bern
3010 Bern, Switzerland

‡Department of Medicine
University of British Columbia
Vancouver, British Columbia
Canada V5Z 4E3

DOI: 10.1016/B978-0-12-375045-7.00027-1

I. Introduction

A. Telomere Length Dynamics

Linear chromosome ends are composed of TTAGGG repeats and associated proteins, which are assembled into a dynamic three-dimensional structure—the telomere. Most of the telomeric tract consists of double-stranded DNA, but the very end of the chromosome contains a short (50–300 nucleotide) single-stranded G-rich 3' overhang. This overhang appears essential for telomere function and the formation of a typical fold-back structure called a *telomere loop* or *t-loop* (Griffith *et al.*, 1999). Current data support the concept that telomeres can exist in at least two states: an uncapped or open form and a capped or closed t-loop form (Blackburn, 2001). It is possible that telomeres reversibly switch between these two states. Such an equilibrium would provide entry for regulation of the cell cycle, programmed cell death, DNA damage responses, and possibly other cellular functions.

In the open or uncapped state, the single-stranded overhang at telomeres can serve as a substrate for telomerase, a cellular reverse transcriptase that can add new telomeric hexanucleotides onto chromosome ends. Telomerase action is required to compensate for the loss of telomere repeats with each cell division because of incomplete DNA replication (Olovnikov, 1971; Watson, 1972), oxidative damage (von Zglinicki, 2002), and possibly other causes. In cells that do not express sufficient compensatory telomerase, telomeres shorten with each cell division *in vitro* and *in vivo*. Critically short telomeres trigger apoptosis or irreversible cell cycle arrest but can also result in chromosome fusions and genomic instability. Important evidence for a link between telomere shortening and cellular senescence is derived from experiments with diverse human cell types showing that ectopic expression of telomerase results in telomere elongation and extension of their proliferative potential and replicative life span (Bodnar *et al.*, 1998; Vaziri and

Benchimol, 1998). Haploinsufficiency for the human telomerase template gene (hTERC), resulting in a modest reduction in telomerase activity, triggers the autosomal dominant disorder dyskeratosis congenita (DKC). Patients with DKC show clinical manifestations in rapidly proliferating tissues such as nail, skin, and bone marrow and typically die before the age of 50 years from marrow failure, immune deficiencies, or cancer (Vulliamy *et al.*, 2001). In patients with DKC, telomeres are critically short and telomere dysfunction during tissue development and regeneration is the most likely explanation for the proliferative defect and related pathology. Patients with DKC highlight the critical importance of adequate telomerase levels in human cells and point to carefully controlled telomere length regulation to sustain the proliferation in stem cells of various organs.

Current data support the idea that telomere shortening with age evolved as a tumor suppressor mechanism in long-lived species. By suppressing immortal growth, younger and healthier (stem) cells are recruited into action as the propagation of older cells (which are more likely to have acquired detrimental mutations) is suppressed. Human cancer cells have to overcome the restrictions in proliferative potential imposed by progressive telomere shortening. Typically, this is achieved by upregulation of telomerase activity, a notion that is supported by the observation that more than 90% of human tumor cells show readily detectable telomerase activity (Kim *et al.*, 1994; Shay and Bacchetti, 1997).

In summary, telomere length dynamics are very important for the regulation of the replicative life span of cells, especially in long-lived species. Telomere shortening and telomerase activity are believed to be important factors in aging and tumorigenesis. To study the telomere length in different cell types, we developed flow-fluorescence *in situ* hybridization (flow-FISH), a sensitive and reproducible technique to measure the telomere length in individual cells.

B. Methods to Measure Telomere Length

We focus on the three most frequently used methods to measure the length of telomere repeats in cells (Table I): Southern blot, Q-FISH (telomere length measurements using digital images of metaphase chromosomes after *in situ* hybridization with fluorescently labeled telomere probe), and flow-FISH (measurements of the average telomere length in cells using flow cytometry after *in situ* hybridization with fluorescently labeled telomere probe). Another promising polymerase chain reaction (PCR)-based method was described to measure telomere length (Baird *et al.*, 2003). This sensitive but laborious method is not discussed in this chapter because of space limitations.

1. Southern Blot Analysis

Telomere length measurement by Southern blot is based on the following steps:

1. Isolation and restriction enzyme digestion of genomic DNA.

Table I
Methods to Measure Telomere Length

	Southern blot	Quantitative Q-FISH	Quantitative flow-FISH
Cell requirement	10^5–10^6	10^5–10^6 for 10–15 metaphases	10^4–10^5
Cell cycle status	Cell cycle independent	Proliferating cells	Cell cycle independent
Processing time needed	3–5 days	3–5 days	1–2 days
Hybridization targets	Telomeric and subtelomeric repeats within a cell population	Telomeric repeats at specific chromosome ends	Telomeric repeats and intrachromosomal telomeric repeats in single cells
Additional parameter	None	None	Forward light scatter and side scatter, DNA fluorescence, limited cell-surface markers

2. Separation of the DNA fragments containing telomeric DNA based on their size using gel electrophoresis.

3. Hybridization and visualization of telomere fragments.

4. Estimation of the average length of the fragments containing telomere repeats from the smear of heterogeneously sized fragments.

Each step has significant drawbacks:

1. There are no restriction enzyme sites directly at the end of the telomeric repeats, so a variable amount of subtelomeric chromosomal DNA will be included in the telomere fragments. As a consequence, telomere length will be invariably overestimated.

2. Very small and very long telomere fragments cannot be efficiently separated on the same gel. Because of the nonlinear relation between fragment size and migration distance in the gel, very short telomeres will run faster (distributed over a wider range) and their size will be underestimated. Long telomeres (>30 kb) may not migrate into the gel very well and often cannot be resolved. These problems are exacerbated by the fact that the number of labeled telomere repeat oligonucleotide molecules that can hybridize to the fragments is related to the size of the fragment. This is yet another factor that makes it difficult, if not impossible, to avoid overestimations of telomere length using Southern blot analysis.

3. To be able to visualize telomere fragments in a gel, high-quality genomic DNA from relatively large numbers of cells (typically corresponding to 10^5–10^6 cells) is required. The resulting telomere length estimate represents the average telomere length in a population of cells and does not allow estimates of the telomere length in minor subpopulations of cells within the sample.

4. Southern blot is a time-consuming method that typically requires radiola-beled probes.

2. Q-FISH

For telomere length measurements by Q-FISH, the following steps are performed:

1. Stimulation of cells in culture and harvest of cells arrested in metaphase.
2. Preparation of slides with fixed metaphase cells.
3. *In situ* hybridization with fluorescently labeled telomere peptide nucleic acid (PNA) probe and DNA counterstaining.
4. Acquisition and analysis of the telomere fluorescence on individual chromosome ends using digital images acquired with a sensitive camera mounted on a fluorescence microscope.

There are two major advantages of this method: Telomere length measurements are highly accurate and sensitive (with appropriate controls) (Martens *et al.*, 1998; Poon *et al.*, 1999) and information on the average telomere length at specific chromosome ends can be obtained. However, each step of the Q-FISH protocol also has its limitations:

1. The requirement for metaphase chromosomes excludes Q-FISH for telomere length analysis in nondividing cells. Presenescent cells or chronic lymphocytic leukemic B cells, which hardly divide, are good examples of cells in which measurement of telomere length by Q-FISH is problematic. Furthermore, to prepare slides with suitable metaphase spreads requires specific technical expertise.

2. Although the nature of telomere-PNA probes allows specific and sensitive hybridization under certain conditions (see Section II), the accessibility of the telomere-PNA probe to the condensed target sequences in metaphase chromosomes remains incompletely known. Thus far, no internal standard to control for possible variation in accessibility during *in situ* hybridization has been developed.

3. Expensive fluorescence microscope and digital camera/computer equipment are required and personnel need to be properly trained to guarantee the quality and standardization of telomere length measurements from captured fluorescence images.

4. Finally, the processes of fluorescence image acquisition and analysis, despite the use of a freely available dedicated computer program (TFL-Telo, available at www.flintbox.com) (Poon *et al.*, 1999), remain cumbersome and time-consuming.

3. Flow-FISH

Telomere length measurements by the basic flow-FISH technique (Rufer *et al.*, 1998) rely on the following steps:

1. Isolation of the nucleated cells from whole blood or tissue.

2. *In situ* hybridization with fluorescently labeled telomere-PNA probe and DNA counterstaining of the cells.

3. Acquisition and analysis of the telomere fluorescence on the flow cytometer. Significant advantages of this method are that telomere length measurements can be performed with individual cells in suspension, that telomere length measurements are not compromised by the presence of cells with very long or short telomeres, that cells can be nondividing, that many cells can be analyzed within a short period, and that in samples of nucleated blood cells, two subpopulations of cells (granulocytes and lymphocytes) can be analyzed separately based on differences in light scatter properties.

Nevertheless, each step also has its problems:

1. Hemoglobin in the cell suspension can interfere with telomere fluorescence in cells and the cell isolation procedures that are required can lead to cell loss and fragmentation.

2. Heat and formamide treatment of cells, though resulting to some extent in the fixation of cells, results in loss of cells and loss of most cell-surface antigens.

3. Binding of the telomere probe to cellular organelles and membranes and incomplete removal of excess probe can result in overestimates of telomere length.

4. Accessibility of the telomere-PNA probe to target sequences may vary with the degree of DNA condensation in the interphase nucleus.

5. The autofluorescence of cells (especially granulocytes) and the DNA fluorescence of the DNA counterstain (used for the gating of cells and for distinction between cells with a diploid [2N] chromosome content and cells in S or G2/M) can interfere with the detection of specific telomere fluorescence.

C. New Method: Automated Multicolor Flow–FISH

Flow-FISH offers many advantages compared to Southern blot and Q-FISH (Table I). However, the first flow-FISH protocols (Rufer *et al.*, 1998, 1999) were still quite time-consuming, especially when more than 10 samples were processed at a time, and further problems were encountered when differences in telomere length were small or when cell counts were (very) low. In addition, no information on the telomere length in subsets of cells (except for granulocytes and lymphocytes) within one sample could be obtained unless such subpopulations were sorted before the procedure.

We optimized and automated the basic flow-FISH technique (Baerlocher *et al.*, 2002) and combined the procedure with limited immunophenotyping (Baerlocher and Lansdorp, 2003; Baerlocher *et al.*, 2006). Many variables influencing the *in situ* hybridization process and the subsequent results were tested in extensive protocol development steps to find the most optimal protocol for telomere length measurements with high sensitivity and good reproducibility. One of the crucial improvements was the automation of most cell-processing steps by a robotic washing station—the Hydra (Robbins Scientific or Apogent Discoveries). The Hydra device allows to aspirate and dispense very small volumes of liquids simultaneously and gently in and from 96 microtubes arranged in a standard 96-well plate format. Again, each step of our original protocol was adjusted to optimally integrate the Hydra device. Furthermore, we included control cells (bovine thymocytes) in each tube as an internal standard to further control for the inevitable method-related variation between tubes. Finally, flow-FISH was combined with limited immunophenotyping.

D. Chapter Outline

In this chapter, we first provide a brief overview of the development of quantitative *in situ* hybridization techniques and we describe specific properties of PNA probes. We then present our latest improvements in the flow-FISH technique for highly sensitive and reproducible measurements of telomere length in interphase cells. We aim to point out the critical parameters of the technique and important aspects of data acquisition on the flow cytometer. Furthermore, we elaborate on *in situ* hybridization combined with immunophenotyping. Finally, we discuss some current applications of the flow-FISH method.

=========   ## II. Background

A. Quantitative Fluorescence *In Situ* Hybridization

At the end of the 1970s, recombinant DNA technology, which allowed production of large quantities of known DNA or RNA sequences in bacteria, paved the way for *in situ* hybridization methods to map mammalian genes and study gene expression in cells and tissue sections. The first FISH experiments used messenger RNA (mRNA) molecules, which were linked to fluorescent latex microspheres to visualize globin genes in human metaphase chromosomes (Cheung *et al.*, 1977). Since then, direct labeling of probes with suitable fluorescent dyes was introduced and many other improvements in FISH methods, and in general, fluorescence detection by fluorescence microscopy, were made (Trask, 1991). As a result, FISH has become a standard *qualitative* tool in cytogenetics and gene mapping.

In the late 1980s, the first reports can be found describing *in situ* hybridization of probes to interphase nuclei in suspension followed by the quantification of bound probe by flow cytometry (Trask *et al.*, 1988). The aim in those studies was to combine FISH with the speed and quantitative analysis provided by flow cytometry. Improvements in FISH probes, antibody labeling, fluorescence labels, and flow cytometers contributed to further advances in FISH techniques using either fluorescence microscopy or flow cytometry. However, one of the fundamental drawbacks of conventional FISH techniques for DNA targets remained that the

DNA or RNA probes that were used always had to compete for hybridization to the denatured target DNA sequences with the original complementary DNA strand. Although high probe concentrations are used to increase the likelihood of probe hybridization, this was not a satisfactory solution for the detection of repeat sequences (as in telomeric DNA) in which a high concentration of short oligonucleotide probes cannot effectively or reproducibly compete with long complementary strands. This shortcoming of conventional FISH was overcome in 1993 when a new sort of molecule was invented: PNAs (Nielsen and Egholm, 1999). Soon, it was found that directly fluorescently labeled PNA probes could be hybridized to target sequences under conditions (low ionic strength) that *did not allow* reannealing of complementary strands (Lansdorp, 1996). These observations allowed the development of the first *quantitative* FISH (Q-FISH) techniques.

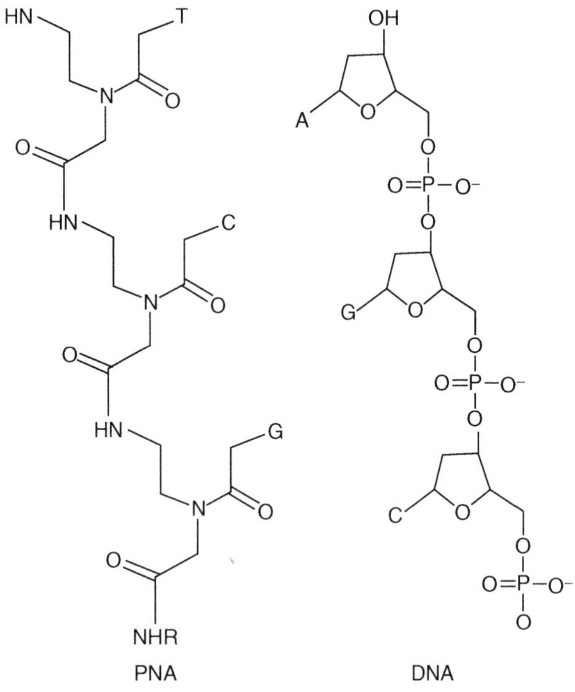

Fig. 1 The general structure of deoxyribose nucleic acid (DNA) oligonucleotides compared to peptide nucleic acid (PNA) is shown. The PNA monomer consists of *N*-(2-amino-ethyl)-glycine units linked by a methylene carbonyl to one of the four bases (adenine, guanine, thymine, or cytosine) found in DNA. Like amino acids, PNA monomers have amino and carboxy termini. Unlike DNA or RNA, PNAs lack pentose sugar phosphate groups.

B. Peptide Nucleic Acids

PNA is a nucleic acid analog in which the sugar phosphate backbone of natural nucleic acid has been replaced by a synthetic peptide backbone formed from *N*-(2-amino-ethyl)-glycine units, resulting in an achiral and uncharged moiety that mimics RNA or DNA oligonucleotides (Fig. 1). PNA is chemically stable, resistant to hydrolytic (enzymatic) cleavage, and cannot be degraded inside living cells. PNA is capable of sequence-specific recognition of DNA and RNA, obeying the Watson-Crick hydrogen bonding scheme, and the hybrid complexes exhibit extraordinary thermal stability and unique ionic strength properties (Ray and Norden, 2000). PNA resulted from efforts during the 1980s and early 1990s by Peter Nielsen and Michael Egholm in Ole Buchardt's laboratory in Denmark. The unique properties of PNA-DNA hybrid complexes were exploited in FISH techniques for the first time in 1995 by Peter Lansdorp during a sabbatical visit in the laboratory of Hans Tanke and Ton Raap at Leiden University in the Netherlands. Although directly labeled PNA probes were already commercially available, their use in FISH techniques required the development of conditions that allowed maximal hybridization at low ionic strength (preventing competition between probe hybridization and DNA renaturation) and low background staining of the labeled probe. The latter was particularly problematic because PNA probes are very hydrophobic and strongly bind, for example, to glass and various hydrophobic sites in cells and chromosomes. Solutions to these problems were found by including blocking protein in the hybridization solution and performing hybridization and wash steps in the presence of high concentrations (70%) of formamide (Lansdorp, 1996). As in all protocol development efforts, countless experiments had to be performed to delineate the importance of each parameter in the technique, distill the most important variables, and establish a procedure that works reproducibly.

III. Methods

This section is based on many experiments performed over the last several years to optimize flow-FISH for *quantitative* measurements of the telomere repeat content in interphase cells. We try to give an overview of the most important conclusions drawn from the findings in many experiments. For additional details, we refer to our two method papers on this topic (Baerlocher and Lansdorp, 2003; Baerlocher *et al.*, 2006; Baerlocher *et al.*, 2002). Table II lists what type of equipment, material, and solutions are needed to set up automated multicolor flow-FISH. In Fig. 2, the main steps of the protocol, the most critical parameters, and our recommended processing steps are presented in a diagram.

Table II
Equipment, Materials, and Supplies Needed for Automated Multicolor Flow-FISH

Equipment	Materials	Solutions
Rotator	Sodium heparin Vacutainer tubes	Ammonium chloride
Shaker	50-ml tubes	Phosphate buffer solution, pH 7.4
Tube racks	14-ml tubes	Dextrose 5%, 252 mOsm, pH 4.0
Centrifuge for 10/14-ml tubes	1.5-ml tubes	Bovine serum albumin 10%
Microcentrifuge for 1.5-ml tubes	Polymerase chain reaction tubes	Hepes 100 mM
Circulating water bath	96 deep-well plates	Trizma base, pH 7.1
Thermometer	Pipette tips	Deionized formamide
Fume hood	Foil	Formamide
Biohazard hood	Parafilm	Sodium chloride 1 M
Pipettes (1000, 100, 20 μl)	Resin beads	Telomere-PNA-fluorescein at 30 μg/ml
Repeater pipette	0.22 μm polyethersulfone	Tween 20 10%
Washing station (e.g., Hydra)	Vacuum filter system	LDS 751 at 0.2 mg/ml
Flow cytometer	Whatman paper	RNase T1 at 100,000 U/ml
	Sieve/strainer	Sodium acid 20%
		FACSFlow
		Dulbecco modified Eagle's medium
		DNAse
		Heparin
		Dimethylsulfoxide
		Paraformaldehyde

A. Optimization of Quantitative *In Situ* Hybridization

1. Preparation/Isolation of Nucleated Cells for Flow-FISH

Ideally, one would like to start the flow-FISH procedure with a homogenous single-cell suspension containing sufficient (2–10 $\times$ 10^5), intact, and stable (not necessarily viable) nucleated cells. In reality, the cells of interest for telomere length measurements typically come in a cell mixture with red blood cells (RBCs) (e.g., whole blood) and/or they are embedded in tissue (e.g., thymus, spleen). Both situations make isolation or separation procedures for the cells of interest necessary. Hemoglobin has fluorescent properties and can act as quencher (absorb the fluorescence from other fluorescent molecules). Therefore, a high concentration of RBCs or hemoglobin (corresponding to a RBC suspension with a hematocrit $\geq$2%) can lead to interference with fluorescence detection from the fluorescein-labeled telomere probe. Second, any cell isolation or separation step can lead to changes in the properties (size, shape, autofluorescence), viability, and recovery of cells. Therefore, the aim is to use gentle techniques for cell preparation and keep the preparation time as short as possible.

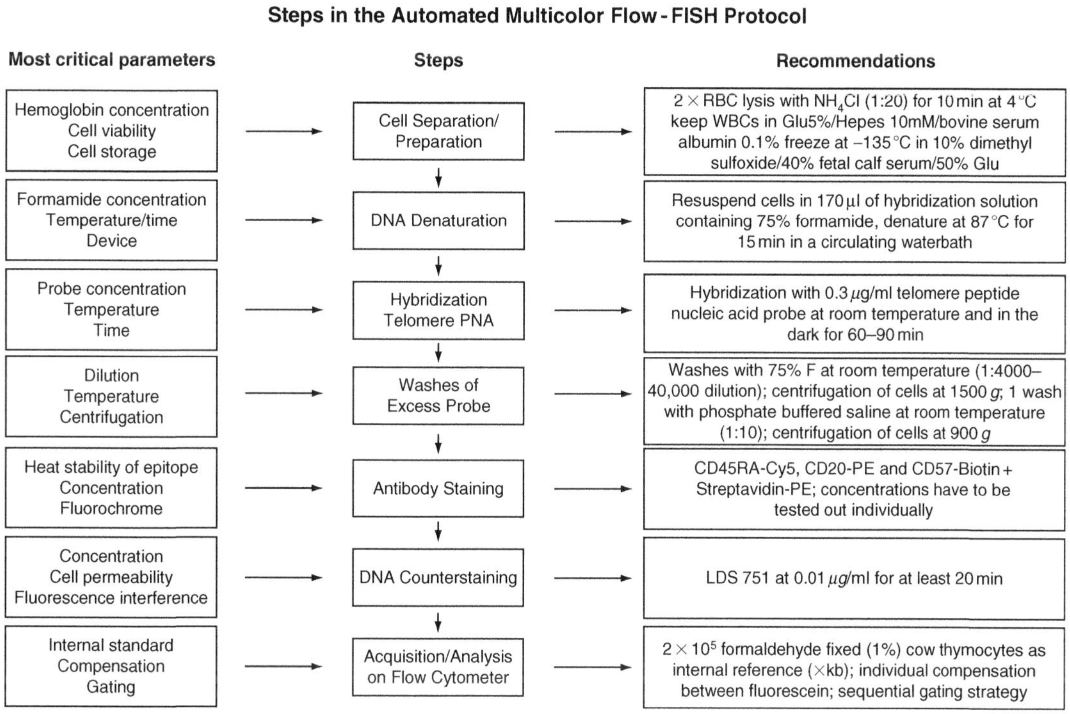

Steps in the Automated Multicolor Flow-FISH Protocol

Most critical parameters	Steps	Recommendations
Hemoglobin concentration Cell viability Cell storage	Cell Separation/ Preparation	$2 \times$ RBC lysis with NH_4Cl (1:20) for 10 min at 4 °C keep WBCs in Glu5%/Hepes 10mM/bovine serum albumin 0.1% freeze at -135 °C in 10% dimethyl sulfoxide/40% fetal calf serum/50% Glu
Formamide concentration Temperature/time Device	DNA Denaturation	Resuspend cells in 170 µl of hybridization solution containing 75% formamide, denature at 87 °C for 15 min in a circulating waterbath
Probe concentration Temperature Time	Hybridization Telomere PNA	Hybridization with 0.3 µg/ml telomere peptide nucleic acid probe at room temperature and in the dark for 60–90 min
Dilution Temperature Centrifugation	Washes of Excess Probe	Washes with 75% F at room temperature (1:4000–40,000 dilution); centrifugation of cells at 1500 g; 1 wash with phosphate buffered saline at room temperature (1:10); centrifugation of cells at 900 g
Heat stability of epitope Concentration Fluorochrome	Antibody Staining	CD45RA-Cy5, CD20-PE and CD57-Biotin + Streptavidin-PE; concentrations have to be tested out individually
Concentration Cell permeability Fluorescence interference	DNA Counterstaining	LDS 751 at 0.01 µg/ml for at least 20 min
Internal standard Compensation Gating	Acquisition/Analysis on Flow Cytometer	2×10^5 formaldehyde fixed (1%) cow thymocytes as internal reference ($\times$kb); individual compensation between fluorescein; sequential gating strategy

Fig. 2 Diagram of the seven steps in the automated multicolor flow-fluorescent *in situ* hybridization (FISH) protocol. The most critical parameters in each step are indicated on the left and our recommendations on how to perform each step are shown on the right.

Unfortunately, there is no good, easy, and fast method to completely separate nucleated cells from RBCs without any loss or destruction of nucleated blood cells. Two approaches to isolate or separate nucleated cells from RBCs have been explored: (1) direct RBC destruction by solutions that lead to swelling and lysis of the RBC via changes in the RBC osmolality (examples are Zap-oglobulin, water, 3% acetic acid in water or ammonium chloride) and (2) separation/isolation of RBCs based on cell-surface antigens that are expressed specifically on RBCs (immunomagnetic selection).

Many RBC lysing reagents are commercially available to prepare leukocyte cell suspensions for use on a cell counter or flow cytometer. Such lysing solutions have been optimized for those specific applications and appear very tempting for the isolation of nucleated cells for flow-FISH. Unfortunately, many lysing reagents contain variable amounts of various cross-linking or noncross-linking fixatives to prevent destruction of the nucleated cells by the lysing reagent (e.g., Zap-oglobulin). Such fixatives may affect the access of the fluorescent probe to target sequences and decrease the efficiency of DNA denaturation by subsequent treatment with heat and formamide (see Section III.A.2). We tested several lysing reagents without fixatives for their efficiency of RBC lysis and recovery of nucleated blood cells from moderate volumes (3–7 ml) of whole blood sufficient for flow-FISH (Baerlocher et al., 2002). Hypotonic lysis with 3% acetic acid in water or water itself was found to be very effective in lysing RBCs but also resulted in poor recovery of white blood cells (WBCs) (typically only 30–50% of the WBCs). The results were better for RBC lysis with ammonium chloride. Interestingly, the first approaches to lyse RBCs with ammonium chloride date back to the 1940s and 1950s. Then, it was found that ammonium chloride can rapidly enter the RBC and change its osmolality via ion exchanges (OH^- or HCO_3^- with Cl^-) over the cell membrane. These processes lead to cell swelling and hemolysis of the RBC within 6–10 min at 4 °C without significant loss of WBCs. Because of differences in membrane ion transport between RBCs and WBCs, the lysis with ammonium chloride occurs in the first minutes primarily in the RBCs. Therefore, leukocytes can be preserved and RBCs highly efficiently eliminated using ammonium chloride. A few essential points, however, have to be taken into account: The plasma should be removed from the cell suspension before lysis; the dilution factor of whole blood to ammonium chloride should be at least 20 times; the mechanism of RBC lysis does exhaust after two rounds of ammonium chloride exposure; and resuspending WBCs after ammonium chloride lysis in a solution containing dextrose greatly increases the recovery of WBCs.

Isolation of cells from tissues (spleen, thymus, tonsils) for flow-FISH typically requires either physical (e.g., dissecting the tissue with a knife or scissors or straining pieces of tissue through a sieve or mesh) or enzymatic digestion (e.g., collagenase) procedures. Such processing steps typically result in a certain amount of cell destruction with release of DNA even with the most gentle cell handling and viable cells can be trapped in the resulting clumps. Addition of DNAse at 1 μg/ml to the cell suspension effectively digests DNA in solution and prevents or reverses such cell clumping. Cell losses occurring during the cell isolation procedures also have to be considered. Isolation of mononuclear cells by Ficoll-Hypaque density centrifugation, for example, results in about a 50% loss of mononuclear cells (as is mentioned in the product sheet by the company). If specific subtypes of cells are of interest for telomere length measurements by flow-FISH, isolation of such cells by immunomagnetic isolation procedures or fluorescence-activated cell sorting (FACS) is an option. However, such cell preparation procedures are time-consuming and labor intensive and typically require a lot of cells to start with and inevitable cell losses have to be taken into account. Besides attempts to minimize such cell losses, it is crucial to keep cells in as good shape as is possible. For this reason, the time for cell isolation and preparation of cells should be as short as possible and cells should be kept at 4 °C in a balanced salt solution with an optimal pH and some protein (e.g., 0.1% BSA or 2% FCS). Cell losses can be further reduced by minimizing centrifugation steps and avoiding transfer of cells between tubes.

After the isolation or separation of cells and performing cell counts, it is often convenient to store cells before subsequent flow-FISH because the preparation already took some time and because samples often arrive in small batches. Fortunately, it is possible to store cell suspensions after processing in 10% (v/v) DMSO, 40% (v/v) fetal calf serum, and 50% (v/v) glucose 5%/Hepes 20 mM at −70 °C for a few weeks and at −135 °C for years. Thawing cells for 1–2 min at 37 °C in a water bath and resuspension in glucose 5%/Hepes 20 mM with 0.1% BSA without DMSO allows quick and efficient recovery of the previously frozen cells for the next steps in the flow-FISH protocol.

2. DNA Denaturation

Many protocols are available for the denaturation of isolated genomic DNA. Optimization of such molecular biology protocols was aimed at complete denaturation of DNA without loss or destruction of the resulting single-stranded DNA. For *in situ* hybridization using whole cells, it is important that complete breakdown of cells and membranes is avoided to allow for subsequent analysis of cells on a flow cytometer and to acquire additional information on the cell type using remaining cellular characteristics. The first experiments with *in situ* hybridization using whole cells were performed with fixatives to preserve nuclear and membrane structures during the harsh conditions required for denaturation of DNA in FISH. Later, it was found that heat and formamide already result in partial fixation of cells, most likely by dehydration and denaturation of proteins. Such fixation can be compared with what is observed when an egg is boiled: Protein denaturation by heat solidifies the liquid egg white. Fixation by heat and formamide does not prevent the PNA-telomere probe to pass through the remains of the cell and nuclear membrane and bind efficiently to complementary target sequences in the denatured DNA. In contrast, prior fixation almost invariably decreases the hybridization efficiency. In our hands, denaturation at 87 °C in a solution containing 75% formamide for 15 min followed by hybridization in the same solution at room temperature for at least 90 min results in the most optimal hybridization with the least loss or disintegration of DNA. Interestingly, the denaturation parameters are somewhat different for lymphocytes and granulocytes, most likely because of their different nuclear structure. Granulocytes need somewhat higher temperatures for optimal denaturation than lymphocytes (Baerlocher *et al.*, 2002). In addition, there are several ways to apply the heat for the denaturation step. A circulating water bath offers the most efficient and reliable method to denature many samples uniformly at the same time. In contrast, denaturation in a heat block does not allow the same degree of temperature control because wells at the periphery typically have lower temperatures than wells in the middle of the block.

Theoretical Hybridization Pattern

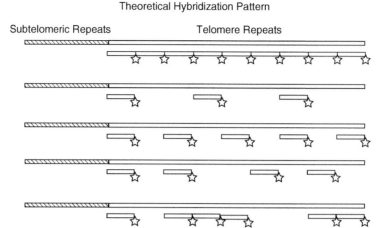

Fig. 3 Possible hybridization patterns for the telomere-peptide nucleic acid (PNA) 18-mer probe. With the assumption that those telomere-PNA probes hybridize with a high specificity and sensitivity and have most optimal accessibility to telomere ends, no bigger unhybridized gaps than 12 nucleotides should exist after hybridization. In theory, at least 70% or more of the telomere repeats will be complexed with PNA-telomere probe.

3. *In Situ* Hybridization

Quantitative *in situ* hybridization became feasible only with the introduction of directly labeled PNA probes. PNA probes and other uncharged oligomimetics have a higher specificity and sensitivity than DNA or RNA oligos and can hybridize under conditions (low salt concentration) that do not favor the renaturation of target sequences with complementary DNA strands. The definition of such conditions is a problem when *in situ* hybridization is performed with whole cells in suspension. In our experience, whole cells can maximally withstand 10–20 mM Tris in the hybridization solution containing 75% formamide, 1% BSA, 0.3 μg/ml PNA probe and water. 20 mM Tris in the hybridization solution is the optimal salt concentration to still preserve the cell structure on the one hand and to achieve favorable conditions for denaturation/hybridization of the telomere-PNA probe to telomeric DNA on the other hand. Generally, synthetic PNA probes do not easily pass cell membranes, which is one of the hurdles for antisense therapy with PNA. However, with heat and formamide treatment, PNA probes enter cells and nuclei without prior permeabilization. We calculated that in theory about 70% of the telomere repeats in cells should be hybridized with PNA when telomere-PNA 18-mers (C-probe) are used (Fig. 3). This may be an underestimate because the hydrophobic nature of PNA may force further alignment of PNA probes along target sequences. Optimal hybridization with the use of the telomere-PNA fluorescein-CCC TAA CCC TAA CCC TAA C-terminus probe is reached at a concentration of 0.3 μg/ml. This is an estimated more than 10^4-fold excess of probe relative to the calculated number of target sites available for hybridization in typical flow-FISH samples.

4. Washes of Excess Probe

The concentration of telomere-PNA probe used for hybridization is lower than that which would typically be used for the hybridization with DNA probes. Nevertheless, an excess amount of telomere-PNA probe is needed to guarantee complete hybridization. Therefore, steps to remove unbound and nonspecifically bound excess fluorescent PNA probes are crucial to only acquire and quantitate the specifically bound probe. When *in situ* hybridization is performed on metaphase chromosomes, one to two wash steps are generally sufficient to dilute the excess telomere-PNA probe to undetectable background levels. However, to eliminate the excess telomere-PNA probe in interphase cells in suspension, more extensive washes are required because the dilution factor with each wash step is much smaller. With our protocol, wash steps need to dilute the excess PNA probe more than 10^4-fold to effectively eliminate most of the background fluorescence, resulting from unbound or nonspecifically bound telomere-PNA probe and allow specific detection of hybridization probe. We perform four wash steps with a wash solution containing 75% formamide at room temperature and one wash step with a phosphate-buffered saline (PBS) solution to achieve the required dilution of unbound probe. Wash steps can also be performed with wash solutions containing PBS at 40–50 °C. Note that cells in suspension have to be spun down to recover them for the next steps in the protocol. Importantly, washes in formamide solution require a higher gravity force ($1500 \times g$) to adequately sediment cells in 5-min centrifugation steps than washes in aqueous solution ($900 \times g$).

5. Antibody Staining

The combination of *quantitative in situ* hybridization for the telomere length measurement with immunostaining has great potential for studies of the replicative histories of cells within one sample and for telomere length measurements in rare cells, which so far have been impossible because of low cell numbers. Unfortunately, the heat and formamide treatment used for the denaturation of the DNA also denatures most membrane and cytoplasmic proteins and epitopes. Most antibodies used for the immunophenotyping of viable or paraformaldehyde-fixed cells fail to react with cells after the flow-FISH procedure. However, a few antibodies (anti-human CD45RA, anti-human CD20, anti-human CD57) are able to specifically bind to cells despite the changes that occur during the heat and formamide treatment. These antibodies are typically used in immunohistochemistry, in which antibody staining is performed on paraffin-embedded tissue sections. We assume that such antibodies recognize peptide or carbohydrate epitopes that are resistant to heat and formamide denaturation. In our hands, any linker or fixative reagent, even in a low concentration, to attach antibodies to the cell membrane before *in situ* hybridization or to preserve epitopes during exposure to heat and formamide, results in a variable decrease in hybridization efficiency that is difficult, if not impossible, to control.

6. DNA Counterstaining

The idea behind counterstaining DNA for flow-FISH is to distinguish diploid (2N) cells from polyploid cells or cells in the S phase or the G_2M (4N) phase of the cell cycle. An optimal DNA dye for flow-FISH should easily enter the flow-FISH-processed cells, bind specifically and tightly to (denatured) DNA, allow discrimination between 2N cells and 4N cells, be optimally excited by a 488-nm laser (available on most flow cytometers), and display a narrow emission spectrum, which is far from the fluorescein isothiocyanate absorbance and emission spectrum (fluorescein-labeled in our protocol to the telomere-PNA). We tested many DNA dyes (propidium iodide, Syto orange dyes, Syto red dyes, 7-AAD, Hoechst H33342, 4,6-diamidino-2-phenylindole dihydrochloride (DAPI), DRAQ5, and LDS 751) and none fulfilled all of the above criteria. LDS 751 at a low concentration (0.01 μg/ml) and DAPI (0.01–1.00 μg/ml) were the only DNA dyes that did not notably interfere with the detection of fluorescein from the telomere-PNA probe and the phycoerythrin (PE)- and Cy5-labeled antibodies that are used in our protocol. However, LDS 751 and Cy5 have largely overlapping fluorescence emission spectra and these dyes require significant cross-compensation.

7. Acquisition on Flow Cytometer

To acquire the telomere fluorescence quantitatively in flow-FISH-processed cells, it is important that the flow cytometer is perfectly set up (warmup, alignment of laser, adjustment of time delay, etc.) according to the manufacturer's recommendations. Before test cells are acquired on the flow cytometer, we usually run a mixture of fluorescein-labeled calibration beads. This mixture of calibration beads contains four subpopulations of beads that are tagged with four defined amounts of molecular equivalents of soluble fluorochrome (MESF). The bead fluorescence is acquired to determine the linear range of the fluorescence detectable in FL1 and to convert telomere fluorescence values into MESF units for comparison of measurements between instruments and from experiment to experiment. Typical dot plots and histograms used for the acquisition of the telomere fluorescence in subtypes of human leukocytes are shown in Fig. 4. For each detector (FL1, FL2, FL3, FL4), optimal settings (values for the voltage of preamplifiers and photo multiplier tubes) have to be determined. Most optimally, the cell populations (granulocytes, lymphocytes, bovine thymocytes) should fall into the center part of the forward and side scatter dot plot and the three cell populations should be easily separable on the side scatter versus DNA dye (FL3) dot plot. In addition, subsets of cells should be easily detectable in forward scatter versus the specific antibody (e.g., antihuman CD45RA-Cy5 in FL4 for naive T cells or antihuman

Gating Strategy for Multicolor Flow - FISH

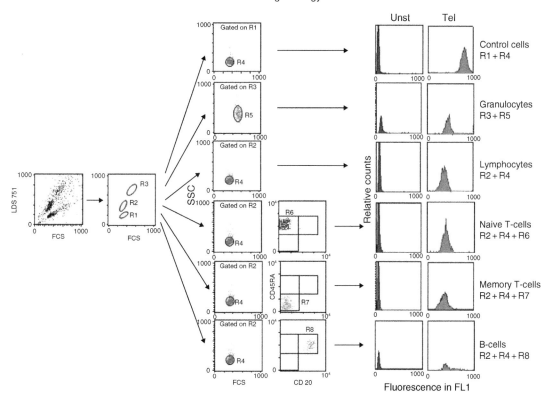

Fig. 4 Flow cytometric dot plots and histograms used for the gating of the internal standard (bovine thymocytes) and the subpopulations of leukocytes and for the analysis of the telomere fluorescence. First, a dot plot with forward scatter channel (FSC) versus LDS 751 is created to gate control cells (R1), lymphocytes (R2), and granulocytes (R3). In a second step, those same populations are gated on FSC versus side scatter channel (SSC). Note that bovine thymocytes and lymphocytes fall into the same gate (R4), whereas granulocytes are somewhat bigger and more granular (R5). Based on CD20 versus CD45RA, three subpopulations of lymphocytes can be gated. CD45RA-positive and CD20-negative naive T cells (R6), CD45RA-negative and CD20-negative memory T cells (R7), and CD45RA-positive and CD20-positive B cells (R8). For each of the subpopulations of leukocytes (bovine thymocytes in gates R1 + R4, granulocytes in gates R3 + R5, lymphocytes in gate R2 + R4, naive T cells in gates R2 + R4 + R6, memory T cells in gates R2 + R4 + R7, and B cells in gates R2 + R4 + R8), the telomere fluorescence shown as a histogram will then be acquired on a unstained sample (Unst = no telomere-PNA probe in the hybridization solution) and on a stained sample (Tel = telomere-PNA probe in the hybridization solution).

CD20-PE in FL2 for B cells) and the peaks for the telomere fluorescence should be in the lowest hundred units (linear scale) for unstained cells and in the highest 200 units (linear scale) for the telomere-fluorescent bovine thymocytes. To optimize the detection of antibody-labeled cells and to guarantee accurate telomere

Table III

Example of Calculation of Telomere Length in kb from Flow–FISH Telomere Fluorescence Values

	Unst	Unst	Mean Unst	Tel	Tel	Mean Tel	Specific Tel	Corr factor (×1.22)	Chromosome no.:120 or:92	Telomere length (kb)
Bovine thymocytes	43	45	44	810	803	806.5	762.5	930.3	7.8	15.0
Leukocytes	67	73	70	392	393	392.5	322.5		3.5	6.8
Granulocytes	112	125	118.5	448	438	443	324.5		3.5	6.8
Lymphocytes	60	65	62.5	372	368	370	307.5		3.3	6.5
Naive T lymphocytes	58	63	60.5	381	382	381.5	321		3.5	6.8
Memory T lymphocytes	62	70	66	351	344	347.5	281.5		3.1	5.9
B lymphocytes	61	65	63	383	375	379	316		3.4	6.6

Note: Channel numbers of the Fl 1 channel (linear scale) of a FACSCalibur flow cytometer following calibration with molecular equivalents of soluble fluorochrome beads are shown.

Unst, sample without telomere-PNA probe; Tel, sample with telomere-PNA probe; Specific Tel, Tel - Unst; Corr factor, to compensate for the lower fluorescence due to fixation; Chromosome #, 120 chromosome ends for bovine thymocytes and 92 chromosome ends for human cells; Telomere length in kb, 15 for this specific aliquot of bovine thymocytes.

fluorescence measurements, compensations have to be performed between the diverse detectors. It is important that the fluorescence from PE detected in FL1 is subtracted by compensation to guarantee accurate telomere fluorescence measurements on cells labeled with PE. Each sample pair is, therefore, individually compensated.

8. Calculation and Presentation of Results

For each sample, subtypes of cells are gated according to the example shown in Fig. 4, and the telomere fluorescence histogram of each subtype of cell is generated. In general, we typically use the median value for the telomere fluorescence. However, when histograms are very asymmetrical, the mean values or percentiles can be more useful. For each sample, duplicate measurements are performed and the specific telomere fluorescence is calculated by subtraction of the (auto-) fluorescence of unstained controls from the telomere fluorescence measured in cells hybridized with the fluorescein-labeled telomere-PNA probe (Table III). The specific telomere fluorescence (in MESF) of the cells of interest is then set in relation to the specific telomere fluorescence (in MESF) of the bovine thymocytes analyzed in the same tube according to the following calculation:

$$\frac{(\text{MESF value test cells} - \text{MESF value unstained test cells}) : 92^{a} \times 15.0 \text{ kb}^{d,e}}{(\text{MESF value control cells} - \text{MESF value unstained control cells}) \times 1.22^{b} : 120^{c}},$$

where [a]number of chromosome ends in a normal diploid human cell; [b]correction factor to compensate for the measured difference in median telomere fluorescence between fixed and unfixed control cells (fluorescence fixed cells was 82% of that in unfixed control cells in two experiments); [c]number of chromosome ends in each control cell (bovine thymocyte); [d]measured telomere restriction fragment size in DNA from control bovine thymocytes; [e]assuming the length of subtelomeric DNA in terminal restriction fragments is similar to human and bovine DNA and that internal T_2AG_3 repeats in human and bovine chromosomes do not contribute significantly to the telomere fluorescence.

Because we measured the telomere restriction fragment size in DNA from the bovine thymocytes, we are able to convert the fluorescence values of the test cells into kilobases (Table III). With duplicate measurements, the average (mean), standard deviation (SD), and coefficient of variation (CV) can be calculated. Duplicates with CVs of more than 10% are excluded from analysis.

There are many ways to present telomere length data. Depending on the data set and the experimental questions, dot plots, histograms, box plots, or simple data tables can be used.

B. Preparation of Internal Standard

Even with the standardization of all steps in the protocol, a certain variation between measurements remains. To reduce the variation in duplicate or triplicate samples from the same individual, and to increase the reproducibility of measurements at different times, we started to add control cells to each tube as an internal standard (Hultdin *et al.*, 1998). We selected bovine thymocytes as internal controls based on their small size, abundant availability, ease of preparation and storage, and low autofluorescence after heat and formamide treatment. In addition, they are mostly diploid cells and have telomeres that are twice as long as humans and up to 2–10 times shorter telomeres than those in mice. Only one characteristic was not optimal: after heat and formamide treatment, bovine thymocytes have very similar forward and side scatter properties as human, baboon, or mice lymphocytes when analyzed on a flow cytometer. Therefore, separation of those cell types was not feasible using light scatter parameters alone. However, we found that fixation of bovine thymocytes with 1% paraformaldehyde after the isolation procedure (details of the protocol are described in Baerlocher and Lansdorp, 2003; Baerlocher *et al.*, 2006) results in weaker staining with the DNA dye LDS 751. Most likely, the modest cross-linking by the paraformaldehyde fixation changes the accessibility of LDS 751 to the chromosomal DNA. This small difference in the DNA counterstaining makes it possible to separate between bovine thymocytes and lymphocytes when they are mixed in the same tube for *in situ* hybridization (Fig. 4, left). As expected, telomere hybridization is compromised by the fixation step, and the resultant telomere fluorescence is lower in fixed cells compared to unfixed cells. The difference in telomere fluorescence was measured (the telomere fluorescence of 1% paraformaldehyde fixed cells is 82% of that in nonfixed cells) and a correction factor ($\times 1.22$) was incorporated into the formula for the calculation of the telomere length in cells of interest (for further details, see Baerlocher and Lansdorp, 2003; Baerlocher *et al.*, 2006).

C. Automation

Nowadays, automation systems are widely used in laboratories to standardize and speed up processes. Many repetitive steps such as aspiration and dispensing of cells and fluids are necessary during the flow-FISH procedure, and we were interested in finding a device that can very accurately aspirate and dispense diverse volumes of solutions to and from many tubes at the same time. The Hydra microdispensers by Robbins Scientific are bench-top instruments that feature 96 or 384 syringes arranged in a standard array format (Baerlocher and Lansdorp, 2003; Baerlocher *et al.*, 2006). The Hydra can be programmed to very accurately pipette or mix solutions into, from, and in 96- or 384-well plates. Most often, this type of device is used in genome analysis and drug discovery programs, and to our knowledge, it has not been used for processing of cell suspensions. We tested the standard 96-well Hydra instrument with 1-ml syringes and somewhat wider needles (inner diameter 0.0155 inches) for use in flow-FISH. Because of the even and gentle aspiration and dispensing of the machine, cells in suspension can easily be transferred or mixed with various solutions. As a result, the cell recovery is better and cells are more evenly distributed than can be achieved manually. We adapted most of the steps in our flow-FISH protocol specifically for integration with the Hydra. For this purpose, the adequate volume, position (height) of the needles, and the number of cycles were determined for each aspiration, dispensing, or mixing step following the instructions in the manual of the manufacturer for the plates and tubes used in flow-FISH. Using the Hydra, we greatly decreased the variation in telomere fluorescence measurements, and we saved up to about 50% of the solutions and reagents needed for most of the steps in the protocol. Furthermore, we processed about five times as many samples as was possible by hand.

IV. Results

A. Variation/Sensitivity/Accuracy

To analyze which improvements could be achieved by optimizing and automating most steps of the basic flow-FISH protocol, we compared the following parameters for samples processed with the basic flow-FISH protocol and with the automated flow-FISH protocol:

1. Variation in telomere fluorescence values within one experiment and from experiment to experiment
2. Cell recovery

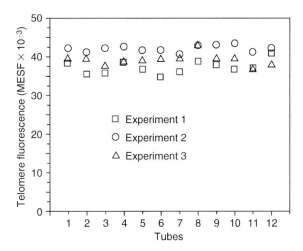

Fig. 5 Variation in telomere fluorescence of control cells within an experiment and from experiment to experiment. Shown are telomere fluorescence values (in molecular equivalents of soluble fluoro-chrome (MESF) $\times$ 10^{-3}) for 1% paraformaldehyde-fixed bovine thymocytes. For each, experiment (experiments 1-3 indicated as different symbols), 12 tubes with the same amount of bovine thymocytes were hybridized and analyzed. Note that the average telomere fluorescence value ($\sim$40 $\times$ 10^{3} MESF per cell) corresponds to about 40% of the calculated maximum fluorescence using the following assumptions: one hybridized PNA-fluorescein molecule yields one MESF unit; bovine cells have 60 chromosomes (120 telomeres per cell); with on average 15 kb of telomere repeats per telomere, each cell has 1.8 Mb of telomere "target" DNA allowing a maximum of 10^{5} PNA molecules (18-mers) to hybridize. The measured value ($\sim$0.4 $\times$ 10^{5} MESF) is even closer to the expected efficiency of about 70% or 0.7 $\times$ 10^{5} MESF (Fig. 3) if the estimated 3 kb of subtelomeric DNA included in the TRF measurements used for calibration is subtracted from the estimated 15 kb of telomere repeats.

3. Time to process samples

4. Amounts of solutions and reagents used

First, we took aliquots of the paraformaldehyde-fixed bovine thymocytes (internal standard cells) and analyzed the variation in telomere fluorescence acquired from tube to tube and from experiment to experiment with the optimized/automated protocol. In Fig. 5, the variation in telomere fluorescence in MESF units from tube to tube (12 tubes for each experiment) is shown. Within the same experiment, the SD ranged from 0.6 MESF (experiment 2) to 1.5 MESF (experiment 3) (CV ranged from 1.5% to 3.7%). From experiment to experiment ($n = 26$), the MESF SD was 2.8 (7.1% coefficient of variation). Note, that the variation in telomere fluorescence of the control cells (2 $\times$ 10^{5}) was measured in the presence of test cells (typically around 0.5 $\times$ 10^{6} human, baboon, or murine hematopoietic cells) to most accu-rately mimic a relevant situation. It is very likely that the variation in telomere fluorescence of control cells is influenced by the presence of other cells in the same test tube (dilution of telomere PNA-probe, binding of DNA to cells, etc.).

Reproducibility of Telomere Length Measurements

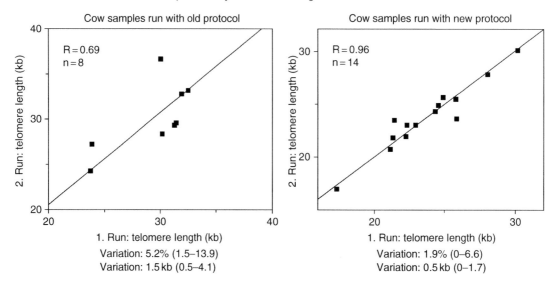

Fig. 6 Comparison of reproducibility in telomere length measurements in various bovine nucleated blood samples analyzed by the old flow-fluorescent *in situ* hybridization (FISH) protocol and the new version. For each sample, two telomere length measurements were performed at different times using each method. The data points on the graphs, where the results from the first run represent measured telomere length values, are plotted against the results from the second run. Correlation analysis was performed for all the data points obtained by the old protocol ($n = 8$, $R = 0.69$) and by the new protocol ($n = 14$, $R = 0.89$).

Two conclusions can be drawn from this experiment: Even with the optimized and most careful execution of each step in the protocol, there still is a certain variation in telomere fluorescence for the same biological sample between tubes within an experiment and from experiment to experiment. The variation within an experiment ranges from 0.3 to 0.7 kb and the variation from experiment to experiment can go up to 1.4 kb. When we looked at unfixed human, baboon, bovine, or murine hematopoietic cells, the variation in telomere fluorescence was in the range of 5–10% within an experiment and 20–30% from experiment to experiment. By using the bovine thymocytes as internal standards, we reduced this variation in telomere fluorescence to 2–5% within one experiment and to 5–10% from experiment to experiment. As a result, the minimum detectable difference in telomere length for human, baboon, bovine, or murine hematopoietic cells was calculated to be in the order of 0.5 kb.

Figure 6 illustrates the improvement in the reproducibility of telomere length measurements using the basic and current flow-FISH methods. In this experiment, the telomere length in hematopoietic cells from various bovine blood samples was calculated from two flow-FISH experiments (two different runs with the same samples at different times). With the basic method, the CV was 5.2% ($n = 8$) and with the new protocol 1.9% ($n = 14$). In addition, the reproducibility of telomere length measurements came into the range of ± 0.5 kb between experiments. Many similar experiments with human, baboon, and murine hematopoietic cells have confirmed these findings.

Because of improvements in the solutions used during the flow-FISH protocol (e.g., inclusion of glucose), the use of an internal standard (to guide the position and identify test cells on the flow cytometry plot more readily), and the more gentle mixing of the cell suspensions by the Hydra, we improved the cell recovery of cells for analysis on the flow cytometer from 1–10% to 25–50% of the initial cell count. As a result, telomere length measurements can now be performed with much lower cell numbers (in the range of 10^4) (Baerlocher and Lansdorp, 2003; Baerlocher *et al.*, 2006; Van Ziffle *et al.*, 2003).

With the Hydra robotic device, 96 tubes or 24 samples (duplicates of unstained and stained samples) can easily be processed at the same time. This is not feasible with the manual basic flow-FISH protocol. For example, the time to perform a wash step with the processes of aspiration, mixing, and addition of wash solution takes maximally 5 min for 96 tubes with the Hydra, whereas aspirating, mixing, and addition of wash solution to 96 tubes by hand (where you have to take each tube out of the rack, open it, and hold it slanted to aspirate fluid, etc.) takes at least 30–60 min. With the five wash steps in the flow-FISH protocol, manual processing of 96 tubes clearly becomes impractical. With automated flow-FISH, 24 duplicate samples can easily be processed within 1 day and, if necessary, still be acquired on the flow cytometer in the later afternoon/evening or the next morning.

Because of the smaller volume and diameter of the tubes in a standard 96-well plate, we tested out and adjusted all the volumes of solutions used during the various steps in the flow-FISH protocol. Most importantly, we reduced the volume for the hybridization solution (from 300 to 170 μl) and for the solutions with the antibodies (100-30 μl). The reduced volumes in the automated flow-FISH protocol have allowed significant cost reductions.

V. Critical Aspects of Methodology

A. PNA Probes

In collaboration with scientists at Applied Biosystems, an extensive survey was made of possible PNA probes that can hybridize with mammalian telomeric $(TTAGGG)_n$ repeats. It was found that targeting the G-rich strand was advantageous over targeting the C-rich strand because PNA probes containing TTAGGG sequences are poorly soluble and give rise to a high nonspecific staining/hybridization. It is possible that PNA probes containing the TTAGGG sequence, similar to DNA oligonucleotides containing the same sequence, form guanine-quadruplex structures, a stable four-stranded structure in which the repeating unit is a plane of four guanine bases (Sen and Gilbert, 1988). Using Q-FISH, it was found that

12-mers (CCCTAA)$_2$ PNA can hybridize but, as expected, at a lower efficiency than 18-mers containing the same sequence. The latter were selected for further studies based on their good solubility and predicted high specificity (the stability of PNA/DNA hybrids containing nucleotide mismatches is very poor (Ratilainen et al., 2000)). Various fluorochromes were explored for labeling of (CCCTAA)$_3$ PNA in Q-FISH and flow-FISH techniques. For Q-FISH, the Cy3 dye appears ideal (good excitation at the 546-nm peak emission of the mercury light source in most fluorescence microscopes, very photostable, and an excellent quantum yield), whereas for flow cytometry the 488-nm argon laser available on most flow cytometers favored the use of fluorescein. This dye has some disadvantages because it is rather hydrophobic (as is PNA) and bleaches relatively fast. For flow cytometry, the exposure to the laser excitation is very short and relatively constant, so bleaching is not so much of a concern, but the hydrophobic nature of the fluorescein-labeled (CCCTAA)$_3$ PNA does impose specific requirements for prevention of nonspecific binding, which we found were best addressed by initial wash steps in 75% formamide. Although labeling of (CCCTAA)$_3$ PNA with two fluorescein molecules increased the fluorescence yield (not by a factor of two, presumably as a result of some quenching), the further increase in hydrophobicity and decreased solubility in aqueous solutions did not favor the use of such probes. The preferred way to label fluorescein to PNA is using succinimidyl esters of fluorescein rather than fluorescein because the former yields a more stable bond. Cy5-labeled telomere PNA probes can also be used, but because of the necessity of a helium-neon laser to excite the Cy5 dye, and because of a somewhat lower sensitivity of Cy5 detection on our flow cytometer, we prefer telomere-PNA probes labeled with fluorescein for the most sensitive telomere length measurements. It has not been possible to hybridize to both C and G strands quantitatively with the use of a C-strand and G-strand telomere PNA probes because the combination of both probes results in the formation of PNA hybrids. Phosphoramidite probes for centromeric repeats were described (O'Sullivan et al., 2002). It is possible that synthetic oligonucleotide probes other than PNA can also be used in Q-FISH for telomere repeats if hybridization conditions for such probes can be developed that prevent annealing of complementary DNA sequences.

B. Fixation

In our hands, any fixative or cross-linking solution (paraformaldehyde, ethanol, methanol/acetic acid, BS3 or bis-[sulfosuccinimyl]-substrate) to preserve nuclear and membrane structures led to a decrease in the hybridization efficiency. Furthermore, such reagents did induce agglutination and clumping of cells, which resulted in a higher loss of cells and a greater variability in telomere fluorescence. Difficulties to control for the increased variability in hybridization efficiency led us to abandon such prefixation steps; however, it remains tempting to continue to explore this approach because the possibility to measure the telomere length in rare cell subpopulations, defined by (combinations of) labile cell surface markers, clearly would be advantageous (Batliwalla *et al.*, 2001; Plunkett *et al.*, 2001; Schmid *et al.*, 2002).

C. Choice of Fluorochromes

For the telomere-PNA probe, it is necessary to use heat-stable fluorochromes (excluding the use of highly fluorescent PE and other fluorescent proteins). In addition, it is optimal to choose a fluorochrome for the telomere-PNA probe that can be used on most flow cytometers (availability of lasers to excite the selected fluorochrome). As mentioned earlier, we decided to use fluorescein-labeled PNA as the telomere probe in our flow-FISH protocol. Initially, we used propidium iodide in a subsaturating concentration (0.06 μg/ml) as a DNA counterstain. However, because of its broad emission spectrum, which can lead to interference with the detection of the weak fluorescence from the fluorescein-labeled telomere-PNA probe (spectral overlap, energy transfer), we switched to LDS 751 at 0.01 μg/ml (Baerlocher *et al.*, 2002). LDS 751 can be excited by the 488-nm argon-ion laser available on most flow cytometers, and its emission spectrum is very distinct from that of fluorescein. Next to telomere-PNA-fluorescein and LDS 751, PE and Cy5, or allophycocyanin (APC) can be used as fluorochromes for antibodies. Because some of the bright fluorescence from PE-labeled antibodies will spill into the detector (FL1) used for the detection of telomere fluorescence, each sample needs to be individually compensated for cross fluorescence of FL2 into FL1. We decided to use the PE-labeled antibody for the detection of epitopes with a low abundance per cell (in our protocol, CD20-PE and CD57-PE). Whereas the signals from Cy5- or APC-labeled antibody (in our protocol, CD45RA-Cy5) are generally less bright than those from PE-labeled antibodies, good discrimination above background is typically possible into the FL4 channel of the FACSCalibur, although significant cross compensation with LDS 751 detected in FL3 is required.

VI. Pitfalls and Misinterpretation of Data

A. DNA Content

The telomere fluorescence in a cell has to be correlated to its DNA content to measure the average length of telomeres. To distinguish cells with 2N or 4N DNA, or to identify cells in S phase, cells are usually stained with a DNA dye at saturating concentrations. Unfortunately, most DNA dyes used at saturating concentration (except for DAPI, which is excited by an ultraviolet (UV) laser and, therefore, not useful on most flow cytometers) interfere notably with the detection of the relatively weak telomere fluorescence from PNA fluorescein (especially in human cells with short telomeres). Therefore, we had to drop the DNA dye concentration to subsaturating levels with the disadvantage of losing much of the resolution in the DNA profile (already compromised by the flow-FISH procedure: denaturation and loss of DNA). As a result, it is at this point not possible to directly correlate cell cycle status with telomere length and telomere length measurements by flow-FISH in rapidly dividing cells such as tumor cells or cultured cells remain problematic. No doubt, future flow-FISH protocols will aim to improve the resolution of both DNA and telomere fluorescence while preserving a linear relationship between telomere length and telomere fluorescence.

B. Background Fluorescence

Another major hurdle for accurate measurements of telomere length by flow-FISH is related to the autofluorescence of cells. Various cellular structures influence the autofluorescence properties of a cell. Such factors include various cell organelles and fluorescent molecules in the cytoplasm (such as lipofuscin, hemoglobin, and cytochromes). Some of these factors only marginally increase the autofluorescence in viable cells. However, with heat and formamide treatment, the autofluorescence may increase dramatically and this may interfere with the detection of specific telomere fluorescence. Cultured fibroblasts with a large volume of cytoplasm are an example of highly autofluorescent cells where it is difficult, if not impossible, to measure telomere length by flow-FISH accurately. Another problem is that the PNA telomere probe may bind nonspecifically to cells in ways that may vary between cell types (e.g., as a function of their size). Unfortunately, there are no good ways to control for this source of variation (different PNA "control" probes yield highly variable probe-specific "background" fluorescence). Extreme caution should, therefore, be used to apply the techniques described in this chapter to cell types other than the ones that have been investigated and validated in extensive experiments. In developing protocols for such novel cell types, inspection of cells under a fluorescence microscope to confirm that fluorescence is derived only from distinct (telomere) spots in the nucleus is of critical importance.

VII. Comparison with Other Methods

There is no "gold standard" or optimal method for comparisons of telomere length measurements by flow-FISH. Several times, we have compared the telomere length measured by flow-FISH to telomere length values obtained by Southern blot. Telomere length measurements with either the basic or the optimized/automated flow-FISH protocol typically showed a high correlation (with R values ≥ 0.80) with measurements obtained by Southern blot. However, improvements made in the sensitivity of telomere length measurements by the optimized/automated flow-FISH protocol do not really show up in this type of comparison because the variation in measurements by Southern blot is typically much higher (with an estimated SD of 1–3 kb for measurements of telomeres in the range of 5–15 kb). In addition, only short telomeres (within a range of 5–30 kb) can be optimally assessed by Southern blot. Therefore, no comparisons between methods can be made using cells with very long telomeres, such as murine cells. Telomere length measurements by flow-FISH were also compared to telomere length measurements by Q-FISH. Such results also correlate very well ($R > 0.8$). However, with Q-FISH, cells have to be able to divide to obtain metaphase spreads. The resulting selection of cells combined with errors in measurements of telomere length using limited numbers of metaphase cells can easily explain the differences in telomere length measurements between the two methods.

VIII. Applications

A. Bovine Samples

The regulation of telomere length in the germline and during early embryogenesis is of great interest. At one point, it was believed that organisms created by nuclear transfer, such as the cloned sheep Dolly, might undergo premature senescence and aging, resulting from insufficient repair of preexisting short telomeres by the cloning procedure (Shiels *et al.*, 1999). In collaborative studies, we looked at the telomere length in hematopoietic cells of cloned calves derived by nuclear transfer of either senescent or "young donor" nuclei from somatic cells into enucleated oocytes (Lanza *et al.*, 2000). Almost all cloned animals showed longer telomere length values in hematopoietic cells than those measured in age-matched control samples or in the cells used for nuclear transfer. These flow-FISH telomere length data clearly demonstrate that telomeres can be extended upon nuclear transfer and cloning. This is just one example in which we were able to assess the telomere length in cells within a very short period and with a higher sensitivity than is possible by Southern blot. Many other examples, most notably in studies of mutant mice, can be found in the literature (Ding *et al.*, 2004).

B. Baboons

Another area of interest is the age-related regulation of telomere length in various subpopulations of cells *in vivo*. Telomere length measurements in lymphocytes and granulocytes of humans at different ages suggest that the decline in telomere length with age is not linear. In humans, a rapid decline of telomere length is observed early in childhood, with a more gradual but steady telomere loss thereafter (Rufer *et al.*, 1999). Because telomere length shows a large variation among individuals at any given age, a highly sensitive and reproducible method to measure telomere length is necessary to exclude method-related variation, and ideally, longitudinal studies of telomere length with age should be performed. Baboons are in many aspects (hematopoietic stem cell turnover, development of immune system, etc.) very similar to humans, and blood samples from baboons are somewhat more easily available for longitudinal studies. For these reasons, we examined the telomere length in lymphocytes and granulocytes of 22 baboons (Papio hamadryas cynocephalus) ranging in age from 2 months to 26 years (Baerlocher et al., 2003a). In addition, we developed multicolor flow-FISH for baboon hematopoietic cells to look at telomere length in diverse subsets of leukocytes (granulocytes, naive T cells, memory T cells, and B cells). Telomere length in granulocytes, B cells, and subpopulations of T cells decreased with age. However, the telomere length kinetics were lineage and cell subset specific. T cells showed the most pronounced overall decline in telomere length, and memory T cells with very short telomeres accumulated in old animals. In contrast, the average telomere length values in B cells remained relatively constant from middle age onward. Individual B cells showed highly variable telomere length, and B cells with very long telomeres were observed after the ages of 1–2 years. To look more specifically at the decline in telomere length after birth inferred from population studies, we also studied the telomere length dynamics in subsets of hematopoietic cells longitudinally every 2 weeks in the first 1 and 1/2, years of four baboons. The telomere length in granulocytes, T lymphocytes, and B lymphocytes was maintained (variation <0.5 kb between serial samples) or even slightly increased in the first 2–3 months for three animals and up to 9 months for one animal. After this period, the telomere length values started to decrease slowly but significantly. In addition, we found that certain B lymphocytes with a much higher fluorescence appeared in the circulation between 9 and 16 weeks of life, and heterogeneity within B cells was clearly detectable after the age of 1 year (Baerlocher et al., 2003b). Again, these telomere length data clearly demonstrate the value of highly sensitive and reproducible measurements of telomere length in combination with immunophenotyping.

C. Humans

A clinically highly interesting and relevant question is whether the replicative demand imposed on transplanted donor stem and progenitor cells may result in substantial telomere shortening, thereby potentially compromising the long-term function of the marrow graft. We looked at a situation in which a 10-year-old male recipient was transplanted with donor cells from his 64-year-old grandmother (Awaya et al., 2002). Histograms for the telomere fluorescence in subtypes of peripheral blood leukocytes from the donor and the recipient are shown in Fig. 7. The telomere length values in subsets of cells obtained from the donor cells are comparable to those of age-matched control individuals. However, values in post-transplantation donor cells harvested from the patient were 0.8–2 kb shorter than those in the donor, and these values were much lower than those in normal individuals age-matched with the recipient. Furthermore, the patient presented similar telomere length values for CD45RA-positive naive T cells and CD45RA-negative memory T cells and showed practically no B cells with longer telomeres as can typically be found in normal individuals. These findings suggest

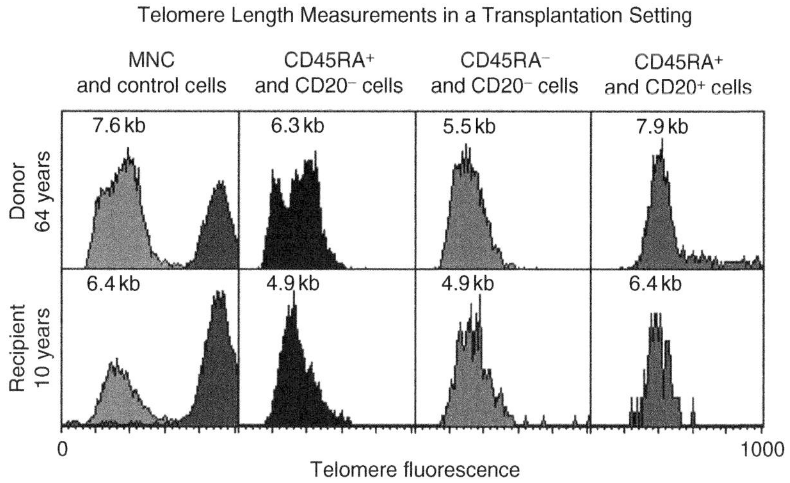

Telomere Length Measurements in a Transplantation Setting

Fig. 7 Telomere length measurements in a transplantation setting. Histograms of telomere fluorescence in control cells and subsets of leukocytes (MNCs, mononuclear cells; CD45RA-positive, CD20-negative naive T cells, CD45RA-negative, CD20-negative memory T cells, CD45RA-positive, CD20-positive B cells) from the donor (64 years old) and the recipient (10 years old) after transplantation. The telomere length in kilobases is calculated from the median telomere fluorescence.

that the telomere erosion with cell division in this patient is not compensated by telomerase activity. The very low telomere length values in various lymphocyte subsets could limit cell replication and compromise immune function as a result. In general, studies of the telomere length in subsets of cells in various (autoimmune) immune disorders are in their infancy and the results of such studies are eagerly awaited.

IX. Future Directions

A. DNA Dyes

Improvements in DNA profiles without affecting the acquisition of quantitative telomere fluorescence is one of the next goals for telomere length measurements. Such an improvement will enable studies of the timing of telomere replication during the cell cycle in various cell types and telomere length measurements in dividing cell populations. One way to optimize the DNA profile without affecting the telomere fluorescence might be to use DAPI. However, this DNA dye will require access to a flow cytometer with a UV laser. Furthermore, there is hope that new DNA dyes that fulfill all the requirements of a DNA counterstain in the flow-FISH protocol will become available.

B. Antibodies

To measure telomere length in diverse subsets of hematopoietic cells (CD4-, CD8-, CD19-positive cells, etc.), additional antibodies that are able to recognize and bind specifically to epitopes following the harsh flow-FISH denaturation step would be of great advantage. Another option is to use already available antibodies with better fixatives or linker molecules, which do not affect the hybridization efficiency. As was discussed, such fixation steps require extremely careful control to exclude variable effects on hybridization efficiency.

Acknowledgments

Work in the laboratory of PML is supported by grants from the Canadian Institutes of Health Research (MOP38075 and GMH-79042), the National Cancer Institute of Canada (with support from the Terry Fox Run). GMB was supported by the Swiss National Foundation and the Bernese Cancer League. Jennifer Mak, Teri Tien, and Irma Vulto are thanked for excellent technical assistance. We thank Jim Coull, Krishan Taneja, Jens Hyldig-Nielsen, and Tish Creasey at Applied Biosystems (Boston) for their generous help in providing various PNA probes. Michael Egholm (Molecular Staging Inc., New Haven, CT) is thanked for initial encouragement and Andre van Agthoven (Beckman Coulter Immunotech, Marseille, France) for generous gifts of CD20-PE.

References

Awaya, N., Baerlocher, G. M., Manley, T. J., Sanders, J. E., Mielcarek, M., Torok-Storb, B., and Lansdorp, P. M. (2002). Telomere shortening in hematopoietic stem cell transplantation: A potential mechanism for late graft failure? *Biol. Blood Marrow Transplant.* **8,** 597–600.

Baerlocher, G. M., and Lansdorp, P. M. (2003). Telomere length measurements in leukocyte subsets by automated multicolor flow-FISH. *Cytometry* **55A,** 1–6.

Baerlocher, G. M., Mak, J., Tien, T., and Lansdorp, P. M. (2002). Telomere length measurements by fluorescence *in situ* hybridization and flow cytometry: Tips and pitfalls. *Cytometry* **47,** 89–99.

Baerlocher, G. M., Mak, J., Roth, A., Rice, K. S., and Lansdorp, P. M. (2003a). Telomere shortening in leukocyte subpopulations from baboons. *J. Leukoc. Biol.* **73,** 289–296.

Baerlocher, G. M., Vulto, I. M., Rice, K. S., and Lansdorp, P. M. (2003b). Longitudinal measurements of telomere length in leukocyte subsets from baboons. *Blood* **102,** 567a (Abstract).

Baerlocher, G. M., Vulto, I., de Jong, G., and Lansdorp, P. M. (2006). Flow cytometry and FISH to measure the average length of telomeres (flow FISH). *Nat. Protoc.* **1,** 2365–2376.

Baird, D. M., Rowson, J., Wynford-Thomas, D., and Kipling, D. (2003). Extensive allelic variation and ultrashort telomeres in senescent human cells. *Nat. Genet.* **33,** 203–207.

Batliwalla, F. M., Damle, R. N., Metz, C., Chiorazzi, N., and Gregersen, P. K. (2001). Simultaneous flow cytometric analysis of cell surface markers and telomere length: Analysis of human tonsillar B cells. *J. Immunol. Methods* **247,** 103–109.

Blackburn, E. H. (2001). Switching and signaling at the telomere. *Cell* **106,** 661–673.

Bodnar, A. G., Ouellette, M., Frolkis, M., Holt, S. E., Chiu, C.-P., Morin, G. B., Harley, C. B., Shay, J. W., Lichtsteiner, S., and Wright, W. E. (1998). Extension of life-span by introduction of telomerase into normal human cells. *Science* **279,** 349–353.

Cheung, S. W., Tishler, P. V., Atkins, L., Sengupta, S. K., Modest, E. J., and Forget, B. G. (1977). Gene mapping by fluorescent *in situ* hybridization. *Cell Biol. Int. Rep.* **1,** 255–262.

Ding, H., Schertzer, M., Wu, X., Gertsenstein, M., Selig, S., Kammori, M., Pourvali, R., Poon, S., Vulto, I., Chavez, E., Tam, P. P., Nagy, A., *et al.* (2004). Regulation of murine telomere length by Rtel: An essential gene encoding a helicase-like protein. *Cell* **117**(7), 873–886.

Griffith, J. D., Comeau, L., Rosenfield, S., Stansel, R. M., Bianchi, A., Moss, H., and de Lange, T. (1999). Mammalian telomeres end in a large duplex loop. *Cell* **97,** 503–514.

Hultdin, M., Gronlund, E., Norrback, K., Eriksson-Lindstrom, E., Just, T., and Roos, G. (1998). Telomere analysis by fluorescence *in situ* hybridization and flow cytometry. *Nucl. Acids Res.* **26,** 3651–3656.

Kim, N. W., Piatyszek, M. A., Prowse, K. R., Harley, C. B., West, M. D., Ho, P. L. C., Coviello, G. M., Wright, W. E., Weinrich, S. L., and Shay, J. W. (1994). Specific association of human telomerase activity with immortal cells and cancer. *Science* **266,** 2011–2015.

Lansdorp, P. M. (1996). Close encounters of the PNA kind. *Nat. Biotechnol.* **14,** 1653.

Lanza, R. P., Cibelli, J. B., Blackwell, C., Cristofalo, V. J., Francis, M. K., Baerlocher, G. M., Mak, J., Schertzer, M., Chavez, E. A., Sawyer, N., Lansdorp, P. M., and West, M. D. (2000). Extension of cell life-span and telomere length in animals cloned from senescent somatic cells. *Science* **288,** 665–669.

Martens, U. M., Zijlmans, J. M., Poon, S. S. S., Dragowska, W., Yui, J., Chavez, E. A., Ward, R. K., and Lansdorp, P. M. (1998). Short telomeres on human chromosome 17p. *Nat. Genet.* **18,** 76–80.

Nielsen, P. E., and Egholm, M. (1999). An introduction to peptide nucleic acid. *Curr. Issues Mol. Biol.* **1,** 89–104.

Olovnikov, A. M. (1971). Principles of marginotomy in template synthesis of polynucleotides. *Dokl. Akad. Nauk. SSSR* **201,** 1496–1499.

O'Sullivan, J. N., Bronner, M. P., Brentnall, T. A., Finley, J. C., Shen, W. T., Emerson, S., Emond, M. J., Gollahon, K. A., Moskovitz, A. H., Crispin, D. A., Potter, J. D., and Rabinovitch, P. S. (2002). Chromosomal instability in ulcerative colitis is related to telomere shortening. *Nat. Genet.* **32,** 280–284.

Plunkett, F. J., Soares, M. V., Annels, N., Hislop, A., Ivory, K., Lowdell, M., Salmon, M., Rickinson, A., and Akbar, A. N. (2001). The flow cytometric analysis of telomere length in antigen-specific CD8 + T cells during acute Epstein-Barr virus infection. *Blood* **97,** 700–707.

Poon, S. S. S., Martens, U. M., Ward, R. K., and Lansdorp, P. M. (1999). Telomere length measurements using digital fluorescence microscopy. *Cytometry* **36,** 267–278.

Ratilainen, T., Holmen, A., Tuite, E., Nielsen, P. E., and Norden, B. (2000). Thermodynamics of sequence-specific binding of PNA to DNA. *Biochemistry* **39,** 7781–7791.

Ray, A., and Norden, B. (2000). Peptide nucleic acid (PNA): Its medical and biotechnical applications and promise for the future. *FASEB J.* **14,** 1041–1060.

Rufer, N., Dragowska, W., Thornbury, G., Roosnek, E., and Lansdorp, P. M. (1998). Telomere length dynamics in human lymphocyte subpopulations measured by flow cytometry. *Nat. Biotechnol.* **16,** 743–747.

Rufer, N., Brummendorf, T. H., Kolvraa, S., Bischoff, C., Christensen, K., Wadsworth, L., Schultzer, M., and Lansdorp, P. M. (1999). Telomere fluorescence measurements in granulocytes and T lymphocyte subsets point to a high turnover of hematopoietic stem cells and memory T cells in early childhood. *J. Exp. Med.* **190,** 157–167.

Schmid, I., Dagarag, M. D., Hausner, M. A., Matud, J. L., Just, T., Effros, R. B., and Jamieson, B. D. (2002). Simultaneous flow cytometric analysis of two cell surface markers, telomere length, and DNA content. *Cytometry* **49,** 96–105.

Sen, D., and Gilbert, W. (1988). Formation of parallel four-stranded complexes by guanine-rich motifs in DNA and its implications for meiosis. *Nature* **334,** 364–366.

Shay, J. W., and Bacchetti, S. (1997). A survey of telomerase activity in human cancer. *Eur. J. Cancer* **33,** 787–791.

Shiels, P. G., Kind, A. J., Campbell, K. H., Waddington, D., Wilmut, I., Colman, A., and Schnieke, A. E. (1999). Analysis of telomere lengths in cloned sheep. *Nature* **399,** 316–317.

Trask, B. (1991). Fluorescence *in situ* hybridization. Application in cytogenetic and gene mapping. *Trends Genet.* **7,** 149–154.

Trask, B., Van Den Engh, G., Pinkel, D., Mullikin, J., Waldman, F., van Dekken, H., and Gray, J. (1988). Fluorescence *in situ* hybridization to interphase cell nuclei in suspension allows flow cytometric analysis of chromosome content and microscopic analysis of nuclear organization. *Hum. Genet.* **78,** 251–259.

Van Ziffle, J. A. G., Baerlocher, G. M., and Lansdorp, P. M. (2003). Telomere length in subpopulations of human hematopoietic cells. *Stem Cells* **21,** 654–660.

Vaziri, H., and Benchimol, S. (1998). Reconstitution of telomerase activity in normal human cells leads to elongation of telomeres and extended replicative life span. *Curr. Biol.* **8,** 279–282.

von Zglinicki, T. (2002). Oxidative stress shortens telomeres. *Trends Biochem. Sci.* **27,** 339–344.

Vulliamy, T., Marrone, A., Goldman, F., Dearlove, A., Bessler, M., Mason, P. J., and Dokal, I. (2001). The RNA component of telomerase is mutated in autosomal dominant dyskeratosis congenita. *Nature* **413,** 432–435.

Watson, J. D. (1972). Origin of concatameric T7 DNA. *Nat. New Biol.* **239,** 197–201.

CHAPTER 28

Sperm Chromatin Structure Assay: DNA Denaturability

Donald Evenson and Lorna Jost

Olson Biochemistry Laboratories
Department of Chemistry
South Dakota State University
Brookings, South Dakota 57007

I. Update

No significant changes in the acridine orange-based methodology of SCSA presented in this chapter have been made since its publication. However, important findings in the area of cytometric assessment of sperm cells in terms of their reproductive capability and relation to the SCSA methodology have been reported. Initially, the extensive DNA fragmentation revealed by the presence of DNA double-strand breaks (DSBs) was detected in abnormal sperm cells by the TUNEL assay; the presence of DSBs strongly correlated with DNA sensitivity to denaturation detected by SCSA (Gorczyca et al., 1993). Since then the TUNEL method developed by Gorczyca et al. (1992) has become widely used either as an alternative or as a complementary approach to identify and quantify reproductively dead sperm cells by flow cytometry (Evenson et al., 2007; Martins et al., 2007). The combination of DNA fragmentation and its increased sensitivity to denaturation is the hallmark of apoptosis, also termed "programmed cell death" (Darzynkiewicz, 1997; Hotz et al., 1994). Other hallmarks of apoptosis assessed by cytometry, namely activation of caspases (Marchetti et al., 2004) and externalization of phosphatidyl-serine (Paasch et al., 2004) were further found to characterize the reproductively dead sperm cells. It became thus apparent that the increased susceptibility of DNA in situ to denaturation detected by the SCSA is one of many reporters of apoptosis that identify both somatic as well as in germ cells. Markers of apoptotic cells have become then widely used to assess integrity of sperm cells by cytometry.

Another significant development in methodology of detection of abnormal and reproductively dead sperm cells is the assay utilizing toluidine blue (Erenpreisa et al., 2003, 2004). The toluidine blue-based assay most likely detects similar apoptosis-associated changes in nuclear chromatin as the SCSA utilizing acridine orange. Because the cells stained with the absorption-dye toluidine blue can visualized be light microscopy, the method complements the SCSA by offering a possibility of simple identification of abnormal sperm cells using light microscopy or quantitative image analysis cytometry.

II. Introduction

A normal mammalian testis in a sexually mature animal or human produces up to hundreds of millions of sperm daily. Spermatogenesis entails the proliferation of stem germ cells and their subsequent steps of cell differentiation including the unique event of meiosis. Following spermatogenesis, spermiogenesis includes the extensive and unique differentiation of meiotic daughter cells in the testis and further maturation during epididymal passage culminating in mature sperm ready for ejaculation.

Many environmental agents can have an effect on one or more of the numerous biochemical/differentiation events resulting in altered kinetics of cell division and differentiation thereby reducing total sperm output. More seriously, the ejaculated sperm may have altered characteristics that reduce fertility potential and/or have damage to the genetic material leading to early embryo death or birth defects.

An increasing number of flow cytometry (FCM) techniques have been developed in recent years to measure abnormalities of germ cells that may be related to decreased reproductive function (see Spano and Evenson, 1993 for review). Early studies concentrated on measurements of ratios of testicular cells obtained by surgical biopsy or fine needle aspirates (Clausen and Abyholm, 1980). Initial studies used univariate analysis of DNA stainability of the various cell types to detect cell-type-specific death and/or altered kinetics of maturation. Univariate analysis has also been used to detect induction of aneuploid or diploid spermatids (Otto and Hettwer, 1990).

In recent years, flow cytometry techniques have been developed to study characteristics of ejaculated sperm related to fertility potential and effects of potential reproductive toxicants. An advantage of these newer studies on ejaculated sperm is that the sample is obtained by noninvasive means and the cell is the finished product prepared for fertilization of the female gamete. Included among these new techniques are measurements of mitochondrial function (Evenson *et al.*, 1982), membrane integrity (Garner *et al.*, 1983), and ratio of X to Y bearing sperm (Johnson *et al.*, 1989).

Studies over the past decade have proven the usefulness of measurements of the integrity of sperm chromatin structure (Evenson, 1989; Evenson *et al.*, 1980; Spano and Evenson, 1993). During spermiogenesis, DNA in round spermatids is complexed with histones which are then exchanged for transition proteins and finally for protamines. The tertiary and quaternary structure of protamine-complexed sperm DNA is likely important for protection of the genetic information and possibly for early genetic events postfertilization (Ward and Coffey, 1991). Studies have shown that chromatin structure is related to fertility potential of sperm (Ballachey *et al.*, 1987, 1988; Evenson, 1986, 1989; Evenson *et al.*, 1980) and also serves as a biomarker for exposure to reproductive toxicants (Evenson and Jost, 1993; Evenson *et al.*, 1985, 1986a, 1989a, 1993b,c).

The FCM measurement of chromatin structure is based on the principle that abnormal sperm chromatin has a greater susceptibility to physical induction of partial DNA denaturation *in situ* (Darzynkiewicz *et al.*, 1975). The extent of DNA denaturation following heat (Evenson *et al.*, 1980, 1985) or acid (Evenson, 1989; Evenson *et al.*, 1985) treatment is determined by measuring the metachromatic shift from green fluorescence [Acridine orange (AO) intercalated into double-stranded nucleic acid] to red fluorescence (AO associated with single-stranded DNA; Darzynkiewicz *et al.*, 1976). Apparently acid conditions that cause partial denaturation of protamine-complexed DNA in sperm with abnormal chromatin structure do not cause denaturation of histone-complexed somatic cell DNA (Evenson *et al.*, 1986a). The FCM measurement of sperm chromatin structure as described here has been termed the sperm chromatin structure assay (SCSA) to distinguish it from other AO staining protocols. This protocol was formerly divided into SCSA$_{acid}$ and SCSA$_{heat}$ to distinguish the physical means of inducing

DNA denaturation. The two methods give essentially the same results but the SCSA$_{acid}$ method is much easier to use and is the method of choice (Evenson *et al.*, 1985). The method is the same as that developed by Darzynkiewicz and colleagues (1975) for the "two-step AO method." However, the SCSA has been developed with numerous additional details for application and data manipulation and is the subject of this chapter.

III. Applications of the SCSA

The primary applications of the SCSA are in the fields of environmental toxicology, animal husbandry, and human infertility. In the field of toxicology, the described techniques provide for rapid, objective assessment of the effects of germ cell toxicants that interfere with chromatin differentiation. Evenson and colleagues have shown that exposure of mice to reproductive toxicants caused changes in the relative ratio of testicular cell types present (Evenson and Jost, 1993; Evenson *et al.*, 1985, 1986a, 1989a, 1993b), presence of abnormal cell types in epididymi (Evenson *et al.*, 1989b), and increased sensitivity of sperm DNA to acid (Evenson and Jost, 1993; Evenson *et al.*, 1985, 1986a, 1989a, 1993b) or heat-induced denaturation (Evenson, 1986; Evenson *et al.*, 1980, 1985). In studies using 10 different chemicals, the dose-response curves of FCM-derived α_t values (see Section VIII) were very similar in shape to the percentage abnormal sperm head morphology curves (Evenson *et al.*, 1985, 1986a, 1989a). Of added interest, sperm cells arising from stem cells exposed to stem cell-specific mutagenic chemicals maintained chromatin structural abnormalities detectable by these FCM methods for at least 45 weeks (Evenson *et al.*, 1989a).

Studies have shown that exposure to chemicals that alkylate free-SH groups on protamine molecules (e.g., methylmethane sulfonate, MMS) in late testicular or early epididymal sperm caused nearly 100% of sperm to have altered chromatin structure by 3 days postexposure (Evenson *et al.*, 1993b). This was 8 days prior to altered sperm head morphology. Of greater interest, the maximum chromatin alterations, measured by the SCSA, corresponded to the temporal pattern of sperm produced that resulted in maximal dominant lethal mutations. Since dominant lethal mutations are caused by chromosomes breaks, the SCSA was likely detecting damage to chromosomes about 8 days prior to maximum dominant lethal mutations. In another experiment (Estop *et al.*, 1993), mouse sperm aged *in vitro* showed alterations of chromatin structure by the SCSA at only 2 h incubation; however, when these aged sperm fertilized mouse oocytes *in vitro*, pronuclear chromosomes from *in vitro* aged sperm did not show chromosome breaks until sperm had been aged for 6 h. The SCSA appears to detect early stages of chromatin alterations that likely lead to whole chromosome breaks. Thus, the SCSA is viewed as a potentially important method to assay for early events of toxicant-induced chromosome damage.

In addition to screening for toxicant-induced damage to sperm, an equally important impact of the SCSA technique may be for assessment of animal and human subfertility (Ballachey *et al.*, 1987, 1988; Evenson, 1986). In both types of studies it is important to know what the normal variation of chromatin structure is over time and how it relates to other typically measured semen parameters.

A longitudinal study of human sperm chromatin structure was made on monthly semen samples from 45 men over 8 consecutive months (Evenson *et al.*, 1991). The study showed that although the SCSA data often differed between donors, there was a remarkable homogeneity within a donor from month to month. In fact, the repeatability of the positions of just a few dots to major clusters of dots on the FCM cytograms from month to month were suggestive that particular stem cells had a consistent abnormality leading to very particular levels of abnormality in their progeny sperm. The SCSA data had higher repeatabilities and lower CVs than the classical measures of semen volume, sperm count, motility, morphology, and viability. Since sperm chromatin structure is a more repeatable feature of sperm and yet is responsive to environmental toxicants, it is looked upon favorably as a valuable biomarker for human toxicology studies.

Data for the relationship between human sperm chromatin structure and fertility are limited. Two major studies are in progress. Both confirm that semen samples from human infertility clinics demonstrate a high degree of chromatin structural heterogeneity within the sperm population (Evenson *et al.*, 1993a). Unresolved yet is the degree of heterogeneity that is compatible with fertility and normal embryo development.

IV. Materials

A. Acridine Orange Staining Solutions

1. *AO stock solution:* Chromatographically purified AO (Polysciences, Warrington, PA) is dissolved in double-distilled water to a final concentration of 1 mg/ml. Nonpurified AO is not acceptable for this technique. AO is a toxic chemical and considerable care should be used when weighing out the powder. Typically a 15-ml glass scintillation vial is tared on a balance pan and about 5–10 mg of AO powder carefully transferred with a cupped spatula into the vial. An equivalent number of milliliters of water are then added to the vial which is then capped, covered with aluminum foil to minimize light exposure, and placed in refrigerator. This solution can be kept at 4 °C for several months.

2. *Acid/detergent treatment solution for step 1 of AO staining procedure:* 0.15 M NaCl, 0.1% Triton X-100 (Sigma Chemical Co., St. Louis, MO), and 0.08 N HCl in double-distilled water. For 500 ml, admix 20 ml 2.0 N HCl, 4.39 g NaCl, 0.5 ml Triton X-100, and 480 ml double-distilled water. This solution will keep up to several months at 4 °C. The working solution is kept in a 16-oz glass amber bottle containing an Oxford adjustable, 0.20- to 0.80-ml automatic dispenser (Lancer Division of Sherwood Medical, St. Louis, MO).

3. *Stock 0.1 M citric acid buffer:* To 21.01 g citric acid monohydrate (FW = 210.14) add double-distilled water to 1 l. Store at 4 °C.

4. *Stock 0.2 M Na2PO4 buffer:* To 28.4 g sodium phosphate dibasic (FW = 141.96) add double-distilled water to 1 l. Store at 4 °C.

5. *Stock AO staining solution:* Mix 370 ml 0.1 M citric acid buffer, 630 ml 0.2 M Na_2PO_4 buffer (due to crystallization at 4 °C, this stock must be warmed prior to use), 372 mg EDTA, and 8.77 g NaCl. Mix well. Adjust to pH 6.0.

6. *AO staining solution:* 0.60 ml AO stock solution is added to each 100 ml of stock AO staining solution. This AO staining solution, kept in a 16-oz glass amber bottle containing an Oxford adjustable, 0.80- to 3.0-ml automatic dispenser, is made fresh biweekly.

B. Buffers and Other Materials

1. *TNE buffer:* 0.01 M Tris, 0.15 M NaCl, and 1 mM EDTA, pH 7.4. Be cautious that this buffer remains free of bacterial contamination as this may cause problems with sample interpretation; it is preferable that this buffer is filtered through a 0.22-μm filter (Corning sterile filter unit No. 25932, Cat. No. 210–963, Curtin Matheson Scientific, Inc., Houston, TX) to remove any debris and bacteria and then stored in a sterile tissue culture flask at 4 °C.

2. *Hanks' Balanced Salt Solution (HBSS):* (Gibco Laboratories, Grand Island, NY).

3. Polypropylene microtubes, 0.5 or 1.0 ml (Sarstedt, Inc., Princeton, NJ).

4. Polystyrene 12 × 75 mm, 4.5 ml, conical tubes (Cat. No. 57.477, Sarstedt, Inc.).

5. Falcon No. 3033, 16 × 125 mm, tissue culture tubes (Becton Dickinson Labware, Lincoln Park, NY).

6. Nylon filters, 153-μm mesh, 1 in. diameter (Tetko, Elmsford, NY).

7. Dental plugger (Henry Schein, Inc., Port Washington, NY).

8. Corning No. 25702 cryogenic vials, with internally threaded caps, 2 ml (Cat. No. 237–347, Curtin Matheson Scientific, Inc.).

9. Tuberculin syringes, 1 cc, Becton Dickinson No. 9602 (Cat. No. 262–247, Curtin Matheson Scientific, Inc.).

V. Cell Preparation

A. Fresh, Frozen, and Fixed Sperm Samples

Human semen samples are obtained by masturbation into plastic clinical specimen jars preferably after 2–3 days abstinence. For safety against potential infectious agents, for example, hepatitis and HIV, samples are handled with disposable gloves in a biological safety cabinet. After 30 min has been allowed for semen liquefaction, aliquots of semen can be frozen directly in an ultracold freezer (-70 to $-110\,°C$) or placed into LN_2. A nonselfdefrosting refrigerator freezer can be used for short-term freezing and storage and then the sample can be transferred preferably that same day to an ultracold freezer. In field situations, an ice chest containing dry ice may be used. Care should be taken to freeze the samples in an upright position using a test tube rack in the freezer. This is especially important if snap cap tubes are used because if the tube is inverted when frozen, the freezing pressure may partially open the snap cap. Furthermore, samples frozen at the bottom of a tube are thawed in the water bath with greater ease and safety. The easiest sample to work with is one that has been diluted into TNE buffer prior to freezing which reduces the viscosity for handling at the time of FCM measurements. Tall, narrow plastic snap cap tubes are preferred to minimize the surface to volume ratio of the sample in order to reduce cell "freezer burn" that may occur during long-term storage (months to years); freezing in an upright position helps to minimize this potential problem. A 1-ml microtube is favorable to store a mixture of 100 μl semen + 400 μl TNE buffer. Since 200 μl of semen/TNE mixture is used for preparation of the sample for FCM measurement, this sample can be used directly if the sperm count is low or diluted to about 1–2×10^6 if the sperm count is too high. If the first sample is too concentrated, the remaining 300 μl can be used for dilution or a repeat measurement. Since the AO/DNA phosphate ratio is very important for appropriate staining (Darzynkiewicz, 1979; practically applied as the number of cells/0.2 ml aliquot sample used for AO staining), it is very useful to know the sperm count in advance so that an aliquot can be diluted very closely to the desired range without a trial and error process. For severe oligospermic samples, undiluted semen can be used directly; the acid/detergent solution used in the fist step dramatically reduces any semen viscosity (Evenson and Melamed, 1983).

Commercial animal semen is often diluted with a variety of extenders which serve as a cryoprotectant and as a diluent for increasing the number of samples from a single ejaculate. These extenders generally do not interfere with the SCSA measurements of sperm chromatin. However, bull semen extended in *nonclarified* egg yolk citrate extender often causes some background noise which may or may not show up as debris noise extending into the sample signal region.

Frozen samples are preferred and the easiest to work with. However, in some cases it may be desirable to fix samples. Data from fixed samples are essentially similar to that obtained on fresh material (Evenson *et al.*, 1986a). In this case, sperm are centrifuged out of semen and resuspended in HBSS at a concentration of about $10^7\ ml^{-1}$, and 1 ml of this suspension is forcefully pipetted into 10 ml cold ($-20\,°C$) 80% ethanol. Prior to analysis, the sperm are again pelleted, washed once in TNE buffer, and then processed for SCSA, always keeping the sample at $4\,°C$.

B. Epididymal or Vas Deferens Sperm

For animal studies, a specific segment of the epididymis can be surgically removed from a killed animal and minced in TNE buffer with a curved pair of scissors in a petri dish set on crushed ice (Evenson *et al.*, 1986a). The mixture is transferred to a 12 × 75-mm tube and the larger fragments are allowed to settle. "Home-made" filtering systems are made by mounting a 1-in. 153-μm nylon mesh between the end of a 1-cc tuberculin syringe and its cap with its end cut off. The supernatant suspension is then passed through this filter. Do not apply pressure to the plunger. The vas deferens may also be excised and placed in a 60-mm petri dish containing TNE buffer and the sperm removed by pressing a blunt probe, a dental plugger works very well, along the length of the organ. In order to easily visualize the white "cord" of sperm being expressed, the petri dish is placed on a black Teflon-coated plate of steel set on the surface of crushed ice.

C. Sonication of Sperm Cells

Our laboratory has measured thousands of animal and human sperm samples by the SCSA. In previous years, there has been concern about whether any residual cytoplasmic droplets potentially containing RNA would add an artifact to the measurement of single-stranded DNA. Thus, in earlier studies, all samples were sonicated or some unsonicated samples were compared with their sonicated counterpart. The results have been so nearly identical (Evenson *et al.*, 1991) that sonication is no longer a routine procedure and this saves a great deal of time and effort. Investigators must remain aware of this potential problem and, if there is a reason to be concerned, then some or all samples should be sonicated. Sonication is preferred over RNase incubation which has the potential of causing incubation-related changes in chromatin structure possibly due to protease digestion of chromatin proteins. As an exception, rat sperm samples often used in toxicology experiments, *must* be sonicated because the long, fibrous tails tend to clog the flow cell. The broken tails in the sonicate can cause a problem in SCSA analysis. This can be corrected by electronic gating (see Section VIII) or purification through a sucrose gradient (Evenson *et al.*, 1985). Mouse sperm, with much shorter tails, do not require sonication relative to the problem with rat sperm.

Sperm suspended in TNE buffer in a Falcon 3033 test tube immersed in an ice water slurry are sonicated for 30 s at a setting of 50 on low power (Bronwill Biosonik IV Sonicator, VWR Scientific, Inc., Minneapolis, MN), cooled for 30 s, and sonicated again for 30 s. The half-inch probe is placed just above the bottom of the tube. Optimal time and power required for sperm head-tail/cytoplasm separation varies between species and needs to be tested for each sonicator to achieve an approximate ≥95% head/tail separation. Allow the sonicate to set 2 min on ice before preparing for the SCSA.

Human semen samples, potentially containing infectious agents such as hepatitis or HIV, must be sonicated only in a closed tube. Our laboratory utilizes a Branson

Sonifier II, Model 450, coupled to a Branson Cup Horn (VWR Scientific, San Francisco, CA). Sample temperature is kept cold by 4 °C water flowing through the cup horn. This is derived by using a Masterflex peristaltic pump (Cole Parmer Instrument Co., Chicago, IL) that drives water (21 ml/min) through approximately 3 ft of copper tube coil (1/4 in. i.d.) set in a 4-l flask containing an ice water slurry. Place 0.5 ml of TNE buffer containing $\leq 1 \times 10^6$ sperm cells into a 2-ml Corning cryogenic screw cap vial. The top end of this capped vial is inserted into the bottom side of a No. 11 rubber stopper which has a 12-mm hole drilled through it that will hold the vial securely. The rubber stopper, holding the vial, is then placed on top of the cup horn with the vial protruding down into the cup horn so that the bottom of the vial is just off the bottom of the cup. Samples are sonicated for 30 s (rats) to 40 s (humans) using 70% of 1-s cycles at a setting of 3.0 output power. The cup horn sonicator is preferred over the probe method for ease of use, uniformity between samples, and safety precautions.

VI. Cell Staining and Measurement

In contrast to many procedures where a large batch of samples can be prepared at a lab distant from the flow cytometer, the SCSA procedure requires that samples are thawed and prepared in the immediate vicinity of the flow cytometer. Elapsed times for various components of the procedure are very important. A frozen sample is held by the top of the test tube which is mostly immersed in a 37 °C water bath, just until the last remnant of ice disappears. After thawing, the sample is either diluted (all buffers and staining solutions are kept on crushed liquid ice) and prepared further or sonicated first.

Prior to measuring experimental samples, the instrument must obviously be checked for alignment. Very importantly, especially if the sample tubing has been bleached clean, an AO buffer mixture (0.4-ml acid detergent solution and 1.2-ml AO staining solution) needs to be passed through the instrument sample lines for at least 30–45 min prior to setting the photomultiplier tubes (PMTs) with the reference sample and measuring samples. This ensures that AO is equilibrated with the sample tubing. In contrast to the rumor spread by representatives of some commercial companies and uninformed flow operators, using AO in a flow cytometer does not ruin it for other purposes! The sample lines do not need to be replaced after measuring AO-stained samples! While the system does need to be equilibrated with AO (i.e., AO adheres to the sample tubing), this is FULLY rectified by rinsing the system for about 10 min with a 50% filtered bleach solution after finishing the AO measurements. Our laboratory has measured many other dyes and sample types after measuring AO-stained sperm without any associated problem.

A 0.20-ml aliquot of sample is placed into a 12×75-mm conical tube. A 0.40-ml aliquot of the first step low pH buffer is added with an automatic dispenser. This dispenser needs to be accurate and to have a maximum capacity near the amount being dispensed. A stopwatch is started immediately after the first buffer is dispensed. Exactly 30 s later the AO staining solution is added. The sample tube is then placed into the flow cytometer sample chamber in a 30-ml beaker containing an ice water slurry. Although it is preferable to have the sample setting in an ice bath, the configuration of some FCM sample chambers may not permit this; since the sample is measured shortly after being removed from ice, this should not cause a significant difference in the data. The sample flow is started immediately after it is placed in the sample holder. Using the same stopwatch that was started with the addition of the first step buffer, the acquisition of the data is started at 3.0 min from the time of step one buffer addition. This allows for equilibration and stabilization of the sample, as well as uniformity between samples, all important points for AO staining. Also, the flow rate is checked during this time and if it is too fast, that is, >300 cells/s, a new sample is made at an appropriate dilution. The original sample cannot be diluted with AO buffer. It is implied that the sample and sheath flow valve settings of the instrument are never changed during these measurements and that the liquid flow rate is constant. Thus, a change in count rate is a function of sperm cell concentration only. A critical part of SCSA measurements is the use of a reference sample to monitor instrument stability throughout any experiment (see Section VIII for more detail).

VII. Instruments

Blue laser light (488 nm) excitation of AO-stained cells at a power of >35 mW is optimal. Fluorescence of individual cells is measured at wavelengths of red (630- to 650-nm long-pass filter) and green (515- to 530-nm band-pass filter). Both green and red fluorescences are processed in *peak* mode of signal rather than area mode. Since mature, AO-stained mammalian sperm have very little red fluorescence, due to lack of RNA and single-stranded DNA, the red PMT gain may need to be set high enough that electronic noise may result with some instruments. Ortho Diagnostics engineers made a slight modification of the red fluorescence preamplifier circuit board to reduce background noise on the Cytofluorograf II.

Aliquots of the same semen samples from humans, stallions, and mice were measured on Cytofluorograf, Becton Dickinson FACScan, and Coulter Elite flow cytometers. The scattergram patterns for all instruments were similar indicating that any of these instruments can be used with the SCSA protocol. However, neither of the latter two instruments is capable of generating the very important ratio of red/red + green fluorescence (α_t). In this case, list-mode files were transferred to an IBM compatible computer and processed using LISTVIEW software (Phoenix Flow, Inc., San Diego, CA) which gave desired SCSA data. It is not appropriate to use a software or hardware configuration that defines α_t as red (>630 nm)/total (515 long-pass) fluorescence; this adds the unknown component of the 530- to 630-nm wavelengths into the denominator.

As an excellent alternative to the more expensive laser-based instruments, the Hg-arc lamp-driven Ortho ICP22A flow cytometer interfaced to the 2150 Data Handling system has been successfully used (Evenson *et al.*, 1993b; Jost and Evenson, 1993). Alternatively, the ICP22A can be interfaced to an IBM compatible 386 or 486 personal computer with the ACQCYTE system (Phoenix Flow, Inc.) installed. Figure 1 compares AO-stained mouse sperm measured by an orthogonal and an epi-illumination flow cytometer. The epi-illumination ICP22A does not produce an optical artifact (see Evenson *et al.*, 1993b for discussion) as seen with orthogonal instruments when measuring sperm. Although the cytograms do not have the same cell fluorescence distribution, the pertinent data are the same and equally useful. Unfortunately, the ICP22A is no longer commercially available but other Hg-arc systems should work. The Argus flow cytometer (Skatron, Bio-Rad, Italy) is known to work satisfactorily (P. De Angelis, O. P. F. Clausen, and D. P., Evenson, unpublished); however, it is hoped that a small, relatively inexpensive Hg-arc system will be produced in the near future that would accommodate the SCSA technique.

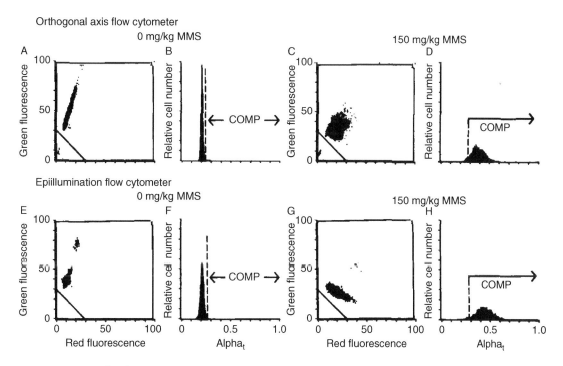

Fig. 1 Green versus red fluorescence cytograms of SCSA data obtained by measuring caudal mouse sperm on a Cytofluorograf (A, C) or an ICP22A (E, G) flow cytometer. Sperm were obtained from mice treated with 0 or 150 mg/kg MMS and killed at 13 days. (B, D, F, H) α_t frequency histograms; cells to the right of the vertical lines are cells with denatured DNA (COMPαt; Evenson *et al.*, 1993b). Reprinted with permission from Wiley-Liss.

VIII. Results and Discussion

A. SCSA Parameters

Two parameters are of particular importance in the evaluation of the SCSA-derived data. The first is the mean green fluorescence (X Green) which is related to the condensation of the sperm chromatin and extent of restricted access of DNA dyes. The sperm nuclear condensation process normally produces a fivefold reduction of DNA stainability relative to round spermatids (Evenson *et al.*, 1986b). Lack of appropriate sperm maturation results in an increased DNA stainability. Studies have shown that patients attending an infertility clinic often have an increased DNA stainability (Engh *et al.*, 1992; Evenson and Melamed, 1983). This can be visualized by univariate analysis (Engh *et al.*, 1992) as well as by the SCSA bivariate analysis. The SCSA has an advantage however in distinguishing debris from sperm signals.

The second parameters of particular importance are those of α_t. Interestingly, AO-stained samples that show high X Green usually do not exhibit extensive DNA denaturation and thus both abnormalities can be studied via the SCSA. SCSA analysis also includes mean total fluorescence (X Total = X Green + X Red), and the α_t parameters of mean α_t ($X\alpha_t$), standard deviation α_t ($SD\alpha_t$), and cells outside the main population α_t ($COMP\alpha_t$). Practically, $COMP\alpha_t$ indicates the percentage of abnormal cells and $SD\alpha_t$ describes the extent of the abnormalities. Note that the defined value range of α_t is from 0.0 to 1.0 (i.e., all green and no red fluorescence to all red and no green fluorescence; Darzynkiewicz *et al.*, 1975), but for practical considerations it is expressed in 1000 channels of fluorescence.

Harsher physical conditions have been applied to sperm with the SCSA (i.e., increasing AO concentration, more acidic conditions, longer incubation times) resulting in a higher percentage of $COMP\alpha_t$ and higher $SD\alpha_t$; however, parallel dose-response curves are observed with the standardized procedure when compared with increased physical conditions for DNA denaturation. Thus, new information is not gained with increased physical conditions for denaturation. The current view is that the abnormalities observed by the $COMP\alpha_t$ and $SD\alpha_t$ are a "tip of the iceberg" effect and perhaps reflect abnormalities that may be present at lesser levels in the main population. Each component has valuable information which sometimes needs to be interpreted with special regards to the results; likewise, a specific component may be of more value when the response differs. For example, if the total sperm population shifts from green to red fluorescence, $X\alpha_t$ and $COMP\alpha_t$ may be the most valuable descriptors. However, in many toxicology experiments, only a small to moderate percentage of cells have shifted to various degrees; in this case, $SD\alpha_t$ has had the highest correlation with dosage of toxicant or some other parameters.

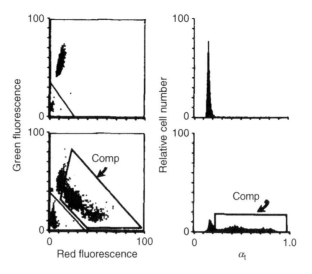

Fig. 2 FCM cytogram of two-parameter green (double-stranded DNA) versus red (single-stranded DNA) fluorescence distribution of sperm from a fertile human and a patient from an infertility clinic. The box marked COMP shows the cells outside the main population with abnormal chromatin structure. The α_t distribution shows the extent of the abnormality. (Evenson, 1990). Reprinted with permission of from Academic Press).

B. Debris Exclusion

Figure 2 compares SCSA-derived data from a fertile male with those from a patient attending an infertility clinic. The distribution for DNA stainability (green fluorescence) of the normal population is broad due to a known optical artifact (see discussion in Evenson *et al.*, 1993b) but which has no effect on the α_t distribution of interest in this technique. Note that the distribution is much more homogeneous for the fertile individual than for the clinic patient. A very important, but sometimes difficult point, is deciding where to make the computer gates to cut out cellular debris and to distinguish between a normal population of sperm and cells with denatured DNA. The problem is accentuated with low sperm count samples and especially those samples derived from patients on chemotherapy that may result in debris from killed cells. For most animal and relatively normal human samples, a box is drawn first very near to the perimeters of the cytogram boundaries to exclude those events that are beyond the full channel limits, for example, channels 2–254 inside a box of 0–256. Next, debris that falls to the lower left hand corner is dealt with in one of two ways. With samples having very little debris a near 45° angle line is drawn just below the bottom of the sperm signal as seen in Fig. 2. The 45° line is based on the premise that cells gain red fluorescence at the expense of green fluorescence at a near 45° angle. Human SCSA data are often more complicated and in some cases an elliptical circle has been used to exclude what was considered to be debris from the data (Evenson *et al.*, 1991). There is no perfect answer on this matter. After inspection of the data set a decision is made on what best fits the experiment. Whatever method is chosen, the important point is to be consistent throughout the entire data set and preferably between current and future experiments.

Because of their long tails and the need for sonication, rat sperm pose a particularly difficult problem when excluding debris. The rat tails break up into many pieces of debris that are seen as fluorescent signals by the flow cytometer. This problem can be overcome by using the flow cytometer's electronic signal processing capabilities. In Fig. 3, cytograms A and D show the regular peak mode, green and red fluorescence signals before debris is gated out. The sperm signal is not resolved from the debris. By gating out the debris in the green area versus peak cytogram (B), the resulting cytograms (C and E) are relatively "debris free" and analyzable. This processing technique may be useful in other species and situations where a large debris to sperm ratio exists.

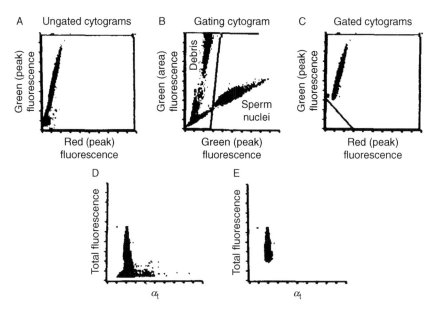

Fig. 3 FCM cytograms of ungated versus gated AO-stained rat sperm nuclei. Cytograms (A) and (D) (debris and sperm signals are not resolved) are prior to gating out the debris using the green fluorescence area versus peak-mode cytogram (B), while cytograms (C) and (E) have the debris gated out and the sperm signal is resolved from the debris.

C. Placement of the COMPα_t Line

The computer gate defining COMPα_t is easy to set if a significant percentage of the cells have fluorescence values equal to a normal population. This was true for all early experiments. However, when the entire population shifts (Evenson *et al.*, 1993b), then COMPα_t cannot be defined as the cells to the right of the main population in that cytogram. In that case, the computer gate must be set to the right side of the *control sample* population and all cells to the right of that line are defined as COMPα_t cells which in some cases may equal 100%. This procedure points out very strongly again the absolute need to have a reference sample to precisely set the instrument variables.

D. Reference Samples

SDα_t has been the most useful and most closely correlated parameter related to known fertility ratings (Ballachey *et al.*, 1987, 1988). Because α_t measures, particularly SDα_t, are very sensitive to small changes in chromatin structure, studies using this parameter require very precise, repeat instrument settings for all comparative measurements whether done on the same or different days. These settings are obtained by using aliquots of a single semen sample that demonstrates heterogeneity of α_t. A semen sample is identified as a reference sample and then diluted with TNE buffer to a working concentration of 2×10^6 cells/ml. Several hundred 250-μl aliquots of this dilution are placed into small snap cap vials and frozen (-70 to -100 °C) immediately. These reference samples are used to set the red and green PMTs to the same X Red, X Green, and α_t values (Xα_t and SDα_t) from day to day. The PMTs are set so that X Green is at about 50/100 channels and X Red at about 13/100 channels. Whatever channel numbers are established by a laboratory should be used consistently thereafter. All reference sample X Red and X Green values should fall within ± 5 channels of a value that the flow operator decides upon. Strict adherence to keeping the reference values in this range should be maintained throughout the experiment. One or two freshly thawed reference samples are typically run after every 5 or 10 regular samples to ensure that there is no instrument drift. If the reference sample values are out of the range, first run a second reference sample, then check focus, and if that does not correct the problem, finally adjust PMT settings. If the first samples run during a day are seen to drift, it usually implies that AO was not equilibrated with the sample tubing.

E. Sampling Order

If the intent of an experiment is to determine the smallest amount of change measurable over time or increasing toxicant dosages, the most ideal situation is to measure all experimental samples of a particular set at one time period. Statisticians often do not like this approach and prefer placing all samples randomly coded in a box to be measured blindly. However, if the best "truth" of the experiment is desired, and recognizing that the flow cytometer randomly and objectively measures the samples, it is preferable to measure sets of samples in a single time frame. Each set is randomized within itself. However, if totally random measurements are preferred, it has been shown that very careful repeat settings of the red and green PMTs allow measurements to be made of compared samples over an extended period of time with nearly identical results. This includes samples that were measured fresh, then frozen, and the frozen/thawed samples measured up to 3 years later.

F. SCSA and Fertility

At this time, it is difficult to define what values are incompatible with normal fertility since the interpretation of α_t parameters is still being explored. However, from measurements of thousands of sperm samples derived from a variety of mammals, some of which had known fertility potential, the evidence strongly suggests that a broadly heterogeneous pattern is indicative of sub- or infertility. It is important to note, however, that fertility may occur with sperm with high α_t values and this may lead to early embryo death (Evenson $et\ al.$, 1993b).

IX. Critical Points

A. Freezing and Thawing

Repeated evidence from various species shows that freezing and thawing a sperm or semen sample $once$ does not cause significantly altered SCSA data relative to fresh samples. However, it is strongly emphasized that the samples must stay frozen in an ultracold freezer or in dry ice ($-70\,^{\circ}$C or colder). Refrigerator freezers with automatic defrosters should be avoided due to the rise and fall of temperatures on a daily basis. Likewise, investigators must be aware that removal of samples from an ultracold freezer into ambient air shifts the sample to an approximate $100\,^{\circ}$C temperature change that may not immediately turn the frozen sample into a liquid sample but the strong physical forces of the temperature increase may damage the chromatin structure and cause an artifact. Thus, when handling any sample, for example, moving from box to box in the freezer, DO NOT pick up the small test tube by the body of the tube; a warm human hand will produce microthawing of the sample very quickly. Use gloves or forceps or at least grasp the tube by the very top lip of the tube. If samples boxes must be manipulated, place them into a deep ice chest containing dry ice that will keep the box and the samples at least dry ice cold. Since ultracold freezers are often shared by a number of personnel, place the sample box in a rack that others are instructed not to remove. It is safest to place the sample box near the bottom of the freezer so that it is not subjected to ambient air when others may keep the door open for extended periods of time. When a group of samples are to be analyzed by flow cytometry, place the samples in an approximate 18-in. deep styrofoam box containing dry ice with a good cover and place it near the flow cytometer where the samples are prepared. Individual samples are removed, placed into a 37 °C water bath to thaw, and then processed immediately.

B. Shipping of Samples

Typically semen samples are shipped by Federal Express (or similar overnight carrier) in 10¾ × 7½ × 10½ (L × W × H in., inside dimensions) commercial insulated shipping containers (i.e., FreezSafe Insulated Containers, Polyfoam Packers Corp., Wheeling, IL; Cat. No. 272–524, Curtain Matheson Scientific, Inc.). Small chunks of dry ice are first placed on the bottom of the shipping container, then the sample box is placed near the center of the shipping box, and then dry ice placed over and around the box. Ten pounds of dry ice added to the shipping container is satisfactory for Priority One overnight shipments from any point in the United States during any season. This amount will keep for at least 2 days. Shipments are made only Monday through Wednesday due to the rare case where the shipment box is "miss-shipped" somewhere else or it is held up for a day in a storm.

C. Reference Samples

Very few FCM protocols are as particular as the SCSA for using a reference sample. Fresh aliquots of this sample are run after approximately every 10 experimental samples to ensure that the instrument has not drifted or lost focus. Strict adherence to keeping the reference values in the range set by the flow operator should be maintained throughout the experiment.

D. Sample Flow Rate

The rate of sperm cell measurement is important and should be about 150 cells/s. Samples that run over 300/s are discarded and a new aliquot is diluted and stained to produce that approximate rate.

Acknowledgments

This work was supported in part by NSF Grant EHR-9108773 and the South Dakota Futures Fund. It is Publication No. 2749 from South Dakota State University Experiment Station. We gratefully acknowledge the skilled collaborative efforts and technical assistance of Donna Gandor and colleagues (Becton Dickinson, Inc. San Jose, CA); Ole Petter Clausen, Kenneth Purvis, and Paula De Angelis (The National Hospital, Oslo, Norway); Richard Coico and Andrew Daley (CUNY, NY); and Barbara Stanton (Coulter Corporation, Hialeah, FL).

References

Ballachey, B. E., Hohenboken, W. D., and Evenson, D. P. (1987). *Biol. Reprod.* **36,** 915–925.

Ballachey, B. E., Saacke, R. G., and Evenson, D. P. (1988). *J. Androl.* **9,** 109–115.

Clausen, O. P. F., and Abyholm, T. (1980). *Fertil. Steril.* **34,** 369–373.

Darzynkiewicz, Z. (1979). *In* "Flow Cytometry and Sorting" (M. R. Melamed, P. F. Mullaney, and M. L. Mendelsohn, eds.), pp. 285–316. Wiley, New York.

Darzynkiewicz, Z., Juan, G., Li, X., Gorczyca, W., Murakami, T., and Traganos, F. (1997). Cytometry in cell necrobiology. Analysis of apoptosis and accidental cell death (necrosis). *Cytometry* **27,** 1–20.

Darzynkiewicz, Z., Traganos, F., Sharpless, T., and Melamed, M. R. (1975). *Exp. Cell Res.* **90,** 411–428.

Darzynkiewicz, Z., Traganos, F., Sharpless, T., and Melamed, M. R. (1976). *Proc. Natl. Acad. Sci. USA* **73,** 2881–2884.

Engh, E., Clausen, O. P. F., Scholberg, A., Tollefsrud, A., and Purvis, K. (1992). *Int. J. Androl.* **15,** 407–415.

Erenpreisa, J., Erenpreis, J., Freivalds, T., Slaidina, M., Krampe, R., Butikova, J., Ivanov, A., and Pjanova, D. (2003). Toluidine blue test for sperm DNA integrity and elaboration of image cytometry algorithm. *Cytometry A* **52A,** 19–27.

Erenpreiss, J., Jepson, K., Giversman, A., Tsarev, I., Eranpresa, J., and Spano, M. (2004). Toluidine blue cytometry test for sperm DNA conformation with the flow cytometric sperm chromatin structure and TUNEL assays. *Hum. Reprod.* **19,** 2277–2282.

Estop, A. M., Munne, S., Jost, L. K., and Evenson, D. P. (1993). *J. Androl.* **14,** 282–288.

Evenson, D. P. (1986). *In* "Clinical Cytometry" (M. Andreeff, ed.), pp. 350–367. New York Academy of Sciences, New York.

Evenson, D. P. (1989). *In* "Flow Cytometry: Advanced Research and Clinical Applications" (A. Yen, ed.), Vol. **1,** pp. 217–246. CRC Press, Boca Raton, FL.

Evenson, D. P. (1990). *In* "Methods in Cell Biology" (Z. Darzynkiewicz and H. A. Crissman, eds.), Vol. 33, pp. 401–410. Academic Press, San Diego, CA.

Evenson, D. P., and Jost, L. K. (1993). *Cell Prolif.* **26,** 147–159.

Evenson, D. P., and Melamed, M. R. (1983). *J. Histochem. Cytochem.* **31,** 248–253.

Evenson, D. P., Darzynkiewicz, Z., and Melamed, M. R. (1980). *Science* **240,** 1131–1133.

Evenson, D. P., Darzynkiewicz, Z., and Melamed, M. R. (1982). *J. Histochem. Cytochem.* **30,** 279–280.

Evenson, D. P., Higgins, P. H., Grueneberg, D., and Ballachey, B. (1985). *Cytometry* **6,** 238–253.

Evenson, D. P., Baer, R. K., Jost, L. K., and Gesch, R. W. (1986a). *Toxicol. Appl. Pharmacol.* **82,** 151–163.

Evenson, D. P., Darzynkiewicz, Z., Jost, L., Janca, F., and Ballachey, B. (1986b). *Cytometry* **7,** 45–53.

Evenson, D. P., Baer, R. K., and Jost, L. K. (1989a). *J. Environ. Mol. Mutagen.* **14,** 79–89.

Evenson, D. P., Janca, F. C., Jost, L. K., Baer, R. K., and Karabinus, D. S. (1989b). *J. Toxicol. Environ. Health* **28,** 67–80.

Evenson, D. P., Jost, L., Baer, R., Turner, T., and Schrader, S. (1991). *Reprod. Toxicol.* **5,** 115–125.

Evenson, D. P., De Angelis, P., Jost, L. K., Purvis, K., and Clausen, O. P. F. (1993a). In *Congr. Int. Soc. Anal. Cytol., 16th,* Colorado Springs, CO. *Cytometry,*Suppl. 6, March 21–26, p. 73.

Evenson, D. P., Jost, L. K., and Baer, R. K. (1993b). *J. Environ. Mol. Mutagen.* **21,** 144–153.

Evenson, D. P., Jost, L. K., and Gandy, J. G. (1993c). *Reprod. Toxicol.* **7,** 297–304.

Evenson, D. P., Kasperson, K., and Wixon, R. L. (2007). Analysis of sperm DNA fragmentation using flow cytometry and other techniques. *Soc. Reprod. Fertil. Suppl.* **65,** 93–113.

Garner, D. L., Gledhill, B. L., Pinkel, D., Lake, S., Stephenson, D., Van Dilla, M. A., and Johnson, L. A. (1983). *Biol. Reprod.* **28,** 312–321.

Gorczyca, W., Bruno, S., Darzynkiewicz, R. J., Gong, J., and Darzynkiewicz, Z. (1992). DNA strand breaks occurring during apoptosis: Their early *in situ* detection by the terminal deoxynucleotidyl transferase and nick translation assays and prevention by serine protease inhibitors. *Int. J. Onc.* **1,** 639–648.

Gorczyca, W., Traganos, F., Jesionowska, H., and Darzynkiewicz, Z. (1993). Presence of DNA strand breaks and increased sensitivity of DNA *in situ* to denaturation in abnormal human sperm cells. Analogy to apoptosis of somatic cells. *Exp. Cell Res.* **207**, 202–205.

Hotz, M. A., Gong, J. P., Traganos, F., and Darzynkiewicz, Z. (1994). Flow cytometric detection of apoptosis. Comparison of the assays of *in situ* DNA degradation and chromatin changes. *Cytometry* **15**, 237–244.

Johnson, L. A., Flook, J. P., and Hawk, H. W. (1989). *Biol. Reprod.* **41**, 199–203.

Jost, L. K., and Evenson, D. P. (1993). *In "Congr. Int. Soc. Anal. Cytology, 16th"*, Colorado Springs, CO. *Cytometry*,Suppl. 6, March 21–26, p. 17.

Marchetti, C., Gallego, M. A., Defossez, A., Formstecher, P., and Marchetti, P. (2004). Staining of human sperm with fluorochrome-labeled inhibitor of caspases to detect activated caspases: correlation with apoptosis and sperm parameters. *Hum. Reprod.* **19**, 1127–1134.

Martins, C. F., Dode, M. N., Bao, S. N., and Rumpf, R. (2007). The use of acridine orange test and the TUNEL assay to assess the integrity of freeze-dried bovine spermatozoa. *Genet. Mol. Res.* **15**, 94–104.

Paasch, U., Grunewald, S., Agarwal, A., and Glandera, H. J. (2004). Activation pattern of caspases in human spermatozoa. *Fertil. Steril.* **91**(Suppl. 1), 802–809.

Spano, M., and Evenson, D. P. (1993). *Biol. Cell* **78**, 53–62.

Ward, W. S., and Coffey, D. S. (1991). *Biol. Reprod.* **44**, 569–574.

PART VI

Cell Physiology Assays

CHAPTER 29

Cell Membrane Potential Analysis

Howard M. Shapiro

283 Highland Avenue
West Newton, Massachusetts 02465-2513

I. Introduction

Cytoplasmic and mitochondrial membrane potential ($\Delta\Psi$) changes may occur during the early stages of surface receptor-mediated activation processes related to the development, differentiated function, and pathology of a large number of cell types, and can play a role in signal transduction between the cell surface and the interior. Investigations in this area have been facilitated by the development of methods for flow cytometric (FCM) estimation of $\Delta\Psi$ in single cells (Shapiro, 1981, 1982, 2003; Shapiro et al., 1979). This chapter discusses the principles, practical aspects, and limitations of such methods, emphasizing measurements of cytoplasmic $\Delta\Psi$.

ESSENTIAL CYTOMETRY METHODS
Reprinted from *Methods in Cell Biology*, Volume 41 (Academic Press, 1994).
DOI: 10.1016/B978-0-12-375045-7.00029-5

A. The Basis of Membrane Potentials

Electrical potential differences normally exist across eukaryotic cell membranes, due in part to concentration gradients of Na^+, K^+, and Cl^- ions across the cell membrane, and in part to the operation of electrogenic pumps. The potential differences across the cytoplasmic membranes of resting mammalian cells range from ~10 to 90 mV, with the cell interior negative with respect to the exterior. There is also a potential difference of ≥ 100 mV across the membranes of energized mitochondria, with the mitochondrial interior negative with respect to the cytosol; this potential is dependent on energy metabolism. In prokaryotes, the enzymes responsible for energy metabolism are located on the inner surface of the cytoplasmic membrane, and most of the 100–200 mV, interior negative, potential difference across this membrane is generated by energy metabolism.

B. Potential Measurement by Indicator Distribution

Membrane potential can be estimated from the distribution of lipophilic cationic indicators or dyes between cells and the suspending medium. Lipophilicity, or hydrophobicity—that is, a high lipid:water partition coefficient—enables dye molecules to pass freely through the lipid portion of the membrane; the concentration gradient of a lipophilic cationic dye C^+ across the membrane is determined by the transmembrane potential difference according to the Nernst equation:

$$[C^+]_i/[C^+]_o = e^{-F\Delta\Psi/RT},$$

where $[C^+]_i$ is the concentration of C^+ ions inside the cell, $[C^+]_o$ is the concentration of C^+ ions outside the cell, $\Delta\Psi$ is the membrane potential, R is the gas constant, T is the temperature in degrees Kelvin, and F is the Faraday. Indicators or dyes used in this fashion are referred to as distributional probes of membrane potential.

Once cells have been equilibrated with a cationic dye, a subsequent electrical depolarization of the cells (i.e., a decrease in $\Delta\Psi$) will cause release of dye from cells into the medium, while a hyperpolarization (i.e., an increase in $\Delta\Psi$) will make cells take up additional dye from the medium. The dye distribution will not adequately represent the new value of $\Delta\Psi$ until equilibrium has again been reached; this process requires periods ranging from a few seconds to several minutes. Thus, although distributional probes may be suitable for detection of slow potential changes, they cannot be used to monitor the rapidly changing action potentials in excitable tissues such as nerve and muscle.

The use of cyanine dyes as $\Delta\Psi$ probes began with the study of Hoffman and Laris (1974), who used 3,3′-dihexyloxacarbocyanine [$DiOC_6(3)$ in the common notation introduced by Sims *et al.* (1974)] to estimate membrane potential in red blood cell (RBC) suspensions based on the partitioning of the dye into the cells. They noted that addition of RBC to a micromolar dye solution in a spectrofluorometer cuvette produced a suspension with lower fluorescence than that of the original solution, indicating that—at these concentrations—the fluorescence of cyanine dyes taken into cells is quenched. When extracellular ion concentrations were manipulated so as to hyperpolarize the cells, increasing cellular uptake of dye, the fluorescence of the suspension decreased further; when the cells were depolarized, releasing dye into the medium, the fluorescence of the suspension increased. Membrane potential measurements of giant *Amphiuma* RBC using cyanine dyes were consistent with results obtained from microelectrode measurements.

Under normal circumstances, the intracellular concentration of K^+ is considerably higher than the extracellular concentration, while the intracellular concentration of Na^+ is considerably lower than the extracellular concentration. Valinomycin (VMC) a lipophilic potassium-selective ionophore, forms complexes with K^+ ions and thus readily transports them across cell membranes, effectively increasing cells' potassium permeability to the point at which membrane potential is determined almost entirely by the transmembrane K^+ gradient. Addition of VMC hyperpolarizes cells if the external K^+ concentration is lower, and depolarizes them if it is higher, than the internal concentration. The ionophore gramicidin A, which forms transmembrane channels passing Na^+, K^+, and other ions, depolarizes cells in solutions with approximately physiologic ionic concentrations.

C. Single-Cell Measurements with Distributional Probes: Principles and Problems

Potential estimation by fluorometry of cell suspensions in cuvets requires dye concentrations sufficiently high that the fluorescence of intracellular dye is largely quenched, so that most of the measured fluorescence comes from free dye in solution. When cells are hyperpolarized, they take up more dye from the solution, decreasing total fluorescence. When cells are depolarized, dye molecules that were quenched when inside the cell are released into solution, increasing total fluorescence.

In cytometric estimation of $\Delta\Psi$, it is the fluorescence of intracellular dye which is measured. The dye concentrations used are lower than those used for bulk measurements, to minimize quenching of intracellular dye. In principle, $\Delta\Psi$ could be calculated accurately from measurements of intracellular and extracellular dye concentrations using the Nernst equation (Ehrenberg *et al.*, 1988). However, the accuracy of the FCM procedure is limited, for several reasons.

The flow cytometer measures the amount, not the concentration, of dye in each cell. To find the intracellular concentration, it would be necessary to divide the fluorescence value for each cell by the cell's volume, obtained by an electronic (Coulter) volume sensor or estimated from forward scatter or extinction measurements. Measuring the extracellular concentration of dye is more problematic; this would require signal processing electronics of a type not normally used in flow cytometers. However, the large variances of fluorescence distributions obtained from conventional FCM measurements of cells at a known $\Delta\Psi$, even when cell size corrections are introduced (Seamer and Mandler, 1992; Shapiro, 1981), suggest that accuracy would not be significantly increased by the instrumental refinements just discussed.

One possible source of fluorescence variance is mitochondrial uptake of dye. Since the mitochondrial $\Delta\Psi$ is typically ≥ 100 mV negative with respect to the cytosol, dye should be present in energized mitochondria at almost 100 times the concentration found in the cytoplasm. The mitochondrial $\Delta\Psi$, in fact, provides the basis for the accumulation in mitochondria of cationic dyes such as rhodamine 123 (Darzynkiewicz *et al.*, 1981; Johnson *et al.*, 1980, 1981), pinacyanol, Janus green, and JC-1 (Reers *et al.*, 1991; Smiley *et al.*, 1991). However, the high concentration of dye in mitochondria should result in substantial quenching. Indeed, in at least some cell types, treatment with metabolic inhibitors and uncouplers which abolish the mitochondrial $\Delta\Psi$ gradient, which should eliminate concentration of dye in mitochondria, reduces neither the intensity nor the variance of cyanine dye fluorescence.

In cells depolarized with gramicidin A, $\Delta\Psi$ should be zero; the intracellular and extracellular dye concentrations predicted by the Nernst equation should, therefore, be equal, and the very fact that the flow cytometer can detect the cells' fluorescence above background requires some explanation. One reason for the cells' increased fluorescence is that the fluorescence of cyanine dyes is enhanced [approximately sixfold in the case of $DiOC_6(3)$, less for other dyes] when the dye is in a hydrophobic environment (Sims *et al.*, 1974) such as the membranous structures in which it can be observed in cells.

Perhaps even more important, the lipophilic, hydrophobic character of cyanine dyes causes them to be concentrated in cells even in the absence of a potential gradient. Similar problems with such "non-Nernstian" probe binding are also encountered when $\Delta\Psi$ is estimated with radiolabeled lipophilic cations such as [^{3}H]triphenylmethylphosphonium (TPMP$^+$) or with less hydrophobic cationic dyes (Ehrenberg *et al.*, 1988). In order to get accurate values of cytoplasmic $\Delta\Psi$ using these indicators, it is necessary to inhibit the mitochondria and to correct for probe uptake by cells in the absence of a potential gradient. The affinity of hydrophobic cyanine dyes for cells is sufficiently high that cells may take up most of the dye molecules even in a suspension in which the cells occupy only a fraction of a percent of the total volume. This makes it necessary to keep cell concentrations relatively constant from sample to sample in order to obtain reproducible results.

FCM of cells exposed to increasing concentrations of cyanine dyes shows that saturation occurs, that is, that, eventually, further increasing the dye concentration does not increase fluorescence in the cells. For $DiOC_6(3)$, this happens when cells at a concentration of 10^6 ml^{-1} are incubated with 2 μM dye. The variance of the fluorescence distribution remains large. It is likely that most of the fluorescence measured in cells comes from dye in hydrophobic regions, which represent both the highest affinity binding sites and the sites in which dye fluorescence yield would be highest; the variance of fluorescence would then be explained by cell-to-cell variations in the number of binding sites. It has been observed (Shapiro, 1981) that, when dye-binding sites are saturated, cellular fluorescence does not change when cells are depolarized or hyperpolarized by ionophores; however, such potential changes would be detectable by bulk fluorometry in a cell suspension exposed to the 2 μM saturating concentration of DiOC$_6$(3). Thus, at this concentration, the cells contain dye which is essentially nonfluorescent due to quenching, as well as dye bound to the hydrophobic sites in which fluorescence is enhanced.

The dye concentration at which saturation of binding sites occurs is determined primarily by the hydrophobicity of the dye; the fluorescence of cells equilibrated (at 10^6 ml^{-1}) with 2 μM diethyloxacarbocyanine [DiOC$_2$(3)], which is less hydrophobic than DiOC$_6$(3), is less than the fluorescence of cells equilibrated with the C$_6$ dye, and does change when $\Delta\Psi$ is changed by ionophore addition or by manipulation of ion concentrations in the medium, indicating that the saturating concentration for the C$_2$ dye is higher than for the C$_6$ dye. When cells in 2 μM DiOC$_2$(3) in NaCl (normal $\Delta\Psi$, higher fluorescence) are mixed with an equal volume of cells in 2 μM DiOC$_2$(3) in KCl (depolarized, lower fluorescence), the cells and dye reequilibrate within a few minutes, yielding a fluorescence distribution that reflects the intermediate value of $\Delta\Psi$ resulting from the ionic composition of the mixed suspending medium. Thus, the fluorescence of cell-associated cyanine dye can provide a reasonably rapid indication of substantial changes in $\Delta\Psi$, even if fluorescence variance limits overall accuracy and precision.

Fluorescence variance also limits the capability of FCM measurements to detect heterogeneous responses within cell populations; two subpopulations of equal size with a mean difference of 50% in fluorescence intensity will be clearly resolved, while a 5% subpopulation with a mean 10% above the population mean might be undetectable.

II. Materials and Methods

A. Dyes and Reagents

1. Cyanine Dyes

The choice of dye is determined primarily by the excitation wavelength(s) available and the emission wavelengths desired. Dihexyl- or dipentyloxacarbocyanine [DiOC$_6$(3) or DiOC$_5$(3)] are both suitable for blue-green (488 nm) excitation. They are green fluorescent and can be used with the same detector/filter combination used for fluorescein. Hexamethylindodicarbocyanine [DiIC$_1$(5); sometimes also known as HIDC] can be excited with a red (633) He-Ne laser and measured through 665 nm long pass filters using a red-sensitive photomultiplier tube. The dyes are available from Molecular Probes (a division of Invitrogen/Life Technologies, Eugene, OR) and other sources.

Stock solutions with a 1 mM concentration of any of the dyes mentioned in dimethyl sulfoxide (DMSO) may be kept for several weeks in the dark at room temperature (RT). Working solutions are made by diluting the stock solution with absolute ethanol or DMSO to allow the desired final dye concentration to be obtained by adding 5 μl of working solution to each 1 ml of cell suspension. For DiOC$_5$(3) and DiOC$_6$(3), a 10 μM working solution and a 50 nM final concentration are appropriate; for DiIC$_1$(5), a 20 μM working solution yields a 100 nM final concentration.

2. Ionophores

VMC is used to hyperpolarize cells in high-Na$^+$, low-K$^+$ media and to depolarize cells in high-K$^+$ media. A 1 mM stock solution in DMSO is stable for several weeks at RT; adding 5 μl of stock solution to a 1 ml cell sample produces a 5 μM concentration of VMC.

Gramicidin D is used to depolarize cells; this material is a mixture containing a variable percentage of gramicidin A, which is the active ingredient. A stock solution of 1 mg/ml gramicidin D in DMSO is stable for several weeks at RT; addition of 5 μl stock solution to 1 ml of cell suspension yields a final concentration of approximately 2 μM gramicidin A.

The proton ionophore carbonyl cyanide chlorophenylhydrazone (CCCP), an uncoupler and mitochondrial inhibitor, is an effective depolarizing agent for both aerobic and anaerobic bacteria. A final concentration of 5 μM is obtained by adding 5 μl of a 1 mM stock solution in DMSO to 1 ml of dilute bacterial suspension.

B. Staining Procedures

Cyanine dye equilibration with cells in protein-free media is usually complete after 15 min at RT. For cells in media containing protein, 30 min incubation at 37 °C generally suffices. The cell concentration should be kept relatively constant from sample to sample, since the high affinity of the dye for cell constituents otherwise produces variations in fluorescence intensity per cell with cell concentration. Concentrations in the range of 10^6 cells/ml give satisfactory results. Cells are not washed prior to analysis.

C. FCM and Data Analysis

The incubation temperature and the interval between dye addition and introduction of samples into the flow cytometer should be kept as nearly constant as possible; it is also advisable to run all samples in an experiment at the same flow rate to avoid artifacts due to dye diffusion between core and sheath, and resultant changes in intracellular dye concentration.

While cyanine dye fluorescence signals are generally strong enough to be used as triggering or gating signals, it is preferable to use a forward scatter and/or a side scatter signal to avoid missing depolarized cells. Since $\Delta\Psi$ is meaningful only in putatively viable cells, whereas cyanine dyes will also stain dead cells and debris, using scatter signals for triggering and gating allows extraneous signals to be excluded during data collection. Pronounced changes in scatter signals following exposure of cells to some stimulus to be tested provide evidence that an apparent $\Delta\Psi$ effect may be secondary to a more obvious change such as membrane lysis.

When using $DiOC_5(3)$ or $DiOC_6(3)$ as a $\Delta\Psi$ probe, it is convenient to add propidium iodide (5–20 μg/ml final concentration) to the cell suspension, allowing dead (i.e., membrane-damaged) nucleated cells to be gated out of analyses on the basis of their strong red fluorescence.

The effects of many agents on $\Delta\Psi$ of homogeneous cell populations may be appreciated by simple comparison of single-parameter fluorescence distributions such as those shown in Fig. 1. In other situations, for example, observation of stimulated lymphocytes, multiparameter analysis that, for example, relates $\Delta\Psi$ to nuclear DNA content in cells vitally stained with Hoechst 33342 (Shapiro, 2003, p. 396), may be preferable. Rapid changes in $\Delta\Psi$ are best appreciated when time is used as a measurement parameter.

D. Controls

When an experiment extends over a period of a few hours, control samples should be run during the course of the experiment as well as at the beginning and end. Controls should include an untreated cell sample, a sample of cells

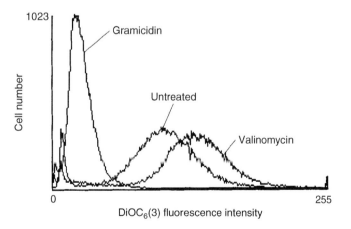

Fig. 1 Distributions of the fluorescence of $DiOC_6(3)$ in CCRF-CEM lymphoblasts in phosphate-buffered saline (pH 7.4), initially incubated with 50 nM dye for 15 min. Fluorescence histograms are shown for untreated cells, for cells hyperpolarized after an additional 10 min incubation with 5 μM valinomycin, and for cells depolarized after a further 10 min exposure to 5 μg/ml gramicidin following the valinomycin treatment. Each histogram represents 25,000 cells; fluorescence measurements were made in a "Cytomutt" flow cytometer (Shapiro, 1988, pp. 211–265) with 21 mW excitation at 488 nm, using forward scatter signals for triggering.

hyperpolarized by addition of 5–10 μM VMC, and another sample depolarized by addition of 5–10 μg/ml gramicidin.

A more dynamic control procedure, which verifies that cells are capable of response to hyper- and depolarizing stimuli, can be implemented by examining one aliquot of a sample of cells after they have equilibrated with dye, then adding VMC and analyzing a second aliquot after 5–10 min, and finally adding gramicidin and analyzing a third aliquot after an additional 5-10 min. The data shown in Fig. 1 are typical of the results of such a procedure. VMC should increase the fluorescence of the cells; gramicidin should decrease the fluorescence well below the initial value (and, incidentally, render cells unresponsive to subsequent treatment with VMC or other hyperpolarizing agents).

Although the large variance (CV values typically $\sim$30%) of cyanine dye fluorescence distributions provides a strong argument against attempting to calibrate flow $\Delta\Psi$ distributions to absolute voltages, the nature and magnitude of changes observed by different investigators in many cell systems have been fairly consistent. On several occasions, cyanine dye-stained lymphocytes from different donors, run on the same flow cytometer on successive days, have yielded distributions with peaks within 5% of one another. This suggests that both $\Delta\Psi$ and the amount of dye-binding sites per cell are relatively well regulated.

The control procedures just described make it possible to establish that the cells under study have a detectable $\Delta\Psi$ and that the dye in use will respond to potential changes from the control value in either direction. All of this is prerequisite to any investigation of the effects of biologic, chemical, and/or physical agents on cell $\Delta\Psi$.

E. Pitfalls and Cautions

Addition of any substantial amount of protein to a cell suspension which has been equilibrated with dye in a protein-free medium will decrease the dye concentration in cells, because the dye will bind to the protein in solution. This artifactual apparent depolarization can be avoided by working in a medium with added protein (e.g., 1–10% albumin) when trying to determine effects of adding specific proteins to cells. Similar cautions apply to studies of the effects of solvents such as DMSO, which alter the partitioning of dye between cells and the medium. The total amount of added solvent should be maintained relatively constant from sample to sample and not exceed 2% of sample volume. Other problems with cyanine dyes have been discussed at length elsewhere (Shapiro, 2003, pp. 386–394). At micromolar concentrations, cyanines have been observed to be toxic to bacteria and mammalian cells. The dyes themselves may perturb $\Delta\Psi$ directly by altering membrane conductivity; their inhibition of energy metabolism might also result in potential changes. When used to monitor neutrophil responses to chemotactic peptides, $DiOC_6(3)$ and $TPMP^+$ were reported to give contradictory results, while the thiacyanine dye $DiSC_3(5)$ was found to be destroyed by oxidation following neutrophil activation.

Toxicity is a liability shared by the cyanines with other families of cationic dyes such as acridines, safranins, oxazines, pyronins, rhodamines, and triarylmethanes, and with lipophilic cations such as TPMP$^+$. When the dyes are used in FCM, at concentrations of 5–100 nM, toxicity is less than when radiolabeled cations or dyes are used at micromolar concentrations for bulk measurements. Different cell types appear to have different degrees of susceptibility to cyanine dye toxicity. Crissman *et al.* (1988) found that simultaneous staining with DiOC$_5$(3) improved Hoechst 33342 staining of live CHO cells, presumably by interfering sufficiently with energy metabolism to block efflux of the dye from cells via a pump mechanism. However, the cyanine dye did not affect cell viability following sorting, which remained ~90%. In this instance, at least, cyanine dye toxicity is evidently entirely reversible.

Oxonols, which are negatively charged, lipophilic dyes, bind to the cytoplasmic membrane and some cellular constituents because of their lipophilicity, but are largely excluded from mitochondrial and cytoplasmic compartments because of their charge. This probably explains why oxonols are much less toxic than cyanines and other cationic dyes and why oxonol fluorescence is less affected by mitochondrial potential changes than are the fluorescence signals from cationic dyes. These desirable characteristics of oxonols are offset somewhat by their weaker fluorescence, as compared to cyanines. As can be seen from Fig. 2, the variance of oxonol fluorescence distributions is no better than that of cyanine fluorescence distributions, and oxonol and cyanine dyes produce comparable results in most cases (e.g., Lazzari *et al.*, 1990).

The oxonol dye most widely used for $\Delta\Psi$ measurements is bis-(1,3-dibutylbarbituric acid) trimethine oxonol [DiBAC$_4$(3)], which is typically excited at 488 nm and emits in the same spectral region as the DiOC$_n$(3) dyes. It is probably now preferable to the cyanines for work with eukaryotic cells; recently, Klapperstück *et al.* (2009), amplifying on earlier work by Krasznai *et al.* (1995), have described a

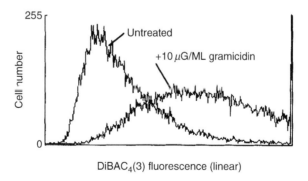

Fig. 2 Oxonol fluorescence distributions from CCRF-CEM cells. Cells in RPMI 1640 medium with 10% fetal calf serum were incubated for 15 min at RT with 100 nM bis-(1,3-dibutylbarbituric acid) trimethine oxonol [DiBAC$_4$(3)] with and without 10 µg/ml gramicidin; fluorescence was excited with 25 mW at 488 nm and measured at 525 nm using a 20 nm bandwidth interference filter, using forward scatter signals for triggering. Each histogram represents 20,000 cells.

procedure for calibrating FCM DiBAC$_4$(3) measurements to yield mean values nearly identical to those obtained by patch clamping. Klapperstück *et al.* emphasize the use of gramicidin-treated rather than fixed controls, and note that the relationship between fluorescence and $\Delta\Psi$ remains linear when the dye concentration is $\leq$100 nM.

The glycoprotein efflux pump responsible for the multidrug-resistance (MDR) phenotype in cells can transport a variety of neutral and positively charged compounds and dyes and, in some experimental situations (Kessel *et al.*, 1991), changes in pump function can be misinterpreted as changes in membrane potential and *vice versa*. Rhodamine 123 and cyanine dyes can be used for demonstration of MDR by FCM observation of loss of fluorescence from washed cells over time, and some observations in the literature relating low rhodamine 123 retention by cells to loss of mitochondrial activity may need to be reexamined in the light of these findings. Efflux pump activity has less effect on the fluorescence of cationic dyes when cells are kept in equilibrium with dye, as in the $\Delta\Psi$ measurement techniques described here, than on fluorescence in washed cells. Since the anionic oxonol dyes are not substrates for the pump, oxonol fluorescence is not affected by its activity.

F. Mitochondrial Staining with Rhodamine 123 and JC-1

The lipophilic cationic dye rhodamine 123 has been used for investigations of mitochondrial structure and function (Darzynkiewicz *et al.*, 1981; Johnson *et al.*, 1980, 1981); it accumulates in energized mitochondria as a result of their $\Delta\Psi$. Cells are equilibrated for 30 min with 10 μg/ml ($\sim$25 μM) dye, then washed and examined; most of the retained dye is found in mitochondria. Safranin and the less hydrophobic cyanine and styryl dyes show similar potential-dependent mitochondrial staining when cells are washed as they are for the rhodamine 123 procedure.

Chen and coworkers (Reers *et al.*, 1991; Smiley *et al.*, 1991) first described the fluorescence properties of the cyanine dye 5,5',6,6'-tetrachloro-1,1',3,3'-tetraethylbenzimidazolocarbocyanine iodide (JC-1), which emits green (527 nm) fluorescence in the monomeric form and orange (590 nm) when aggregates form. Although it has been suggested that JC-1 detects local differences in mitochondrial $\Delta\Psi$ even within individual mitochondria, based on spectral differences in fluorescence emission, this dye is no less likely than other cyanines to be affected by local variations in the number and availability of binding sites; it has, however, become widely used as a ratiometric probe of mitochondrial membrane potential in relation to studies of apoptosis (Troiano *et al.*, 2007).

G. Ratiometric Measurement of $\Delta\Psi$ in Bacteria

Bacterial $\Delta\Psi$s are primarily dependent on metabolic activity, of which they are a sensitive indicator. Membrane potential changes substantially and rapidly in response to the availability or lack thereof of suitable energy sources, and is rapidly dissipated when the organism is killed by drugs or other agents. Cyanine dye fluorescence in bacteria is illustrated in Fig. 3. It is possible to exploit potential-sensitive dyes in rapid FCM procedures for bacterial detection, identification, and antibiotic susceptibility testing. This application is facilitated by the availability of ratiometric potential probes (Shapiro, 2003, pp. 400-402).

Results of FCM measurements of JC-1 fluorescence in *Staphylococcus aureus* stained with 100 nM dye are shown in Fig. 4. When the organisms are depolarized with CCCP, orange fluorescence from the aggregates is greatly diminished, although green fluorescence is not changed significantly. The differences between untreated and CCCP-treated cells in Fig. 4 appear greater than those in Fig. 3, but

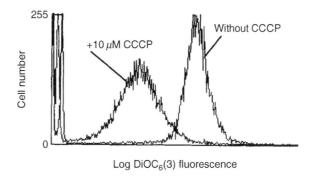

Fig. 3 Fluorescence distributions from *S. aureus* in a Tris buffer containing glucose and EDTA. Cells were incubated for 2 min at RT with 50 nM $DiOC_6(3)$ with and without 10 μM CCCP, which depolarizes the bacteria. The instrument was set up as described in the legend of Fig. 2; fluorescence was measured on a logarithmic scale.

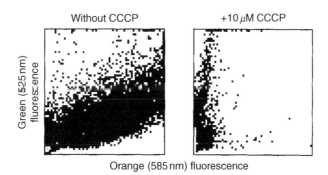

Fig. 4 Two-parameter distributions of orange (585 nm, 20 nm bandwidth) and green (525 nm, 20 nm bandwidth) from *S. aureus* in a Tris buffer containing glucose and EDTA. Cells were incubated for 15 min at RT with 100 nM JC-1 (Molecular Probes), with and without 10 μM CCCP. Fluorescence was excited with 25 mW at 488 nm and measured on linear scales; the green fluorescence signal was used for triggering. Each distribution represents 20,000 cells.

this is due primarily to the use of a linear fluorescence scale in Fig. 4 rather than the logarithmic scale of Fig. 3.

A more accurate and precise ratiometric measurement of $\Delta\Psi$ in bacteria than can be obtained with JC-1 utilizes diethyloxacarbocyanine [DiOC$_2$(3)] (Novo *et al.*, 1999, 2000; Shapiro, 2008). Organisms are equilibrated with 30 μM dye. At this relatively high concentration, the green (488 nm excitation, 530 nm emission) fluorescence is largely dependent on cell size and independent of $\Delta\Psi$. Red (610 nm) fluorescence, probably due to the formation of dye aggregates, also appears; this is dependent on both size and $\Delta\Psi$. A quantity proportional to the logarithm of the ratio of red to green fluorescence is calculated for each cell by subtracting the green fluorescence channel value from the red fluorescence channel value (both are on a logarithmic scale) and adding a constant to keep values of the calculated parameter on scale. Figure 5 shows ratiometric $\Delta\Psi$ measurements in untreated *S. aureus* and in organisms depolarized with CCCP; a comparison with Fig. 3 demonstrates that the ratiometric method gives much better separation.

Values of the calculated fluorescence ratio parameter can be calibrated to values of $\Delta\Psi$. Cells are measured in buffers containing 5 μM VMC and various concentrations of potassium; the concentration of sodium ion is also varied to keep the combined molarity of potassium and sodium at 300 mM. Measurements of $\Delta\Psi$ in *S. aureus* using the ratiometric method are accurate over the range from -50 to -120 mV; the normal $\Delta\Psi$ of the organism was determined to be -113 mV. The technique should be readily applicable to Gram-positive species; however, while it can be used to discriminate metabolically active and CCCP-treated Gram-negative

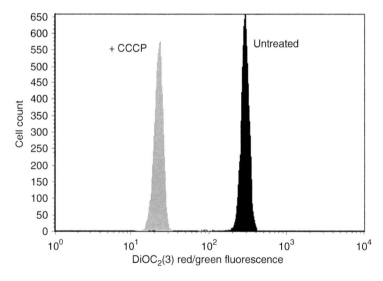

Fig. 5 Red/green fluorescence ratio measurements of $\Delta\Psi$ in CCCP-treated and untreated *S. aureus*. Comparison with Fig. 3 shows the superior separation of populations achieved using the ratiometric method.

organisms after EDTA permeabilization of the outer membrane, it has not, thus far, been possible to construct calibration curves for Gram-negative species. The $DiOC_2(3)$ ratio measurement of $\Delta\Psi$ should also be usable for isolated mitochondria. We have also combined the fluorescence ratio measurement with a measurement of fluorescence of the impermeant red-excited nucleic acid dye TO-PRO-3 to provide a precise ratiometric measurement of membrane permeability (Novo *et al.*, 2000; Shapiro, 2008). Finally, improved measurements of $\Delta\Psi$ in bacteria have been reported by others using the ratio of oxonol fluorescence to side scatter (Amor *et al.*, 2002; Papadimitriou *et al.*, 2007).

III. Improving Cytometry of Membrane Potentials

The limits of accuracy and precision with which cell $\Delta\Psi$s and changes can be measured by FCM are due largely to the population variance of fluorescence intensity, presumably reflecting cell-to-cell differences in numbers of dye-binding sites. This problem is largely eliminated when repeated measurements of the fluorescence of a cyanine dye or other distributional probe of $\Delta\Psi$ are made of the same cell at intervals using a static low-resolution or image cytometer.

Even under the best of circumstances, however, distributional probes could only be used to monitor relatively slow $\Delta\Psi$ changes, occurring over periods of seconds to minutes. Better results would be expected using faster responding dyes, which sense $\Delta\Psi$ by different mechanisms. A number of dyes developed for use by neurophysiologists respond to $\Delta\Psi$ changes by changing their position and/or orientation in the membrane; most do not penetrate to the cell interior. Since all of the transmembrane potential difference is developed across the thickness of the membrane, the electric field strength in the membrane itself can be quite high. The fluorescence or absorption changes seen with fast response dyes are typically small, in the order of a few percent for a 60-mV change in potential, but such changes, which are difficult to detect reliably in a flow cytometer, are readily measured in a static apparatus.

At present, interest among investigators studying transmembrane signaling in eukaryotic cells has shifted from cytometry of $\Delta\Psi$ to cytometry of intracellular pH and calcium. This is due in part to the relative inconstancy and variable magnitude of $\Delta\Psi$ changes associated with surface ligand-receptor interactions, as compared to changes in calcium concentration and distribution and pH, and in part to better probe technology. Both pH and calcium probes suitable for ratiometric measurements of fluorescence using two excitation or emission wavelengths are now available; a ratiometric procedure eliminates the contribution of cell-to-cell differences in dye binding to measurement variance, yielding narrow distributions and making it easier to identify heterogeneity in populations.

In the particular area of lymphocyte activation, calcium flux measurements have proved more useful than $\Delta\Psi$ measurements for analysis of changes occurring within the first minutes following stimulation by mitogens or antigens.

Measurements of cytoplasmic or mitochondrial $\Delta\Psi$ may be useful by 5–12 h, when changes are sufficiently large to detect relatively small activated subpopulations, but measurements of the expression of early activation antigens can also be done at this time, and do not require that the cells be maintained alive and in good condition until FCM analyses are done. Recent work by Waggoner and his colleagues (Hahn *et al.*, 1993) has, however, made it feasible to analyze mitochondrial $\Delta\Psi$ in cells after fixation; they have developed a cyanine derivative which can be covalently bound to mitochondria by photo-induced cross-linking after live cells are incubated with dye and washed.

Loew and his coworkers have developed electrochromic probes of $\Delta\Psi$, which undergo spectral changes in responses to changes in the electric field in the membrane. One such dye, di-4-ANEPPS, has been used for dual excitation beam ratiometric $\Delta\Psi$ measurements in an image cytometer; it yields a 10% change in fluorescence ratio for a 90-mV potential change (Montana *et al.*, 1989). A flow cytometer with helium-cadmium (441 nm) and argon ion (515 nm) laser excitation could be used for di-4-ANEPPS measurements, but it is not clear that the precision needed to detect $\Delta\Psi$ changes of a few tens of millivolts could be readily attained.

Gonzalez and Tsien (1997) introduced another ratiometric method that specifically senses the fast potential response of oxonol molecules in the membrane lipid bilayer using fluorescence resonance energy transfer between the oxonol dye and a fluorescently labeled lectin or phospholipid, allowing fairly precise measurement of cytoplasmic $\Delta\Psi$. Changes in fluorescence ratio exceed 50% for a 100-mV change in $\Delta\Psi$, and responses can be detected in fractions of a millisecond in static cytometers; the technique is also applicable to bulk fluorometry and FCM. However, there is some question as to its suitability for use in bacteria, because the number of oxonol molecules bound within a single bacterial membrane is quite small.

Most recently, it has been possible to engineer voltage-responsive fluorescent proteins (Baker *et al.*, 2008); these have been of interest primarily to investigators working on cells of the nervous system but might be applicable to other cell types.

Even if practical probes usable for ratiometric $\Delta\Psi$ measurements become available, it is, obviously, more appropriate to assess effects of stimuli on cell $\Delta\Psi$ by measuring the same cell at different times, as is done in a static system, than by measuring different cells at different times and assuming their behavior to be similar, a compromise forced on the experimenter by the nature of FCM. It, thus, seems likely that further progress in understanding transmembrane signaling in living cells will come from the refinement of static cytometric techniques for measurement of $\Delta\Psi$ and of other functional parameters such as intracellular pH, calcium flux, and redox state.

References

Amor, K. B., Breeuwer, P., Verbaarschot, P., Rombouts, F. M., Akkermans, A. D., De Vos, W. M., and Abee, T. (2002). Multiparametric flow cytometry and cell sorting for the assessment of viable, injured, and dead bifidobacterium cells during bile salt stress. *Appl. Environ. Microbiol.* **68**, 5209–5216.

Baker, B. J., Mutoh, H., Dimitrov, D., Akemann, W., Perron, A., Iwamoto, Y., Jin, L., Cohen, L. B., Isacoff, E. Y., Pieribone, V. A., Hughes, T., and Knöpfel, T. (2008). Genetically encoded fluorescent sensors of membrane potential. *Brain Cell Biol.* **36**, 53–67.

Crissman, H. A., Hofland, M. H., Stevenson, A. P., Wilder, M. E., and Tobey, R. A. (1988). Use of DiO-C5-3 to improve Hoechst 33342 uptake, resolution of DNA content, and survival of CHO cells. *Exptl. Cell Res.* **174**, 388–396.

Darzynkiewicz, Z. D., Staiano-Coico, L., and Melamed, M. R. (1981). Increased mitochondrial uptake of rhodamine 123 during lymphocyte stimulation. *Proc. Natl. Acad. Sci. USA* **78**, 6696–6698.

Ehrenberg, B., Montana, V., Wei, M.-D., Wuskell, J. P., and Loew, L. M. (1988). Membrane potential can be determined in individual cells from the Nernstian distribution of cationic dyes. *Biophys. J.* **53**, 785–794.

Gonzalez, J. E., and Tsien, R. Y. (1997). Improved indicators of cell membrane potential that use fluorescence resonance energy transfer. *J. Chem. Biol.* **4**, 269–277.

Hahn, K. M., Conrad, P. A., Chao, J. C., Taylor, D. L., and Waggoner, A. S. (1993). A photocross-linking fluorescent indicator of mitochondrial membrane potential. *J. Histochem. Cytochem.* **41**, 631–634.

Hoffman, J. F., and Laris, P. C. (1974). Determination of membrane potentials in human and *Amphiuma* red cells by means of a fluorescent probe. *J. Physiol. (London)* **239**, 519–552.

Johnson, L. V., Walsh, M. L., and Chen, L. B. (1980). Localization of mitochondria in living cells with rhodamine 123. *Proc. Natl. Acad. Sci. USA* **77**, 990–994.

Johnson, L. V., Walsh, M. L., Bockus, B. J., and Chen, L. B. (1981). Monitoring of relative mitochondrial membrane potential in living cells by fluorescence microscopy. *J. Cell Biol.* **88**, 526–535.

Kessel, D., Beck, W. T., Kukuraga, D., and Schulz, V. (1991). Characterization of multidrug resistance by fluorescent dyes. *Cancer Res.* **51**, 4665–4670.

Klapperstück, T., Glanz, D., Klapperstück, M., and Wohlrab, J. (2009). Methodological aspects of measuring absolute values of membrane potential in human cells by flow cytometry. *Cytometry A* **75**, 593–608.

Krasznai, Z., Márián, T., Balkay, L., Emri, M., and Trón, L. (1995). Flow cytometric determination of absolute membrane potential of cells. *J. Photochem. Photobiol. B.* **28**, 93–99.

Lazzari, K.G., Proto, P., and Simons, E.R. (1990). Simultaneous measurement of stimulus-induced changes in cytoplasmic Ca^{2+} and in membrane potential of human neutrophils. *J. Biol. Chem.* **265**, 10959–10967.

Montana, V., Farkas, D. O., and Loew, L. M. (1989). Dual-wavelength ratiometric fluorescence measurements of membrane potential. *Biochem.* **28**, 4536–4539.

Novo, D., Perlmutter, N. G., Hunt, R. H., and Shapiro, H. M. (1999). Accurate flow cytometric membrane potential measurement in bacteria using diethyloxacarbocyanine and a ratiometric technique. *Cytometry.* **35**, 55–63.

Novo, D. J., Perlmutter, N. G., Hunt, R. H., and Shapiro, H.M. (2000). Multiparameter flow cytometric analysis of antibiotic effects on membrane potential, membrane permeability, and bacterial counts of *Staphylococcus aureus* and *Micrococcus luteus.* Antimicrob. *Agents. Chemother.* **44**, 827–831.

Papadimitriou, K., Pratsinis, H., Nebe-von-Caron, G., Kletsas, D., and Tsakalidou, E. (2007). Acid tolerance of Streptococcus macedonicus as assessed by flow cytometry and single-cell sorting. *Appl. Environ. Microbiol.* **73**, 465–476.

Reers, M., Smith, T. W., and Chen, L. B. (1991) J-aggregate formation of a carbocyanine as a quantitative fluorescent indicator of membrane potential. *Biochem.* **30**, 4480–4486.

Seamer, L. C., and Mandler, R. N. (1992). Method to improve the sensitivity of flow cytometric membrane potential measurements in mouse spinal cord cells. *Cytometry* **13**, 545–552.

Shapiro, H. M., Natale, P. J., and Kamentsky, L. A. (1979). Estimation of membrane potentials of individual lymphocytes by flow cytometry. *Proc. Natl. Acad. Sci. USA* **76**, 5728–5730.

Shapiro, H. M. (1981). Flow cytometric probes of early events in cell activation. *Cytometry* **1**, 301–312.

Shapiro, H. M. (1982). *USPat.* No. 4,343,782.

Shapiro, H. M. (1988). "Practical Flow Cytometry", Second Edition. Liss/Wiley, New York.

Shapiro, H. M. (2003). "Practical Flow Cytometry", Fourth Edition. Wiley, Hoboken (NJ). A printable. pdf file may be downloaded free at http://coulterflow.com/bciflow/research01.php.

Shapiro, H. M. (2008). Flow cytometry of bacterial membrane potential and permeability. *Methods Mol. Med.* **142,** 175–186.

Sims, P. J., Waggoner, A. S., Wang, C.-H., and Hoffman, J. F. (1974). Studies on the mechanism by which cyanine dyes measure membrane potential in red blood cells and phosphatidylcholine vesicles. *Biochem.* **13,** 3315–3330.

Smiley, S. T., Reers, M., Mottola-Hartshorn, C., Lin, M., Chen, A., Smith, T. W., SteeleJr., G. D., and Chen, L.B. (1991). Intracellular heterogeneity in mitochondrial membrane potentials revealed by a J-aggregate-forming lipophilic cation JC-1. *Proc. Natl. Acad. Sci. USA* **88,** 3671–3675.

Troiano, L., Ferraresi, R., Lugli, E., Nemes, E., Roat, E., Nasi, M., Pinti, M., and Cossarizza, A. (2007). Multiparametric analysis of cells with different mitochondrial membrane potential during apoptosis by polychromatic flow cytometry. *Nat. Protoc.* **2,** 2719–2727.

CHAPTER 30

Measurement of Intracellular pH

Michael J. Boyer* and David W. Hedley[†]

*Department of Medical Oncology
Royal Prince Alfred Hospital
Sydney, New South Wales 2050
Australia

[†]Departments of Medicine and Pathology
Ontario Cancer Institute/Princess Margaret Hospital
Toronto, Ontario
Canada M4X 1K9

ESSENTIAL CYTOMETRY METHODS
Reprinted from *Methods in Cell Biology*, Volume 41 (Academic Press, 1994).
Copyright © 1994 by Academic Press, Inc.,
DOI: 10.1016/B978-0-12-375045-7.00030-1

I. Introduction

A. Maintenance of Intracellular pH

The level of intracellular pH (pH_i) is of considerable importance to the viability and normal function of mammalian cells. The passive diffusion of ions based on their electrochemical gradients is predicted by the Donnan equilibrium; if these calculations are performed for H^+, the predicted level of pH_i when extracellular pH (pH_e) is 7.4 is approximately 6.4, considerably lower than the measured pH_i under these conditions. It follows, therefore, that mechanisms exist which regulate pH_i and maintain it at a level above that predicted by the electrochemical gradient.

Two major mechanisms allow cells to regulate their pH_i. These are the buffering capacity of the cytosolic and organellar contents and membrane-based transport systems, including the Na^+/H^+ exchanger and anion exchangers. The buffering capacity of a cell is its ability to buffer a change in pH_i following the addition (or removal) of H^+ and is comprised of both bicarbonate-dependent and nonbicarbonate (mainly protein) components (Boron, 1989; Roos and Boron, 1981). Intracellular buffering provides substantial protection for the cell against the effects of an acid load, with most cells capable of buffering millimolar concentrations of H^+ (compared to the submicromolar concentrations that are normally present) (Roos and Boron, 1981).

The Na^+/H^+ exchanger is a membrane-based transport mechanism that is ubiquitous in mammalian cells. The exchanger is a 110-kDa protein whose gene has been cloned from several tissues in different species (Fliegel *et al.*, 1991; Hildebrandt *et al.*, 1991; Reilly *et al.*, 1991; Sardet *et al.*, 1989; Tse *et al.*, 1991). It uses the inwardly directed Na^+ gradient to pump H^+ out of cells, with a 1:1 stoichiometry. Its actions are not directly dependent on the utilization of ATP, although energy is required to maintain the 10-fold inwardly directed Na^+ gradient, which allows the pump to operate. The Na^+/H^+ antiport is reversibly inhibited by amiloride and its more potent analogues such as ethyliso-propylamiloride and hexamethylene amiloride (HMA) (Grinstein and Rothstein, 1986; L'Allemain *et al.*, 1984).

At levels of pH_i close to the normal resting value, the Na^+/H^+ exchanger is relatively quiescent. It becomes activated as pH_i falls below ~7, and there is a rapid increase in its rate of activity at levels of pH between 7 and 6.6. Its activity is also modulated by several other stimuli including hormones, mitogens, and chronic exposure of cells to acidic environments (Horie *et al.*, 1990).

Anion exchange also contributes to regulation of pH_i and at levels of pH_i close to neutrality, it may be more active than Na^+/H^+ exchange in some cell lines. Two different anion exchangers exist, the cation-independent Cl^-/HCO_3^- exchanger and the Na^+-dependent Cl^-/HCO_3^- exchanger. Under physiological conditions, the cation-independent Cl^-/HCO_3^- exchanger allows Cl^- ions to enter the cell in exchange for HCO_3^- ions and therefore acts to decrease pH_i following cytoplasmic alkalination (Cassel *et al.*, 1988; Reinertsen *et al.*, 1988). By contrast, the Na^+-dependent Cl^-/HCO_3^- exchanger acts to protect cells from cytoplasmic acidification (Cassel *et al.*, 1988; Reinertsen *et al.*, 1988). Both of these exchangers are inhibited by the stilbene derivative DIDS.

B. Measurement of pH$_i$

1. Fluorescent pH$_i$ Probes

The measurement of pH$_i$ using fluorescent probes generated *in situ* was first carried out by Thomas *et al.* (1979) using fluorescein and carboxyfluorescein diacetate. These early pH$_i$ probes suffered from considerable limitations including difficulties with cellular uptake, rapid leakage out of cells, and pK_a that was too acidic to make accurate measurements of pH$_i$ at values close to neutrality. In

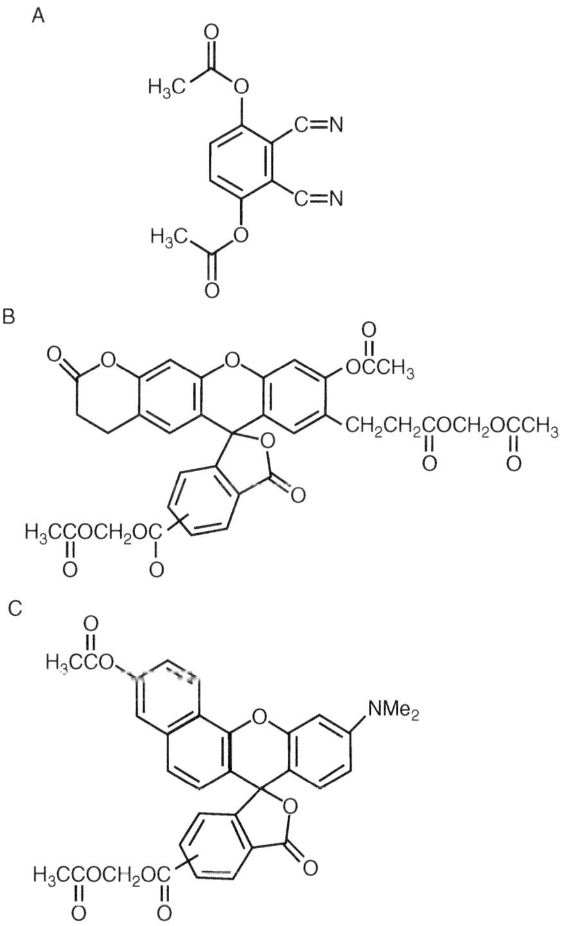

Fig. 1   Structures of (A) 1,4-diacetoxy-2,3-dicyanobenzene (the esterified form of 2,3-dicyanohydroquinone), (B) BCECF-AM, and (C) carboxy-SNARF-1.

addition to the flourescein derivatives, 2,3-dicyanohydroquinone (DCH) (Fig. 1) has been used to measure pH$_i$ using a ratiometric technique (Musgrove *et al.*, 1987; Valet *et al.*, 1981). DCH has a shift in emission wavelength and a decrease in emission intensity with increasing pH. Use of this probe, however, is associated with significant problems including the need for UV excitation, rapid leakage from cells, and the fact that it enters organelles, so that the values of pH$_i$ obtained include a component of organellar pH. Finally, its pK of 8 is not ideal for measurements of pH$_i$ in mammalian cells.

A derivative of carboxyfluorescein, biscarboxyethylcarboxyfluorescein (BCECF) (Fig. 1) overcomes many of these problems. It is well retained within cells for up to 2 h (Musgrove *et al.*, 1987) and has a pK_a of 6.98. After uptake into cells as the acetoxymethyl (AM) ester, BCECF is confined to the cytoplasmic compartment (Paradiso *et al.*, 1984), providing a measure of the pH of this compartment only and not some average value that includes organelles. Although originally introduced for use in fluorometers, BCECF has been used widely in the flow cytometric measurement of pH$_i$.

The acetoxymethyl ester of BCECF (BCECF-AM) is a nonfluorescent molecule that enters cells easily. Once within the cytoplasm the AM groups are cleaved by the action of nonspecific esterases, yielding the highly fluorescent molecule BCECF. Following excitation at 480–500 nm, the emission intensity of BCECF at 525–535 nm is pH dependent with greater intensity at higher pH. The excitation source can be either an argon laser or a mercury arc lamp. In order to make measurements of pH$_i$, a ratio is usually taken between a pH-dependent-emission intensity (e.g., 525 nm) and a pH-independent-emission intensity (e.g., 640 nm). The value obtained is independent of such factors as the amount of dye that was loaded into or leaked from cells.

New probes have become available for the measurement of pH$_i$. The carboxy-seminaphthofluoresceins (SNAFLs) and the carboxyseminaphthorhodafluors (SNARFs) (Fig. 1) were designed to be long wavelength, dual-emission pH indicators with a pK_a near neutrality. The most promising of these for flow cytometry is SNARF-1 which has a pK_a of 7.4 (Haugland, 1992). They are loaded into cells as AM esters in a manner analogous to BCECF. These probes have dual-emission wavelengths, one (587 nm for SNARF-1) which is maximal under acidic conditions and the other (636 nm for SNARF-1) which is maximal under alkaline conditions. This provides the theoretical advantage of increased sensitivity.

2. Calibration

Calibration of fluorescence ratio to pH is carried out by the use of the H$^+$/K$^+$ ionophore, nigericin. Nigericin causes the ratios of intracellular to extracellular hydrogen ion concentration ([H$^+$]$_i$ and [H$^+$]$_e$, respectively) to be equal to the ratios of intracellular to extracellular potassium ion concentration ([K$^+$]$_i$ and [K$^+$]$_e$) (Thomas *et al.*, 1979):

$$\frac{[K^+]_i}{[K^+]_e} = \frac{[H^+]_i}{[H^+]_e}$$

If $[K^+]_i$ and $[K^+]_e$ are equal, then $[H^+]_i$ will be equal to $[H^+]_e$, and hence pH_i can be estimated simply by measureing pH_e. The accuracy of this method of calibration depends on the equality of intra- and extracellular potassium concentrations which typically are not determined experimentally but are assumed to be of the order of 130–140 mM. This is a valid assumption for most cell types, and the method gives values of pH_i which are close to those obtained using other techniques. However, if experiments are to be performed with cells whose $[K^+]_i$ is variable (e.g., excitable cells) or under conditions which might alter $[K^+]_i$, the value must be measured and the appropriate concentrations used for calibration.

3. Measurement Accuracy

The sensitivity of measurements of pH_i is the ability to detect differences in the mean of a population distributed normally in pH_i. The theoretical limit of sensitivity is determined by the magnitude of difference in pH_i represented by one channel number (i.e., the slope of the calibration curve). Calibration curves with BCECF obtained in this laboratory typically have slopes of $\sim$250 channels per pH unit. Although SNARF-1 would be expected to have greater sensitivity than BCECF, because it has an obvious isosbestic point, in practice we have found it to have identical sensitivity (see calibration curves in Fig. 2). These values correspond to a maximum sensitivity of $\sim$0.004 pH units.

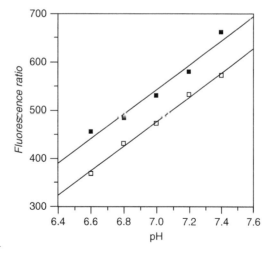

Fig. 2 Calibration curves obtained with BCECF and carboxy-SNARF-1, in EMT-6 cells, using a Coulter Epics Elite cytometer.

The maximum sensitivity described above is not achieved in practice because of lack of resolution. The resolution of measurements of pH_i is the ability to resolve subpopulations of different pH_i within the measured sample and can be estimated by examining the overlap between the ratio histograms of populations differing in pH_i. The breadth of the ratio histogram is usually measured as a coefficient of variation (CV). The CV is a measure of the cell-to-cell variation in intracellular pH, together with variation introduced by the instrument and uncertainties inherent to the technique. The cell-to-cell variation can be virtually eliminated by the use of nigericin in high $[K^+]$ buffer, allowing the resolution of the technique to be assessed.

II. Application

These techniques are widely applicable and have been used to measure pH_i in cell lines, as well as in cells derived from multicellular spheriods and solid tumors. In addition to their use to measure steady-state values of pH_i, these techniques can be adapted to measure the effect of various stimuli on pH_i. They have also been employed to measure the operation of the Na^+/H^+ exchanger. The major limitation to the use of fluorescent pH probes is that some cell lines fail to retain the probes, although this is a relatively uncommon problem.

III. Materials

A. Chemicals

1. BCECF

BCECF-AM is available from Molecular Probes (Eugene, OR). Stock solutions at a concentration of 1 mg/ml in dimethyl sulfoxide are stored at $-20\,°C$.

2. SNARF

cSNARF-AM is available from Molecular Probes (Eugene, OR). Stock solutions at a concentration of 1 mM in dimethyl sulfoxide are stored at $-20\,°C$.

3. DCH

The esterified form of DCH, 1,4-diacetoxy-2,3-dicyanobenzene (ADB) is obtained from Boehringer-Mannheim (Indianapolis, IN). Stock solutions are made at 2 μl/ml in anhydrous dimethyl formamide and stored at $-20\,°C$.

4. Other Chemicals

Nigericin is available from Sigma Chemical Co. (St. Louis, MO). A stock solution of 1 mg/ml in absolute ethanol is stored at −20 °C. Hexamethylene amiloride is available from Research Biochemicals, Inc. (Natick, MA) and a stock solution at a concentration of 1 mM in dimethyl sulfoxide is stored at 4 °C.

B. Buffers

1. Physiological Saline Solution (PSS)

- 140 mM NaCl
- 5 mM KCl
- 5 mM glucose
- 1 mM $CaCl_2$
- 1 mM $MgCl_2$
- 20 mM Tris/MES

Adjust to pH 7.2–7.4 by mixing appropriate amounts of MES-buffered and Tris-buffered solution.

2. NMG Buffer

NMG buffer is identical to PSS, but with isoosmotic replacement of NaCl with N-methyl-D-glucamine (NMG). In preparing this buffer, the NMG should be dissolved first and the pH adjusted to ~7.4 with 10 N HCl. Other components can be added, and the final pH adjusted as described above.

3. NH_4 Buffer

NH_4 buffer is prepared as NMG buffer, but with 130 mM NMG and 10 mM NH_4Cl.

4. High [K^+] Buffer

Solution 1:

- 130 mM KH_2PO_4
- 20 mM NaCl

Solution 2:

- 110 mM K_2HPO_4
- 20 mM NaCl

Mix solutions 1 and 2 to give buffers with a range of pH between 6 and 7.5.

IV. Cell Preparation and Staining: BCECF or SNARF

Following harvest, cells are rinsed once with PSS and then resuspended in the same solution at their final concentration (usually 10^6 cells/ml). Add BCECF-AM to give a final concentration of 2 μl/ml (add 2 μl stock solution per 1 ml of cells[+] buffer). Phosphate-buffered saline or serum-free medium can be used in place of PSS; serum must not be present at this stage since the activity of esterases in serum will cleave BCECF-AM extracellularly, resulting in poor loading.

Incubate cells at 37 °C for 20–30 min to allow cleavage of the AM ester. Although shorter periods can be used because the method is independent of the amount of dye loaded into the cells, we have obtained better results with incubations of this duration.

Remove aliquots of 10^6 cells and centrifuge. Discard the supernatant and resuspend the pellet (which will appear yellow) in 1 ml of PSS (or in another test solution as appropriate). For calibration samples, the pellet should be resuspended in high $[K^+]$ buffer of differing pH. Nigericin 1 μl/ml (1 μl stock solution per 1 ml of cells[+] buffer) should be added 2–3 min prior to measurement of pH_i.

The procedure for SNARF is identical except that SNARF-AM is added rather than BCECF-AM. The final concentration of SNARF should be 5 μM (5 μl stock per 1 ml of cells[+] buffer). Cleavage of SNARF-AM seems to take a little longer than that of BCECF-AM so a minimum incubation of 30 min should be used. The cell pellet will appear pink.

V. Instruments

A flow cytometer capable of generating a ratio signal either in hardware or in software is needed to carry out these measurements. In addition, since measurements of the activity of the Na^+/H^+ exchanger is actually a measurement of rate, time becomes a component of these assays. This is most conveniently done on flow cytometers that are able to use time as a parameter, although it is possible to carry out these experiments by using multiple samples, if the software is not capable of collecting events over time.

A. Measurement of Steady–State pH_i

1. BCECF

Excitation of BCECF is provided by the 488-nm line of an argon laser. When used with an Epics Elite cytometer, power as low as 15–20 mW is adequate for excitation. The resulting fluorescence is separated into high- and low-wavelength components by a 550-nm dichroic filter. These components are further narrowed by passing through a 640-nm band pass filter (10 nm bandwidth) and 525-nm band pass filter (20 nm bandwidth) respectively. The ratio of 525/640 nm fluorescence is measured; this ratio increases with increasing pH_i (Fig. 2).

Measurements of pH$_i$ should be possible with a benchtop flow cytometer equipped with filters for both fluorescein and red wavelengths, provided that the ratio of these two signals can be obtained. However, we have not performed such measurements ourselves.

2. SNARF

As with BCECF, excitation of SNARF is provided by the 488-nm line of an argon laser, at a power of 20–25 mW. The emitted light is passed through a 625 dichroic filter, and the resultant beams are narrowed by passage through 640-nm band pass (10 nm bandwidth) and 580-nm band pass (10 nm bandwidth) filters. The ratio of 640/580 nm fluorescence is measured; this ratio increases with increasing pH (Fig. 2).

3. DCH

Although DCH has been largely superseded by more recent pH$_i$ indicators capable of excitation at 488 nm, the following procedure based on Cook and Fox (1988) can be used.

Cells are loaded with the esterified form of DCH, ADB. This is obtained from Boehringer-Mannheim and made up at 2 mg/ml in anhydrous dimethyl formamide. Add 5 μl to each milliliter of cells in buffer, giving a final concentration of 10 μg/ml (41 μM). Incubate at room temperature for 20 min, and then run on flow cytometer immediately. Instrument setup requires UV excitation, using either a mercury arc or the UV doublet of an argon ion laser, and band pass filters centered around 430 and 480 nm, separated by a suitable dichroic mirror. Calibration is as for BCECF or SNARF-1.

B. Measurement of Na$^+$/H$^+$ Exchange Activity

In order to make these measurements, a flow cytometer which allows the measurement of time as a parameter is desirable (though not essential). Cells are loaded with either BCECF or SNARF-1 using the procedure described above, but NMG buffer is substituted for PSS. Three minutes prior to the end of loading with fluorochrome, nigericin, at a final concentration of 2 μg/ml, is added (2 μl of stock solution per 1 ml of cells$^+$ buffer). This will produce intracellular acidification to a level of ~6.5 in most cell lines; the absence of extracellular Na$^+$ prevents activity of the Na$^+$/H$^+$ exchanger. Following the conclusion of incubation, the sample is centrifuged and resuspended in a small volume (100–200 μl) of NMG buffer. The action of the Na$^+$/H$^+$ exchanger is commenced by the addition of 1 ml of PSS, at which time measurement should begin, using the technique described above. Inhibition of the exchanger can be produced by adding HMA at a final concentration of 1–10 μM.

C. Measurement of Intracellular Buffering Capacity

Cells are loaded with fluorochrome as described above. Following centrifugation, the pellet is resuspended in 1 ml NMG buffer, and measurement of pH_i is commenced. After a sufficient number of cells have been measured, and (ideally) while the sample is still running, 1 ml of NH_4 buffer is added to the sample. A boost should be applied to the sample so that measurement of pH_i continues immediately after the addition of NH_4. Continue making measurements until sufficient events have been collected, but not for longer than 1 min.

Intrinsic (nonbicarbonate) buffering (β_I) capacity is given by the formula $\beta_I = \Delta[NH_4^+]_i/\Delta pH_i$, where $\Delta[NH_4^+]_i$ and ΔpH_i are the changes which occur in intracellular $[NH_4^+]$ and intracellular pH, respectively, following the addition of NH_4 buffer. The change in pH_i is measured. If it is assumed that $[NH_4^+]_i$ prior to the addition of NH_4 buffer is 0, then $\Delta[NH_4^+]_i$ is simply the value of $[NH_4^+]_i$ after the change in buffer and is given by the equation (Roos and Boron, 1981):

$$[NH_4^+]_i = \frac{[NH_4Cl]_e \times 10^{(pH_e - pK)}}{(1 + 10^{(pH_e - pK)}) \times 10^{(pH_i - pK)}}.$$

The pK of NH_4^+ is 9.3.

D. Calibration

Cells are loaded with BCECF or SNARF in PSS, as described above. At the conclusion of incubation, the cell samples are centrifuged and the pellet is resuspended in high $[K^+]$ buffers at several different levels of pH. At least four high $[K^+]$ calibration solutions should be prepared with pH increasing in 0.2–0.3 pH unit increments; ideally these should have values of pH centered around the values expected to be measured in the experimental samples. Following addition of nigericin (to equilibrate pH_i and pH_e) fluorescence is measured.

The calibration curve is obtained by plotting the mean fluorescence ratio of samples measured in high $[K^+]$ buffer and nigericin against pH (Fig. 2). At levels of pH_i close to the pK_a of the probe, the calibration curve is linear, and its equation may be obtained by linear regression. Outside this range, the calibration curve is sigmoidal. Although computerized curve-fitting programs are available which will provide the equation of such a curve, the appropriate selection of pH probes based on knowledge of their pK can prevent this problem.

VI. Critical Aspects

The success of techniques to measure pH_i depends upon obtaining a fluorescence signal of adequate strength. The major causes of weak signal strength are poor intracellular loading or rapid leakage of the fluorescent probe. Poor loading of esterified probes may be due to extracellular deesterification which may occur if serum (which may contain esterases) is present in the buffer during the exposure of cells to the probes. The use of serum-free buffers should overcome this problem. Another cause of a weak signal is poor intracellular hydrolysis of esterified probes if experiments are carried out at temperatures below 37 °C. If it is necessary to carry out experiments at low temperature, increasing the concentration of BCECF-AM or SNARF-AM may overcome the problems of slow or poor loading.

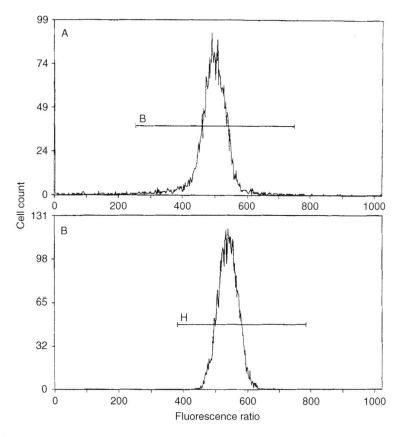

Fig. 3 Fluorescence ratio histograms for a population of EMT-6 cells, suspended in physiological saline buffer at extracellular pH 7.30, obtained with the use of (A) BCECF or (B) carboxy-SNARF-1.

The accuracy of the values of pH$_i$ obtained is dependent upon the calibration procedure. As pointed out above, calibration is based on the assumption that intra- and extracellular [K$^+$] are equal. Under some conditions (e.g., in excitable cells, or in the presence of inhibitors of ion transporter) this assumption may not be valid, and use of the calibration procedure as described will result in a systematic error in the values of pH$_i$. This difficulty can be overcome by measuring the intracellular concentration of [K$^+$] and adjusting the calibration buffers appropriately.

The methods described in this chapter outline the measurement of pH$_i$ under bicarbonate-free conditions. The presence of bicarbonate may result in changes in the steady-state level of pH$_i$ as a result of the activation of additional pH$_i$ regulating systems (Ganz *et al.*, 1989). Whether measurements of pH$_i$ should be made in the presence of bicarbonate will depend on the purpose of the experiment; if an

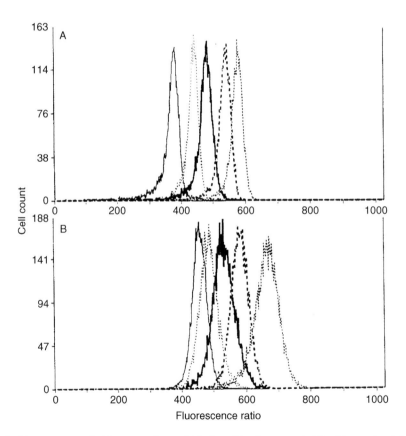

Fig. 4 Fluorescence ratio histograms from which the calibration curve in Fig. 2 is constructed. The peaks in each figure (from left to right) are at pH 6.6, 6.8, 7, 7.2, and 7.4, and are shown for (A) BCECF and (B) carboxy-SNARF-1.

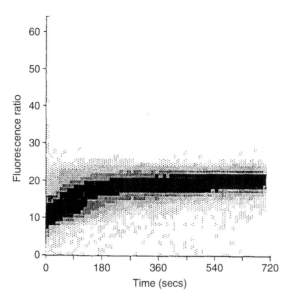

Fig. 5 Change in fluorescence ratio (which is proportional to intracellular pH) in acidified EMT-6 cells following the addition of Na$^+$. The slope of the curve is a measure of the activity of the Na$^+$/H$^+$ exchanger.

estimate of pH$_i$ is desired under physiological conditions, experiments should be carried out with bicarbonate. In order to carry out experiments in the presence of bicarbonate, the composition of buffers must be altered accordingly. This will also have an impact on buffering capacity (Boron, 1989). Care should be taken when working with bicarbonate containing solutions to prevent its loss into the atmosphere, thus lowering the concentration of bicarbonate in the buffer.

VII. Results and Discussion

The fluorescence ratio histogram shown in Fig. 3A is from a population of EMT-6 cells stained with BCECF and suspended in PSS. The mean pH$_i$ (in the absence of bicarbonate) is 7.07 and the CV is 4.12%. These are typical values for mammalian cells, although under different environmental conditions (e.g., reduced extracellular pH) different values are obtained. The use of SNARF-1 to make the same measurements is shown in Fig. 3B; pH$_i$ is 6.99 and the CV is 4.39%.

Figure 4 shows histograms obtained during the construction of a calibration curve with BCECF (a), SNARF-1 (b); the calibration curve is shown in Fig. 2. Two points are of note. First, the slope of the calibration curves for BCECF and SNARF-1 are almost identical (253 and 254 channels/pH unit, respectively). Based on these data, the maximum sensitivity of the measurements are of the order of 0.003 pH units. This compares well with previously reported values.

Second, the decrease in CV due to nigericin is apparent (compare with Fig. 3A and B). The resolution based on these data is dependent on the level of pH. The resolution with BCECF for values of pH_i below 7 is ~0.1 pH units, while for values above 7 it is ~0.2 pH units. For SNARF-1, the resolution above pH_i 7 is ~0.1 pH units, while below 7 it is ~0.2 pH units. Therefore, BCECF appears to be the probe of choice for when values of pH_i below 7 are being measured, while above this level, SNARF-1 is a better choice.

An application of the technique is shown in Fig. 5. This is a plot of fluorescence ratio (pH_i) against time for EMT-6 cells following experimental acidification in Na^+-free medium and then recovery as a result of the activity of the Na^+/H^+ exchanger following addition of Na^+. The slope of the curve is a measure of the activity of this exchanger. The slope can be obtained by measuring the mean fluorescence ratio during several time intervals.

References

Boron, W. F. (1989). *In* "The Regulation of Acid-Base Balance" (D. W. Seldin and G. Giebisch, eds.), pp. 33–56. Raven Press, New York.

Cassel, D., Scharf, O., Rotman, M., Cragoe, E. J., Jr., and Katz, M. (1988). *J. Biol. Chem.* **263**, 6122–6127.

Cook, J. A., and Fox, M. H. (1988). *Cytometry* **9**, 441–447.

Fliegel, L., Sardet, C., Pouyssegur, J., and Barr, A. (1991). *FEBS Lett.* **279**, 25–29.

Ganz, M. B., Boyarsky, G., Sterzel, R. B., and Boron, W. F. (1989). *Nature (Lond.)* **337**, 648–651.

Grinstein, S., and Rothstein, A. (1986). *J. Membr. Biol.* **90**, 1–12.

Haugland, R. P. (1992). "Handbook of Fluorescent Probes and Research Chemicals," 2nd edn., Molecular Probes, Eugene, OR.

Hildebrandt, F., Pizzonia, J. H., Reilly, R. F., Reboucas, N. A., Sardet, C., Pouyssegur, J., Slayman, C. W., Aronson, P. S., and Igarashi, P. (1991). *Biochim. Biophys. Acta* **1129**, 105–108.

Horie, S., Moe, O., Tejedor, A., and Alpern, R. J. (1990). *Proc. Natl. Acad. Sci. USA* **87**, 4742–4745.

L'Allemain, G., Franchi, A., Cragoe, E. J., Jr., and Pouyssegur, J. (1984). *J. Biol. Chem.* **259**, 4313–4319.

Musgrove, E., Rugg, C., and Hedley, D. (1987). *Cytometry* **7**, 347–355.

Paradiso, A. M., Tsien, R. Y., and Machen, T. E. (1984). *Proc. Natl. Acad. Sci. USA* **81**, 7436–7440.

Reilly, R. F., Hildebrandt, F., Biemesderfer, D., Sardet, C., Pouyssegur, J., Aronson, P. S., Slayman, C. W., and Igarashi, P. (1991). *Am. J. Physiol.* **261**, F1088–F1094.

Reinertsen, K. V., Tonnessen, T. I., Jacobsen, J., Sandvig, K., and Olsnes, S. (1988). *J. Biol. Chem.* **263**, 11117–11125.

Roos, A., and Boron, W. F. (1981). *Physiol. Rev.* **61**, 296–434.

Sardet, C., Franchi, A., and Pouyssegur, J. (1989). *Cell (Cambridge, MA)* **56**, 271–280.

Thomas, J. A., Buchsbaum, R. N., Zimniak, A., and Racker, E. (1979). *Biochemistry* **18**, 2210–2218.

Tse, C. M., Ma, A. I., Yang, V. W., Watson, A. J., Levine, S., Montrose, M. H., Potter, J., Sardet, C., Pouyssegur, J., and Donowitz, M. (1991). *EMBO J.* **10**, 1957–1967.

Valet, G., Raffael, A., Moroder, L., Wunsch, E., and Ruhenstroth-Bauer, G. (1981). *Naturwissenschaften* **68**, 265–266.

CHAPTER 31

Intracellular Ionized Calcium

Carl H. June* and Peter S. Rabinovitch†

*Department of Immunobiology
Naval Medical Research Institute
Bethesda, Maryland 20889

†Department of Pathology
University of Washington
Seattle, Washington 98121

ESSENTIAL CYTOMETRY METHODS
Reprinted from *Methods in Cell Biology*, Volume 41 (Academic Press, 1994).
Copyright © 1994 by Academic Press, Inc.,
DOI: 10.1016/B978-0-12-375045-7.00031-3

I. Introduction

Measurement of intracellular ionized calcium concentration ($[Ca^{2+}]_i$) in living cells is of considerable interest to investigators over a broad range of cell biology. Calcium has an important role in a number of cellular functions and, perhaps most interestingly, can transmit information from the cell membrane to regulate diverse cellular functions. An optimal indicator of $[Ca^{2+}]_i$ should span the range of calcium concentrations found in quiescent cells ($\sim$100 nM) to levels measured in stimulated cells (micromolar free Ca^{2+}), with greatest sensitivity to small changes at the lower end of that range. The indicator should freely diffuse throughout the cytoplasm and be easily loaded into small cells. The response of the indicator to transient changes in $[Ca^{2+}]_i$ should be rapid. Finally, the indicator itself should have little or no effect upon $[Ca^{2+}]_i$ itself or on other cellular functions.

Until 1982, it was not possible to measure $[Ca^{2+}]_i$ in small intact cells, and attempts to measure cytosolic free calcium were restricted mostly to large invertebrate cells where the use of microelectrodes was possible. Bioluminescent indicators such as aequorin, a calcium-sensitive photoprotein, are well suited for certain applications (Blinks *et al.*, 1982). Their greatest limitation is the necessity for loading into cells by microinjection or other forms of plasma membrane disruption. $[Ca^{2+}]_i$ was first measured in populations of small nonadherent cells with the development of quin2 (Tsien *et al.*, 1982). The indicator was easily loaded into intact cells using a chemical technique developed by Tsien (1981). Cells are incubated in the presence of the acetoxymethyl ester of quin2. This uncharged form is cell permeant and diffuses freely into the cytoplasm where it serves as a substrate for esterases. Hydrolysis releases the tetraanionic form of the dye which is trapped inside the cell. Unfortunately, quin2 has several disadvantages that limit its application to flow cytometry (Ransom *et al.*, 1986). A relatively low extinction coefficient and quantum yield have made detection of the dye at low concentrations difficult; at higher concentrations, quin2 itself buffers the $[Ca^{2+}]_i$. Grynkiewicz *et al.* (1985) have described a family of highly fluorescent calcium chelators which overcome most of the above limitations. One of these dyes, indo-1, has spectral properties which make it especially useful for analysis with flow cytometry. In particular, indo-1 exhibits large changes in fluorescent emission wavelength upon calcium binding (Fig. 1). As described below, use of the ratio of intensities of fluorescence at two wavelengths allows calculation of $[Ca^{2+}]_i$ to be made independent of variability in cellular size or intracellular dye concentration. Although a single visible light-excited ratiometric dye has yet to be developed, several alternatives have been developed that have proven especially useful for analysis requiring visible excitation (Table I). Nonratiometric dyes that exhibit *increased* fluorescence upon binding of calcium include fluo-3 (Minta *et al.*, 1989) and a series of dyes developed by Molecular Probes: calcium green, calcium orange, and calcium crimson. Fura-Red exhibits *decreased* fluorescence intensity upon calcium binding and, as described below, can be combined with simultaneous

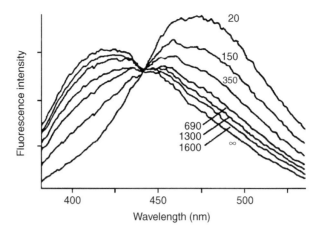

Fig. 1 Emission spectra for indo-1 as a function of ionized calcium. Indo-1 (3 μM) was added to a buffer consisting of 21.64 mM K_2H_2EGTA, 100 mM KCl, 20 mM Hepes, pH 7.20. Small aliquots of a buffer that contained equimolar calcium and EGTA as $K_2CaEGTA$ and was otherwise identical to the first buffer were added. Fluorescence excited at 356 nm was measured with a spectrofluorimeter.

Table I
Calcium Indicator Dyes Useful for Flow Cytometric Applications

Indicator	Emission response to elevated calcium	Excitation wavelength (nm)	Emission wavelength(s) (nm)	Calcium affinity, K_d (nm)	
				22 °C	37 °C
Indo-1	ΔRatio	325–360	390/520		~250
Fluo-3	Increase	488	530	~400	~860
Calcium green-1	Increase	488	530	~250	
Calcium orange	Increase	550	575	~330	
Calcium crimson	Increase	590	610	~200	
Fura-Red	Decrease	488	660	~400	
Fluo-3/Fura-Red	ΔRatio	488	530/660	~400	

cellular loading of fluo-3 to provide a sensitive form of ratiometric analysis using 488-nm excitation. Improvement in available calcium probes that exhibit different fluorescence properties now permits single-cell measurements of $[Ca^{2+}]_i$ in large number of cells with flexibility that was not possible previously. The use of a flow cytometer allows the correlation of $[Ca^{2+}]_i$ with other cell parameters such as surface antigen expression and cell cycle and, furthermore, allows one to electronically sort cells based upon $[Ca^{2+}]_i$.

II. Flow Cytometric Assay with Indo-1

A. Loading of Cells with Indo-1

Uptake and retention of indo-1 ([1-[2 amino-5-[6-carboxylindol-2-yl]-phenoxy]-2-[2'-amino-5'-methylphenoxy]ethane N,N,N',N'-tetraacetic acid] are facilitated by the use of the pentaacetoxymethyl ester of indo-1, using the scheme described above. Approximately 20% of the total dye is trapped in this manner during typical loadings. After loading, the extracellular indo-1 should be diluted 10- to 100-fold before flow cytometric analysis (Rabinovitch et al., 1986). One incidental benefit of this loading strategy is that this procedure, like the more familiar use of fluorescein diacetate or carboxyfluorescein diacetate, allows one to distinguish between live and dead cells. The latter will not retain the hydrophilic impermeant dye and can be excluded during subsequent analysis.

The lower limit of useful intracellular loading concentrations of indo-1 is determined by the sensitivity of fluorescence detection of the flow cytometer and the upper limit is determined by avoidance of buffering of $[Ca^{2+}]_i$ by the presence of the calcium chelating dye itself. In practice, one should use the least amount of indo-1 that is necessary to reliably quantitate the fluorescence signal. Fortunately, indo-1 has excellent fluorescence characteristics (30-fold greater quantum yield than quin2 (Grynkiewicz et al., 1985)) and useful ranges of indo-1 loading are much lower than the millimolar amounts required with quin2. For human peripheral blood T cells, we have found adequate detection at or above 1 μM indo-1 ester, under conditions that achieve ~5 μM intracellular indo-1. Buffering of $[Ca^{2+}]_i$ in human T cells was observed as a slight delay in the rise in $[Ca^{2+}]_i$ and a retarded rate of return of $[Ca^{2+}]_i$ to baseline values when loading concentrations above 3 μM (22 μM intracellular concentration) were used. A reduction in peak $[Ca^{2+}]_i$ occurred at even higher indo-1 concentrations (Rabinovitch et al., 1986). Chused et al. (1987) have observed slightly greater sensitivity of murine B cells to indo-1 buffering, recommending a loading concentration of no greater than 1 μM. In side-by-side comparisons, we have found that calcium transients in B cells are much more sensitive to the effects of buffering by indo-1 than T cells. For human platelets, a 2 μM loading concentration has been reported (Davies et al., 1988). Rates of loading of the indo-1 ester can be expected to vary between cell types, perhaps as a consequence of variations in intracellular esterase activity, as well as varying abilities of cells to extrude the cell-impermeant form. In peripheral human blood, more rapid rates of loading are seen in platelets and monocytes than in lymphocytes. Even within one cell type, donor or treatment-specific factors may affect loading; for example, lower rates of indo-1 loading were seen in splenocytes from aged than from young mice (Miller et al., 1987).

Indo-1 has been found to be remarkably nontoxic to cells subsequent to loading. Analysis of the proliferative capacity of human T lymphocytes loaded with indo-1 (Table II) has shown no adverse effects on the ability of cells to enter and complete three rounds of the cell cycle. Similar results have been obtained with murine B

Table II
Proliferation of Peripheral Blood Lymphocytes Loaded with Indo-1[a]

Day of culture	Indo-1/AM (μM)	Percentage of cells			
		Noncycling	1st cycle	2nd cycle	3rd cycle
2	0	69.8 ± 1.3	28.4 ± 0.8	1.9 ± 0.7	–
	3	70.7 ± 1.8	27.5 ± 1.2	1.5 ± 0.8	–
3	0	58.1 ± 2.5	14.3 ± 0.5	26.5 ± 2.7	1.2 ± 0.4
	3	59.5 ± 2.6	14.9 ± 1.1	24.7 ± 3.4	0.9 ± 0.3
4	0	43.8 ± 0.7	8.4 ± 0.5	29.5 ± 1.0	18.4 ± 1.7
	3	44.6 ± 0.8	9.5 ± 0.9	28.6 ± 1.4	17.3 ± 3.0

[a]Peripheral blood lymphocytes were either loaded or not loaded with indo-1 ester and cultured with PHA (10 μg/ml) and BrdU (1×10^{-4} M). The cells were harvested on the indicated days, stained with Hoechst 33258, and analyzed by flow cytometry, and the percentage of cells (mean ± S.E.M., $n = 4$) in each cell cycle was quantitated as described (Kubbies et al., 1985).

lymphocytes (Chused et al., 1987). This is especially pertinent to the sorting of indo-1 loaded cells based on [Ca^{2+}]$_i$ as described subsequently.

B. Instrumental Technique

The absorption maximum of indo-1 is between 330 and 350 nm, depending upon the presence of calcium (Grynkiewicz et al., 1985); this is well-suited to excitation at either 351–356 nm from an argon ion laser or 337–356 nm from a krypton ion laser. Laser power requirements depend upon the choice of emission filters and optical efficiency of the instrument; however, it is our experience that although 100 mW is often routinely employed, virtually identical results can be obtained with 10 mW. The use of helium-cadmium lasers has been compared to argon laser excitation, and little difference observed for analysis in the case of indo-1 (Goller and Kubbies, 1992). Stability of the intensity of the excitation source is less important in this application than in many others, because of the use of the ratio of fluorescence emissions. Analysis with indo-1 has also been performed using excitation by a mercury arc lamp (FACS Analyzer, Becton Dickinson). It is instructive to consult previous work using quin2 for flow cytometry in order to appreciate how much the current probes have improved over previous generations (Ransom et al., 1987).

The cellular [Ca^{2+}]$_i$ signaling response is an active process and is highly temperature dependent in most cell types. Thus, the sample chamber must usually be kept warm, and the time that cells spend in cooler tubing in transit to the flow cell must be kept minimal (<10 s), or else the sample tubing must be kept warm also. The agonist is ordinarily introduced into the sample by quickly ceasing flow, removing the sample container, adding agonist, and restarting flow, "boosting" the new sample to the flow cell quickly. With practice, this procedure can be completed in less than 20 s. If more rapid $[Ca^{2+}]_i$ transients after agonist addition must be analyzed, then various agonist injection methods may be utilized so that disruption of sample flow is not required. A commercial device is available for on-line addition of agonist using a syringe (Cytek Development, Fremont, CA).

An increase in $[Ca^{2+}]_i$ is detected with indo-1 as an increase in the ratio of fluorescence intensity at a lower to a higher emission wavelength. The optimal strategy is to select band pass filters so that one minimizes the collection of light near the isosbestic point and maximizes collection of fluorescence that exhibits the largest variation in calcium-sensitive emission. The choice of filters used to select these wavelengths is dictated by the spectral characteristics of the shift in indo-1 emission upon binding to calcium (Fig. 1). The original spectral curves published for indo-1 (Grynkiewicz et al., 1985) did not depict the large amounts of indo-1 emission in the blue-green and green wavelengths; in practice, on flow cytometers, we find that there is more light available in the blue region than in the violet region, although the "information" content of the light from the violet region exceeds that of the blue region (Fig. 1). Commercially available band pass filters for analysis with indo-1 are usually centered on the "violet" emission of the calcium-bound indo-1 dye (405 nm) and calcium-free indo-1 dye "blue" emission (485 nm). However, we have found these wavelengths to be suboptimal and that a larger dynamic range in the ratio of wavelengths is obtained if "blue" emission below 485 nm is not collected and the center of the blue emission band pass filter is instead moved upward. Similarly, the violet band pass filter should be chosen to minimize the collection of wavelengths above 405 nm. Thus, in order to optimize the calcium signal, it is important to exclude light from analysis that is near or at the isosbestic point. This effect is summarized in Table III by values R_{max}/R, which indicates the range of change in indo-1 ratio observed from resting intracellular calcium to saturated calcium.

Table III
Effect of Wavelength Choice on Calcium–Sensitive Indo–1 Ratio Shifts[a]

Wavelength pair (nM)	R_{max}	R_{min}	R	R_{max}/R_{min}	R_{max}/R
475/395	2.33	0.040	0.352	58.2	6.62
475/405	2.38	0.100	0.410	23.8	5.80
495/395	3.51	0.048	0.429	73.1	8.18
495/405	3.59	0.119	0.501	30.2	7.17
515/395	5.75	0.070	0.644	82.1	8.93
515/405	5.88	0.176	0.752	33.4	7.82
530/395	**9.68**	**0.117**	**1.073**	**82.7**	**9.02**
530/405	9.89	0.292	1.252	33.9	7.90

[a]By spectrofluorimetry, 2-nm slit width; uncorrected fluorescence. The wavelength choice shown in bold demonstrated the best ratio shift.

C. Calibration of Ratio to $[Ca^{2+}]_i$

Prior to the development of indo-1, $[Ca^{2+}]_i$ determination with quin2 fluorescence was sensitive to cell size and intracellular dye concentration as well as $[Ca^{2+}]_i$. This made necessary calibration at the end of each individual assay by determination of the fluorescence intensity of the dye at zero and saturating Ca^{2+}. In contrast, with indo-1 use of the Ca^{2+}-dependent shift in dye emission wavelength allows the ratio of fluorescence intensities of the dye at the two wavelengths to be used to calculate $[Ca^{2+}]_i$:

$$[Ca^{2+}]_i = K_d \frac{(R - R_{min})S_{f2}}{(R_{max} - R)S_{b2}}, \tag{1}$$

where K_d is the effective dissociation constant (250 nM); R, R_{min}, and R_{max} are the fluorescence intensity ratios of violet/blue fluorescence at resting, zero, and saturating $[Ca^{2+}]_i$, respectively; and S_{f2}/S_{b2} is the ratio of the blue fluorescence intensity of the calcium-free and bound dye, respectively (Grynkiewicz et al., 1985). Because this calibration is independent of cell size and total intracellular dye concentration as well as instrumental variation in efficiency of excitation or emission detection, it is not necessary to measure the fluorescence of the dye in the calcium-free and saturated states for each individual assay. In principle, it is sufficient to calibrate the instrument once after setup, and tuning by measurement of the constants R_{max}, R_{min}, S_{f2}, and S_{b2}, after which only R is measured for each subsequent analysis on that occasion. It is important to note that the apparent K_d of indo-1 was measured at 37 °C at an ionic strength of 0.1 at pH 7.08; this value will change significantly at different temperatures, pH, and ionic strength. For example, the K_d of Fura-2 changes from 225 to 760 nM when ionic strength is changed from 0.1 to 0.225. Similar values for indo-1 have not been published.

One strategy to obtain the R_{max} and R_{min} values for indo-1 is to lyse cells in order to release the dye to determine fluorescence at zero and saturating $[Ca^{2+}]_i$, as is carried out in fluorimeter-based assays with quin2 and Fura-2. However, this is not possible with flow cytometry, due to the loss of cellular fluorescence. Another strategy is the use of an ionophore to saturate or deplete $[Ca^{2+}]_i$ in order to allow approximation of the true end points without cell lysis. For this approach the ionophore ionomycin is best suited, due to its specificity for calcium and low fluorescence. When flow cytometric quantitation of fluorescence from intact cells treated with ionomycin or ionomycin plus EGTA was compared with spectrofluorimetric analysis of lysed cells in medium with or without EGTA, the indo-1 ratio of unstimulated cells (R) and the ratio at saturating amounts of Ca^{2+}, R_{max}, were similar by both techniques (Rabinovitch et al., 1986). The latter indicates that ionomycin-treated cells reach near-saturating levels of $[Ca^{2+}]_i$. The value of R_{min} which is obtained by treatment of intact cells with ionomycin in the presence of EGTA, however, is substantially higher than either that predicted from the spectral emission curves (Fig. 1) or that obtained by cell lysis and spectrofluorimetry. Spectrofluorimetric quantitation with either quin2 or indo-1 indicates that $[Ca^{2+}]_i$

remains at approximately 50 nM in intact cells treated with ionomycin and EGTA. Thus, due to the inability to obtain a valid flow cytometric determination with calcium-free dye, we have used for calibration the values of R_{min} and S_{f2} or S_{b2} derived from spectrofluorimetry, of either the indo-1 pentapotassium salt or, after lysis, cells loaded with indo-1 acetoxymethyl ester in the presence of EGTA. It is essential that the same optical filters be used for flow cytometry and spectrofluorimetry, since the standardization is very sensitive to the wavelengths chosen. Typical values of R_{max}, R_{min}, R, and S_{f2}/S_{b2} are shown for different emission wavelength combinations in Table III. Even if careful calibration of the fluorescence ratio to $[Ca^{2+}]_i$ is not being performed for a particular experiment, ordinary quality control can include a determination of the value of R_{max}/R. A limited range of R_{max}/R values should be obtained since unperturbed cells are routinely found to have a reproducible value of $[Ca^{2+}]_i$ and day-to-day optical variations in the flow cytometer are usually minimal (with the same filter set). The R_{max}/R values obtained on the Ortho Cytofluorograph and on Coulter instruments are typically in the range of 6–8; for unexplained reasons, several groups have reported that the value obtained on Becton-Dickinson instruments is only about four.

Chused *et al.* (1987) have suggested that metabolic inactivation of cells by the use of a cocktail of nigericin, high concentrations of potassium, 2-deoxyglucose, azide, and carbonyl cyanide *m*-chloro-phenylhydrazone and the calcium ionophore ionomycin can collapse the calcium gradient to zero and, therefore, that $[Ca^{2+}]_i = [Ca^{2+}]_o$. To avoid the apparent impossibility of assessing R_{min} in intact cells, the calibration is based upon a regression formula that relates R to ionomycin-treated cells suspended in a series of precisely prepared calcium buffers. Thus, this technique allows one to estimate $[Ca^{2+}]_i$ without the need to determine R_{min}, S_{f2}, S_{b2}, although it is subject to the limitation that calcium concentrations that are less than those found in resting cells cannot be quantitated and, further, by the precision with which one can prepare a series of calcium buffers that yield known and reproducible free calcium concentrations.

Accuracy of prediction of ionized Ca^{2+} concentration in buffer solutions is dependent upon a variety of interacting factors, so that care must be exercised in formulating Ca^{2+} standards. The ionized calcium concentration in an EGTA buffer system is dependent upon the magnesium concentration; other metals such as aluminum, iron, and lanthanum also bind avidly to EGTA. In addition, the dissociation constant of Ca^{2+}-EGTA is a function of pH, temperature, and ionic strength (Blinks *et al.*, 1982; Harafuji and Ogawa, 1980). For example, in an EGTA buffer (total EGTA 2 mM, total Ca^{2+} 1 mM, ionic strength 0.1 at 37 °C), changing the pH from 7.4 to 7 can result in the ionized calcium increasing by more than 200 nM, a change that is approximately twice the magnitude of that found in resting cells, and is easily measured on a flow cytometer. Finally, it is important to prepare the buffers using the "pH metric technique," in part because of the varying purity of commercially available EGTA. Buffer solutions suitable for calibration are available (Molecular Probes, Eugene, OR).

As an alternative to the above approaches, there are several potential schemes for use of the flow cytometer to directly analyze the fluorescence of dye in solution; the fluorescence measurement must be converted to a brief pulse for processing by the flow cytometer, either by strobing the exciting laser beam or by chopping the PMT output signals (Kachel *et al.*, 1990).

D. Display of Results

It is possible to display the data as a bivariate plot of the inversely correlated "violet" versus "blue" signals derived from each cell. Thus, the increase in ratio seen with increased $[Ca^{2+}]_i$ will be observed as a rotation about the axis through the origin. This method of ratio analysis is cumbersome, and fortunately, commercial flow cytometers all have some provision for a direct calculation of the value of the fluorescence ratio itself, either by analog circuitry or by digital computation.

Plotted as a histogram of the ratio values, quiescent cell populations show narrow distributions of ratio, even when cellular loading with indo-1 is very heterogeneous, and coefficients of variation of less than 10% are not uncommon (Fig. 2). The effects of perturbation of $[Ca^{2+}]_i$ by agonists can be noted by changes in the ratio histogram profiles by storing histograms sequentially with subsequent analysis of data. The above approach results in a linear display of indo-1 blue and violet fluorescence. If cellular indo-1 loading is extremely heterogenous, it may be desirable to work with a logarithmic conversion of "violet" and "blue" emission intensities in order to observe a broader range of cellular fluorescence. In this case, the hardware must permit the logarithm of the ratio to be calculated by *subtraction* of the log blue from the log violet signals (Rabinovitch *et al.*, 1986).

A more informative and elegant display is obtained by a bivariate plot of ratio versus time. The bivariate histogram can be stored and the data displayed as "dot plots" on which the indo-1 ratio or each cell (proportional to $[Ca^{2+}]_i$) is plotted on the y-axis versus time on the x-axis (Fig. 3A). Alternatively, the data can be subjected to further analysis for presentation as "isometric plots" in which the x-axis represents time; the y-axis, $[Ca^{2+}]_i$; and the z-axis, number of cells (Fig. 3B). In these bivariate plots, kinetic changes in $[Ca^{2+}]_i$ are seen with much greater resolution, limited only by the number of channels on the time axis, the interval of time between each channel, and the rate of cell analysis. Changes in the fraction of cells responding, in the mean magnitude of the response, and in the heterogeneity of the responding population are best observed with these displays. For example, it can be seen in Fig. 3 that not all murine thymocytes respond to stimulation by anti-CD3, and of those that do, the values of $[Ca^{2+}]_i$ are quite heterogeneous.

Calculation of the mean y-axis value for each x-axis time interval allows the data to be presented as mean ratio versus time (Rabinovitch *et al.*, 1986). Calibration of the ratio to $[Ca^{2+}]_i$ allows data to be presented in the same manner as traditionally displayed by spectrofluorimetric analysis, that is, mean $[Ca^{2+}]_i$ versus time. While this presentation yields much of the information of interest in an easily displayed format, data relating to heterogeneity of the $[Ca^{2+}]_i$ response is lost. Some of this

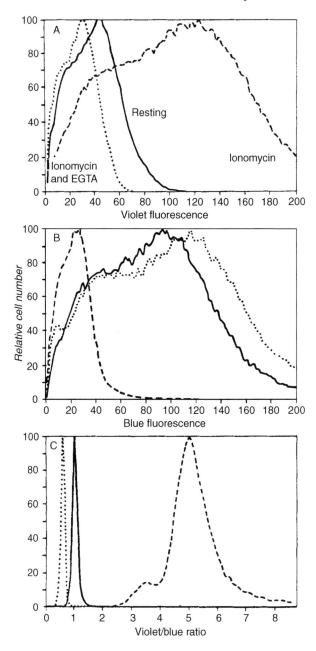

Fig. 2 Histograms of (A) violet (405 nm) fluorescence, (B) blue (485 nm) fluorescence, and (C) the ratio of violet/blue fluorescence of peripheral blood lymphocytes loaded with indo-1. Cells were analyzed in a basal state (solid lines), in the presence of 3 μM ionomycin (dashed lines), or in the presence of ionomycin and EGTA, used to reduce free calcium in the medium to ∼20 nM (dotted lines). Data are plotted on linear scales; the violet/blue ratio was normalized to unity for resting cells. Reproduced from Rabinovitch *et al.* (1986), by copyright permission of the American Association of Immunologists.

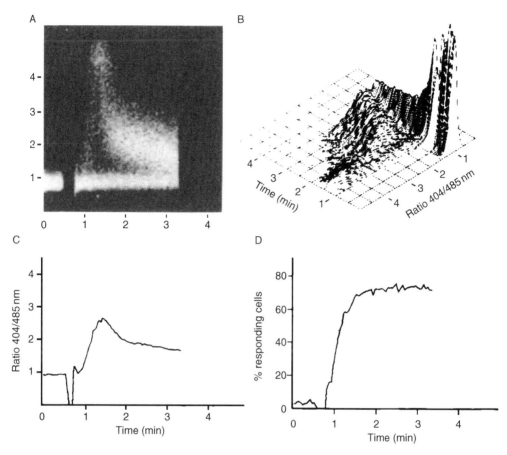

Fig. 3 Methods to express calcium signaling measured in single cells as a function of time. Thymocytes from a C57B1/6 mouse were loaded with indo-1 and stimulated with anti-CD3 antibody 145–2C11. The cells were analyzed at 300 cells/s. (A) Results were displayed on a "dot plot" in which $[Ca^{2+}]_i$ is plotted for each cell analyzed on a $100\times$ 100-pixel grid. The number of cells per pixel is displayed by intensity which ranges over 16 levels. (B) Isometric plots of the same experiment as in panel A are shown. Sequential histograms are plotted in which the x-axis represents time, the y-axis $[Ca^{2+}]_i$, and the z-axis number of cells. (C and D) Mean calcium versus time and percentage cells responding to two standard deviations above the mean of the cells before antibody stimulation are plotted for the same experiment depicted in panels A and B.

information can be displayed by a calculation of the "proportion of responding cells." If a threshold value of the resting ratio distribution is chosen, for example, one at which only 5% of control cells are above, the proportion of cells responding by ratio elevations above this threshold versus time yields a presentation informative of the heterogeneity of the response (Fig. 3C and D). In some instances, however, plots of mean response and percentage responding cells versus time do not provide an adequate representation of the complexity of the data. For example, in

Fig. 3, it can be seen from the "dot plot" and the "isometric" plot that the early response consists of a small population ($\sim$10%) of cells with an extremely high magnitude response of brief duration while the later response is comprised $\sim$50% of cells that have a low-magnitude response. Thus in this example, the time of the occurrence of the peak response is dissociated from the time of the maximal mean response (Fig. 3A and B vs. C). Note that this heterogeneity of the pattern of response is not apparent in the displays of mean $[Ca^{2+}]_i$ versus time and the percentage responding cells versus time.

E. Simultaneous Analysis of $[Ca^{2+}]_i$ and Other Fluorescence Parameters

Although the broad spectrum of indo-1 fluorescence emission will likely preclude the simultaneous use of a second UV-excited dye, the use of two or more excitation sources allows additional information to be derived from visible light-excited dyes. Perhaps the most usual application will be determination of cellular immunophenotype simultaneously with the indo-1 assay, allowing alterations in $[Ca^{2+}]_i$ to be examined in, and correlated with, specific cell subsets.

Combining the use of FITC and PE-conjugated antibodies with indo-1 analysis allows determination of $[Ca^{2+}]_i$ in complex immunophenotypic subsets. On instruments without provision for analysis of four separate fluorescence wavelengths, detection of both FITC and the higher indo-1 wavelength with the same filter element may allow successful implementation of these experiments. Gating the analysis of indo-1 fluorescence upon windows of FITC versus PE fluorescence allows information relating to each identifiable cellular subset to be derived from a single sample.

Using other probes excited by visible light, it may be possible to analyze additional physiologic responses in cells simultaneously with $[Ca^{2+}]_i$. The simultaneous analysis of membrane potential and $[Ca^{2+}]_i$ has been accomplished by several groups (Ishida and Chused, 1988; Lazzari et al., 1986). Similarly, several investigators have measured pH_i and $[Ca^{2+}]_i$ simultaneously (Van Graft et al., 1993).

F. Sorting on the Basis of $[Ca^{2+}]_i$ Responses

The ability of the flow cytometric analysis with indo-1 to observe small proportions of cells with $[Ca^{2+}]_i$ responses different than the majority of cells suggests that the flow cytometer may be useful to identify and sort variants in the population for their subsequent biochemical analysis or growth. Results of artificial mixing experiments with Jurkat (T cell) and K562 (myeloid cell) leukemia lines indicated that subpopulations of cells with variant $[Ca^{2+}]_i$ comprising <1% of total cells could be accurately identified (Rabinovitch et al., 1986). Goldsmith and Weiss (1987, 1988) have reported the use of sorting on the basis of indo-1 fluorescence to identify mutant Jurkat cells that fail to mobilize $[Ca^{2+}]_i$ in response to CD3 stimulus, despite the expression of structurally normal CD3/Ti complexes. The basis for impaired signal transduction was found to be the absent expression of the *lck* protein tyrosine kinase (Straus and Weiss, 1992). These experiments suggest that sorting on the basis of indo-1 fluorescence can be an important tool for the selection and identification of genetic variants in the biochemical pathways leading to Ca^{2+} mobilization and cell growth and differentiation. The potential of this technique is not limited to lymphoid cells, as a specific calcium response was used to sort a rare subpopulation of gastrointestinal cells that secrete cholecystokinin (Liddle *et al.*, 1992).

III. Use of Flow Cytometry and Fluo-3 to Measure [Ca^{2+}]$_i$

Fluo-3 is a fluorescein-based, calcium-sensitive probe developed by Minta and coworkers (1989). This was the first calcium indicator that did not require UV excitation, and, therefore, it could be used on all flow cytometers that have the capability to measure fluorescence emission from fluorescein. Fluo-3 may be less sensitive than indo-1 at detecting small changes in [Ca^{2+}]$_i$, in part because the K_d is higher. An advantage with fluo-3 is that it can be used with other probes such as caged calcium chelators that may themselves require UV excitation (Tsien, 1989). Because fluo-3 emission characteristics are virtually identical to those of FITC, the use of fluo-3 can be combined with almost all of the dyes used with multicolor analysis in addition to FITC, including simultaneous determination of cell cycle.

Fluo-3 has been successfully adapted for flow cytometry by several laboratories (Table IV). Loading of fluo-3 is performed similarly to the protocol described above for indo-1. Calibration must be performed either by use of spectrofluorimetry and nonratiometric equations as originally described for quin2 (Tsien *et al.*, 1982) or by reference to a series of known calcium buffers (Chused *et al.*, 1987). A direct comparison of fluo-3 to indo-1 was done by loading both dyes in T cells and analyzing the signals simultaneously (Rabinovitch and June, 1990). Both dyes showed a readily measurable response to agonist, although the change in indo-1 fluorescence ratio was greater than the increase in fluo-3 fluorescence intensity. Due to the heterogeneity in intracellular concentration of fluo-3 in loaded cells, the determination of [Ca^{2+}]$_i$ in any given cell is less accurate, and discrimination of heterogeneity in the cellular response to an agonist is less clear using fluo-3. These differences are not unexpected, given the nonratiometric properties of fluo-3. However, fluo-3 can be combined with SNARF-1 (Rijkers *et al.*, 1990), or, even better, with Fura-Red, to obtain a ratiometric measurement that helps to overcome these difficulties.

IV. Use of Flow Cytometry and Fluo-3/Fura-Red for Ratiometric Analysis of [Ca^{2+}]$_i$

Indo-1 has been the ratiometric calcium indicator dye most commonly used in flow cytometry because of its shift in emission frequency when excited at a single wavelength. A significant drawback to the use of indo-1, however, is the

Table IV
Studies Using Fluo-3 or Fura-Red to Measure $[Ca^{2+}]_i$ by Flow Cytometry

Investigators	Study aims/conclusions	Comments
Vandenberghe and Ceuppens (1990)	Initial application of fluo-3 for flow cytometry. Human PBL assayed after fluo-3 loading using 488-nm excitation	Leakage of dye was rapid, and this was a significant limitation, given absence of ratiometric properties with fluo-3
Yee and Christou (1993)	Human PMNs assayed after fluo-3 loading using 488-nm excitation. Cells were primed with LPS and stimulated with FMLP	Subpopulations of cells detected after FMLP stimulation
Sei and Arora (1991)	Human PBL were stained with PE-conjugated antibodies and loaded with fluo-3	Demonstration that simultaneous analysis of surface antigen expression and $[Ca^{2+}]_i$ is possible with single-beam excitation at 488 nm
Van Graft *et al.* (1993)	Develop a method to measure $[Ca^{2+}]_i$ and pH_i using fluo-3 and SNARF-1 in killer cell-target cell conjugates	Conjugates were clearly distinguished from single cells. Killers and targets developed increased $[Ca^{2+}]_i$; only killer cells had changes in $[pH]_i$
Rijkers *et al.* (1990)	Determine if simultaneous loading of cells with SNARF-1 and fluo-3 would improve $[Ca^{2+}]_i$ signal	SNARF-1 emission collected at isobestic point. Ratio of fluo-3 to SNARF-1 reduces variation in fluorescence intensity caused by variable fluo-3 loading. Caveat: SNARF-1 can be protein bound
Rabinovitch and June (1990)	Compare signals from T cells loaded with both indo-1 and fluo-3	Resting CV smaller with indo-1; the quality of the agonist response was similar with fluo-3 and indo-1
Nolan *et al.* (1992)	Determine if $[Ca^{2+}]_i$ can be measured in bull sperm with fluo-3	Sperm exocytosis successfully studied
Kozak and Yavin (1992)	PC12 cells studied with indo-1 and fluo-3 to determine if cell density affects $[Ca^{2+}]_i$	Pronounced effects of cell density on $[Ca^{2+}]_i$ homeostasis suggested by both dyes
Elsner *et al.* (1992)	Study characteristics of PMN response to fMLP, C5a, and IL-8	Heterogeneity of $[Ca^{2+}]_i$ response suggests functional subsets of neutrophils
Sanchez Margalet *et al.* (1992)	Determine if pancreastatin affects $[Ca^{2+}]_i$ in RINm5f cells	Both pancreastatin and ATP caused a fourfold increase in $[Ca^{2+}]_i$

requirement for UV excitation, which is not as widely available as 488-nm excitation. A ratiometric analysis with visible excitation is possible using two dyes simultaneously: Fura-Red (Haugland, 1992) and fluo-3, a fluorescein-based, calcium-sensitive probe (Minta *et al.*, 1989). As seen in Table I, both dyes are excited at 488 nm, but fluo-3 fluoresces in the green region with increasing intensity when bound to calcium while Fura-Red exhibits inverse behavior, fluorescing most intensely in the red region when not calcium bound. The spectral characteristics of the two dyes mixed together (Fig. 4) are reminiscent in shape of the UV-excited

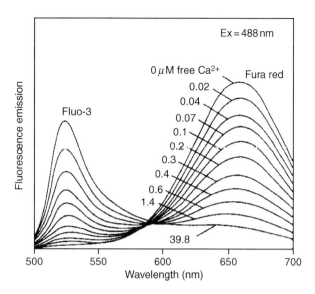

Fig. 4 Fluorescence emission spectra of a mixture of fluo-3 and Fura-Red in the presence of buffers containing various concentrations of free calcium, as noted. The fura red concentration is approximately 10 times that of fluo-3. (Courtesy of Molecular Probes, Inc., with copyright permission.)

emission of indo-1 (Fig. 1). Fura-Red emissions are dimmer than emissions from cells loaded with indo-1 and fluo-3 at the same concentration, and approximately 2–3.5 times as much Fura-Red as fluo-3 may be required to produce emissions of optimal intensity; loading concentrations of 10 and 4 μM, respectively, were optimal for T cells (Novak and Rabinovitch, 1994). As Fig. 5 shows, the fluo-3/Fura-Red ratio provides a better resolved picture of the increase in $[Ca^{2+}]_i$ associated with cellular response to antibody stimulus than does either fluo-3 or Fura-Red alone. Qualitatively, the fluo-3/Fura-Red ratio (Fig. 5C) achieves results similar to those for indo-1. Compared to fluo-3 alone, there is a larger magnitude response, with less variability in measurements from different cells. The narrower distribution of resting ratios from nonstimulated cells can be observed graphically by comparing the prestimulated regions of Fig. 5A and B with the region in Fig. 5C. The fluo-3/Fura-Red ratio can also be used simultaneously with analysis of PE-labeled immunofluorescence markers. The emission characteristics of this dye combination (Fig. 4) have a "window" between the fluo-3 and Fura-Red emission peaks that allows 488-nm-excited PE emission to be observed (Novak and Rabinovitch, 1994). When T lymphocytes, for example, are loaded with 10 μM Fura-Red, emission from Fura-Red is approximately 35% as intense as PE-CD4$^+$ emission in the region of 562–588 nm. Using spectral crossover compensation, this degree of overlap allows bright and moderately bright PE-labeled antibody probes to be used. If additional lasers are available, further label combinations can be used.

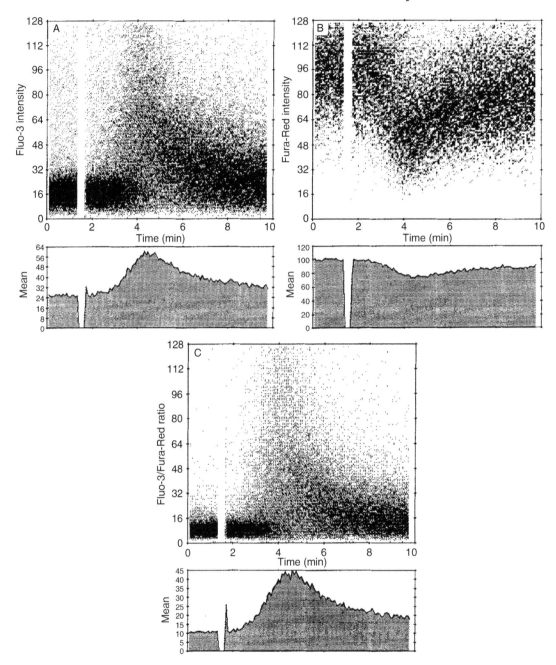

Fig. 5 Kinetic displays of fluorescence intensity/ratio and mean values of (A) fluo-3 intensity, (B) Fura-Red intensity, and (C) fluo-3/Fura-Red ratio of cells simultaneously loaded with 4 μM fluo-3 + 10 μM Fura-Red esters, gated for CD4$^+$ PE-labeled cells. Cells were stimulated by addition of 10 μg/ml anti-CD3 antibody at time 1 min. Data analysis and display performed using the software program MultiTime (Phoenix Flow Systems, San Diego, CA).

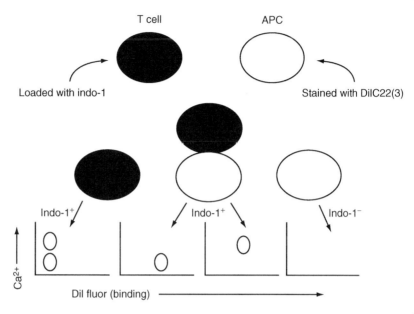

Fig. 6 Schematic display of cell conjugate assay combined with simultaneous measurement of $[Ca^{2+}]_i$. One population of cells (e.g., T cells) is loaded with indo-1, and another population (e.g., antigen presenting (APC) cells) with a cell marker dye such as DilC22(3). The cells are mixed, pelleted, and assayed. Indo-1 and DilC22(3) signals are detected by triggering on the indo-1 signal so that, in this case, either T cells or T cells conjugated to APC are analyzed. APC alone do not trigger analysis. T cells which have reacted with the APC and exhibit elevated $[Ca^{2+}]_i$ can be sorted for further analysis (for details see Abe *et al.*, 1992 and Alexander *et al.*, 1992).

V. Cell Conjugate Assays Combined with Calcium Analysis and Flow Cytometry

The realization that many forms of signal transduction are initiated by intercellular contact combined with increasing interest in a number of cell adhesion molecules has led to the description of flow cytometric assays capable of measuring conjugate formation and calcium elevations. Abe and coworkers have devised a powerful technique to analyze adhesive and signaling interactions between bone marrow-derived cells (Fig. 6). This technique permitted the analysis of signal transduction initiated by viral "superantigens" that induce clonal deletion of self-reactive thymocytes (Abe *et al.*, 1992). Others have used a similar approach to measure calcium flux induced by tumor bearing cells from populations of T cells (Alexander *et al.*, 1992), permitting, in principle, the ability to isolate antigen-specific T cells. As was noted before, a conjugate assay was used to measure $[Ca^{2+}]_i$ and $[pH]_i$ simultaneously in natural killer cells bound to target cells (Van Graft *et al.*, 1993).

VI. Pitfalls and Critical Aspects

A. Instrumental Calibration and Display of Data

In some instruments there may be difficulty in displaying the ratio of indo-1 fluorescence so that increases in Ca^{2+} are depicted as increased ratio values (Breitmeyer *et al.*, 1987). In particular, the analog ratio circuits of older Coulter instruments were limited in their range of acceptable inputs; for example, that the "violet" signal never be greater than the "blue," yielding a ratio of greater than one. Rather than reversing the ratio (blue/violet) so that a rise in calcium results in a counterintuitive decline in ratio, the violet/blue ratio can be used as long as the signal gains are initially set such that subsequent rises in the ratio will not exceed the permitted value. Alternatively, several list-mode programs allow the ratio signal to be calculated with software subsequent to analysis.

Some instruments may have nonlinearity in signal amplification or introduce errors into the calculation of the indo-1 ratio. By either analog or digital calculation, it is important that no artifactual offset be introduced in the ratio; this can be quickly tested by altering the excitation power over a broad range of values in an analysis of a nonperturbed indo-1-loaded cell population—a correctly calculated ratio will not show any dependence upon excitation intensity. It can similarly be shown that loading of cells with a broad range of indo-1 concentrations results in a constant value of the violet/blue ratio (Rabinovitch *et al.*, 1986).

B. Difficulties with Loading Indo-1

Several problems may be encountered in cells loaded with indo-1. These include compartmentalization, leakage, secretion, quenching by heavy metals, and incomplete deesterification of indo-1 ester. The analysis of $[Ca^{2+}]_i$ using indo-1 is predicated upon achieving uniform distribution of the dye within the cytoplasm. In several cell types, the related dye Fura-2 has been reported to be compartmentalized within organelles (Di Virgilio *et al.*, 1989; Malgaroli *et al.*, 1987). In bovine aortic endothelial cells, Fura-2 has been reported to be localized to mitochondria; however, under those conditions, indo-1 remained diffusely cytoplasmic (Steinberg *et al.*, 1987). We have observed that indo-1 may become compartmentalized in cells. Some cell types, such as neutrophils and monocytes, and some cell lines rather than primary cells are more susceptible to compartmentalization. In addition, compartmentalization is enhanced by prolonged incubation of cells at 37 °C. Thus, it is possible that there will be fewer problems with compartmentalization of indo-1 than with Fura-2; however, it is advisable to examine the cellular distribution of indo-1 microscopically, and in each new application to confirm the expected behavior of the dye. This is done by determining the ratio of R_{max} to R as a control for each experiment as described below. In addition, one should store indo-1-loaded cells at room temperature after loading and use the cells promptly after loading. In prolonged experiments, it is preferable to discard cells after several hours and to reload fresh batches of cells.

To reduce indicator quenching by heavy metals, the use of the membrane-permeant chelator diethylenetriaminepentaacetic acid (TPEN) for cell lines that contain increased amounts of heavy metals has been described (Arslan et al., 1985). Probenecid, a blocker of organic anion transport may be useful in cells that actively secrete Fura-2, indo-1, or other indicators (Di Virgilio et al., 1989). In some circumstances, the use of Pluronic F-127, a nonionic, high-molecular-weight surfactant, may aid in the loading of probe into cells that are otherwise difficult to load (Cohen et al., 1974; Poenie et al., 1986). Pluronic F-127 may be obtained from Molecular Probes (Eugene, OR).

If, for a particular cell type loaded with a calcium-sensitive dye, the magnitude of change between R and R_{max} are in good agreement with the values predicted from spectral curves of the probe in a cell-free buffer, then it would be unlikely that the dye is in a compartment inaccessible to cytoplasmic Ca^{2+}, in a form unresponsive to $[Ca^{2+}]_i$ (e.g., still esterified) or in a cytoplasmic environment in which the spectral properties of the dye were altered. With regard to the second condition, it has been proposed that since indo-1 fluorescence, but not that of the indo-1 ester, is quenched in the presence of millimolar concentrations of Mn^{2+}, Mn^{2+} in the presence of ionomycin can be used as a further test of complete hydrolysis of the indo-1 ester within cells (Luckhoff, 1986). It has been suggested that both Fura-2 and indo-1 may be incompletely deesterified within some cell types (Luckhoff, 1986; Scanlon et al., 1987). Since the fluorescence of the ester has little spectral dependence upon changes in Ca^{2+}, the presence of this dye form could lead to false estimates of $[Ca^{2+}]_i$. Again, results of calibration experiments are helpful in excluding this possibility.

C. Unstable Baseline

Under typical conditions, the baseline indo-1 ratio should show little (<3%) variation from sample to sample. Some cell lines may have altered mean values of "resting" $[Ca^{2+}]_i$, which can often be ascribed to a subpopulation of cells with elevated $[Ca^{2+}]_i$. This may result from impaired viability of some cells or, presumably, may be due to cells traversing certain phases of the cell cycle. In circumstances where the baseline is not stable from sample to sample, the following situations should be considered. The cells should be equilibrated to 37 °C for 3–5 min before analysis. Regulation of the temperature of the cell sample is essential, as transmembrane signaling and calcium mobilization are temperature dependent and active processes. Most applications will require analyses at 37 °C. If cells are allowed to cool before they flow past the laser beam, calcium signals will often become impaired, so that either the sample input tubing should be warmed or narrow-gauge tubing and high flow rates should be used to keep transit times from warmed sample to flow cell minimized. As noted above, it is necessary to maintain the cells at room temperature and to warm the cells just prior to the assay. It is not clear as to what mechanism the variation in basal $[Ca^{2+}]_i$ with temperature variation may be ascribed. It is possible that the changes reported by indo-1 are real and reflect strict temperature requirements of the cell for the maintenance of calcium homeostasis. Alternatively, the changes of calcium reported by the indicator dye may in part be due to temperature-dependent changes in the effective dissociation constant of the dye for calcium.

Sometimes the baseline will start at a normal level and then rise with time. This may be due to the failure to completely remove an agonist from the sample lines from a previous experiment. The most common problem has been residual calcium ionophore; this can be efficiently removed by first washing the sample lines with dimethyl sulfoxide and then scavenging residual ionophore by washing with a buffer containing 2% bovine serum albumin.

D. Sample Buffer

The choice of medium in which the cell sample is suspended for analysis can be dictated primarily by the metabolic requirements of the cells, subject only to the presence of millimolar concentrations of calcium (to enable calcium agonist-stimulated calcium influx) and reasonable pH buffering. The use of phenol red as a pH indicator does not impair the flow cytometric detection of indo-1 fluorescence signals. Although the new generation of Ca^{2+} indicator dyes are not highly sensitive to small fluctuations of pH over the physiologic range (Grynkiewicz et al., 1985), unbuffered or bicarbonate-buffered solutions can impart large and uncontrolled pH shifts. Finally, if analysis of release of Ca^{2+} from intracellular stores is desired, independent of extracellular Ca^{2+} influx, addition of 5 mM EGTA to the cell suspension (final concentration) will reduce Ca^{2+} from several millimolars to <20 nM, thus abolishing the usual extracellular to intracellular gradient.

E. Poor Cellular Response

When one encounters cells that are poorly responsive to various treatments, it is necessary to first determine if there is difficulty with the cells or with the instrument. The cells should be stimulated with the calcium ionophore ionomycin and the magnitude of R_{max} to R determined. If the ratio increases by the expected magnitude (approximately sixfold for indo-1), then the instrument is functioning properly. If the increase is less than expected, then one should obtain an independent preparation of cells, such as murine thymocytes or human peripheral blood lymphocytes. Aliquots of cryopreserved cells are convenient for this purpose. If these cells load properly and also respond poorly, then the instrument alignment should be checked. For ratiometric analyses, one of the two signals may not be properly focused, or perhaps there is interference from a second laser. This problem can be pinpointed by analyzing separately each of the two wavelength signals after ionophore treatment; for indo-1, the violet signal should increase at least threefold and the blue signal should decrease approximately twofold (Figs. 1 and 2).

If the instrument is functioning properly, then the problem may be in the cells. The cells must be loaded with sufficient dye to be easily detected. This should be checked independently with fluorescence microscopy. If the cells are too dim or excessively bright, or if the probe is compartmentalized, the ability to detect calcium signals will be impaired. For unknown reasons, the calcium signaling of B cells and not T cells is particularly sensitive to overloading with probes (Chused *et al.*, 1987; Rabinovitch *et al.*, 1986). The cells must be suspended in a medium that contains calcium; responses can appear blunted due to the inadvertant resuspension of cells in medium that contains no added calcium. In the simultaneous analysis of $[Ca^{2+}]_i$ and immunofluorescence, consider that the use of the antibody probe can itself alter the cellular $[Ca^{2+}]_i$. It is becoming increasingly clear that binding of monoclonal antibodies (mAbs) to cell-surface proteins can alter $[Ca^{2+}]_i$, even when these proteins are not previously recognized as part of a signal transducing pathway. For example, antibody binding to CD4 will reduce CD3-mediated $[Ca^{2+}]_i$ signals; if the anti-CD4 mAb is cross-linked to the CD3 complex, as with a goat-anti-mouse mAb, the CD3 signals are augmented (Ledbetter *et al.*, 1987, 1988). Antibody binding to CD8 has similar effects.

As a consequence of these concerns, a reciprocal staining strategy should be used whenever possible, so that the cellular subpopulation of interest is unlabeled while undesired cell subsets are identified by mAb staining. The CD4$^+$ subset in PBL may be identified, for example, by staining with a combination of CD8, CD20, and CD11 mAbs (Rabinovitch *et al.*, 1986), and the CD5$^+$ subset can be identified by staining with CD16, CD20, and HLA-DR mAbs (June *et al.*, 1987). Finally, it is important when staining cells with mAbs for functional studies that the antibodies be azide free, in order that metabolic processes be uninhibited. Commercial antibody preparations may thus require dialysis before use.

As with many assays, artifacts can be encountered from diverse causes. When analyzing specimens of bone marrow, it is possible that fat droplets may be included in the sample preparation. This may pose a problem, as calcium probes can label fat droplets as well as cells (Bernstein *et al.*, 1989). Thimerosal, a commonly used preservative that is included in many drug preparations, can elevate calcium in cells (Gericke *et al.*, 1993). Finally, ethanol can inhibit the mitogen-induced initial increase in $[Ca^{2+}]_i$ in mouse splenocytes (Sei *et al.*, 1992), so that diluent controls must be performed diligently.

VII. Limitations

There are several limitations to the flow cytometric assay of cellular calcium concentration. First, certain problems attributable to the use of fluorescent indicators have been mentioned. In some cells, indo-1 will not load uniformly into cells, or may not be uniformly hydrolyzed to the calcium-sensitive moiety. Quin2 at

with the HIV-1 retrovirus was found to impair signal transduction in CD4 cells (Linette *et al.*, 1988). T cells from tumor bearing animals have been found to have abnormal calcium responses (Mizoguchi *et al.*, 1992). There are many potentially exciting clinical applications of the flow cytometric assay of cellular calcium concentration (Rabinovitch *et al.*, 1993). Demonstration of such heterogeneity in $[Ca^{2+}]_i$ signals would have been impossible to discern in conventional assays carried out in a fluorimeter where only the mean calcium response is recorded.

Acknowledgments

We thank Ryo Abe and Thomas Chused for the material adapted for Fig. 6, W. C. Gause and J. Bluestone for the reagents used for the experiment shown in Fig. 3, and R. J. Hartzman and N. Hensel for use of the spectrofluorimeter. This work was supported in part by National Institutes of Health Grant AG01751 and by the Naval Medical Research and Development Command, Research Task No. M0095.003-1402. The opinions and assertions expressed herein are those of the authors and are not to be construed as official or reflecting the views of the Navy Department or the naval service at large.

References

Abe, R., Ishida, Y., Yui, K., Katsumata, M., and Chused, T. M. (1992). *J. Exp. Med.* **176,** 459–468.

Alexander, R. B., Bolton, E. S., Koenig, S., Jones, G. M., Topalian, S. L., June, C. H., and Rosenberg, S. A. (1992). *J. Immunol. Methods* **148,** 131–141.

Allen, T. J., and Baker, P. F. (1985). *Nature (Lond.)* **315,** 755–756.

Ambler, S. K., Poenie, M., Tsien, R. Y., and Taylor, P. (1988). *J. Biol. Chem.* **263,** 1952–1959.

Arslan, P., Di Virgilio, F., Beltrame, M., Tsien, R. Y., and Pozzan, T. (1985). *J. Biol. Chem.* **260,** 2719–2727.

Bernstein, R. L., Hyun, W. C., Davis, J. H., Fulwyler, M. J., and Pershadsingh, H. A. (1989). *Cytometry* **10,** 469–474.

Blinks, J. R., Wier, W. G., Hess, P., and Prendergast, F. G. (1982). *Prog. Biophys. Mol. Biol.* **40,** 1–114.

Breitmeyer, J. B., Daley, J. F., Levine, H. B., and Schlossman, S. F. (1987). *J. Immunol.* **139,** 2899–2905.

Chused, T. M., Wilson, H. A., Greenblatt, D., Ishida, Y., Edison, L. J., Tsien, R. Y., and Finkelman, F. D. (1987). *Cytometry* **8,** 396–404.

Cobbold, P. H., and Rink, T. J. (1987). *Biochem. J.* **248,** 313–328.

Cohen, L. B., Salzberg, B. M., Davila, H. V., Ross, W. N., Landowne, D., Waggoner, A. S., and Wang, C. H. (1974). *J. Membr. Biol.* **19,** 1–36.

Davies, T. A., Drotts, D., Weil, G. J., and Simons, E. R. (1988). *Cytometry* **9,** 138–142.

Di Virgilio, F., Steinberg, T. H., and Silverstein, S. C. (1989). *In* "Methods in Cell Biology" (A. Tartakoff, ed.), Vol. 31, pp. 453–462. Academic Press, San Diego, CA.

Elsner, J., Kaever, V., Emmendorffer, A., Breidenbach, T., Lohmann Matthes, M. L., and Roesler, J. (1992). *J. Leukocyte Biol.* **51,** 77–83.

Gericke, M., Droogmans, G., and Nilius, B. (1993). *Cell Calcium* **14,** 201–207.

Goldsmith, M. A., and Weiss, A. (1987). *Proc. Natl. Acad. Sci. USA* **84,** 6879–6883.

Goldsmith, M. A., and Weiss. A. (1988). *Science* **240,** 1029–1031.

Goller, B., and Kubbies, M. (1992). *J. Histochem. Cytochem.* **40,** 451–456.

Grynkiewicz, G., Poenie, M., and Tsien, R. Y. (1985). *J. Biol. Chem.* **260,** 3440–3450.

Harafuji, H., and Ogawa, Y. (1980). *J. Biochem. (Tokyo)* **87,** 1305–1312.

Haugland, R. P. (1992). *In* "Handbook of Fluorescent Probes and Research Chemicals" (K. D. Larison, ed.), pp. 117–118. Molecular Probes, Eugene, OR.

Hesketh, T. R., Smith, G. A., Moore, J. P., Taylor, M. V., and Metcalfe, J. C. (1983). *J. Biol. Chem.* **258,** 4876–4882.

31. Intracellular Ionized Calcium 711

Ishida, Y., and Chused, T. M. (1988). *J. Exp. Med.* **168,** 839–852.

June, C. H., Ledbetter, J. A., Rabinovitch, P. S., Martin, P. J., Beatty, P. G., and Hansen, J. A. (1986). *J. Clin. Invest.* **77,** 1224–1232.

June, C. H., Rabinovitch, P. S., and Ledbetter, J. A. (1987). *J. Immunol.* **138,** 2782–2792.

Kachel, V., Kempski, O., Peters, J., and Schodel, F. (1990). *Cytometry* **11,** 913–915.

Kozak, A., and Yavin, E. (1992). *J. Mol. Neurosci.* **3,** 203–212.

Kubbies, M., Schindler, D., Hoehn, H., and Rabinovitch, P. S. (1985). *Cell Tissue Kinet.* **18,** 551–562.

Lazzari, K. G., Proto, P. J., and Simons, E. R. (1986). *J. Biol. Chem.* **261,** 9710–9713.

Ledbetter, J. A., June, C. H., Grosmaire, L. S., and Rabinovitch, P. S. (1987). *Proc. Natl. Acad. Sci. USA* **84,** 1384–1388.

Ledbetter, J. A., June, C. H., Rabinovitch, P. S., Grossmann, A., Tsu, T. T., and Imboden, J. B. (1988). *Eur. J. Immunol.* **18,** 525–532.

Liddle, R. A., Misukonis, M. A., Pacy, L., and Balber, A. E. (1992). *Proc. Natl. Acad. Sci. USA* **89,** 5147–5151.

Linette, G. P., Hartzman, R. J., Ledbetter, J. A., and June, C. H. (1988). *Science* **241,** 573–576.

Luckhoff, A. (1986). *Cell Calcium* **7,** 233–248.

Malgaroli, A., Milani, D., Meldolesi, J., and Pozzan, T. (1987). *J. Cell Biol.* **105,** 2145–2155.

Miller, R. A., Jacobson, B., Weil, G., and Simons, E. R. (1987). *J. Cell Physiol.* **132,** 337–342.

Minta, A., Kao, J. P., and Tsien, R. Y. (1989). *J. Biol. Chem.* **264,** 8171–8178.

Mizoguchi, H., O'Shea, J. J., Longo, D. L., Loeffler, C. M., McVicar, D. W., and Ochoa, A. C. (1992). *Science* **258,** 1795–1798.

Nolan, J. P., Graham, J. K., and Hammerstedt, R. H. (1992). *Arch. Biochem. Biophys.* **292,** 311–322.

Novak, E. J., and Rabinovitch, P. S. (1994). *Cytometry* (in press).

Owen, C. S. (1988). *Cell Calcium* **9,** 141–147.

Poenie, M., Alderton, J., Steinhardt, R., and Tsien, R. (1986). *Science* **233,** 886–889.

Poenie, M., Tsien, R. Y., and Schmitt-Verhulst, A. M. (1987). *EMBO J.* **6,** 2223–2232.

Rabinovitch, P. S., and June, C. H. (1990). *In* "Flow Cytometry and Sorting" (M. R. Melamed, T. Lindmo, and M. L. Mendelsohn, eds.), pp. 651–668. Wiley-Liss, New York.

Rabinovitch, P. S., June, C. H., Grossmann, A., and Ledbetter, J. A. (1986). *J. Immunol.* **137,** 952–961.

Rabinovitch, P. S., June, C. H., and Kavanagh, T. J. (1993). *Ann. NY Acad. Sci.* **677,** 252–264.

Ransom, J. T., DiGiusto, D. L., and Cambier, J. C. (1986). *J. Immunol.* **136,** 54–57.

Ransom, J. T., DiGiusto, D. L., and Cambier, J. (1987). *In* "Methods in Enzymology" (P. Conn and A. Means, eds.), Vol. 141, pp. 53–63. Academic Press, Orlando, FL.

Rijkers, G. T., Justement, L. B., Griffioen, A. W., and Cambier, J. C. (1990). *Cytometry* **11,** 923–927.

Sanchez Margalet, V., Lucas, M., and Goberna, R. (1992). *Mol. Cell Endocrinol.* **88,** 129–133.

Scanlon, M., Williams, D. A., and Fay, F. S. (1987). *J. Biol. Chem.* **262,** 6308–6312.

Sei, Y., and Arora, P. K. (1991). *J. Immunol. Methods* **137,** 237–244.

Sei, Y., McIntyre, T., Skolnick, P., and Arora, P. K. (1992). *Life Sci.* **50,** 419–426.

Steinberg, S. F., Bilezikian, J. P., and Al-Awqati, Q. (1987). *Am. J. Physiol.* **253,** C744–C747.

Straus, D. B., and Weiss, A. (1992). *Cell (Cambridge, MA)* **70,** 585–593.

Tsien, R. Y. (1981). *Nature (Lond.)* **290,** 527–528.

Tsien, R. Y. (1989). *In* "Methods in Cell Biology" (A. Tartakoff, ed.), Vol. 30, pp. 127–156. Academic Press, San Diego, CA.

Tsien, R. Y., Pozzan, T., and Rink, T. J. (1982). *J. Cell Biol.* **94,** 325–334.

Vandenberghe, P. A., and Ccuppens, J. L. (1990). *J. Immunol. Methods* **127,** 197–205.

Van Graft, M., Kraan, Y. M., Segers, I. M., Radosevic, K., De Grooth, B. G., and Greve, J. (1993). *Cytometry* **14,** 257–264.

Ware, J. A., Smith, M., and Salzman, E. W. (1987). *J. Clin. Invest.* **80,** 267–271.

Wilson, H. A., Greenblatt, D., Poenie, M., Finkelman, F. D., and Tsien, R. Y. (1987). *J. Exp. Med.* **166,** 601–606.

Yee, J., and Christou, N. V. (1993). *J. Immunol.* **150,** 1988–1997.

CHAPTER 32

Oxidative Product Formation Analysis by Flow Cytometry

J. Paul Robinson,★ Wayne O. Carter,† and Padma Kumar Narayanan★,‡

★Department of Basic Medical Sciences
School of Veterinary Medicine
Purdue University
West Lafayette, Indiana 47907

†Hill's Pet Food
Topeka, KS

‡Amgen Inc., Bothell
Washington

ESSENTIAL CYTOMETRY METHODS
Copyright 2009, Elsevier Inc. All rights reserved.

DOI: 10.1016/B978-0-12-375045-7.00032-5

I. Introduction

Prior to flow cytometric methods, measurement of oxidative function was labor intensive and often very difficult to quantitate. The *Staphylococcus aureus* killing methods (Alexander *et al.*, 1968) required a large volume of blood and at least 48 h to complete. More recent methods such as chemiluminescence, while quantitative and rapid, have been difficult to interpret (Cheung *et al.*, 1983; DeChatelet *et al.*, 1982). The advantages of the flow-based methods are that a significantly reduced cell number is required, the procedure is relatively easy, and, if a flow cytometer is already available, the procedure is inexpensive to perform. A significant feature of the flow-based methods not available using any other technique is the capability to determine heterogeneity of response (Neil *et al.*, 1985; Taga *et al.*, 1985). Subpopulations defined by either light scatter or fluorescence intensity can be identified and quantified. Thus, unresponsive or poorly responsive cell populations are easily discerned.

II. Oxidative Burst

A. Application

A variety of reactive oxygen species (ROS) is produced by several cells. Our principal application for measuring oxidative systems is for functional evaluation of "oxidative or respiratory bursting" in neutrophils. The respiratory burst results from activation of the membrane-bound NADPH oxidases via an electron transfer reaction. Two electrons are transferred from NADPH to oxygen through an FAD-flavoprotein utilizing cytochrome b_{-245}. Superoxide anion is produced and then dismutates to hydrogen peroxide (H_2O_2) either spontaneously or by superoxide dismutase (SOD). The ROS and the hydrogen peroxide produced are necessary for normal bactericidal mechanisms in the neutrophil. These oxidative mechanisms exist in many cells, and we have used these techniques to evaluate a variety of cell types including neutrophils, macrophages, HL-60 cells (human leukemia-60 cells, see Vol. 42, Chapter 25), and endothelial cells.

A calibration curve can be generated based upon spectrophotometric data and flow cytometric measurements. This allows for conversion of flow cytometry fluorescence channels into quantitative estimations of H_2O_2, if necessary (Bass *et al.*, 1983).

The assays described below utilize two dyes. In the first assay, 2′,7′-dichlorofluorescin diacetate (DCFH-DA), which is freely permeable, is incorporated into hydrophobic lipid regions of the cell (Bass *et al.*, 1983). The acetate moieties are cleaved off leaving the nonfluorescent 2′,7′-dichlorofluorescin (DCFH). Hydrogen peroxide and peroxidases produced by the cell oxidize DCFH to 2′,7′-dichlorofluorescein (DCF), which is fluorescent (530 nm). The green fluorescence is thus proportional to the H_2O_2 produced.

The second assay utilizes hydroethidine (HE), which can be directly oxidized to ethidium bromide by superoxide anions produced by the cell. The third assay incorporates both dyes and has been used to detect selective defects in phagosomal oxidation following lysosomal degranulation as has been reported to occur in sepsis (Rothe and Valet, 1990). There are several advantages of HE over the DCF assay.

B. Materials

1. Hanks' balanced salt solution (HBSS)

Stock HBSS 10× concentration: NaCl, 40 g; KCl, 2 g; Na_2HPO_4, 0.5 g; $NaHCO_3$, 0.5 g; q.s. to 500 ml with distilled water.

Stock Tris 1 M: Tris base, 8 g; Tris-HCl, 68.5 g; q.s. to 500 ml with distilled water, pH to 7.3.

Preparation of 100 ml HBSS:

Stock HBSS, 10× : 10 ml
Distilled water: 80 ml
Tris 1 M: 2.75 ml
$CaCl_2$ 1.1 M: 170 μl
$MgSO_4$ 0.4 M: 200 μl
Dextrose: 220 mg

Adjust pH to 7.4 and q.s. to 100 ml with distilled water.

2. PBS gel

- Stock PBS gel

   EDTA (disodium salt) 0.2 M: 7.604 g
   Dextrose 0.5 M: 9 g
   10% Gelatin (Difco): 10 g
   Distilled water: 100 ml

- Heat water to 45–50 °C and slowly add gelatin while mixing with a magnetic stirrer. Continue stirring and add EDTA and dextrose. Do not exceed 55 °C because gelatin and glucose will "caramelize." Store in 1.2-ml aliquots at −20 °C.

- Working solution PBS gel (make daily as needed). Warm 1 ml gel to 45 °C. Add 95 ml warm PBS (phosphate buffered saline) and mix. Adjust pH to 7.4 and q.s. to 100 ml with distilled water.

3. Erythrocyte lysing solution (preparation of 100 ml):

NH_4Cl 0.15 M: 0.8 g
Na HCO_3 10 mM: 0.084 g
EDTA (disodium) 10 mM, 0.037 g
Distilled water: 95 ml

Adjust pH to 7.4 and q.s. to 100 ml with distilled water.

4. DCFH-DA (MW 487.2) [Molecular Probes, Inc., Eugene, OR], 20 mM solution:

   • Weigh 2–9 mg of DCFH-DA and place in a foil-covered 12×75-mm tube.
   • Add absolute ethanol in a volume equivalent to the weight in milligrams of the DCFH-DA divided by 9.74.
   • Cap the tube, mix, cover in foil, and store at $4\,^\circ$C until use.

5. HE (MW 315) (Molecular Probes Inc., Eugene, OR), 10 mM solution:

   • Stock solution, 10 mM in dimethylformamide (3.15 mg/ml).

6. PMA (phorbol 12-myristate 13-acetate) [Sigma Chemical Co., St. Louis, MO]: PMA is toxic and carcinogenic; additionally dimethyl sulfoxide (DMSO) is readily absorbed through the skin. Wear gloves while handling solutions, prepare solutions in a hood, and be extremely cautious!

Stock PMA (2 mg/ml in DMSO): Mix well and aliquot 15–20 μl stock PMA in small capped polypropylene bullets. Store at $-20\,^\circ$C.

Working PMA solution (make daily as needed): 5 μl PMA stock in 10 ml PBS gel (1000 ng/ml PMA solution). A final PMA concentration of 100 ng/ml will predictably result in maximal cell stimulation (e.g., 900 μl cells in solution and 100 μl working PMA solution).

C. Instrumentation

Excitation is at 488 nm for both the above probes. Emission filters should be 525 nm for DCF and 590 nm for HE. Collection of forward light scatter and 90° scatter, as well as both fluorescence wavelengths, is necessary. Wherever possible collect list-mode data for further analysis. If performing a kinetic assay, time may be required.

D. Methods

DCFH-DA Assay (using whole blood):

1. Place 2 ml preservative-free heparinized whole blood in a 50-ml conical tube.
2. Add 48 ml erythrocyte lysing solution.
3. Gently mix solution 10 min at $25\,^\circ$C on a hematology rotator.
4. Centrifuge 10 min at $350 \times g$ and $4\,^\circ$C.
5. Decant supernatant and resuspend in 5 ml PBS gel (working solution).
6. Centrifuge 10 min at $350 \times g$ and $4\,^\circ$C. Decant supernatant and resuspend in 2.5 ml HBSS.
7. Count leukocytes and adjust cell suspension to 2×10^6 cells/ml.
8. Add 1 μl 20 mM DCFH-DA per ml of cell suspension to be loaded.
9. Incubate loaded cells 15 min at $37\,^\circ$C.

10. Stimulate cells with PMA: add 100 μl PMA (working solution) to 900 μl cell suspension (final PMA concentration 100 ng/ml). Reserve some loaded, unstimulated cell suspension for a control.

11. Maintain cell sample at 37 °C and run stimulated and unstimulated samples every 10 min on the cytometer for a total of 40 min.

HE Assay: Procedures 1–7 are identical (as above).

8. Add 1 μl HE per ml of cell suspension to be loaded.

9. Incubate loaded cells 15 min at 37 °C.

Procedures 10 and 11 are identical (as above).

Combined DCFH-DA and HE Assay: Procedures 1–7 are identical (as above)

8. Add 1 μl 20 mM DCFH-DA per ml of cell suspension to be loaded.

9. Incubate loaded cells 5 min at 37 °C.

10. Add 1 μl HE per ml of cell suspension to be loaded.

11. Incubate loaded cells an additional 15 min at 37 °C.

12. Stimulate cells with PMA: add 100 μl PMA (working solution) to 900 μl cell suspension (final PMA concentration 100 ng/ml). Reserve some loaded, unstimulated cell suspension for a control.

13. Maintain cell sample at 37 °C and run stimulated and unstimulated samples every 10 min on the cytometer for a total of 40 min.

E. Critical Aspect

Clumped cells are obviously detrimental to the assay and may potentially plug the flow cell. Additionally, clumped cells cannot be measured as "functionally normal" and must be eliminated. Clumping may be related to highly activated cells and may, therefore, provide important information regarding the context of the assay. Keeping the cell sample at 4 °C (on ice) and using PBS gel will help prevent clumping of these reactive cells. Additionally, the dextrose, calcium, and magnesium in the HBSS are necessary for optimal cell function. Preservative-free heparin is also important to prevent any alteration in normal cell function. Dimethylformamide, used for preparation of the HE solution, will dissolve plastics and should be stored in glass. Alternatively DMSO can be used to prepare the HE solution. Another crucial aspect of these assays is that an unstimulated cell sample must be used for comparison at all time points.

There is a steady increase in oxidation of the DCFH-DA and HE even in unstimulated cells. This spontaneous oxidation varies with the cell type and with the dye and is primarily mitochondrial in origin. Mitochondrial oxidation can be blocked with azide or cyanide. In addition to measuring an increase in red-fluorescent ethidium bromide, a decrease in blue-fluorescent HE (excitation 350–380 nm, emission 418–500 nm) can be simultaneously measured in cells. One of the advantages of HE is the measurement of one ROS, superoxide anion, which occurs earlier in the cascade of oxidation events than hydrogen peroxide.

An advantage of the combined assay is the identification of different subpopulations of neutrophils with disparate oxidative function.

F. Results

Cells must be preloaded for the correct time before activation. The time period varies with the cell type. Neutrophils are usually fully "loaded" within 15 min at 37 °C.

Figures 1–3 show typical histograms of an unstimulated and a stimulated population of human neutrophils. A 4- to 20-fold increase in green fluorescence is expected and evident.

III. Phagocytosis

A. Application

Phagocytosis involves a series of stages by a cell progressing from particle attachment to ingestion. Any step in the phagocytic process can potentially fail. By using both opsonized and unopsonized particles in the assay we can determine whether abnormal phagocytosis is due to defective ingestion or opsonization. The

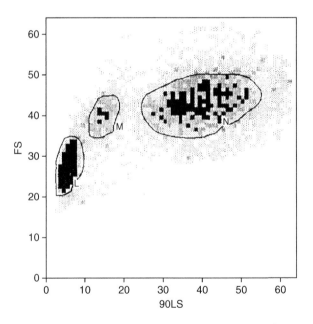

Fig. 1 Histogram with forward angle light scatter (FS) on the *y*-axis and 90° light scatter (90LS) on the *x*-axis. The cell sample was whole blood (human) prepared by erythrocyte lysis with ammonium chloride. Three different cell populations are readily identified and labeled: L, lymphocytes; M, monocytes; N, neutrophils. The gate (N) drawn around the neutrophil population was applied to additional histograms to identify the fluorescent changes in only the neutrophils.

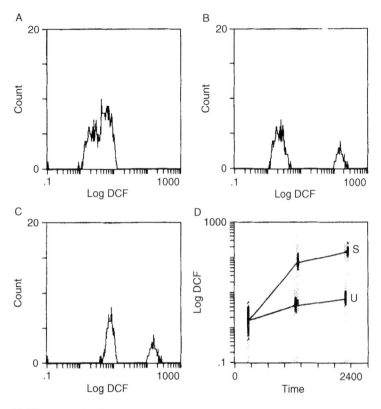

Fig. 2 (A) Histogram showing the increase in green fluorescence (log DCF) from 0 to 40 min in unstimulated neutrophils. (B) Histogram similar to A except the neutrophils are stimulated with PMA. (C) Histogram indicating the difference in fluorescence (log DCF) at 40 min between unstimulated and PMA-stimulated neutrophils (a 4- to 20-fold increase in fluorescence is expected in normal neutrophils). (D) Overview histogram of three time points (time in seconds) versus green fluorescence (log DCF on *y*-axis) for both unstimulated (U) and PMA-stimulated (S) neutrophils. The lines are drawn through the mean channel fluorescence of each time point. As indicated in the text, unstimulated neutrophils will oxidize (primarily mitochondrial oxidation) some DCFH-DA to DCF and increase the intensity of green fluorescence. Nevertheless, the change in fluorescence is much greater in PMA-stimulated neutrophils.

technique described below employs the use of opsonized and unopsonized bacteria (Bjerknes and Bassøe, 1984). Alternatively, fluorescent beads may be used or FITC (fluorescein isothiocyanate)-labeled yeast. After phagocytosis of the FITC-labeled yeast, ethidium bromide can be added to stain the extracellular yeast red so they can be differentiated on the flow cytometer.

A related version of this assay can also be used for evaluating phagocytosis in macrophage populations. This has particular importance in evaluating lavage fluids such as peritoneal lavage or bronchoalveolar lavage.

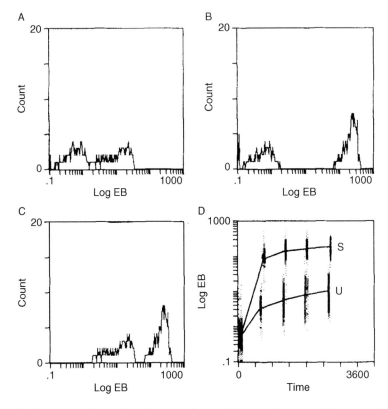

Fig. 3 (A) Histogram similar to that of 2A except hydroethidine is oxidized to ethidium bromide (EB) in unstimulated neutrophils. (B) Histogram similar to A except the neutrophils are stimulated with PMA. The change in red fluorescence (log EB) is evident from 0 to 40 min in PMA-stimulated neutrophils. (C) Difference in red fluorescence (log EB) between unstimulated and PMA-stimulated neutrophils at 40 min. (D) Overview histogram showing five time points (time in seconds) versus EB fluorescence for unstimulated and PMA-stimulated neutrophils. The lines are drawn through the mean channel fluorescence of each time point. Similar to the changes with DCF, an increase in EB fluorescence is evident in unstimulated neutrophils. However, the fluorescence change is much greater for PMA-stimulated neutrophils. More time points were collected for the HE assay than the DCF assay (Fig. 2), but the result is the same.

B. Materials

        Bacterial culture
        Blood agar plate
        Brain-heart infusion broth
        Brain-heart infusion agar slant
        Neutrophil isolation medium (NIM) (Cardinal Associates, Santa Fe, NM)
        HBSS (as above)

Pooled human serum
Glycerin
Carbonate/bicarbonate buffer (pH 9.5):

Na_2CO_3, 0.5 M (5.3 g/100 ml) 1 volume
$NaHCO_3$, 0.5 M (4.2 g/100 ml) 3 volumes

FITC: 0.02 mg/ml in carbonate/bicarbonate buffer (make a concentrated solution and dilute).

0.9% NaCl-0.02% EDTA:

NaCl: 900 mg
EDTA disodium: 20 mg
Distilled water: q.s. to 100 ml

Trypan blue (Gibco Laboratories, Grand Island, NY):
Stock: 4 mg/ml in saline
Working solution: 3 mg/ml (dilute stock 3:4 with saline).

C. Methods

1. Bacterial Culture

Bacteria are cultured on a blood agar plate and then subcultured to obtain discrete colonies. A discrete colony is then cultured on a brain-heart infusion agar slant and again subcultured. Using a sterile loop, some of the colonies are transferred into broth and cultured overnight in a 37 °C incubator.

2. Labeling Bacteria with FITC

The bacteria are washed with HBSS and centrifuged 10 min at $10,000 \times g$ and the supernatant is decanted. The bacterial slurry is heat killed for 1 h at 60 °C. Enumeration of the bacteria is achieved by serial dilutions and subsequent plating and correlation using spectrophotometry so that the final concentration is approximately 10^9 ml^{-1}. The bacteria are resuspended in carbonate/bicarbonate buffer such that the absorbance is, for example, 0.35 at 620 nm (this may differ depending upon the bacterial species and the media composition). The volume of the bacterial suspension is doubled with 0.02 mg/ml FITC in carbonate/bicarbonate buffer and incubated 30 min at 37 °C with end-over-end rotation. The bacteria are washed three times with HBSS, counted, and resuspended at a final concentration of 1×10^8 ml^{-1}. The bacteria are aliquoted into sterile Eppendorf microcentrifuge tubes at a volume of 1 ml/vial; a drop of sterile glycerin is added to each vial and the vials are frozen at −70 °C. The labeled bacteria will last for several years at −70 °C.

3. Phagocytosis Assay (Using a Purified Population of Neutrophils)

1. Overlay 5 ml EDTA-treated or preservative-free heparinized whole blood on NIM.

2. Centrifuge 30 min at $400 \times g$ and 25 °C.

3. Remove neutrophil layer and wash in PBS gel; centrifuge 10 min at $250 \times g$ and 4 °C.

4. Remove supernatant and resuspend neutrophil pellet in HBSS.

5. Count neutrophils and adjust cell suspension to 1×10^6 cells/ml.

6. Thaw one vial of bacteria; sonicate five times for 10-s intervals at approximately 75% power. Cool bacteria on ice between cycles.

7. For opsonization, dilute 1 ml pooled human serum with 3 ml HBSS (1:4 dilution).

8. Mix 1 ml bacterial solution with 4 ml diluted human serum and incubate 15 min at 37 °C with end-over-end rotation.

9. Mix 5 ml neutrophil solution with opsonized bacteria (adjust volumes to approximate a 20:1 bacteria:neutrophil ratio) and immediately remove 1 ml of the mixture and place into a 12×75–mm tube containing 1 ml 0.9% NaCl-0.02% EDTA solution at 4 °C to stop phagocytosis.

10. Maintain the remaining neutrophil/bacteria mixture at 37 °C and remove 1-ml aliquots at 10, 15, 30, and 60 min, each time mixing the aliquot with 1 ml cold 0.9% NaCl-0.02% EDTA solution at 4 °C in a 12×75–mm tube to stop phagocytosis.

11. For a control with unopsonized bacteria, mix 100 μl bacterial solution with 400 μl HBSS. Incubate 15 min at 37 °C, then add 500 μl neutrophil solution and incubate the tube 30 min at 37 °C. Add 1 ml cold 0.9% NaCl–0.02% EDTA solution at 4 °C in a 12×75–mm tube to stop phagocytosis.

12. Run on cytometer, measuring green fluorescence emission at 525 nm.

13. Immediately after measuring the green fluorescence of each tube, add 1 ml trypan blue (3 mg/ml) and repeat measurements on the cytometer. The trypan blue quenches the fluorescence of the extracellular bacteria and thus allows for measurement of only the intracellular bacteria.

D. Critical Aspects

As with the oxidative burst, proper cell handling and care are necessary to prevent clumping. Since phagocytosis is an active functional assay, dextrose, calcium, and magnesium in the HBSS are necessary for normal cell function. This assay can be performed with a leukocyte mixture instead of a pure neutrophil suspension. The neutrophil population is relatively easy to separate on the cytometer using forward-angle and 90° light scatter. However, it may be difficult to separate the neutrophil phagocytosis from the monocyte phagocytosis.

The optimal ratio of bacteria to neutrophils is 20:1. Small particles such as bacteria may be difficult to see using light scatter. It is necessary to trigger using green fluorescence to detect these small particles (1 μm or less).

IV. Controls and Standards

It is important to establish a standard procedure for running oxidative burst assays. This can be achieved by finding a fluorescent bead which falls generally within the range of fluorescence of activated cells. This bead is then used to set up the flow cytometer, each time setting the high voltage of the photomultiplier tubes based upon the bead fluorescence. If a full calibration is performed, the mean channel fluorescence can then be equated with the quantity of H_2O_2 formed per cell. Finally, it is not possible to perform kinetic assays for oxidative metabolism without proper background controls. The fluorescent probes discussed will become oxidized over the period of time with or without any stimulation. Thus, it is always necessary to measure responses compared to one of these controls. This relative stimulation value is the only measure that accurately compares a cellular response based on a true metabolic reaction and nonspecific oxidation reactions.

Acknowledgment

Funding for these studies was provided by Grants P42 ES04911 and GM38827 from the National Institutes for Health.

References

Alexander, J. W., Windhorst, D. B., and Good, R. A. (1968). Improved tests for the evaluation of neutrophil function in human disease. *J. Lab. Clin. Med.* **72**, 136–148.

Bass, D. A., Parce, J. W., DeChatelet, L. R., Szejda, P., Seeds, M. C., and Thomas, M. (1983). Flow cytometric studies of oxidative product formation by neutrophils: a graded response to membrane stimulation. *J. Immunol.* **130**, 1910–1917.

Bjerknes, R., and Bassøe, C.-F. (1984). Phagocyte mediated attachment and internalization: flow cytometric studies using a fluorescence quenching technique. *Blut* **49**, 315–323.

Cheung, K., Archibald, A., and Robinson, M. (1983). The origin of chemiluminescence produced by neutrophils stimulated by opsonized zymosan. *J. Immunol.* **130**, 2324–2329.

DeChatelet, L. R., Long, G. D., Shirley, P. S., Bass, D. A., Thomas, M. J., Henderson, F. W., and Cohen, M. S. (1982). Mechanism of the luminol-dependent chemiluminescence of human neutrophils. *J. Immunol.* **129**, 1589–1593.

Neill, M. A., Henderson, W. R., and Klebanoff, S. J. (1985). Oxidative degradation of leukotriene C4 by human monocytes and monocyte-derived macrophages. *J. Exp. Med.* **162**, 1634–1644.

Rothe, G., and Valet, G. (1990). Flow cytometric analysis of respiratory burst activity in phagocytes with hydroethidine and 2',7'-dichlorofluorescin. *J. Leukoc. Biol.* **47**, 440–448.

Taga, K., Seki, H., Miyawaki, T., Sato, T., Taniguchi, N., Shomiya, K., Hirao, T., and Usui, T. (1985). Flow cytometric assessment of neutrophil oxidative metabolism in chronic granulomatous disease on small quantities of whole blood: heterogeneity in female patients. *Hiroshima J. Med. Sci.* **34**, 53–60.

CHAPTER 33

Phagocyte Function

Gregor Rothe and Mariam Klouche

Laborzentrum Bremen
Friedrich-Karl-Strasse 22
D-28205 Bremen, Germany

ESSENTIAL CYTOMETRY METHODS
© 2009 Published by Elsevier Inc.

DOI: 10.1016/B978-0-12-375045-7.00033-7

I. Update

Flow cytometry in the past years has further firmly established itself as the method of choice for the functional characterization of phagocytic cells. This is due to the high specificity of probes and assays that has been reached after more than 25 years of research on neutrophil and monocyte function. Furthermore, the analysis of these cells without the need for cell isolation and in a physiological environment enables correlations to *in vivo* conditions which cannot be achieved with conventional biochemical assays for phagocyte function.

There has been only a very limited methodological development regarding the flow cytometric analysis of phagocyte function. Thus, the repertoire of fluorogenic probes for intracellular analysis of calcium has been continuously increasing. Indo-1 and fluo-3 which are described in the technical protocols are, however, still among the first choices of fluorochromes. Phagocytosis assays typically have been difficult to standardize and not very sensitive to changes of the *in vivo* function. Therefore, their use has been restricted mainly to the analysis of severe inborn defects of neutrophil function. Here as an example for new developments an interesting new approach has been recently published by Bicker *et al.* (2008). In their study enhanced green fluorescent protein has been used to label viable *Escherichia coli* and a kinetic assay on diluted whole blood is used to determine the efficiency of the ingestion as well as the degradation of bacteria. A decrease of bacterial degradation was observed in immunosuppressed patients.

Furthermore, the mechanisms of neutrophil function and especially the central role of the generation of reactive oxidants have been further elucidated. In this context, animal models have pointed to the central role of phagosomal pH changes generated by the activation of NADPH oxidase in conjunction with neutral proteinases in bacterial killing (Behe and Segal, 2007). On the other hand, reactive oxygen species have been identified as important modulators of intracellular signaling which includes the modulation of protein and lipid kinases as well as transcription factors (Fialkow *et al.*, 2007). The interpretation of the biological effects of an altered oxidative function thus is becoming increasingly complex.

II. Introduction

A. Biological Functions and Mechanisms of Activation

Professional phagocytes, including neutrophil, eosinophil, and basophil granulocytes, and monocytes/macrophages play a major role in innate host defense, and function as scavengers of debris removing damaged, senescent, and apoptotic cells, as well as foreign substances, such as latex particles (Djaldetti *et al.*, 2002). Moreover, phagocytes participate in inflammatory reactions and exert tumoricidal activity. Besides their principal function in the first-line defense, monocytes have a critical function regulating hemostasis and wound repair via expression of tissue factor and through their functional capacity to interact with platelets and endothelial cells (Osterud, 1998). Phagocytes may directly engulf microorganisms, particles, or altered endogenous substances; however, their activity is greatly enhanced by opsonizing factors, such as complement, particularly C3b and C3bi, immunoglobulins, or the acute-phase reactants C-reactive protein and pentraxin-3 (Hart *et al.*, 2004; Mantovani *et al.*, 2003). While granulocytes primarily degrade phagocytosed materials until completion, ultimately leading to cell death, the principal role of monocytes/macrophages comprises the specific degradation to antigenic peptides for presentation via the major histocompatibility complex, thus linking the innate to the specific immune response.

The functional reaction cascade of the phagocytes to external microorganisms, foreign substances, or endogenous decayed material, such as cellular debris or apoptotic cells, encompasses a complex series of fine-tuned responses from chemotaxis, actin assembly, and migration to cell surface receptor recognition, receptor assembly, adhesion, aggregation, phagocytosis, degranulation, reactive oxygen species production, and intracellular pH changes (Fig. 1; Aderem and Underhill, 1999; Kwiatkowska and Sobota, 1999).

Chemotaxis and migration are the initial steps of phagocyte function, which are largely governed by adhesion events, chemokine receptors, and ultimate reorganizations of the actin cytoskeleton. The binding of opsonized and free particles to specific cell surface receptors limited to phagocytes initiates phagocytosis. Phagocytes are equipped with at least two sets of receptors for recognition of these different kinds of presentation of particles, the first recognizes primarily host opsonin molecules, while the second consists of receptors recognizing integral surface or membrane components of particles. The best characterized opsonin-dependent receptors include the immunoglobulin superfamily receptors FcγRI, FcγRII, FcγRIII, the complement receptor-3, CR3, and the integrin receptors $\alpha_5\beta_1$ (VLA-5) and $\alpha_v\beta_3$ (Kwiatkowska and Sobota, 1999). The opsonin-independent receptors encompass receptors for *N*-formylated peptides such as *N*-formyl-Met-Leu-Phe (fMLP), the lectin class mannose receptor, the β-glucan receptor, the class A scavenger receptors I and II, CD36 and MARCO, as well as CD68 (macrosialin), several toll-like receptors and CD14 (Ozinsky *et al.*, 2000; Palecanda and Kobzik, 2001; Savill, 1998).

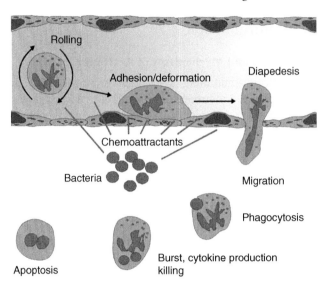

Fig. 1 Stepwise process of neutrophil activation during the extravasation to sites of infection. An upregulation of adhesion is induced by chemoattractants followed by actin polymerization, diapedesis through the endothelium and extravasation. Following migration along a chemotactic gradient, the phagocytosis of bacteria is accompanied by the release of reactive oxidants, secretion of granule-associated proteases, and microbicidal proteins as well as the autocrine synthesis of proinflammatory cytokines or chemokines such as IL-8. Apoptosis finally represents an important regulatory step under the control of interleukin-10 which limits the inflammatory reaction.

The first event triggered by the binding of ligands, irrespective of the nature of the phagocyte receptors, is the clustering of the cell surface membrane receptors (Allen and Aderem, 1996; Jongstra-Bilen *et al.*, 2003; Pfeiffer *et al.*, 2001). In fact, receptor clustering as a salient feature of phagocyte activation transmits several signal transduction pathways, including tyrosine kinases, small GTPases of the Rho and ARF families, and phosphatidylinositol 3-kinase (Kwiatkowska and Sobota, 1999). Particularly for the phosphatidylinositol 3-kinase, a crucial and nonredundant role determining the responsiveness of phagocytes to chemoattractants and the process of chemotactic leukocyte recruitment has been demonstrated (Wymann *et al.*, 2000). This action is probably based on the capacity of phosphatidylinositol 3-kinase to modulate the assembly of the submembranous actin filament system locally, leading to directed migration, but also to particle internalization.

These stimulus-dependent activation steps can be greatly enhanced by phagocyte "priming," a process involving several mediators, which at higher concentrations may also act as direct stimuli, such as interleukin-8 (IL-8), and others, which act primarily as facilitators of subsequent activators, such as tumor-necrosis factor-α (TNF-α). Currently, several classes of priming substances have been identified (Fig. 2), including proinflammatory cytokines (TNF-α, IL-1), chemokines such as IL-8, granulocyte-monocyte colony-stimulating factor (GM-CSF)

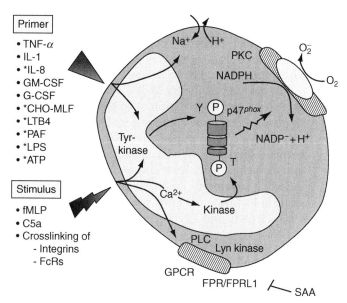

Fig. 2 Cellular signals triggered by priming of phagocytes. Ligand-specific activation may be substantially amplified by priming of phagocytes with multiple activators, including proinflammatory cytokines, colony-stimulating factors, lipid mediators, and ATP. Several interactive priming mechanisms have been discovered, including limited phosphorylation of p47phox, enhanced expression and microdomain redistribution of the NADPH oxidase complex at the plasma membrane, increased expression of triggering receptors, including G-protein coupled receptors redistribution of these receptors and second messengers as well as adaptor molecules and kinases in rafts. Flow cytometric quantification of responses revealed a homozygous alkalinization as crucial events during the oxidative burst cascade. (*) Designates ligands involved in priming at substimulatory concentrations, which may act as direct stimulants of the oxidative response at higher concentrations. The abbreviations not yet mentioned in the text include: CHO-MLF, N-formyl-methionyl-leucyl-phenylalanine; ATP, adenosine trisphosphate; SAA, serum amyloid A.

and granulocyte colony-stimulating factor (G-CSF), lipid mediators, such as lipopolysaccharide (LPS), leukotriene B_4 (LTB4), and platelet-activating factor (PAF), as well as formylated peptides (Dewas *et al.*, 2003; Hallett, 2003; Hallett and Lloyds, 1995; Wittmann *et al.*, 2004). The mechanisms of neutrophil priming encompass a complex and increasing array of cross talking cellular events, which ultimately amplify oxidant production and cytoskeletal rearrangements (Fig. 2). Priming may result from increases of formylated receptor (FPR) expression and enhancement of NADPH oxidase assembly and expression at the plasma membrane. This assumption has recently been supported by the finding of priming-induced redistribution of FPRs with G-protein-coupled receptors (GPCRs), Lyn kinase, and phospholipase C (PLC), as well as recruitment of the NADPH oxidase system and its effector protein kinase C (PKC) in specialized membrane microdomains or rafts (Keil *et al.*, 2003; Shao *et al.*, 2003). In fact, kinetic analyses showed that depletion of rafts delayed the onset of NADPH oxidase activation;

supporting the role of rafts as bundeling sites for functional receptor clustering and colocalization of signal transduction components (Shao *et al.*, 2003). Moreover, quantitative flow cytometry has revealed a central role of alkalinization for promoting neutrophil activation.

The final step in the phagocyte activation cascade, intracellular killing of microorganisms, and degradation of foreign substances, also involves a rearrangement of plasma membrane composition. Upon ligand binding, several resting dispersed cytosolic and membrane-bound enzymatic components of the NADPH oxidase become rapidly assembled in the plasma membrane. NADPH oxidase and the granule-associated enzyme myeloperoxidase generate reactive oxygen species including superoxide anion, hydroxyl radical, singlet oxygen, and hydrogen peroxide in the oxidative burst reaction (Babior *et al.*, 2003; Dahlgren and Karlsson, 1999; Forman and Torres, 2002).

B. Phagocyte Functions and Disease

Primary or acquired phagocyte dysfunctions or quantitative defects are associated with increased susceptibility to infections, leading to recurrent and often life-threatening infections, particularly with encapsulated bacteria, fungi, or parasites. Clinically, cutaneous infections, for example, abscesses, furuncles, carbuncles; respiratory infections, for example, sinusitis, pneumonia; and systemic purulent infection, for example, liver abscesses, are prevalent (Malech and Nauseef, 1997). The WHO classification of primary immunodeficiencies is expanded every 5 years, and defects have been described for all stages of phagocyte function, including (1) deficient chemotaxis, mobility, and adhesion, for example, actin dysfunction, Chédiak-Higashi syndrome, Shwachman disease, and gp110 deficiency, as well as (2) defective endocytosis and killing, for example, X-linked or autosomal recessive chronic granulomatous disease (CGD), myeloperoxidase deficiency, GP150 and neutrophil G6-PD deficiency (WHO, 1999). One of the best characterized groups of rare inherited disorders of phagocyte function encompasses CGD, which is characterized by defects in the phagocyte-specific NADPH-oxidase complex resulting in absent or reduced respiratory burst (Dinauer and Orkin, 1992; Segal and Holland, 2000).

Acquired dysfunctions of phagocytes are by far more frequent, and constitute important causes of morbidity and mortality particularly from infections in chronically ill patients, in patients with diabetes mellitus, autoimmune disease, renal or hepatic failure, alcoholism, as well as in immunosuppression in the context of trauma, surgery, burns, and viral infections (Engelich *et al.*, 2001). The mechanisms leading to phagocyte suppression are complex and often less well characterized compared to primary defects of phagocyte function and include chemical toxins (e.g., ethanol), metabolic disturbances (e.g., uremia), or pathologic activation in the circulation (e.g., burns, bypass). An impairment of monocytes/macrophage phagocytosis following HIV infection has been well documented, and forms the basis of increased susceptibility to opportunistic pathogens (Kedzierska *et al.*, 2003).

While reactive oxygen derivatives and proteases normally serve a microbicidal function, excessive or inappropriate release of these products contribute to inflammatory reactions and tissue injury (Babior, 2000). This increased activation of phagocytes with concomitant tissue destruction is believed to contribute to the pathogenesis of the adult respiratory distress syndrome (ARDS), as well as the systemic inflammatory response syndrome (SIRS), which may occur in absence of infection (Djaldetti *et al.*, 2002; Hasleton and Roberts, 1999). The expression of tissue factor by monocytes/macrophages contributes to their pathophysiological roles in disseminated intravascular coagulation linked to sepsis, postoperative thrombosis, unstable angina, atherosclerosis, chronic inflammation, and cancer (Osterud, 1998). Tissue factor is now considered to be the primary physiological activator of the blood coagulation system, and altered expression of tissue factor by monocytes has been recognized as a trigger for pathologic intravascular coagulation and has been observed in general inflammatory disorders, such as sepsis and hypoxic shock syndromes.

A further mechanism interfering with phagocyte function includes iatrogenic suppression by several drugs, such as general and regional anesthesia, particularly after using inhalative anesthetics and, paradoxically, many antibiotics (Djaldetti *et al.*, 2002). Antimicrobial drugs may adversely influence the interaction between neutrophils and microorganisms, and conversely, neutrophils may also interfere with the action of antimicrobial drugs (Pallister and Lewis, 2000). Moreover, a number of drugs has been reported not only to suppress functional activation, but also to trigger reduction of phagocyte numbers up to agranulocytosis, these include antiviral substances, for example, zidovudine, gancyclovir, and antibiotics, for example, sulfonamides, the dihydrofolate reductase inhibitors: trimetroprim, pyrimethamine, as well as antiparasitic drugs, such as pentamidine (Pisciotta, 1990).

C. Flow Cytometry as a Tool to Study Phagocyte Function

Numerous methods have been developed which allow the analysis of phagocyte function by flow cytometry at the single cell level. Phagocytosis of bacteria or latex particles can be quantitated per cell by flow cytometry (Steinkamp *et al.*, 1982). Furthermore, cell-bound and internalized pathogens can be discriminated (Bjerknes and Bassoe, 1983). Quantitative studies of bacterial uptake in this context allow the characterization of mechanisms of opsonization (Bassoe and Bjerknes, 1984), while the decay of microbe-associated fluorescence represents a correlate of bacterial degradation (Bassoe and Bjerknes, 1985).

The specific assessment of the functional activation of phagocytic cells became possible when fluorogenic substrates were induced, which allow the quantitation of the phagocyte-specific generation of reactive oxidants in the oxidative burst response (Bass *et al.*, 1983). Subsequently, probes were induced which can be used to detect the oxidative response to soluble receptor ligands such as *N*-formylated bacterial peptides (Rothe *et al.*, 1988) and which allow the differential analysis of membrane-associated superoxide anion generation and lysosomal peroxidase activity (Rothe and Valet, 1990).

These tools now allow to characterize receptor-specific cellular activation processes. Thus, cross-linking of receptors by antibodies combined with analysis of calcium fluxes and the oxidative burst response was applied for the characterization of cellular activation processes by glycosyl-phosphatidylinositol (GPI) anchored surface receptors (Lund-Johansen *et al.*, 1990). Flow cytometry similarly has been used to characterize the antagonism of the acute-phase protein serum amyloid A and fMLP in stimulating the oxidative burst (Linke *et al.*, 1991) which later was identified to be due to binding to the same receptor FPRL1 (He *et al.*, 2003). The interaction of different activation processes such as the priming of the cellular response to fMLP by proinflammatory cytokines (Fig. 3) can also be analyzed through the flow cytometric oxidative burst analysis at the single cell level (Wittmann *et al.*, 2004). Finally, regulatory processes in the conformational activation of receptor clusters such as the clustering of the LPS receptor CD14 with integrins or Toll-like-receptors and other receptors mediating signal transduction have been addressed using fluorescence resonance energy transfer (FRET; Pfeiffer *et al.*, 2001; Zarewych *et al.*, 1996). The anti-inflammatory downregulation of activation processes by IL-10, finally, has been addressed by the analysis of apoptosis (Keel *et al.*, 1997) and has been linked to intracellular recruitment of IL-10 receptors (Elbim *et al.*, 2001).

The methods now allow direct assessment of basal and activated cell surface receptor expression, measurement of intracellular pH changes by the use of pH-sensitive fluorescent colors, detection of phagocytic activity based on incubation of phagocytic cells with fluorescent conjugated particles, and FRET, using suitable labeled antibody pairs, to visualize functional receptor clustering.

This review focusses on several of the assays which can be used to characterize the graded response of neutrophils and monocytes to activation. These assays can be grouped in four categories as shown in Table I.

1. Cytosolic Ca^{2+} Transients

Cytosolic free Ca^{2+} is a universal second messenger, which is increased within seconds following cellular activation (Berridge *et al.*, 2000; da Silva and Guse, 2000; Lewis, 2001). In resting cells, a low cytosolic free Ca^{2+} concentration is actively maintained at 100–150 nM by a Ca^{2+} ATPase, compared to the 10^4-fold higher free Ca^{2+} concentration (1.3 mM) in the extracellular space. Upon surface receptor stimulation an increase in the cytosolic free Ca^{2+} concentration in nonmuscle cells typically occurs in three phases. First, receptor-coupled PLC activity leads to an initial liberation of calcium from calciosomes, the intracellular calcium stores of nonmuscle cells. In the second phase, an influx of calcium from the extracellular space is induced. This is followed within a few minutes by reuptake of Ca^{2+} by the calciosomes and export through the plasma membrane-bound Ca^{2+} ATPase leading to the termination of the generalized increase in cytosolic Ca^{2+}.

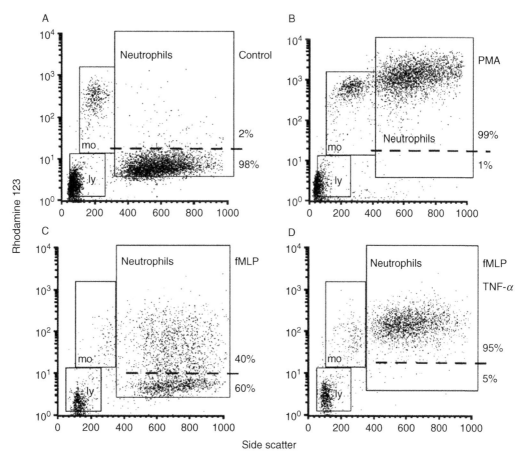

Fig. 3 Oxidative burst response of neutrophils and monocytes (mo) as detected by intracellular oxidation of dihydrorhodamine to rhodamine 123. (A) In the absence of stimulation only monocytes show rhodamine 123 fluorescence resulting from spontaneous intracellular oxidation of the substrate; (B) PMA induces a homogeneous and maximal response of neutrophils and monocytes while lymphocytes (ly) which do not express NADPH oxidase remain negative; (C) fMLP as a physiological stimulus induces an oxidative burst response in a subpopulation of neutrophils, only; (D) preincubation with proinflammatory cytokines such as TNF α leads to an enhanced response to stimulation with fMLP.

A sustained activation of calcium-dependent cellular responses is then maintained by locally increased calcium fluxes in the submembrane space only.

Ca^{2+} signaling, in general, is regulated by a number of tools (Berridge *et al.*, 2000): (1) Ca^{2+}-mobilizing signals including the intracellular messengers inositol-1,4,5-triphosphate, cyclic ADP ribose, nicotinic acid dinucleotide phosphate, and sphingosine-1-phosphate; (2) channels in the plasma membrane or the endoplasmic reticulum membrane; (3) Ca^{2+} sensors, including the ubiquitous Ca^{2+}-binding protein calmodulin and a wide variety of Ca^{2+}-calmodulin-activated proteins;

Table I

Applications of Flow Cytometry in the Study of Neutrophil and Monocyte Activation

Method	Labeling
Analysis of signaling and priming and the phagocytic defense cascade	
Ca^{2+} transients	Indo-1, Fluo-3
F-actin generation	Phalloidin, phallicidin
Cytosolic alkalinization	SNARF-1
The oxidative burst response	Dihydrorhodamine 123, hydroethidine
Lysosomal proteinases	$(Z-Ala-Ala)_2$-R110, $(Z-Arg-Arg)_2$-R110
Analysis of phagocytosis and an adhesive phenotype	
Phagocytosis of bacteria	FITC-conjugated bacteria
Expression of adhesion antigen and granule proteins	CD11b, CD62L, CD66b
Characterization of ligands and receptors	
Recruitment of receptors involved in signal transduction	FLPEP, Interleukin-10-Receptor
Coassembly of heteromeric receptor complexes	CD14/CD11b, CD14/TLR-4
Autocrine secretion of cytokines	IL-8
Analysis of the anti-inflammatory termination of activation	
Induction of apoptosis	Annexin V

and (4) mechanisms which reduce the free cytoplasmic Ca^{2+} concentration by sequestering/buffering or pumping it out of the cytoplasm.

2. Cytosolic Alkalinization

Intracellular pH is closely regulated in the range of 7–7.4 in eukaryotic cells. The regulation of the pH is tightly related to the function of leukocytes (Hackam *et al.*, 1996) and especially phagocytic cells which maintain a distinct pH within their phagocytic organelles dependent on cellular activation (Bernardo *et al.*, 2002; Jankowski *et al.*, 2002). pH regulation also is closely related to cell survival and apoptosis (Ishaque and Al-Rubeai, 1998). The most common pH-regulatory mechanism is an amiloride-sensitive, electroneutral Na^+/H^+-antiport which uses the inward-directed Na^+ gradient for H^+ extrusion. Alkalinization of intracellular pH is induced by upregulation of the antiport following stimulation by hormones or growth factors. In addition, pH-regulatory anion antiport mechanisms such as a Cl^-/HCO_3^--antiport seem to be present in most cells. The regulation of the intracellular pH is of major importance for cellular activation as increases in intracellular pH are associated with an increased metabolic activity of cells. In the case of neutrophils activation of the Na^+/H^+-antiport, for example, during spreading, is tightly linked to NADPH oxidase activity and the oxidative burst response as the intracellular alkalinization induced by the antiport directly compensates for the acidification induced by the transmembrane transport of electrons during superoxide anion generation (Demaurex *et al.*, 1996).

3. The Oxidative Burst Response

Professional phagocytes are well equipped with specific enzyme complexes that raise the production of reactive oxygen species, and of ozone following ligand-induced assembly in cell membranes (Babior, 2004; Babior et al., 2003; Clark, 1999; Dahlgren and Karlsson, 1999; Forman and Torres, 2002). Reactive oxidants are released into phagosomes and into the extracellular space during the oxidative burst reaction, which contributes to their microbicidal activity. The NADPH oxidase constitutes one of the best characterized phagocyte-specific enzyme complexes, which in resting cells is disassembled in cytosolic and membrane-bound complexes. Upon activation, the enzyme incorporates into the plasma or lysosomal membrane and catalyzes the transmembrane transport of electrons from NADPH to molecular oxygen. Superoxide anion, the product of the one-electron-reduction of molecular oxygen, dismutates to hydrogen peroxide, which finally is converted by myeloperoxidase into long-lasting oxidants such as hypochlorous acid or chlorinated tauramines.

$$\underset{\text{NADPHoxidase}}{NADPH + 2O_2 \rightarrow NADP^+ + H^+ + 2O_2^-}$$

$$\underset{\text{Superoxidedismutase}}{2O_2^- + 2H^+ \rightarrow H_2O_2 + O_2}$$

$$\underset{\text{Myeloperoxidase}}{H_2O_2 + X^- + H^+ \rightarrow HOX + H_2O} \quad (X^- = Cl^-, Br^-, I^-, SCN^-)$$

Two other oxidants, hydroxyl radical originating either from an iron-catalyzed reaction of O_2^- with H_2O_2, a reaction of O_2^- with HOCl, or a reaction of O_2^- with nitric oxide, and singlet oxygen, generated from a water-induced dismutation of O_2^-, or in a myeloperoxidase-dependent reaction, may also play a role in the respiratory burst reaction. Moreover, activated granulocytes may produce an oxidant with the chemical signature of ozone, a particularly powerful oxidant (Babior et al., 2003). Whether a production of hydroxyl radical or singlet oxygen by neutrophils occurs under physiological reactions is unclear.

4. Lysosomal Proteinases

Cellular endopeptidases are comprised of four classes of enzymes defined by the action of class-specific inhibitors, (1) serine proteinases inhibitable by diisopropyl-phosphofluoridate (DFP), (2) cysteine proteinases inhibitable by E-64, (3) aspartic proteinases inhibitable by pepstatin, and (4) metalloproteinases inhibitable by phenanthroline (Hooper, 2002; Kirschke and Barrett, 1987). Due to a wide substrate specificity, these enzymes are involved in intracellular protein turnover as well as extracellular protein degradation during inflammation. Cell lineage-dependent

depend on carbon dioxide in case of incubation in ambient air. Phagocytic cells tend to adhere to surfaces when incubated at 37 °C at low protein concentrations. Polypropylene tubes are preferable to polystyrene or glass tubes due to a less adhesive surface.

B. Ligands

The bacterial peptide fMLP represents a ligand which is capable of inducing calcium transients, activation of the Na^+/H^+-antiport leading to cytosolic alkalinization and the oxidative burst response in a dose-dependent manner at concentrations ranging from 10^{-9} to 10^{-5} M. Cytokines such as TNF-α, GM-CSF, and IL-8 can be employed for *in vitro* induction of neutrophil priming similar to the reactivity of neutrophils in inflamed tissues. Phorbol 12-myristate 13-acetate can be used to induce a maximal activation of neutrophils or monocytes through direct activation of PKC.

C. Analysis of Ca^{2+} Transients

The cytosolic-free calcium concentration can be measured intracellularly by the spectral shift of indo-1 from green fluorescence in the absence of Ca^{2+} to blue fluorescence in the presence of Ca^{2+}. This allows precise determination of the calcium concentration independently of the cell size and cellular dye loading. First, indo-1 is accumulated intracellularly through intracellular cleavage of the membrane-permeable derivative indo-1/acetoxymethyl ester (AM) by esterases. The cellular blue to green fluorescence ratio is then analyzed as a measure of the intracellular free calcium concentration. A molar calibration curve can be obtained from the analysis of cells incubated in buffers of defined Ca^{2+} concentration in the presence of the Ca^{2+} ionophore ionomycin. Intracellular buffering of Ca^{2+} may lead to difficulties in calibrating concentrations of Ca^{2+} lower than the 100–150 nM typically found in resting cells. In addition, protein interaction can modify the Kd of Ca^{2+} chelating fluorophores. The spectral response of the flow cytometer when flushed with indo-1 and buffers of known Ca^{2+} concentrations, therefore, also may be recorded to compare the spectral Ca^{2+} calibration curve of the flow cytometer with the Ca^{2+} calibration curve of the cells.

The analysis of the kinetic response of intracellular Ca^{2+} to stimulation is achieved in three steps (do Céu Monteiro *et al.*, 1999; Réthi *et al.*, 2002). First, cells loaded with indo-1 are analyzed at 37 °C without stimulation. Then a stimulus is added, and the increase and decline in intracellular Ca^{2+} concentration are measured kinetically until a stable level is reached again. This typically occurs in 2–5 min. Finally, ionomycin is added to record the maximum indo-1 ratio in the presence of Ca^{2+} as a positive control.

If fluorescence excitation at 365 nm is not possible, for example, using argon laser based instruments which allow excitation at 488 nm only, the Ca^{2+}-sensitive fluorescein derivative fluo-3 may be used alternatively. Binding of Ca^{2+} does not change the fluorescence emission spectrum of fluo-3 but increases the green fluorescence intensity by 40-fold (Minta *et al.*, 1989) or more dependent on the source of the reagent. This allows sensitive detection of the kinetically fast relative increases in cytosolic free Ca^{2+} following stimulation. Absolute Ca^{2+} concentrations, however, are difficult to measure with fluo-3 as the cellular fluorescence intensity is substantially affected by differences in the extent of enzymatic dye loading, cell size, compartmentalization of the dye, and effects of protein binding on the Kd of the probe.

D. Cytosolic Alkalinization

The cytoplasmic pH value can be measured intracellularly through the spectral shift of the 488-nm excitable SNARF-1 to a longer wavelength fluorescence emission with increasing pH value. In the first step, SNARF-1 is accumulated intracellularly by intracellular cleavage of the membrane-permeable derivative SNARF-1/AM by esterases. The cellular red to orange (SNARF-1) fluorescence ratio is then analyzed as a measure of the intracellular pH value. A pH calibration curve can be obtained from the analysis of cells incubated in buffers of defined pH value with equilibration of the intracellular pH by addition of the ionophore nigericin in the presence of a high K^+ concentration (Thomas *et al.*, 1979).

In addition to differences in resting intracellular pH value between cells, the response of the intracellular pH to stimulation may be analyzed. The stimulation-dependent activation of the Na^+/H^+ antiport typically leads to prolonged cytoplasmic alkalinization. This response can be analyzed by addition of the stimulus during the SNARF-1 loading procedure at 37 °C.

E. The Oxidative Burst Response

Formation of the reactive oxidants superoxide anion and hydrogen peroxide during the oxidative burst is measured intracellularly by the oxidation of membrane-permeable fluorogenic substrates such as DHR or hydroethidine (HE) to fluorescent products. The intracellular oxidation of nonfluorescent DHR to green fluorescent rhodamine 123 by hydrogen peroxide and peroxidases is the most sensitive method for the analysis of the oxidative burst response (Lund-Johansen *et al.*, 1990; Rothe *et al.*, 1988, 1991; Vowells *et al.*, 1995). This high sensitivity is reached by the intracellular retention of the positively charged fluorescent product rhodamine 123 at mitochondrial binding sites. The blue fluorescent HE, in contrast to DHR, can be oxidized to red fluorescent ethidium bromide (EB) already by superoxide anion, the primary product of the oxidative burst response (Rothe and Valet, 1990; Walrand *et al.*, 2003). Cells are first preincubated with the fluorogenic substrates at 37 °C. An oxidative burst response is then induced by addition of stimulatory ligands such as fMLP with or without response modifiers such as TNF-α. PMA may be used as a positive control or for the analysis of the cellular expression of the oxidative burst enzymes. The incubation is stopped by transferring cells onto ice. The cellular accumulation of the fluorescent oxidation products is then analyzed on the flow cytometer. As the cellular fluorescence is stable for about 90 min on ice, cells may be subsequently labeled with mAbs. Fixation with paraformaldehyde retains the cellular localization of the fluorescent product.

F. Lysosomal Proteinases

Lysosomal proteinase activity can be analyzed intracellularly by the conversion of nonfluorescent and membrane-permeable, specifically bis-substituted R110-peptide derivatives to green fluorescent monosubstituted R110 and free R110 in a two-step reaction (Leytus et al., 1983a,b; Rothe et al., 1992). Cells are incubated kinetically at 37 °C with appropriate substrates such as $(Z-Arg-Arg)_2-R110$ for the cysteine proteinases cathepsin B and L of monocytes or $(Z-Ala-Ala)_2-R110$ for the serine proteinase elastase of neutrophils. The specificity of the reaction is shown by the inhibition of the reaction through preincubation with specific inhibitors such as $Z-Phe-Ala-CHN_2$ for cathepsin B and L (Green and Shaw, 1981) or DFP for neutrophil elastase (Powers, 1986). As the cellular fluorescence is stable for about 90 min on ice, cells may be subsequently labeled with mAbs.

G. Analysis of fMLP-Binding

Quantification of binding sites for fMLP can be performed by staining with the fluoresceinated peptide formyl-Nle-Leu-Phe-Nle-Tyr-Lys (FLPEP; Sklar et al., 1981). The binding of FLPEP is quantitated by flow cytometry after incubation at 4 °C in order to avoid cellular stimulation (Allen et al., 1992). Nonspecific binding is assessed by incubation with an excess of fMLP in parallel.

H. Coassembly of Heteromeric Receptor Complexes

Receptor colocalization can be assessed by flow cytometry using FRET for the determination of the proximity of antibodies directed at different components of a heteromeric receptor clusters (Szöllösi, this volume). In short, in case of R-phycoerythrin as the donor and Cy5 as the acceptor, spatially resolved dual laser excitation at 488 and 635 nm is used for the independent quantification of both fluorochromes. Energy transfer is calculated as described by Szöllösi (this volume) based on the determination of the increased Cy5-specific fluorescence excited at 488 nm after simultaneous staining with donor and acceptor in comparison to the fluorescence of donor-only or acceptor-only stained samples.

A change in the spatial distance to CD11b can be used for the determination of the activation of the pattern recognition receptor CD14 by LPS or one of its alternative bacterial or inflammatory ligands (Pfeiffer et al., 2001; Zarewych et al., 1996). Pairs of antibodies to the same pairs of proteins, for example CD14 and CD11b, differ in their efficiency for showing FRET due to the relative proximity of epitopes. Indirect staining can improve transfer efficiency due to an increased radius of fluorochromes around epitopes (Koksch et al., 1995).

======== **IV. Methods**

A. Preparation of Cells

1. Isolation of Leukocytes Through Sedimentation of Erythrocytes on Ficoll

Blood is depleted of erythrocytes by a method which avoids contact of leukocytes to any separation media. Undiluted heparinized (10 U/ml) whole blood (3 ml) is layered on top of 3 ml lymphocyte separation medium (ficoll, density 1.077). Erythrocytes will aggregate at the interface and sediment at room temperature without centrifugation. After 40 min, the upper 800 μl supernatant plasma is withdrawn avoiding contact to the plasma fraction near the interface to the separation medium and stored on ice. This will contain platelets and approximately 2×10^7 ml^{-1} unseparated leukocytes in autologous plasma.

B. Analysis of Ca^{2+} Transients

1. Measurement of Cytosolic Free Calcium with Indo-1

1. Incubate 5×10^6 cells/ml in HEPES-buffered medium with 0.5–5 μM indo-1/AM at 37 °C. An equilibrium of dye loading is reached within 20–30 min of incubation. Depending on the cell type, select the lowest concentration of substrate which results in a homogeneous fluorescence ratio prior to stimulation.

2. Add propidium iodide (PI) at 30 μM final concentration and analyze the blue fluorescence of indo-1/Ca^{2+} complexes using a 390- to 440-nm bandpass filter, and the blue-green fluorescence of Ca^{2+}-free indo-1 and the red PI fluorescence of dead cells using a 490-nm long-pass filter with excitation at 351/363 nm by an argon laser or a high pressure mercury arc lamp with a 360- to 370-nm bandpass filter.

3. For the analysis of stimulation-dependent changes in intracellular Ca^{2+}, add an appropriate stimulus, for example, fMLP (10^{-8} M) in the case of neutrophils, following an initial measurement for 30 s. Continue kinetic data acquisition for approximately 3 min.

4. Add ionomycin (2 μM) as a positive control for the maximal Ca^{2+} response of the cells.

2. Measurement of Cytosolic Free Calcium with Fluo-3

1. Incubate 5×10^6 cells/ml in HEPES-buffered medium with 0.5–2 μM fluo-3/AM at 37 °C. An equilibrium of dye loading is reached within 20–30 min of incubation. Depending on the cell type, select the lowest concentration of substrate which results in a homogeneous fluorescence prior to stimulation.

2. Add PI and at 30 μM final concentration analyze the green fluorescence of fluo-3 using a 515- to 535-nm bandpass filter, and the red PI fluorescence of dead cells using a 620-nm long-pass filter with excitation by the 488-nm line of an argon laser.

3. For the analysis of stimulation-dependent changes in intracellular Ca^{2+}, add an appropriate stimulus, for example, fMLP (10^{-8} M) in the case of neutrophils, following an initial measurement for 30 s. Continue kinetic data acquisition for approximately 3 min.

4. Add ionomycin (2 μM) as a positive control for the maximal Ca^{2+} response of the cells.

C. Cytosolic Alkalinization

1. Incubate 5 $\times$ 10^6 cells/ml in HEPES-buffered medium with 0.2–1 μM SNARF-1/AM at 37 °C. An equilibrium of dye loading is reached within 20-30 min of incubation. Depending on the cell type, select the lowest concentration of substrate which results in a homogeneous fluorescence ratio in resting cells.

2. For the analysis of stimulation-dependent changes in intracellular pH, incubate for the final 15 min in the presence of appropriate stimuli, for example, fMLP (10^{-8} M) or TNF-α (1 ng/ml) in the case of neutrophils.

3. Add PI at 30 μM final concentration and analyze the orange fluorescence of acidic SNARF-1 using a 575- to 595-nm bandpass filter, and the red fluorescence of basic SNARF-1 and the PI fluorescence of dead cells using a 620-nm long-pass filter with excitation by the 488-nm line of an argon laser or a high pressure mercury arc lamp with a 470–500 nm bandpass filter.

4. For the generation of a calibration curve, incubate cells with SNARF-1/AM as above. Divide the sample into six aliquots and spin down at 60 $\times$ g and 4 °C for 5 min. Resuspend for 5 min at room temperature in the high K^+ calibration buffers (pH 6.40, 6.80, 7.20, 7.60, 8, 8.40) supplemented with 10 μM nigericin followed by flow cytometric analysis.

D. The Oxidative Burst Response

1. Incubate 5 $\times$ 10^6 cells/ml in HEPES-buffered medium with 1 μM DHR (stock 1 mM in N,N-dimethylformamide, DMF) or 10 μM HE (stock 10 mM in DMF) at 37 °C for 5 min.

2. Incubate for 15–30 min with appropriate stimuli, for example, fMLP (10^{-6} M) or PMA (10^{-7} M) in the case of neutrophils. The cellular "priming" by incomplete stimuli such as TNF-α (10 ng/ml) may be analyzed by 5 min preincubation with the cytokine followed by incubation for 15 min a low concentration of fMLP (10^{-7} M).

3. Add PI at 30 μM final concentration and analyze the green fluorescence of rhodamine 123 using a 515- to -535-nm bandpass filter, and the PI red fluorescence of dead cells or the EB red fluorescence of HE-stained cells using a 620 nm long-pass filter using excitation with the 488 nm line of an argon laser.

E. Lysosomal Proteinases

1. Preincubate 5×10^6 cells/ml in HEPES-buffered medium with or without specific proteinase inhibitor such as 100 μM Z-Phe-Ala-CHN$_2$ (stock 100 mM in DMSO) or 1 mM DFP (stock 1 M in DMSO) at 37 °C for 10 min.

2. Incubate for 20 min with appropriate proteinase substrate such as 4 μM (Z-Arg-Arg)2-R110 for the cathepsin B/L of monocytes or 4 μM (Z-Ala-Ala) 2-R110 for the elastase of neutrophils.

3. Add PI at 30 μM final concentration and analyze the green fluorescence of R110 using a 515- to 535-nm bandpass filter, and the red PI fluorescence of dead cells using a 620 nm long-pass filter using excitation with 488-nm line of an argon laser.

F. Analysis of fMLP-Binding

1. Preincubate 5×10^6 cells/ml in HEPES-buffered medium with or without specific agonists for 30 min at 37 °C for the study of the upregulation of binding sites. Stop the incubation by cooling to 4 °C.

2. Incubate for 20 min with 20 nM FLPEP and wash twice. Coincubate with fMLP (10^{-4} M) in a parallel set of tubes for determination of nonspecific binding of FLPEP.

3. Analyze the green fluorescence of FLPEP using a 515- to 535-nm bandpass filter and excitation with 488-nm line of an argon laser.

G. Coassembly of Heteromeric Receptor Complexes

1. Preincubate whole blood with or without specific agonists, for example, LPS from *Salmonella minnesota* for 15 min at 37 °C. Wash cells with cold PBS containing 0.1% NaN$_3$.

2. Incubate the washed whole blood samples for 15 min at 4 °C with saturating concentrations of the fluorochrome-conjugated or biotinylated antibodies, for example, CD11b (clone D12) as a R-PE-conjugate and CD14 (clones x8) in biotinylated form. A separate sample is incubated with the biotinylated antibody only. Lyse the cells with a one-step fix and lyse solution and wash twice.

3. Divide the samples that have been stained with both antibodies into two parts and stain one part of each sample with a saturating amount of streptavidin-Cy5 for 15 min on ice.

4. Wash the cells and analyze the cells using spatially resolved dual laser excitation at 488 and 635 nm. For all samples, independent of staining, the orange fluorescence of R-PE is analyzed at with a bandpass filter centered at 585 nm with 488 nm excitation and the red fluorescence of Cy5 is analyzed above 650 nm with separate excitation by both lasers.

5. Energy transfer is calculated separately both as the decrease in donor fluorescence and as the increase in donor-dependent and 488-nm excited acceptor fluorescence as described in detail elsewhere (Pfeiffer *et al.*, 2001; Szöllösi, this volume).

V. Results

The analysis of the oxidative burst response can serve as a typical result for the activation-dependent characterization of neutrophil function. As shown in Fig. 3A, neutrophils do not show intracellular oxidation of dihydrorhodamine 123 to rhodamine 123 in absence of stimulation as indicated by a similar fluorescence to lymphocytes which do not express NADPH oxidase. Monocytes, however, typically show a significant oxidative burst activity representing spontaneous generation of oxidants. Stimulation with PMA induces a more than 100-fold and homogenous increase of rhodamine 123 fluorescence in neutrophils and a more than threefold increase of fluorescence in monocytes indicating a maximal induction of the respiratory burst response.

fMLP as a physiological agonist of neutrophil activation when applied at a low concentration, for example, 10^{-7} M typically induces an oxidative burst response in a subpopulation of neutrophils, only (Fig. 3C). Increasing the amount of fMLP, for example, incubation with 10^{-6} M fMLP, results in an increase of the size of the population of reactive neutrophils indicating that the cellular heterogeneity corresponds to the graded induction of an all-or-none-response in all cells rather than to a constitutive heterogeneity of cells. Preincubation with an "incomplete" stimulus, for example, 1 ng/ml TNF-α which does not directly induce an oxidative burst response in cells in suspension, results in a significantly enhanced response to fMLP-stimulation (Fig. 3D) indicating "priming" of neutrophils by the cytokine (Fig. 2).

A dose-dependent priming of the oxidative burst response of neutrophils is obtained for priming with TNF-α, GM-CSF, or IL-8 (Wittmann *et al.*, 2004). fMLP-receptor expression and an intracellular alkalinization are induced with a similar potency by these cytokines and can serve as a direct indicator of the "priming" of cells without a secondary stimulation, for example, by fMLP.

Transient increases of cytosolic Ca^{2+} concentrations are typically induced by lower concentrations of fMLP, for example, 10^{-9} M, compared to those necessary for the induction of an oxidative burst response. The assay is particularly useful for the determination of the reagibility of cells to a given ligand or the cross-linking of receptors (Lund-Johansen *et al.*, 1990, 1993), but can be hardly used for a quantitative comparison of the potency of ligands.

The determination of intracellular elastase activity in neutrophils and of cathepsin B activity in monocytes or macrophages allows the identification of an altered maturation and differentiation of cells, for example, in inflammatory disorders or neoplastic diseases of the hematopoietic systems. An increased cathepsin B activity is closely related to the *in vivo* activation of monocytes. Degranulation of cells which is observed *in vivo* at sites of inflammation, however, is hardly observed in cells circulating in peripheral blood suggesting that this process is linked to cellular adhesion.

A dose-dependent clustering of CD14 and CD11b *in vitro* occurs in monocytes following stimulation with LPS or alternative ligands of CD14 such as lipoteichoic acid (Pfeiffer *et al.*, 2001). Stimulation with agonists such as fMLP or PMA which differ in their mechanism of cell activation does not result in spatial proximity of the receptors as detected by FRET analysis. Spontaneous clustering of CD14 and CD11b can be observed in the *ex vivo* analysis of blood from patients with inflammatory conditions.

VI. Pitfalls and Misinterpretation of the Data

An artificial activation of cells as indicated by a spontaneous oxidative burst response or nonspecific staining which occurs through redistribution of fluorescent products following incubation, are the two most critical problems in the functional analysis of phagocytic cells. Both problems can be addressed by an undelayed processing of fresh samples as well as a timely analysis of cells after staining.

A. Analysis of Ca^{2+} Transients

• If only low cellular accumulation of fluorescent indo-1 is reached, dye loading may be improved by addition of the nonionic dispersing agent pluronic F-127 (Poenie *et al.*, 1986).

• Low changes in the cellular indo-1 fluorescence ratio even after addition of ionomycin may be due to incomplete hydrolysis of indo-1/AM. This may occur in some cell types resulting in Ca^{2+} insensitive fluorescence. A more complete hydrolysis may be reached by washing the cells and incubating for a prolonged time in a substrate-free medium prior to analysis.

• A lack of a cellular response to stimulation despite a normal response following addition of ionomycin may be due to a low Ca^{2+} concentration extracellularly, for example, when media such as RPMI 1640 are used, or due to intracellular buffering of Ca^{2+} by the fluorescent indicator. This can occur due to high intracellular concentrations of the dyes in the micromolar range compared with the 100-fold lower Ca^{2+} concentration and the low Kd values of the dyes for Ca^{2+}, for example, 250 nM in the case of indo-1. Therefore, the lowest substrate concentration should be used which results in a homogeneous fluorescence ratio of resting cells. No interaction of the dye loading procedure with the cellular sensitivity for stimulation may be shown by independent cellular responses such as chemotaxis, depolarization of the membrane potential, or an oxidative burst response in phagocytic cells.

B. Cytosolic Alkalinization

Cytosol has a high buffering capacity for intracellular pH. Therefore, the additional buffering of the intracellular pH value by the intracellular indicator usually does not significantly affect the pH measurement.

- A lack of a cellular response to stimulation despite pH-sensitive cellular fluorescence as observed by calibration with nigericin may be due to the simultaneous activation of H^+ extruding antiports and metabolic activity which leads to the intracellular generation of H^+. Na^+/H^+ antiport activity can be shown in this case by pH changes following preincubation with the specific Na^+/H^+ antiport inhibitor amiloride.

C. The Oxidative Burst Response

- Only low cellular responses to stimulation may be due to the presence of high [>0.1% (v/v)] concentrations of organic solvents such as dimethylsulfoxide (DMSO) or DMF used in the reagent stock solutions, which both act as scavenger for reactive oxidants and interfere with cellular stimulation.

- High fluorescence in nonphagocytic cells such as lymphocytes indicates spontaneous oxidation of the fluorogenic substrates, which are unstable when exposed to excessive light or stored at higher temperature.

- Specificity of the intracellular accumulation of the fluorescent products for the oxidative burst response may be shown by preincubating the cells with 100 μM diphenyl iodonium as a specific inhibitor of the NADPH oxidase of phagocytic cells (Cross, 1987). Hydrogen peroxide (1 mM) may be used as positive control to show sensitivity of the substrates for intracellular oxidation.

- Cellular staining is usually stable for at least 2 h at 4 °C due to the intracellular trapping of rhodamine 123 at mitochondrial binding sites. If prolonged stability of the assay samples is required, fixation with 1% paraformaldehyde (w/v) in phosphate buffered saline may be performed to maintain cellular staining.

D. Lysosomal Proteinases

- Selective cell death occurring in samples incubated with the proteinase substrates but not in samples preincubated with the inhibitors prior to incubation with the proteinase substrates may be a result of the local accumulation of high amounts of the product R110 inside cellular lysosomes. This can be avoided by reducing the substrate concentration or time of incubation.

- The susceptibility of fluorogenic peptide substrates for cleavage by different enzymes typically strongly depends on the precise assay conditions such as pH or buffer as well as the fluorophore (Assfalg-Machleidt *et al.*, 1992). It is, therefore, critical to prove the specificity of the intracellular cleavage of substrates using, for example, specific cell permeant inhibitors or cellular models which selectively do or do not express a specific enzymes such as transfected cells.

E. Coassembly of Heteromeric Receptor Complexes

- A significant energy transfer between antibodies directed against CD14 and CD11b when analyzing blood samples obtained from healthy individuals in the absence of stimulation indicates artifactual activation of cells. The time after sample drawing as well as contamination of buffers is the most critical components of the assay.

VII. Comparison with Other Methods

The intracellular oxidation of 2′,7′-dichlorofluorescin (DCFH) to green fluorescent 2′,7′ dichlorofluorescein was used in an earlier flow cytometric assay for the oxidative burst response of phagocytic cells (Bass *et al.*, 1983). DCFH oxidation in comparison to the intracellular oxidation of DHR results in a far lower cellular fluorescence of stimulated phagocytic cells, making the analysis of cellular responses to low amounts of physiological stimuli impossible (Rothe *et al.*, 1991). The assay is further complicated by a high background fluorescence in cells which are not capable of oxidative burst activity, such as lymphocytes and the need to accumulate the fluorogenic substrate in a first step through a hydrolytic intracellular cleavage of the membrane-permeable derivative 2′,7′ dichlorofluorescin diacetate. In addition to H_2O_2 and peroxidases, further mechanisms such as reactive nitrogen species are involved in DCFH oxidation (Walrand *et al.*, 2003).

In the analysis of phagocytosis by neutrophils or monocytes flow cytometry can be used to assess the attachment and internalization of ligands depending as a result of their opsonization as well as secondary reactions such as the phagosomal pH and the oxidative burst response induced. These techniques are reviewed, for example, by Lehmann *et al.* (2000). As a major drawback phagocytosis is a rather strong stimulus for cells which can be hardly applied in a graded manner. Defects in intracellular killing of microbes are not specifically addressed by the currently available methodologies. These techniques are, therefore, not very sensitive to detect changes of phagocyte function in disease.

VIII. Applications and Biomedical Information

A. Applications in Identification of Carriers in Primary Immunodeficiency Screening

While the role of flow cytometry in the diagnosis of primary immunodeficiencies is undisputed, only few reports address the potential role in detecting heterozygous carriers in affected families. One important example is CGD, which is caused by different mutations in at least five components of the phagocytic NADPH oxidase. The most frequent mutations, such as the 25% affecting the NCF1 or p47[phox] gene, causing complete deletions, may be addressed by sequencing-based methods (Dekker *et al.*, 2001). However, the majority of the CGD-causing mutations cannot be addressed by gene-scan methods. By contrast, flow cytometric methods determining the phagocyte oxidative burst in response to phagocytosis of bacteria or activation by PMA, are suitable as a very sensitive diagnostic screening (Emmendörffer *et al.*, 1994). Flow cytometry, using dihydrorhodamine 123 oxidation, clearly allows to distinguish heterozygous individuals with one NCF1 gene (carriers) from controls and from NCF1-deficient patients (Atkinson *et al.*, 1997; Crockard *et al.*, 1997; Emmendörffer *et al.*, 1994). Moreover, it permits the detection of novel defects of phagocyte function, prior to molecular identification. Thus, flow cytometry may serve as an adjunct for genetic counseling, particularly in sisters of affected patients.

B. Application in Inflammation and Sepsis

Professional phagocytes in the circulation respond weekly to many external ligands. Under conditions of stress, particularly in inflammation, cytokines may prime phagocytes to dramatically increase their responsiveness to bacterial ligands (Figs. 2 and 3). Flow cytometry allowed the characterization of cytokine-priming of phagocytes to fMLP-induced oxidative burst and actin polymerization in the presence of rapid changes of fMLP-receptor cycling (Elbim *et al.*, 1993; Wittmann *et al.*, 2004). In sepsis and SIRS flow cytometry may be a tool for the identification of individuals with a hyperresponsive inflammation which may profit from immunomodulatory therapy. Since a dysregulation of fMLP-induced L-selectin shedding and oxidative burst in cytokine-primed granulocytes of asymptomatic HIV patients could be demonstrated, flow cytometric follow-up may contribute to the prediction of the susceptibility to infections (Elbim *et al.*, 1994).

A central feature of flow cytometry in the characterization of the inflammatory transition from resting, over an intermediate primed state with restrained neutrophil activity to full-blown activation of neutrophils is the capability to identify regulatory Ca^{2+} fluxes (Hallett, 2003; Hallett and Lloyds, 1995), as well as actin polymerization, intracellular alkalinization, upregulation of fMLP-receptors, and the heterogeneity of the cytokine-primed oxidative burst response in a closely correlated way in the same cell (Rothe and Valet, 1994; Wittmann *et al.*, 2004; Yamashiro *et al.*, 2001). This has already led to the identification of a number of mechanisms involved in the altered function of phagocytic cells in inflammation (Fig. 2) and should be of value to identify potential mechanisms for the modulation of the altered function of these cells.

C. Immunopharmacological Applications

Several drugs compromise phagocyte function, including such different chemical compounds as general and inhalative anesthetics, antibiotics, chemotherapeutics, or antiparasitic drugs, increase the risk of infections or counteract the beneficial effect of antimicrobials. By using flow cytometry, a direct dose-dependent reduction of complement or fMLP-induced oxidative response of neutrophils in patients during general anesthesia nitrous oxide could be demonstrated

(Fröhlich *et al.*, 1997). The mechanism leading to functional depression of granulocytes differed with the chemical compounds, and involved inhibition of cellular signaling upstream from PKC (Fröhlich *et al.*, 2002). In fact, distinctive inhibition of the neutrophil oxidative function was shown for different anesthetic substances (Fröhlich *et al.*, 2002). Since suppression of granulocyte function following general anesthesia may contribute to the postoperative risk of infection, rapid determination of the extent of the reduction of the oxidative response by flow cytometry may add to an optimized and individualized patient care.

D. Application for Characterization of Ligands

Ligand-specific interaction with professional phagocytes as well as with platelets may affect distinct functional activation profiles, including increased calcium fluxes in the absence of an oxidative burst, or, conformational changes of receptors without general platelet degranulation. Discriminatory flow cytometric analyses allowed the identification of selective induction of cytoplasmic calcium fluxes and oxidative burst by cross-linking of GPI-anchored receptors using anti-CD14 or anti-CD55 in unprimed myeloid cells, while cross-linking of a number of non-GPI-anchored antigens (e.g., CD11a, CD18, CD45) had no effect (Lund-Johansen *et al.*, 1993). Similarly, in platelets a selective alteration of the resting and the activated $\alpha IIb\beta 3$ (GPIIb/IIIa) receptor conformation in the absence of general platelet activation by the GPIIb/IIIa antagonist tirofibran could be visualized by flow cytometry (Barlage *et al.*, 2002). Moreover, ligand-specific activation mechanisms based on distinctive recruitment of coreceptors, for example, of TLR-2 and TLR-4 was identified using FRET, which allowed the characterization of ligand-specific clusters of pattern-recognition receptors (Pfeiffer *et al.*, 2001). This may be coupled to the analysis of alterations in the spatial organization of membranes into rafts and the lipid composition and microfluidity of membranes in general as determinants of receptor mobility (Triantafilou *et al.*, 2002). Thus, flow cytometry using receptor cross-linking, clustering or conformational changes, may serve to characterize mechanisms of receptor-, and ligand-specific signal transduction in conditions of altered phagocyte reactivity in disease, and during treatment with drugs and specific inhibitors.

IX. Future Directions

Flow cytometry is increasingly used to assess the functional status of leukocytes and leukocyte-subpopulations, including phagocytes. Going further from multi-parameter immunophenotyping, cell function-based flow cytometry has provided many new insights into the ligand-specific functional receptor clustering, intracellular processes, such as actin polymerization, cytokine production, and protein phosphorylation. Direct visualization and quantification of distinct phagocytic cells allows an estimation of the quality of the innate immune response. Since all steps of the phagocytosis cascade are accessible to flow cytometric study, the method is particularly suited to assess the functional status of phagocytes in a variety of human diseases. Flow cytometry may serve both as a simple and distinguished method to control therapeutic regimens addressed at stimulating phagocytosis, but also to monitor therapies with potential adverse effects on phagocytes.

Besides the ultimate diagnosis of primary phagocyte defects, the monitoring of secondary deficiencies of phagocyte function during chronic and acute diseases, including diabetes or SIRS/sepsis, as well as the control in the course of phagocyte-suppressive therapies provides an important field of application of diagnostic flow cytometry. Particularly for the prediction of the risk of infections associated with chronic diseases, such as diabetes, flow cytometry may gain additional impact. Similarly, depressed phagocytosis in immunosuppressive infectious diseases, in particular with AIDS, predisposes to opportunistic infections. Here, flow cytometry could not only contribute to the assessment of the remaining phagocyte function, but could also serve to monitor the efficiency of therapeutic measures to reactivate phagocyte activity with interferon gamma (IFNγ) or GM-CSF. Similarly, the functional capacity of transfused granulocyte concentrates in neutropenic patients may gain importance in predicting the therapeutic outcome. Further potentially very useful applications include testing of adverse effects of antibiotics on phagocyte function, which may counteract antimicrobial effectivity, thus predicting *in vivo* treatment failure.

References

Aderem, A., and Underhill, D. M. (1999). Mechanisms of phagocytosis in macrophages. *Annu. Rev. Immunol.* **17,** 593–623.

Allen, L. A., and Aderem, A. (1996). Molecular definition of distinct cytoskeletal structures involved in complement- and Fc receptor-mediated phagocytosis in macrophages. *J. Exp. Med.* **184,** 627–637.

Allen, C. A., Broom, M. F., and Chadwick, V. S. (1992). Flow cytometry analysis of the expression of neutrophil fMLP receptors. *J. Immunol. Methods* **149,** 159–164.

Assfalg-Machleidt, I., Rothe, G., Klingel, S., Banati, R., Mangel, W. F., Valet, G., and Machleidt, W. (1992). Sensitive determination of cathepsin L activity in the presence of cathepsin B using rhodamine-based fluorogenic substrates. *Biol. Chem. Hoppe-Seyler* **373,** 433–440.

Atkinson, T. P., Bonitatibus, G. M., and Berkow, R. L. (1997). Chronic granulomatous disease in two children with recurrent infections: Family studies using dihydrorhodamine-based flow cytometry. *J. Pediatr.* **130,** 488–491.

Babior, B. M. (2000). Phagocytes and oxidative stress. *Am. J. Med.* **109,** 33–44.

Babior, B. M. (2004). NADPH oxidase. *Curr. Opin. Immunol.* **16,** 42–47.

Babior, B. M., Takeuchi, C., Ruedi, J., Gutierrez, A., and Wentworth, P. Jr. (2003). Investigating antibody-catalyzed ozone generation by human neutrophils. *Proc. Natl. Acad. Sci. USA* **100,** 3031–3034.

Barlage, S., Wimmer, A., Pfeiffer, A., Rothe, G., and Schmitz, G. (2002). MK-383 (tirofiban) induces a GPIIb/IIIa receptor conformation which differs from the resting and activated receptor. *Platelets* **13,** 133–140.

Bass, D. A., Parce, J. W., Dechatelet, L. R., Szejda, P., Seeds, M. C., and Thomas, M. (1983). Flow cytometric studies of oxidative product formation by neutrophils: A graded response to membrane stimulation. *J. Immunol.* **130,** 1910–1917.

Bassoe, C. F., and Bjerknes, R. (1984). The effect of serum opsonins on the phagocytosis of *Staphylococcus aureus* and zymosan particles, measured by flow cytometry. *Acta Pathol. Microbiol. Immunol. Scand. [C]* **92**, 51–58.

Bassoe, C. F., and Bjerknes, R. (1985). Phagocytosis by human leukocytes, phagosomal pH and degradation of seven species of bacteria measured by flow cytometry. *J. Med. Microbiol.* **19**, 115–125.

Behe, P., and Segal, A. W. (2007). The function of the NADPH oxidase of phagocytes, and its relationship to other NOXs. *Biochem. Soc. Trans.* **35**, 1100–1103.

Bernardo, J., Hartlaub, H., Yu, X., Long, H., and Simons, E. R. (2002). Immune complex stimulation of human neutrophils involves a novel $Ca^{++}/H+$ exchanger that participates in the regulation of cytoplasmic pH: Flow cytometric analysis of Ca^{++}/pH responses by subpopulations. *J. Leukoc. Biol.* **72**, 1172–1179.

Berridge, M. J., Lipp, P., and Bootman, M. D. (2000). The versatility and universality of cal-cium signaling. *Nat. Rev. Mol. Cell. Biol.* **1**, 11–21.

Bicker, H., Höflich, C., Wolk, K., Vogt, K., Volk, H. D., and Sabat, R. (2008). A simple assay to measure phagocytosis of live bacteria. *Clin. Chem.* **54**, 911–915.

Bjerknes, R., and Bassoe, C. F. (1983). Human leukocyte phagocytosis of zymosan particles measured by flow cytometry. *Acta Pathol. Microbiol. Immunol. Scand. [C]* **91**, 341–348.

Boyum, A. (1968). Isolation of leucocytes from human blood. Further observations. Methylcellulose, dextran, and ficoll as erythrocyteaggregating agents. *Scand. J. Clin. Lab. Invest. Suppl.* **97**, 31–50.

Clark, R. A. (1999). Activation of the neutrophil respiratory burst oxidase. *J. Infect. Dis.* **179**(Suppl. 2), S309–S317.

Crockard, A. D., Thompson, J. M., Boyd, N. A., Haughton, D. J., McCluskey, D. R., and Turner, C. P. (1997). Diagnosis and carrier detection of chronic granulomatous disease in five families by flow cytometry. *Int. Arch. Allergy Immunol.* **114**, 144–152.

Cross, A. R. (1987). The inhibitory effects of some iodonium compounds on the superoxide generating system of neutrophils and their failure to inhibit diaphorase activity. *Biochem. Pharmacol.* **36**, 489–493.

Dahlgren, C., and Karlsson, A. (1999). Respiratory burst in human neutrophils. *J. Immunol. Methods* **232**, 3–14.

Dalpiaz, A., Spisani, S., Biondi, C., Fabbri, E., Nalli, M., and Ferretti, M. E. (2003). Studies on human neutrophil biological functions by means of formyl-peptide receptor agonists and antagonists. *Curr. Drug Targets Immune Endocr. Metabol. Disord.* **3**, 33–42.

da Silva, C. P., and Guse, A. H. (2000). Intracellular Ca^{++} release mechanisms: Multiple pathways having multiple functions within the same cell type? *Biochim. Biophys. Acta* **1498**, 122–133.

Dekker, J., de Boer, M., and Roos, D. (2001). Gene-scan method for the recognition of carriers and patients with p47(phox)-deficient autosomal recessive chronic granulomatous disease. *Exp. Hematol.* **29**, 1319–1325.

Demaurex, N., Downey, G. P., Waddell, T. K., and Grinstein, S. (1996). Intracellular pH regulation during spreading of human neutrophils. *J. Cell Biol.* **133**, 1391–1402.

Dewas, C., Dang, P. M., Gougerot-Pocidalo, M. A., and El-Benna, J. (2003). TNF-alpha induces phosphorylation of p47(phox) in human neutrophils. Partial phosphorylation of p47phox is a common event of priming of human neutrophils by TNF-alpha and granulocyte-macrophage colony-stimulating factor. *J. Immunol.* **171**, 4392–4398.

Dinauer, M. C., and Orkin, S. H. (1992). Chronic granulomatous disease. *Annu. Rev. Med.* **43**, 117–124.

Djaldetti, M., Salman, H., Bergman, M., Djaldetti, R., and Bessler, H. (2002). Phagocytosis—The mighty weapon of the silent warriors. *Microsc. Res. Tech.* **57**, 421–431.

do Céu Monteiro, M., Sansonetty, F., Gonçalves, M. J., and O'Connor, J. E. (1999). Flow cytometric kinetic assay of calcium mobilization in whole blood platelets using Fluo-3 and CD41. *Cytometry* **35**, 302–310.

Elbim, C., Chollet-Martin, S., Bailly, S., Hakim, J., and Gougerot-Pocidalo, M. A. (1993). Priming of polymorphonuclear neutrophils by tumor necrosis factor alpha in whole blood: Identification of two polymorphonuclear neutrophil subpopulations in response to formyl-peptides. *Blood* **82**, 633–640.

Elbim, C., Prevot, M. H., Bouscarat, F., Franzini, E., Chollet-Martin, S., Hakim, J., and Gougerot-Pocidalo, M. A. (1994). Polymorphonuclear neutrophils from human immunodeficiency virus-infected patients show enhanced activation, diminished fMLP-induced L-selectin shedding, and an impaired oxidative burst after cytokine priming. *Blood* **84,** 2759–2766.

Elbim, C., Reglier, H., Fay, M., Delarche, C., Andrieu, V., El Benna, J., and Gougerot-Pocidalo, M. A. (2001). Intracellular pool of IL-10 receptors in specific granules of human neutrophils: Differential mobilization by proinflammatory mediators. *J. Immunol.* **166,** 5201–5207.

Emmendörffer, A., Nakamura, M., Rothe, G., Spiekermann, K., Lohmann-Matthes, M. L., and Roesler, J. (1994). Evaluation of flow cytometrical methods for diagnosis of chronic granulomatous disease (CGD)-variants under routine laboratory conditions. *Cytometry* **18,** 147–156.

Engelich, G., Wright, D. G., and Hartshorn, K. L. (2001). Acquired disorders of phagocyte function complicating medical and surgical illnesses. *Clin. Infect. Dis.* **33,** 2040–2048.

Fialkow, L., Wang, Y., and Downey, G. P. (2007). Reactive oxygen and nitrogen species as signaling molecules regulating neutrophil function. *Free Radical Biol. Med.* **42,** 153–164.

Forman, H. J., and Torres, M. (2002). Reactive oxygen species and cell signaling: Respiratory burst in macrophage signaling. *Am. J. Respir. Crit. Care Med.* **166,** S4–S8.

Forsyth, K. D., and Levinsky, R. J. (1990). Preparative procedures of cooling and re-warming increase leukocyte integrin expression and function on neutrophils. *J. Immunol. Methods* **128,** 159–163.

Fröhlich, D., Rothe, G., Schwall, B., Schmid, P., Schmitz, G., Taeger, K., and Hobbhahn, J. (1997). Effects of volatile anaesthetics on human neutrophil oxidative burst response to the bacterial peptide fMLP. *Br. J. Anaesth.* **78,** 718–723.

Fröhlich, D., Wittmann, S., Rothe, G., Schmitz, G., and Taeger, K. (2002). Thiopental impairs neutrophil oxidative response by inhibition of intracellular signalling. *Eur. J. Anaesthesiol.* **19,** 474–482.

Green, D. J., and Shaw, E. (1981). Peptidyl diazomethyl ketones are specific inactivators of thiol proteinases. *J. Biol. Chem.* **256,** 1923–1928.

Hackam, D. J., Grinstein, S., and Rotstein, O. D. (1996). Intracellular pH regulation in leukocytes: Mechanisms and functional significance. *Shock* **5,** 17–21.

Hallett, M. B. (2003). Holding back neutrophil aggression; the oxidase has potential. *Clin. Exp. Immunol.* **132,** 181–184.

Hallett, M. B., and Lloyds, D. (1995). Neutrophil priming: The cellular signals that say 'amber' but not 'green'. *Immunol. Today* **16,** 264–268.

Hart, S. P., Smith, J. R., and Dransfield, I. (2004). Phagocytosis of opsonized apoptotic cells: Roles for 'old-fashioned' receptors for antibody and complement. *Clin. Exp. Immunol.* **135,** 181–185.

Hasleton, P. S., and Roberts, T. E. (1999). Adult respiratory distress syndrome—An update. *Histopathology* **34,** 285–294.

He, R., Sang, H., and Ye, R. D. (2003). Serum amyloid A induces IL-8 secretion through a G protein-coupled receptor, FPRL1/LXA4R. *Blood* **101,** 1572–1581.

Honey, K., and Rudensky, A. Y. (2003). Lysosomal cysteine proteases regulate antigen presentation. *Nat. Rev. Immunol.* **3,** 472–482.

Hooper, N. M. (2002). Proteases: A primer. *Essays Biochem.* **38,** 1–8.

Ishaque, A., and Al-Rubeai, M. (1998). Use of intracellular pH and annexin-V flow cytometric assays to monitor apoptosis and its suppression by bcl-2 over-expression in hybridoma cell culture. *J. Immunol. Methods* **221,** 43–57.

Jankowski, A., Scott, C. C., and Grinstein, S. (2002). Determinants of the phagosomal pH in neutrophils. *J. Biol. Chem.* **277,** 6059–6066.

Jongstra-Bilen, J., Harrison, R., and Grinstein, S. (2003). Fcγ-receptors induce Mac-1 (CD11b/CD18) mobilization and accumulation in the phagocytic cup for optimal phagocytosis. *J. Biol. Chem.* **278,** 45720–45729.

Kedzierska, K., Azzam, R., Ellery, P., Mak, J., Jaworowski, A., and Crowe, S. M. (2003). Defective phagocytosis by human monocyte/macrophages following HIV-1 infection: Underlying mechanisms and modulation by adjunctive cytokine therapy. *J. Clin. Virol.* **26,** 247–263.

Keel, M., Ungethum, U., Steckholzer, U., Niederer, E., Hartung, T., Trentz, O., and Ertel, W. (1997). Interleukin-10 counterregulates proinflammatory cytokine-induced inhibition of neutrophil apopto-sis during severe sepsis. *Blood* **90**, 3356–3363.

Keil, M. L., Solomon, N. L., Lodhi, I. J., Stone, K. C., Jesaitis, A. J., Chang, P. S., Linderman, J. J., and Omann, G. M. (2003). Priming-induced localization of G(ialpha2) in high density membrane microdomains. *Biochem. Biophys. Res. Commun.* **301**, 862–872.

Kirschke, H., and Barrett, A. J. (1987). Chemistry of lysosomal proteases. *In* "Lysosomes: Their Role in Protein Breakdown" (H. Glaumann and F. J. Ballard, eds.), pp. 193–238. Academic Press, London.

Koksch, M., Rothe, G., Kiefel, V., and Schmitz, G. (1995). Fluorescence resonance energy transfer as a new method for the epitope-specific characterization of anti-platelet antibodies. *J. Immunol. Methods* **187**, 53–67.

Kono, H., Suzuki, T., Yamamoto, K., Okada, M., Yamamoto, T., and Honda, Z. (2002). Spatial raft coalescence represents an initial step in Fc gamma R signaling. *J. Immunol.* **169**, 193–203.

Kwiatkowska, K., and Sobota, A. (1999). Signaling pathways in phagocytosis. *Bioessays* **21**, 422–431.

Le, Y., Murphy, P. M., and Wang, J. M. (2002). Formyl-peptide receptors revisited. *Trends Immunol.* **23**, 541–548.

Lehmann, A. K., Sørnes, S., and Halstensen, A. (2000). Phagocytosis: Measurement by flow cytometry. *J. Immunol. Methods* **243**, 229–242.

Lewis, R. S. (2001). Calcium signaling mechanisms in T lymphocytes. *Annu. Rev. Immunol.* **19**, 497–521.

Leytus, S. P., Melhado, L. L., and Mangel, W. F. (1983a). Rhodamine-based compounds as fluorogenic substrates for serine proteinases. *Biochem. J.* **209**, 299–307.

Leytus, S. P., Patterson, W. L., and Mangel, W. F. (1983b). New class of sensitive and selective fluorogenic substrates for serine proteinases. Amino acid and dipeptide derivatives of rhodamine. *Biochem. J.* **215**, 253–260.

Linke, R. P., Bock, V., Valet, G., and Rothe, G. (1991). Inhibition of the oxidative burst response of *N*-formyl peptide-stimulated neutrophils by serum amyloid-A protein. *Biochem. Biophys. Res. Com-mun.* **176**, 1100–1105.

Lund-Johansen, F., Olweus, J., Aarli, A., and Bjerknes, R. (1990). Signal transduction in human monocytes and granulocytes through the PI-linked antigen CD14. *FEBS Lett.* **273**, 55–58.

Lund-Johansen, F., Olweus, J., Symington, F. W., Arli, A., Thompson, J. S., Vilella, R., Skubitz, K., and Horejsi, V. (1993). Activation of human monocytes and granulocytes by monoclonal antibodies to glycosylphosphatidylinositol-anchored antigens. *Eur. J. Immunol.* **23**, 2782–2791.

Malech, H. L., and Nauseef, W. M. (1997). Primary inherited defects in neutrophil function: Etiology and treatment. *Semin. Hematol.* **34**, 279–290.

Mantovani, A., Garlanda, C., and Bottazzi, B. (2003). Pentraxin 3, a non-redundant soluble pattern recognition receptor involved in innate immunity. *Vaccine* **21**, S43–S47.

McCarthy, D. A., and Macey, M. G. (1996). Novel anticoagulants for flow cytometric analysis of live leucocytes in whole blood. *Cytometry* **23**, 196–204.

Minta, A., Kao, J. P. Y., and Tsien, R. Y. (1989). Fluorescent indicators for cytosolic calcium based on rhodamine and fluorescein chromophores. *J. Biol. Chem.* **264**, 8171–8178.

Moraes, T. J., Chow, C. W., and Downey, G. P. (2003). Proteases and lung injury. *Crit. Care Med.* **31**, S189–S194.

Osterud, B. (1998). Tissue factor expression by monocytes: Regulation and pathophysiological roles. *Blood Coagul. Fibrinolysis* **9**, S9–S14.

Ozinsky, A., Underhill, D. M., Fontenot, J. D., Hajjar, A. M., Smith, K. D., Wilson, C. B., Schroeder, L., and Aderem, A. (2000). The repertoire for pattern recognition of pathogens by the innate immune system is defined by cooperation between toll-like receptors. *Proc. Natl. Acad. Sci. USA* **97**, 13766–13771.

Palecanda, A., and Kobzik, L. (2001). Receptors for unopsonized particles: The role of alveolar macrophage scavenger receptors. *Curr. Mol. Med.* **1**, 589–595.

Pallister, C. J., and Lewis, R. J. (2000). Effects of antimicrobial drugs on human neutrophil-microbe interactions. *Br. J. Biomed. Sci.* **57**, 19–27.

Pfeiffer, A., Böttcher, A., Orsó, E., Kapinsky, M., Nagy, P., Bodnár, A., Spreitzer, I., Liebisch, G., Drobnik, W., Gempel, K., Horn, M., and Holmer, S. (2001). Lipopolysaccharide and ceramide docking to CD14 provokes ligand specific receptor clustering in rafts. *Eur. J. Immunol.* **31**, 3153–3164.

Pisciotta, A. V. (1990). Drug-induced agranulocytosis. Peripheral destruction of polymorphonuclear leukocytes and their marrow precursors. *Blood Rev.* **4**, 226–237.

Poenie, M., Alderton, J., Steinhardt, R., and Tsien, R. (1986). Calcium rises briefly and throughout the cell at the onset of anaphase. *Science* **233**, 886–889.

Powers, J. C. (1986). Serine proteases of leukocyte and mast cell origin: Substrate specificity and inhibition of elastase, chymases, and tryptases. *Adv. Inflammation Res.* **11**, 145–157.

Reeves, E. P., Lu, H., Jacobs, H. L., Messina, C. G., Bolsover, S., Gabella, G., Potma, E. O., Warley, A., Roes, J., and Segal, A. W. (2002). Killing activity of neutrophils is mediated through activation of proteases by K^+ flux. *Nature* **416**, 291–297.

Réthi, B., Detre, C., Gogolák, P., Kolonics, A., Magócsi, M., and Rajnavölgyi, E. (2002). Flow cytometry used for the analysis of calcium signaling induced by antigen-specific T-cell activation. *Cytometry* **47**, 207–216.

Rothe, G., Assfalg-Machleidt, I., Machleidt, W., Klingel, S., Zirkelbach, C. h., Banati, R., Mangel, W. F., and Valet, G. (1992). Flow cytometric analysis of protease activities in vital cells. *Biol. Chem. Hoppe-Seyler* **373**, 547–554.

Rothe, G., Emmendörffer, A., Oser, A., Roesler, J., and Valet, G. (1991). Flow cytometric measurement of the respiratory burst activity of phagocytes using dihydrorhodamine 123. *J. Immunol. Methods* **138**, 133–135.

Rothe, G., Oser, A., and Valet, G. (1988). Dihydrorhodamine 123: A new flow cytometric indicator for respiratory burst activity in neutrophil granulocytes. *Naturwissenschaften* **75**, 354–355.

Rothe, G., and Valet, G. (1990). Flow cytometric analysis of respiratory burst activity in phagocytes with hydroethidine and 2′,7′-dichlorofluorescin. *J. Leukoc. Biol.* **47**, 440–448.

Rothe, G., and Valet, G. (1994). Flow cytometric assays of oxidative burst activity in phagocytes. *In* "Methods in Enzymology: Oxygen Radicals in Biological Systems, Part C" (L. Packer, ed.), Vol. 233, pp. 539–548. Academic Press, Orlando.

Savill, J. (1998). Apoptosis. Phagocytic docking without shocking. *Nature* **392**, 442–443.

Segal, B. H., and Holland, S. M. (2000). Primary phagocytic disorders of childhood. *Pediatr. Clin. North Am.* **47**, 1311–1338.

Shao, D., Segal, A. W., and Dekker, L. V. (2003). Lipid rafts determine efficiency of NADPH oxidase activation in neutrophils. *FEBS Lett.* **550**, 101–106.

Simons, K., and Ehehalt, R. (2002). Cholesterol, lipid rafts, and disease. *J. Clin. Invest.* **110**, 597–603.

Sklar, L. A., Oades, Z. G., Jesaitis, A. J., Painter, R. G., and Cochrane, C. G. (1981). Fluoresceinated chemotactic peptide and high-affinity antifluorescein antibody as a probe of the temporal characteristics of neutrophil stimulation. *Proc. Natl. Acad. Sci. USA* **78**, 7540–7544.

Steinkamp, J. A., Wilson, J. S., Saunders, G. C., and Stewart, C. C. (1982). Phagocytosis: Flow cytometric quantitation with fluorescent microspheres. *Science* **215**, 64–66.

Thomas, J. A., Buchsbaum, R. N., Zimniak, A., and Racker, E. (1979). Intracellular pH measurements in Ehrlich ascites tumor cells utilizing spectroscopic probes generated *in situ*. *Biochemistry* **18**, 2210–2218.

Triantafilou, M., Miyake, K., Golenbock, D. T., and Triantafilou, K. (2002). Mediators of innate immune recognition of bacteria concentrate in lipid rafts and facilitate lipopolysaccharide-induced cell activation. *J. Cell Sci.* **115**, 2603–2611.

Vowells;, S. J., Sekhsaria, S., Malech, H. L., Shalit, M., and Fleisher, T. A. (1995). Flow cytometric analysis of the granulocyte respiratory burst: A comparison study of fluorescent probes. *J. Immunol. Methods* **178**, 89–97.

Walrand, S., Valeix, S., Rodriguez, C., Ligot, P., Chassagne, J., and Vasson, M. P. (2003). Flow cytometry study of polymorphonuclear neutrophil oxidative burst: A comparison of three fluorescent probes. *Clin. Chim. Acta* **331**, 103–110.

WHO Consensus Group. (1999). Primary immunodeficiency diseases. *Clin. Exp. Immunol.* **118**, 1–28.

Wittmann, S., Rothe, G., Schmitz, G., and Fröhlich, D. (2004). Cytokine upregulation of surface antigens correlates to the priming of the neutrophil oxidative burst response. *Cytometry* **57A,** 53–62.

Wymann, M. P., Sozzani, S., Altruda, F., Mantovani, A., and Hirsch, E. (2000). Lipids on the move: phosphoinositide 3-kinases in leukocyte function. *Immunol. Today* **21,** 260–264.

Yamashiro, S., Kamohara, H., Wang, J. M., Yang, D., Gong, W. H., and Yoshimura, T. (2001). Phenotypic and functional change of cytokine-activated neutrophils: Inflammatory neutrophils are heterogeneous and enhance adaptive immune responses. *J. Leukoc. Biol.* **69,** 698–704.

Zarewych, D. M., Kindzelskii, A. L., Todd, R. F., and Petty, H. R. (1996). LPS induces CD14 association with complement receptor type 3, which is reversed by neutrophil adhesion. *J. Immunol.* **156,** 430 433.

CHAPTER 34

Analysis of RNA Synthesis by Cytometry

Peter Østrup Jensen,★ **Jacob Larsen,**† **and Jørgen K. Larsen**‡

★Department of Clinical Microbiology
Rigshospitalet
DK-2100 Copenhagen, Denmark

†Department of Clinical Pathology
Næstved Hospital
4700 Næstved, Denmark

‡Borgergade 30III
DK-1300 Copenhagen K, Denmark

I. Update

This manuscript describes a method of immunochemical detection of nascent RNA by cross-reacting antibromodeoxyuridine antibodies following an *in vitro* or *in vivo* bromouridine (BrUrd) labeling. The method has found use in various

DOI: 10.1016/B978-0-12-375045-7.00034-9

biological problems like pharmacologic screening (Chiu and Yang, 2007), mapping of the cell nucleus (Casafont *et al.*, 2006), apoptosis research (Halicka *et al.*, 2000), and gene expression analysis (Ohtsu *et al.*, 2008). The passive uptake of BrUrd and its conversion to BrUTP *in vivo* is less labor intensive than direct introduction of BrUTP by, for example, microinjection or lipofection. Evaluation by fluorescence microscopy or electron microscopy is seen more often than flow cytometry.

The ABDM antibody referred to in our manuscript is no longer being produced. However, other antibodies have been reported to detect BrUrd-substituted RNA: Clone BU-33 (B8434, Sigma-Aldrich, St. Louis, MO) (Boisvert *et al.*, 2000) and clone PRB1-U (Phoenix Flow Systems Inc., San Diego, CA) (Halicka *et al.*, 2000).

Other halogenated nucleosides (fluorouridine) can be used alternatively to BrUrd (Boisvert *et al.*, 2000; Casafont *et al.*, 2006).

II. Introduction

The biological significance of RNA has been studied by different cytometric approaches, which can be divided into "static" approaches, describing the momentary distribution of the RNA content in the cell population, and "dynamic" approaches, describing the distribution of the content of RNA precursors that have been incorporated during a preset time period. In this chapter, we will focus on the dynamic approach, with emphasis on the flow cytometric analysis limited to applications with mammalian cells.

III. Background

The RNA content of a cell can be estimated by direct staining with fluorochromes such as pyronin Y and acridine orange, the latter binding metachromatically to single-stranded RNA and DNA (Darzynkiewicz, 1994). The content of double-stranded RNA can be estimated by staining of DNase-pretreated cells with propidium iodide (Frankfurt, 1990). These static staining techniques only enable an indirect estimation of the net amount of synthesized RNA, as may be revealed by comparison of cell samples taken at different time points.

Dynamic measurements of RNA synthesis are based on a different approach. The object of interest, cell cultures or living organisms, must be labeled *in vitro* or *in vivo* with a detectable RNA precursor during a given time period. Thus, RNA synthesis has been traditionally investigated by incorporation of [3H]uridine and counting the silver grains developed in an autoradiographic film over the individual cells (Fakan, 1986). As an alternative to this time-consuming and laborsome method, a faster and nonradioactive technology has emerged based on immunochemical detection. This method utilizes the incorporation of brominated RNA precursors, either 5′-bromouridine (BrUrd) (Hozák *et al.*, 1994; Jensen *et al.*, 1993a) or 5′-bromouridine triphosphate (BrUTP) (Jackson *et al.*, 1993), followed by immunofluorescent staining using anti-5′-bromodeoxyuridine (BrdUrd) antibodies. Alternatively, nascent RNA can be detected by fluorescein-UTP incorporation (LaMorte *et al.*, 1998).

For studies of RNA synthesis in experimental animals, *in vivo* labeling with BrUrd (or [3H]uridine) is required. It is thought that the BrUrd enters the cell and that phosphate groups are attached by the cell's own anabolic machinery to produce BrUTP, which is suitable for incorporation into nascent RNA during transcription. For the incorporation of BrUTP, biotin-UTP, or fluorescein-UTP into nascent RNA by the RNA polymerases, access to the transcription sites in the cells has to be facilitated. Thus, stripping of the plasma membrane (Jackson *et al.*, 1993), permeabilization (Dundr and Raska, 1993), microinjection (Carmo-Fonseca *et al.*, 1996; LaMorte *et al.*, 1998; Wansink *et al.*, 1993), or fusion with precursor-containing liposomes (Haukenes *et al.*, 1997) have been applied. The BrUrd-substituted RNA may be detected by permeabilizing the cells and staining with certain anti-BrdUrd antibodies. Indeed, several of the antibodies marketed as anti-BrdUrd antibodies are produced by hybridomas that are derived from rodents that have been immunized with halogenated uridine, and not deoxyuridine. A fluorescein isothrocyanate (FITC)-conjugated anti-BrdUrd antibody may be used for visualization of the BrUrd-substituted RNA, or this may be indirectly stained with a fluorochrome-conjugated or gold-labeled secondary antibody. The secondary structure of RNA may influence the accessibility for the antibodies to the BrUrd-labeled RNA, in analogy with the well-known experience from staining of BrdUrd-labeled DNA, where staining with anti-BrdUrd antibodies requires that the BrdUrd is exposed in a denaturated, single-stranded form (Dolbeare, 1995). Thus, it is to be expected that BrUrd-substituted single-stranded RNA is far more accessible to the anti-BrdUrd antibodies than is BrUrd-substituted double-stranded RNA.

Owing to their relatively short lifetime in comparison to other types of RNA molecules, specific RNA transcripts may provide estimates of the cellular expression of the corresponding genes at the RNA level. Specific mRNA molecules may be cytometrically recognized by hybridization with specific oligonucleotide probes. The bound probes may then be visualized by addition of fluorochrome-conjugated probes (fluorescence *in situ* hybridization, FISH) or by incorporation of fluorochrome-conjugated nucleotides where the specific oligonucleotide probes serve as primers (primed *in situ* labeling, PRINS).

IV. Methods for Analysis of 5′-Bromouridine Incorporation and DNA Content

The described techniques have been developed for flow cytometric measurements of BrUrd incorporation, but they will also allow for investigations by fluorescence microscopy.

A. 5'-Bromouridine Labeling

Cell cultures of, for example, activated lymphocytes, leukemia cell lines, or adherent cell cultures are labeled by incubation for 1 h with 1 mM BrUrd (850187, Sigma-Aldrich Co., St. Louis, MO). After harvest, samples of 1.000.000 cells are washed in cold PBS (Ca^{2+}- and Mg^{2+}-free Dulbecco's phosphate-buffered saline, pH 7.2).

For *in vivo* labeling of rats, 200 mg BrUrd dissolved in 2 ml PBS is injected intraperitoneally. After 1 h the rats are anesthetized, and organ samples are quickly excised and placed on ice. A single cell suspension is prepared mechanically and stored at −20 °C in a freezing buffer [250 mM sucrose, 40 mM trisodium citrate dihydrate, 5% (v/v) dimethyl sulfoxide, pH 7.6].

B. Cell Preparation

The cells must be fixed and permeabilized before staining. With methods A and B suspensions of nuclei are produced, and with methods C and D suspensions of entire cells.

For method A, according to Landberg and Roos (1991) and Jensen *et al.* (1993a), the cells are lysed for 15 min on ice with 0.5 ml of buffer [0.5% Triton X-100, 1% bovine serum albumin (BSA), and 0.2 μg/ml EDTA in PBS]. The nuclei are subsequently fixed by addition of 3 ml methanol at −20 °C. Higher temperature increases the risk for clotting of the BSA. Store at −20 °C.

For method B, modified after Otto (1990), the cells are lysed for 15 min on ice with 1 ml of buffer (2.1% citric acid and 0.5% Tween 20 in distilled water, pH 2). The nuclei are washed in PBS and fixed by addition of 1 ml methanol at −20 °C. This method is suitable for release of nuclei from monolayer cultures. Store at −20 °C.

For method C, modified after Carayon and Bord (1992) and according to Jensen *et al.* (1993b), the cells are fixed and permeabilized in 1 ml of buffer (1% formaldehyde and 0.05% Nonidet P-40 in PBS) at room temperature, slowly agitated for 15 min, and then stored at 4 °C.

For method D, according to Li *et al.* (1994), the cells are fixed in 1 ml of 1% formaldehyde in PBS for 15 min. The fixed samples are washed in PBS and permeabilized by addition of 1 ml ethanol at −20 °C. Store at −20 °C.

C. Staining

Samples of approximately 5 × 105 fixed cells are washed in PBS, and 50 μl of monoclonal antibromodeoxyuridine antibody (ABDM, Partec, Münster, Germany), diluted 1:10 in PBS with 0.01% (v/v) Nonidet P-40, is added for 60 min. After washing once in PBS, a secondary, FITC-conjugated antimouse immunoglobulin antibody (e.g., F0313, Dako) is added for 60 min. Finally, after washing once in PBS, 100 μl of 50 μg^{-1} ml^{-1} propidium iodide (P-4170, Sigma-Aldrich Co.,

St. Louis, MO) in PBS is added for at least 15 min before flow cytometry. Samples are kept on an ice bath during and after staining.

In addition to the ABDM antibody, several antibodies exist that will bind to BrUrd in RNA: clone B-44 (347580, Becton Dickinson, San Jose, CA), and clone BMC9318 (11 170 376 001, Roche, Mannheim, Germany), both from mouse, and clone BU1/75 (ab7384, Abcam plc, Cambridge, UK), derived from rats. Interestingly, the clone BU20a anti-BrdUrd antibody (M0744, Dako, Copenhagen, Denmark), which is derived from mice immunized with BrdUrd, shows no cross-reactivity to BrUrd.

D. Flow Cytometry

Flow cytometric analysis of the BrUrd/DNA distribution is performed on a flow cytometer using laser excitation at 488 nm and collection of FITC fluorescence at 515–545 nm and of propidium iodide fluorescence at >615 nm. Recording of 10–20.000 cells is triggered by the propidium iodide signal. The instrument is adjusted to minimal coefficient of variation (CV) for the DNA measurement, using, for example, propidium iodide-stained chicken erythrocytes.

E. Fluorescence Microscopy

The subcellular distribution of the incorporated BrUrd is investigated by fluorescence microscopy of cells that are processed as described, but without the final staining with propidium iodide. Aliquots of the cell suspensions are either pipetted onto slides and air dried for 30 min, or spun onto microscope slides for 5 min at $800 \times g$ using 1-ml, 30-mm two cytocentrifuge containers (Zytokammer, Hettich Zentrifugen, Tuttingen, Germany). The slides are finally mounted with Fluoromount-G (Southern Biotechnology Associates, Birmingham, AL), containing 0.5 μg/ml 4′,6-diamidino-2-phenylindole (DAPI) (Serva Feinbiochemica, Heidelberg, Germany).

F. Critical Aspects of the Methodology

For control of the specificity of the immunofluorescence signal, a comparison with a BrUrd-unlabeled sample as well as a BrUrd-labeled and ribonuclease (RNase)-treated sample is recommended. For this purpose, it is important to apply the RNase treatment before immunochemical staining of the BrUrd (Jensen et al., 1993a). It has been reported that a BrUTP-specific cytoplasmic fluorescence can occur. This fluorescence could not be inhibited with actinomycin D and was only partially removed by treatment with RNase (Haukenes et al., 1997). This problem might be omitted by using preparations of cell nuclei. Because the BrUrd signal is quite stable against RNase treatment after immunostaining, addition of RNase to the propidium iodide staining solution (e.g., 5 mg/ml ribonuclease A,

Sigma R-4875) is advantageous for increasing the specificity and resolution of the DNA measurement (Jensen *et al.*, 1993a).

In our experience, the flow cytometric detection level for BrUrd incorporation is at about 15 min incubation with 100 μM BrUrd. Application of BrUrd over extended periods of time (>-50 μM for >-24 h) may induce cell cycle perturbation and apoptotic cell death, as shown in HL-60 and MOLT-4 cells by Li *et al.* (1994).

Limited access for the anti-BrdUrd antibody to bind to the incorporated BrUrd, which may be masked by proteins due to fixation with formaldehyde or not accessible due to incorporation into double-stranded RNA, may be an important factor in interpretation of the results. The problems due to masking of incorporated BrUrd may be omitted by alternative labeling with FITC-UTP (LaMorte *et al.*, 1998). Care should be taken to avoid contamination with RNases during handling of the cells. The importance of the labeling efficiency of the artificial RNA precursors into the various types of RNA has not been completely investigated. It should be mentioned that BrUrd is incorporated unbiased in intact cells (Jackson *et al.*, 1998). Biotin-11-UTP and digoxigenin-11-UTP are less efficiently incorporated into RNA than BrUTP (Jackson *et al.*, 1993). For microscopic detection of BrUTP incorporation, labeling with 2 mM BrUTP for 2.5 min is sufficient (Jackson *et al.*, 1993).

Compared to [3H]uridine incorporation, exposure to BrUrd has little effect initially on transcription rates (Jackson *et al.*, 1998), but the chemically modified uridines are known to inhibit the subsequent processing of RNA transcripts (Wansink *et al.*, 1994a). This implies that the very fast RNA processing is delayed. However, in studies of the morphology of transcription this may turn out to be a practical advantage, as it provides the researcher the needed time to conduct the experiment.

V. Results of Labeling RNA with 5′-Bromouridine

A. Specificity

In exponentially growing cultures of HL-60 cells, RNA synthesis can be demonstrated in the majority of the cells by flow cytometric analysis of nuclear BrUrd incorporation (Fig. 1). The specificity of the FITC fluorescence is indicated by the fact that RNase treatment before the immunochemical staining (Fig. 1C) induced a reduction of the signal from the labeled nuclei (Fig. 1A) to the level of the unlabeled nuclei (Fig. 1B). Attempts to measure the effect of ribonucleotide reductase activity, which might convert BrUrd to BrdUrd leading to labeling of DNA during replication, had no effect, as deliberately BrdUrd-labeled nuclei stained negatively (Fig. 1D), which is supported by Dundr and Raska (1993). Also, treatment of the cell culture with the RNA polymerase inhibitor actinomycin D prior to incubation with BrUrd decreased the labeling efficiency, in agreement with results from Dundr and Raska (1993) and Jackson *et al.* (1993).

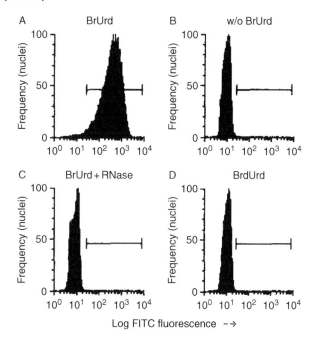

Fig. 1 Flow cytometric analysis of RNA synthesis according to BrUrd incorporation in exponentially growing HL-60 cells, demonstrating the specificity of the immunochemical staining. The cells were lysed and fixed according to method A before staining with ABDM anti-BrdUrd antibody (Partec) and secondary FITC-conjugated antibody. Flow cytometric analysis was performed on a FACS IV (Becton Dickinson), measuring the incorporated BrUrd by the log FITC fluorescence of fixed nuclei. Markers indicate histogram regions of BrUrd positive nuclei. (A) Cells incubated with 1 mM BrUrd for 1 h (98% positive nuclei). (B) Cells incubated without BrUrd (0.3% positive nuclei). (C) Cells incubated with 1 mM BrUrd for 1 h, but the fixed nuclei were treated with RNase before staining with the antibody (0.3% positive nuclei). (D) Cells incubated with 10 μM BrdUrd instead of BrUrd (0.4% positive nuclei). Reprinted from Jensen *et al.* (1993b) with permission from the publisher.

B. RNA Synthesis and the Cell Cycle

Dual parameter analysis of FITC (BrUrd) and propidium iodide (DNA) fluorescence enables the correlation of transcriptional activity and cell cycle phase distribution, as shown for HL-60 cells in Fig. 2A. Nuclei with S phase DNA content show a high level of BrUrd incorporation, whereas those with G1 and G2/M phase DNA content show a low or intermediate level. The horseshoe shaped distribution, similar to the distribution found when labeling with the DNA precursor BrdUrd, indicates that the major part of RNA synthesis in this exponentially growing cell line is related to replication. *In vivo* labeling of rat liver cells reveals a somewhat different distribution, shown in Fig. 2B. These largely noncycling and partially polyploid cells show a substantial labeling in the G1 and G_2/M phases, thus indicating RNA synthesis in the noncycling state.

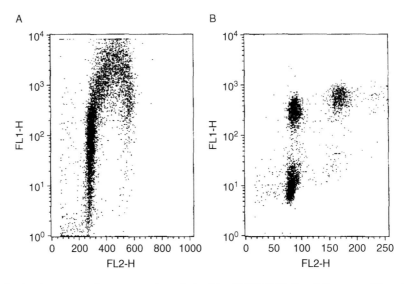

Fig. 2 Dual parameter flow cytometric analysis of log FITC (BrUrd) and linear propidium iodide (DNA) fluorescence. (A) Exponentially growing HL-60 cells were incubated with 1 mM BrUrd for 1 h and prepared according to method B. (B) Rat liver cells harvested after *in vivo* labeling for 1 h with 200 mg BrUrd and prepared according to method A.

C. Subcellular Distribution of Incorporated 5′-Bromouridine

In Fig. 3, the FITC staining pattern of BrUrd incorporated after a labeling period of 1 h may be compared with the localization of nuclei and nucleoli according to simultaneous staining of the DNA with DAPI. When cells were prepared according to method B, the nuclei showed intense staining of the nucleoli as well as several extranucleolar smaller foci in a darker nuclear matrix (Fig. 3A). This staining pattern is similar to labeling with [3H]uridine and FITCUTP. In contrast, cells prepared with the milder extraction method A showed only extra-nucleolar staining. With methods C and D, the FITC staining in the nucleus was similarly confined to the extranucleolar portion, and in addition the cytoplasm contained FITC-stained loci.

VI. Applications

The measurement of incorporated BrUrd by flow cytometry has provided a basis for correlating the overall transcriptional activity to the cell cycle by simultaneous measurement of DNA content (Haider *et al.*, 1997; Jensen *et al.*, 1993a; Li *et al.*, 1994), as well as to the phenotype by simultaneous measurement of cell surface markers (Jensen *et al.*, 1993b). In toxicological studies, it was shown that the RNA synthesis was affected at a lower concentration of 5-azacytidine than necessary for affecting the DNA synthesis (Murakami *et al.*, 1995). Using

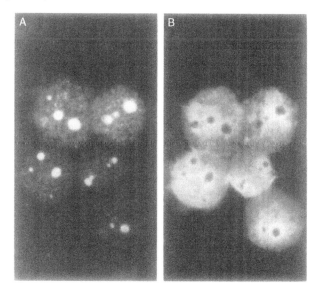

Fig. 3 Fluorescence microphotographs of exponentially growing HL-60 cells that were incubated with 1 mM BrUrd for 1 h and prepared according to method B: (A) FITC-stained BrUrd, (B) DAPI stained DNA.

fluorescence or electron microscopy, several studies have utilized the BrUrd technique. In this way, approximately 2000 extranucleolar transcription foci were detected in the nuclei of HeLa cells (Iborra *et al.*, 1996; Jackson *et al.*, 1998). The transcription loci have been studied by simultaneous staining of the splicing factors and nascent RNA (Jackson *et al.*, 1993; Pombo and Cook, 1996). When BrUTP incorporation into RNA was combined with biotin-dUTP or digitonin-dUTP incorporation into DNA, distributions of transcription sites could be compared to replication sites in the nuclei (Hassan *et al.*, 1994; Jackson *et al.*, 1993; Wansink *et al.*, 1994b). Implementing the BrUrd method in animal cells, nucleolar transcription sites were mapped to the dense fibrillar component (Dundr and Raska, 1993; Wansink *et al.*, 1994b) and additionally to the nucleolar fibrillary centers and vacuoles in plants (Melcák *et al.*, 1996).

Cytometric analysis in itself, only reporting the amount of recognized Br-substituted RNA, does not provide any information of the primary structure of the synthesized RNA. However, by exploiting the knowledge on BrUrd antigen-antibody reactivity in the field of methods that are applicable for cell lysates, the detection of RNA synthesis from specific genes is possible. Using immunoseparation of the RNA labeled with brominated precursors during a short labeling period, conditioned by specific external stimuli, and followed by reverse transcriptase polymerase chain reaction (RT-PCR) and Northern blotting or sequencing for relating the signal to specific genes, seems a promising path to follow (Haider *et al.*, 1997).

References

Boisvert, F. M., Hendzel, M. J., and Bazett-Jones, D. P. (2000). Promyelocytic leukemia (PML) nuclear bodies are protein structures that do not accumulate RNA. *J. Cell Biol.* **148,** 283–292.

Carayon, P., and Bord, A. (1992). Identification of DNA-replicating lymphocyte subsets using a new method to label the bromodeoxyuridine incorporated into the DNA. *J. Immunol. Methods* **147,** 225–230.

Carmo-Fonseca, M., Cunha, C., Custódio, N., Carvalho, C., Jordan, P., Ferreira, J., and Parreira, L. (1996). The topography of chromosomes and genes in the nucleus. *Exp. Cell Res.* **229,** 247–252.

Casafont, I., Navascues, J., Pena, E., Lafarga, M., and Berciano, M. T. (2006). Nuclear organization and dynamics of transcription sites in rat sensory ganglia neurons detected by incorporation of 5′-fluorouridine into nascent RNA. *Neuroscience* **140,** 453–462.

Chiu, S. C., and Yang, N. S. (2007). Inhibition of tumor necrosis factor-alpha through selective blockade of pre-mRNA splicing by shiokonin. *Mol. Pharmacol.* **71,** 1640–1645.

Darzynkiewicz, Z. (1994). Simultaneous analysis of cellular RNA and DNA content. *Methods Cell Biol.* **41A,** 401–420.

Dolbeare, F. (1995). Bromodeoxyuridine: A diagnostic tool in biology and medicine, Part 1: Historical perspectives, histochemical methods and cell kinetics. *Histochem. J.* **27,** 339–369.

Dundr, M., and Raska, I. (1993). Nonisotopic ultrastructural mapping of transcription sites within the nucleolus. *Exp. Cell Res.* **208,** 275–281.

Fakan, S. (1986). Structural support for RNA synthesis in the cell nucleus. *Methods Achiev. Exp. Pathol.* **12,** 105–140.

Frankfurt, O. S. (1990). Flow cytometric analysis of double stranded RNA content distributions. *Methods Cell Biol.* **33,** 299–304.

Haider, S. R., Juan, G., Traganos, F., and Darzynkiewicz, Z. (1997). Immunoseparation and immunodetection of nucleic acids labeled with halogenated nucleotides. *Exp. Cell Res.* **234,** 498–506.

Halicka, D. H., Bedner, E., and Darzynkiewicz, Z. (2000). Segregation of RNA and separate packaging of DNA and RNA in apoptotic bodies during apoptosis. *Exp. Cell Res.* **260,** 248–256.

Hassan, A. B., Errington, R. J., White, N. S., Jackson, D. A., and Cook, P. R. (1994). Replication and transcription sites are colocalized in human cells. *J. Cell Sci.* **107,** 425–434.

Haukenes, G., Szilvay, A. M., Brokstad, K. A., Kanestrøm, A., and Kalland, K. H. (1997). Labeling of RNA transcripts of eukaryotic cells in culture with brutp using a liposome transfection reagent (DOTAP). *BioTechniques* **22,** 308–312.

Hozák, P., Cook, P. R., Schöfer, C., Mosgöller, W., and Wachtler, F. (1994). Site of transcription of ribosomal RNA and intranucleolar structure in hela cells. *J. Cell Sci.* **107,** 639–648.

Iborra, F. J., Pombo, A., Jackson, D. A., and Cook, P. R. (1996). Active RNA polymerases are localized within transcription factories in human nuclei. *J. Cell Sci.* **109,** 1427–1436.

Jackson, D. A., Hassan, A. B., Errington, R. J., and Cook, P. R. (1993). Visualization of focal sites of transcription within human nuclei. *EMBO J.* **12,** 1059–1065.

Jackson, D. A., Iborra, F. J., Manders, E. M. M., and Cook, P. R. (1998). Numbers and organization of RNA polymerases, nascent transcripts, and transcription units in hela nuclei. *Mol. Biol. Cell* **9,** 1523–1536.

Jensen, P. O., Larsen, J., Christiansen, J., and Larsen, J. K. (1993a). Flow cytometric measurement of RNA synthesis using bromouridine labelling and bromodeoxyuridine antibodies. *Cytometry* **14,** 455–458.

Jensen, P. O., Larsen, J., and Larsen, J. K. (1993b). Flow cytometric measurement of RNA synthesis based on bromouridine labelling and combined with measurement of DNA content or cell surface antigen. *Acta Oncol.* **32,** 521–524.

LaMorte, V. J., Dyck, J. A., Ochs, R. L., and Evans, R. M. (1998). Localization of nascent RNA and CREB binding protein with the PML-containing nuclear body. *Proc. Natl. Acad. Sci. USA* **95,** 4991–4996.

Landberg, G., and Roos, G. (1991). Antibodies to proliferating cell nuclear antigen (PCNA) as S-phase specific probes in flow cytometrie cell cycle analysis. *Cancer Res.* **51,** 4570–4575.

Li, X., Patel, R., Melamed, M. R., and Darzynkiewicz, Z. (1994). The cell cycle effects and induction of apoptosis by 5-bromouridine in cultures of human leukemic MOLT-4 and HL-60 cell lines and mitogen stimulated normal lymphocytes. *Cell Prolif.* **27,** 307–320.

Melcák, I., Risueño, M. C., and Raska, I. (1996). Ultrastructural nonisotopic mapping of nucleolar transcription sites in onion protoplasts. *J. Struct. Biol.* **116,** 253–263.

Murakami, T., Li, X., Gong, J., Bhatia, U., Traganos, F., and Darzynkiewicz, Z. (1995). Induction of apoptosis by 5-azacytidine: Drug concentration-dependent differences in cell cycle specificity. *Cancer Res.* **55,** 3093–3098.

Ohtsu, M., Kawate, M., Fukuoka, M., Gunji, W., Hanaoka, F., Utsugi, T., Onoda, F., and Murakami, Y. (2008). Novel DNA microarray system for analysis of nascent mrnas. *DNA Res.* **15,** 241–251

Otto, F. (1990). DAPI staining of fixed cells for high-resolution flow cytometry of nuclear DNA. *Methods Cell Biol.* **33,** 105–110.

Pombo, A., and Cook, P. R. (1996). The localization of sites containing nascent RNA and splicing factors. *Exp. Cell Res.* **229,** 201–203.

Wansink, D. G., Schul, W., van der Kraan, I., van Steensel, B., van Driel, R., and de Jong, L. (1993). Fluorescent labeling of nascent RNA reveals transcription by RNA polymerase II domains scattered throughout the nucleus. *J. Cell Biol.* **122,** 283–293.

Wansink, D. G., Nelissen, R. L. H., and de Jong, L. (1994a). *In vitro* splicing of pre-mrna containing bromouridine. *Mol. Biol. Rep.* **19,** 109–113.

Wansink, D. G., Manders, E. E., van der Kraan, I., Aten, J. A., van Driel, R., and de Jong, L. (1994b). RNA polymerase II transcription is concentrated outside replication domains throughout S-phase. *J. Cell Sci.* **107,** 1449–1456.

CHAPTER 35

Analysis of Mitochondria by Flow Cytometry

Martin Poot★ and Robert H. Pierce[†]

★Department of Pathology
University of Washington
Seattle, Washington 98195

†Department of Pathology
Wright-Patterson Medical Center
Wright-Patterson Air Force Base
Dayton, Ohio 45433

ESSENTIAL CYTOMETRY METHODS
Reprinted from *Methods in Cell Biology*, Volume 64 (Academic Press, 2001).
Copyright © 2001 by Academic Press, Inc.,
DOI: 10.1016/B978-0-12-375045-7.00035-0

I. Introduction

A. Principles of Fluorescent Detection of Mitochondria

Most of the energy (ATP) needed for proper functioning of the cell is generated by the mitochondrion. This process involves oxidation of reduced nucleotides (e.g., NADH and FADH) and concomitant generation of a negative inside membrane potential. The complex biochemical reactions that ultimately lead to formation of ATP are not fully understood. However, the finding that mitochondrial dysfunction may lead to neuromuscular diseases, neurodegenerative disorders, and apoptosis (Green and Reed, 1998; Kroemer *et al.*, 1998; Wallace, 1995) spurred intense investigation of this organelle.

For study by flow cytometry, cells and their organelles have to be (made) fluorescent. Reduced nucleotides, such as NADH and FADH, emit blue and blue-green fluorescence after excitation with light in the ultraviolet (UV) range and 488 nm, respectively (Thorell, 1983). As a function of their negative inside membrane potential, mitochondria will take up fluorescent cations. Initially, fluorescent probes for mitochondria have been devised based on this reasoning. Meanwhile, two dyes have been described that stain mitochondria based on unique properties that do not depend on the mitochondrial membrane potential. Nonyl acridine orange (NAO) specifically stains the mitochondrial membrane lipid cardiolipin (Petit *et al.*, 1992). The MitoTracker Green FM (MTG) dye was found to specifically stain mitochondria of cells in which the mitochondrial membrane potential was dissipated after cell fixation (Hollinshead *et al.*, 1997). Table I lists the response of some of these dyes to perturbation of mitochondrial function.

B. Xanthylium Dyes: Rhodamine 123 and CMXRosamine

The first dye described as a specific probe for the mitochondrion was the xanthylium dye rhodamine 123 (Johnson *et al.*, 1981). The amount of fluorescence obtained with this dye responded to the physiological state of the mitochondrion, as was predicted from the assumption that the mitochondrial membrane potential was the driving force behind dye accumulation inside the cell. Its poor photostability notwithstanding, rhodamine 123 has found widespread application, and its use was described by Chen in an earlier volume of this series (Chen, 1989). To improve photostability of mitochondrial staining, the MitoTracker Red dye CMXRosamine (CMXRos) was developed. The bright and relatively photostable red fluorescence of CMXRos (excitation 594 nm, emission 608 nm) showed strong sensitivity toward manipulation of the mitochondrial membrane potential and colocalized with cytochrome *c* oxidase (Poot *et al.*, 1996). The enhanced dye retention after cell fixation has been linked to the presence of a chloromethyl moiety (Poot *et al.*, 1996), with which CMXRos can covalently modify reduced thiol groups of mitochondrial membrane proteins. In addition to improved photostability, CMXRos showed better retention in stained cells during washing with

Table I
Sensitivity to Mitochondrial Poisons[a]

Dye	Rotenone	Antimycin A	CCCP
Rhodamine 123	97 ± 15	65 ± 12	71 ± 5
CMXRos	100 ± 6	37 ± 3	82 ± 7
MTG	108 ± 8	94 ± 7	105 ± 3
NAO	107 ± 4	98 ± 8	92 ± 11
JC-1	103 ± 12	58 ± 8	66 ± 7

[a]Logarithmically growing cultures of lymphoblastoid cells were pretreated with rotenone, antimycin A, and carbonyl cyanide m-chlorophenyl-hydrazone (CCCP) and stained with 100 nM of rhodamine 123, CMXRos, or MTG or with 500 nM of nonyl acridine orange (NAO) or JC-1 according to the methods described by Poot *et al.* (1996). With rhodamine 123, MTG, and NAO, green fluorescence was recorded (525 ± 20 nm); with CMXRos and JC-1, red fluorescence (above 640 nm) was recorded. All data are fluorescence intensities as percentages of untreated controls from three independent experiments each performed in triplicate.

phosphate-buffered saline than rhodamine 123 (Poot *et al.*, 1996). The latter may be due to the increased lipophilicity of CMXRos. Because of this characteristic, CMXRos fluorescence is likely to be less sensitive to mitochondrial swelling than in rhodamine 123. Changes in fluorescence after rhodamine 123 staining may represent the sum of changes in the mitochondrial membrane potential and in the volume of the intramitochondrial sap (Vander Heiden *et al.*, 1997). Fluorescence after CMXRos staining, on the other hand, may be sensitive to the amount of mitochondrial protein. Rhodamine 123, CMXRos, and other cationic xanthylium dyes share a critical feature: they are fluorescent in any medium, which leads to a certain level of background fluorescence even in a cell that does not contain mitochondria.

C. Symmetrical Carbocyanine Dyes: DiOC₆(3) and JC-1

To overcome the drawbacks of xanthylium dyes and to obtain stains that exhibit much less background fluorescence, symmetrical lipophilic carbocyanine dyes such as 3,3′-dihexyloxacarbocyanine iodide [DiOC₆(3)] and 5,5′,6,6′-tetrachloro-1,1′,3,3′-tetraethylbenzimidazoly iodide (JC-1) (Reers *et al.*, 1991) have been developed. The DiOC₆(3) dye turned out to be sensitive to the membrane potential of both the plasma membrane, and the mitochondrial membrane, and it showed strong fluorescence enhancement in a hydrophobic versus a hydrophilic environment (Sims *et al.*, 1974). In an earlier volume of this series Shapiro described the use of DiOC₆(3) to determine distributions of plasma membrane potential by flow cytometry (Shapiro, 1994). It is therefore no surprise that DiOC₆(3) fluorescence does not accurately monitor changes in the mitochondrial membrane potential (Salvioli *et al.*, 1997).

The JC-1 dye exhibits green fluorescence (excitation 490 nm, emission 527 nm) if present in low concentrations, and red fluorescence (excitation 490 nm, emission 590 nm) if accumulating at higher concentrations. The intramitochondrial concentration of the dye, and thus its fluorescence maximum, depends on the mitochondrial membrane potential (Smiley *et al.*, 1991). In addition to accumulating inside the mitochondrion, carbocyanine dyes, such as $DiOC_6(3)$ and JC-1, were found to stain the endoplasmic reticulum (Chen, 1989; Terasaki *et al.*, 1984). In addition, JC-1 forms J-aggregates that can result in nonspecific speckled staining of the cytoplasm (Poot *et al.*, 1996). These findings cast doubt on the alleged specificity of these carbocyanine dyes for mitochondria. Thus, the use of these dyes for studies of mitochondrial physiology by flow cytometry does not appear to be warranted. In particular, the contention that during apoptosis the mitochondrial membrane potential decreases cannot be based on the observed changes in $DiOC_6(3)$ and JC-1 fluorescence (Cossarizza *et al.*, 1994; M'etivier *et al.*, 1998).

D. Asymmetric Carbocyanine Dyes: MitoTracker Green FM and MitoFluor

To overcome the numerous drawbacks associated with symmetric carbocyanine dyes, a novel family of asymmetric carbocyanine dyes, including MitoTracker Green FM (MTG) and MitoFluor, have been developed. Both MTG and Mito-Fluor dyes are well excited by the 488 nm line of the argon laser and emit in the green region of the spectrum with little fluorescence emission in the orange-red region. Since the fluorescence emission range of the MTG and MitoFluor dyes is much narrower than that of rhodamine 123 (which emits significantly at wavelengths up to 620 nm), these dyes are more suitable for multicolor applications. The MTG dye is equipped with a chloromethyl moiety, which causes it to be retained after fixative treatment, whereas MitoFluor fluorescence vanishes after cell fixation. Both dyes do not respond to alterations in the mitochondrial membrane potential (see Table I). Moreover, the MTG dye has been used to specifically stain mitochondria in fixed cells (Hollinshead *et al.*, 1997) and is believed to monitor possible changes in mitochondrial protein level (Poot and Pierce, 1999).

E. The Cardiolipin Dye: Nonyl Acridine Orange

Nonyl acridine orange (NAO) (Maftah *et al.*, 1989) accumulates in mitochondria due its specific, high affinity binding to cardiolipin (Petit *et al.*, 1992). On binding to monoacidic phospholipids NAO emits green fluorescence, whereas in the presence of the diacidic phospholipid cardiolipin additional red fluorescence arises (Gallet *et al.*, 1995). Since cardiolipin is situated at the inner mitochondrial membrane, the fluorescence of NAO depends on the amount of cardiolipin. In other words, the fluorescence intensity of NAO can be taken as a direct measure of the amount of mitochondrial membrane lipid in a cell (Petit *et al.*, 1992).

F. Reduced Dyes

Reduced forms of some dyes (e.g., dihydrorhodamine 123, H_2-CMXRos), which yield only a fluorescent response after they are oxidized in intact mitochondria, have become available (Whitaker *et al.*, 1991). These dyes allow monitoring the rate of oxidant formation in mitochondria (Poot and Pierce, 1999).

G. Subcellularly Targeted Green Fluorescent Protein

The fluorescence emission of the green fluorescent protein (GFP) from the jellyfish *Aequorea victoria* is sensitive to the calcium level (Rizzuto *et al.*, 1992) and the pH of its environment (Kreen *et al.*, 1998; Llopis *et al.*, 1998). Via recombinant DNA techniques, fusion proteins containing GFP and subcellular targeting signals have been prepared. These fusion proteins have been shown to localize to the mitochondrion (Rizzuto *et al.*, 1992), the endoplasmic reticulum (Kendall *et al.*, 1992), and the nucleus (Rizzuto *et al.*, 1994). Thus, a set of reagents has been devised that can be used to detect rapid changes in subcellular calcium levels (Rizzuto *et al.*, 1994). Although this approach proved to be successful in image analysis, no flow cytometric protocol has been published yet.

By introducing amino acid substitutions, forms of the GFP with different pK_a values and fluorescence emission spectra have been created (Llopis *et al.*, 1998). By simultaneously detecting the fluorescence emission from the "cyan" and least pH-sensitive protein with the yellow and most pH-sensitive protein, changes in intracellular pH can be deduced from the ratio of fluorescence intensity in the two wavelength domains (Llopis *et al.*, 1998). Thus, intracellular pH can be measured by a fluorescence ratioing method. The genes for the cyan and yellow proteins were then fused with targeting sequences for the Golgi system and the mitochondrion. These fusion proteins were then expressed in HeLa cells. By confocal microscopy, it was shown that the modified proteins localized to the organelles of interest (Llopis *et al.*, 1998). This novel approach may be of great promise, but it has as yet not been implemented in flow cytometry.

In this chapter, protocols to determine changes in NADH level, mitochondrial membrane potential as normalized for possible differences in mitochondrial protein content, mitochondrial oxidative turnover, mitochondrial protein, and cardiolipin content will be described.

II. Materials and Methods

A. Dyes

Inside the mitochondrion NADH and FADH exist as natural fluorophores. Stock solutions of 2.5 or 5.0 mM of rhodamine 123 and NAO are prepared in phosphate-buffered saline and stored at 4 °C in the dark. The MTG, MitoFluor, and CMXRos dyes are dissolved at 0.2 mM in dimethyl sulfoxide and stored at −20 °C in the dark. Immediately before use, these dyes are thawed at room temperature in the dark. In our experience these dyes do not suffer from repeated freeze-thaw cycles. Reduced dyes, such as dihydrorhodamine 123 and H$_2$-CMX-Ros, are dissolved in dimethyl sulfoxide and flushed with an inert gas (e.g., nitrogen) before storage at −20 °C. At room temperature, reduced dyes will oxidize readily. It is therefore recommended to divide dye solutions into aliquots and not to refreeze them. Key features of the dyes used in the protocols below are displayed in Table II.

B. Cell Preparation

Cells are harvested by standard procedures. In case mitochondrial function during cell death by apoptosis is to be studied, care should be taken that all (attached and detached) cells are included. After harvesting, cells are resuspended in regular cell culture medium and warmed to 37 °C.

C. Reduced Nucleotides (NADH and FADH)

In unstained cells, the UV-excited (360 nm) blue autofluorescence (around 450 nm) is proportional to the mitochondrial NADH content and can easily be detected by flow cytometry (Thorell, 1983). The NADH content of cells proved to be a very sensitive parameter for changes in mitochondrial metabolism during apoptosis (Poot and Pierce, 1999). Excitation with 488 nm light of isolated mitochondria gives green autofluorescence that is inversely proportional to their FADH content. In intact cells, FADH fluorescence is overshadowed by other sources of green autofluorescence (e.g., lipofuscin). Therefore, it is not possible to quantify FADH levels in intact cells based on green autofluorescence.

Table II
Features of Mitochondrial Dyes

Dye	Excitation maximum (nm)	Emission maximum (nm)	Fixability with aldehydes
Rhodamine 123	506	530	−
CMXRos	594	608	−
MTG	480	516	+
MitoFluor	480	516	−
NAO	497	519	−
JC-1	514	529, 590	+/−[a]

[a]The JC-1 dye is partially retained after formaldehyde fixation, though a covalent bond with cellular macromolecules appears unlikely (Poot et al., 1996).

D. Normalized Mitochondrial Membrane Potential

Flow cytometry measures the total amount of fluorescence obtained from each individual cell. When cells are stained with dyes of which the fluorescence intensity is proportional to the mitochondrial membrane potential, the total cellular fluorescence will reflect the average of the mitochondrial membrane potential per mitochondrion multiplied by the amount of mitochondrial mass per cell. Thus, a cell with more mitochondrial mass will be more fluorescent than a cell with less mitochondria. To correct for possible differences in mitochondrial mass, it is necessary to have a measure for the amount of mitochondrial mass in each individual cell. This measure is provided by counterstaining the cells with a mitochondria-specific dye that is not sensitive to the mitochondrial membrane potential. This dye has to emit at wavelengths that are different from those of the mitochondrial membrane sensitive dye. That means that if the mitochondrial membrane sensitive dye emits in the green, the mitochondrial membrane insensitive dye has to emit in the red and vice versa. By dividing the fluorescence intensity of the membrane potential sensitive dye by the fluorescence intensity from the insensitive dye, the normalized mitochondrial membrane of each individual cell is obtained (Poot and Pierce, 1999). A dye pair that has worked successfully in this way is MTG (insensitive) and CMXRos (membrane potential sensitive) (Poot and Pierce, 1999).

To 1-ml aliquots of prewarmed cell suspensions, CMXRos and MTG are added to obtain a final concentration of maximally 100 nM. Cell suspensions are incubated for 30 min at 37 °C in the dark. After incubation cell suspensions are placed in a melting ice bath and immediately assayed by flow cytometry. Both dyes are excitable with the 488-nm line of an argon laser; MTG emits fluorescence around 530 nm, whereas CMXRos is best detected at wavelengths above 610 nm. To minimize possible interference from MTG fluorescence, a 640-nm-long-pass filter is preferred.

The ratio of CMXRos to MTG fluorescence intensity for each individual cell represents its normalized mitochondrial membrane potential. Since the MTG and the CMXRos dyes emit green and red fluorescence after excitation with 488 nm laser light, it is possible to combine this assay with quantification of NADH by UV-excited blue fluorescence as described earlier. Thus, a combined assay for normalized mitochondrial membrane potential and normalized NADH content is obtained. Figure 1 shows a typical result of such an assay.

E. Mitochondrial Oxidative Turnover

To 1-ml aliquots of prewarmed cell suspensions, H_2-CMXros is added at a final concentration of 100 nM. Cell suspensions are incubated for 30 min at 37 °C in the dark. After incubation cell suspensions are placed in a melting ice bath and immediately assayed by flow cytometry. After oxidation by mitochondrial metabolism H_2-CMXRos becomes CMXRos and can be excited with the 488-nm line of an

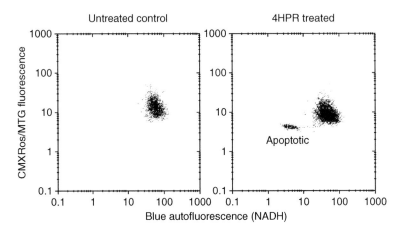

Fig. 1 Human breast cancer cells cultured for 4 days with 1 μM N-(4-hydroxyphenyl)retinamide (4HPR) and stained with CMXRos and MTG as described in the protocol for normalized mitochondrial membrane potential. Abscissa: UV-excited blue autofluorescence (which represents the NADH level); ordinate: the ratio of CMXRos to MTG fluorescence for each individual cell (which represents the normalized mitochondrial membrane potential). The signal dots labeled apoptotic represent cells that show a simultaneous decrease in NADH level and normalized mitochondrial membrane potential.

argon laser; CMXRos is best detected at wavelengths above 610 nm. Since the amount of fluorescence generated by mitochondrial metabolism is usually low, it is not recommended to combine H_2-CMXros staining with MTG or any other dye that may generate some yellow and red fluorescence (e.g., NAO). Generation of fluorescence from H_2-CMXRos can be abolished by preincubation with antimycin A, an inhibitor of mitochondrial electron flux (Poot and Pierce, 1999). Fluorescence generated in cells incubated with dihydrorhodamine 123 was found to be less sensitive toward inhibition of mitochondrial metabolism (M. Poot and R. H. Pierce, 1997, unpublished observation). Therefore, H_2-CMXros is now the preferred dye for measuring mitochondrial oxidative turnover.

F. Mitochondrial Protein

In fixed cells, the MTG dye stains mitochondria specifically (Hollinshead *et al.*, 1997). In flow cytometric assays the amount of fluorescence after MTG staining appears to correlate with cell volume (Poot and Pierce, 1999). In addition, mitochondrial proteins can be resolved as green bands by SDS-polyacrylamide gel electrophoresis (M. Poot, 1997, unpublished observation). Therefore, it is believed that the amount of MTG fluorescence represents the mitochondrial protein level. Incubation of cell suspensions with 100 nM of MTG for 30 min at 37 °C followed by flow cytometric analysis is adequate to compare mitochondrial protein levels in different cell samples.

G. Mitochondrial Membrane Phospholipid (Cardiolipin)

Nonyl acridine orange (Maftah *et al.*, 1989) stains specifically the mitochondrial membrane lipid cardiolipin (Petit *et al.*, 1992). Cell suspensions are stained with 1 μM NAO during 30 min at 37 °C, placed in a melting ice bath, and assayed immediately. NAO can be excited with the 488-nm line of an argon laser. The emission spectrum of NAO covers the green, yellow, and red domain. Because the red emission appears to most accurately reflect the mitochondrial cardiolipin level (Gallet *et al.*, 1995) a 640-nm long-pass filter is preferred (Pierce *et al.*, 2000).

III. Critical Aspects

A. Staining Parameters

After harvesting and before staining, cells are resuspended in regular cell culture medium and warmed to 37 °C. The latter is critical, because parameters of mitochondrial function, such as the mitochondrial membrane potential, and thus dye uptake, depend on the temperature. In order to not disturb mitochondrial function, regular cell culture medium should be used. If effects of certain environmental components are to be investigated, it is also possible to use special buffers or media. The staining protocols outlined are based on saturation of the target with dye. The recommended staining time is intended to be sufficient for this purpose. In case of doubt, it is recommended to test a series of staining times.

To avoid nonspecific staining it is important to keep the dye concentration as low as possible. With most cell types and with most flow cytometers, a sufficiently strong signal can be obtained with 100 nM of both CMXRos and MTG. If enough signal can be obtained, a lower dye concentration (<100 nM) may be preferable. A similar reasoning applies to the use of dihydrorhodamine 123, H_2-CMXRos, and NAO.

B. Cell Fixation

Some of the fluorescence of MTG and CMXRos is retained after cell fixation. However, the level of retained fluorescence may reflect the amount of protein thiols to which the dyes have bound covalently. This inference, which is derived from the behavior of chloromethyl moieties of other dyes (Poot *et al.*, 1991), has been supported experimentally (Gilmore and Wilson, 1999). One should not confuse the fluorescence intensities retained after cell fixation with the original fluorescence intensities, which are proportional with the mitochondrial membrane potential. It is clearly in error to assume that the retained fluorescence after cell fixation represents the mitochondrial membrane potential, as has been done in the literature (Macho *et al.*, 1996).

C. Combinations of Dyes

Since dyes to monitor mitochondrial parameters show different emission spectra, it is possible to use combinations of dyes in a single sample. This serves two purposes: first, concordant and discordant changes in metabolic parameters can be detected. Thus, a concordant change in mitochondrial membrane potential and cellular reduced glutathione level was elicited by exposing leukemia cells to 1-β-D-arabinofuranosylcytosine. This change preceded a high rate of reactive oxygen generation and an increase in intracellular calcium (Backway *et al.*, 1997). Second, it is possible to use the fluorescence emission of one dye to calibrate the fluorescence of other fluorochromes. Thus, ratios of fluorescence intensities on a per cell basis will represent "normalized" values for the parameters under investigation. Protocols for normalized mitochondrial membrane potential and normalized NADH levels were previously described. Such protocols give valid results only if the dyes used either do not interact with each other or if their interaction is constant. Owing to the small size of the mitochondrion, it is possible that dye molecules will bind in such a way that fluorescence resonance energy transfer may take place. Thus, it is conceivable that part of the fluorescence emission of MTG may be absorbed by CMXRos. It is therefore recommended to include samples stained with a single dye to verify that the differences between samples observed do not result from differences in efficiency of energy transfer between pairs of dyes.

Using a combination of NAO and CMXRos, we have been able to document a change in the structure of the mitochondrial membrane during apoptosis (Poot and Pierce, 1999). Owing to the broad emission spectrum of NAO, combinations of NAO with other dyes require the use of control samples stained with a single dye or careful electronic compensation to eliminate cross-talk of NAO into the red fluorescence emission channel.

Acknowledgments

The authors are supported by a grant from the U.S. Army Medical Research and Material Command (DAMD: 17-94-J-4028) (M.P.), by the National Institute for Aging Research (Grant PO1 AG 01751) (M. P.), by the Nathan Shock Center of Excellence for the Basic Biology of Aging (Grant P30 AG 13240) (M. P.), by a Seed Money grant as part of an Institutional Research Grant of the American Cancer Society to the Fred Hutchinson Cancer Research Center (M.P.), and by grants from the National Institute of Environmental Health Sciences Environmental Pathology/Toxicology Training Program (Grant 5 T32 ES07032) and American Liver Foundation Irwin Arias Postdoctoral Research Fellowship (R.H.P.). We kindly acknowledge advice by W. T. Shen and Dr P. S. Rabinovitch regarding the MPLUS software package. The views expressed in this article are those of the authors and do not reflect the official policy or position of the United States Air Force, Department of Defense, or the U.S. government.

References

Backway, K. L., McCulloch, E. A., Chow, S., and Hedley, D. W. (1997). Relationships between the mitochondrial permeability transition and oxidative stress during ara-C toxicity. *Cancer Res.* **57**, 2446–2451.

Chen, L. B. (1989). Fluorescent labeling of mitochondria. *Methods Cell Biol.* **29,** 103–123.

Cossarizza, A., Kalashnikova, G., Grassilli, E., Chiapelli, F., Salvioli, S., Capri, M., Barbieri, D., Troiano, L., Monti, D., and Franceschi, C. (1994). Mitochondrial modifications during rat thymocyte apoptosis: A study at the single cell level. *Exp. Cell Res.* **214,** 323–330.

Gallet, P. F., Maftah, A., Petit, J. M., Denis-Gay, M., and Julien, R. (1995). Direct cardiolipin assay in yeast using red fluorescence emission of 10-*N*-nonyl acridine orange. *Eur. J. Biochem.* **228,** 113–119.

Gilmore, K., and Wilson, M. (1999). The use of chloromethyl-X-rosamine (MitoTracker Red) to measure loss of mitochondrial membrane potential in apoptotic cells is incompatible with cell fixation. *Cytometry* **36,** 355–358.

Green, D. R., and Reed, J. C. (1998). Mitochondria and apoptosis. *Science* **281,** 1309 1312.

Hollinshead, M., Sanderson, J., and Vaux, D. J. (1997). Anti-biotin antibodies offer superior organelle-specific labeling of mitochondria over avidin or streptavidin. *J. Histochem. Cytochem.* **45,** 1053–1057.

Johnson, L. V., Walsh, M. L., Bockus, B. J., and Chen, L. B. (1981). Monitoring of relative mitochondrial membrane potential in living cells by fluorescence microscopy. *J. Cell Biol.* **88,** 526–535.

Kendall, J. M., Dormer, R. L., and Campbell, A. K. (1992). Targeting aequorin to the endoplasmic reticulum of living cells. *Biochem. Biophys. Res. Commun.* **189,** 1008–1016.

Kreen, M., Farinas, J., Li, Y., and Verkman, A. S. (1998). Green fluorescent protein as a noninvasive intracellular pH indicator. *Biophys. J.* **74,** 1591–1599.

Kroemer, G., Dallaporta, B., and Resche-Rigon, M. (1998). The mitochondrial death/life regulator in apoptosis and necrosis. *Annu. Rev. Physiol.* **60,** 619–642.

Llopis, J., McCaffery, M., Miyawaki, A., Farquhar, M. G., and Tsien, R. Y. (1998). Measurement of cytosolic, mitochondrial, and Golgi pH in single living cells with green fluorescent proteins. *Proc. Natl. Acad. Sci. USA* **95,** 6803–6808.

M'etivier, D., Dallaporta, B., Zamzami, N., Larochette, N., Susin, S. A., Marzo, I., and Kroemer, G. (1998). Cytofluorometric detection of mitochondrial alterations in early CD95/Fas/APO-1-triggered apoptosis of Jurkat T lymphoma cells. Comparison of seven mitochondrion-specific fluorochromes. *Immunol. Lett.* **61,** 157–163.

Macho, A., Decaudin, D., Castedo, M., Hirsch, T., Susin, S. A., Zamzami, N., and Kroemer, G. (1996). Chloromethyl-X-Rosamine is an aldehyde-fixable potential-sensitive fluorochrome for the detection of early apoptosis. *Cytometry* **25,** 333–340.

Maftah, A., Petit, J. M., Ratinaud, M. H., and Julien, R. (1989). 10 *N*-Nonyl-acridine orange: A fluorescent probe which stains mitochondria independently of their energetic state. *Biochem. Biophys. Res. Commun.* **164,** 185–190.

Petit, J.-M., Maftah, A., Ratinaud, M.-H., and Julien, R. (1992). 10 *N*-Nonyl-acidine orange interacts with cardiolipin and allows the quantification of this phospholipid in isolated mitochondria. *Eur. J. Biochem.* **209,** 267–273.

Pierce, R. H., Campbell, J. S., Stephenson, A. B., Franklin, C. C., Chaisson, M., Poot, M., Kavanagh, T. J., and Fausto, N. (2000). Disruption of redox homeostasis in tumor necrosis factor induced apoptosis in murine hepatocytes. *Am. J. Pathol.* **157,** 221–236.

Poot, M., and Pierce, R. H. (1999). Detection of changes in mitochondrial function during apoptosis by simultaneous staining with multiple fluorescent dyes and correlated multiparameter flow cytometry. *Cytometry* **35,** 311–317

Poot, M., Kavanagh, T. J., Kang, H.-C, Haugland, R. P., and Rabinovitch, P. S. (1991). Flow cytometric analysis of cell cycle-dependent changes in cell thiol level by combining a new laser dye with Hoechst 33342. *Cytometry* **12,** 184–187.

Poot, M., Zhang, Y.-Z., Krämer, J., Wells, K. S., Jones, L. J., Hanzel, D. K., Lugade, A. G., Singer, V. L., and Haugland, R. P. (1996). Analysis of mitochondrial morphology and function with novel fixable fluorescent stains. *J. Histochem. Cytochem.* **44,** 1363–1372.

Reers, M., Smith, T. W., and Chen, L. B. (1991). J-aggregate formation of a carbocyanine as a quantitative fluorescent indicator of membrane potential. *Biochemistry* **30,** 4480 4486.

Rizzuto, R., Simpson, A. W. M., Brini, M., and Pozzan, T. (1992). Rapid changes of mitochondrial Ca^{2+} revealed by specifically targeted recombinant aequorin. *Nature* **358,** 325–327.

Rizzuto, R., Brini, M., and Pozzan, T. (1994). Targeting recombinant aequorin to specific intracellular organelles. *Methods Cell Biol.* **40,** 339–358.

Salvioli, S., Ardizzoni, A., Franceschi, C., and Cossarizza, A. (1997). JC-1, but not DioC$_6$(3) or rhodamine 123, is a reliable fluorescent probe to assess $\Delta\Psi$ changes in intact cells: Implications for studies on mitochondrial functionality during apoptosis. *FEBS Lett.* **411,** 77–82.

Shapiro, H. M. (1994). Cell membrane potential analysis. *Methods Cell Biol.* **41,** 121–133.

Sims, P. J., Waggoner, A. S., Wang, C. H., and Hoffman, J. F. (1974). Studies on the mechanism by which cyanine dyes measure membrane potential in red blood cells and phosphatidylcholine vesicles. *Biochemistry* **13,** 3315–3330.

Smiley, S. T., Reers, M., Mottola-Hartshorn, C., Lin, M., Chen, A., Smith, T. W., Steele, G. D., Jr., and Chen, L. B. (1991). Intracellular heterogeneity in mitochondrial membrane potentials revealed by a J-aggregate-forming lipophilic cation JC-1. *Proc. Natl. Acad. Sci. USA* **88,** 3671–3675.

Terasaki, M., Song, J., Wong, J. R., Weiss, M. J., and Chen, L. B. (1984). Localization of endoplasmic reticulum in living and glutaraldehyde-fixed cells with fluorescent dyes. *Cell* **38,** 101–108.

Thorell, B. (1983). Flow-cytometric monitoring of intracellular flavins simultaneously with NAD(P)H levels. *Cytometry* **4,** 61–65.

Vander Heiden, M. G., Chandel, N. S., Williamson, E. K., Schumacker, P. T., and Thompson, C. B. (1997). Bcl-$_{xL}$ regulates the membrane potential and volume homeostasis of mitochondria. *Cell* **191,** 627–637.

Wallace, D. C. (1995). Mitochondrial DNA variation in human evolution, degenerative disease, and aging. *Am. J. Hum. Genet.* **57,** 201–223.

Whitaker, J. E., Moore, P. L., Haugland, R. L., and Haugland, R. P. (1991). Dihydrotetramethylrosamine: A long wavelength, fluorogenic peroxidase substrate evaluated *in vitro* and in a model phagocyte. *Biochem. Biophys. Res. Commun.* **175,** 387–393.

CHAPTER 36

Analysis of Platelets by Flow Cytometry

Kenneth A. Ault and Jane Mitchell

Maine Medical Center Research Institute
South Portland, Maine 04106

I. Introduction

Platelets are anucleate cellular fragments which circulate in large numbers in blood (2–$4 \times 10^5 \, \mu l^{-1}$). They are primarily responsible for maintaining the integrity of the vasculature by plugging leaks as they occur. In order to do this they have the capability to respond to changes in blood flow (shear stress) and to the subendothelial matrix of vessels. When they detect an abnormality they are able

ESSENTIAL CYTOMETRY METHODS
Reprinted from *Methods in Cell Biology*, Volume 42 (Academic Press, 1994).
Copyright © 1994 by Academic Press, Inc.,
All rights of reproduction in any form reserved.

DOI: 10.1016/B978-0-12-375045-7.00036-2

to respond in basically three ways. They can undergo adhesion to the vessel wall, they can aggregate with other platelets to form a "platelet plug," and they can release the contents of their cytoplasmic granules which result in the recruitment of more platelets, the initiation of an inflammatory response, and the initiation of the coagulation cascade. Interestingly, this entire functional repertory of platelets can be evaluated by flow cytometry. Platelet dysfunction can result in an increased tendency to bleeding and may very well play an important role in a variety of cardiovascular diseases such as myocardial infarction or stroke.

Only in the past 5 years has flow cytometry assumed any importance in the evaluation of platelets. However, it has now become clear that flow cytometric techniques are extremely useful in the study of platelets. The flow cytometer can overcome some very difficult problems that had slowed progress in understanding platelets. For this reason, flow cytometric techniques for the study of platelets have achieved considerable and still rapidly growing importance.

This chapter attempts to outline the methods used for analysis of platelets by flow cytometry. Because there are a number of different flow cytometric applications associated with platelets, we do not attempt to detail each of the applications but rather cover the general principles that are important whenever platelets are being studied by flow cytometry. The biology and physical properties of platelets are sufficiently different from those of other cell types in that they frequently require different methods of preparation, labeling, and analysis than are used for flow cytometric evaluation of other cell types. These special considerations are dealt with in detail. The specific applications will be mentioned and whatever unique aspects they entail will be outlined, but the reader is referred to more detailed accounts of specific applications in the literature for discussion of the rationale, alternative methods, and interpretation of individual tests. There have been several recent reviews which cover the biological and medical issues rather than strictly technical issues (Ault, 1992; Ault and Mitchell, 1993; Ault *et al.*, 1991).

The flow cytometric evaluation of platelets has a historical component which is both of some interest and relevant to some of the discussion to follow. Much of the early work with flow cytometry was directed toward the study of nucleated cells of the blood, particularly lymphocytes. Until the past 5 years, no one had used flow cytometers for the study of platelets, and in fact most workers referred to the small particles found in blood as "debris" with little thought that the debris consisted almost entirely of platelets. Thus, one can find extensive discussions of methods to eliminate debris and of the confusion created when debris contaminated the lymphocyte gates.

It has now become clear that the study of platelets by flow cytometry is a fruitful area of research and clinical applications. They have been elevated above the category of debris. However, as we discuss below, there is now a new category of debris which those of us interested in platelets take care to eliminate from our platelet gates. It is likely that there is much useful information in this new debris which includes fragments of red cells, white cells, and platelets as well as large immune complexes and perhaps other interesting objects. Thus, this chapter might be subtitled 'The Flow Cytometric Analysis of Debris'.

II. Applications

There are several distinct applications of flow cytometry to the study of platelets which differ to some extent in the methods of platelet preparation, labeling, and analysis, and of course in the specific markers used. These are summarized in Table I and described briefly below.

Phenotyping: Platelets can be characterized according to their expression of a considerable variety of surface glycoproteins that are identified by monoclonal antibodies. The surface of the platelet is at least as well understood as that of the lymphocyte. In the case of most of the major platelet surface glycoproteins, their function is known and monoclonal antibodies are readily available. Unlike lymphocytes, however, there are very few subsets of platelets that have been identified. Thus, the major utility of platelet phenotyping is to determine unambiguously that one is studying platelets. Markers such as CD41 and CD42 can be used to reliably distinguish platelets from "debris." In addition, there are a few, relatively rare diseases in which there are abnormalities of platelet surface structures. For example, in Glanzman's thrombasthenia the GPIIb/IIIa molecule is absent or markedly reduced and this can be easily detected using a CD41 monoclonal.

Platelet-associated immunoglobulin (PAIg): There are abundant data suggesting that the ability to measure the amount of antibody on the surface of platelets provides useful information in diagnosing the relatively common disorders in which antibodies directed against platelets result in thrombocytopenia. This has been a controversial area due in large part to technical problems associated with the measurement of Palg. Recently, flow cytometry has been used to an increasing extent to reliably measure the amount of Ig (and sometimes C3) on the surface of platelets. The flow cytometric method has several advantages, such as the ability to perform the measurement in patients who are severely thrombocytopenic with relatively small volumes of blood; the ability to be absolutely certain that one is measuring platelet-associated Ig rather than Ig on other types of "debris" such as red cell fragments and immune complexes, and the ability to restrict the measurement to the platelet surface thus avoiding the problem of a large amount of Ig inside the platelet which may not be relevant to the process of immune platelet destruction.

Platelet activation and aggregation: Numerous investigators have recently demonstrated that it is possible to elegantly study various aspects of platelet function through the use of activation-specific monoclonal antibodies and the use of light scatter to measure platelet aggregation. These methods of studying platelet function have the advantages that they are applicable to whole blood, thus avoiding artifactual platelet activation during preparation of washed platelets; they require small volumes of blood; they permit correlated measurement of two

Table I

Applications of Flow Cytometry to the Study of Platelets

Application	Platelet preparation	Labeling	Comments	References
Phenotyping	Whole blood or washed platelets	Platelet specific glycoproteins, for example, CD37, 41, 42.	Can be used to diagnose diseases associated with loss of specific markers	Adelman et al. (1985), Ault (1988), Marti et al. (1988)
Platelet-associated immunoglobulin	Washed platelets	CD41 and anti-IgG, IgM, IgA, and C3	Used to aid in the diagnosis of immune thrombocytopenia	Corash and Rheinschmidt (1986), Lazarchick and Hall (1986), Rosenfield et al. (1987)
Platelet activation	Whole blood	CD41 and an activation marker such as CD62, 63 or an epitope of GPIIb/IIIa	Used to measure spontaneous or stimulated platelet function. May be combined with measure of platelet aggregation	Abrams and Shattil (1991), Ault et al. (1989), Berman et al. (1986), Carmody et al. (1990), Coller (1985), Corash (1990), Corash et al. (1986), Ginsberg et al. (1990), Johnston et al. (1987) Nieuwenhuis et al. (1987), Rinder et al. (1991a,b,c), Shattil et al. (1987),
Reticulated platelets	Whole blood or washed platelets	Thiazole-orange	Defines a subset of platelets which are newly released into the circulation	Ault et al. (1992), Ingram and Coopersmith (1969), Kienast and Schmitz (1990), Rinder et al. (1993)
Platelet-leukocyte interactions	Whole blood	CD41 and a leukocyte specific marker	Used to study the interactions between platelets and leukocytes	de Bruijne-Admiraal et al. (1992) Rinder et al. (1991d,e, 1992),
Platelet microparticles	Whole blood	CD41 or 42	Measures platelet fragmentation	Abrams et al. (1990), Sims et al. (1988)
Platelet calcium flux	Washed platelets		Measures platelet response to agonists	Davies et al. (1988), Johnson et al. (1985)

or more different aspects of platelet function, such as activation and aggregation or more than one activation-specific marker; and they appear to be more sensitive than conventional platelet function tests to increased platelet reactivity. Traditional methods for evaluating platelet function were geared to measuring platelet dysfunction rather than hyperactivity. For these reasons there is a considerable interest in the possibility that the flow cytometric methods may lead to useful clinical tests for platelet hyperfunction or *in vivo* platelet activation which almost certainly plays a role in some cardiovascular diseases.

Reticulated platelets: This subset of platelets is characterized by an increased content of nucleic acid which can be readily identified by the flow cytometer. There is now good evidence that these are newly released platelets and that their measurement may be useful in evaluating thrombocytopenia in the same way that a reticulocyte count is useful in anemia, that is, as an estimate of the rate of new platelet production.

Platelet-leukocyte interactions: By using a combination of platelet-specific and leukocyte-specific markers, platelet leukocyte interactions can be readily measured in the flow cytometer. In whole blood systems, it has been possible to demonstrate that significant numbers of leukocytes circulate in association with platelets. The possible physiological consequences of this observation and the possible diagnostic uses of such a measurement are only beginning to be explored.

Platelet microparticles: These particles have the immunological properties of platelets, that is, they contain platelet-specific glycoproteins and are labeled by platelet-specific monoclonal antibodies, but they are physically smaller than normal platelets. These particles have been most clearly defined using flow cytometric techniques and there are published data showing that these particles are produced when platelets are activated by some agonists. It is thought that these particles represent fragmentation or vesiculation of platelets. There is some suggestion that these particles may have more activity in activating the coagulation cascade than do normal platelets. Their clinical significance has not been established, but they are likely to be the subject of considerably more study in the near future.

III. Materials

Most of the reagents used in the preparation and labeling of platelets are not unusual. These include standard anticoagulants used in the collection of blood. If platelets are to be fixed, the universally accepted fixative is paraformaldehyde made up in phosphate-buffered saline. Although this is a standard reagent, its use in the fixation of platelets requires more care than is usually necessary. Paraformaldehyde solutions should be checked for pH (should be near 7.0) and careful attention has to be paid to the concentration. Overfixation is a common cause of poor labeling.

Monoclonal antibodies directed against platelet-specific markers are becoming more widely available. The most complete catalog of these reagents (at the time of this writing) is that of AMAC Inc. (160B Larrabee Road, Westbrook, ME 04092; Fax: 1-800-458-5060).

There are a number of reagents used to either activate platelets (agonists) or to prevent their activation. Some of these are listed in Table II with their usual concentrations and stability. Sources for these reagents include standard chemical supply companies and the manufacturers of hematology instruments used to evaluate platelet function.

IV. Cell Preparation

There are three basic sources for preparation of platelets for flow cytometry. These are whole blood, platelet-rich plasma, and washed platelets. The use of whole blood is recommended because it greatly reduces the artifactual platelet activation that is inevitable in any process which manipulates the platelets. Platelets can be studied in whole-blood samples which have been fixed using paraformaldehyde. The protocol used in our laboratory is to place 50 μl of fresh blood directly into a tube containing 1 ml of 2% (w/v) paraformaldehyde in phosphate-buffered saline. This results in rapid fixation of the platelets and preserves them in a state of activation and aggregation that accurately reflects their status in the blood (Ault *et al.*, 1989). The great advantage to fixation is that the effects of subsequent artifactual changes in the platelets are removed. This procedure lends itself particularly well to clinical studies in which the samples cannot be analyzed immediately. The major disadvantage of fixation is that it may destroy some immunological determinants that are of interest. For example, the very interesting activation marker on GPIIb/IIIa which is identified by the monoclonal antibody PAC-1 is altered by fixation. Alternatively, whole blood may be labeled without fixation by placing a small amount of blood into tubes containing premeasured amounts of the antibodies (Shattil *et al.*, 1987). This results in rapidly labeling the platelets, which are then diluted to a final volume

Table II
Reagents Used to Activate Platelets and to Prevent Activation

	Agent	Concentration	Stability	Comments
Agonists	Adenosine diphospate	5–20 μM	Stable	Works poorly on washed platelets
	Epinephrine	100 μM	Stable	A very weak agonist, frequently combined with ADP
	Thrombin	0.5–1 unit/ml	Unstable, freeze aliquots	A strong agonist, difficult to use in whole blood due to clotting
	Arachidonic acid	500 μg/ml	Unstable, freeze aliquots	
	Collagen	0.2 mg/ml	Stable	
Antagonists	Prostoglandin E1	1–10 μg/ml	Unstable, freeze aliquots	The best single antagonist
	Theophyllin	1 mM	Stable	
	Heparin	10–100 μg/ml	Stable	Blocks thrombin-induced activation
	Hirudin	0.5–25 μg/ml	Stable	Blocks thrombin-induced activation

of 0.5–1 ml. The result is that the effective concentration of the antibodies is reduced so that the pace of subsequent labeling is slowed. When such samples can be analyzed within a short period of time, perhaps an hour or less, the results are excellent. However, aggregation is not observed in such protocols due to the dilution of the platelets and deaggregation which occurs.

Platelet-rich plasma (PRP) is prepared from blood by a slow centrifugation which removes most of the erythrocytes and results in plasma containing large numbers of platelets. Because the platelets have not been pelleted there is not a great deal of artifactual activation. However, there is a time penalty in that there is progressive activation during the time required for the centrifugation. Nevertheless, the use of PRP is the customary method for platelet function studies and flow cytometric techniques based on PRP are still widely used. In studies which do not depend upon platelet activation, such as PAIg or reticulated platelets, PRP may be an excellent preparative step. It has the advantage that the platelets are still in their normal physiologic plasma.

Finally, washed platelets may be prepared in a variety of ways, most of which require centrifugation to pellet the platelets. Unless inhibitors of activation are used this will always result in considerable platelet activation. In our hands, it is not possible to obtain washed platelets with less than 20–40% expression of CD62 unless one uses inhibitors such as PGE1. Even in the presence of such inhibitors the levels of activation obtained are highly variable. Recently, it has been suggested that it is possible to obtain washed platelets without pelleting them using a discontinuous density gradient. This approach has not been explored extensively to our knowledge. Washed platelets are best used in protocols which are not affected by platelet activation. For example in measuring PAIg it is necessary to wash the platelets out of the plasma in order to label with anti-immunoglobulin reagents (Ault, 1988).

A. Sample Protocol for Preparation of Fixed Whole-Blood Samples for Measuring Platelet Activation and Aggregation

- Draw blood using minimal application of a tourniquet and a clean venipuncture.
- Use a heparinized syringe or a green top (heparinized) Vacutainer tube.
- Transfer 50 μl of blood immediately into a tube containing 1 ml of 2% paraformaldehyde in PBS.
- Transfer 0.45 ml of blood into a tube containing 50 μl of 200 μM ADP.
- Mix the tube and incubate 5 min at room temperature.
- Remove 50 μl of blood from the tube and transfer to a tube containing 1 ml of 2% paraformaldehyde in PBS.
- Allow both fixed samples to fix for 1 h at room temperature.

- Wash the cells three times with 1 ml of sterile filtered Tyrode's buffer. Centrifuge at 1800 × g for 5 min. During washing there may be considerable lysis of erythrocytes.
- Resuspend the cells in 1 ml of Tyrode's buffer and store at 4 °C until labeled.

B. Sample Protocol for Preparation of Platelet–Rich Plasma Containing Resting and Activated Platelets

- Draw blood in purple top Vacutainer tubes (EDTA anticoagulant).
- Centrifuge tubes 200 × g for 10 min.
- Remove the supernatant platelet-rich plasma from each tube and pool (the volume of PRP will be about 25% of the blood volume).
- Transfer 4.5 ml of PRP to a tube containing 0.5 ml of 100 μM PGE 1 (to keep platelets resting).
- Transfer 4.5 ml of blood to another tube containing 0.5 ml of 200 μM ADP (to induce activation). Incubate 5 min at room temperature.
- The PRP can be used immediately for labeling or can be fixed in an equal volume of 2% paraformaldehyde in PBS.
- Wash fixed platelets and store as above.

C. Sample Protocol for Preparation of Washed Platelets for Measurement of PAIg

- Collect blood in purple top Vacutainer tubes (EDTA anticoagulant).
- Centrifuge blood at 200 × g for 10 min.
- Remove supernatant platelet-rich plasma.
- Centrifuge PRP at 1800 × g for 10 min.
- Resuspend platelet pellet in 1 ml Tyrode's buffer containing 0.3% EDTA and 0.5% bovine serum albumin.
- Wash platelets again with Tyrode's EDTA/BSA.
- Add washed platelets to an equal volume of 2% paraformaldehyde in PBS.
- Wash fixed platelets and store as above.

V. Staining

Labeling fresh or fixed platelets with fluorescent monoclonal antibodies follow exactly the same protocol used for labeling other cells with only a few additional considerations. If fixed platelets are being labeled it is critically important in preliminary experiments to confirm that the epitope detected by the antibody is still detectable after fixation. Thus, any new labeling procedure should be first tested on both fresh and fixed platelets. In our experience, the commercially available antibodies for CD41, CD42, and CD62 all work well on fixed platelets. One should obviously perform a titration of a new antibody on a fixed number of platelets in order to determine the level of antibody which is saturating. It is theoretically possible for an antibody to cause agglutination of the platelets which will remove them from the normal platelet region of the flow cytometer. We have encountered this difficulty only when using antiplatelet antiserum, not with monoclonal antibodies.

Platelets have an Fc receptor (FcRII, CD32) and can thus bind antibodies nonspecifically. However, this is not, in our experience, a sufficient problem to require the use of blocking for the Fc receptors. In general, the nonspecific labeling of activated platelets will be somewhat higher than that of resting platelets, perhaps due to increased expression of the Fc receptor.

Labeling platelets with thiazole orange to detect RNA can be done with either fresh or fixed platelets. In our experience, the dye penetrates fixed platelets more rapidly and reaches equilibrium within 15 min at room temperature (Ault *et al.*, 1992). Fresh platelets may require longer incubation to achieve full labeling; we generally allow 1 h at room temperature. The labeling of fixed platelets is stable for about 1 h at room temperature but will then gradually decrease. The labeling of fresh platelets is more stable. The dye should not be washed out as it will leak back out of the platelets.

A. Sample Protocol for Labeling of Fixed Whole–Blood Samples for Platelet Activation and Aggregation

- If the samples have not been washed out of fixative before, the fixative must be removed before labeling.
- If the platelet count is normal it should be possible to divide one sample containing 50 μl of fixed whole blood into at least five tubes for labeling and still have enough platelets in each tube to analyze, (50 μl of normal blood contains 1.2×10^7 platelets).
- Pellet the platelets by centrifugation at $1800 \times g$ for 5 min.
- Add 10 μl of FITC-CD41 at a dilution determined in preliminary experiments.
- Add 10 μl of biotinylated-CD62 at a dilution determined in preliminary experiments.
- Mix and incubate 15 min at room temperature.
- Wash once with Tyrode's buffer.
- To the pelleted platelets add 10 μl of phycoerythrin-streptavidin at the manufacturer's recommended dilution.
- Incubate 15 min at room temperature.
- Wash once and resuspend in 0.5 ml Tyrode's buffer for analysis.

B. Sample Protocol for Labeling of Fixed Washed Platelets for PAIg

- Transfer approximately 1×10^6 fixed platelets to each of five tubes.
- Add the following reagents (using the antibodies at predetermined dilutions):

Tube	Purpose	Add	Add
1	Control	10 μl Biotin CD41	Nothing
2	Two-color anti-Ig	10 μl Biotin CD41	FITC-anti-human Ig
3	IgG	10 μl Biotin CD41 (optional)	FITC-anti-human IgG
4	IgM	10 μl Biotin CD41 (optional)	FITC-anti-human IgM
5	C3	10 μl Biotin CD41 (optional)	FITC-anti-human C3

We have not found it cost effective to add CD41 to every tube. The anti-immunoglobulin reagents are all $F(ab')_2$ rabbit anti-human antibodies. The use of IgM, C3, and IgA are optional, see Ault (1988).

After 15 min at room temperature the tubes are washed once with Tyrode's buffer and phycoerythrin-streptavidin is added to each tube containing biotin CD41.

After an additional 15 min the tubes are washed again and the pellet is resuspended in 0.5 ml Tyrode's buffer for analysis.

C. Sample Protocol for Labeling of Fixed Platelets for Reticulated Platelets

- Place 1×10^6 fixed platelets in a tube in a volume of 50 μl.
- Resuspend the platelets in 0.5 ml of thiazole orange solution (50 μg/ml in saline).
- Incubate 15 min to 1 h at room temperature.
- Analyze.

VI. Critical Aspects

One of the most critical aspects for successful flow cytometric analysis of platelets is the resolving power of the flow cytometer. Even when platelets are properly prepared and labeled one will not obtain good data unless the platelets can be clearly resolved. The size distribution of normal platelets is from 1 to 5 μm. Thus a good test of the flow cytometer is the ability to clearly resolve 1-μm beads using forward light scatter. Platelet microparticles are smaller, in the range of 0.1–1 μm, thus the instrument must be capable of resolving smaller particles if they are of interest. It is particularly important to realize that there are a large number of particles in this size range other than platelets. Cellular debris, as discussed in details below, can be easily confused with platelets. Even more problematic is the fact that bacteria which may contaminate reagents or buffers can fall exactly in the forward scatter region of platelets. If there are even small numbers of bacteria in a buffer used for washing they will be concentrated with every wash step and may contaminate the sample in very large numbers making resolution of platelets impossible.

Thus, it is critical that all reagents and buffers used in the preparation and labeling of platelets be filtered through a 0.2-μm filter and that they be stored and used under sterile conditions.

As an additional guarantee that the particles in the size range of platelets really are platelets it is advisable to include in every procedure a sample labeled with at least one platelet-specific marker such as CD41 or CD42. In some cases, this may be used simply to qualify the gate used for analysis of platelets, in other cases it may be desirable to trigger the instrument on the fluorescent signal for the platelet-specific marker thus excluding nonplatelet particles.

Another critical aspect of successful analysis of platelets has to do with the wide size distribution of normal platelets. They vary in size under normal conditions from 1 to 5 μ. In some disease states they can be considerably larger, and of course they have a marked propensity to form aggregates which may range in size up to 20–30 μm. This is much greater size heterogeneity than is seen in other cell types such as lymphocytes. For this reason, it is customary for platelets to span about one decade on a log light scatter scale and more than a decade on log fluorescent scales. The result is a unique signature for platelets on the flow cytometer which allows the experienced operator to be fairly certain one is dealing with platelets. However, regardless of what label is used with platelets there will always be strong correlation between fluorescence and size which must be taken into account. Thus, it is usually desirable when studying platelets to use log amplification of all parameters, including light scatter, and to present results in a two-parameter fashion with light scatter versus fluorescence (see examples below). The definition of a platelet as "positive" for a particular marker must frequently take into account the size of the platelet as well as its fluorescence intensity. Analysis of platelets using single-parameter histograms will result in very wide distributions with poorly resolved "positive" populations.

VII. Standards

There are no standards that are specific to the study of platelets. It is advisable to check the ability of the flow cytometer to resolve small particles with the routine use of 1-μm beads. Since fluorescent signals from platelets are generally of considerably lower magnitude than those from larger cells, good fluorescence sensitivity should also be assured through the use of calibrated fluorescent beads.

In studies of platelet activation it is necessary to have negative (resting) controls and positive (activated) controls. Fresh blood platelets fixed immediately should have a level of CD62 labeling that is less than 5% of the population. The most common problem is the inability to obtain good resting samples. The use of PGE1 to help ensure quiescence was described above. Platelets activated with thrombin or high doses of ADP should be greater than 90% positive for CD62. Samples of fixed resting and activated platelets can be kept in the laboratory for many weeks and can even be kept frozen with little change in their light scatter or labeling properties.

VIII. Instrument

The setup of a flow cytometer for the analysis of platelets is similar to that for other types of immunofluorescence measurements. All parameters, including light scatter, should be log amplified. The forward scatter gain should be set high enough to clearly resolve 1-μm particles, and the alignment and "tuning" of the instrument should be such that instrument noise falls well below the platelet cluster. Gains for fluorescence channels should be set higher than is normal for analysis of lymphocytes. This can only be determined by running positive and negative controls.

The setting of compensation for FITC and PE is identical to that for any immunofluorescence protocol and can be adequately done with beads designed for the purpose.

If the platelets have been enriched in the process of preparation (PRP or washed platelets) the instrument will probably be triggered on light scatter with the threshold set just below the level of the platelet cluster. If the platelets are in whole blood, or if platelet microparticles are of interest, the samples will have been labeled with a platelet-specific antibody and the instrument must be triggered on the fluorochrome corresponding to that antibody, with the threshold set below that of the platelet cluster. This is best done by looking at a dot display of forward scatter versus fluorescence while adjusting the threshold. It is usually not possible to exclude all of the leukocytes with a single threshold. Variable numbers of leukocytes will fall above the threshold due to either increased autofluorescence or binding of platelets to the leukocytes. Platelet microparticles are seen as particles that are positive for the platelet-specific marker but distinctly smaller than normal platelets. They are best resolved on a display of forward or side scatter versus fluorescence.

IX. Results and Discussion

Figure 1 shows the typical platelet cluster with a strong positive correlation between the two light scatter measurements. This cluster is usually easily distinguished from the leukocyte and erythrocyte clusters which do not show a strong size correlation. Figure 1 also illustrates that it may be difficult to accurately separate platelets from leukocytes using light scatter gating alone. This is especially

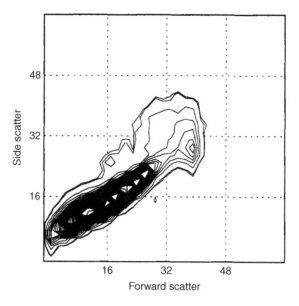

Fig. 1 The typical platelet cluster in whole blood. A fixed whole-blood sample labeled with FITC-CD41 analyzed by triggering on FL1 so as to include only those particles that express CD41. The result is the typical platelet cluster showing a more than one decade variation in light scatter. A small cluster of leukocytes can be seen to the right of the platelet cluster. Both parameters are four-decade log scales.

true if the platelets have become aggregated in the preparation process. Thus, in Fig. 2, a more useful display of forward scatter versus fluorescence is shown. Here the platelet cluster is much easier to distinguish from that of leukocytes, and it is on this display that platelet gating is best performed.

Figure 3 shows an analysis of resting platelets gated as shown in Fig. 2. There is minimal expression of CD62. The control is not shown here; however in our laboratory control labeling is always within the first decade (below channel 16 in the figure). Figure 4 shows an analysis of platelets which were intentionally activated and allowed to aggregate. There are several features of interest. First is the strong expression of CD62 both on normal-sized (single) platelets and on the aggregates which extend up to sizes that somewhat exceed that of leukocytes. Of interest is the fact that in this whole-blood system, the streak of aggregates does not extend to infinity. Thus we do not observe the appearance of macroscopic aggregates such as would be formed in PRP. It is likely that the size of the aggregates is self-limited by the marked fall in the effective concentration of platelets and the presence of large numbers of erythrocytes which inhibit further aggregation. Also of note is the presence of a small number of platelets that do not express CD62 even with maximal stimulation. These apparently defective platelets have not been well studied.

A difficulty arises when one attempts to quantitate platelet aggregation in this system. We have used the percentage of platelet particles that are larger than resting platelets as judged by forward light scatter (Ault *et al.*, 1989). This is a

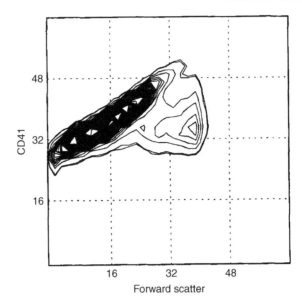

Fig. 2 Use of CD41 to identify platelets in whole blood. The same data as shown in Fig. 1 illustrating the CD41 labeling of the platelets. The trigger threshold is located at the boundary of the first and second decade. The labeling of the leukocytes is to a large extent due to the binding of platelets to monocytes and granulocytes. These parameters are the best ones for placing a gate around the platelet cluster for subsequent analysis of platelet activation.

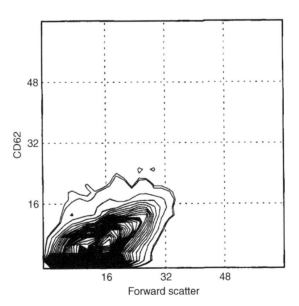

Fig. 3 Analysis of platelet activation and aggregation in resting platelets. Shown here are the same data shown in Figs. 1 and 2, now illustrating the labeling of the gated platelets with CD62. In resting blood the proportion of platelets labeled above control levels (approximately the boundary of the first decade) is less than 5%.

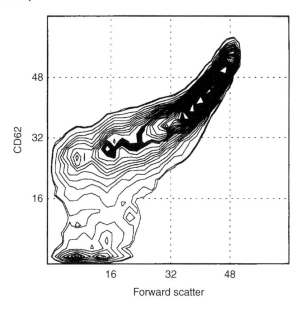

Fig. 4 Measurement of platelet activation and aggregation in stimulated whole blood. This sample was prepared from blood which had been stimulated with 20 μM ADP prior to fixation. There is now extensive positive labeling with CD62 as well as dramatic change in the forward scatter distribution. The peak of large CD62-positive particles isare platelet aggregates approximately the size of leukocytes.

useful parameter which correlates fairly well with the degree of platelet aggregation estimated in a traditional platelet aggregometer, which also uses light scatter to detect aggregation (Carmody *et al.,* 1990). However, it grossly underestimates the proportion of platelets that are actually participating in the aggregation. The maximal value for the percentage of platelet particles larger than resting platelets is usually about 50–70%. At that level of aggregation we estimate that in excess of 99% of the platelets are in aggregates. Each leukocytesized aggregate contains at least several hundred platelets as judged by their staining characteristics. Most platelet aggregation in whole blood is reversible, thus unfixed samples which are not analyzed for more than 30 min after preparation will not show much if any platelet aggregation.

Typical results for the measurement of PAIg are shown in Figs. 5 and 6. Figure 5 shows a sample which has normal levels of PAIg. In this sample, nearly all of the particles gated as platelets (using light scatter gating in this case) were indeed CD41 positive. It can be seen that there is some heterogeneity in the level of anti-Ig labeling of these normal platelets. Sometimes in patients and occasionally in normals it is possible to resolve two distinct populations of platelets having considerably different levels of Ig labeling. The explanation for this phenomenon has not been clearly defined. It most likely reflects differences in permeability of the platelets to the anti-Ig antibody, perhaps due to differences in activation status. It does not appear to correlate with platelet size nor with the presence of reticulated

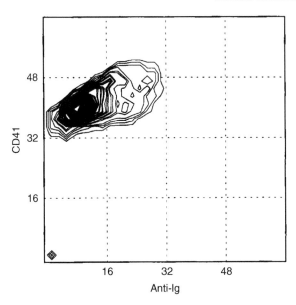

Fig. 5   Labeling of platelets for immunoglobulin. Illustrated here is the result of an assay for PAIg which is within the normal range. The platelets are identified by the presence of CD41 and the mean level of PAIg is determined from the anti-Ig staining.

platelets. Figure 6 shows data from a patient with immune thrombocytopenia. The platelet gate now contains several distinct populations of particles. Those that are CD41 positive have a clearly increased level of anti-Ig labeling. The mean fluorescence intensity of this population can be used as measure of PAIg. In this particular example, there are three additional populations of particles that do not express CD41. It is important to realize that all of these populations copurify with platelets by centrifugation, and all fall within the same "platelet" light scatter gate on the flow cytometer. In our experience, the particles which are negative for Ig frequently label with anti-glycophorin and probably represent red cell fragments or "microspherocytes." Nonplatelet particles which label brightly for Ig can usually be shown to also label very brightly for IgG, IgM, and C3 and, we feel, represent large immune complexes. The intermediate population in this particular sample is unexplained.

The frequent presence of these nonplatelet particles in patient samples raises several important issues. First is the one discussed above, that many particles fall in the platelet size range. Thus, light scatter gating alone cannot be relied upon to identify platelets, especially in abnormal samples. Second, the presence of significant numbers of nonplatelet particles in preparations of "washed platelets" from patients could have a major impact on the measurement of PAIg. We feel that this is one of the strongest arguments in favor of the flow cytometric approach to PAIg measurement. As shown in Fig. 6 the flow cytometer permits one to reliably restrict the measurement to platelets, ignoring particles which are not platelets and which

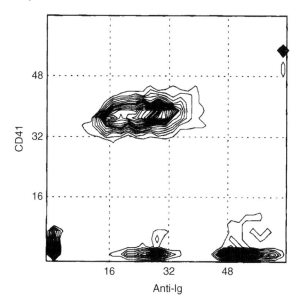

Fig. 6 A sample with elevated PAIg and nonplatelet particles. A sample from a patient with elevated PAIg was processed the same as that in Fig. 5. Here the platelets which are CD41 positive have elevated levels of PAIg. In addition to the platelets the light scatter gate contains three additional populations of CD41-negative particles which have no immunoglobulin, low levels, and high levels, respectively. Although not illustrated here, further workup usually suggests that these are red cell fragments and immune complexes.

may have 10–100 times more or less immunoglobulin per particle than the platelets. It is likely that this is one of the difficulties which has plagued previous attempts to measure PAIg by other techniques.

Figure 7 shows a typical analysis of reticulated platelets. This sample was obtained by Thiazole orange labeling of whole blood followed by light scatter gating on platelets. Similar results can be obtained by Thiazole orange labeling of fixed washed platelets. All of the platelets label to some extent with Thiazole orange; however, a subset can usually be resolved which has increased labeling and which, on average, is slightly larger than the average normal platelet. This is a good example of a situation in which a population can be fairly clearly identified in two-parameter, light scatter versus fluorescence display but is very poorly resolved in a one-parameter fluorescence histogram. The normal size heterogeneity of platelets can easily mask a small population with increased fluorescence for their size. In order to enumerate these reticulated platelets we have used a line which is parallel to the platelet cluster and defines the upper bound of the cluster (Ault *et al.*, 1992). Platelets falling above this line are considered "positive." Thus the definition of "positive" includes consideration of the size of the platelet.

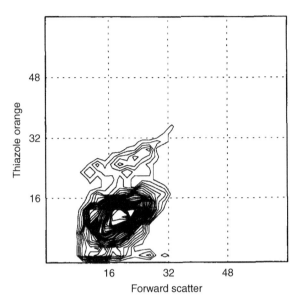

Fig. 7 Reticulated platelets. This shows results from a whole-blood sample labeled with Thiazole orange and gated on platelets. A subpopulation of platelets, about 11% in this case, have increased thiazole orange labeling and have been referred to as reticulated platelets. They have the properties of platelets newly released into the circulation.

References

Abrams, C. S., and Shattil, S. J. (1991). *Thromb. Haemostasis* **65**, 467–473.

Abrams, C. S., Ellison, N., Brudzynski, A. Z., and Shattil, S. J. (1990). *Blood* **75**, 128–138.

Adelman, B., Michelson, A. D., Handin, R. I., and Ault, K. A. (1985). *Blood* **66**, 423.

Ault, K. A. (1988). *Pathol. Immunol. Res.* **7**, 395–408.

Ault, K. A. (1992). *In* "Clinical Flow Cytometry" (A. L. Landay, K. A. Ault, K. D. Bauer, and P. S. Rabinovitch, eds.), pp. 387–403. New York Academy of Sciences, New York.

Ault, K. A., and Mitchell, J. (1993). *In* "Flow Cytometry of the Platelet-Megakaryocyte System" (R. E. Scharf and K. J. Clemeson, eds.). Elsevier, Amsterdam (in press).

Ault, K. A., Rinder, H. M., Mitchell, J. G., Rinder, C. S., Lambrew, C. T., and Hillman, R. S. (1989). *Cytometry* **10**, 448–455.

Ault, K. A., Mitchell, J. G., Rinder, H. M., Rinder, C., and Hillman, R. S. (1991). *In* "Flow Cytometry in Hematology" (O. D. Laerum and R. Bjerknes, eds.), pp. 153–163. Academic Press, London.

Ault, K. A., Rinder, H. M., Mitchell, J., Carmody, M. B., Vary, C. P. H., and Hillman, R. S. (1992). *Am. J. Clin. Pathol.* **93**, 637–646.

Berman, C. L., Yeo, E. L., Wencel-Drake, J. D., Furie, B. C., Ginsberg, M. H., and Furie, B. (1986). *J. Clin. Invest.* **78**, 130–137.

Carmody, M., Ault, K. A., Mitchell, J. G., Rote, N. S., and Ng, A. (1990). *Hybridoma* **9**, 631–641.

Coller, B. S. (1985). *J. Clin. Invest.* **76**, 101.

Corash, L. (1990). *Blood Cells* **160**, 97–108.

Corash, L., and Rheinschmidt, M. (1986). *In* "Manual of Clinical Laboratory Immunology" (N. R. Rose, H. Friedman, and J. L. Fahey, eds.), pp. 254–257. American Society of Microbiology, Washington, DC.

Corash, L., Mok, Y., and Rheinshmidt, M. (1986). *In* "Applications of Fluorescence in the Biomedical Sciences," pp. 567–584. Liss, New York.

Davies, T. A., Drotts, D., Weil, G. J., and Simons, E. R. (1988). *Cytometry* **9**, 138–142.

de Bruijne-Admiraal, L. G., Modderman, P. W., von dem Borne, A. E. G. K., and Sonnenberg, A. (1992). *Blood* **80**, 134–142.

Ginsberg, M. H., Frelinger, A. L., Lam, S. C., Forsyth, J., McMillan, R., Plow, E. F., and Shattil, S. J. (1990). *Blood* **76**, 2017–2023.

Ingram, M., and Coopersmith, A. (1969). *Br. J. Haematol.* **17**, 225–228.

Johnson, P. C., Ware, J. A., Cliveden, P. B., Smith, M., Dvorak, A. M., and Salzman, E. W. (1985). *J. Biol. Chem.* **260**, 2069 2076.

Johnston, G. I., Pickett, E. B., McEver, R. P., and George J. N. (1987). *Blood* **69**, 1401–1403.

Kienast, J., and Schmitz, G. (1990). *Blood* **75**, 116–121.

Lazarchick, J., and Hall, S. A. (1986). *J. Immunol. Methods* **87**, 257–265.

Marti, G. E., Magruder, L., Schuette, W. E., and Gralnick, H. R. (1988). *Cytometry* **9**, 448–455.

Nieuwenhuis, H. K., van Oosterhout, J. J. G., Rozemuller, E., van Iwaarden, F., and Sixma, J. J. (1987). *Blood* **70**, 838–845.

Rinder, C. S., Bohnert, J., Rinder, H. M., Mitchell, J., Ault, K. A., and Hillman, R. S. (1991a). *Anesthesiology* **75**, 388–393.

Rinder, C. S., Mathew, J. P., Rinder, H. M., Bonan, J., Ault, K. A., and Smith, B. R. (1991b). *Anesthesiology* **75**, 563–570.

Rinder, H. M., Murphy, J., Mitchell, J. G., Stocks, J., Hillman, R. S., and Ault, K. A. (1991c). *Transfusion (Philadelphia)* **31**, 409–414.

Rinder, H. M., Bonan, J., Rinder, C. S., Ault, K. A., and Smith, B. R. (1991d). *Blood* **78**, 1730–1737.

Rinder, H. M., Bonan, J., Rinder, C. S., Ault, K. A., and Smith, B. R. (1991e). *Blood* **78**, 1760–1769.

Rinder, C. S., Bonan, J. L., Rinder, H. M., Mathew, J., Hines, R., and Smith, B. R. (1992). *Blood* **79**, 1201–1205.

Rinder, H. M., Munz, U., Smith, B. R., Bonan, J. L., and Ault, K. A. (1993). *Arch. Pathol. Lab. Med.* **117**, 606–610.

Rosenfield, C. S., Nichols, G., and Bodensteiner, D. C. (1987). *Am. J. Clin. Pathol.* **87**, 518–522.

Shattil, S. J., Cunningham, M., and Hoxie, J. A. (1987). *Blood* **70**, 307–315.

Sims, P. J., Faioni, E. M., Wiedmer, T., and Shattil, S. J. (1988). *J. Biol. Chem.* **263**, 18205–18212.

PART VII

Detection of Microorganisms
and Pathogens

CHAPTER 37

Detection of Specific Microorganisms in Environmental Samples Using Flow Cytometry

Graham Vesey,★ Joe Narai,† Nicholas Ashbolt,‡ Keith Williams,★ and Duncan Veal★

★School of Biological Sciences
Macquarie University
Sydney, New South Wales 2109
Australia

†Commonwealth Centre for Laser Applications
Macquarie University
Sydney, New South Wales 2109
Australia

‡Australian Water Technologies
Science and Environment
Sydney, New South Wales 2114
Australia

ESSENTIAL CYTOMETRY METHODS
Reprinted from *Methods in Cell Biology*, Volume 42 (Academic Press, 1994).

DOI: 10.1016/B978-0-12-375045-7.00037-4

I. Introduction

Flow cytometers are technologically advanced instruments which combine laser interrogation of a fluid stream with sophisticated data handling technology for obtaining information about, and potentially isolating, particles that pass through the laser beam. Traditionally, they are among the most expensive of laboratory instruments and require highly skilled personnel to operate them. This combined with the fact that samples must be particulate and of reasonably uniform size has limited their application in biology to well-funded laboratories in biomedical research where they are used to analyze blood cell subpopulations (immunology, AIDS, cancer) or separate chromosomes.

The reduced cost and ease of operation of analytical flow cytometers (which collect data but do not sort particles) means that applications in other areas of biology are now envisaged. The applications of flow cytometry to clinical microbiology have been described by Shapiro (1990). In this chapter, we discuss the use of flow cytometry within the environmental microbiology laboratory. In particular, we focus on flow cytometric methods for the detection of low numbers of, and even single, specific microorganisms within environmental samples.

Flow cytometric analysis performed in environmental microbiology laboratories is often more stringent than that required for the analysis of mammalian cells and can push sensitivities close to limits of operation. This is because the volume, nucleic acid and protein content of bacteria are approximately $1000\times$ less than that in mammalian cells. Since detection involves identification of light scatter, specific proteins, or DNA, the signals produced by bacteria are generally several orders of magnitude lower than those from eukaryotic cells. However, recently flow cytometers have been used to great effect for microbiological diagnosis and even more recently they have been applied in environmental microbiology.

Developments in both biological techniques and instrumentation, described in this chapter, will considerably increase the range of applications of flow cytometry within environmental microbiology laboratories. Furthermore, these developments result in greatly simplified protocols allowing not only research laboratories but also routine environmental testing laboratories to perform these analyses. We envisage that within the foreseeable future small, robust, relatively cheap, and simple to operate flow cytometers will be available for the detection of a vast range of microorganisms in environmental samples.

A. The Instrument

The principles of flow cytometry have been described by Shapiro (1988) and here we provide only the basic features and terminology.

Flow cytometry is used to quantitatively measure physical or chemical characteristics of cells as they are presented, in single file, into a focused light beam. The flow rate of fluid through the focused beam is usually around 10 m/s which allows approximately 2000–10,000 cells/s to be analyzed. The light source can be either a high-pressure mercury lamp or one of an assortment of different lasers. Various parameters of an individual cell, such as fluorescence (FL), forward angle light scatter (FALS), and side (90°) angle light scatter (SALS), can be measured simultaneously. Forward angle light scatter provides information on the size of a cell while side angle light scatter correlates with cell refractibility and, thus, it is thought, provides information on the surface properties and internal structure. Fluorescence can be used to detect autofluorescence in the cell emanating from cellular components such as flavin nucleotides, pyridine, chlorophyll, and other photosynthetic pigments. In natural environments, FALS, SALS, and autofluorescence can provide sufficient information to enable clustering of cells into particular types (Yentsch, 1990). However, alone, these parameters normally do not provide sufficient information to identify specific microorganisms in environmental samples. To identify specific microorganisms, it is usually necessary to label cells of interest with fluorescent molecules such as fluorescein isothiocyanate (FITC). Cells can be tagged using specific antibodies, lectins, or nucleic acid probes conjugated to particular fluorochromes. In addition to analyzer flow cytometers described above, some machines have the capability of physically sorting cells using information from the various detectors as discriminators. Sorting enables purification of a particular cell type from a mixture. To sort cells, the fluid stream is broken into droplets by the vibration of a piezoelectric crystal. If a particle is to be collected, the droplet containing the particle is electrically charged and then deflected, using charged plates, from the main stream into a collection vessel.

A major hurdle limiting to the use of flow cytometry in environmental microbiology has been the initial cost of the instrument. A basic analyzer cytometer (such as the Coulter XL) costs around $100,000 (US). The XL is an excellent instrument and will certainly find applications in the area of routine monitoring in environmental microbiology. However, to sort cells doubles the cost of the basic instrument. Instruments with the ability to sort include the Coulter Elite and the Becton-Dickinson FACStar plus or FACS Vantage. Further, costs are involved with cell sorting since the instrument takes considerably longer to set up and requires a highly skilled operator. Nevertheless, the ability to sort cells is an important feature for environmental microbiology since it enables the collection of the organisms of interest and confirmation of results. We regard the ability to sort as essential to most research and development applications in environmental microbiology. However, one aim of this research and development should be to develop protocols which can be used reliably in routine situations on analysis only instruments.

B. Environmental Microbiology

Despite the high cost of the instrumentation, flow cytometers are now found in a few environmental microbiology laboratories around the world. This preparedness to invest in sophisticated instrumentation is due, in part, to an increasing appreciation of the potential applications of flow cytometry in environmental microbiology (Button and Robertson, 1993; Edwards *et al.*, 1992) and awareness of the limitations of traditional techniques (Mills and Bell, 1986). The availability of funds to purchase flow cytometers for environmental microbiology also reflects growing awareness and concern for environmental issues. Certainly, environmental microbiology is crucial to the major global environmental issues and flow cytometry is now beginning to play its role in the development of this discipline. For example, global warming has been described as a most significant environmental issue. In the oceans, two groups of microorganisms are essential for carbon cycling and, therefore, have important impacts on future trends in global warming. These are the phytoplankton, which are responsible for about half of the total primary production on the earth, and the heterotrophic bacteria. Of the phytoplankton, picophytoplankton (including cyanobacteria, prochlorophytes, diatoms, and cryotophytes), due to their autofluorescence, these are well suited to flow cytometric analysis, as are their predators (Balfoort *et al.*, 1992; Olson *et al.*, 1993). On the other hand, flow cytometric quantification of heterotrophic aquatic bacteria is in its infancy (Button and Robertson, 1993; Robertson and Button, 1989), although methods based on fluorescent ribosomal probes for aquatic bacteria (Lim *et al.*, 1993) could be adapted for flow cytometry (Amann *et al.*, 1990b). The latter approach is already leading to a better understanding of the identification and biodiversity of the predominant bacteria in environments (Manz *et al.*, 1993; Wagner *et al.*, 1993).

Pollution is another major environmental issue. Flow cytometry has been used to study the hydrocarbon degrading bacteria in Resurrection Bay after the Exon Valdez grounding (Button *et al.*, 1992). In these studies, microbial biomass was computed from the population and mean cell volume (obtained from forward scatter histograms). Significantly, this study concluded that turnover times for toluene were decades longer than expected and that dissolved spill components such as toluene should enter the world ocean pool of hydrocarbons rather than biooxidize in place. There is surprisingly little information on flow cytometry for human or animal health microbiology. The analysis of foods, body fluids, or culture media should, however, pose fewer problems than environmental samples and in fact the challenge to apply flow cytometry to the bioreactor has been taken up by the fermentation industry (Betz *et al.*, 1984; Kell *et al.*, 1991; Scheper *et al.*, 1987). Also, pathogens in food or body fluid will be present at relatively high concentrations (Humphreys *et al.*, 1993; Pinder and McClelland, 1993). Other environmental areas awaiting flow cytometry study include monitoring the fate of probiotics or nuisance species in aquaculture.

C. Environments

1. Water

Water is routinely tested for microorganisms on a large scale by water utilities and government agencies to ensure that the water is safe for consumption, recreation, and shell fish production. The detection of microorganisms in water depends largely on the use of selective and differential media to isolate, culture, and identify specific types of microorganisms. These culture-based methods are limited because they rely on the growth of the microorganisms of interest on laboratory medium. For most microorganisms this culturing takes between a few hours and a few days, thus results are never immediately available. Furthermore, in our experience, only between 0.001 and 10% of bacteria from natural environments are culturable, including introduced pathogens.

Due to the resources required to screen for a wide range of water-borne pathogens (including viruses, bacteria, protozoa, nematodes, and flukes), the fact that many pathogens are currently nonculturable or are very slow growing, and the potential risks associated with culturing pathogens to laboratory workers, microbial water quality is routinely assessed by enumerating the presence of certain indicator bacteria, generally fecal coliforms and enterococci (Ashbolt and Veal, 1994). These bacteria are used to indicate recent fecal contamination and depend on the indicator organism not growing in the environment, being present at greater numbers than any pathogen, and dying in the environment at a rate no faster than that of any pathogen. Unfortunately, it is now well established that dependence on these indicator bacteria can lead to false conclusions as all of these assumptions can fail some of the time; for example, human viruses and pathogenic protozoa of human origin have been reported in waters when indicator bacteria were not detected (Ashbolt and Veal, 1994; Badenoch, 1990; Smith and Rose, 1990). Thus, there is renewed interest in detection of the specific pathogens themselves in water.

Work in our laboratories has concentrated on the detection of the enteric protozoan parasites *Giardia* and *Cryptosporidium* in water. These protozoa are among the most frequent causative agents of diarrheal disease in humans (Adam, 1991; Current, 1986). No effective chemotherapy is available for *Cryptosporidium* and this organism can be life threatening in immunosuppressed individuals (Current, 1986). Although the most common route of infection of *Cryptosporidium* and *Giardia* is direct person to person, data from the United Kingdom and the United States indicate that the zoonotic water-borne route is important (Badenoch, 1990; Smith and Rose, 1990).

Cryptosporidium and *Giardia* form robust oocysts and cysts, respectively, which are shed in the feces. These oocysts and cysts are very infective with as few as 10 cysts or oocysts leading to the development of disease (Miller *et al.*, 1986; Rendtorff, 1954).

One of the features of the oocysts and cysts of *Cryptosporidium* and *Giardia* is that they are robust and survive for extended periods in the environment. Bacterial indicators, such as fecal coliforms are an unsatisfactory standard by which to evaluate the health risk of the protozoan parasites, since bacterial indicators survive for days or at most weeks in aquatic environments, while cysts and oocysts may survive for several months. Establishment of the health risks of these organisms in drinking water using indicator organisms is further complicated because indicator organisms are inactivated by levels of chlorine used routinely in drinking water, whereas cysts and oocysts are not (Adam, 1991; Korich *et al.*, 1990). Several water-borne outbreaks of *Cryptosporidium* and *Giardia* have been reported in situations where the water meets all statutory requirements of microbial quality (Badenoch, 1990; Smith and Rose, 1990). Some of these outbreaks have resulted in hundreds of thousands of individuals becoming infected (e.g., cryptosporidiosis in Milwaukee in 1993).

Flow cytometry has the potential for rapid detection of nonculturable microorganisms in water samples at high levels of sensitivity. Vesey and his coworkers (Vesey *et al.*, 1993a,c, 1994) have described the use of a sorting flow cytometer to detect *Cryptosporidium* oocysts and *Giardia* cysts in concentrated water samples. This flow cytometric detection method involves flocculation with calcium carbonate to concentrate cysts and oocysts from large volumes (1–40 l) of water (Vesey *et al.*, 1993b). The flocculant is dissolved and then stained with FITC-labeled monoclonal antibodies and analyzed by flow cytometry. Presumptive cysts and oocysts are then sorted on to microscope slides and confirmed by epifluorescence microscopy.

The method described by Vesey and coworkers is a significant advance on the American Society for Testing Materials (ASTM) method evaluated by LeChevallier *et al.* (1991). The method is, however, far from perfect. At present, to achieve the level of sensitivity required for detecting *Cryptosporidium* oocysts, the method is restricted to an instrument fitted with cell sorting facility. Nonsorting cytometers, however, are less expensive, much simpler to operate, and better suited to routine analysis within a water microbiology laboratory. Features such as an automatic sample loader enable samples to be loaded and left to run unattended. Within our laboratory the development of a nonsorting detection method is underway. We believe that in the near future a flow cytometric method which can achieve the sensitivity required for the routine monitoring of *Cryptosporidium* oocysts in water will be developed. Already we have developed a reliable method for the detection of *Giardia* cysts in water that does not depend on cell sorting (Vesey *et al.*, in preparation).

2. Foods and Beverages

In the food and beverage industries, routine microbiological testing is performed on a large scale to ensure freedom from pathogens and acceptable levels of spoilage organisms. Most of this testing is still performed using traditional culture-based techniques. These techniques are laborious and relatively expensive in terms of culture media and reagents, and it is generally several days before results are available. Flow cytometry has considerable potential in the food and beverage industry since it can be used to detect nonculturable organisms, and in the analytical mode it is readily automated and can be used to rapidly enumerate microorganisms in a sample. Immediate microbial assessment has both stock control and production advantages as microbiological problems can be immediately identified and rectified, reducing the amount of lost product and negating the need to store products while awaiting results of microbiological analyses.

Despite the potential applications of flow cytometry in food microbiology there have been few reports detailing its use. Pettipher (1991) has evaluated the Chem-Flow flow cytometer system (Chemunex S. A.) for the rapid detection of spoilage yeasts in soft drinks. Minimum sensitivity achieved using this system varied from 50 to 14,000 yeasts/ml depending on the product being tested. Pettipher (1991) concluded that the ChemFlow system looks promising for the rapid detection of relatively low numbers of yeast in soft drinks, but without some concentration or preincubation it was not sufficiently sensitive for use on site by a production laboratory. Using 24-h preincubation with a ChemFlow Autosystem II sensitivities as low as 1 viable yeast/g of product have been achieved (Bankes and Richard, 1993). Without preincubation these authors achieved sensitivities $<10^3$ cells/ml of product within 1 h of sampling.

Infections with wild yeast are a severe problem for breweries, with spoilage of the beer occurring at infection levels as low as 1 wild yeast in 10^3–10^4 culture yeast. Here, the problem is to detect the presence of relatively low numbers of wild yeast against a high background of culture yeast. Jespersen et al. (1993) have described a flow cytometric procedure based on preincubation on wild yeast-selective medium followed by staining with the Fluorassure Substrate B. W. (Chemunex S. A., Paris). By examining fluorescence intensity, differentiation of the wild yeast from culture yeast was possible. Using this method the authors claimed to be able to detect 1 wild yeast in the presence of 10^6 culture yeast after 48–72 h of incubation.

Donnelly and Baigent (1986) have described the use of flow cytometry for the detection of *Listeria monocytogenes* in milk using FITC-labeled polyclonal antibodies. This work is of particular significance because food-borne listeriosis can be fatal and because culturing of *L. monocytogenes* is tedious, may take several days of enrichment, and can be unreliable as cultures are frequently overgrown by competing microorganisms. However, Haslett et al. (1991) have reported problems with the application of flow cytometry to the detection of this organism since commercially available antibodies cross-react with other *Listeria* species and various other Gram-positive organisms.

3. Soils and Sediments

A rapidly growing area of biotechnology is in the development of microbial plant and soil inoculants. Microorganisms are used to break down pollutants in soil, control pests and disease of plants, and promote plant growth through N_2 fixation or by producing plant growth stimulating factors. In developing new products in this area, it is important to be able to monitor the inoculated micro-oganisms, to determine their fate, and to assess the success or failure of inoculation (Ryder *et al.*, 1994). Many microbial inoculants which are being developed for environmental applications contain recombinant DNA. The ability to monitor the survival, growth, and migration of such genetically modified products is an essential prerequisite for their licensing as products (Ford and Olsen, 1988). Methods for monitoring genetically modified microorganisms need to be sensitive and able to detect both culturable and viable but nonculturable cells (Colwell *et al.*, 1985).

The ability to enumerate bacteria introduced into soil using flow cytometry has been investigated by Page and Burns (1991). When compared to plate counts and direct counts, flow cytometry was found to be more accurate at low numbers. These authors concluded that flow cytometry is a more rapid and objective method for bacterial enumeration in soil than direct counting.

An unresolved problem with soils and sediments is the difficulty of separating microorganisms from particulates (Hopkins *et al.*, 1991). Enumeration of particle-bound pathogens is important as they act as reservoirs readily resuspended into the water column by a number of forces including storms, dredging, bioturbation, current, and wave actions, impacting on recreational use and shellfish culture (Grimes *et al.*, 1986; Lewis *et al.*, 1986). Virtually all methods for enumerating bacteria in soils or sediments, including flow cytometry, are dependent on separating the microorganisms from the particulates. Generally, the highest yields of desorbed bacteria are achieved by sonicating sediments, often in the presence of a surfactant (McDaniel and Capone, 1985). Once in suspension, microorganisms may be separated from detritus by differential centrifugation, such as sedimention field-flow fractionation which may recover 59–87% of a seeded bacterial population (Sharma *et al.*, 1993). However, in our experience of protozoan parasites differential centrifugation of cysts results in poor quantitation and sonication can lead to the removal of the identifying epitopes required for monoclonal antibody binding (Vesey and Slade, 1990). Hence, we are investigating the chemical encrusting of iron and humic salts that may bind environmental cysts.

D. Current Technologies for Rapid Detection of Specific Microorganisms in Environmental Samples

1. Antibody–Based Methods

Most rapid techniques used to detect specific microorganisms in environmental samples depend on either antibody- or nucleic acid-based methodologies. Most of the antibody-based techniques employ some form of colorimetric detection system such as immunofluorescence, ELISA, or chemiluminescence. These techniques normally involve immobilization of the sample onto a solid substance and then probing with an antibody labeled with the detection reagent. In the case of immunofluorescence, the detection reagent is normally FITC and the sample is visualized using epifluorescence microscopy. The entire sample is scanned for fluorescing particles with the morphological characteristics of the target organism. The microscopic examination requires highly skilled operators and is tedious, time-consuming, and, with dirty samples, may result in low detection efficiencies (Vesey *et al.*, 1994).

ELISA techniques often employ an antibody bound to a solid substrate to capture the target organism and then a horseradish peroxidase-labeled antibody, which also identifies the target organism, is attached to the captured sample. The horseradish peroxidase is then developed using hydrogen peroxide and tetra-methylbenzidine and the sample is examined for color development either visually or with an ELISA reader.

The sensitivity of colorimetric detection systems can be increased by employing chemiluminescence. In this technique, light emitting substrates are utilized in conjunction with enzyme-linked antibodies which allow the detection of the target organism to be performed by exposing the sample to photographic film.

The advantage of ELISA and chemiluminescence techniques is that they are simple to perform and enable the analysis of multiple samples simultaneously. However, these techniques often lack the specificity required to detect low numbers of specific microorganisms within environmental samples (Campbell *et al.*, 1993). This is because detection is based on a single paramater, namely the binding of an antibody to the organism of interest. In our experience, antibodies always bind to some interfering particles found in environmental samples, no matter how specific the antibody is and how good the blocking agents used. Flow cytometry can overcome this problem because additional detection parameters, such as light scatter and autofluorescence can be used to be certain that the antibody is binding an authentic target organism.

2. Gene Probes

Methods that employ specific sequences of nucleic acids to probe samples utilize either colorimetric enzyme-labeled probes or radioactively labeled probes. These techniques are potentially extremely sensitive and have been successfully used to detect pathogens in environmental samples (Abbaszadegan *et al.*, 1991). However, radioactively labeled probes are often required to achieve the required sensitivity. The facilities needed to handle radioactive materials are not generally available in environmental microbiology laboratories. As with antibody-based techniques, the detection relies on a single parameter. The use of fluorescently labeled nucleic acid probes with flow cytometry allows additional parameters and, therefore, increased selectivity. The use of these probes is discussed later.

3. Polymerase Chain Reaction

The polymerase chain reaction (PCR) is a technique which can detect single microorganisms within environmental samples with high specificity (Brooks *et al.*, 1992; Mahbubani *et al.*, 1991; Tsai and Olson, 1992). These techniques have been successfully used for the detection of specific microorganisms within a range of samples including the detection of lactic acid bacteria in beer (Dimichele and Lewis, 1993), *Escherichia coli* in sewage and sludge (Tsai *et al.*, 1993) and raw milk (Keasler and Hill, 1992), *L. monocytogenes* in food (Furrer *et al.*, 1991; Niederhauser *et al.*, 1993; Wernars *et al.*, 1991), *Carnobacterium* spp. in meat (Brooks *et al.*, 1992), and *Campylobacter jejuni* in raw milk and dairy products (Wegmuller *et al.*, 1993). These techniques often enable inexpensive and rapid analysis of multiple samples. However, there are problems with the use of these techniques in many environmental samples. Substances present in some samples interfere with the polymerase enzyme and inhibit the reaction (Herman and De-Ridder, 1993; Tsai and Olson, 1992). Procedures for removing the inhibitors such as Sephadex columns have been developed for some types of samples (Abbas-zadagen *et al.*, 1991; Tsai and Olson, 1992). Unfortunately, these procedures are often tedious and labor intensive and are not applicable to many microorganisms. The use of magnetic-activated cell sorting to remove inhibitors may prove to be successful.

II. Preparation of Water Samples for Flow Cytometric Analysis

A. Concentration

Most microbial pathogens occur in water at very low concentrations. Thus, the detection of specific pathogens in water by flow cytometry requires some concentration step prior to analysis. The degree of concentration depends on the infective dose of the organism and its frequency of occurrence. For the detection of many organisms, such as *Legionella*, which needs to occur in waters in high numbers to cause a serious health threat (Badenoch, 1987), sample concentration simply involves a 10- to 50-fold concentration using centrifugation. With organisms that have a very low infectious dose, such as *Cryptosporidium*, much higher levels of concentration are required, and typically 10–40 l of water must be processed. Because it is impractical to routinely centrifuge tens of liters of water, alternative concentration methods have been developed. These include filtration using wound cartridge filters (Rose *et al.*, 1986), tangentinal flow filtration (Issac-Renton *et al.*, 1986), continuous flow centrifugation (Bee *et al.*, 1991), and vortex filtration. However, most of these techniques do not solve the basic problems associated with concentrating microorganisms from environmental samples, that is, achieving good recovery without damaging the target microorganism and with the minimum of labor and expense.

We now routinely use the following calcium flocculation method for concentrating parasites from samples up to 20 l in volume. The method is inexpensive and simple and results in high recoveries of *Cryptosporidium* oocysts and *Giardia* cysts (Vesey *et al.*, 1993b).

1. Calcium Flocculation (Vesey *et al.*, 1993b)

1. Collect a 10–l sample in a flat-bottomed plastic container.
2. Add 100 ml of 1 M sodium bicarbonate and 100 ml of 1 M calcium chloride.
3. Adjust the pH to a value of 10.0 (±0.1) by the addition of sodium hydroxide and allow the sample to stand for at least 4 h.
4. Remove the supernatant by aspiration without disturbing the calcium deposit.
5. Dissolve the calcium carbonate residue by adding sulfamic acid.
6. Decant the sample and centrifuge.
7. Resuspend the pellet in 50 ml of distilled water and centrifuge. Wash a second time in phosphate-buffered saline (PBS), pH 7.4, and centrifuge.
8. Resuspend the pellet in a small volume of PBS and analyze (e.g., by flow cytometry).

Many organisms, including *Cryptosporidium* oocysts, are very sticky and will adhere to any containers used. To improve recoveries, all containers should be rinsed with 0.01% Tween 80 and the washings combined with the sample.

Samples over 40 l in volume are too large to be conveniently flocculated in any number. To overcome this logistical problem, we use a membrane filtration method (Ongerth and Stibbs, 1987) to concentrate samples up to 5000 l in volume on site.

2. Membrane Filtration

1. Place a 293- or 142-mm polycarbonate membrane of appropriate pore size (2 μm for *Cryptosporidium* oocysts, 5 μm for *Giardia* cysts, 0.22 μm for bacteria) in a stainless steel housing and filter the sample. Ensure that the pressure across the filter does not exceed 15 psi or the membrane may be damaged.
2. Once filtration is complete and no water is left in the housing, carefully remove the membrane and place it onto a perspex plate slightly larger in size than the membrane.
3. Hold the perspex plate upright above a narrow plastic box. Use a spray bottle a squirt 0.01% (v/v) Tween 80 onto the membrane. Allow the washings to drip into the plastic box.
4. While squirting Tween 80 at the membrane, rub the entire surface with a squeegee (we use a modified windshield wiper). Start at the top of the membrane and work downward.

5. Continue the process until the membrane appears clean. Use a minimum amount of Tween 80 (<100 ml).

6. Transfer the washings to a centrifuge pot and centrifuge at 3000 × g for 10 min.

7. Resuspend in a small volume of PBS and analyze (e.g., by flow cytometry).

The methods described here were developed for concentrating *Cryptosporidium* oocysts and *Giardia* cysts. However, with slight modifications they should be applicable to a range of other microorganisms.

Prior to staining and analysis, all samples should be prefiltered through a small disk of stainless steel with 100-μm pores.

B. Magnetic–Activated Cell Sorting

Concentration by magnetic-activated cell sorting is an area which has received considerable attention within environmental microbiology. In particular, the technique has been used extensively within the food industry to selectively enrich food-borne pathogens (Patel and Blackburn, 1991). The technique has some of the advantages of fluorescence-activated cell sorting, but is much cheaper to set up. Two magnetic separation systems are commercially available. The first, MACS, is produced by Becton-Dickinson. In this system, very small magnetic particles (50 nm diameter) are conjugated with a ligand specific to the target organism. The magnetic particle-conjugated ligands are then reacted with the sample containing the target cells. The cell suspension is passed through a column, filled with steel wool, contained within a magnetic field. Magnetically labeled material is retained in the column until the magnetic field is removed and then the labeled fraction is eluted. The second magnetic system is produced by Dynal. The technique employs 2- or 4-μm magnetic beads and a simple sample tube holder with a rare earth magnet in the base. The ligand-conjugated beads are reacted with the sample and then placed in the tube holder. Within seconds the magnetic particles are concentrated in the bottom of the tube allowing the supernatant to be discarded. Thus, the sample is concentrated to a small (<1 ml) volume which is an ideal volume for flow cytometric analysis. Furthermore, although the Dynal beads change the optical properties of the target organism, Dynal now markets a reagent for detaching the beads from their target. The MACS system has the advantage that the small magnetic particles used do not aggregate, do not change the optical properties of the target organism, and provide a stronger attachment to the target organism. However, the MACS system is expensive [about $5000 (US)] compared to the Dynal system [<$1000 (US)]. Also, the steel wool columns have a pore size of 50 μm which may limit their application in environmental microbiology.

Although considerable literature is available on the use of magnetic-activated cell sorting within food microbiology, very little is available within water or soil microbiology. A method has been described by Bifulco and Schaerfer (1993) to purify *Giardia* cysts from water which involves the use of a custom-made magnetic capture system. However, losses averaged 18% and very turbid samples could not be analyzed.

Research in our laboratory on the use of magnetic-activated cell sorting for purification of microorganisms from water samples has been hampered by the large numbers of magnetic particles present in environmental samples. These particles are sufficiently magnetic to interfere with the method but not magnetic enough to be easily removed prior to addition of the magnetically conjugated ligands. Further details on magnetic cell sorting are provided in Chapter 6.

III. Staining of Organisms from Water Samples for Flow Cytometric Analysis

Detection of specific single organisms within the diverse array of other particles normally found in environmental samples requires a high level of discrimination. To achieve this, highly specific labels are used which have spectral properties dissimilar to the autofluorescent material found in environmental samples. Described below are a range of staining methods including gene probes, lectins, and various DNA stains.

At present, the simplest and most reliable method of staining organisms within environmental samples is with antibodies.

A. Antibody–Based Staining

1. Polyclonal and Monoclonal Antibodies

Commerical companies now produce antibodies to a range of environmental microorganisms. Armed with catalogs from these antibody companies, Molecular Probes Inc., (Eugene, OR) "Handbook of Fluorescent and Research Chemicals," and a reasonable quantity of dollars the environmental microbiologist can label a range of microorganisms with almost whatever color takes their fancy.

In our experience, monoclonal antibodies are far superior to polyclonal antibodies for staining microorganisms within environmental samples. Considerably, more binding of polyclonal antibody to debris occurs than with monoclonal antibodies. Figure 1 illustrates the superior staining qualities of monoclonal antibodies in an environmental water sample seeded with *Cryptosporidium* oocysts and stained with a FITC-labeled monoclonal antibody specific to the oocyst wall. Figure 1B shows the same sample stained with a FITC-labeled polyclonal antibody. Note, the higher fluorescence of the background debris in Fig. 1B. Similar results were obtained when comparing indirect immunofluorescence staining of microorganisms within environmental samples and staining with directly conjugated antibodies. Again, we have found that directly conjugated antibodies are far superior.

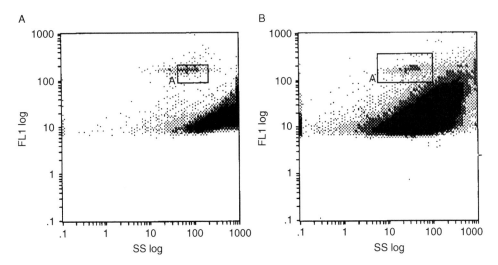

Fig. 1 Results from analyzing a river water sample seeded with *Cryptosporidium* oocysts and stained with monoclonal (A) or polyclonal FITC-labeled antibodies (graph B). The areas labeled "A" show the position of oocysts. Note the higher fluorescence of the debris in (B).

The concentration and the incubation temperature at which antibodies are used are critical to good staining. Concentrations used for environmental samples are often considerably higher than those for other applications. The antibody should be titered out to determine the optimal concentration. These titrations need to be performed using typical environmental samples seeded with realistic numbers of the target organism. Environmental samples from different locations may require different titers of antibody.

2. Blocking Agents

The use of blocking agents to prevent nonspecific binding of antibodies to debris in environmental samples is essential. Even when highly specific, directly conjugated monoclonal antibodies are used, considerable nonspecific binding may occur unless adequate blocking agents are employed. This is demonstrated in Fig. 2 which represents a river water sample seeded with *Cryptosporidium* oocysts and stained with a FITC-conjugated mAb with blocking agents (Fig. 2A) and without blocking agents (Fig. 2B). The choice of blocking agent and concentration to use must be determined empirically for each antibody that is used. Table I gives examples of some blocking agents and the range of concentrations at which we use them. Determining the optimal blocking agent(s) is performed by staining and analyzing seeded environmental samples using different combinations of blocking agents. However, results may differ from one type of sample to another. The easiest way to overcome this is to prepare a cocktail sample by combining all

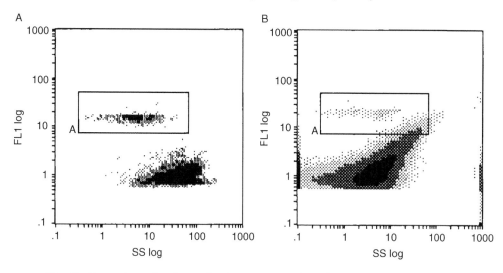

Fig. 2 Comparison of analyses of a river water sample seeded with *Cryptosporidium* oocysts and stained with monoclonal antibodies labeled with FITC with (A) and without (B) blocking agents. The areas labeled "A" show the position of oocysts. Note the higher fluorescence of the debris in (B).

Table I
Examples of Blocking Agents and the Concentrations at Which They are Employed

Blocking agent	Range of concentrations
Bovine serum albumin	0.1–5% (w/v)
Skim milk	0.1–5% (w/v)
Fetal calf serum	0.1–10% (v/v)
Glycine	0.01–1% (w/v)
Tween 20	0.001–1% (v/v)

types of samples that will be analyzed. The method used to concentrate the sample may also influence the choice of blocking agents.

3. Loss of Epitopes

Commercially available antibodies are not normally developed or prepared for staining environmental samples for flow cytometric analysis. Instead, they are developed for clinical samples using epifluorescence microscopy or ELISA. This can create problems when they are used in novel applications. For example, the antigens which the antibodies recognize may be stable and always present on the organism in clinical samples but this may not be true for organisms which have been in the environment for some time. We have found that the carbohydrate antigen on *Cryptosporidium* oocyst walls which is recognized by all commercially available antibodies is slowly shed by oocysts in the aquatic environment (data not presented). Furthermore, this antigen is removed by oxidizing agents, such as chlorine, and is not essential to viability (Moore *et al.*, in preparation). Therefore, oocysts which have been treated with high levels of chlorine may still be infectious but they cannot be detected using the commercially available antibodies.

4. Dual Staining

The use of two different antibodies to stain a microorganism increases the sensitivity of a detection method to such a degree that nonsorting detection is often possible (see Section IV). The antibodies must not compete with each other. Staining can be performed with both antibodies simultaneously. If uneven staining is seen then stain with one antibody, fix in 10% formalin, wash in PBS, and stain with the second antibody.

5. Fluorescence Amplification

The sensitivity of a detection method can often be increased by using a mAb to the fluorochrome bound to the antibody. These antibodies, such as an anti-FITC mAb, can be conjugated with a second fluorochrome such as R-phycoerythrin (RPE) or allophycocyanine (APC) (a HeNe laser is required for APC). This technique can help to discriminate the target organism from autofluorescent particles. However, any particle to which the first antibody is attached will be stained and, thus, amplified, with the second fluorochrome.

The use of biotinylated probes with avidin or streptavidin as secondary reagents is, in our experience, not suitable for staining environmental samples. We have experienced considerable increases in background staining when using these techniques. This is probably attributable to the fact that avidin's positively charged residues and its oligosaccharide component can interact nonspecifically with negatively charged cell surfaces and nucleic acids. Similarly, streptavidin binds to integrins and related cell-surface molecules (Alon *et al.*, 1990) causing background problems. However, Molecular Probes, Inc., now markets a new form of avidin, NeutraLite avidin, which appears to be less susceptible to background staining problems.

B. DNA and Vital Stains

Fluorescence methods are available for accurate, reproducible determinations of cellular viability and proliferation. These methods normally rely on a metabolic process, such as enzymatic activity or membrane potential, to identify living cells and employ a membrane impermeable nuclear stain to identify dead cells. Considerable research has been performed into the use of these techniques on bacteria (Diaper *et al.*, 1993; Kaprelyants and Kell, 1992, 1993). However, all of these studies involve the analysis of pure cultures of bacteria. There is very little literature available on the detection of specific microorganisms and simultaneous evaluation of their viability within environmental samples. Donnelly and Baigent (1986) combined the DNA stain propidium iodide (PI) with immunofluorescence staining for the detection of *Listeria* in milk. However, the authors did not use PI as an indicator of viability but fixed the cells prior to staining so that PI would enter all cells. The authors then used the PI signal in conjunction with the immunofluorescence FITC signal to detect the *Listeria* cells. Pettipher (1991) reported on the use of the ChemFlow system for detecting yeasts in soft drinks. The ChemFlow system is a flow cytometer and bacterial stain developed by Chemunex (France) specifically for microbiology laboratories. The fluorochrome used is a propriety viability stain. The system is capable of enumerating viable microorganisms but is not able to distinguish between different types of microorganisms.

To be able to detect specific microorganisms and assess their viability simultaneously, separate fluorochromes for viability and detection must be used which do not spectrally overlap. If two different fluorochromes are being used for detection then the choice of fluorochrome for viability is limited. This type of three-color cytometry normally requires a second laser beam operating at a different wavelength. A UV laser enables the use of stains such as 4',6-diamidino-2-phenylindole (DAPI) used in combination with two detection fluorochromes such as FITC and APC.

We have found that Molecular Probes new DNA stains YOPRO-1 and TOTO-1 are far superior to other DNA stains. These stains are considerably brighter than PI or ethidium bromide. We routinely use the stains at concentrations of 1 in 10,000 in conjunction with immunofluorescent staining.

The tetrazolium salt iodonitrotetrazolium violet (INT) is a useful nonfluorescent viability stain. Respiring cells oxidize the stain with their electron transport system and deposit the red formazan product (Vesey *et al.*, 1990; Zimmerman *et al.*, 1978). These formazan deposits are clearly visible as pink spots within cells when examined with light microscopy. Furthermore, the formazan deposits cause a large increase in SALS during flow cytometric analysis. We routinely use INT at a concentration of 0.02% (w/v) in PBS, pH 9.2.

A fluorescent tetrazolium dye 5-cyano-2,3-ditolyl tetrazolium chloride (CTC) has been reported for the direct visualization of actively respiring bacteria (Rodriguez *et al.*, 1992). The dye is similar in principle to INT except that the CTC-formazan that accumulates intracellularly fluoresces at 602 nm. However, in our experience, fluorescence is often considerably less than when other fluorescent viability dyes are used.

The use of fluorescent dyes to determine the viability of *Cryptosporidium* oocysts and *Giardia* cysts is an area which is receiving considerable attention at present. Schupp and Erlandsen (1987) reported on the use of fluorescein diacetate (FDA) and PI to distinguish between live and dead *Giardia* cysts. Live cysts stained with FDA and dead cysts with PI. By staining and flow cytometrically sorting cysts the authors were able to show that FDA-positive cysts caused infection in a mouse model whereas PI-stained cysts did not. However, the authors occasionally observed cysts which did not stain with FDA or PI. Further, evaluation of PI as an indicator of cyst viability was performed by Sauch *et al.* (1991). The authors demonstrated that PI staining is not satisfactory for determining the viability of cysts exposed to commonly employed water disinfection methods.

A viability assay for oocysts of *Cryptosporidium* based on the inclusion or exclusion of the dyes DAPI and PI has been reported by Campbell *et al.* (1993). The authors report that oocysts which stain with PI are nonviable and oocysts which stain with only DAPI are viable. The authors found that not all oocysts stained with the two dyes unless a pretreatment involving incubation in acid was performed. Korich *et al.* (1993) investigated the use of a range of methods for determining oocyst viability and in contrast to Campbell *et al.* (1993) the authors reported that the frequency of DAPI-stained nonviable oocysts was 18 times higher than stained viable oocysts. The authors state that the ability of DAPI to stain nuclei within sporozoites of oocysts was not sufficient to establish oocyst viability and that the most promising technique involved the use of a mAb which is specific for partially opened oocysts. The mAb will only bind to nonviable oocysts but in our opinion it is very unlikely that the mAb will stain all nonviable oocysts.

C. Lectins

The use of fluorescently conjugated lectins for staining environmental samples is a simple and inexpensive method to attach an additional probe to the target organism. A range of fluorescently conjugated lectins is commercially available. On their own, lectins are not specific enough to enable detection of individual strains or species of microorganisms. However, by combining lectins with antibodies or gene probes the sensitivity of a detection method can be increased. A lectin does not have to stain the target organism to be useful. When screening a range of fluorescently conjugated lectins against samples seeded with *Giardia* cysts we found that several lectins stained the cysts, several stained nothing, and some stained the debris but not the cysts. Further, investigations revealed that the lectins which stained cysts were not specific enough to be of use as an additional probe in a detection method. However, by combining several of the lectins which stained the debris particles we were able to stain virtually all debris particles without staining the cysts. By combining this lectin staining with cyst-specific antibodies labeled with a different fluorochrome to the lectins we were able to significantly increase the sensitivity of our detection method.

D. Fluorescent *In Situ* Hybridization

The use of fluorescent *in situ* hybridization (FISH) is a relatively new and exciting method to label specific microorganisms. The technique relies on identification of a specific sequence of DNA or RNA within the target organism. Thousands of copies of a complementary sequence are then chemically synthesized and labeled with a fluorochrome. The fixation procedure leads to permeation of the wall and membranes of microorganisms allowing the probe to penetrate (Amann *et al.*, 1990a). The probe is added to the sample and incubated at a temperature which allows hybridization to the organism. The probe will only attach to complementary sequences within the target organism. However, these specificities are strongly dependent on the hybridization temperatures, concentrations of monovalent cations, and denaturing agents.

Considerable research has been performed on the use of ribosomal RNA-targeted oligonucleotide probes with flow cytometry by Amann *et al.* (1990a,b). Molecules of rRNA are ideal targets for fluorescently labeled nucleic acid probes. High sensitivity can be achieved due to the target being present in very high numbers (normally more than 10,000 per cell) and a denaturation step is not required as the target is single stranded. Also, in active cells the degree of fluorescence depends on the integrity of the organism and an organism which is nonviable will have little or no fluorescence.

We have successfully used the following method for staining both bacteria and protozoa (Fig. 3) with fluorescently labeled rRNA-targeted probes.

1. Staining with FISH

a. Fixation

1. Add three volumes of 4% (w/v) paraformaldehyde in PBS to one volume of sample.

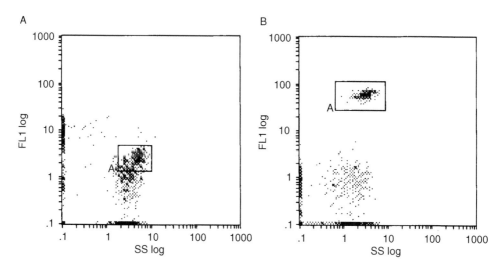

Fig. 3 Results from analyzing *Cryptosporidium* oocysts labeled with a FISH probe specific to bacteria (A) and a probe specific to Eukaryote rRNA (B). The areas labeled "A" show the position of oocysts. Note the high fluorescence of oocysts in (B) and the high fluorescence of the contaminating bacteria in (A) (the bacteria are next to the *y*-axis).

2. Hold for 2 h at 4 °C and then wash twice in PBS.

3. Add one volume of ice-cold ethanol. Samples can be stored at this stage for several months at −20 °C.

b. Hybridization

1. Add to a 1.5-ml Eppendorf tube:
   - 50 μl fluorescent probe (25 ng/μl in distilled water),
   - 500 μl hybridization buffer (0.1% SDS, 0.9 M sodium chloride, 20 mM Tris/HCl, pH 7.2) prewarmed to 50 °C,
   - 50 μl of sample.

2. Incubate at 50 °C for 2 h.

3. Wash cells in prewarmed (50 °C) hybridization buffer.

4. Wash cells in PBS and analyze immediately.

One of the major attractions of FISH is that probes of desired specificity can be simply and inexpensively produced. Compared to the production of monoclonal antibodies where numerous fusions and screening of hundreds of cell lines is often required before a suitable antibody is found, the technique appears very attractive. FISH is still in its infancy and there are still many practical problems associated with the application of this technique. Often, the amount of fluorescence achieved is considerably less than that with antibodies. We are currently trying to increase the fluorescence signal by increasing the amount of fluorochrome attached to each probe.

Unfortunately, at present, when more probe is attached the fluorescence increases but so does the background. We are confident that this and similar problems will be overcome in the near future and the technique of FISH will become a routine procedure.

IV. Flow Cytometric Analysis of Water Samples

A. Instrumentation

The type of samples which are of interest to environmental microbiologists are very different from the cell suspensions which flow cytometers are designed to analyze. A water or sediment sample may contain up to 10^{11} particles/ml of sufficient size to be detected by a flow cytometer. Against a diverse background composed of bacterial, algal, plant, animal, and mineral particles the analyst may be trying to detect a single microorganism. These types of samples create problems which have not been encountered previously in flow cytometry.

A major problem with flow cytometric analysis of environmental samples is blockages. Blockages within the flow cell and within the sample tubing can result in hours of frustration. Extensive prefiltering of samples is often not possible due to losses of organism of interest. We have overcome blockages on a Coulter Elite cytometer by using a quartz flow cell with a 140-μm orifice and the shortest possible length of sample tubing from the sample tube to the flow cell. Problems were encountered with unstable droplet breakoff due to the large size of the droplets. However, by replacing the pietzoelectric crystal wafer with a cylindrical bimorph and replacing the flow cell body with a prototype flow cell body made from thinner plastic and which has a finer sample insertion rod permanently glued in position, the problem was overcome. The voltage across the deflection plates was increased to 3000 V to enable deflection of the large droplets.

The Becton-Dickinson (BD) Facstar Plus has the potential to solve blockage problems with their optional extra the Macro-Sort. The Macro-Sort enables the use of air flow cells with an orifice size of up to 400 μm. However, we have not yet successfully used a flow cell of greater than 100 μm diameter on the Facstar plus. We have also encountered problems due to the turbid and viscous nature of environmental samples causing disturbances within the sample stream. The Facstar plus analyzes the sample in air after it has left the flow cell. The disturbance of the sample stream caused by the injection of viscous and turbid samples into the stream causes noise in both the light scatter and fluorescence signals.

The two instruments discussed above are both state of the art flow cytometers which can sort. Analysis only instruments, such as the Coulter XL or the BD Facscan, offer considerable advantages, over the sorters, to the environmental microbiologist. They are half the cost of sorters, they are simple to operate, they can be fully automated, they have a large orifice which is seldom blocked, and they require little or no alignment or maintenance. However, because they cannot sort their level of sensitivity may limit their application, as discussed later in this chapter.

B. Discriminators

Once an environmental microbiologist has managed to persuade a flow cytometer to analyze a sample without blocking up, decisions need to be made on which particles the cytometer should collect data from, which particles it should ignore, and what signals should be processed. The cytometer is not capable of collecting data from all particles present in a typical sample since there are too many. Therefore, a discriminator is used to reduce the amount of data to be collected to a level within the capabilities of the instrument.

During flow cytometric analysis a particle enters the laser intersection point and measurements of one parameter, the parameter chosen as the discriminator, are made. If the signal for this parameter is higher than the level set for the discriminator then the acquisition of data for all other parameters is performed. When analyzing environmental water samples for the presence of low numbers of fluorescently labeled particles the fluorescent signal is the most appropriate parameter to use as the discriminator. By setting the discriminator at a level of fluorescence slightly lower than that of the particles to be detected, the number of measurements the cytometer needs to make per second (the data rate) is typically less than 500 s^{-1}. If FALS or SALS were used as the discriminator for the same sample the data rate would be $>10,000 \text{ s}^{-1}$. The reason for this is that environmental samples contain large numbers of particles of which only a small proportion have high fluorescence.

The next question is what type of fluorescence signal should be used as the discriminator: a log, a linear, or a peak signal? The quick answer is if recovery is important, then use a peak signal as the discriminator because a peak signal is processed much faster than a log or linear signal. The reason for wanting to use the fastest signal as the discriminator is explained below.

C. Signals and Interference

We have experienced problems with the analysis of very turbid river samples and sediment slurries that contain large quantities of small particulate matter. In these samples, a significant reduction in the recovery of the target organism was observed. When fluorescently labeled beads were seeded into a very turbid sample and flow cytometric analysis was performed, a high proportion of the beads produced light scatter signals quite different from those obtained for pure beads (Fig. 4).

Further investigations revealed that when analyzing turbid environmental samples, considerable reductions in recovery can occur due to interference caused by small particles. These losses can be completely overcome by careful control of the

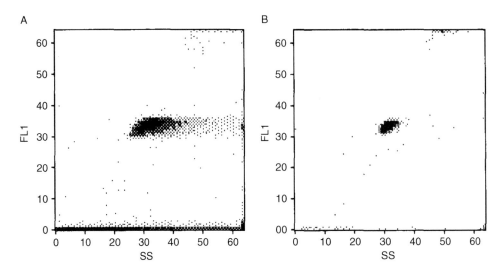

Fig. 4 Analysis of fluorescently labeled beads seeded into a turbid river water sample (A) and the same beads alone (B). Note how the population in graph A has spread due to interference caused by small particulate matter present in the sample.

length of time a particle remains within the laser intersection point and by using peak signals for the discriminator and for analysis. We have observed that by adjusting the laser spot size and the acquisition start delay value we can still operate the Elite at the sheath pressure which results in optimal sorting. Using this set up we can obtain 100% recovery of organisms seeded into our most turbid samples.

D. Analysis

The two major concerns while analyzing a sample are that the cytometer is detecting or sorting all the target organisms present in the sample and that there is no contamination of the target organism from a previous sample.

To confirm that the cytometer is detecting or sorting all target organisms present, two quality control procedures should be performed. The first procedure involves the analysis of a sample seeded with a known number of the target organism within each batch of samples analyzed. The second procedure involves seeding samples with a known number of fluorescent beads. If sorting is being performed then the beads should be sorted along with the target organism. If the correct number of beads is not detected or sorted then the sample should be reanalyzed.

When analyzing environmental samples for low numbers or single microorganisms it is essential to ensure that carryover of target organisms from one sample to the next does not occur. The best way to achieve this is to run a sample of 20% sodium hypochlorite. Sodium hypochlorite removes the antibody from the target organism, so even if an organism is carried over into the next sample it will not be fluorescing. A 2-min run of sodium hypochlorite followed by a 30-s run of Coulter Cleanze is the protocol we use routinely.

E. Detection

In an ideal situation, flow cytometric enumeration of specific microorganisms in environmental samples simply involves recording the number of particles within a predefined area on a scatter plot. The area should be defined by analysis of real samples seeded with the target organism. A scatter plot or scatter plots are then set up and an area is defined which encloses all of the organism of interest. The ability of the software to allow you to define amorphous regions is a considerable aid which actually increases the sensitivity of the detection method.

Defining the correct area in which the target organism will appear is the most important stage in developing a detection method. As an example, we examine the development of a method for the detection of an organism which can be stained with a FITC-labeled antibody. The first step is to analyze a seeded environmental sample and to set up a scatter plot of FALS or SALS versus FITC. The choice between FALS or SALS depends on which produces the tightest population as well as separation from debris. Once the graph of choice has been created it is worth running the sample at several different laser power and PMT high-voltage settings to optimize the separation between the organisms and the debris. The autofluorescence of the debris and the fluorescence of the FITC have different fluorescence lifetimes. By adjusting the laser power and the high-voltage settings it is possible to obtain the maximum amount of FITC emissions without increasing the autofluorescence of the debris. To increase the sensitivity of the method, it is important to use as much information as possible to define the target organisms. Although we only have one label on the organism, FITC, it is worth looking at fluorescence at different wavelengths. The organism will have an autofluorescent signal different from other particles. It does not have to have a high autofluorescent signal to be useful. For example, the organism may have very low autofluorescence at 600 nm, whereas much of the autofluorescent material may be considerably higher. To be able to use this information we need to perform backgating. This is simply performed by setting up histograms of the fluorescent parameter (e.g., FL3) against FALS or SALS (use the light scatter that is not used in the FL1 histogram) and gating the histogram on the area defined in the first histogram. After analyzing a sample or playing list-mode data, define an area on the new histogram which encloses all of the target organisms. Remove the gate from the second histogram and gate the first histogram on the new area. Some wavelengths will be useful and some will not, but by trial and error and using a range of different wavelengths it is possible to increase the sensitivity of the detection method considerably.

Another technique which we have found useful is to look at the ratio of fluorescence from one wavelength to that of another wavelength. Most flow cytometers enable this to be performed by assigning the chosen fluorescence parameters to the ratio parameter. The ratio can then be used on a two-dimensional histogram. For example, one of the limitations in sensitivity is the presence of very bright autofluorescent mineral and algal particles. Because these particles normally autofluoresce across a very wide spectrum it is possible to discriminate between them and the target organism, which only fluoresces across a narrow spectum, by using a backgating method.

The target organism used to seed must be as similar as possible to the organism which is actually present in environmental samples. For example, if the method which is being developed is for the detection of *Legionella* in environmental samples then preparing the seed by simply using *Legionella* growth from an agar plate is not sufficient. Cells of *Legionella* in the environment are considerably smaller and more rounded than freshly cultured cells. In our experience, these types of differences between freshly cultured bacteria and the same organism occurring in the environment are very common. Differences may also occur in organisms which have been purified from fecal material and the same organism once it has been in the environment for a period of time. For example, *Cryptosporidium* oocysts which have been freshly isolated from feces fluoresce around the entire oocyts wall when stained with a specific mAb, whereas oocysts which have been in an aquatic environment for more than a few days show patchy and considerably reduced fluorescence. To overcome these differences when setting up our detection method we seeded oocysts into various types of water and stored them at a range of temperatures for periods up to 3 months (Vesey *et al.*, 1993a). These samples as well as fresh samples were then used to define the area in which oocysts would appear. We recommend that similar experiments are performed before defining an area in which the organism to be detected will appear.

Now that we have defined an area in which our target organism will always appear, we need to determine the sensitivity of the method. This is performed by seeding environmental samples with known numbers of the test organism. When a negative sample is analyzed no particles should appear in the defined area and a sample seeded with 10 organisms should have 10 particles within the defined area. However, at present it is extremely difficult to obtain this level of sensitivity and often a negative sample will contain some particles within the defined area (Vesey *et al.*, 1991). These particles are typically autofluorescent mineral or algal particles or particles which cross-react with the antibody. If the level of sensitivity required is higher than the number particles present in the negative sample then there is no problem. For example, if a method for detecting *Legionella* in cooling tower water is being developed and a negative sample contains 100 particles in the defined region and the level of sensitivity required is 1000 cells/l [this is a realistic figure for *Legionella* because below this level there is considered to be little health risk (Badenoch, 1987)] then the method is satisfactory. However, for organisms such as *Cryptosporidium*, for which 1 oocyst in 1 l may be a potential health risk (Badenoch, 1990), then the sensitivity is not satisfactory. To increase the sensitivity, we need to be able to obtain more information from the target organism so that we can distinguish it from the interfering particles. To achieve this level of discrimination, one option is to label the organism with a second fluorochrome. This second label can be achieved with a second antibody which does not compete with the first antibody, a gene probe or a simple stain. In our experience, a second antibody is the most successful method. The fluorescent signal from the second probe is used as an additional gate for the first histogram.

Figure 5 demonstrates the use of two probes to achieve good sensitivity. The figure shows results of analyzing concentrated river water samples for the presence of seeded *Dictyostelium* spores. Samples were stained with two different antibodies specific to the spore wall. One antibody was labeled with FITC and the other with RPE. Figure 5A and B are the results from analyzing a sample of river water which contained no spores. The area within the RPE graph (Fig. 5B) is the defined area in which spores will appear. This area is gated on a second area within the FITC graph (Fig. 5A). Note that the area on the RPE graph contains no particles. Figure 5C and D represent the analysis of a similar sample seeded with a single spore. Note that the area on the RPE graph contains a single particle.

If a second probe is not available the way to increase the sensitivity to the single organism level is to sort all particles in the defined area. By using three-droplet sorting virtually 100% recoveries of the target organism can be achieved. Furthermore, the sorted sample contains very little contaminating debris which enables rapid examination using microscopy, gene probes, or PCR.

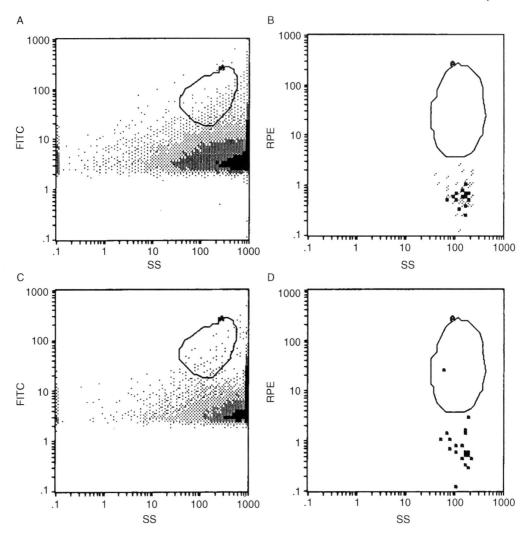

Fig. 5 Analysis of a river water sample for Dictyostelium spores. Samples were stained with two different spore coat antibodies labeled with FITC and RPE. (A, B) A negative sample. The area (A) is used to gate (B). All spores will appear in the area in (B). Note that there are no particles in this area in (B). (C, D) The same analysis performed on a similar sample seeded with a single spore. Note the single particle in the area in (D).

If an analysis only method is used, that is, no sorting and confirmation is performed, then we strongly recommend that, initially, some form of confirmation is performed on positive samples.

V. Instrumentation Developments for Environmental Applications

Current cytometers only give peak, integral, or log intensity for each parameter and in our experience, even with fluorescently labeled antibodies in environmental samples, this is not sufficient to discriminate many of the background particles from the target. Several detection techniques currently being developed to increase the amount of information that can be collected from each particle in a flow cytometer are discussed below.

A. Array Detection

Intensified diode arrays and charge coupled device (CCD) arrays have been used extensively for both spectroscopy and imaging for many years (Messenger, 1991). The main reason they have not been suitable for flow cytometry until now is their readout speed. Generally, spectroscopic and imaging detectors read out at "video" type rates, up to about 50 Hz. Flow cytometry demands data rates (and therefore readout rates) of up to 10,000 particles/s or more. Recent advances in array detector have seen extremely fast CCD arrays become available; for example, 128-element linear CCD arrays are now available which have readout at rates over 100 kHz. This means that we can collect either 128 spectral channels or a 128-element line image for each particle passing through the cytometer.

B. Spectral Fingerprinting

As it is now possible to read out from an array detector at sufficient speeds we can apply the array detector in a variety of ways to increase the amount of information obtained from each particle passing through the cytometer. If we remove all conventional filters and photomultiplier tubes and replace them with a dispersing prism or grating and a high-speed array detector we have the equivalent of 128 spectral channels! [CCD array detectors have sensitivities approaching those of PMTs (Anon, 1992)]. We are examining the possibility of characterizing the autofluorescence of both the background particles found in the environmental samples and also the various fluorochromes. The concept of spectral fingerprinting may then allow the discrimination of particles solely from autofluorescence spectra or allow enhanced discrimination of fluorescent tags. This is demonstrated in Fig. 6 where the spectral fingerprint of three dyes and the light scatter from milk are compared. The three dyes can be easily separated.

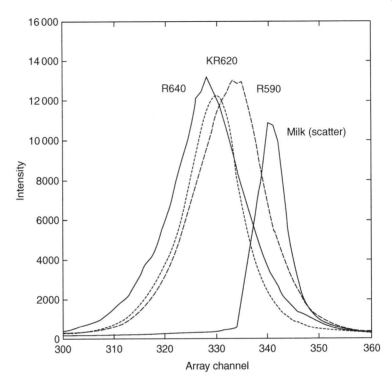

Fig. 6 Spectral fingerprinting of three fluorescent dyes: kiton red 620 (KR620), rhodamine 590 (R590), and rhodamine 640 (R640) and the light scatter from milk. The sample was illuminated with a pulsed, frequency-doubled, diode-pumped laser producing at 532 nm.

C. Pulsed Lasers

Most current laser flow cytometers employ countinuous wave lasers as the excitation source. Generally, these are continuous wave argon lasers, operating in the green, blue, and ultraviolet. The use of pulsed lasers in flow cytometry systems allows the possibility of "in-flow" particle imaging and the use of time-resolved fluorescence techniques. Also, solid-state diode-pumped lasers have the potential for reducing flow cytometry costs as they have high efficiencies, low power consumption, and long lifetimes [estimated 50,000 h (Byer, 1988; Fan, 1990)]. They also have the potential for the generation of new wavelengths in pulse mode by nonlinear processes, as well as temporally multiplexing different excitation wavelengths to the interaction region. Another advantage of solid-state lasers is that they are very small and robust thus enabling the development of portable machines which would be ideal for environmental monitoring.

D. Particle Imaging

One- or two-dimensional imaging can now be realized with a high repetition rate (kHz) pulsed laser source and a high-speed array detector. This technique can be applied in two ways: using a conventional cw laser and detection system to trigger the pulsed laser and an array detector to capture either a forward scatter or fluorescence image. This effectively "automates" the visual confirmation step of the current methodology. The other method uses only the pulsed laser source with an illumination method developed in our laboratory (Grant *et al.*, 1993; Narai *et al.*, 1993), which ensures illumination of all particles passing through the interaction region of the cytometer. We then replace the conventional foward/side scatter detector with a linear or area array. This effectively views the interaction region of the cytometer and forward/side scatter images can be obtained.

E. Time-Resolved Fluorescence

A pulsed laser source is ideal for the technique of time-resolved fluorescence, where the detection of the fluorescence signal is gated with time to remove any autofluorescent background. Currently in microscopy, the technique commonly uses electronic or mechanical choppers with a continuous source to create a pulsed illumination of the sample. As the short pulse duration of many high repetition rate pulsed lasers reduces the number of excitation/relaxation cycles of the fluorochrome, conventional probes do not yield as many photons when illuminated by a pulsed laser source. However, probes with lifetimes longer than around 1 μs (the time it takes a particle to pass through the laser beam in a conventional cytometer) will yield the same number of photons when illuminated by a pulsed laser source, which has the advantage that the detectors can be gated so that the illuminating laser light is not detected, improving our signal to noise ratio. Suitable fluorochromes are the europium chelates and other rare earth chelates (Leif *et al.*, 1993).

VI. The Future of Flow Cytometry within Environmental Microbiology

Flow cytometry has the potential to revolutionize detection methods within environmental microbiology. At present, flow cytometry enables the rapid detection of low numbers and even single specific microorganisms within extremely turbid samples. These methods, once set up, can be performed on easy to use flow cytometers which require little calibration and enable the analysis of multiple samples to be made without an operator being present. However, unless two different probes, which are highly specific to the target organism (such as monoclonal antibodies), are available the sensitivity required to detect single microorganisms cannot be achieved using this type of cytometer. To achieve this level of sensitivity without two probes a sorter is required to sort the suspect particles for further examination. Unfortunately, cytometers capable of sorting are expensive and complex instruments which are not totally suitable for examining environmental samples.

Current developments within the technique of FISH promise to enable the simple and inexpensive development of highly specific probes to a vast range of microorganisms. If the current problems with insufficient probe fluorescence are overcome and the hybridization method is simplified and more rapid, the detection of single specific microorganisms within environmental samples, on basic analyzer flow cytometers, will be possible for many microorganisms.

Alongside the biological advances which are currently occurring are developments within cytometer hardware. Flow cytometers have always been developed for the analysis of mammalian cells. The types of samples which the environmental microbiologist wishes to analyze are very different from suspensions of mammalian cells. Currently, in our laboratory developments in cytometry hardware are underway which are specifically designed for aiding the analysis of environmental samples. These developments include the use of array detectors to image particles and to enable spectral fingerprinting to characterize the autofluorescence of microorganisms and fluorochromes. The use of modern solid-state lasers may also enable the development of small, robust cytometers which can be used for field sampling. Such instrument developments will facilitate major advances in our understanding of the role of microorganisms in the environment.

We foresee that, in the future, flow cytometers that are designed specifically for the analysis of environmental samples will be available to environmental microbiologists. Coupled with the advancements in probe technology, we believe that flow cytometric detection will replace culture methods for the analysis of environmental samples. Portable flow cytometers will enable much of the analysis to be performed on site; at water treatment plants cytometers will be on-line instruments continually analyzing the end product.

Acknowledgments

The authors thank Günter Wallner without whose expertise the work on FISH would not have been possible; Graham Chapman of Coulter Electronics for his enthusiastic support; and the Sydney Water Board for funding this work.

References

Abbaszadegan, M., Gerba, C. P., and Rose, J. B. (1991). *Appl. Environ. Microbiol.* **57,** 927–931.
Adam, D. A. (1991). *Microbiol. Revs.* **55,** 706–732.
Alon, M., Bayer, E. A., and Wilchek, M. (1990). *Biochem. Biophys. Res. Commun.* **170,** 1236.
Amann, R. I., Binder, B. J., Olson, R. J., Chisholm, S. W., Devereux, R., and Stahl, D. (1990a). *Appl. Environ. Microbiol.* **56,** 1919–1925.
Amann, R. I., Krumholz, L., and Stahl, D. A. (1990b). *J. Bacteriol.* **172,** 762–770.
Anon (1992). "CCD Image Sensors Data Book." Dalsa, Inc., Waterloo, Ont., Canada.
Ashbolt, N. J., and Veal, D. A. (1994). *Today's Life Sci.*, **6**(6), 28–29.

Badenoch, J. (1987). "Second Report of the Committee of Inquiry into the Outbreak of Legionnaires' Disease in Stafford in April 1985." HMSO, London.

Badenoch, J. (1990). "*Cryptosporidium* in Water Supplies." HMSO, London.

Balfoort, H. W., Berman, T., Maestrini, S. Y., Wenzel, A., and Zohary, T. (1992). *Hydrobiologia* **238,** 89–97.

Bankes, P., and Richard, F. (1993). *In* "The Society for Applied Bacteriology". 62nd Annual Meeting and Summer Conference. University of Nottingham. 13–16 July 1993. p. xviii.

Bee, C. A., Christy, P. E., and Robinson, B. S. (1991). "Methods for Detection of Giardia and Cryptosporidium in Water: A Preliminary Assessment." Report Number 25, Urban Water Research Association of Australia, Melbourne. (Published for the Urban Water Research Association of Australia by the Melbourne and Metropolitan Board of Works.)

Betz, J. W., Aretz, W., and Härtel, W. (1984). *Cytometry* **5,** 145–150.

Bifulco, J. M., and Schaerfer, III, F. W. (1993). *Appl. Environ. Microbiol.* **59,** 772–776.

Brooks, J. L., Moore, A. S., Pratchett, R. A., Collins, M. D., and Kroll, R. G. (1992). *J. Appl. Bacteriol.* **72,** 294–301.

Button, D. K., and Robertson, B. R. (1993). *In* "Handbook of Methods in Aquatic Microbial Ecology" (P. F. Kemp, B. F. Sherr, E. B. Sherr, and J. J. Cole, eds.), pp. 163–173. Lewis, Boca Raton, FL.

Button, D. K., Robertson, B. R., McIntosh, D., and Juttner, F. (1992). *Appl. Environ. Microbiol.* **58,** 243–251.

Byer, R. L. (1988). *Science* **239,** 742–747.

Campbell, A. T., Robertson, L. J., and Smith, H. V. (1993). *Appl. Environ. Microbiol.* **58,** 3488–3493.

Colwell, R. R., Brayton, P. R., Grimes, D. J., Roszak, D. B., Huq, S. A., and Palmer, L. M. (1985). *Biotechnology* **3,** 269–277.

Current, W. L. (1986). *Crit. Rev. Environ. Control* **17,** 21–51.

Diaper, J. P., Thither, K., and Edwards, C. (1993). *Appl. Microbiol. Biotechnol.* **38,** 268–272.

Dimichele, L. J., and Lewis, M. J. (1993). *J. Am. Soc. Brewing Chem.* **51,** 63–66.

Donnelly, C. W., and Baigent, G. J. (1986). *App. Environ. Microbiol.* **52,** 689–695.

Edwards, C., Porter, J., Saunders, J. R., Diaper, J., Morgan, J. A. W., and Pickup, R. W. (1992). *SGM Q.* **19**(4), 105–108.

Fan, T. Y. (1990). *Lincoln Lab. J.* **3,** 413–425.

Ford, S., and Olsen, B. H. (1988). *Adv. Microb. Ecol.* **10,** 45–79.

Furrer, B., Candrian, U., Hoefelein, C., and Luethy, J. (1991). *J. Appl. Bacteriol.* **70,** 372–379.

Grant, K. J., Piper, J. A., Ramsay, D. J., and Williams, K. L. (1993). *Appl. Optics* **32,** 416–417.

Grimes, D. J., Atwell, R. W., Brayton, P. R., Palmer, L. M., Rollins, D. M., Roszak, D. B., Singleton, P. L., Tamplin, M. L., and Colwell, R. R. (1986). *Microbiol. Sci.* **3,** 324–329.

Haslett, N. G., Adams, M. R., Cordier, J. I., and Cox, L. J. (1991). *J. Appl. Bacteriol.* **71,** XXIII–XXIV.

Herman, L., and De-Ridder, H. (1993). *Netherlands Milk Dairy J.* **47,** 23–29.

Hopkins, D. W., O'Donnell, A. G., and Macnaughton, S. J., (1991). *Soil Biol. Biochem.* **23,** 227–232.

Humphreys, M. J., Allman, R., and Lloyd, D. (1993). *In* "Proceedings of the Society for Applied Bacteriology 62nd Annual Meeting and Summer Conference." University of Nottingham, 13–16 July, 1993. Society for Applied Bacteriology, Nottingham. p. xix.

Issac-Renton, J. L., Fury, C. P., and Lochen, A. (1986). *Appl. Environ. Microbiol.* **52,** 400–402.

Jespersen, L., Lassen, S., and Jakobsen, M. (1993). *Int. J. Food Microbiol.* **17,** 321–328.

Kaprelyants, A. S., and Kell, D. B. (1992). *J. Appl. Bacteriol.* **72,** 410–422.

Kaprelyants, A. S., and Kell, D. B. (1993). *Appl. Environ. Microbiol.* **59,** 3187–3196.

Keasler, S. P., and Hill, W. E. (1992). *J. Food Protect.* **55,** 382–384.

Kell, D. B., Ryder, H. M., Kaprelyants, A. S., and Westerhoff, H. V. (1991). *Antonie van Leeuwenhoek* **60,** 145–158.

Korich, D. G., Mead, J. R., Madore, M. S., Sinclair, M. A., and Sterling, C. R. (1990). *Appl. Environ. Microbiol.* **57,** 2610–2616.

Korich, D. G., Yozwiak, M. L., Marshal, M. M., Sinclair, N. A., and Sterling, R. S. (1993). Report to the American Water Works Association Research Foundation.

LeChevallier, M. W., Norton, W. D., and Lee, R. G. (1991). *In* "Monitoring Water in the 1990's: Meeting New Challenges" (J. R. Hall and D. Glysson, eds.), pp. 483–498. American Society for Testing and Materials.

Leif, R. C., Vallarino, L., Harlow, P. M., and Cayer, M. L. (1993). *Cytometry*, **6**(Suppl.), Abstract 404B.

Lewis, G., Loutit, M. W., and Austin, F. J. (1986). *N. Z. J. Mar. Freshwater Res.* **20,** 431–437.

Lim, E. L., Amaral, L. A., Caron, D. A., and DeLong, E. F. (1993). *Appl. Environ. Microbiol.* **59,** 1647–1655.

Mahbubani, M. H., Bej, A. K., Perlin, M., Schaefer, F. W., Jakubowski, W., and Atlas, R. M. (1991). *Appl. Environ. Microbiol.* **57,** 3456–3461.

Manz, W., Szewzyk, U., Ericsson, P., Amann, R., Schleifer, K. H., and Stenstrm, T. A. (1993): *Appl. Environ. Microbiol.* **59,** 2292–2298.

McDaniel, J. A., and Capone, D. G. (1985). *J. Microbiol. Methods* **3,** 291–302.

Messenger, H. W. (1991). *Laser Focus World* **27,** 77–82.

Miller, R. A., Bronsdon, M. A., and Morton, W. R. (1986). *Ann. Meeting Am. Soc. Microbiol. (Washington)*, 48.

Mills, A. L., and Bell, P. E. (1986). *In* "Microbial Autoecology: A Method for Environmental Studies" (R. L. Tate, ed.) pp. 27–60. Wiley, New York.

Narai, J., Vesey, G., Champion, A. C., Piper, J., and Williams, K. (1993). *In* "Proceedings of the Australian Flow Cytometry Group Annual Symposium," 30 June, Melbourne, Australia.

Niederhauser, C., Candrian, U., Hoflein, C., Jermini, M., Buhler, H. P., and Luthy, J. (1992). *Appl. Environ. Microbiol.* **58,** 1564–1568.

Olson, R. J., Zettler, E. R., and DuRand, M. D. (1993). *In* "Handbook of Methods in Aquatic Microbial Ecology" (P. F. Kemp, B. F. Sherr, E. B. Sherr, and J. J. Cole, eds.), pp. 163–173. Lewis, Boca Raton, FL.

Ongerth, J. E., and Stibbs, H. H. (1987). *Appl. Environ. Microbiol.* **53,** 672–676.

Page, S., and Burns, R. G. (1991). *Soil Biol. Biochem.* **23,** 1025–1028.

Patchett, R. A., Back, J. P., Pinder, A. C., and Kroll, R. G. (1991). *Food Microbiol.* (*Lond.*) **8**(2), 119–126.

Patel, P. D., and Blackburn, C. (1991). *In* "Magnetic Separation Techniques Applied to Cellular and Molecular Biology" (J. T. Kemshead, ed.), pp. 93–106. Wordsworths' Conference Publications, Bristol, UK.

Pettipher, G. L. (1991). *Lett. Appl. Microbiol.* **12,** 109–112.

Pinder, A. C., and McClelland, R. G. (1993). *In "Proceedings of the Society for Applied Bacteriology 62nd Annual Meeting and Summer Conference."* pp. ix. University of Nottingham, 13–16 July, 1993. The Society for Applied Bacteriology, London.

Rendtorff, R. C. (1954). *Am. J. Hyg.* **59,** 209–220.

Robertson, B. R., and Button, D. K. (1989). *Cytometry* **10,** 70–76.

Rodriguez, G. C., Phipps, D., Ishoguro, K., and Ridgeway, H. F. (1992). *Appl. Environ. Microbiol.* **58,** 1801–1808.

Rose, J. B., Madore, M. S., Riggs, J. L., and Gerba, C. P. (1986). *In* "Proceedings of the AWWA Water Quality Technology Conference." pp. 417–424. American Water Works Association, Denver, CO.

Ryder, M. H., Pankhurst, C. E., Rovira, A. D., and Correll, R. L. (1994). *In* "Microbiol Ecology of the Rhizosphere" (O'Garra, F., Dowling, D., and Boestein, B., eds.).Weinheim, Germany, VCH.

Sauch, J. F., Flanagan, D., Galvin, M. L., Berman, D., and Jakubowski, W. (1991). *Appl. Environ. Microbiol.* **57,** 3243–3247.

Scheper, T., Hitzmann, B., Rinas, U., and Schugerl, K. (1987). *J. Biotechnol.* **5,** 139–148.

Schupp, D. G., and Erlandsen, L. S. (1987). *App. Environ. Microbiol.* **53,** 704–707.

Shapiro, H. M. (1988). "A Practical Guide to Flow Cytometry." A.R. Liss, New York.

Shapiro, H. M. (1990). *Am. Soc. Microbiol. News* **56,** 584–588.

Sharma, R. V., Edwards, R. T., and Beckett, R. (1993). *Appl. Environ. Microbiol.* **59,** 1864–1875.

Smith, H. V., and Rose, J. B. (1990). *Parasitol. Today* **6,** 8–12.

Tsai, Y. L., and Olson, B. H. (1992). *Appl. Environ. Microbiol.* **58,** 2292–2295.

Tsai, Y. L., and Palmer, C. J., and Sangermano, L. R. (1993). *Appl. Environ. Microbiol.* **59,** 353–357.

Venkateswaran, K., Shimada, A., Maruyama, A., Higashihara, T., Sakou, H., and Maruyama, T. (1993). *Can. J. Microbiol.* **319,** 506–512.

Vesey, G., and Slade, J. S. (1990). *Water Sci. Technol.* **24,** 165–167.

Vesey, G., Nightingale, A., James, D., Hawthorne, D. L., and Colbourne, J. S. (1990). *Lett. Appl. Microbiol.* **10,** 113–116.

Vesey, G., Slade, J. S., and Fricker, C. R. (1991). *Lett. Appl. Microbiol.* **13,** 62–65.

Vesey, G., Slade, J. S., Bryne, M., Shepherd, K., Dennis, P. J., and Fricker, C. R. (1993a). *J. Appl. Bacteriol.* **75,** 87–90.

Vesey, G., Slade, J. S., Byrne, M., Shepherd, K., and Fricker, C. R. (1993b). *J. Appl. Bacteriol.* **75,** 82–86.

Vesey, G., Slade, J. S., Byrne, M., Shepherd, K., Dennis, P. J. L., and Fricker, C. R. (1993c). *In* "Application of New Technology in Food and Beverage Microbiology." Society of Applied Bacteriology Technical Series No. 31. Blackwell Scientific, Oxford, UK.

Vesey, G., Hutton, P. E., Champion, A. C., Ashbolt, N. J., Williams, K. L., Warton, A., and Veal, D. A. (1994). *J. Cytometry*, **16,** 1–6.

Wagner, M., Amann, R., Lemmer, H., and Schleifer, K. H. (1993). *Appl. Environ. Microbiol.* **59,** 1520–1525.

Wegmuller, B., Luthy, J., and Candrian, U. (1993). *Appl. Environ. Microbiol.* **59,** 2161–2165.

Wernars, K., Heuvelman, C. J., Chakraborty, T., and Notermans, S. H. W. (1991). *Listeria J. Appl. Bacteriol.* **70,** 121–126.

Yentsch, C. M. (1990). *In* "Methods in Cell Biology" (Z. Darzynkiewicz and H. A. Crissman, eds.), Vol. **33,** pp. 572–612. Academic Press, San Diego, CA.

Zimmerman, R., Iturriaga, R., and Becker-Birch, J. (1978). *Appl. Environ. Microbiol.* **36,** 926–935.

CHAPTER 38

Flow Cytometric Analysis of Microorganisms

J. Paul Robinson★,†

★Department of Basic Medical Sciences
School of Veterinary Medicine
Purdue University Cytometry Laboratories
West Lafayette, Indiana 47907

†Weldon School of Biomedical Engineering
Purdue University
West Lafayette, Indiana 47907

I. Introduction

Conventional techniques (i.e., growth on laboratory media) employed for the detection and enumeration of microbes in clinical and environmental samples require time (24–48 h), and they have a strong bias in that these methods detect only organisms that grow under a selected set of conditions. Problems with the current technology for microbial cell analysis led to development of alternative techniques that include flow cytometry. Flow cytometry allows rapid, multiparameter data acquisition, and analysis of individual cells.

Although flow cytometry was rapidly accepted into hospital pathology and immunology laboratories, microbiology laboratories have remained essentially oblivious to the use of this technology. With few exceptions (Dubelaar *et al.*, 1999, Steen, 1980, 1983; Steen and Boye, 1980), flow cytometers were not designed to measure microorganisms, but rather mammalian cells in the range of 5–15 μm, the general size of blood cells. In practice, the measurement of smaller particles, while possible, often required modifications to the instrument or a greater understanding of and interest in the technological aspects of cytometry than generally possessed by those with expertise in clinical microbiology. In addition, clinical microbiologists generally found the technology expensive and inappropriate for their cells of interest.

Improvements in the sensitivity and specificity of flow cytometric instrumentation have made possible a wide range of techniques to rapidly characterize microbial populations. More importantly, microbiologists have started to recognize the potential of flow cytometry to study the responses of individual cells in environmental and clinical samples and to report their findings. An excellent review describing applications of flow cytometry in the field of microbiology has been published (Davey and Kell, 1996).

This chapter discusses experimental approaches that have been or could be used to study individual microbial cells using flow cytometry and key factors that may impact these studies, including instrument setup, instrument operation, and sample preparation. A brief discussion of flow cytometric applications to the field of medical and food microbiology is also included.

A. Instrument Setup for Microbes

Flow cytometers designed specifically for small particles (i.e., Bio-Rad Bryte HS, Hercules, CA; Skatron, Oslo, Norway) are no longer commercially available, and technical support for existing instruments is limited. An ordinary flow cytometer optimized for mammalian cells can be adapted for microbial cell analysis with a few simple changes in instrument setup and operation. For example, sheath fluid, sample buffer, media used to grow bacteria, and other reagents (i.e., dyes, antibodies) must be filtered (0.2-μm filter or smaller) to remove any particles that could interfere with bacterial measurements. The laboratory water system used to prepare sample buffer and sheath fluid should also be rigorously cleaned and maintained to prevent bacterial contamination.

Daily quality control procedures should include the instrument alignment beads recommended by the manufacturer and latex beads of size similar to that of the microbe of interest (1, 1.5, 2, 4, 6 μm). Because small latex beads can give a scatter signal quite different from that of bacteria of similar size, ethanol- or heat-fixed vegetative cells (e.g., *Escherichia coli*) or unfixed spores in water (e.g., *Bacillus subtilis*) should also be used as an internal laboratory standard to check light scatter parameters. Fixed cells or spores can be stored at 4 °C for up to 6 months.

Initial light-scatter parameters should be established using target microorganisms spiked with latex beads. For example, in Fig. 1 *B. subtilis* cells were spiked with a small number of 1-μm beads. Bacteria and beads in the spiked sample were separated using a dual-parameter histogram of log forward scatter (FS) and log side scatter (SS). A region was established for the bacterial population and used as a gating parameter to exclude cell aggregates and debris from further analysis. Sterile-filtered sample buffer was used to set the discriminator or threshold on forward light scatter to eliminate background particles.

In order to reduce the risk associated with analyzing potentially hazardous microorganisms, certain protective measures should be followed and strictly enforced. In particular, the protective doors that shield the instrument sample probe should be kept closed to reduce aerosolization of bacterial particles, bleach should be added to the waste container to kill any harmful organisms, and personal protective gear (i.e., gloves, mask, laboratory coat) should be worn at all times. Laboratory personnel should also avoid contaminating the computer keyboard and mouse with bacteria. Instrument maintenance should include frequent flushing of the system between samples to reduce instrument carryover of bacteria and dyes and rigorous shutdown/cleaning procedures.

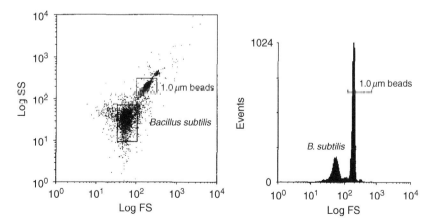

Fig. 1 Light-scatter measurements of a mixture of *Bacillus subtilis* cells and latex beads. Fluorescent microspheres (1 μm) were added as an internal standard.

B. Sample Preparation

Biological characteristics of bacteria such as size, shape, DNA, RNA, and protein content can change depending on growth conditions and cell source. For example, exponentially growing cells are larger than dormant or starved cells and contain considerably higher levels of nucleic acids. Growing cells have a wide light scatter distribution with a comet-like tail in the direction of increasing scatter (Thomas *et al.*, 1997). Prior to flow cytometric analysis, growing cells should be washed with sterile-filtered buffer to remove debris and reduce cell clumping (Fig. 2). A washing step will also remove medium that may interfere with staining. Bacteria can grow as single cells or in pairs, chains, or clusters. Gentle pipetting or vortexing may be necessary to disrupt the chains or clusters and form a single-cell suspension.

Some bacteria have considerable permeability barriers (i.e., cell walls, endospores, capsules, efflux pumps) to fluorescent dyes or DNA probes and may require use of fixatives or EDTA. However, sample preparation methods necessary for efficient penetration of a fluorochrome into target cells may significantly affect light scatter profiles. For example, alcohol fixation can cause considerable cell shrinkage and a reduction in cell size.

II. Experimental Approaches

The basic problem in developing flow cytometric protocols for microbial cell analysis is the assumption that procedures developed and optimized for mammalian cells will work for bacteria. In some cases, ignorance of traditional flow methods is an advantage; however, the fundamentals of microbiology must always be understood. In this chapter, we have outlined a few experimental approaches for using flow cytometry to study microbial cells. These approaches included generic

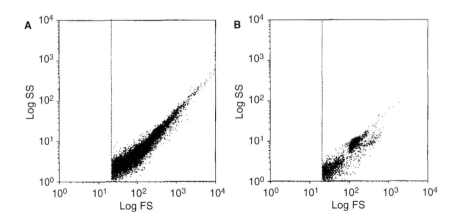

Fig. 2 Changes in light-scatter profiles of bacteria due to sample preparation: (A) *Bacillus subtilis* spore slurry in water and (B) washed spores.

detection of microorganisms, specific identification of target organisms, cell viability determinations, and Gram staining.

A. Detection of Microbes

Nucleic acid dyes can be combined with light-scatter measurements to detect bacteria using flow cytometry. A detailed discussion of bacterial DNA appears in this volume. The dye selected for a detection assay should have high specificity for DNA binding, high extinction coefficient, and high quantum yield. Depending on the available excitation source, 4',6-diamidino-2-phenylindole (DAPI), Hoechst 33258, propidium iodide, YO-PRO-1, or YOYO-1 could be used for a rapid detection assay. YOYO-1 (Molecular Probes, Inc., Eugene, OR) is a membrane-impermeant cyanine dye (excitation 491 nm, emission 509 nm) that is essentially nonfluorescent unless bound to nucleic acids. Dyes that are membrane impermeant will stain only cells that are dead or have compromised membranes. Live cells must be fixed for the dye to pass through the membrane. Rapid fixation with ice-cold 70% ethanol will ensure that the selected dye will enter all cells in the sample and bind to nucleic acids. Alcohol fixation will cause some cell shrinkage and prevent further studies regarding cell viability.

Figure 3 is an example of a rapid detection assay. A "bacteria" region (region F) was created using *E. coli* cells fixed 5 min with ice-cold 70% ethanol. Cells were washed briefly with filtered 0.8% NaCl, stained 5 min with 0.1 μM YOYO-1 in the dark, and then analyzed using flow cytometry (Sincock *et al.*, 1996a). YOYO–1–stained *E. coli* cells were gated on region F; the fluorescence of the gated

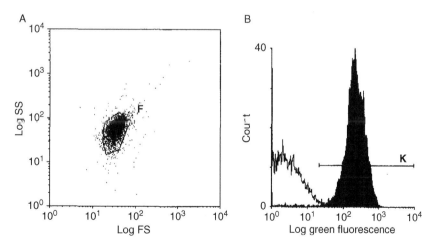

Fig. 3 Detection of *E. coli* cells in environmental samples using YOYO-1 nucleic acid stain. (A) Light scatter measurements of ethanol-fixed *E. coli* cells (region F). (B) Fluorescence histogram overlay of YOYO-1 stained (■) and unstained *E. coli* cells (□).

population was then measured and displayed as a histogram with fluorescence intensity on the *x*-axis and the number of cells on the *y*-axis. The background noise was determined using 0.2 μm-filtered 0.8% NaCl.

Samples containing dust, pollen, fungal spores, or unknown bacteria were tested with this assay. Particles in the test samples that met the light scatter requirements (bacteria region) and stained positive for nucleic acids were classified as bacteria. Although pollen, mold, and fungal spores contain nucleic acids and will stain with YOYO-1, they do not meet the light scatter gating requirements owing to their large size (20–100 μm) and therefore can be excluded. YOYO-1 will not stain dust particles that fall within the bacteria region because they do not contain nucleic acids.

This assay can be used for generic detection of bacteria within a heterogeneous sample but cannot be used for specific identification. In mixed populations of bacteria (i.e., *E. coli* combined with *Staphylococcus aureus*), it was not possible to discriminate between bacteria using light-scatter or using differences in relative staining.

B. Identification of Specific Microorganisms

During the 1990s, experimental approaches to identifying microorganisms in liquid samples using flow cytometry included light-scatter profiles, DNA content, immunoassays, neural nets, and rRNA probes, with varying degrees of success. A brief introduction to this material follows.

1. Light Scatter Measurements

Light-scattering profiles are a function of cellular size, shape, and refractive index of a cell. Morphological features of bacteria that can influence light-scatter profiles include shape (rods, cocci, vibrios, spirilla, spirochetes), flagella, pili, and capsules. Growth conditions, cell source, and responses to stress (e.g., starvation, antimicrobial exposure) can also influence light-scatter profiles.

Light-scatter profiles are a useful first step in characterizing microorganisms. Identifying specific organisms within mixed populations is difficult. Allman *et al.* (1993) collected dual-parameter contour plots of forward versus side scatter for artificial mixtures of clinically relevant microorganisms using an arc lamp–based cytometer. Mixtures of vegetative cells (i.e., *Salmonella typhimurium, Legionella pneumophila, S. aureus*) had overlapping light-scatter profiles. However, light-scatter profiles could be used to resolve spore-forming bacteria (e.g., *Clostridium perfringens*) from vegetative cells. Spores give a forward light-scatter signal that is out of proportion to their size, which may be explained on the basis of a high value for their refractive index (Allman *et al.*, 1993). Using a cytometer specifically designed for small particles (Bio-Rad Bryte HS), light-scatter profiles could also be used to resolve populations of closely related Gram-positive spores

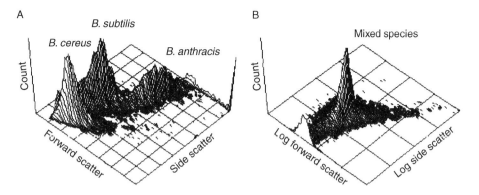

Fig. 4 Isometric plots of forward scatter versus side scatter for artificial mixtures of bacterial species: (A) mixture of three closely related *Bacillus* spores (Bio-Rad Bryte flow cytometer) and (B) mixture of the same three spores (Coulter EPICS XL flow cytometer).

(Sincock *et al.*, 1996b) (Fig. 4A). A laser-based commercial cytometer (Coulter EPICS XL, Hialeah, FL) could not resolve a mixture of *E. coli, S. aureus,* and *B. subtilis* spores using light scatter measurements (Fig. 4B).

2. DNA Content

DNA base composition of individual cells within a bacterial sample can be determined without extraction of DNA using flow cytometry. Van Dilla *et al.* (1983) used a combination of DNA-specific fluorochromes to analyze six species of ethanol-fixed bacteria with differing DNA base composition. Using a combination of Hoechst 33258, a fluorochrome that binds preferentially to the regions of DNA rich in AT base pairs, and chromomycin A3, a fluorochrome that binds preferentially to regions of DNA rich in GC base pairs, this group established a direct relationship between the fluorescence dye ratio calculated by flow cytometry and the %guanine + cytosine (%[G + C]) content. This method was able to resolve individual species within an artificial mixture of *S. aureus, E. coli,* and *Pseudomonas aeruginosa* based on differences in DNA content. Each species formed a distinct cluster within the dual-parameter fluorescence histogram. However, further studies using this method suggested that flow cytometric determination of % [G + C] be limited to samples containing only one bacterial species (Sanders *et al.*, 1990).

3. Immunofluorescence Approach

Measurement of fluorescently labeled antibodies combined with light scatter can be used for the specific identification of microorganisms. Microbes that have been identified using a flow cytometric immunoassay include pathogenic microorganisms found in food, water, sewage, and aerosols (Table I). An example of a direct flow

Table I
Examples of Microbes Identified Using a Flow Cytometric Immunoassay

Microbe	References
Food	
Escherichia coli O157:H7	Seo *et al.* (1998a,b), Tortorello *et al.* (1998)
Listeria monocytogenes	Donnelly and Baigent (1986), Pinder and McClelland (1994),
Salmonella typhimurium	Clarke and Pinder (1998), McClelland and Pinder (1994b),
	Pinder and McClelland (1994),
Salmonella serotypes	McClelland and Pinder (1994a)
Oral bacteria	
Streptococcus mutans and	Barnett *et al.* (1984)
Actinomyces viscosus	
Streptococcus pyogenes	Sahar *et al.* (1983)
Dental plaque	Obernesser *et al.* (1990)
Aerosols	
Francisella tularensis	Henningson *et al.* (1998)
Water and sewage	
Legionella pneumophila	Ingram *et al.* (1982)
Nitrosomonas serotypes	Volsch *et al.* (1990)
Salmonella spp.	Desmonts *et al.* (1990)
Fecal bacteria	Apperloo-Renkema *et al.* (1992), van der Waaij *et al.* (1994)
Cryptosporidium parvum	Arrowood *et al.* (1995), Valdez *et al.* (1997), Vesey *et al.* (1993, 1997)
Giardia spp.	Bruderer *et al.* (1994), Dixon *et al.* (1997)
	Heyworth and Pappo (1989)
Biowarfare agents	
Bacillus anthracis	Phillips and Martin (1983, 1988), Sincock *et al.* (1996b)
Cell surface polysaccharides or proteins	
Bacteroides fragilis	Lutton *et al.* (1991)
E. coli lipopolysaccharide expression	Nelson *et al.* (1991)
Myxococcus virescens	Martinelli *et al.* (1995)
Pseudomonas aeruginosa outer membrane protein	Hughes *et al.* (1996)
Microsphere-based immunoassays	
Helicobacter pylori	Best *et al.* (1992)
E. coli O157:H7	Seo *et al.* (1998a,b)

cytometric immunoassay can be found in Fig. 5. *E. coli* O157:H7 cells at a concentration of 10^6 cells/ml were incubated with a fluorescein isothiocyanate (FITC)-conjugated rabbit anti-*E. coli* O157:H7 polyclonal antibody 5 min at room temperature and analyzed by flow cytometry. For this assay, the desired population of cells was selected by gating on light-scatter signals. A discriminator was set on forward scatter and used to resolve bacteria from noncellular material and electronic noise. The fluorescence of the gated population was then measured and displayed as a histogram with fluorescence intensity on the *x*-axis and the number of cells on the *y*-axis (5000 counts). Target *E. coli* O157:H7 cells were identified and enumerated within a few minutes of obtaining the sample. Culturing of the target organism was not necessary for identification in this direct immunoassay.

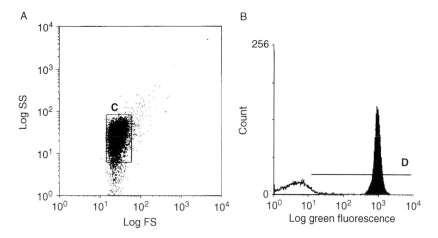

Fig. 5 (A) Dual-parameter histogram of forward versus side scatter for live *E. coli* O157:H7 cells stained with a FITC-labeled anti-*E. coli* O157:H7 polyclonal antibody. (B) Comparison of stained (■) and unstained (□) *E. coli* O157:H7 cells. Region D represents bacteria stained positive with FITC-labeled antibody. Cells were gated on region C (5000 events).

Low numbers of target organisms can be identified in the presence of large numbers of nontarget organisms or high levels of background particulate material using nonselective media. Unknown samples are incubated in nonselective media for a short period of time, washed with buffer, and then stained with a specific antibody. Flow cytometry can be used to discriminate target organisms from nontarget organisms by means of specific antibody binding. Enrichment media specific to the nutritional requirements of the target organism can also be used. Ideally, only the target organism will grow. Nutritional supplements can also be used to facilitate expression of specific polysaccharides on bacterial cell surfaces that can be used to discriminate between closely related species. For example, viable *Bacillus anthracis* spores were identified after a brief incubation (20 min, 37 °C) in media selected to stimulate the outgrowth and expression of specific polysaccharides on the surface of vegetative cells. After exposure to the food source, *B. anthracis* spores were able to sporulate and transition to vegetative cells. Cells were stained with FITC-conjugated monoclonal antibody specific for *B. anthracis* cell-wall polysaccharide and analyzed using flow cytometry (Fig. 6) (Sincock *et al.*, 1996b).

Because cell fixation is not necessary for antibody binding, the immunofluorescence approach can be combined with certain stains to identify viable target organisms. For example, the survival ratio of *Francisella tularensis,* the causative agent of tularemia, was determined before and after aerosolization using a specific anti–*F. tularensis* monoclonal antibody to identify the target organisms together with rhodamine 123 to count the number of viable or metabolically active cells (Henningson *et al.*, 1998). In a second example, Red613-conjugated

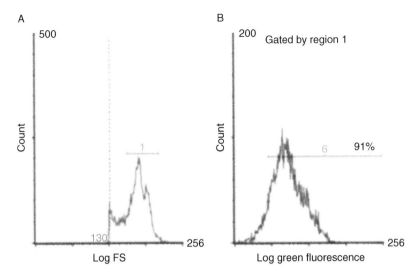

Fig. 6 Flow cytometric analysis of *B. anthracis* spores incubated 30 min in polysaccharide medium at 37 °C. A specific FITC-conjugated anti-cell wall polysaccharide MAb stained emerging vegetative *B. anthracis* cells (region 6) (viable cells) but did not stain encapsulated *B. anthracis* cells, dormant *B. anthracis* spores, or other *Bacillus* species. Forward light scatter was used to identify vegetative cells population in the sample (region 1). (See Plate no. 22 in the Color Plate Section.)

anti-*S. typhimurium* monoclonal antibody combined with Chemchrome, a live cell stain, was used to detect viable *S. typhimurium* cells in the presence of large number of nontarget and dead organisms (Clarke and Pinder, 1998).

Flow cytometric immunoassays are rapid (usually <10 min), sensitive (lower limit of 10^3 cells/ml), specific, require very little sample preparation, and need no cell fixation. Immunoassay-based methods rely on the specificity and sensitivity of the selected antibody to identify the target organism. Unfortunately, very few good antibodies for microbes are commercially available, and in-house antibody development can be time-consuming and expensive. High levels of nonbacterial particles, bacterial debris, and antibody aggregates in the test sample can also produce sensitivity problems.

4. Automated Classification and Identification Techniques

Flow cytometry can be used to generate multiparameter data for individual cells. However, the vast quantity of information generated can make data analysis difficult. Artificial neural networks are computing technologies that can be used to discriminate between different cell types based on flow cytometry data (Boddy and Morris, 1993). A computer is "taught" how to recognize data patterns (i.e., staining profiles of different organisms) and to analyze cell populations using examples. Eventually, the neural network can identify specific cell types in real time and adapt to changing conditions (Frankel *et al.*, 1989). Artificial neural networks have been developed for chromosome classification (Errington and Graham, 1993), leukemia subsets (Maguire *et al.*, 1994a,b), and phytoplankton populations (Frankel *et al.*, 1989). Davey *et al.* (1999) developed an artificial neural network for detection and identification of *Bacillus globigii* spores against a background of other microorganisms (*E. coli, Micrococcus luteus, Saccharomyces cerevisiae*). Datasets were collected for microorganisms stained with six cocktails of fluorescent stains. These stains included Tinopal CBS-X, Nile Red, propidium iodide, FITC, $DiSC_2(5)$, Oxonol V, SYTO 17, and TOPRO-3. Forward scatter, side scatter, and autofluorescence measurements were also included in the datasets. Careful selection of the staining cocktail and data analysis method allowed accurate identification of the target organism (*Bacillus* spores). Trained neural networks may be useful in identifying specific organisms against a high background of particulate matter or discriminating between closely related organisms in real time. Applications could include food analysis, clinical microbiology samples, and identification of biowarfare agents.

C. Cell Viability

Fluorescent dyes have been successfully used as indicators of cell viability in fluorescence microscopy and flow cytometry. With these dyes, live and dead cells within a heterogeneous sample population can be identified and counted within a few minutes. Traditional methods employed to detect and enumerate bacteria (such as growth on laboratory media) require time (24–48 h) and may underestimate the number of viable bacteria. Therefore, direct methods for the assessment of microbial viability are of increasing importance. Because each technique has its limitations, each investigator must choose the experimental approaches that are best suited for the test organisms and the specific questions being asked.

1. Membrane Integrity

Membrane-integrity analysis is based on the capacity of bacterial cells to exclude certain compounds. Stains that are commonly used to determine membrane integrity include ethidium bromide, propidium iodide, and SYTOX Green dead-cell stain. These dyes passively enter stressed, injured, or dead cells via damaged membranes and intercalate into DNA and RNA. The fluorescence indicates a loss of viability or membrane integrity. Flow cytometry can be used to quantify the fluorescence associated with dead or injured cells. Because the influx of the dye can be correlated with the extent of the bacterial wall permeability, the number of fluorescent cells counted using flow cytometry is inversely proportional to the number of viable cells. These dye-exclusion methods have been successfully used to monitor antibiotic-induced changes in bacterial membrane permeability (Gant *et al.*, 1993). For example, the oral pathogen *Streptococcus mutans* was treated with the antibiotic clindamycin and then stained with SYTOX Green (Fig. 7). After 2 h

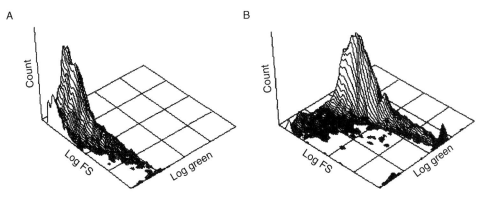

Fig. 7 Isometric plots of log green fluorescence versus log forward scatter for *Streptococcus mutans* cells stained with SYTOX Green after exposure to clindamycin (10× MIC) for (A) 0 h and (B) 2 h.

of exposure to the antibiotic, a significant number of cells were dead, as indicated by strong green fluorescence.

Membrane-integrity analysis is not suitable for all cell types because some bacteria can rapidly pump out dyes using an efficient efflux pump (Jernaes and Steen, 1994). In this case, damaged or injured cells would not fluoresce and would be counted as viable.

2. Membrane Potential

Membrane-potential analysis is based on the selective permeability and active transport of charged molecules through intact membranes. Cells with a membrane potential actively take up lipophilic cationic dyes or actively exclude lipophilic anionic dyes. Using flow cytometry, any particle in the approximate size range of bacteria that is found to have a membrane potential can be identified as a viable organism. However, organisms can show considerable variation in dye uptake owing to differences in membrane potential (Allman *et al.*, 1993).

Using the lipophilic cation rhodamine 123, which preferentially accumulates within viable cells, several groups have been able to discriminate between live, dead, and dormant cells in culture. Viable and nonviable cells have been enumerated using flow cytometry (Kaprelyants and Kell, 1992, 1993a,b; Kaprelyants *et al.*, 1993). Studies using this dye have determined that dye uptake is variable both between species and among cells from the same culture (Porter *et al.*, 1995). Furthermore this dye can be used for Gram-negative bacteria only after they have been treated with EDTA (Diaper *et al.*, 1992).

In contrast to rhodamine 123, the lipophilic oxonol dyes are anionic and preferentially accumulate within dead bacteria; they have been used to assess bacterial antibiotic susceptibility (Deere *et al.*, 1995; Mason *et al.*, 1995b) and cell viability (Jepras *et al.*, 1995; Mason *et al.*, 1995a) by flow cytometry. In these studies, either heat or bactericidal antibiotics were used to kill cells prior to oxonol staining, and comparisons were made with untreated cells. Figure 8 is a fluorescence histogram overlay of *E. coli* cells treated with gentamicin at 10 times the mimimum inhibitory concentration (10 × MIC) and then stained with bis (1,3-dibutylbarbituric acid) trimethine oxonol (DiBAC$_4$(3)). Over time, the number of dead cells increased, as indicated by an overall shift in green fluorescence.

3. Enzymatic Activity

Flow cytometric detection of intracellular enzymatic activity utilizes lipophilic, uncharged, nonfluorescent derivatives such as fluorescein diacetate (FDA) that readily diffuse across cell membranes. Once inside the cell, the derivative is hydrolyzed by nonspecific esterases to release the highly fluorescent parent compound. Because the parent compound is polar and charged, it is retained inside cells with intact membranes. Dead or dying cells with compromised membranes rapidly leak the dye.

Flow cytometry can be used to detect the number of viable bacteria and to verify the metabolic activity of these cells (Diaper and Edwards, 1994; Diaper *et al.*, 1992). However, FDA does not efficiently penetrate some types of membranes and the fluorescein product tends to leak from cells or can be actively pumped out (Edwards, 1996). Other related fluorescent compounds such as carboxyfluorescein diacetate (CFDA) and sulfofluorescein diacetate (SFDA) exhibit similar problems (Tsuji *et al.*, 1995).

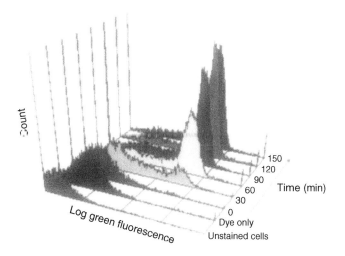

Fig. 8 Fluorescence histogram overlays of *E. coli* cells stained with DiBAC$_4$(3) (oxonol) after exposure to 10× MIC gentamicin. Cells were stained with oxonol after 0, 30, 60, 90, 120, or 150 min of drug incubation.

4. Bacterial Respiration

The redox dye 5-cyano-2,3-ditolyl tetrazolium chloride (CTC) was first employed for the direct microscopic enumeration of respiring bacteria in environmental samples. CTC is readily reduced via electron transport activity to insoluble, highly fluorescent, and intracellularly accumulated CTC-formazan through bacterial respiration. Actively respiring bacteria (red fluorescence) can be distinguished from nonrespiring bacteria and abiotic material. More recent studies have used flow cytometry to enumerate respiring bacteria in lakes (del Giorgio et al., 1997), marine systems (López-Amorós et al., 1998), and after exposure to antibiotics (Suller and Lloyd, 1999).

However, several problems associated with using CTC have been identified. The CTC assay may not be sensitive enough to detect low respiration rates of microorganisms, especially in very small bacteria. In addition, not all bacteria are able to reduce tetrazolium salts. It is also thought that CTC may have an inhibitory effect on bacterial metabolism (Ullrich et al., 1996; Yu et al., 1995).

D. Identification of Viable Bacteria with Fluorescent *In Situ* Hybridization

Staining with membrane-integrity or membrane-potential fluorochromes offers limited information on the numbers of viable bacterial cells within a sample and none at all about their identity. Staining with fluorochromes that preferentially bind to specific DNA base pairs offers limited information on species identification but not cell viability. Fluorescent *in situ* hybridization can be used to label specific nucleic acid sequences inside intact, viable cells, and identify species of bacteria present in the sample. Probe binding to ribosomal RNA (rRNA) is perhaps the best target for bacterial cells.

rRNA can be found in all bacteria and consists of both highly conserved and variable regions. Synthetic probes have been developed that can target sections of the rRNA based on the amount of conserved and variable regions. Appropriate probes can be composed of oligonucleotide sequences that distinguish between the primary kingdoms (eukaryotes, eubacteria, archaebacteria) and between closely related organisms (DeLong et al., 1989). Probes that target very conserved regions can be used as universal probes to measure total rRNA within a sample (Amann et al., 1990).

The rRNA content of microorganisms is proportional to the growth rate in pure culture. Using microfluorimetry, DeLong et al. (1989) quantified the binding of a universal rRNA probe to *E. coli* cells grown in media that support different growth rates. The fluorescence intensity of single cells owing to hybridization with the universal probe varies linearly with growth rate and can be used to estimate the growth rate of that particular organism in a natural population. Further studies conducted by Wallner et al. (1993) demonstrated that 16S rRNA probe conferred fluorescence is directly proportional to ribosome content. Because the amount of fluorescence can be correlated with cellular rRNA content, it is possible to obtain information on the physiological state (i.e., growth rate, activity, viability) of specific bacterial cells (Manz *et al.*, 1993; Wallner *et al.*, 1993). Owing to the abundance of cellular ribosomes in rapidly growing cells (approximately 10^4–10^5 per cell), the binding of fluorescent probes to individual cells can be readily visualized (DeLong *et al.*, 1989).

After appropriate selection, rRNA-targeted oligonucleotides can be sequenced, labeled with an appropriate fluorochrome, and used as probes in hybridization experiments. After hybridization, the fluorescence conferred by rRNA-targeted oligonucleotide probes can be analyzed by flow cytometry (Amann *et al.*, 1990; Lange *et al.*, 1997; Rice *et al.*, 1997; Simon *et al.*, 1995; Thomas *et al.*, 1997; Wallner *et al.*, 1993, 1995) or confocal microscopy (Amann *et al.*, 1996).

E. Gram Stain

Gram staining is the most commonly used procedure in clinical microbiology laboratories. Specimens are smeared on glass slides, heat fixed, Gram stained, and examined microscopically. Based on the outcome of the Gram reaction, bacteria are divided into two taxonomic groups. Cells stained purple-blue are Gram-positive; cells stained red are Gram-negative. This technique is relatively simple, albeit messy. However, some organisms can show gram variability (i.e., *Acinetobacter* species), particularly anaerobes.

Sizemore *et al.* (1990) reported on the use of a fluorescently labeled lectin as an alternative Gram staining technique. Lectin isolated from *Triticum vulgaris,* or wheat germ agglutinin (WGA), will bind specifically to *N*-acetylglucosamine in the outer peptidoglycan layer of Gram-positive bacteria. Gram-negative bacteria have an outer membrane covering the peptidoglycan layer that prevents lectin binding. Using this method, heat-fixed bacterial smears were covered with a small aliquot of FITC-conjugated WGA (100 μg/ml), washed briefly with phosphate buffer, and observed using fluorescence microscopy. Unlike the Gram staining method, culture age did not affect lectin binding, suggesting that this technique can be used directly on samples without culturing and may offer an alternative method to classify fastidious, slowly growing, or viable but nonculturable organisms. In theory, flow cytometry could be used to extend this technique.

Flow cytometry has been used to determine the Gram stain of unfixed cells using $DiIC_1(5)$ (Shapiro, 1995) or rhodamine 123 (Allman *et al.*, 1993). More recently, Mason *et al.* (1998) developed a two-color flow assay for mixed populations of bacteria in suspension. Bacterial strains isolated from clinical specimens were cultured overnight, washed, and then stained with a combination of fluorescent nucleic acid–binding dyes: hexidium iodide (excitation 488 nm, emission 605 nm) and SYTO 13 (excitation 488 nm, emission 509 nm). Hexidium iodide (HI) preferentially penetrates Gram-positive bacteria, whereas SYTO 13 enters both Gram-positive and Gram-negative bacteria. When used in combination, these dyes allow differential labeling of unfixed Gram-positive bacteria (HI and SYTO 13, red-orange fluorescence) and Gram-negative bacteria (SYTO 13 only, green fluorescence) in suspension (Mason *et al.*, 1998). Using this method, artificial mixtures of *E. coli* and *S. aureus* cells analyzed using flow cytometry were clearly separated using fluorescence. Total time needed for this assay was 15 min.

III. Applications in Medical and Food Microbiology

To date, the most frequent application of flow cytometry to the study of microorganisms is the field of environmental microbiology, where rapid assessment of bacterial viability in natural samples is important. The rapid methods first described in these studies have been adapted for use in medical and food microbiology. In these areas, flow cytometry can significantly shorten the analysis time required for detection and identification of bacteria compared with conventional detection procedures and provide additional information on responses of individual cells.

A. Antimicrobial Agents

Flow cytometry permits rapid analysis of individual bacterial, fungal, or protozoan responses to antimicrobial agents. Antimicrobial agents such as antibiotics, disinfectants, and antiseptics are used to reduce the number of microorganisms to a level that is insufficient to transmit infection. Antibiotics are products of the metabolism of a microorganism that are inhibitory to other microorganisms. Disinfectants are chemical or physical agents used to kill pathogenic microorganisms on nonliving objects (e.g., sink, table); antiseptics are chemicals used to kill microbes on a living object (e.g., skin, mouth). Flow cytometry can be used to investigate physiological and morphological changes that can occur after drug exposure, even if little is known about a particular antimicrobial agent.

1. Exposure to Antibiotics

Clinical microbiology laboratories devote a great deal of resources to antibiotic susceptibility testing. Routine analysis is limited to growth inhibition assays using fast growing, nonfastidious bacteria. Flow cytometry can supply valuable additional information on the response of individual cells to antibiotic exposure within a short period of time and provide an indication of population dynamics within the heterogeneous test sample. For example, gentamicin was added to early exponential-phase *E. coli* cells in broth and incubation was allowed to continue for 5 h. Untreated *E. coli* cells were used as controls. At timed intervals, aliquots of treated and untreated cells were removed, stained, and analyzed. Membrane perturbation was assessed using the membrane potential–sensitive dye $DiBAC_4(3)$ and the membrane-integrity dye propidium iodide. Dual-parameter histograms of log forward scatter versus log fluorescence suggest that membrane potential of the treated cells collapsed after 5 h; however, a subpopulation of treated cells maintained membrane integrity (Fig. 9).

Table II is a brief summary of work using flow cytometry to investigate the effect of antibiotic and antifungal agents on target organisms. Procedures for antibiotic susceptibility testing using flow cytometry are described in detail in this volume.

2. Exposure to Disinfectants or Antiseptics

Traditional assessment of disinfectant efficacy involves the incubation of microbes in liquid or on solid media for 24–48 h. Most bacteria will not grow in the presence of low concentrations of disinfectants. To avoid this inhibitory effect, disinfectant compounds must be inactivated or neutralized before treated cells are incubated in media or plated. In addition, some cells will experience a lag of regrowth, similar to the postantibiotic effect, after exposure to disinfectants. For example, chlorhexidine delays regrowth after exposure for more than 2 h.

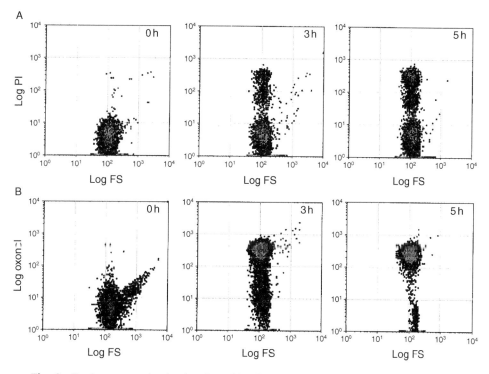

Fig. 9 Dual-parameter density dot plots of log forward scatter versus log fluorescence for *E. coli* cells stained with (A) propidium iodide (PI) or (B) DiBAC₄(3) (oxonol), after exposure to gentamicin for 0, 3, and 5 h.

Table II
Rapid Antimicrobial Susceptibility Testing Using Flow Cytometry

Bacteria: dye	Species	Antibiotic	References
Ethidium bromide (EtBr)	E. coli, P. aeruginosa, S. aureus, P. mirabilis, S. pyogenes	Amikacin	Cohen and Sahar (1989)
EtBr + mithramycin	E. coli	Ceftazidime, ciprofloxacin, gentamicin	Walberg et al. (1997a) Walberg et al. (1997b)
	E. coli, K. pneumoniae	Ampicillin	Mason and Lloyd (1997)
Acridine orange	E. coli	Gentamicin	Gant et al. (1993)
Propidium iodide (PI)	E. coli	Gentamicin, mecillinam, cefotaxime, ampicillin, ciprofloxacin	
	E. coli, P. aeruginosa	Ampicillin, ceftriaxone, ciprofloxacin, rifampin, imipenem	Gottfredsson et al. (1998)
Live/Dead BacLight	Propionibacterium acnes	Lymecycline, minocycline	Arrese et al. (1998)
	L. monocytogenes	Bacteriocin	Swarts et al. (1998)
SYTOX Green	E. coli, S. aureus, P. aeruginosa	Ceftazidime, ampicillin, vancomycin	Suller and Lloyd (1999)
	E. coli, S. aureus, B. cereus	Ampicillin, amoxicillin, penicillin G, vancomycin	Roth et al. (1997)
Rhodamine 123	E. coli, S. aureus, P. aeruginosa	Ceftazidime, ampicillin, vancomycin	Suller and Lloyd (1999)
	E. coli, P. fluorescens, E. aerogenes, A. globiformis	Valinomycin	Porter et al. (1995)
DiBAC₄(3) (oxonol)	E. coli	Azithromycin, cefuroxime, ciprofloxacin	Jepras et al. (1997)
	E. coli, S. aureus, P. aeruginosa	Gramicidin S	Jepras et al. (1995)
	S. aureus	Methicillin	Suller and Lloyd (1998), Suller and Lloyd (1997)
	E. coli, S. aureus	Ampicillin, gentamicin, ciprofloxacin	Mason et al. (1994)
	Aeromonas salmonicida	Gentamicin	Deere et al. (1995)
	E. coli, S. aureus, P. aeruginosa	Ceftazidime, ampicillin, vancomycin	Suller and Lloyd (1999)
Fluorescein diacetate (FDA)	Mycobacterium tuberculosis	Ethambutol, isoniazid, rifampin	Kirk et al. (1998)
CTC	S. aureus	Methicillin	Suller and Lloyd (1998)
	E. coli	Ciprofloxacin	Mason et al. (1995b)
	E. coli, S. aureus, P. aeruginosa	Ceftazidime, ampicillin, vancomycin	Suller and Lloyd (1999)
FITC	E. coli	Amoxycillin, mecillinam, chloramphenicol, ciprofloxacintrimethoprim	Durodie et al. (1995)
Yeast: dye	**Species**	**Antifungal agent**	**References**
Propidium iodide	Candida albicans, S. cerevisiae, Cryptococcus neoformans	Amphotericin B, fluconazole, cilofungin	Green et al. (1994)
	C. albicans, C. krusei, C. parapsilosis	Amphotericin B, fluconazole	Ramani et al. (1997)
Ethidium bromide	Candida spp.	Amphotericin B	O'Gorman and Hopfer (1991)
DiOC₅(3) (carbocyanine dye)	Candida spp., T. glabrata	Amphotericin B	Peyron et al. (1997)
	C. albicans, C. tropicalis	Amphotericin B	Ordóñez and Wehman (1995)
FUN-1	C. albicans	Amphotericin B, flucytosine, fluconazole, ketoconazole	Wenisch et al. (1997)

Flow cytometry combined with fluorescent probes allows the activity of disinfectant compounds on target organisms to be ascertained within a few minutes and provides information on the heterogeneity of the sample population.

Sheppard *et al.* (1997) used oxonol and propidium iodide to monitor chlorhexidine-induced membrane damage in stationary and log-phase *E. coli* cells. Their results indicated that membrane potential (oxonol) of cells collapsed prior to loss of membrane integrity (propidium iodide). Increased light-scattering properties of organisms exposed to higher chlorhexidine concentrations suggest that there are also major changes to internal cellular structure. Comas and Vives-Rego (1997) used rhodamine 123, bis-oxonol, propidium iodide, SYTO-13, and SYTO-17 to assess the effect of formaldehyde and surfactants (e.g., sodium dodecyl sulfate (SDS), benzalkonium chloride) on *E. coli*.

Paul *et al.* (1996) used oxonol to determine the effectiveness of oral antiseptics found in mouthwash and toothpaste to kill bacteria such as *S. mutans, Streptococcus sanguis,* and *Streptococcus oralis* that cause tooth decay and gum disease. Membrane-potential damage after 30-s exposure to triclosan, chlorhexidine, or cetylpyridinium chloride at $5\times$ MIC was assessed using flow cytometry and compared to plate-count data. Flow cytometry provided information within minutes on the immediate effect of oral antiseptics on target bacteria; plate-count data required 24–48 h.

In our laboratory, we have developed a rapid flow cytometric assay to evaluate alternative disinfectant processes. Outbreaks of cryptosporidiosis have been attributed to the inability of chlorine to inactivate the oocyst form of *Cryptosporidium parvum*. Gamma (γ) irradiation may be a viable alternative to conventional chlorine-based wastewater disinfection processes.

Purified *Cryptosporidium parvum* oocysts were exposed in batch reactors to γ-irradiation from a ^{60}Co source. Exposures to γ-irradiation ranged from 50 to 800 krad. Untreated oocysts, heat-killed (70 °C for 30 min) control oocysts, and irradiated oocysts were stained with SYTOX Green dead cell stain (10 μM final concentration), incubated 1 h at 37 °C, and counted using flow cytometry (Fig. 10). Differences in light-scattering properties were used to differentiate oocysts from sporozoites, ghosts (oocyst shells), and debris. After exposure to γ-irradiation, the oocysts were morphologically intact, but the process damaged the oocyst wall and allowed SYTOX Green, a membrane-integrity stain, to enter and bind to nucleic acids. Nonviable oocysts with damaged but intact walls fluoresced bright green; viable oocysts and ghosts did not stain. Flow cytometry was used to count the number of damaged or inactivated oocysts after disinfectant exposure (Sincock *et al.*, 1998).

B. Food and Drink

Flow cytometry has been used to detect and identify pathogenic microorganisms in food samples and to monitor food and drink products for spoilage microorganisms. Food pathogens that can be detected and identified using flow cytometry include

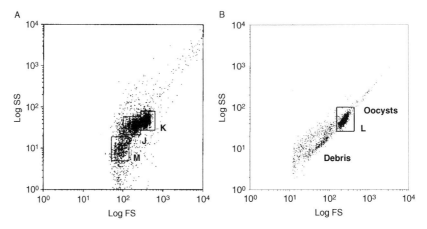

Fig. 10 Flow cytometric analysis of *Cryptosporidium parvum* oocysts exposed to γ-irradiation from a
^{60}Co source: (A) 200-krad dose, (B) 800-krad dose. Intact cysts (region L, K) can be differentiated from
sporozoites (region J), ghosts (region M), and debris using light-scatter measurements. High doses of
γ-irradiation prevented excystation of sporozoites from oocysts.

Listeria monocytogenes in raw milk (Donnelly and Baigent, 1986), *S. typhimurium*
in eggs and milk (McClelland and Pinder, 1994b; Pinder and McClelland, 1994),
and *E. coli* O157:H7 in ground beef, apple juice, and milk (Seo *et al.*, 1998a,b;
Tortorello *et al.*, 1998). In general, flow cytometry requires specific monoclonal
antibodies to detect and identify food pathogens.

Spoilage microorganisms are not necessarily harmful but can interfere with the
quality of a food or drink product and can cause delays in product releases at great
economic cost to the manufacturer. To guarantee that food or drink products
conform to specifications, flow cytometry has been used to detect spoilage caused
by yeast in soft drinks (Pettipher, 1991), yogurt, and fruit juice (Mulard, 1995), and
to monitor the viability of yeast used for beer (Jespersen and Jakobsen, 1994;
Jespersen *et al.*, 1993), wine (Bruetschy *et al.*, 1994), and cider (Lloyd *et al.*, 1996;
Willetts *et al.*, 1997).

In order to identify sources of food contamination and spoilage, a large number
of samples need to be tested. Flow cytometry allows the rapid and semiautomated
analysis of heterogeneous food samples; however, extensive sample preparation is
needed to isolate target organisms from high background levels of nonpathogenic
microflora and particulate matter found in food samples. Sample preparation may
include homogenization of solid food using a stomacher, filtering of large food
particles, serial dilutions, or special reagent addition (e.g., clearing solution to
remove micelles in milk and egg samples). After the cells have been isolated from
the food sample, enrichment media can be used to increase the number of target
organisms and to allow recovery of stressed or injured cells. Increasing the number
of target organisms in food samples is extremely important because the infective
dose for some foodborne illnesses can be as low as 10 cells.

IV. Conclusion

It is clear that there are a tremendous number of excellent uses of flow cytometry in the field of microbiology; however, there are some valid problems in implementing this technology. Listed below are what are considered to be the main advantages and disadvantages in the application of flow cytometry to microbial systems.

1. Advantages of using flow cytometry to analyze microbes

   a. Technology is clinically proven in areas such as leukemia/lymphomas, HIV monitoring, platelet studies, and functional studies. Most hospitals and research centers have already purchased the instrument.

   b. Assays can be performed rapidly, usually taking less than 1 min with little to no sample preparation.

   c. Sensitivity is high, as low as 10^3 cells/ml reported.

   d. Direct detection and identification can be achieved without elaborate or time-consuming culturing of microbes.

   e. Cost per test, after initial investment in instrument, is low.

   f. Automation/walkaway capability is available in most instruments as well as report generation for clinicians.

   g. Flow cytometry is user friendly once the initial protocol is developed.

   h. Instrument maintenance (daily/monthly quality control) is minimal.

2. Disadvantages to using flow cytometry to analyze microbes

   a. Most microbiologists are not comfortable with using nontraditional, high-technology procedures to run routine tests.

   b. Initial cost of instrument is high.

   c. Few microbial reagents or kits are commercially available for use with flow cytometry. Reagents (i.e., antibodies, DNA probes, control cells) must be developed in-house.

   d. Protocols for microbes need to be developed in-house. Because most instruments were designed for mammalian cells, instrument setup and operation must also be modified.

   e. Few, if any, instruments are designed specifically for small particles.

   f. Little support is available from instrument manufacturers. Service technicians and technical support personnel are not familiar with procedures or methods utilized in microbial flow cytometry.

Clearly, the application of flow cytometry to the field of microbiology involves many unresolved problems; however, the continuing development of detailed protocols, appropriately designed instruments, and fluorescent probes will enable flow cytometry to solidify its position as the technology of preference.

Acknowledgments

S. A. Sincock contributed to the original published chapter.

References

Allman, R., Manchee, R., and Lloyd, D. (1993). Flow cytometric analysis of heterogeneous bacterial populations. *In* "Flow Cytometry in Microbiology" (D. Lloyd, ed.), pp. 27–48. Springer-Verlag, New York.

Amann, R. I., Binder, B. J., Olson, R. J., Chisholm, S. W., Devereux, R., and Stahl, D. A. (1990). Combination of 16S rRNA-targeted oligonucleotide probes with flow cytometry for analyzing mixed microbial populations. *Appl. Environ. Microbiol.* **56,** 1919–1925.

Amann, R., Snaidr, J., Wagner, M., Ludwig, W., and Schleifer, K.-H. (1996). In situ visualization of high genetic diversity in a natural microbial community. *J. Bacteriol.* **178,** 3496–3500.

Apperloo-Renkema, H. Z., Wilkinson, M. H. F., and van der Waaij, D. (1992). Circulating antibodies against faecal bacteria assessed by immunomorphometry: Combining quantitative immunofluorescence and image analysis. *Epidemiol. Infect.* **109,** 497–506.

Arrese, J. E., Goffin, V., Avila-Camacho, M., Greimers, R., and Piérard, G. E. (1998). A pilot study on bacterial viability in acne. Assessment using dual flow cytometry on microbials present in follicular casts and comedones. *Int. J. Dermatol.* **37,** 461–464.

Arrowood, M. J., Hurd, M. R., and Mead, J. R. (1995). A new method for evaluating experimental cryptosporidial parasite loads using immunofluorescent flow cytometry. *J. Parasitol.* **81,** 404–409.

Barnett, J. M., Cuchens, M. A., and Buchanan, W. (1984). Automated immunofluorescent speciation of oral bacteria using flow cytometry. *J. Dent. Res.* **63,** 1040–1042.

Best, L. M., Veldhuyzen van Zanten, S. J. O., Bezanson, G. S., Haldane, D. J. M., and Malatjalian, D. A. (1992). Serological detection of Helicobacter pylori by a flow microsphere immunofluorescence assay. *J. Clin. Microbiol.* **30,** 2311–2317.

Boddy, L., and Morris, C. W. (1993). Neural network analysis of flow cytometry data. *In* "Flow Cytometry in Microbiology" (D. Lloyd, ed.), pp. 159–170. Springer-Verlag, New York.

Bruderer, T., Niederer, E., and Köhler, P. (1994). Separation of a cysteine-rich surface antigen-expressing variant from a cloned *Giardia* isolate by fluorescence-activated cell sorting. *Parasitol. Res.* **80,** 303–306.

Bruetschy, A., Laurent, M., and Jacquet, R. (1994). Use of flow cytometry in oenology to analyse yeasts. *Lett. Appl. Microbiol.* **18,** 343–345.

Clarke, R. G., and Pinder, A. C. (1998). Improved detection of bacteria by flow cytometry using a combination of antibody and viability markers. *J. Appl. Microbiol.* **84,** 577–584.

Cohen, C. Y., and Sahar, E. (1989). Rapid flow cytometric bacterial detection and determination of susceptibility to amikacin in body fluids and exudates. *J. Clin. Microbiol.* **27,** 1250–1256.

Comas, J., and Vives-Rego, J. (1997). Assessment of the effects of gramicidin, formaldehyde, and surfactants on Escherichia coli by flow cytometry using nucleic acid and membrane potential dyes. *Cytometry* **29,** 58–64.

Davey, H. M., and Kell, D. B. (1996). Flow cytometry and cell sorting of heterogeneous microbial populations: The importance of single-cell analyses. *Microbiol. Rev.* **60,** 641–696.

Davey, H. M., Jones, A., Shaw, A. D., and Kell, D. B. (1999). Variable selection and multivariate methods for the identification of microorganisms by flow cytometry. *Cytometry* **35,** 162–168.

Deere, D., Porter, J., Edwards, C., and Pickup, R. (1995). Evaluation of the suitability of *bis*-(1,3-dibutylbarbituric acid)trimethine oxonol, (diBAC4(3)($^-$)), for the flow cytometric assessment of bacterial viability. *FEMS Microbiol. Lett.* **130,** 165–169.

del Giorgio, P. A., Prairie, Y. T., and Bird, D. F. (1997). Coupling between rates of bacterial production and the abundance of metabolically active bacteria in lakes, enumerated using CTC reduction and flow cytometry. *Microb. Ecol.* **34,** 144–154.

DeLong, E. F., Wickham, G. S., and Pace, N. R. (1989). Phylogenetic stains: Ribosomal RNA-based probes for identification of single cells. *Science* **243**, 1360–1362.

Desmonts, C., Minet, J., Colwell, R., and Cormier, M. (1990). Fluorescent-antibody method useful for detecting viable but nonculturable *Salmonella* spp. in chlorinated wastewater. *Appl. Environ. Microbiol.* **56**, 1448–1452.

Diaper, J. P., and Edwards, C. (1994). The use of fluorogenic esters to detect viable bacteria by flow cytometry. *J. Appl. Bacteriol.* **77**, 221–228.

Diaper, J. P., Tither, K., and Edwards, C. (1992). Rapid assessment of bacterial viability by flow cytometry. *Appl. Microbiol. Biotechnol.* **38**, 268–272.

Dixon, B. R., Parenteau, M., Martineau, C., and Fournier, J. (1997). A comparison of conventional microscopy, immunofluorescence microscopy and flow cytometry in the detection of *Giardia lamblia* cysts in beaver fecal samples. *J. Immunol. Methods* **202**, 27–33.

Donnelly, C. W., and Baigent, G. J. (1986). Method for flow cytometric detection of Listeria monocytogenes in milk. *Appl. Environ. Microbiol.* **52**, 689–695.

Dubelaar, G. B. J., Gerritzen, P. I., Beeker, A. E. R., Jonker, R. R., and Tangen, K. (1999). Design and first results of CytoBuoy: A wireless flow cytometer for *in situ* analysis of marine and fresh waters. *Cytometry* **37**, 247–254.

Durodie, J., Coleman, K., Simpson, I. N., Loughborough, S. H., and Winstanley, D. W. (1995). Rapid detection of antimicrobial activity using flow cytometry. *Cytometry* **21**, 374–377.

Edwards, C. (1996). Assessment of viability of bacteria by flow cytometry. *In* "Flow Cytometry Applications in Cell Culture" (M. Al-Rubeai and A. N. Emery, eds.), pp. 291–310. Dekker, New York.

Errington, P. A., and Graham, J. (1993). Application of artificial neural networks to chromosome classification. *Cytometry* **14**, 627–639.

Frankel, D. S., Olson, R. J., Frankel, S. L., and Chisholm, S. W. (1989). Use of a neural net computer system for analysis of flow cytometric data of phytoplankton populations. *Cytometry* **10**, 540–550.

Gant, V. A., Warnes, G., Phillips, I., and Savidge, G. F. (1993). The application of flow cytometry to the study of bacterial responses to antibiotics. *J. Med. Microbiol.* **39**, 147–154.

Gottfredsson, M., Erlendsdottir, H., Sigfusson, A., and Gudmunsson, S. (1998). Characteristics and dynamics of bacterial populations during postantibiotic effect determined by flow cytometry. *Antimicrob. Agents Chemother.* **42**, 1005–1011.

Green, L., Petersen, B., Steimel, L., Haeber, P., and Current, W. (1994). Rapid determination of antifungal activity by flow cytometry. *J. Clin. Microbiol.* **32**, 1088–1091.

Henningson, E. W., Krocova, Z., Sandström, G., and Forsman, M. (1998). Flow cytometric assessment of the survival ratio of Francisella tularensis in aerobiological samples. *FEMS Microbiol. Ecol.* **25**, 241–249.

Heyworth, M. F., and Pappo, J. (1989). Use of two-colour flow cytometry to assess killing of *Giardia muris* trophozoites by antibody and complement. *Parasitology* **99**, 199–203.

Hughes, E. E., Matthews-Greer, J. M., and Gilleland, H. E., Jr. (1996). Analysis by flow cytometry of surface-exposed epitopes of outer membrane protein F of Pseudomonas aeruginosa. *Can. J. Microbiol.* **42**, 859–862.

Ingram, M., Cleary, T. J., Price, B. J., Price, R. L., and Castro, A. (1982). Rapid detection of *Legionella pneumophila* by flow cytometry. *Cytometry* **3**, 134–137.

Jepras, R. I., Carter, J., Pearson, S. C., Paul, F. E., and Wilkinson, M. J. (1995). Development of a robust flow cytometric assay for determining numbers of viable bacteria. *Appl. Environ. Microbiol.* **61**, 2696–2701.

Jepras, R. I., Paul, F. E., Pearson, S. C., and Wilkinson, M. J. (1997). Rapid assessment of antibiotic effects on *Escherichia coli* by *bis*-(1,3-dibutylbarbituric acid) trimethine oxonol and flow cytometry. *Antimicrob. Agents Chemother.* **41**, 2001–2005.

Jernaes, M. W., and Steen, H. B. (1994). Staining of *Escherichia coli* for flow cytometry: Influx and efflux of ethidium bromide. *Cytometry* **17**, 302–309.

Jespersen, L., and Jakobsen, M. (1994). Use of flow cytometry for rapid estimation of intracellular events in brewing yeasts. *J. Inst. Brew.* **100**, 399–403.

Jespersen, L., Lassen, S., and Jakobsen, M. (1993). Flow cytometric detection of wild yeast in lager breweries. *Int. J. Food Microbiol.* **17**, 321–328.

Kaprelyants, A. S., and Kell, D. B. (1992). Rapid assessment of bacterial viability and vitality by rhodamine 123 and flow cytometry. *J. Appl. Bacteriol.* **72**, 410–422.

Kaprelyants, A. S., and Kell, D. B. (1993a). The use of 5-cyano-2,3-ditolyl tetrazolium chloride and flow cytometry for the visualisation of respiratory activity in individual cells of *Micrococcus luteus. J. Microbiol. Methods 17, 115–122.*

Kaprelyants, A. S., and Kell, D. B. (1993b). Dormancy in stationary-phase cultures of *Micrococcus luteus*: Flow cytometric analysis of starvation and resuscitation. *Appl. Environ. Microbiol.* **59**, 3187–3196.

Kaprelyants, A. S., Gottschal, J. C., and Kell, D. B. (1993). Dormancy in non-sporulating bacteria. *FEMS Microbiol. Rev.* **104**, 271–286.

Kirk, S. M., Schell, R. F., Moore, A. V., Callister, S. M., and Mazurek, G. H. (1998). Flow cytometric testing of susceptibilities of Mycobacterium tuberculosis isolates to ethambutol, isoniazid, and rifampin in 24 hours. *J. Clin. Microbiol.* **36**, 1568–1573.

Lange, J. L., Thorne, P. S., and Lynch, N. (1997). Application of flow cytometry and fluorescent in situ hybridization for assessment of exposures to airborne bacteria. *Appl. Environ. Microbiol.* **63**, 1557–1563.

Lloyd, D., Moran, C. A., Suller, M. T. E., and Dinsdale, M. G. (1996). Flow cytometric monitoring of rhodamine 123 and a cyanine dye uptake by yeast during cider fermentation. *J. Inst. Brew.* **102**, 251–259.

López-Amorós, R., Comas, J., García, M. T., and Vives-Rego, J. (1998). Use of the 5-cyano-2,3-ditolyl tetrazolium chloride reduction test to assess respiring marine bacteria and grazing effects by flow cytometry during linear alkylbenzene sulfonate degradation. *FEMS Microbiol. Ecol.* **27**, 33–42.

Lutton, D. A., Patrick, S., Crockard, A. D., Stewart, L. D., Larkin, M. J., Dermott, E., and McNeill, T. A. (1991). Flow cytometric analysis of within-strain variation in polysaccharide expression by Bacteroides fragilis by use of murine monoclonal antibodies, *J. Med. Microbiol.* **35**, 229–237.

Maguire, D., King, G. B., Kelley, S., and Robinson, J. P. (1994a). Neural network classification of acute leukemias using flow cytometry analysis data. *Proc. 13th Southern Biomed. Eng. Conf.* pp. 645–648.

Maguire, D. J., King, G. B., and Robinson, J. P. (1994b). A comparison of hard and soft boundaries of intensity regions for the statistical classification of acute leukemias. *Cytometry 7(Suppl.), 48 (Abstract).*

Manz, W., Szewzyk, U., Ericsson, P., Amann, R., Schleifer, K.-H., and Stenström, T.-A. (1993). *In situ* identification of bacteria in drinking water and adjoining biofilms by hybridization with 16S and 23S rRNA-directed fluorescent oligonucleotide probes. *Appl. Environ. Microbiol.* **59**, 2293–2298.

Martinelli, F., Pizzi, R., Cabibbo, E., Licenziati, S., Dima, F., Canaris, A. D., Crea, G., Ravizzola, G., Caruso, A., and Turano, A. (1995). Monoclonal antibodies against antigens exposed on the surface of vegetative forms and spores of *Myxococcus virescens. Microbiologica 18, 399–407.*

Mason, D. J., and Lloyd, D. (1997). Acridine orange as an indicator of bacterial susceptibility to gentamicin. *FEMS Microbiol. Lett.* **153**, 199–204.

Mason, D. J., Allman, R., Stark, J. M., and Lloyd, D. (1994). Rapid estimation of bacterial antibiotic susceptibility with flow cytometry. *J. Microsc.* **176**, 8–16.

Mason, D. J., López-Amorós, R., Allman, R., Stark, J. M., and Lloyd, D. (1995a). The ability of membrane potential dyes and calcafluor white to distinguish between viable and non-viable bacteria. *J. Appl. Bacteriol.* **78**, 309–315.

Mason, D. J., Power, G. M., Talsania, H., Phillips, I., and Gant, V. A. (1995b). Antibacterial action of ciprofloxacin. *Antimicrob. Agents Chemother.* **39**, 2752–2758.

Mason, D. J., Shanmuganathan, S., Mortimer, F. C., and Gant, V. A. (1998). A fluorescent gram stain for flow cytometry and epifluorescence microscopy. *Appl. Environ. Microbiol.* **64**, 2681–2685.

McClelland, R. G., and Pinder, A. C. (1994a). Detection of low levels of specific *Salmonella* species by fluorescent antibodies and flow cytometry. *J. Appl. Bacteriol.* **77**, 440–447.

McClelland, R. G., and Pinder, A. C. (1994b). Detection of *Salmonella typhimurium* in dairy products with flow cytometry and monoclonal antibodies. *Appl. Environ. Microbiol.* **60**, 4255–4262.

Mulard, Y. (1995). Flow cytometry: Real time microbiology testing. *Food Technol. Eur.* **2**, 72–76.

Nelson, D., Bathgate, A. J., and Poxton, I. R. (1991). Monoclonal antibodies as probes for detecting lipopolysaccharide expression on *Escherichia coli* from different growth conditions. *J. Gen. Microbiol.* **137**, 2741–2751.

O'Gorman, M. R. G., and Hopfer, R. L. (1991). Amphotericin B susceptibility testing of *Candida* species by flow cytometry. *Cytometry* **12**, *743–747.*

Obernesser, M. S., Socransky, S. S., and Stashenko, P. (1990). Limit of resolution of flow cytometry for the detection of selected bacterial species. *J. Dent. Res.* **69**, 1592–1598.

Ordóñez, J. V., and Wehman, N. M. (1995). Amphotericin B susceptibility of *Candida* species assessed by rapid flow cytometric membrane potential assay. *Cytometry* **22**, *154–157.*

Paul, F., Jepras, R., Hynes, D., Smith, A., and Marken, B. (1996). Activity of common oral antiseptics against bacteria assessed using the oxonol DiBAC$_4$(3). *Cytometry* **8** *(Suppl.), 117 (Abstract).*

Pettipher, G. L. (1991). Preliminary evaluation of flow cytometry for the detection of yeasts in soft drinks. *Lett. Appl. Microbiol.* **12**, 109–112.

Peyron, F., Favel, A., Guiraud-Dauriac, H., el Mzibri, M., Chastin, C., Duménil, G., and Regli, P. (1997). Evaluation of a flow cytofluorometric method for rapid determination of amphotericin B susceptibility of yeast isolates. *Antimicrob. Agents Chemother.* **41**, 1537–1540.

Phillips, A. P., and Martin, K. L. (1983). Immunofluorescence analysis of Bacillus spores and vegetative cells by flow cytometry. *Cytometry* **4**, *123–131.*

Phillips, A. P., and Martin, K. L. (1988). Limitations of flow cytometry for the specific detection of bacteria in mixed populations. *J. Immunol. Methods* **106**, *109–117.*

Pinder, A. C., and McClelland, R. G. (1994). Rapid assay for pathogenic salmonella organisms by immunofluorescence flow cytometry. *J. Microsc.* **176**, 17–22.

Porter, J., Pickup, R., and Edwards, C. (1995). Membrane hyperpolarisation by valinomycin and its limitations for bacterial viability assessment using rhodamine 123 and flow cytometry. *FEMS Microbiol. Lett.* **132**, 259–262.

Ramani, R., Ramani, A., and Wong, S. J. (1997). Rapid flow cytometric susceptibility testing of Candida albicans. *J. Clin. Microbiol.* **35**, 2320–2324.

Rice, J., Sleigh, M. A., Burkill, P. H., Tarran, G. A., O'Connor, C. D., and Zubkov, M. V. (1997). Flow cytometric analysis of characteristics of hybridization of species-specific fluorescent oligonucleotide probes to rRNA of marine nanoflagellates. *Appl. Environ. Microbiol.* **63**, 938–944.

Roth, B. L., Poot, M., Yue, S. T., and Millard, P. J. (1997). Bacterial viability and antibiotic susceptibility testing with SYTOX Green nucleic acid stain. *Appl. Environ. Microbiol.* **63**, 2421–2431.

Sahar, E., Lamed, R., and Ofek, I. (1983). Rapid identification of *Streptococcus pyogenes* by flow cytometry. *Eur. J. Clin. Microbiol.* **2**, 192–195.

Sanders, C. A., Yajko, D. M., Hyun, W., Langlois, R. G., Nassos, P. S., Fulwyler, M. J., and Hadley, W. K. (1990). Determination of guanine-plus-cytosine content of bacterial DNA by dual-laser flow cytometry. *J. Gen. Microbiol.* **136**, 359–365.

Seo, K. H., Brackett, R. E., and Frank, J. F. (1998a). Rapid detection of *Escherichia coli* O157:H7 using immunomagnetic flow cytometry in ground beef, apple juice, and milk. *Int. J. Food Microbiol.* **44**, 115–123.

Seo, K. H., Brackett, R. E., Frank, J. F., and Hilliard, S. (1998b). Immunomagnetic separation and flow cytometry for rapid detection of Escherichia coli O157:H7. *J. Food Protect.* **61**, 812–816.

Shapiro, H. M. (1995). "Practical Flow Cytometry," 3rd edn. Wiley-Liss, New York.

Sheppard, F. C., Mason, D. J., Bloomfield, S. F., and Gant, V. A. (1997). Flow cytometric analysis of chlorhexidine action. *FEMS Microbiol. Lett.* **154**, 283–288.

Simon, N., LeBot, N., Marie, D., Partensky, F., and Vaulot, D. (1995). Fluorescent in situ hybridization with rRNA-targeted oligonucleotide probes to identify small phytoplankton by flow cytometry. *Appl. Environ. Microbiol.* **61**, 2506–2513.

Sincock, S. A., Anderson, P. E., and Stopa, P. J. (1996a). New fluorescent stains for flow cytometric detection of bacteria. *Cytometry* **8***(Suppl.), 128 (Abstract).*

Sincock, S. A., Anderson, P. E., Stopa, P. J., and Ezzell, J. (1996b). Rapid detection of pathogenic bacteria using flow cytometry. *Cytometry* **8***(Suppl.), 117 (Abstract).*

Sincock, S. A., Thompson, J. E., Blatchley, III, E. R., Ragheb, K. E., and Robinson, J. P. (1998). Flow cytometry can detect inactivation of *Cryptosporidium parvum* oocysts by gamma irradiation. *Cytometry Suppl.* **9***. 46–47 (Abstract).*

Sizemore, R. K., Caldwell, J. J., and Kendrick, A. S. (1990). Alternate gram staining technique using a fluorescent lectin. *Appl. Environ. Microbiol.* **56**, 2245–2247.

Steen, H. B. (1980). Further developments of a microscope-based flow cytometer: Light scatter detection and excitation intensity compensation. *Cytometry* **1***, 26–31.*

Steen, H. B. (1983). A microscope-based flow cytophotometer. *Histochem. J.* **15**, 147–160.

Steen, H. B., and Boye, E. (1980). Bacterial growth studied by flow cytometry. *Cytometry* **1***, 32–36.*

Suller, M. T. E., and Lloyd, D. (1998). Flow cytometric assessment of the postantibiotic effect of methicillin on *Staphylococcus aureus. Antimicrob. Agents Chemother.* **42**, 1195–1199.

Suller, M. T. E., and Lloyd, D. (1999). Fluorescence monitoring of antibiotic-induced bacterial damage using flow cytometry. *Cytometry* **35**, 235–241.

Suller, M. T. E., Stark, J. M., and Lloyd, D. (1997). A flow cytometric study of antibiotic-induced damage and evaluation as a rapid antibiotic susceptibility test for methicillin-resistant Staphylococcus aureus. *J. Antimicrob. Chemother.* **40**, 77–83.

Swarts, A. J., Hastings, J. W., Roberts, R. F., and von Holy, A. (1998). Flow cytometry demonstrates bacteriocin-induced injury to *Listeria monocytogenes. Curr. Microbiol.* **36**, 266–270.

Thomas, J.-C., Desrosiers, M., St.-Pierre, Y., Lirette, P., Bisaillon, J.-G., Beaudet, R., and Villemur, R. (1997). Quantitative flow cytometric detection of specific microorganisms in soil samples using rRNA targeted fluorescent probes and ethidium bromide. *Cytometry* **27***, 224–232.*

Tortorello, M. L., Stewart, D. S., and Raybourne, R. B. (1998). Quantitative analysis and isolation of *Escherichia coli* O157:H7 in a food matrix using flow cytometry and cell sorting. *FEMS Immunol. Med. Microbiol.* **19**, 267–274.

Tsuji, T., Kawasaki, Y., Takeshima, S., Sekiya, T., and Tanaka, S. (1995). A new fluorescence staining assay for visualizing living microorganisms in soil. *Appl. Environ. Microbiol.* **61**, 3415–3421.

Ullrich, S., Karrasch, B., Hoppe, H.-G., Jeskulke, K., and Mehrens, M. (1996). Toxic effects on bacterial metabolism of the redox dye 5-cyano-2,3-ditolyl tetrazolium chloride. *Appl. Environ. Microbiol.* **62**, 4587–4593.

Valdez, L. M., Dang, H., Okhuysen, P. C., and Chappell, C. L. (1997). Flow cytometric detection of *Cryptosporidium* oocysts in human stool samples. *J. Clin. Microbiol.* **35**, 2013–2017.

van der Waaij, L. A., Mesander, G., Limburg, P. C., and van der Waaij, D. (1994). Direct flow cytometry of anaerobic bacteria in human feces. *Cytometry* **16***, 270–279.*

Van Dilla, M. A., Langlois, R. G., Pinkel, D., Yajko, D., and Hadley, W. K. (1983). Bacterial characterization by flow cytometry. *Science* **220***, 620–622.*

Vesey, G., Slade, J. S., Byrne, M., Shepherd, K., Dennis, P. J., and Fricker, C. R. (1993). Routine monitoring of *Cryptosporidium* oocysts in water using flow cytometry. *J. Appl. Bacteriol.* **75**, 87–90.

Vesey, G., Deere, D., Weir, C. J., Ashbolt, N., Williams, K. L., and Veal, D. A. (1997). A simple method for evaluating Cryptosporidium-specific antibodies used in monitoring environmental water samples. *Lett. Appl. Microbiol.* **25**, 316–320.

Volsch, A., Nader, W. F., Geiss, H. K., Nebe, G., and Birr, C. (1990). Detection and analysis of two serotypes of ammonia-oxidizing bacteria in sewage plants by flow cytometry. *Appl. Environ. Microbiol.* **56**, 2430–2435.

Walberg, M., Gaustad P., and Steen, H. B. (1997a). Rapid assessment of ceftazidime, ciprofloxacin, and gentamycin susceptibility in exponentially-growing *E. coli* cells by means of flow cytometry. *Cytometry* **27***, 169–178.*

Walberg, M., Gaustad, P., and Steen, H. B. (1997b). Rapid discrimination of bacterial species with different ampicillin susceptibility levels by means of flow cytometry. *Cytometry* **29***, 267–272.*

Wallner, G., Amann, R., and Beisker, W. (1993). Optimizing fluorescent in situ hybridization with rRNA-targeted oligonucleotide probes for flow cytometric identification of microorganisms. *Cytometry* **14,** *136–143.*

Wallner, G., Erhart, R., and Amann, R. (1995). Flow cytometric analysis of activated sludge with rRNA-targeted probes. *Appl. Environ. Microbiol.* **61,** 1859–1866.

Wenisch, C., Linnau, K. F., Parschalk, B., Zedtwitz-Liebenstein, K., and Georgopoulos, A. (1997). Rapid susceptibility testing of fungi by flow cytometry using vital staining. *J. Clin. Microbiol.* **35,** 5–10.

Willetts, J. C., Seward, R., Dinsdale, M. G., Suller, M. T. E., Hill, B., and Lloyd, D. (1997). Vitality of cider yeast grown micro-aerobically with added ethanol, butan-1-ol or *iso*-butanol. *J. Inst. Brew.* **103,** 79–84.

Yu, W., Dodds, W. K., Banks, M. K., Skaisky, J., and Strauss, E. A. (1995). Optimal staining and sample storage time for direct microscopic enumeration of total and active bacteria in soil with two fluorescent dyes. *Appl. Environ. Microbiol.* **61,** 3367–3372.

CHAPTER 39

Flow Cytometry in Malaria Detection

Chris J. Janse and Philip H. Van Vianen

Laboratory of Parasitology
University of Leiden
2300 RC Leiden, The Netherlands

ESSENTIAL CYTOMETRY METHODS
Reprinted from *Methods in Cell Biology*, Volume 42 (Academic Press, 1994).
Copyright © 1994 by Academic Press, Inc.,

DOI: 10.1016/B978-0-12-375045-7.00039-8

I. Introduction

Malaria is a parasitic disease found in tropical and subtropical regions which is caused by unicellular organisms of the genus *Plasmodium*. In man, four different species are responsible for the disease, *Plasmodium falciparum*, *P. vivax*, *P. ovale*, and *P. malariae*.

A large part of the life cycle of these parasites takes place in the blood circulation, where these organisms invade red blood cells in which they grow and multiply. Failure of existing methods to control malaria, the lack of an effective vaccine, and increasing drug resistance of the parasites are factors which play a role in the increase of malaria cases. Both for laboratory research aimed at the development of new control strategies and for monitoring the effects of existing control projects in the field, the availability of rapid, sensitive, and reproducible techniques for the detection and analysis of blood infection are required. The blood-stage infection is the most relevant part of the life cycle; the blood stages cause the clinical symptoms and are targets for a number of drugs. Moreover, the demonstration of the presence of parasites in the blood is used for diagnosis and treatment of the disease.

Flow cytometry has proven to be a useful tool for the analysis of blood infection by malaria parasites. Analysis of blood-stage development (Janse *et al.*, 1987; Mons and Janse, 1992) and determination of susceptibility to drugs by flow cytometry (van Vianen *et al.*, 1990a,b) are reproducible and rapid and detection of blood-stage parasites appears to be sensitive and reproducible (van Vianen *et al.*, 1993; P. H. van Vianen, unpublished results).

The analysis and detection of malaria infection by flow cytometry makes almost exclusive use of fluorescent dyes which are specific for nucleic acids (for review, see Mons and Janse, 1992). DNA-specific dyes are especially useful since the parasites multiply inside the red blood cell (RBC) population of the blood cells. Since RBC do not contain DNA, DNA-specific fluorescence from infected RBC can only be due to fluorescence of dyes bound to parasite DNA. Consequently, infected cells can be discriminated from noninfected cells based on their fluorescence intensity. In addition, since parasites multiply within the RBC by several mitotic divisions, the fluorescence intensity of stained parasites increases during development of the parasites. This can be analyzed by flow cytometry and used to determine the developmental stage of the parasite. The total DNA content of a parasite is 100–200 times less than that of nucleated blood cells. Therefore, the nucleated blood cells can easily be distinguished from parasites on the basis of the difference in fluorescence intensity.

The vast majority of the reported flow cytometric studies on malaria parasites have been carried out with the A/T-specific DNA dyes, Hoechst 33258 and Hoechst 33342. These dyes give a strong fluorescence with parasite DNA after fixation of infected blood cells (Bianco *et al.*, 1986; Myler *et al.*, 1982) or free parasites and after vital staining of parasites (Franklin *et al.*, 1986). Moreover, the relative fluorescence intensity of different blood stages after Hoechst staining corresponds closely to the relative DNA content of these stages (Janse *et al.*, 1987). A/T-specific dyes are particularly suited since the A/T content of DNA of malaria parasites is extremely high, ranging from 70% to 82% in different species. These dyes are now routinely used in studies to determine parasite development and DNA synthesis in parasites and for determination of the level of drug resistance in parasites obtained from patients. Moreover, it has been shown that these dyes can be used for sensitive detection of parasites in blood samples obtained from patients in clinical and in epidemiological studies (van Vianen *et al.*, 1993; P. H. van Vianen, unpublished results).

Other dyes have been used for parasite detection (for review, see Mons and Janse, 1992), such as acridine orange (Hare, 1986; Whaun *et al.*, 1983), propidium iodide (Pattanapanyasat *et al.*, 1992; Saul *et al.*, 1982), and thiazole orange (Makler *et al.*, 1987). These dyes have the disadvantage that they bind both to DNA and to RNA. The reticulocyte population of blood cells contains RNA and the fluorescence of parasite-infected cells can be in the same range as that of noninfected reticulocytes which hampers the discrimination between infected cells and noninfected cells.

Below we describe flow cytometric methods for (1) measurement of parasite development and DNA synthesis by parasites, (2) determination of susceptibility of parasites to drugs, and (3) detection of low numbers of parasite-infected cells in blood samples from patients.

II. Applications

A. Flow Cytometry and the Developmental Cycle of the Parasite

Malaria parasites have a complex life cycle which alternates between two different hosts: mosquitoes and vertebrates such as reptiles, birds, rodents, nonhuman primates, and humans. An infection starts with a bite of an infected mosquito, which injects parasites into the blood of the vertebrate host. These parasites penetrate liver cells and, after one developmental cycle, the parasites are released in the blood and invade RBC.

For flow cytometry, the blood stages (see Fig. 1) are the most important (see below). After entering the RBC, the small haploid parasites (called ring forms) develop and grow until they nearly fill the RBC (these growing stages are the so-called trophozoites). After this growth phase most of the parasites synthesize DNA and enter mitosis. In each parasite, now called schizont, three to five rapid mitotic divisions follow each other, resulting in the production of 8–32 haploid merozoites. After rupture of the RBC, these merozoites can penetrate new RBC and the multiplication cycle starts again. One multiplication cycle takes 24–72 h, depending on the species. This mitotic multiplication results in a rapid increase in the number of infected red blood cells.

A small percentage of the merozoites do not continue the asexual multiplication but differentiate within the RBC into precursor cells of the gametes, the so-called gametocytes. These stages develop into gametes when they are ingested by mosquitoes, in which further development takes place. No flow cytometric studies on the mosquito stages of the parasite have been reported.

B. Flow Cytometry and Determination of Parasite Development, DNA Synthesis, and Drug Susceptibility

The blood stages of several species of malaria parasites can grow and multiply under culture conditions. These *in vitro* cultures allow the effect of new drugs to be studied on parasite development and DNA synthesis under standardized

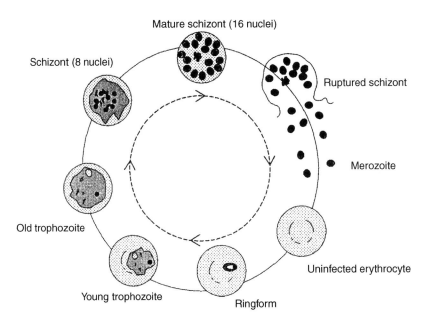

Fig. 1 Schematic representation of blood-stage development of malaria parasites. Only part of the life cycle, the asexual erythrocytic development, is shown.

conditions. The effect of new drugs on parasite development in culture is routinely monitored by microscopic counting of the number of infected cells or by measuring the incorporation of radioactive precursors into the nucleic acids of the parasites. Flow cytometry is a very good alternative for the determination of the effect of new drugs on parasite development in culture. Using DNA-specific dyes the increase/ decrease of the number of infected cells can be measured rapidly and reproducibly. In addition, the amount of DNA synthesis can be determined precisely which is a reliable characteristic of the development of the parasites (Janse *et al.*, 1987, 1989; van Vianen *et al.*, 1990a). The advantages of flow cytometry are the speed of measurement, the accuracy, reproducibility, and the large number of parasites analyzed.

For the determination of drug susceptibility of parasites from field isolates the World Health Organization (WHO) (1982) has developed a standard *in vitro* microtest. In these tests haploid, small blood stages (ring forms, trophozoites) are cultured for short periods in the presence of different concentrations of drugs. Giemsa-stained slides are made from these cultures to monitor development of the parasites from ring forms into the DNA synthesizing stages (schizonts) by light microscopy. Determination of parasite development by light microscopy is time consuming and the results can easily be influenced by human errors. Since development and DNA synthesis of parasites can accurately be assessed by flow cytometry using DNA-specific dyes, flow cytometry is therefore very useful for determination of drug susceptibility. Recently, a method has been developed for the fully automated reading of the WHO microtests by flow cytometry (van Vianen *et al.*, 1990b).

C. Flow Cytometry and Detection of Blood Stages

The demonstration of the presence of blood stages is used for diagnosis and treatment of malaria. The "gold standard" is the microscopic detection of blood stages in thin or thick blood smears, which are stained with Giemsa. However, sensitive detection in large numbers of blood samples collected from patients under primitive field conditions using this method poses problems. The sensitivity is highly influenced by local circumstances and working conditions and depends on the availability of experienced people for recognition of parasites. The quality of microscopic slides is not always optimal, due to improper preparation of the smears, applying contaminated staining solution, or suboptimal use of the staining procedure. The sensitivity is very much dependent on the experience of the microscopist and the time spent reading the slides. Because microscopy is labor intensive, human factors such as loss of concentration, especially when large numbers of samples need to be screened with a low percentage of positives, can account for misreading of samples (Payne, 1988).

In the past few years, alternatives to microscopic detection of malaria parasites have been investigated, such as immunological methods to demonstrate antibodies or antigens (Tharavanij, 1990), detection of parasites by fluorescence microscopy (Kawamoto and Billingsley, 1992; Rickman *et al.*, 1989), and the use of malaria-specific radioactively labeled DNA and RNA probes (Tharavanij, 1990). At present none of these techniques appears to be superior to microscopic examination of Giemsa-stained blood smears. Flow cytometry has been shown to be potentially suited to overcome most of the above-mentioned problems with parasite detection (van Vianen *et al.*, 1993). Although several fluorescent dyes have been reported to be useful for detection of parasite-infected cells, most studies have been performed with the DNA-specific Hoechst dyes (Mons and Janse, 1992). Detection and counting of the number of infected cells can be performed using Hoechst 33258-stained RBCs which are fixed by glutaraldehyde. However, this method is not sensitive enough to detect very low numbers of infected RBC (<0.1%). It appears to be necessary to first free the parasites from the RBC by lysing these blood cells (van Vianen *et al.*, 1993). This reduces the sample volume and the number of cells to be analyzed.

III. Materials

Materials for the *in vitro* culture of *P. berghei* have been described by Janse *et al.* (1989). For materials for *in vitro* cultures of *P. falciparum*, see references in Trigg (1985).

Phosphate-buffered saline (PBS) tablets (Flow Laboratories) are dissolved in double-distilled deionized (demi) water and HCl is used to adjust the pH at 7.2. PBS for fixation solution or staining solution is filtered through a 0.22-μm filter to remove small particles. PBS is stored at 4 °C.

Hoechst 33258 (Janssen Chimica) is dissolved in demi-water at a stock concentration of 500 μM. The stock is stored at −20 °C. Final concentration for cell staining is 2 μM in PBS.

Propidium iodide (Sigma) is dissolved in demiwater at a stock concentration of 1 mg/ml, which is stored at 4 °C. Final concentration for cell staining is 1 μg/ml in PBS.

Glutaraldehyde (Zeiss, high grade, 70%) is diluted with PBS at a stock concentration of 25% and stored at −20 °C. Final solution is made by diluting the stock 1/100 with filtered PBS (to 0.25%) and stored at 4 °C.

Lysis solution (Becton-Dickinson Immunocytometry Systems, 10 × stock) is stored at room temperature. Final solution is made by diluting the stock 1/10 with demiwater and filtering through a 0.22-μm filter. Storage is at 4 °C.

IV. Cell Preparation and Staining

A. Collection of Blood Samples from Infected Humans and Laboratory Animals

For flow cytometric analysis only small blood samples are required. Samples from human patients can be taken from blood collection tubes treated with heparin or EDTA as anticoagulants, which are routinely used for collection of blood. Alternatively, small samples can be drawn using heparinized capillaries from the finger after a finger prick.

Malaria parasites which infect nonhuman primates and rodents are regularly used as models for the study of malaria. Small blood samples (20–200 μl) from infected rodents, such as mice and rats, are usually collected from the veins at the end of the tail using heparinized capillaries and resuspended in PBS or culture medium. Cells are collected by centrifugation for 5 s (15,000 $\times$ g) in an Eppendorf centrifuge or at 450 $\times$ g for 10 min. When larger amounts of blood are needed (e.g., for cultures of the blood stages; see below), a cardiac puncture under ether anesthesia is performed and blood is collected either in PBS containing heparin (20 IU/ml) or in culture medium RPMI 1640 (see below) containing heparin (20 IU/ml). Heparin is added as an anticoagulant. Cells are collected by centrifugation at 450 $\times$ g for 10 min.

B. Collection of Samples from *In Vitro* Cultures of the Blood Stages of Malaria Parasites

In vitro cultures of the blood stages of two species are regularly used for the study of parasite development and drug susceptibility. These are the human parasite *P. falciparum* and a parasite which infects rodents, *P. berghei*. Culture methods for both species have been described extensively (Janse *et al.*, 1989; Trigg, 1985). Here, the methods are described very briefly.

Infected RBCs are obtained either directly from humans and rodents or from liquid nitrogen storage. Cultures are normally started with young stages of the parasite, the ring forms. Infected RBCs are incubated in culture medium RPMI 1640 containing Hepes buffer (5.94 g/l) $NaHCO_4$, serum (10–20%), and antibiotics. This cell suspension at an RBC concentration ranging from 0.5% to 10% is incubated at 37 °C in culture plates, flasks, petri dishes, or Erlenmeyers and gassed with a mixture of 10% CO_2, 5% O_2, and 85% N_2. In these cultures parasites develop from ring forms into schizonts, after which invasion of new RBC takes place. For both parasite species methods have been described to synchronize the asexual development of the blood stages, so that all parasites are at the same stage of development during the complete cycle. Samples from the culture are centrifuged for 5 s (15,000 $\times$ g) in Eppendorf centrifuges or at 450 $\times$ g for 10 min to remove culture medium.

C. Collection of Samples from Standard Drug Susceptibility Tests

For determination of drug susceptibility of *P. falciparum* parasites obtained from patients, the WHO (1982) has developed a standardized microtest. Infected blood, obtained from patients, is incubated in complete culture medium (RPMI 1640 + 10% human serum) in standard 96-well microtiter plates for 26–30 h, according to the WHO procedure, at an RBC concentration of 10%. The plates are predosed with different concentrations of drugs. Samples from the microtest can be prepared as described for the *in vitro* cultures of the blood stages of the parasites. However, for flow cytometric analysis the samples can remain in the culture wells at the end of the culture period and be fixed and stained in the wells after removal of the culture medium (see below).

D. Fixation of Infected Erythrocytes

Infected RBC can be fixed with glutaraldehyde, paraformaldehyde, or a combination of these two. Both glutaraldehyde and (para)formaldehyde have the disadvantage that they crosslink cell membrane components, which hampers the penetration of high-molecular-weight molecules such as monoclonal antibodies, DNA probes, or large fluorochromes. In addition, these fixatives can have a significant quenching effect on the emission of fluorescence from certain DNA-bound fluorochromes (Crissmann et al., 1979). Despite these disadvantages aldehyde-type fixatives appear to be very useful for fixation and staining of infected RBC with Hoechst dyes. They induce no significant cell aggregation or lysis, which are frequently observed when for example ethanol and methanol are used as fixative.

The usual procedure for fixation is as follows: Samples of infected blood cells obtained from humans, laboratory animals, or cultures are washed once in PBS before fixation. Typically, these samples consist of 1 ml of blood cell suspension of 0.5–10% (10^7–10^9 cells), collected in Eppendorf tubes. The blood cells are centrifuged for 5 s at 15,000 $\times$ g and the supernatant is removed. Subsequently, 1 ml of 0.25% glutaraldehyde in PBS is added and the sample is mixed vigorously. Fixation is done at 4 °C for 15 min. The cell suspension in the glutaraldehyde solution ranges between 0.5% and 10%. After fixation, cells are washed twice with PBS. Fixed cells are stored in PBS at 4 °C until being stained for flow cytometry. Cells can be kept for more than a year at 4 °C without deteriorating. We have found that the washing steps with PBS both before and after fixation are not necessary for accurate flow cytometric readings. Blood cells can be added directly to the fixative and can be stored without removal of the fixation solution.

E. Fixation of Free Parasites

To detect very low numbers of parasites in blood samples from patients it is beneficial to lyse the RBC before fixation in order to reduce the sample volume and the number of cells to be analyzed (van Vianen et al., 1993). For this method samples of 50 μl of blood (1–5 $\times$ 10^8 cells) are collected in Eppendorf tubes with 1 ml of FACS lysing solution (Becton-Dickinson Immunocytometry Systems, San Jose, CA) containing 1.5% formaldehyde as a fixative. For proper lysis of the RBC, the blood cells are added directly to the lysis solution and the sample is mixed well in the solution. This treatment ruptures red blood cells, releasing the malaria parasites which are subsequently fixed by the formaldehyde. The white blood cells (WBCs) remain intact and are fixed as well. The samples are lysed and fixed for 30 min at room temperature and are stored in the lysing solution at 4 °C. We found that samples can be stored up to a year in this way.

F. Fixation of Infected Blood Cells from Drug Susceptibility Microtests

Microtests are performed in 96-well microtiter plates. To fix the cells in the wells, culture medium is carefully removed from the cells using a micropipette leaving the blood cells at the bottom of the wells. The cells are fixed immediately by adding 200 μl 0.25% glutaraldehyde in PBS. The plates containing the fixed samples can be sealed and stored at 4 °C until analysis. In this way, material can be kept in fixing solution or in PBS for over 6 months at 4 °C without significant deterioration.

G. Staining of Infected Blood Cells and Free Parasites with Hoechst 33258 After Fixation

Samples of fixed infected blood cells or free parasites are stained for 1 h at 37°C in the dark in 1–2 μM Hoechst 33258 in PBS. In case of the samples containing fixed RBC, part of the sample is diluted with PBS to a volume of 1 ml (approximately 10^6–10^8 RBCs/ml). To this suspension 2–4 μl of a 500 μM stock solution of Hoechst 33258 is added. In case of samples containing free parasites, 200–500 μl of the sample is centrifuged for 1 min in an Eppendorf centrifuge (15,000 × g) and the supernatant is carefully removed to prevent loss of cells. To the pellet 0.2–0.5 ml of a staining solution containing 1 μM Hoechst 33258 in PBS is added, and the suspension is mixed. In the final cell suspension at least 10^5 WBCs/ml should be present. The cells remain in staining solution until analysis, which is usually performed within 0–3 h after staining.

The same procedure can also be applied to samples from drug susceptibility tests. Fixation solution in the wells is carefully removed using a micropipette and the cells are resuspended in 200 μl staining solution which contains 2 μM Hoechst 33258 in PBS. Cells are stained in the plates at 37 °C for 1 h in the dark.

H. Staining of Free Parasites with Hoechst 33258 in Combination with Propidium Iodide

When low numbers of infected cells are present in blood samples (<0.1%), red blood cells are lysed before staining and analysis by flow cytometry. We have found that staining of the parasites with Hoechst 33258 in combination with propidium iodide improves the capability to distinguish parasites from background fluorescence (see Section VIII) (P. H. van Vianen, unpublished results).

Fixed samples of free parasites (approximately 10^5 WBCs) are centrifuged for 1 min (15,000 × g) in an Eppendorf centrifuge. The supernatant is removed and replaced with 0.3 ml staining solution containing 2 μM Hoechst 33258 in PBS. After the samples are stained for 1 h at 37 °C in the dark, 0.3 ml propidium iodide solution in PBS is added to a final concentration of 1 μg/ml. The cells remain in this solution at room temperature for 30 min until analysis. Analysis will be performed within 0–3 h after staining.

V. Critical Aspects of the Preparation and Staining Procedures

In general, for the study of malaria infection, some of the most critical aspects are the preparation and culture of blood stages of the parasites. These procedures have been described in detail elsewhere and do not fall within the scope of this chapter.

Because cell collection and preparation procedures will also occur under primitive conditions in field research in developing countries, these methods must be simple and straightforward. Other prerequisites are that sample handling can be minimized or automated, especially with large numbers of samples, and that samples can be stored for long periods, which is convenient in epidemiological studies. Both cell preparation and staining procedures described here are simple and easy to perform. Washing steps before or after fixation are not essential for reproducible results. The cells can be stored for long periods either in PBS or in fixative.

Fixation of infected RBC with glutaraldehyde (GA) is fast and very easy. However, some problems can occur. Especially, when fixing cells in 96-well micro-titer plates after removing the culture medium, care must be taken to add the fixative before the cells deteriorate. In all cases, the cells must be mixed vigorously with the fixative by shaking, whirl mixing, or using the pipette. Because GA fixation is quick, it can be replaced by PBS after 10–15 min and cells can be stored in PBS. When samples are stored in GA, storage should be at 4 °C. When stored in GA for long periods at higher temperatures (ambient temperatures in the tropics), GA can cause an increase in background fluorescence of uninfected RBC, causing overlap with infected RBC.

For optimal lysis of RBC in the preparation of free parasites, the RBC should ideally be suspended directly in the lysis solution after being collected. Preparing and handling of the free parasites after lysis of RBC must be carefully performed. Loss of cells must be prevented during the steps in which the lysing solution is removed and replaced with the staining solution. Since these methods are used to prepare cells for detection and quantitation of low number of parasites, small losses of free parasites or WBCs could significantly influence the reliability and reproducibility of the results.

The length of the staining period of fixed cells is not very critical: during analysis which can take several hours, cells are normally kept in the staining solution at room temperature. This appears not to affect the fluorescence intensity of the cells.

VI. Standards

Young blood stages (ring forms and trophozoites) are non-DNA synthesizing haploid organisms, which show a narrow symmetrical distribution of fluorescence values after being stained with Hoechst dyes. With flow cytometric analysis using fixed laser power and fixed amplifier settings, haploid parasites fall within a small region in the fluorescence histogram. In most experiments using fixed RBC, samples containing these haploid stages are used as a standard and for determining the initial settings of the flow cytometer (see also Section VII).

Similar to what is described for the analysis of parasites in intact RBC, samples containing free haploid parasites can be used as standard and to determine the initial setting of the flow cytometer (see Section VII).

VII. Instruments

A. Analysis of Samples Containing Hoechst-Stained Infected Erythrocytes

To determine the percentage of infected RBC and DNA synthesis by parasites, samples have been analyzed with a FACS analyzer and with a FACStar (Becton-Dickinson, San Jose, CA). The FACS analyzer is equipped with a mercury arc lamp. Standard filter sets for UV excitation are used: a BP 360 and SP 375 for excitation and a SP 375 as dichroic mirror, and the blue Hoechst fluorescence is selected using a BP 490 and two LP 400 filters. Because the FACStar has a better sensitivity and higher discriminative properties for the light scatter this instrument is preferred and used for most studies reported here.

The FACStar is equipped with a Coherent Innova 90 laser tuned to UV excitation (351 nm, 50 mW). The blue Hoechst fluorescence is selected with a BP 485/22 optical filter. By setting an electronic threshold in the forward angle light scatter (FSC), debris is eliminated from analysis. Tuning and calibration of the FACStar is done using calibration beads containing defined amounts of fluorescent dye (Hoechst). A Hoechst-stained sample containing infected RBC is then used for the initial settings of the machine. These settings are monitored on a two-dimensional dot plot of Hoechst fluorescence and FSC, similar to what is shown in Fig. 4. Since uninfected RBC and infected RBC with single-haploid parasites, as well as schizonts containing more than 30 nuclei, need to be presented in the same histogram, the fluorescence gain setting is set in a logarithmic scale. The lower threshold in the FSC is set so that free merozoites, which are much smaller than the RBC, are still included.

The fluorescence intensity and FSC of 10,000–50,000 cells per sample are measured, collected in list mode, and stored using the standard BD Consort 30 software. Data analysis can be performed using the Consort 30 software, but for analysis of malaria parasite development in culture, specialized software is developed (see also Section VIII).

B. Analysis of Samples Containing Free Parasites Stained with Hoechst 33258 in Combination with Propidium Iodide

To detect and count free parasites, samples are analyzed using a FACStar equipped with a Coherent Innova 90 laser tuned to UV excitation (351 nm, 50 mW). A dichroic mirror (DM560), a BP 485/22 for the blue Hoechst fluorescence, and a LP 620 for the red propidium iodide are selected as emission filters. Calibration of the FACStar is done using calibration beads containing defined amounts of fluorescent dye (Hoechst). A lysed blood sample containing sufficient numbers of parasites is used for the initial settings of the machine. Because both parasites and WBCs have to be included in the same fluorescence histograms, the fluorescence gain settings are set in the logarithmic scale. To eliminate small weakly fluorescing particles which are not of interest, an electronic threshold is set in the red fluorescence signal, just below the fluorescence signal of parasites from the control sample. All samples are analyzed using these fixed settings. Of each sample, 5000 events are analyzed and the data are stored using the standard Becton-Dickinson Consort 30 software. For data analysis, parasite and WBC populations are identified and selected by setting a gate in a two-dimensional dot plot of FSC and blue fluorescence. This gate is set using the data from the control sample.

C. Analysis of Samples from Microtests Containing Hoechst-Stained Parasites

The analysis of samples from microtests is as described for the analysis of Hoechst-stained infected RBC by the FACStar as described above. However, the samples are not fed into the flow cytometer by hand, but sampling from the wells is performed automatically using an AutoMATE (Becton-Dickinson). This enables the fully automatic analysis and storage of samples from a complete 96-well microtiter plate. The acquisition is performed using the AutoMATE Control Program (ACP). With the ACP, samples of interest can be selected. Furthermore information for each sample can be added and stored together with the flow cytometric data. This can be used for identification purposes and for data processing by the program described below (Reinders *et al.*, 1994).

The program selects and identifies the events of interest, such as uninfected RBC, infected RBC, and free parasites by their FSC and fluorescence characteristics. This is used to determine the percentage of infected RBC whereas parasite development is calculated from the fluorescence distribution of the infected RBC which can be expressed as the increase in the mean number of nuclei per parasite (growth) and as the total number of nuclei synthesized, respectively. Results from several samples from the same culture series are combined in graphs and tables (see Fig. 5). Alternatively, results from single samples can also be displayed.

VIII. Results and Discussion

A. Blood Stages: Development and DNA Synthesis

The nuclei of parasites stained with Hoechst dyes show a strong specific fluorescence. Therefore, infected RBCs are clearly separated from uninfected cells on the basis of Hoechst-DNA fluorescence intensity (Fig. 2). The assessment of the

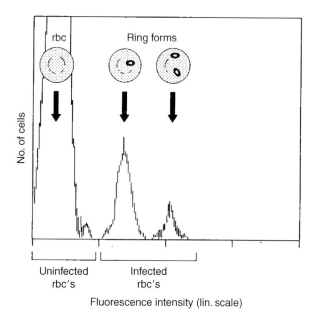

Fig. 2 Schematic representation of a frequency distribution showing the fluorescence distribution of a blood sample containing uninfected RBC and malaria-infected RBC containing haploid non-DNA synthesizing parasites, the ring forms. The RBCs were fixed with glutaraldehyde before being stained with Hoechst 33258. Two peaks represent the infected cells: the first peak consists of RBC with one haploid ring form and the second peak consists of RBC infected with two ring forms.

percentage of infected cells by flow cytometry based on this difference in fluorescence intensity corresponds closely to the assessments by microscopic examination of Giemsa-stained slides (Janse *et al.*, 1987).

Frequency distributions of the fluorescence values of young ring forms and merozoites show narrow, symmetrical Gaussian distributions (see Fig. 2). These stages of the parasite are haploid and non-DNA synthesizing. In the fluorescence histograms of these stages often a small second peak is observed of cells with a double-fluorescence intensity. This peak represents infected cells containing two ring forms (double infected RBC) or is caused by the simultaneous measurement of two infected cells or two free parasites.

During development of the ring forms into the old trophozoites, parasites increase in size but do not synthesize DNA. Old trophozoites show about 10% higher fluorescence intensity than ring forms, which is due to an increase of a nonspecific background fluorescence of the cytoplasm of the parasite (Janse *et al.*, 1987). In the schizont stage of development a rapid increase in DNA content and number of nuclei occurs as the result of three to five mitotic divisions, resulting in the production of 8–32 merozoites per parasite. The increase in the number of nuclei is proportional to the increase of the fluorescence intensity of Hoechst-stained schizonts (Fig. 3). Therefore, the frequency distributions of the fluorescence intensities

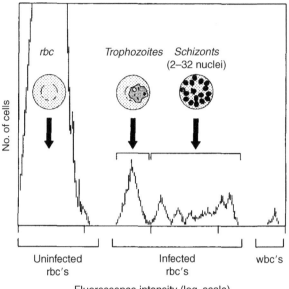

Fig. 3 Schematic representation of a frequency distribution showing the fluorescence intensity of a sample containing uninfected RBC and infected RBC. Parasites range from haploid non-DNA synthesizing trophozoites to mature schizonts containing 16–32 merozoites with immature schizonts in between. The RBCs were fixed with glutaraldehyde before being stained with Hoechst 33258. The fluorescence gain setting is set in a logarithmic scale. The small peak with the highest fluorescence intensity represents the white blood cells (WBCs).From van Vianen *et al.* (1990b), with permission.

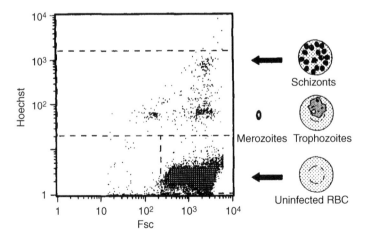

Fig. 4 Two-parameter dot plot representation of fluorescence intensity and FSC from a sample containing uninfected RBC, free merozoites and infected RBC containing trophozoites, and immature and mature schizonts. The cells were fixed with glutaraldehyde before being stained with Hoechst 33258.

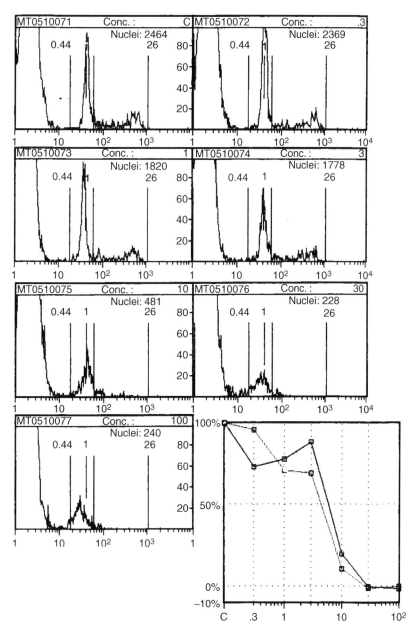

Fig. 5 Results from the computer program which calculates parasite inhibition by an antimalarial drug in a series of cultures. Each histogram represents the fluorescence distribution of parasites cultured without drug (control) or in the presence of different concentrations of the drug ranging from 0.3 to 100 ng/ml. In this experiment, *P. berghei* ring forms were cultured for 22 h in the presence of sodium artesunate after which samples of the cultures were fixed with glutaraldehyde and stained with Hoechst 33258. From the histograms it is clear that ring forms still develop into (mature) schizonts at low

of a population of (dividing) stages at different time points are representative of the development and degree of DNA synthesis of the parasites (Janse *et al.*, 1987). These frequency distributions can be used to determine the inhibition of parasite development and DNA synthesis by antimalarial drugs (see below).

B. Determination of Drug Susceptibility of Parasites

Determination of antimalarial activity of drugs *in vitro* which inhibit development of the blood stages can be performed routinely using flow cytometry. Here, we describe results from experiments using two different species, *P. falciparum* and *P. berghei*.

1. *P. berghei*

Young ring forms are cultured for 24 h under standard culture conditions in RPMI 1640 medium to which different concentrations of the drugs are added. Blood-stage development of *P. berghei* from haploid ring forms to the mature schizonts containing 8–24 nuclei takes 22–24 h. RBC containing the mature schizonts do not burst spontaneously in culture, but remain intact and viable for several hours. Samples for flow cytometry are taken from these cultures, before the culture is started and after 24 h. Samples at the start of the cultures contain ring forms which show a narrow symmetrical frequency distribution of their fluorescence intensities (Fig. 2). The mean fluorescence intensity of ring forms/young trophozoites represents the haploid DNA content and can be used as an internal standard to calculate the number of nuclei in the schizonts. In the 24-h sample from cultures without antimalarial drugs parasites show fluorescence values between 1 and 24 times the haploid amount. This comprises mature schizonts with 8–24 nuclei, immature schizonts in the process of DNA synthesis, and some degenerated parasites and free merozoites which are liberated from the RBC during handling of the samples. Figure 3 shows a schematic representation of a histogram of the fluorescence distribution of a sample containing trophozoites and schizonts. Figure 4 shows a two-parameter dot plot representation of flow cytometric data showing fluorescence intensity and FSC of a sample containing trophozoites, schizonts, and free merozoites.

Based on the mean fluorescence intensity of the cells, the software developed for this purpose calculates the percentage of infected cells and the total number of nuclei and the average number of nuclei per parasite present. Additionally, the parasite growth and DNA synthesis of the whole series are calculated and

concentrations up to 3 ng/ml. No schizonts can be detected at 30–100 ng/ml. The first peak in the histograms contains ring forms/free merozoites and trophozoites and the mean fluorescence intensity of these cells represents the haploid amount of DNA (one nucleus). This value is used to calculate the total number of nuclei in the different samples which is shown in the upper right corner of the histograms. The graph shows the inhibition of growth (...) and DNA synthesis (—) (see also Section VII.B).

presented in a graph and table. Parasite growth is defined as the average number of nuclei/parasite in a sample divided by the maximum average number of nuclei/parasite in the cultures. DNA synthesis is defined as the total number of nuclei in a sample divided by the maximum total number of nuclei in the cultures. Figure 5 shows an example of the calculation of parasite growth and DNA synthesis in samples obtained from cultures containing different concentration of an antimalarial drug. Figure 6 gives an example of the flow cytometric comparison of inhibition of parasite growth/DNA synthesis by three related antimalarial drugs. The results obtained by flow cytometry are comparable with results obtained by microscopic examination of parasites in Giemsa-stained slides (Janse *et al.*, 1987; van Vianen *et al.*, 1990a) or by measurement of the incorporation of radioactive precursors into RNA/DNA (C. J. Janse, unpublished results).

2. *P. falciparum*

Different *in vitro* tests have been described for determination of drug susceptibility of *P. falciparum* parasites. Parasites, often ring forms or young trophozoites, are cultured for prolonged periods (48–96 h) in the presence of different concentrations of drugs. The length of the culture periods depends on the process studied (e.g., reinvasion or schizont development) and on the method used to determine parasite development. In Giemsa-stained slides differences in the number of

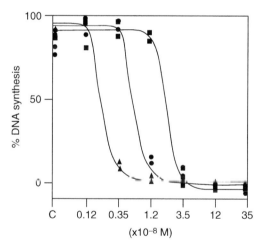

Fig. 6 An example of comparing the antimalarial activity of three related drugs, artemisinin (■), dihydroartemisinin (▲), and sodium artesunate (●) as measured using flow cytometry. In this experiment, ring forms of *P. berghei* were cultured for 22 h in the presence of different concentrations of the three drugs. At the start of the cultures and after 22 h, samples were taken from the cultures, fixed with glutaraldehyde, and stained with Hoechst 33258. The total number of parasite nuclei in the samples is calculated as described in the text of Section VII.B. DNA synthesis is defined as the increase in number of parasite nuclei during the culture period.

infected cells compared to the control culture are counted or the number of DNA synthesizing parasites (schizonts) is determined as parameters for parasite development. Alternatively, the incorporation of radioactive precursors into the DNA of parasites can be measured. The development of ring forms into mature schizonts takes 48 h, after which schizonts burst spontaneously and the free merozoites enter new RBC. In cultures without drugs this will result in the increase of the number of infected cells. Figure 7 gives an example of determination of the chloroquine susceptibility of *P. falciparum* in the "extended 72-h test" by flow cytometry.

A standardized test to determine drug susceptibility is the WHO microtest. Here, parasites are cultured for 26–30 h in 96-well microtiter plates, which are predosed with different concentrations of antimalarial drugs. During this culture period ring forms or young trophozoites develop into (immature) schizonts. The percentage of schizonts after culture is used for the assessment of susceptibility. Since flow cytometry can rapidly and reproducibly determine the increase of the number of nuclei a method has been developed for automated flow cytometric analysis of drug susceptibility of *P. falciparum* in microtests. For this purpose, the Auto-MATE is used for automatic sampling from the plates for flow cytometric measurements. Automatic sampling and flow cytometric analysis of a complete 96-well plate takes 2 h. Data analysis is performed with the described software which

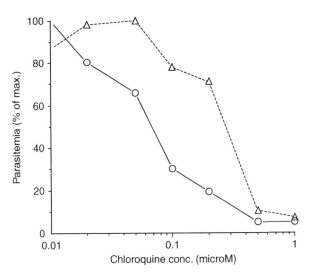

Fig. 7 Comparison of the chloroquine susceptibility of *P. falciparum* isolates (○, clone T9/94; △, isolate TM 152) in the "extended 72-h test" using flow cytometry. In this test, parasites are cultured for 72 h in the presence of different concentrations of chloroquine and inhibition of development is determined by measurement of the increase/decrease of the number of infected cells during the culture period. Here, the number of infected cells is determined by flow cytometry after fixed (infected) cells are stained with Hoechst 33258. In 63 tests, it was found that the results obtained by flow cytometry closely corresponded with results obtained by microscopic examination of Giemsa-stained slides (van Vianen *et al.*, 1990b).From van Vianen *et al.* (1990b), with permission.

calculates the percentage of infected cells, the total number of nuclei per sample, and the average number of nuclei per parasite in each sample. Parasite growth and DNA synthesis for a complete culture series is calculated as described above. Figure 8 shows an example of the determination of chloroquine susceptibility of *P. falciparum* in microtests by flow cytometry compared to results by microscopic examination. We have shown in several experiments that automatic reading of a microtest by flow cytometry gives results comparable with those of microscopic examination of Giemsa-stained slides (van Vianen *et al.*, 1990b; P. H. van Vianen, unpublished results).

C. Detection and Counting of Low Numbers of Parasites

Since infected cells show a higher fluorescence intensity than noninfected cells, the percentage of infected cells can be established by flow cytometry. When this percentage is higher than 0.1%, reproducible counts are obtained in samples where the blood cells are fixed before staining and measurement. However, less reproducible results are obtained with fixed RBC when the percentage is lower than 0.1%, due to the presence of low numbers of RBC or reticulocytes in the blood samples, which show a specific background fluorescence. Their fluorescence intensity is in the same range as that of infected cells and they may interfere with the measurement.

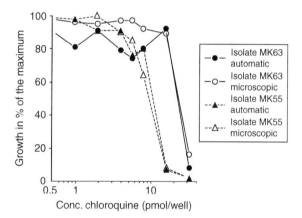

Fig. 8 Comparison of chloroquine susceptibility of two *P. falciparum* isolates in standard drug susceptibility microtests. Inhibition of development was determined either by flow cytometric analysis or by microscopic examination of Giemsa-stained thick smears. Parasites were incubated in 96-well microtiter plates under standard culture conditions for a period of 30 h. These plates were predosed with different concentrations of chloroquine. After 30 h samples were taken to prepare slides for microscopic examination. The rest of the culture material was fixed in the plates with glutaraldehyde and stained with Hoechst 33258. Flow cytometric reading of the plates was performed using the AUTOmate and parasite growth and DNA synthesis was calculated as described in Section VII.B. Calculation of parasite growth by microscopy is done by dividing the number of schizonts counted by the total number of parasites counted. From van Vianen *et al.* (1990b), with permission.

Therefore, to detect low numbers of parasites in blood samples, the RBCs are first lysed before fixation of the parasites. Figure 9 shows the fluorescence intensity and the FSC of a blood sample, which is lysed before fixation and staining with Hoechst 33258. Three populations can be distinguished: WBCs, platelets, and parasites. From dilution experiments, in which infected blood was diluted with uninfected blood, we have found that *P. falciparum* parasites were reproducibly detected at a percentage of about 0.005% (Fig. 10). Using this method, the detection of *P. berghei* is somewhat less sensitive due to interference by residual bodies of the nucleus in a low percentage of rodent RBC. These residual bodies do not lyse and can show fluorescence intensities comparable to that of parasites (see the relative high number of "parasites" in noninfected blood samples in Fig. 10).

The reproducible detection of lower numbers of parasites is hampered by the fact that the fluorescence intensity and FSC of a low percentage of platelets fall in the same range as those of the parasites. We found that the combination of Hoechst 33258 and propidium iodide staining of the lysed blood samples allows a better separation of platelets and parasites. When the samples are stained with propidium iodide alone parasites cannot be separated from platelets on the basis of the red fluorescence. In combination with Hoechst 33258 staining, however, the

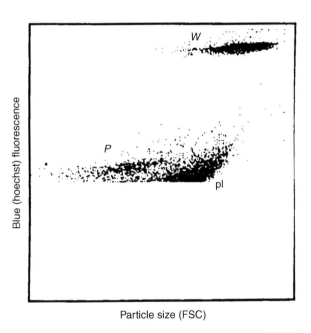

Particle size (FSC)

Fig. 9 Two-parameter dot plot representation of fluorescence intensity and FSC from a blood sample after lysis of the RBC in FACS lysing solution, fixation with formaldehyde, and staining with Hoechst 33258. The blood sample is from a patient infected with *P. falciparum* with a percentage of infected RBC of about 0.01%. Three populations of cells can be distinguished: white blood cells (W), platelets (pl), and parasites (P). From van Vianen *et al.* (1993), with permission.

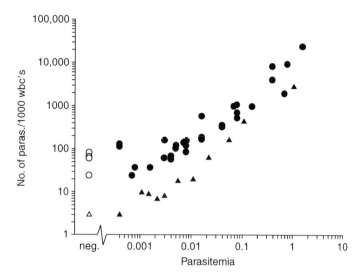

Fig. 10 Relationship between the number of parasites counted by flow cytometry and the expected percentage of infected cells (parasitemia) in blood samples. In these experiments, infected blood was serially diluted with noninfected blood after which RBCs were lysed and the parasites fixed and stained with Hoechst as described in Fig. 9. Parasite numbers counted by flow cytometry (in parasites per 1000 WBCs) are compared with the expected parasitemia calculated from the starting parasitemia and the dilution factor (*P. berghei*, O; *P. falciparum*, △; open symbols are from blood samples containing no parasites).From van Vianen *et al.* (1993), with permission.

red fluorescence of the parasites significantly increases while the fluorescence intensity of the platelets remains the same, allowing a better separation of those two populations. It is suggested that energy transfer from DNA-bound Hoechst 33258 to propidium iodide takes place, by which the red fluorescence of the parasites is more enhanced than that of platelets which do not contain DNA. This staining method has been used in clinical and epidemiological studies to detect low numbers of parasites in blood samples from patients. We found that as few as 20 parasites per μl of blood (equivalent to 0.0004% infected RBC) could be detected (P. H. van Vianen, unpublished results).

IX. Comparison of Methods

A. Drug Susceptibility of Parasites

The use of an instrument such as a flow cytometer for the analysis of samples collected under primitive field conditions in developing countries seems contradictory to the necessity to keep field procedures uncomplicated. However, the collection of samples in the field is very simple. For drug susceptibility tests only the addition of fixative to the culture wells is essential. In fact the procedure is more easy than making smears of the culture material. Compared to other techniques, flow cytometric analysis of microtests provides more information. With microscopic examination of thick smears only the percentage of schizonts is counted and with the use of radioactive precursors only the total amount of precursor in the nucleic acids per sample is used to establish parasite development. Flow cytometry enables the combination of a number of different parameters. Information on the percentage of infected cells, the schizogonic development of individual parasites, extra- versus intraerythrocytic parasites, and invasion of new RBC can be generated in one analysis. An additional improvement is the automation of the analysis and processing of the data. All data are processed in a standardized way, maintaining objectivity, and the data remain available for reexamination. Data from large numbers of tests can be combined.

B. Detection and Counting of Parasites

Routine diagnosis of malaria is generally based on examination of Giemsa-stained blood smears for blood-stage parasites. Since this disease requires a rapid treatment, ideally within several hours after attending the hospital, a rapid sensitive and simple diagnostic method is required. This is provided by the Giemsa-stained smear examination. Since flow cytometric analysis as described in this chapter requires a FACStar flow cytometer, the method is less suitable for direct diagnosis of malaria, especially in hospitals in developing countries. However, flow cytometric analysis has proven to be a sensitive, rapid, and reproducible method when large numbers of samples have to be screened for malaria parasites in epidemiological studies or clinical studies. It allows for the reproducible detection of less than 20 parasites/μl of human blood (0.0004% infected RBC). This is comparable to the sensitivity obtained under optimal laboratory conditions by the use of radioactively labeled DNA probes specific for malaria parasites. Nonradioactive methods are less sensitive. The sensitivity of microscopic examination is reported to be better. In optimal circumstances, one parasite/μl blood (0.00002% parasitemia) can be detected, although 10–20 parasites/μl is probably more realistic. When both the probe and the microscopic method were tested under field conditions their sensitivities appeared to be much lower. In these circumstances, both radioactive DNA probes and microscopy were reported to approach their limits of reproducible detection at parasitemias of 0.016% in the field (Barker et al., 1989). Flow cytometric analysis is not affected by the circumstances in which the samples are collected and handled, and the same sensitivities have been found in the laboratory and with field samples. Furthermore, flow cytometry allows for quantitation of the number of parasites. This is not easily performed with other methods. In conclusion, flow cytometry seems to be the best technique available for the detection and quantitation of parasites in large numbers of blood samples. Especially, in view of the recent interest in vaccine development and the use of newly developed drugs, this method is suitable for the follow-up of patients after treatment or the follow-up of vaccine trials and for large-scale epidemiological research in malaria.

Acknowledgments

This research was financed in part by the Netherlands' Ministry for Development Cooperation (Project ID.OSAM-08775). Flow cytometry was performed within the Department of Haematology, AZL, Leiden by M. van der Keur and P. P. Reinders. We thank Dr H. J. Tanke (Department of Cytochemistry and Cytometry, Medical Faculty, University of Leiden) for critical reading of the manuscript and stimulating discussions and P. P. Reinders for developing the software for the calculation of parasite growth.

References

Barker, R. H., Jr., Suebsang, L., Rooney, W., and Wirth, D. F. (1989). *Am. J. Trop. Med. Hyg.* **41,** 266–272.

Bianco, A. E., Battye, F. L., and Brown, G. V. (1986). *Exp. Parasitol.* **62,** 275–282.

Crissmann, H. A., Stevenson, A. P., Kissane, R. J., and Tobey, R. A. (1979). *In* "Flow Cytometry and Sorting" (M. R. Melamed, P. F. Mullaney, and M. L. Mendelsohn, eds.), pp. 234–262. Wiley, New York.

Franklin, R. M., Brun, R., and Grieder, A. (1986). *Z. Parasitenkd.* **72,** 201–212.

Hare, J. D. (1986). *J. Histochem. Cytochem.* **34**(12), 1651–1658.

Janse, C. J., van Vianen, P. H., Tanke, H. J., Mons, B., Ponnudurai, T., and Overdulve, J. P. (1987). *Exp. Parasitol.* **64,** 88–94.

Janse, C. J., Boorsma, E. G., Ramesar, J., Grobbee, M. J., and Mons, B. (1989). *Int. J. Parasitol.* **19,** 509–514.

Kawamoto, F., and Billingsley, P. F. (1992). *Parasitol. Today* **8,** 81–83.

Makler, M. T., Lee, L. G., and Recktenwald, D. (1987). *Cytometry* **8,** 568–570.

Mons, B., and Janse, C. J. (1992). *In* "Flow Cytometry in Hematology" (O. D. Laerum and R. Bjerkness, eds.), pp. 197–211. Academic Press, London.

Myler, P., Saul, A., Mangan, T., and Kidson, C. (1982). *Aust. J. Exp. Biol. Med. Sci.* **60,** 83–89.

Pattanapanyasat, K., Webster, H. K., Udomsangpetch, R., Wanachiwanawin, W., and Yongvanitchit, K. (1992). *Cytometry* **13,** 182–187.

Payne, D. (1988). *Bull. WHO* **66,** 621–626.

Reinders, P. P., van Vianen, P. H., van der Keur, M., van Engen, A., Mons, B., Tanke, H. J., and Janse, C. J. (1994). *Cytometry* (in press).

Rickman, L. S., Long, G. W., Oberst, R., Cabanban, A., Sangalang, R., Smith, J. I., Chulay, J. D., and Hoffman, S. L. (1989). *Lancet* **1,** 68–71.

Saul, A., Myler, P., Mangan, T., and Kidson, C. (1982). *Exp. Parasitol.* **54,** 64–71.

Tharavanij, S. (1990). *Southeast Asian J. Trop. Med. Public Health* **21,** 3–16.

Trigg, P. I. (1985). *Bull. WHO* **63,** 397–398.

van Vianen, P. H., Klayman, D. L., Lin, A. J., Lugt, C. B., van Engen, A. L., van der Kaay, H. J., and Mons, B. (1990a). *Exp. Parasitol.* **70,** 115–123.

van Vianen, P. H., Thaithong, S., Reinders, P. P., van Engen, A. L., van der Keur, M., Tanke, H. J., van der Kaay, H. J., and Mons, B. (1990b). *Am. J. Trop. Med. Hyg.* **43,** 602–607.

van Vianen, P. H., van Engen, A., Thaithong, S., van der Keur, M., Tanke, H. J., van der Kaay, H. J., Mons, B., and Janse, C. J. (1993). *Cytometry* **14,** 276–280.

Whaun, J. M., Rittershaus, C., and Ip, S. H. C. (1983). *Cytometry* **4,** 117–122.

World Health Organization (WHO) (1982). "WHO/MAP/82.1." WHO, Geneva.

INDEX

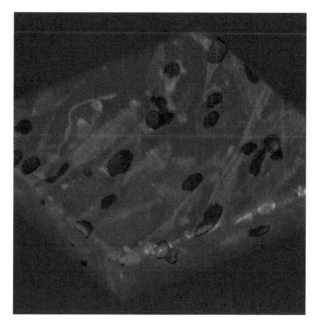

Plate 1 (Figure 1.4 on page 19 of this volume)

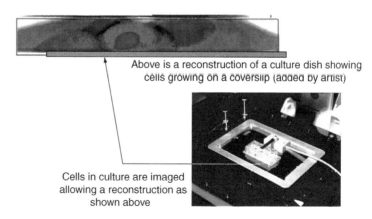

Above is a reconstruction of a culture dish showing cells growing on a coverslip (added by artist)

Cells in culture are imaged allowing a reconstruction as shown above

Plate 2 (Figure 1.5 on page 19 of this volume)

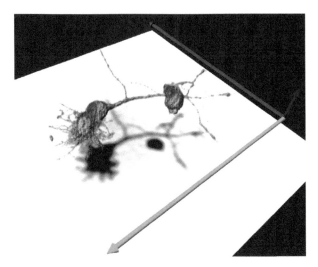

Plate 3 (Figure 1.8 on page 24 of this volume)

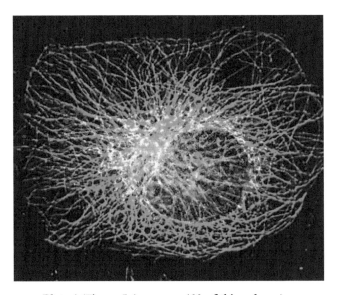

Plate 4 (Figure 5.1 on page 133 of this volume)

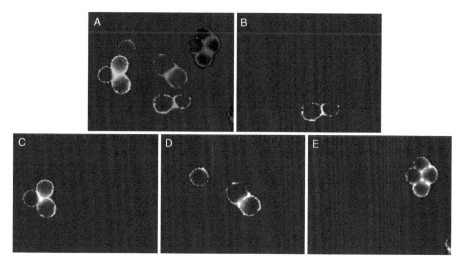

Plate 5 (Figure 5.2 on page 135 of this volume)

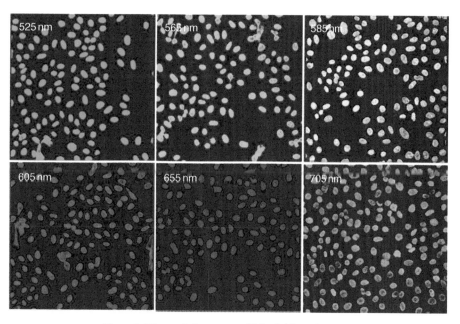

Plate 6 (Figure 5.3 on page 136 of this volume)

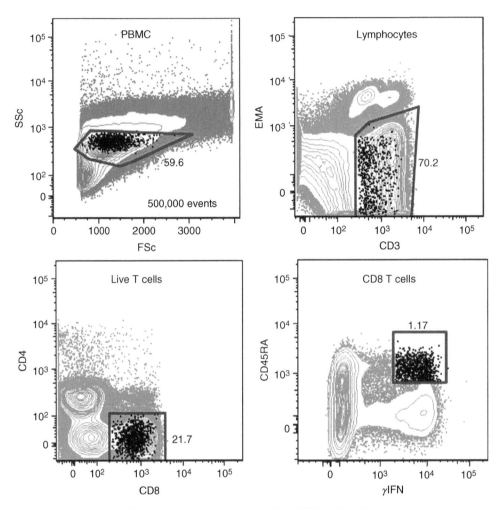

Plate 7 (Figure 9.2 on page 212 of this volume)

Plate 8 (Figure 20.8 on page 434 of this volume)

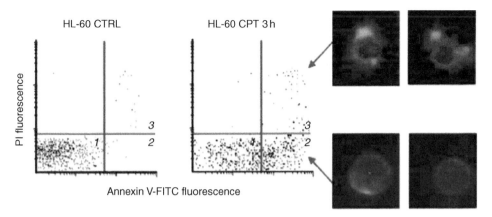

Plate 9 (Figure 22.1 on page 472 of this volume)

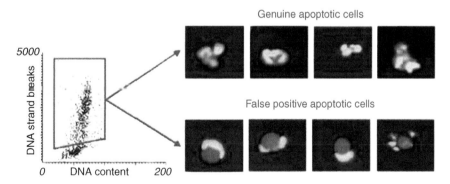

Plate 10 (Figure 22.2 on page 475 of this volume)

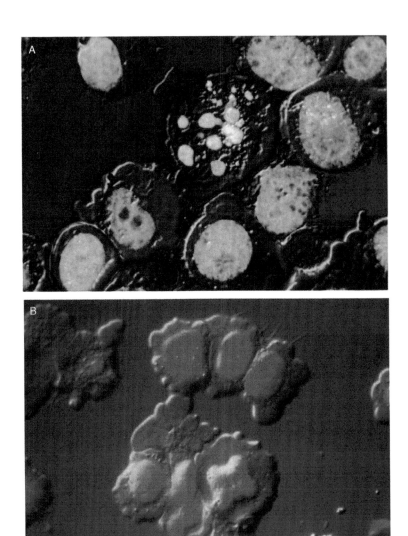

Plate 11 (Figure 22.3 on page 477 of this volume)

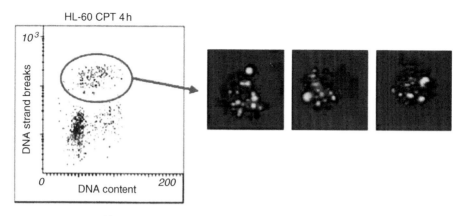

Plate 12 (Figure 22.4 on page 479 of this volume)

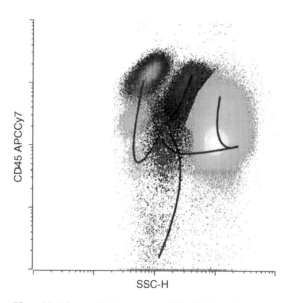

Plate 13 (Figure 24.1 on page 525 of this volume)

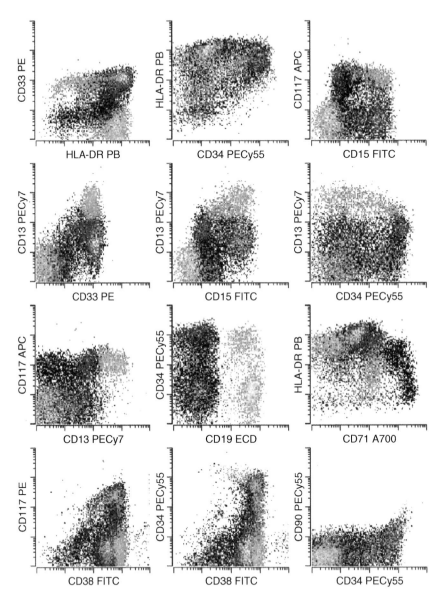

Plate 14 (Figure 24.2 on page 527 of this volume)

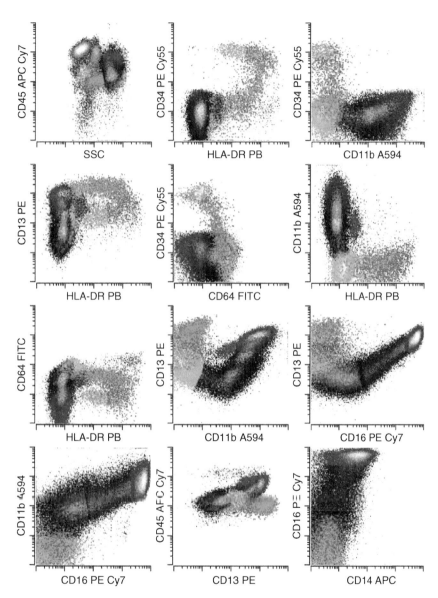

Plate 15 (Figure 24.3 on page 529 of this volume)

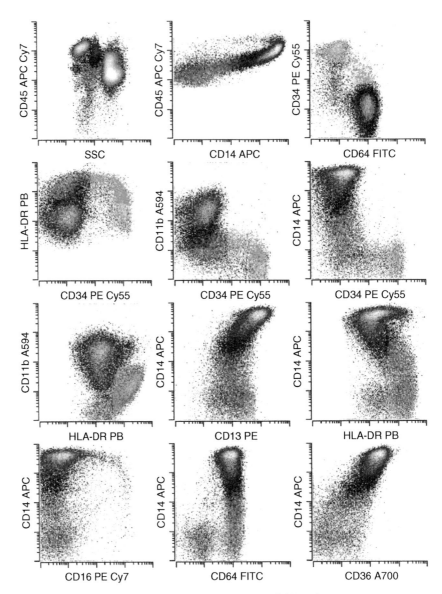

Plate 16 (Figure 24.4 on page 531 of this volume)

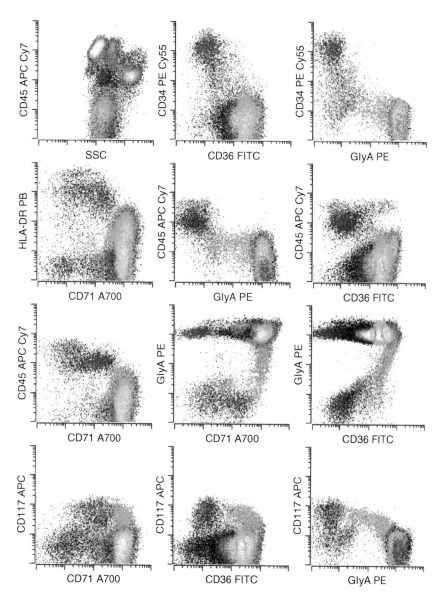

Plate 17 (Figure 24.5 on page 533 of this volume)

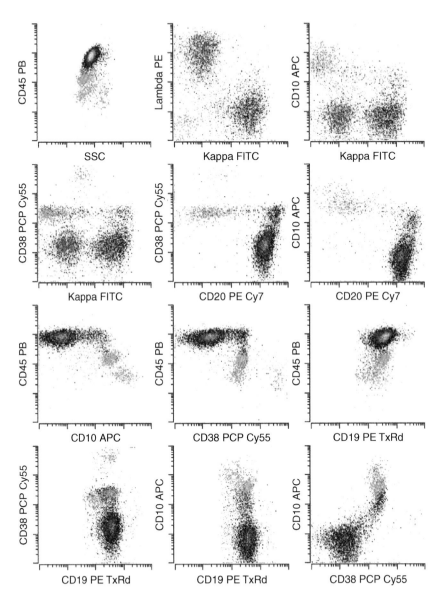

Plate 18 (Figure 24.6 on page 535 of this volume)

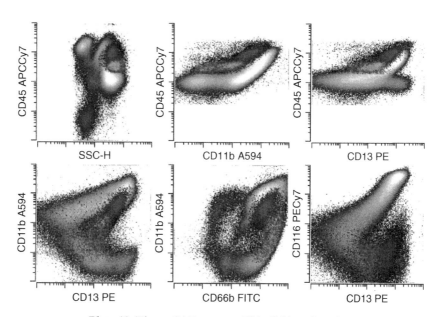

Plate 19 (Figure 24.7 on page 536 of this volume)

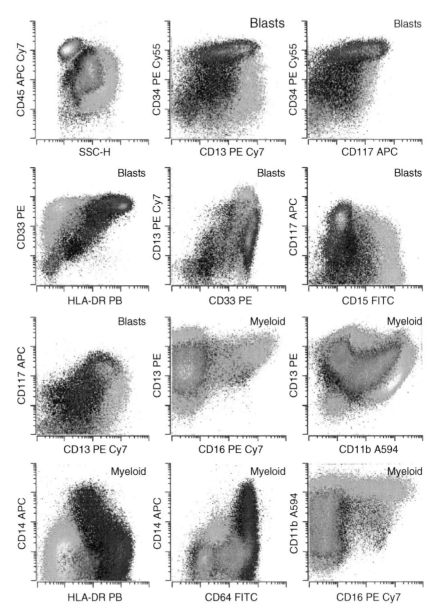

Plate 20 (Figure 24.8 on page 537 of this volume)

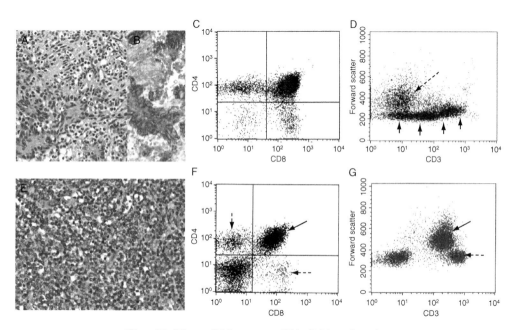

Plate 21 (Figure 25.3 on page 554 of this volume)

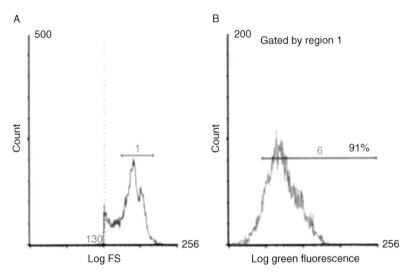

Plate 22 (Figure 38.6 on page 846 of this volume)

Printed and bound by CPI Group (UK) Ltd, Croydon, CR0 4YY
03/10/2024
01040318-0005